ADVANCES IN NANOSTRUCTURED COMPOSITES

Volume 2: Applications of Nanocomposites

Editor

Mahmood Aliofkhazraei

Department of Materials Engineering
Tarbiat Modares University
Tehran, Iran

CRC Press is an imprint of the
Taylor & Francis Group, an **informa** business
A SCIENCE PUBLISHERS BOOK

Cover credit: Paul Fleet/Shutterstock.com

CRC Press
Taylor & Francis Group
6000 Broken Sound Parkway NW, Suite 300
Boca Raton, FL 33487-2742

First issued in paperback 2021

Version Date: 20190306

ISBN 13: 978-0-367-77941-2 (pbk)
ISBN 13: 978-0-367-07631-3 (hbk)

Library of Congress Cataloging-in-Publication Data

Names: Aliofkhazraei, Mahmood, editor.
Title: Applications of nanocomposites / editor Mahmood Aliofkhazraei, Department of Materials Engineering, Tarbiat Modares University, Tehran, Iran.
Description: Boca Raton, FL : CRC Press, Taylor & Francis Group, [2018] | Series: Advances in nanostructured composites ; volume 2 | "A science publishers book.» | Includes bibliographical references and index.
Identifiers: LCCN 2018041058 | ISBN 9780367076313 (hardback : acid-free paper)
Subjects: LCSH: Nanocomposites (Materials)
Classification: LCC TA418.9.N35 A665 2018 | DDC 620.1/18--dc23
LC record available at https://lccn.loc.gov/2018041058

Visit the Taylor & Francis Web site at
http://www.taylorandfrancis.com

and the CRC Press Web site at
http://www.crcpress.com

Preface

The second volume of this book aims to provide a guide for different applications of modern nanocomposites especially those fabricated by carbon nanotubes and graphene. The book makes a comparative study of fiber-reinforced composites which have been embedded into the matrix with nanocomposites containing nanotubes in place of fibers. The main topics of this volume are: Electrochemical Properties of Nanoporous Based Materials, Fabrication and Application of Graphene Oxide-based Metal and Metal Oxide Nanocomposites, Electrochemical Sensors/Biosensors Based on Carbon Aerogels/Xerogels, Advances in Nanobiocatalysis: Strategies for Lipase Immobilization and Stabilization, Metal Oxide Based Heterojunction Nanoscale Materials for Chemiresistive Gas Sensors, Recent Advances in Polymer Nanocomposite Coatings for Corrosion Protection, Recent Advances in the Design of Nanocomposite Materials via Laser Techniques for Biomedical Applications, Carbonaceous Nanostructured Composites for Electrochemical Power Sources: Fuel Cells, Supercapacitors and Batteries, Bismuth Vanadate Based Nanostructured and Nanocomposite Photocatalyst Materials for Water Splitting Application.

Editor

Contents

Section III

Recent Applications of Nanocomposites

Electrochemical Properties of Nanoporous Based Materials

Tebogo P. Tsele,[1,2] *Abolanle S. Adekunle,*[1,2,3] *Omolola E. Fayemi,*[1,2] *Lukman O. Olasunkanmi*[1,2,3] and *Eno E. Ebenso*[1,2,*]

Introduction

Nanoparticles/Nanocomposites

Nanotechnology is the study of the controlling parameters of materials in atomic and molecular scale. Generally, nanotechnology deals with structures of 100 nanometers or less in at least one dimension, and involves developing materials or devices of that size (Sharma et al. 2004). Nanotechnology encompasses the production and application of physical, chemical, and biological systems. The convergence of scientific disciplines (chemistry, biology, electronics, physics, engineering, etc.) is leading to a multiplication of applications in materials manufacturing, computer chips, medical diagnosis and health care, energy, biotechnology, space exploration and security (Taniguchi 1974).

Theoretical background

Metal oxides nanoparticles

Metal and metal oxides nanoparticles are of particular interest for their mechanical, electrical, magnetic, optical, chemical and other special properties (Valden et al. 1998, Gracias et al. 2000, Huang et al. 2001,

[1] Department of Chemistry, Faculty of Natural and Agricultural Sciences, North-West University (Mafikeng Campus), Private Bag X2046, Mmabatho 2735, South Africa, South Africa.

[2] Material Science Innovation and Modelling (MaSIM) Research Focus Area, Faculty of Natural and Agricultural Sciences, North-West University (Mafikeng Campus), Private Bag X2046, Mmabatho 2735, South Africa, South Africa.

[3] Department of Chemistry, Obafemi Awolowo University, Ile-Ife, Nigeria.
Emails: tebogo.palesa@yahoo.com; sadekpreto@gmail.com; omololaesther12@gmail.com; waleolasunkanmi@gmail.com

* Corresponding author: Eno.Ebenso@nwu.ac.za

Pan et al. 2001, Shi et al. 2001, Johnson 2002, Aizpurua et al. 2003, Wang 2004, Song et al. 2005, Karkare 2008). They have better malleability and ductility compared to bulk materials. These are made of tight cluster of very small particles resulting in overlapping electron clouds that induce quantum effect resulting in more efficient conduction of light or electricity (Sammes et al. 1999). In technological applications, metal oxides nanoparticles are used in the fabrication of microelectronic circuits, sensors, fuel cells, as catalysts and biomedicine. The physicochemical and optoelectronic properties of metallic nanoparticles are strongly dependent on the size, shape and its distribution (Metraux et al. 2003, Shankar et al. 2004). Particle size is expected to influence three important groups of basic properties in any material. The first one comprises the structural characteristics, namely the lattice symmetry and cell parameters (Moens et al. 1997).

In order to display mechanical or structural stability, a nanoparticle must have a low surface free energy. As a consequence of this requirement, phases that have a low stability in bulk materials can become very stable in nanostructures. This structural phenomenon has been detected in TiO_2, VO_x, Al_2O_3 and MoOx oxides (Zhang and Bandfield 1998, Samsonov et al. 2003, Song et al. 2003). There are several synthesis procedures for the preparation of ultrafine oxide nanoparticles such as sol-gel, hydrothermal, solvothermal, flame combustion and emulsion precipitation (Kwon et al. 2002, Hambrock et al. 2003, Ge et al. 2007, Kim and Park 2007, Vaezi and Sadrnezhaad 2007, Shokuhfar et al. 2008, Moghaddam et al. 2009).

Titanium oxide (TiO_2)

The performance of TiO_2 based photocatalyst is largely influenced by the size of the nanometric TiO_2 building units because of the increased surface-to-volume ratio that facilitates reaction/interaction between them and the interacting media, which mainly occurs on the surface or at the interface (Macwan et al. 2011). In particular, it has been found that TiO_2 nanoparticles enhanced the total photoreactivity of TiO_2 due to their larger active surface area and capability to reduce the recombination rate of electron-hole pairs. Moreover, the size of the TiO_2 particles is known to alter the width of the band gap and the band bending at the interfaces, and thus influence their photochemical properties (Sugimoto et al. 2003).

Titanium oxide and its application

Titanium oxide (TiO_2) occurs in nature in three different polymorphs which, in order of abundance, are rutile (most abundant), anatase, and brookite which consist of octahedrally coordinated Ti cations arranged in edge sharing chains, but differ in the number of shared edges and corners (Stathato et al. 1997). Additional synthetic phases are called TiO_2(B), TiO_2(H) and TiO_2(R) (Marchand et al. 1980). Traditionally, the rutile phase has been investigated most extensively due to availability of good single crystals for characterization and also because of the comparatively simple crystal structure. In TiO_2 materials, the so-called "quantum-confinement" or "quantum-size effect" is restricted to very low sizes, below 10 nm, due to their rather low excitation Bohr radii (Fernández-García et al. 2004). Various synthesis techniques such as hydrolysis (Guan et al. 2007), solgel route (Selvaraj et al. 1992, Sugimoto et al. 1997), micro emulsion or reverse micelles (Lee et al. 2005) and hydrothermal (Cheng et al. 1995) have been used to prepare the nanoparticles of titanium dioxide.

Titanium dioxide (TiO_2) is an attractive electrode material and hence it is widely used in a number of electrochemical applications. Due to its good biocompatibility and catalytic properties, TiO_2 has been predominantly employed as an electrode material in electrochemical biosensors for the detection of various analytes, even at their lower concentration (Chen et al. 2010, Wang et al. 2012a, 2013b). It is relatively inexpensive, nontoxic, and biocompatible in nature and exhibits a high photostability in adverse environments. The biocompatibility of TiO_2 is highly important in medicinal biology because the TiO_2 layers on Ti and Ti alloys are in direct contact with biological tissue in hip or dental implants (Brunette et al. 2001, Macak et al. 2008).

The desirable surface properties of TiO_2 make it a potential interface for the immobilization of biomolecules and its application (Topoglidis et al. 1998) as a food additive (Phillips and Barbeno 1997), in cosmetics (Selhofer et al. 1999) and as a potential tool in cancer treatment (Fujishima et al. 2000). TiO_2 has been used for the destruction of toxic organic compounds and microorganisms such as bacteria and viruses and hence has been used in purification of polluted air and wastewaters (Bahnemann 1991, Berry and Mueller 1994). The most actively pursued applied research on TiO_2 is its use for photo-assisted degradation of organic molecules.

Graphene oxide and multiwall carbon nanotubes

Graphene oxide (GO) is the product of chemical exfoliation of graphite (Staudenmaier 1898, Lerf et al. 1998, Rao et al. 2009). The oxidation of graphite breaks up the extended two-dimensional (2D) π-conjugation of the stacked graphene sheets into nanoscale graphitic sp^2 domains surrounded by disordered, highly oxidised sp^3 domains as well as defects of carbon vacancies (Berger et al. 2006, Gao et al. 2009). This severe functionalization of the conjugated network renders GO sheets insulating. However, conductivity may be partially restored conveniently by thermal (Schniepp et al. 2006) or chemical treatment (Stankovich et al. 2007), producing chemically modified graphene sheets.

The ease of synthesizing GO and its solution processability have made it a very attractive precursor for large scale production of graphene in applications including transparent conductors (Eda et al. 2008, Hu et al. 2010, Wu et al. 2010, Huang et al. 2011), chemical sensors (Wang et al. 2009a, Sengupta et al. 2011), biosensors (Wang et al. 2009b), polymer composites (Qi et al. 2010, Sun et al. 2011), battery and ultra-capacitors (Iijima 1991).

Carbon nanotubes (CNTs) are seamless cylinders of one or more layers of graphene (denoted single-wall, SWCNT, or multiwall, MWCNT), with open or closed ends (Peng et al. 2008, Harris 2009). Perfect CNTs have all carbons bonded in a hexagonal lattice except at their ends, whereas defects in mass-produced CNTs introduce pentagons, heptagons, and other imperfections in the sidewalls that generally degrade desired properties. CNT lengths range from less than 100 nm to several centimeters, thereby bridging molecular and macroscopic scales (Wei et al. 2001).

MWCNTs are typically metallic and can carry currents of up to 109 A cm^{-2} (Wei et al. 2001). Individual CNT walls can be metallic or semiconducting depending on the orientation of the graphene lattice with respect to the tube axis, which is called the chirality (Pop et al. 2006). For example, the electrical properties of CNT may be tuned by mechanical deformation. Such properties are of great interest for applications such as sensors or smart materials. The study of these properties is multidisciplinary and involves various branches of science and engineering (Sinha and Yeow 2005).

Graphene oxide, carbon nanotubes and application

Graphene oxide (GO) is one of the main precursors for the preparation of various graphene-based photocatalytic materials due to its large-scale production and facile synthetic method (Williams et al. 2008). However, the microstructures of the GO are completely different from that of the pristine graphene owing to the presence of large amount of oxygen-containing functional groups such as hydroxyl and epoxide, resulting in a great decrease of the electronic properties (Berger et al. 2006, Park and Ruoff 2009).

Graphene oxide is considered as a promising material for biosensors due to excellent electrochemical properties, biocompatibility, high defect density and the presence of pendant organic functional groups (–OH, –COOH, –CHO) (Pumera et al. 2010). GO can be electrostatically suspended in water due to the presence of carboxylic (–COOH) groups (Li et al. 2008). The –COOH groups allows easy attachment of various biomolecules, such as protein, enzyme and nucleic acids onto the GO sheets that warrants its use as electrode in the development of immunosensors.

Since graphene oxide and its reduced forms have an extremely high surface area, these materials are under consideration for usage as electrode material in batteries and double-layered capacitors, as well as in studies of hydrogen storage, fuel cells, and solar cells (Zhu et al. 2010a). Graphene oxide was found to be fluorescent, which opened a route for applications in bio-sensing, early disease detection, and even assisting in carrying cures for cancer. Graphene oxide has been successfully used in fluorescent-based biosensors for the detection of DNA and proteins, with a promise of better diagnostics of HIV. Furthermore, graphene oxide is tested as a drug carrier as well (Yang et al. 2011).

Multiwall carbon nanotubes (MWCNT) production today is used in bulk composite materials and thin films, which rely on unorganized CNT architectures having limited properties. Organized MWCNT architectures such as vertically aligned forests, yarns, and sheets show promise to scale up the properties of individual MWCNTs and realize new functionalities, including shape recovery (Cao et al. 2005), dry adhesion (Qu et al. 2008), high damping (Xu et al. 2010, De Volder et al. 2012), terahertz polarization (Ren et al. 2009), large-stroke actuation (Aliev et al. 2009, Lima et al. 2012), near-ideal black-body absorption (Mizuno et al. 2009), and thermoacoustic sound emission (Xiao et al. 2008). Chemical vapour deposition (CVD) is the dominant mode of high-volume CNT production and typically uses fluidized bed reactors that enable uniform gas diffusion and heat transfer to metal catalyst nanoparticles (Endo et al. 2006).

In the automotive industry, conductive CNT plastics have enabled electrostatic-assisted painting of mirror housings, as well as fuel lines and filters that dissipate electrostatic charge. Other products include electromagnetic interference (EMI)–shielding packages and wafer carriers for the microelectronics industry. For load-bearing applications, CNT powders mixed with polymers or precursor resins can increase stiffness, strength, and toughness (Chou et al. 2010). Additionally, engineering nanoscale stick-slip among CNTs and CNT-polymer contacts can increase material damping (Suhr et al. 2005), which is used to enhance sporting goods, including tennis racquets, baseball bats, and bicycle frames.

Dopamine

Dopamine (DA) is a neurotransmitter associated with proper functioning of several organs such as the heart, brain, and suprarenal glands. The determination of dopamine is a subject of great significance for investigating its physiological functions and diagnosing nervous diseases resulting from dopamine abnormal metabolism, such as epilepsy, Parkinsonism and senile dementia (Zou 1999). The fact that compound makes their detection possible by electrochemical methods is based on anodic oxidation (Yavich and Tiihonen 2000). Dopamine has been determined using various electrochemical methods (Wang et al. 2001, Chicharroa et al. 2004, Xue et al. 2005).

Dopamine is one of the most important neurotransmitters and is present in mammalian central nervous system (Sawa and Snyder 2002). Neurotransmitters are chemical messengers that transmit a message from one neuron to the next (Michael and Wightman 1999). Dopamine is a catecholamine in the form of the large organic cations and belongs to the family of excitatory chemical cardiovascular, renal and hormonal system (Velasco and Luchsinger 1998, Mo and Ogorevc 2001). In humans, a deficiency of the neurotransmitter dopamine in the basal ganglia of the brain has been known well to play a critical role in Parkinson's disease (Agid et al. 1987). Parkinson's disease is a degenerative disease of the nervous system associated with trembling of the arms and legs, stiffness and rigidity of the muscle and slowness of movement.

Dopamine acts like a brain chemical to transmit message to parts of the brain for coordination of body movements (Owen et al. 1998). A significant problem for dopamine determination is the fouling effects due to accumulation of reaction product which forms electron-polymerized films on the electrode surface (Lane and Blaha 1990). Thus, one promising approach to overcome problem arising from fouling of the biological substrate is the use of chemical modified electrodes (Ardakani et al. 2009, Manjunatha et al. 2009, Mazloum-Ardakani et al. 2009). Since DA is an oxidisable compound, it can be easily detected by electrochemical methods based on anodic oxidation (Xu et al. 1992, Raj et al. 2001). DA regulates food intake by modulating food reward and motivation (Kennedy and Mitra 1963, Espelin and Done 1968, Leibowitz 1975, Knappertz et al. 1998, Jones et al. 2006, Epstein et al. 2007, Atkinson 2008) without crossing the blood-brain barrier. DA is measured in the body using electrochemical sensors for monitoring HIV infection (Cai et al. 2014, Thien et al. 2014).

Small scale studies on the use of intermittent dopamine infusions and oral levodopa provided initial evidence for a symptomatic benefit in patients with severe heart failure (Sol et al. 1987). Ibopamine, an orally active dopaminergic compound, which as a prodrug is hydrolysed to the active metabolite epinine, was registered some ten years ago in several European countries for the treatment of mild congestive heart failure in combination with diuretics, or for moderate to severe heart failure in combination with diuretics, ACE inhibitors or digoxin. The drug has dopaminergic effects on both D1 and D2 dopamine receptors (Nichols et al. 1987). It has mild renal and diuretic effects, and it reduces norepinephrine plasma levels in patients with heart failure, and in some conditions inhibits reninangiotensin-aldosterone system activation (Lieverse et al. 1995, Van Veldhuisen et al. 1995).

Dopamine is widely used in acute renal failure, in spite of the publication of several reviews advising caution (Thompson and Cockrill 1994). Amantadine was developed as a prophylaxis and treatment for influenza and serendipitously was found to be useful for treatment of Parkinson's disease. Among its actions, it is considered to enhance dopaminergic activity. In one unblinded, uncontrolled study, amantadine in doses ranging from 100–300 mg/day (taken 1 to 3 times a day as needed) was evaluated as an add-on treatment for 21 adult RLS patients who were not adequately treated by their current medications (Evidente et al. 2000).

GO-TiO_2/MWCNT-TiO_2 nanocomposites and applications

A two-step, direct synthesis of TiO_2 nanocrystals on graphene oxide (GO) and advanced photocatalytic properties of the resulting hybrid material has been reported. Due to low cost, high stability and efficient photo-activity, TiO_2 has been widely used for photoelectrochemical and photocatalytic applications (Chen and Mao 2007). Several graphene/TiO_2 composites have been reported recently for lithium ion battery (Wang et al. 2009a), photocatalysis (Zhang et al. 2010) and dye sensitized solar cells (Yang et al. 2010), using surfactant assisted growth. Inorganic nanoparticles such as Ag, Au, TiO_2, ZnO, and carbon nanotubes (CNTs) exhibited antibacterial activity towards several multidrug resistant bacteria (Kumar et al. 2008). Another carbon nanomaterial, graphene was found to be biocompatible for growth, adhesion, and proliferation of many cells such as L-929 cells (Chen et al. 2008), neuron cells, and osteoblasts (Agarwal et al. 2010), and graphene oxide was used as an effective nanocargo to deliver water-insoluble drugs into cells (Liu et al. 2008, Sun et al. 2008).

Experimental procedure

Materials and reagents

Gold electrode (Au, 2 mm diameter) was obtained from CH instrument (USA). Graphite powder (99.9%, 325 mesh, Alfa aesar), sulfuric acid (H_2SO_4), potassium permanganate ($KMnO_4$), hydrogen peroxide (H_2O_2), hydrochloric acid (HCl), sodium nitrate ($NaNO_3$), potassium ferricyanide ($K_3Fe(CN)_6$), potassium phosphate monobasic (KH_2PO_4), sodium dihydrogen phosphate ($NaH_2PO_4.2H_2O$) titanium dioxide (TiO_2) precursor powder, sodium hydroxide (NaOH) were of analytical grade and obtained from Sigma-Aldrich chemicals, Merck chemicals and LabChem. Phosphate buffer solution (PBS, pH 7.0) was prepared with appropriate amounts of $NaH_2PO_4.2H_2O$ and $Na_2HPO_4.2H_2O$ and the pH monitored with already calibrated pH meter.

Apparatus and equipment

Petri dishes, conical flasks, beakers, volumetric flasks, measuring cylinder, Buchner funnel and Buchner flask used were washed in detergent solution and rinsed several times with distilled water. Other apparatus and equipment used included oven, magnetic stirrer, magnetic bar, Fourier transformed infrared spectrometer (Agilent Technology, Cary 600 series FTIR spectrometer, USA) and UV-visible spectrophotometer (Agilent Technology, Cary series UV-vis spectrometer, USA). Electrochemical experiments were carried out using an Autolab Potentiostat PGSTAT (Eco Chemie, Utrecht, and The Netherlands) driven by the GPES software version 4.9. Electrochemical impedance spectroscopy (EIS) measurements were performed with Autolab Frequency Response Analyser (FRA) software between 10 kHz and 1 Hz using a 5 mV rms sinusoidal modulation with the solution of the analyte at their respective peak potential of oxidation (vs. Ag|AgCl in sat'd KCl). A Ag|AgCl in saturated KCl and platinum wire were used as reference and counter electrodes, respectively. All experiments were performed at 25 ± 1°C while the solutions were de-aerated before every electrochemical experiment.

Synthesis of GO

5 g of graphite flakes and 2.5 g of $NaNO_3$ were mixed with 108 mL H_2SO_4 and H_3PO_4 and stirred in an ice bath for 10 minutes. Next, 15 g of $KMNO_4$ were slowly added so that temperature of the mixture remained below 5°C. The suspension was then reacted for 2 hour in an ice bath and stirred for 60 minutes before being stirred in a 40°C water bath for 60 minutes. The temperature of the mixture was adjusted to a constant 98°C for 60 minutes while water was added continuously. Deionized water was further added so that the volume of the suspension was 400 mL. 15 mL of H_2O_2 was added after 5 minutes. The reaction product was washed with deionized water and 5% HCl solution repeatedly. Finally, the product was dried at 60°C (Zhu et al. 2010b, Guo and Dong 2011, Rao et al. 2011, Perera et al. 2012).

Synthesis of functionalized multiwall carbon nanotube

An approximately 100 mg of pristine MWCNTs was first treated with a (v/v 3:1) mixture of concentrated H_2SO_4 and HNO_3 acid (120 mL). This mixture was then sonicated for 3 h at 40°C in an ultrasonic bath to introduce carboxylic acid groups on the MWCNT surface. After cooling to room temperature, the carboxylated MWCNTs (MWCNT–COOH) were added dropwise to 300 mL of cold deionised water and then vacuum-filtered through a 0.05 mm pore size PTFE filter paper; the filtrant was washed with deionised water until pH was neutral. The sample was then dried in a vacuum oven at 80°C for 8 h (Marcano et al. 2010). The obtained MWCNT-COOH product is simply represented as MWCNT in this report.

Synthesis of TiO_2 nanoparticles

2 g of the TiO_2 precursor powder (bulk) was dispersed in 10 mL NaOH (100 mL) and was subjected to hydrothermal treatment at 150°C for 24 hours in furnace. When the reaction had completed, a white solid was obtained and collected. The white solid was washed with 0.1 M HCl, followed by distilled water. The white solid was separated and collected from solution and subsequently dried at 80°C for 24 hours. After drying, the obtained powder was heated at 500°C in static air for 2 hours to produce TiO_2 nanoparticles (Armarego and Chai 2003).

Preparation of GO-TiO_2 and MWCNT-TiO_2 nanocomposite

About 10 mg of GO or CNT and 10 mg of TiO_2 nanoparticles were dissolved in 5 mL DMF. The mixture was stirred with magnetic stirrer overnight. Then it was filtered and dried in the oven at 50°C to obtain the GO-TiO_2 or MWCNT-TiO_2 nanocomposite.

Electrode modification procedure

Gold electrode surface was cleaned by gentle polishing in aqueous slurry of alumina nanopowder (LabChem) on a SiC-emery paper. The electrode was then subjected to ultrasonic vibration for 5 minutes in distilled water, and then in absolute ethanol to remove residual alumina particles that might be trapped on the surface. Three solutions (GO or MWCNT, TiO_2, GO or MWCNT-TiO_2) were prepared and in each solution 5 mg of each synthesized samples (GO, TiO_2, GO-TiO_2) were dissolved in 0.5 mL DMF and ultrasonicated for 15 minutes. Au-GO, Au-MWCNT, Au-TiO_2, Au-MWCNT-TiO_2 and Au-GO-TiO_2 were prepared by a drop-dry method. About 5 μL drops of the GO (or MWCNT), TiO_2-GO or TiO_2-MWCNT and TiO_2 solutions were dropped on the bare Au electrode and dried in an oven at 50°C for 5 minutes.

Characterization of synthesized nanocomposites

Successful synthesis of the nano-materials and the nanocomposite was confirmed using techniques such as transmission electron microscopy (TEM), Fourier transformed infrared spectroscopy (FTIR), UV-visible spectroscopy, Raman spectroscopy and x-ray diffraction spectroscopy (XRD); the results are presented in Chapter four. Successful modification of electrode was confirmed using cyclic voltammetry (CV) and electrochemical impedance spectroscopy (EIS). Electrochemical characterisation was carried out in both 5 mM Ferri/Ferro ($[Fe(CN)_6]^{3-/4-}$) redox probe and 0.1 M pH 7.0 phosphate buffer solution (PBS) to study the electron transport properties of the modified electrodes. Results obtained are presented in Chapter four. Electrocatalytic oxidation of 10^{-4} M dopamine (in pH 7.0 PBS) solution was carried out using both the bare Au, Au-GO, Au-MWCNT, Au-TiO_2, Au-MWCNT-TiO_2 and Au-GO-TiO_2 as the working electrode, a platinum wire as the auxiliary electrode, and an Ag/AgCl sat'd KCl as the reference electrode, respectively. All experiments were performed at 25 ± 1°C while the solutions were de-aerated before every electrochemical experiment.

Results and discussion

Characterization of synthesized compounds

Characterization of synthesized GO, TiO_2 and GO-TiO_2 with UV-VIS spectroscopy

The UV-vis spectra of GO, TiO_2 precursor and GO-TiO_2 nanocomposite are represented in Figure 1a. Graphene oxide shows a typical characteristic absorbance peak at 230 nm, attributed to p-p* transitions of C-C in amorphous carbon systems, and broad absorption in the visible region (Razali et al. 2013). A shoulder peak weakly appears at 300 nm due to n-π* transitions of aromatic C-C bonds (Eda et al. 2010). Figure 1b shows a spectrum of titanium dioxide with two characteristic absorption peaks at around 220 and 290 nm. The first peak corresponds to the direct recombination between electrons in the conduction band and holes in the valence band of TiO_2 (Xu and Wu 2013). Figure 1c is the GO-TiO_2 spectrum with broad peak around 227 nm which represents the partial restoration of the pi-conjugation network of the sample as a result of the hydrothermal and chemical reduction of the titanium dioxide and the graphene oxide chemical process. The disappearance of TiO_2 peaks at around 220 and 290 nm, and appearance of a new peak at 227 nm, confirms successful formation of the GO-TiO_2 nanocomposite. In the spectrum of MWCNT-TiO_2 composite, the presence of the TiO_2 band is at 260 nm and the broad spectrum of MWCNT is between 280 and 300 nm.

Characterization of synthesized compounds with Raman spectroscopy

Figure 2a shows a Raman spectrum of GO-TiO_2 nanocomposite and it appears that there is reduction of Ti^{4+} to Ti^{3+}. However, if there is a reduction of Ti^{4+}, somewhere oxidation occurred, and it is probably on the graphene skeleton in free positions of Π bonds. Following oxidation of graphene due to reduction of Ti^{4+} corresponds to an increase of intensity of Raman bands' GO in the nano-composite GO-TiO_2. The specific vibration modes are located at 144 cm^{-1} (Eg), 396 cm^{-1} (B1g), 512 cm^{-1} (B1g + A1g) and 631 cm^{-1} (Eg) indicating the presence of the anatase phase in all of these samples (Liqiang et al. 2004). The Raman spectrum of MWCNT-TiO_2 in Figure 2b shows well defined D band peak at 1353 cm^{-1}, due to the sp^3 defects and another peak of G band at 1597 cm^{-1}, which can be ascribed to the in plane vibrations of sp^2 carbon atoms and a doubly degenerated phonon mode (E2g symmetry) at the Brillouin zone centre (Ookubo et al. 1990).

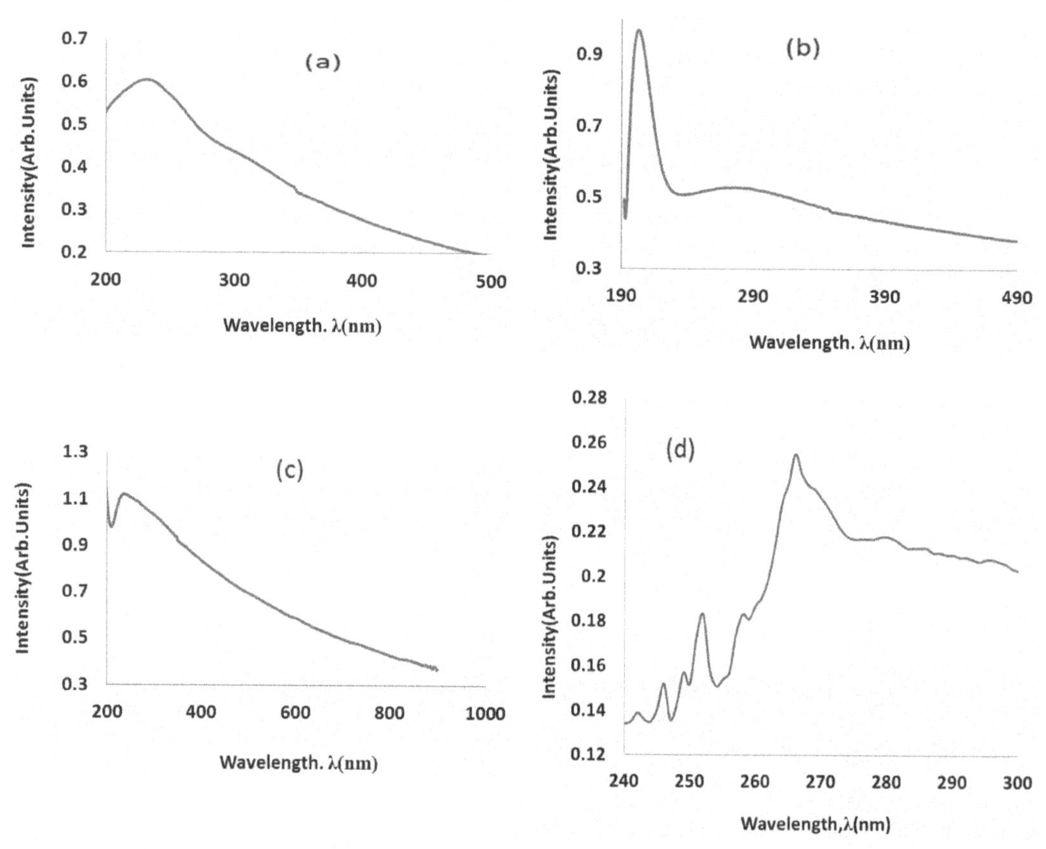

Figure 1. UV-vis spectra of (a) GO (b) TiO_2 (c) GO-TiO_2 nanocomposite and (d) MWCNT-TiO_2 nanocomposite.

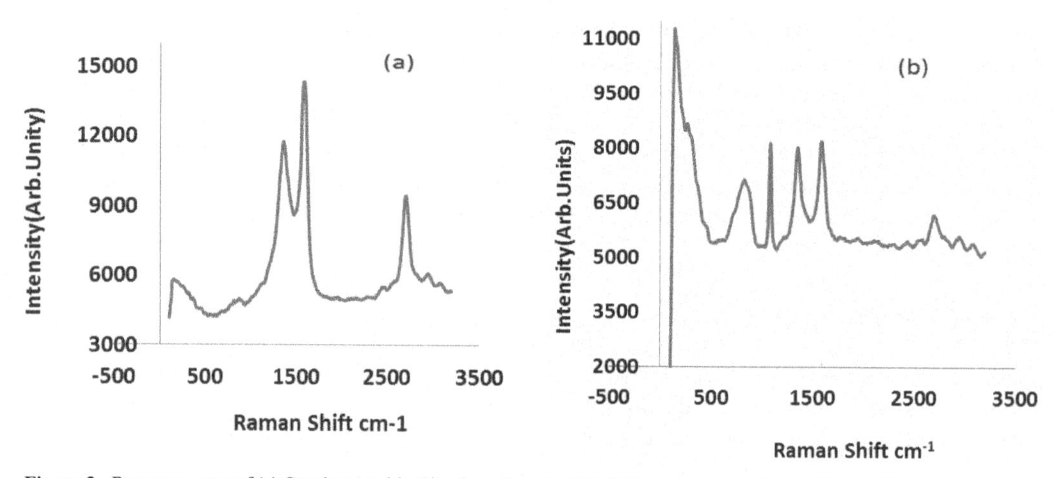

Figure 2. Raman spectra of (a) Graphene oxide-Titanium dioxide (GO-TiO_2) nanocomposite (b) Multiwall carbon nanotube-Titanium dioxide (MWCNT-TiO_2) nanocomposite.

Characterisation of synthesized TiO_2 and GO-TiO_2 with FTIR

The FTIR spectrum of TiO_2 in Figure 3a shows a broad band at 3400 cm^{-1} and the band at 1425 cm^{-1} originates from the surface-adsorbed water, and this indicates the presence of −OH groups on the surface

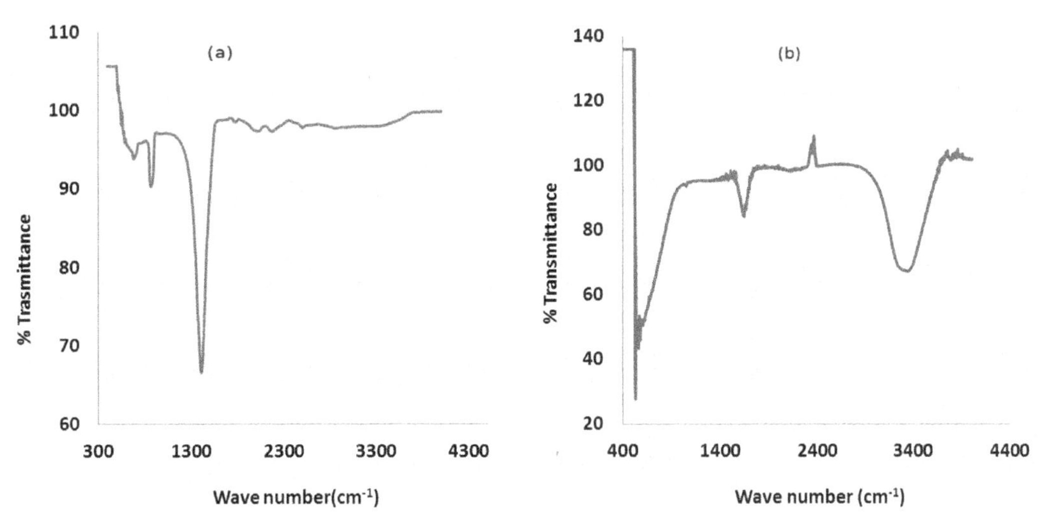

Figure 3. FTIR spectra of (a) titanium dioxide and (b) TiO_2-GO nanocomposite.

of titania (Naumenko et al. 2012). It can be seen from Figure 3b that most of these oxygen functionalities had been removed in the synthesized TiO_2-GO composite. A new band at 1580 cm^{-1}, which may be attributed to the skeletal vibrations of the graphene sheets (Nethravathi and Rajamathi 2008) and a strong band at 500 cm^{-1}, which is due to the Ti–O–Ti vibration, confirms the presence of both graphene oxide and TiO_2 in the synthesized composite. The band at 2900–3400 cm^{-1} of TiO_2-GO is broader, thus indicating the presence of hydrogen bonding in the synthesized composite (Wang et al. 2007a).

Electrochemical characterisation

Comparative cyclic voltammetric evolutions of modified electrodes in 5 mM $[Fe(CN)6]^{4-}/[Fe(CN)6]^{3-}$ and in pH 7.0 PBS (scan rate = 25 mVs^{-1}) are shown in Figures 4 and 5 for GO and MWCNT nanoporous based supporting materials, respectively. The purpose of this study was to determine the electron transport properties of the modified electrodes. In these Figures, the redox peaks around 0.2 V are attributed to the $[Fe(CN)_6]^{3-/4-}$ redox process while the oxidation peak around 1.0 V is attributed to Au in form of Au^{3+} or Au^{4+}. The cyclic voltammograms obtained for the Au modified electrodes in $[Fe(CN)_6]^{3-/4-}$ and PBS electrolyte solution, using GO as supporting nanoporous materials are shown in Figure 4. In 5 mM $[Fe(CN)_6]^{3-/4-}$ redox probe, it was very clear that there was no significant difference in the current response of Au-GO-TiO_2 and the bare Au. In fact, presence of GO on Au electrode (Au-GO) leads to drastic decrease in Au current response probably because of the less conductive nature of the GO nanoporous materials (Liu et al. 2009). However, an increased current response was observed in 0.1 M pH 7.0 PBS for Au-GO-TiO_2 as compared with other electrodes studied. This result is expected for chemically modified electrodes where addition of electroactive materials like metal oxides (MO) increases the conductivity and current response of the bare electrode (Musameh et al. 2005, Adekunle and Ozoemena 2010). It is evident that the employment of the nanomaterial in the biosensor construction significantly enhances the analytical response of the resulting device (Zhao et al. 2007).

However, a significant increase in current response was observed in 0.1 M pH 7.0 PBS and 5 mM Ferri/Ferro ($[Fe(CN)_6]^{3-/4-}$) redox probe when functionalised MWCNT was used as the supporting nanoporous support (Figure 5). The observed cyclic voltammograms were similar to those reported earlier (Liu et al. 2009) for MWCNT modified Au. Remarkably, current enhancement was observed when TiO_2 nanoparticles were coupled with MWCNT as in Au-MWCNT-TiO_2 composite modified Au as compared with that of MWCNT modified Au alone. The redox couple of $[Fe(CN)_6]^{3-}/[Fe(CN)_6]^{4-}$ appears quite reversible at the Au-MWCNT-TiO_2 electrode, with an increase in the peak currents and a slight decrease in the peak potential separation (ΔEp) for the $[Fe(CN)_6]^{3-/4-}$ couple compared to those of

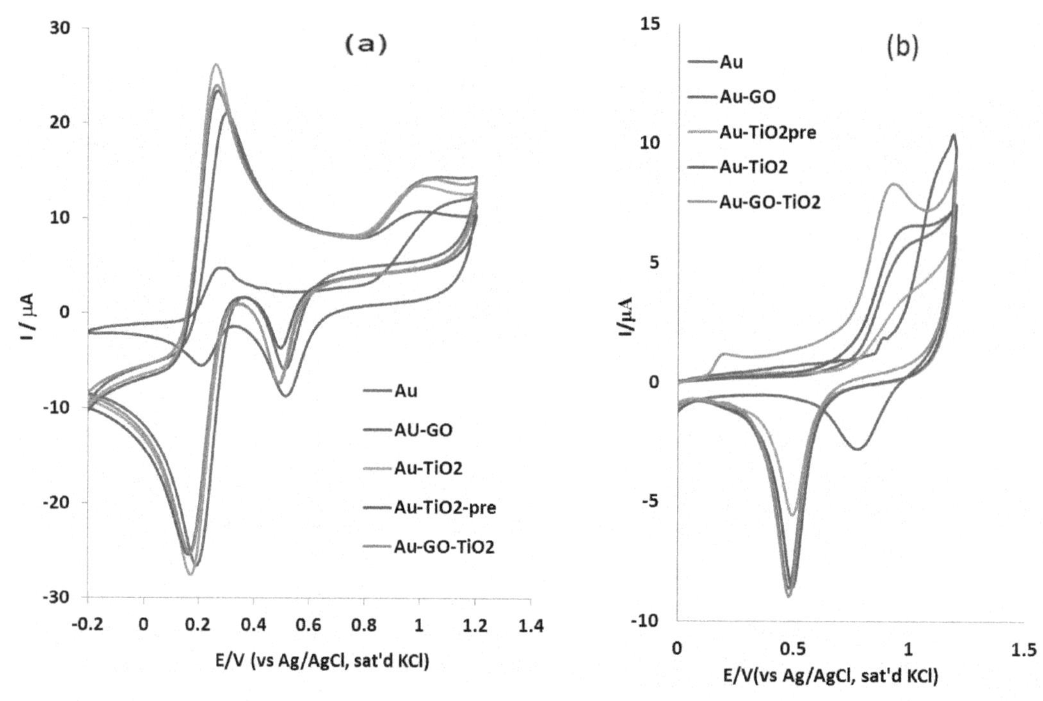

Figure 4. Comparative cyclic voltammetric evolutions of GO nanoporous based modified electrodes in (a) 5 mM $[Fe(CN)6]^{4-}/[Fe(CN)6]^{3-}$ (b) in 0.1 M pH 7.0 PBS (scan rate = 25 mVs^{-1}).

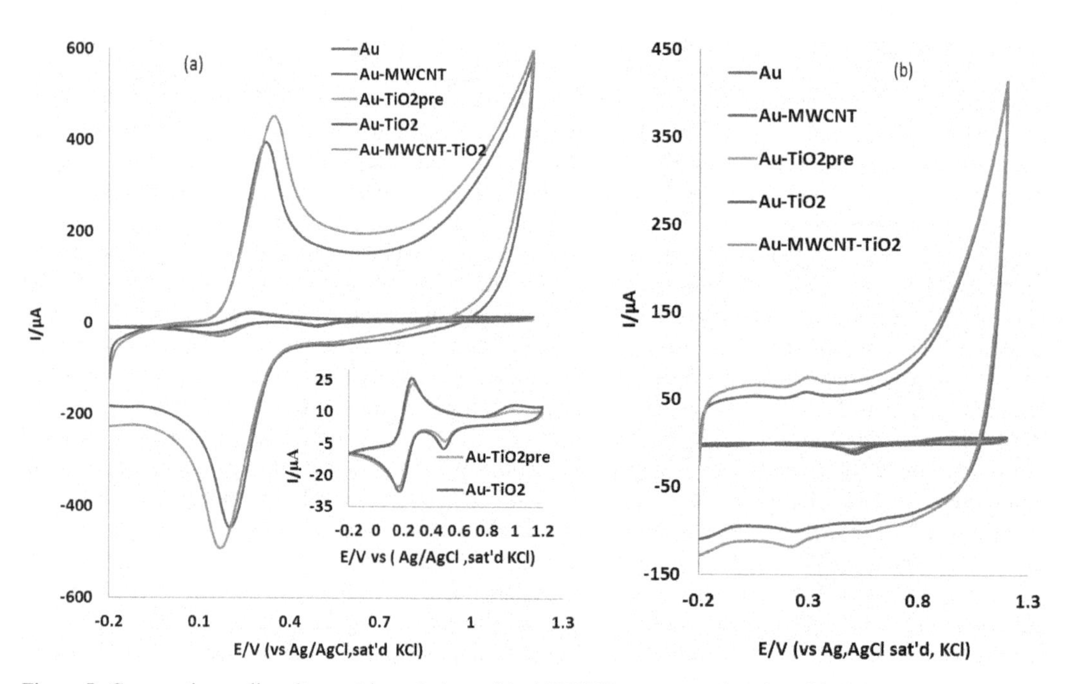

Figure 5. Comparative cyclic voltammetric evolutions of the MWCNT nanoporous based modified electrodes in (a) 5 mM $[Fe(CN)_6]^{4-}/[Fe(CN)_6]^{3-}$ (b) in 0.1 M pH 7.0 PBS (scan rate 25 mVs^{-1}).

the bare Au electrode (Figure 5a). This result suggests that MWCNT and TiO_2 nanoparticles act as an electron transfer medium and enhance electron transfer of bare Au during an electrochemical reaction. The introduction of TiO_2 nanoparticles on the Au electrode surface facilitates the conduction pathway at the modified electrode surface (Kaniyoor et al. 2012). Similar current response was obtained for the Au-MWCNT-TiO_2 electrode in PBS (Figure 5b). The improved electrochemical response of Au-MWCNT-TiO_2 electrode compared with Au-GO-TiO_2 electrode in this study could be attributed to the electrical conductive nature of MWCNT, as has already been reported in literature (Musameh et al. 2005, Adekunle and Ozoemena 2010).

Electrocatalytic oxidation of dopamine

Figure 6 presents the cyclic voltammograms showing the electrocatalytic behaviour of bare Au and Au modified electrodes towards dopamine (DA) oxidation in 0.1 M PBS solution (pH = 7) at a scan rate of 25 mVs^{-1}. A significant DA oxidation current was recorded at Au-MWCNT-TiO_2 modified electrode compare with bare Au and other electrodes studied, suggesting that the Au-MWCNT-TiO_2 modified electrode possessed faster electron transfer kinetics (Wang et al. 2009). After background current subtraction (Figure 6b), DA oxidation current at the electrodes follows the trend: Au-MWCNT-TiO_2 (0.12 mA) > Au-MWCNT (0.25 mA) > Au-TiO_2 (0.5 μA) > Au-TiO_2 pre (0.65 μA) > Au (0.64 μA). Factors such as presence of porous MWCNT, increase electrode surface area and the electrical conductive nature of the MWCNT nanoparticles could be responsible for the improved response of DA oxidation peak around (0.2 V) at Au-MWCNT-TiO_2 electrode. The result obtained in this study agreed with similar observation for DA oxidation on gold nanoparticles choline chloride glassy carbon modified electrode (nano-Au/Ch/GCE), which was reported to be better compared to the bare gold, or nano-Au/GCE or Ch/GCE alone (Hou et al. 2010). MWCNT is a conductive material, and TiO_2 nanoparticles can promote electron exchange. The composited MWCNT and TiO_2 nanoparticles offer synergistic effects in electric catalytic applications (Zhang et al. 2004). On the other hand and contrary to expectation, there is no

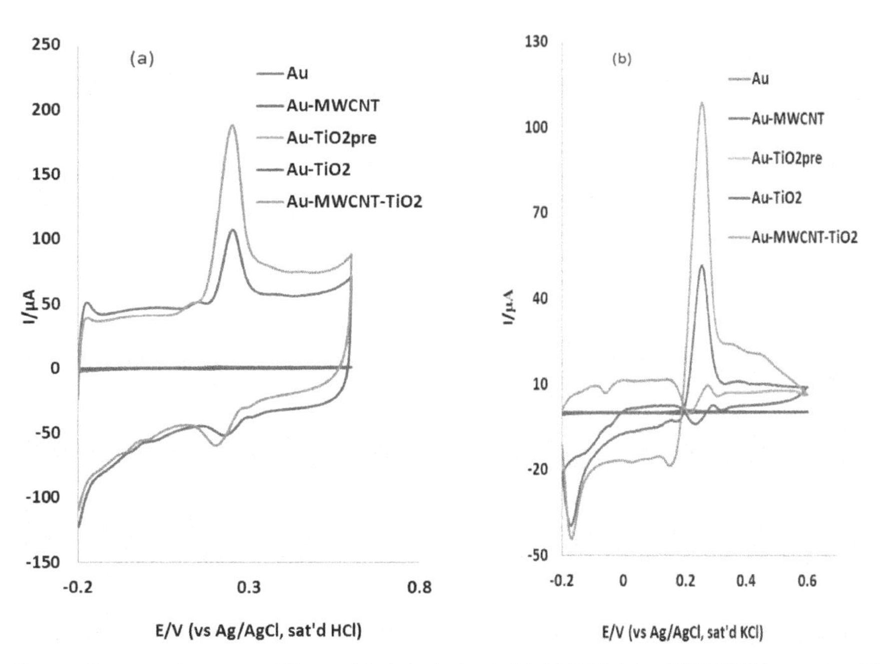

Figure 6. Comparative current response of Au modified electrodes in (a) 10^{-4} M DA in pH 7.0 PBS (scan rate = 25 mVs^{-1}) using CNT support, (b) 10^{-4} M DA in pH 7.0 PBS (scan rate = 25 mVs^{-1}) using CNT support (after background current subtraction).

significant difference in DA oxidation current on bare Au electrode compared with Au-TiO_2 modified electrodes with and without GO nanoporous support (Figure 5c).

Electrochemical impedance studies

Electrochemical impedance spectroscopic (EIS) measurements were performed to investigate the electrochemical behaviour (electron transfer properties) of the different modified electrodes in the presence of dopamine. Figure 7 shows Nyquist plots obtained for the various modified electrodes. Among these, the bare Au, Au-TiO_2pre, Au-TiO_2, Au-GO and Au-GO-TiO_2 showed large semicircles (Figure 7a), which were due to the large charge-transfer resistance (R_{ct}) at the electrode/electrolyte interface due to the sluggish electron transfer kinetics (Wen et al. 2013) during dopamine oxidation.

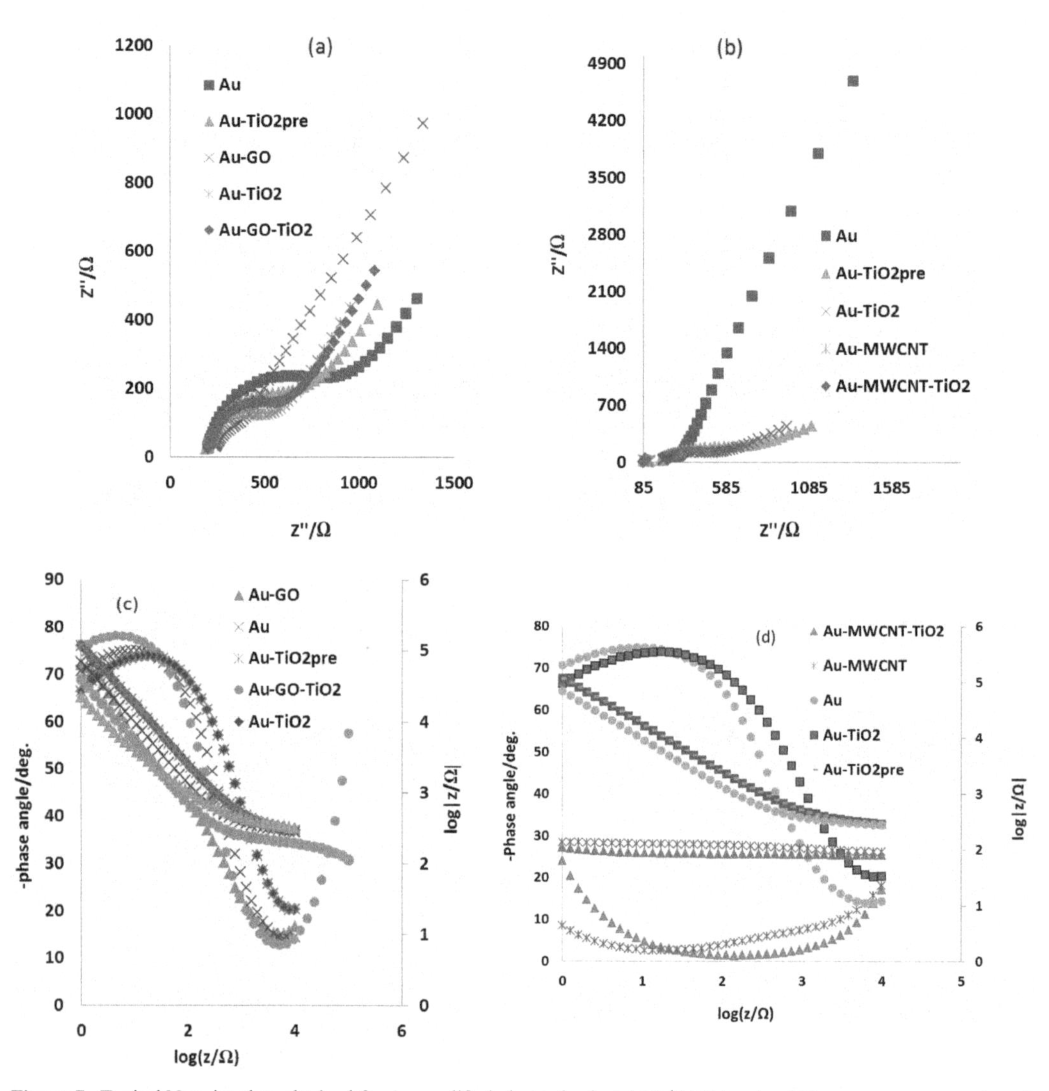

Figure 7. Typical Nyquist plots obtained for Au modified electrodes in (a) 10^{-4} M DA using GO nanoporous support, and (b) 10^{-4} M DA solutions at a fixed potential of 0.8 V (vs Ag|AgCl, sat'd KCl). (c) and (d) are the Bode plots obtained for Au modified electrode in DA using GO and CNT nanoporous support, respectively, showing the plots of -phase angle/deg. vs log (f/Hz), and the plot of log |Z/Ω| vs log (f/Hz).

The TiO_2-CNT nanocomposite on the other hand shows much lower impedance or charge transfer resistance (smaller semicircle) in the lower frequency region compared to that of other electrodes. Therefore, Au-MWCNT-TiO_2 nanohybrids demonstrated lower charge-transfer resistance, signifying remarkable enhancement of the catalytic reaction and electron transfer (Xu et al. 2005); hence, higher DA oxidation current was recorded for its CV in Figure 6b above.

From the Bode plots of -phase angle vs. log (f/Hz) (Figures 7c and d), all the electrodes including the bare Au showed a phase angle lower than –90° expected for a pure capacitive behaviour, therefore suggesting the pseudocapacitive behaviour of the electrodes towards DA electrocatalytic oxidation. However, Au-MWCNT-TiO_2 demonstrated very low phase angle (< –30°) indicating faster electron transport and higher DA oxidation current, as already reported above for the electrode. The lower capacitance, improved electron transport and catalytic behaviour of this electrode can be attributed to the conductive nature of the MWCNT and TiO_2 nanoparticles.

Since Au-MWCNT-TiO_2 electrode demonstrated excellent catalytic properties towards DA oxidation in this study, further studies in this work were carried out using this electrode, unless otherwise stated.

Stability study

The stability of the Au-MWCNT-TiO_2 modified electrode towards dopamine fouling effects was studied by running the electrode (30 scans) in 10^{-4} M DA solution at a scan rate of 25 mVs^{-1}. From the CV present in Figure 8, a current drop of *ca* 30% was observed suggesting the stability of the electrode towards DA oxidation poisoning effect. The result also suggests some levels of adsorption of the analyte, or their oxidation intermediates at the electrode surface. However, after repetitive cycling the electrode in PBS 7.0, a current recovery of > 90% was obtained suggesting that the adsorption procuress is physiosorption, and that the electrode can be reused. The adsorptive nature of the electrodes towards the analyte can be attributed to the porous MWCNT in the composite (Zhang et al. 2004).

Effect of scan rate

The effect of the scan rate on the DA oxidation was studied for the Au-MWCNT-TiO_2 modified electrode by varying the scan rate from 25–1000 mVs^{-1} in 0.1 M PBS (pH = 7) containing 10^{-4} M DA solution. Increasing the scan rate increased the peak separation (ΔE) because of the chemical interaction between the DA and the modified electrode (Figure 9). Simultaneous increases in the anodic and cathodic current intensities were also observed when increasing the scan rate. A plot of the anodic peak current (I_{pa}) versus the square root of the scan rate (not shown) was linear ($R^2 = 0.9931$), which indicates a diffusion controlled redox process (Xu et al. 2005, Wang et al. 2007b, Jo et al. 2008, Wen et al. 2013).

Concentration study

Linear sweep voltammetry

The effect of varying concentrations of DA on DA oxidation current at Au-MWCNT-TiO_2 electrode was carried out using linear sweep voltammetry (LSV) technique. The linear sweep voltammograms Figure 10a were obtained after stirring the mixture thoroughly. Figure 10b represents the calibration curve for the plot of peak current (I_p) versus DA concentration. The measured peak currents were found to be linear with increasing concentrations. The detection limit was calculated based on the relationship LoD = 3.3 δ/m (Christian 2004) where δ is the relative standard deviation of the intercept of the y-coordinates from the line of best fit, and m the slope of the same line. The limit of detection LoD was estimated to be 6.4 μM. The detection limit is comparable with other results obtained in literature using other electrodes (Wang et al. 2012b, Huang et al. 2013, Wang et al. 2013b). This low limit of detection and high sensitivity of Au-MWCNT-TiO_2 electrode will enable its numerous applications in sensor or biosensor systems.

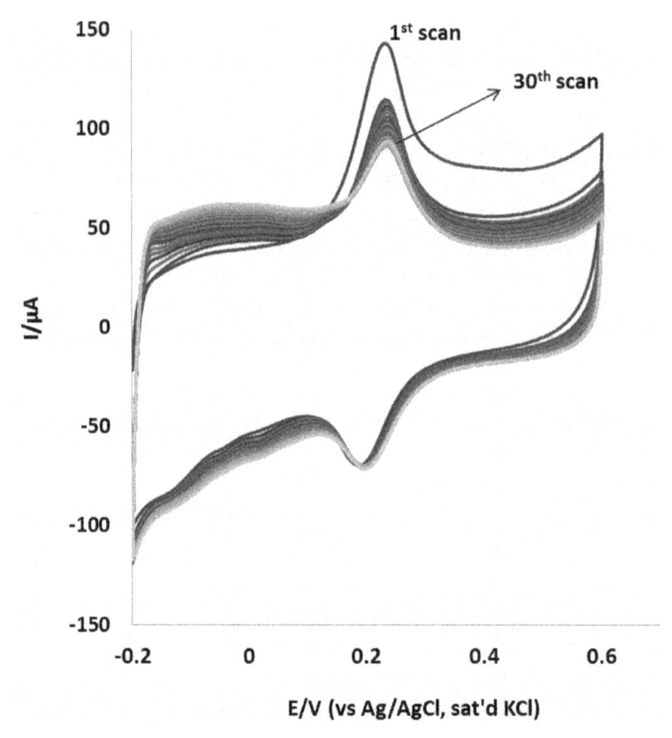

Figure 8. Current response (30 scans) of Au-MWCNT-TiO_2 electrode in (a) 0.1 M (pH 7.0) PBS containing 10^{-4} M DA (Scan rate: 25 mVs^{-1}).

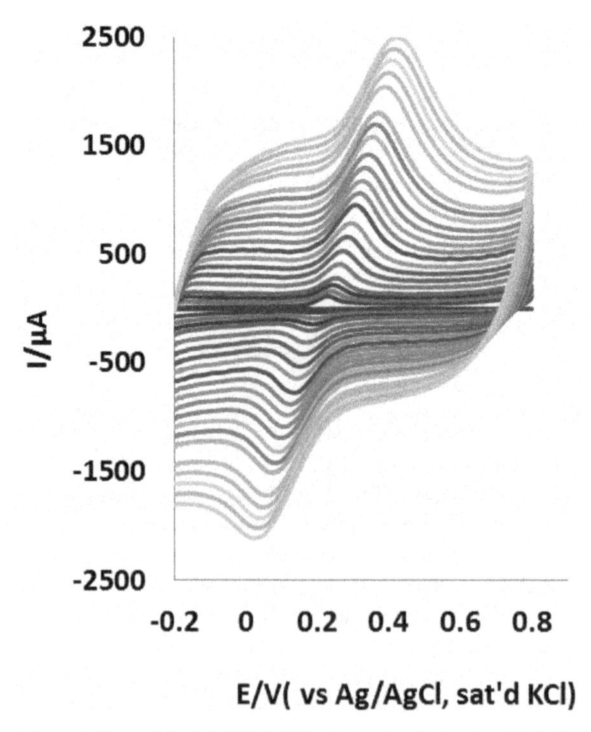

Figure 9. Cyclic voltammetric evolutions of Au-MWCNT-TiO_2 electrode obtained in 0.1 M PBS (pH 7.0) containing 10^{-4} M DA (scan rate range 25–1000 mVs^{-1}; inner to outer).

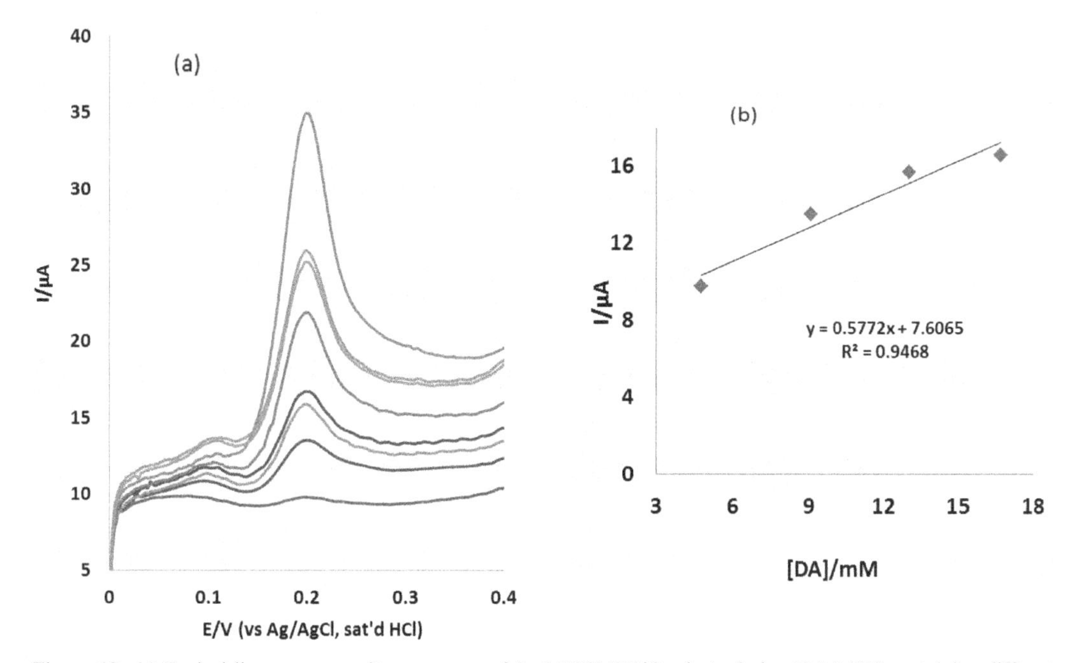

Figure 10. (a) Typical linear sweep voltammograms of Au-MWCNT-TiO_2 electrode in pH 7.0 PBS containing different concentrations of DA (0.0, 4.76, 9.09, 13.0, 16.7, 20.0, 23.1, 25.9 and 28.6 μM). (b) Plot of current (I_p) versus dopamine concentration [DA].

Chronoamperometric detection of dopamine

The chronoamperometric technique was also used to determine the applicability of the Au-MWCNT-TiO_2 modified electrode for DA sensing. Figure 11 shows the chronoamperogram using different DA concentrations. A calibration curve ($R^2 = 0.9914$) showing linear relationship between dopamine concentration and the current response was obtained. The limit of detection (LoD) was calculated using same relationship (LoD = $3\sigma/m$) as 3.62 μM. The LoD value is lower compared to the value obtained using LSV. The difference can be attributed to sensitivity and the different experimental conditions adapted for the two techniques.

Selective determination of DA in the presence of AA

To evaluate the selectivity of electrochemical sensor, the influence of interfering species, 10^{-1} M ascorbic acid (AA) was examined in 0.1 M PBS containing 10^{-4} M DA. During the simultaneous detection of DA and AA at the bare Au (not shown), the oxidation peak potentials of AA and DA are obviously overlapping and indefinable due to poor selectivity. AA directly interferes with DA, so the bare Au fails to determine the individual electrochemical redox peaks for DA. Figure 12 shows the cyclic voltammogram, the modification of the Au with the MWCNT-TiO_2 nanocomposite, and the good separation potential between the AA and DA at about 0.2 V (Wu et al. 2012). Hence, the good selectivity of the modified Au can be successfully achieved which can be attributed to the different charge properties of DA and AA (positively-charge DA, negatively-charged AA) occurring at the MWCNT-TiO_2 surface (Wang et al. 2012a). The result suggests that the Au-MWCNT-TiO_2 has excellent electrocatalytic activity towards the oxidation of DA than AA and exerts no interference in the selective determination of DA. Similar report on modified electrode performance towards DA detection without interference from AA has been reported. MWCNT-TiO_2 nanocomposite could potentially be used for electrochemical sensing applications (Wu et al. 2012).

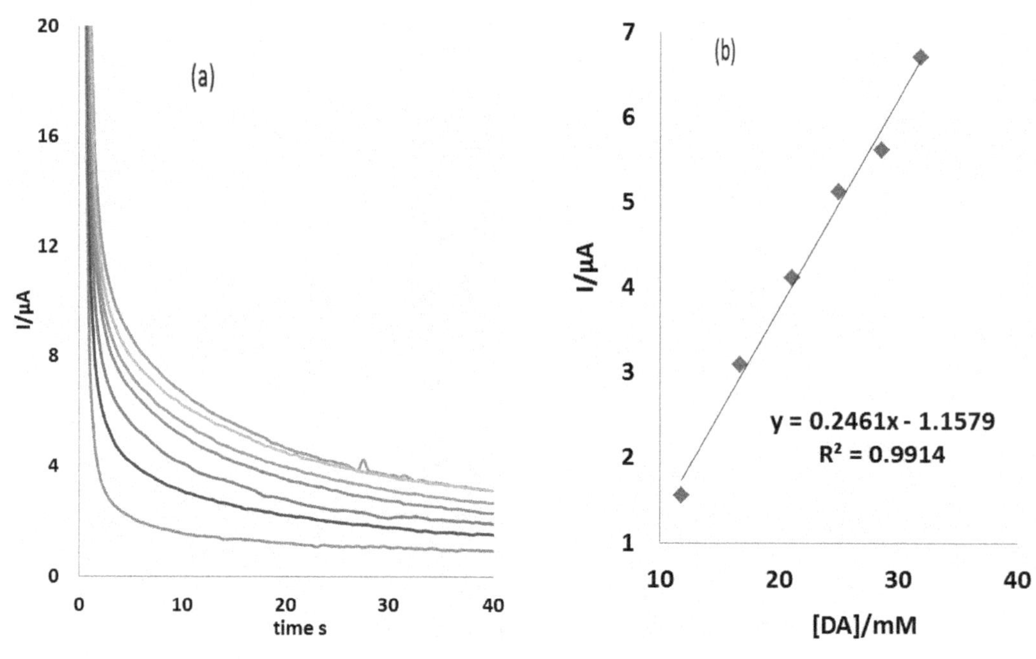

Figure 11. (a) Typical chronoamperograms of Au-MWCNT-TiO_2 electrode in pH 7.0 PBS containing different concentrations of DA (0.0, 11.8, 16.7, 21.1, 25.0, 28.6 31.8 and 34.8 µM). (b) Plot of current (I_p) versus dopamine concentration [DA].

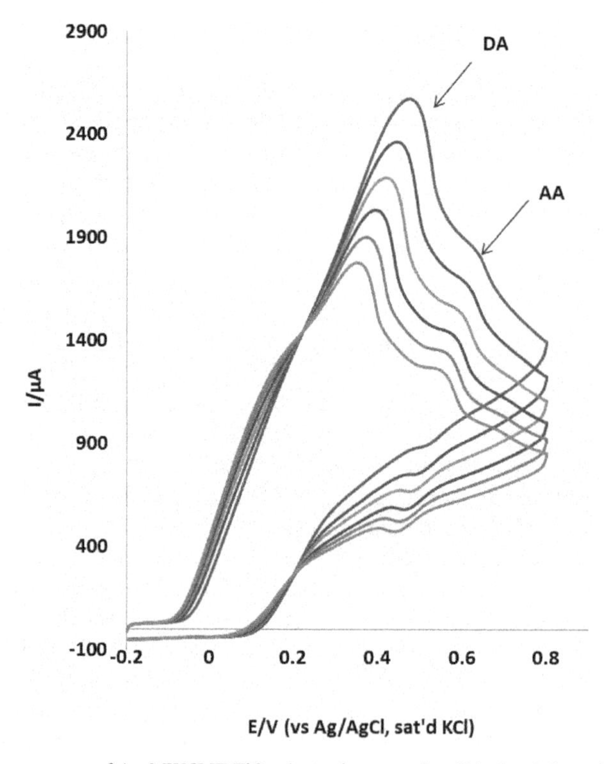

Figure 12. Cyclic voltammograms of Au-MWCNT-TiO_2 electrode separating DA signal from AA signal at different DA concentration.

Conclusion

The study has described successful modification of Au electrode with GO, CNT and TiO_2 nanoparticles. Au-GO-TiO_2 and Au-MWCNT-TiO_2 nanocomposite modified electrode were also developed and described using different spectroscopic techniques. Au-MWCNT-TiO_2 electrode demonstrated faster electron transport and better catalytic behaviour towards dopamine (DA) oxidation at pH 7.0 compared with other electrodes studied. Au-GO-TiO_2 performed poorly towards the analyte probably because of the less conductive nature of GO compared with the functionalised MWCNT which has assumed metallic behaviour, and hence is very conductive and catalytic. The result suggests that MWCNT would have better electrochemical properties when used as electrode nanoporous support compared with GO. Electrocatalytic oxidation of dopamine on the Au-MWCNT-TiO_2 electrode is diffusion controlled. The electrode was also characterised with some degree of adsorption attributed to DA oxidation intermediates products. The electrode showed good resistance to electrode poisoning, and gave μM detection limit that agreed favourably with values previously reported in literature. The electrode can conveniently detect dopamine DA without interferences from ascorbic acid (AA).

Acknowledgements

This project was supported by the North-West University (Mafikeng Campus), Material Science Innovation & Modelling (MaSIM) Focus Area, Faculty of Agriculture, Science and Technology, North-West University (Mafikeng Campus). ASA and LOO thank the North-West University for post-doctoral fellowship and Obafemi Awolowo University, Nigeria for the research leave visit to NWU. EEE acknowledges the National Research Foundation of South Africa for Incentive funding for Rated Researchers.

References

Adekunle, A.S. and K.I. Ozoemena. 2010. Voltammetric and impedimetric properties of nano-scaled γ-Fe_2O_3 catalysts supported on multi-walled carbon nanotubes: catalytic detection of dopamine. Int. J. Electrochem. Sci. 5: 1726–1742.

Agarwal, S., X. Zhou, F. Ye, Q. He, G.C. Chen, J. Soo et al. 2010. Interfacing live cells with nanocarbon substrates. Langmuir 26: 2244–2247.

Agid, Y., F. Javoy-Agid and M. Rubberg. 1987. Biochemistry of neurotransmitters in Parkinson's disease. pp. 166–230. *In*: C.D. Marsden and S. Fahn (eds.). Movement Disorders. Neurology London: Butterworths.

Aizpurua, J., P. Hanarp, S.K. Sutherland, M. K¨all, G.W. Bryant and F.J.G. de Abajo. 2003. Optical properties of gold nanorings. Phys. Rev. Lett. 90: 057401–057404.

Aliev, A.E., J. Oh, M.E. Kozlov, A.A. Kuznetsov, S. Fang, A.F. Fonseca et al. 2009. Giant-stroke, superelastic carbon nanotube aerogel muscles. Science 323: 1575–1578.

Ardakani, M.M., A. Talebi, H. Naeimi, M.N. Barzoky and N. Taghavinia. 2009. Fabrication of modified TiO_2 nanoparticle carbon paste electrode for simultaneous determination of dopamine, uric acid, and l-cysteine. J. Solid State Electrochem. 13: 1433–1440.

Armarego, W.L.F. and C.L.L. Chai. 2003. Purification of Laboratory Chemicals. Elsevier Science Burlington, MA.

Atkinson, T. 2008. Central and peripheral neuroendocrine peptides and signalling in appetite regulation: considerations for obesity pharmacotherapy. Obes. Rev. 9: 108–120.

Bahnemann, D.W. 1991. Mechanisms of organic transformations on semiconductor particles. pp. 251–276. *In*: E. Pellizzetti and M. Schiavello (eds.). Photochemical Conversion and Storage of Solar Energy. Kluwer Academic Publishers, The Netherlands.

Berger, C., Z. Song, X. Li, X. Wu, N. Brown, C.C. Naud et al. 2006. Electronic confinement and coherence in patterned epitaxial graphene. Science 312: 1191–1196.

Berry, R.J. and M.R. Mueller. 1994. Photocatalytic decomposition of crude oil slicks using TiO_2 on a floating substrate. Microchem. J. 50: 28–32.

Brunette, D.M., P. Tengvall, M. Textor and P. Thomsen. 2001. Titanium in Medicine. Springer.

Cai, W., H. Du, J. Ye and T. Lai. 2014. Electrochemical determination of ascorbic acid, dopamine and uricacid based on an exfoliated graphite paper electrode: a high performance flexible sensor. Sens. Actuators, B Chem. 193: 492–500.

Cao, A.Y., P.L. Dickrell, W.G. Sawyer, M.N. Ghasemi-Nejhad and P.M. Ajayan. 2005. Super-compressible foamlike carbon nanotube films. Science 310: 1307–1310.

Chen, H., M.B. Müller, K.J. Gilmore, G.G. Wallace and D. Li. 2008. Mechanically strong, electrically conductive, and biocompatible graphene paper. Adv. Mater. 20: 3557–3561.

Chen, J.S., Y.L. Tan, C.M. Li, Y.L. Cheah, D. Luan, S. Madhavi et al. 2010. Constructing hierarchical spheres from large ultrathin anatase TiO_2 nanosheets with nearly 100% exposed (001) facets for fast reversible lithium storage. J. Am. Chem. Soc. 132: 6124–6130.

Chen, X.B. and S.S. Mao. 2007. Titanium dioxide nanomaterials: synthesis, properties, modifications, and applications. Chem. Rev. 107: 2891–2959.

Cheng, H., J. Ma, Z. Zhao and L. Qi. 1995. Hydrothermal preparation of uniform nanosize rutile and anatase particles. Chem. Mater. 7: 663–671.

Chicharroa, M., A. Sancheza, A. Zapardiel, M.D. Rubianesc and G. Rivasc. 2004. Capillary electrophoresis of neurotransmitters with amperometric detection at melanin-type polymer-modified carbon electrodes. Anal. Chim. 523: 185–191.

Chou, T.W., L. Gao, E.T. Thostenson, Z. Zhang and J.H. Byun. 2010. An assessment of the science and technology of carbon nanotube-based fibers and composites. Compos. Sci. Technol. 70: 1–19.

Christian, G.D. 2004. Analytical Chemistry, 6th ed., Wiley, New York.

De Volder, M.F.L., J. De Coster, D. Reynaerts, C.V. Hoof and S.G. Kim. 2012. High-damping carbon nanotube hinged micromirrors. Small 8: 2006–2010.

Eda, G., G. Fanchini and M. Chhowalla. 2008. Large-area ultrathin films of reduced graphene oxide as a transparent and flexible electronic material. Nat. Nanotechnol. 3: 270–274.

Eda, G., Y.Y. Lin, C.Y. Mattevi, H. Amaguchi, H.A. Chen, I.S. Chen et al. 2010. Blue photoluminescence from chemically derived graphene oxide. Adv. Mater. 22: 505–509.

Endo, M., T. Hayashi and Y.A. Kim. 2006. Large-scale production of carbon nanotubes and their applications. Pure Appl. Chem. 78: 1703–1713.

Epstein, L.H., J.L. Temple, B.J. Neaderhiser, R.J. Salis, R.W. Erbe and J.J. Leddy. 2007. Food reinforcement, the dopamine D2 receptor genotype, and energy intake in obese and nonobese humans. Behav. Neurosci. 121: 877–886.

Espelin, D.E. and A.K. Done. 1968. Amphetamine poisoning: Effectiveness of chlorpromazine. N. Engl. J. Med. 278: 1361–1365.

Evidente, V.G., C.H. Adler, J.N. Caviness, J.G. Hentz and K. Gwinn-Hardy. 2000. Amantadine is beneficial in restless legs syndrome. Movement Disorders 15: 324–327.

Fernández-García, M., A. Martínez-Arias, J.C. Hanson and J.A. Rodríguez. 2004. Nanostructured oxides in chemistry: characterization and properties. Chem. Rev. 104: 4063–4104.

Fujishima, A., T.N. Rao and D.A. Tryk. 2000. Titanium dioxide photocatalysis. J. Photochem. Photobiol. C: Photochem. Rev. 1: 1–20.

Gao, W., L.B. Alemany, L. Ci and P.M. Ajayan. 2009. New insights into the structure and reduction of graphite oxide. Nat. Chem. 1: 403–408.

Ge, M.Y., H.P. Wu, L. Niu, J.F. Liu, S.Y. Chen, P.Y. Shen et al. 2007. Nanostructured ZnO: from monodisperse nanoparticles to nanorods. J. Cryst. Growth 305: 162–166.

Gracias, D.H., J. Tien, T.L. Breen, C. Hsu and G.M. Whitesides. 2000. Forming electrical networks in three-dimensions by self-assembly. Science 289: 1170–1172.

Guan, B., W. Lu, J.F. Richard and B. Cole. 2007. Characterization of synthesized titanium oxide nanoclusters by MALDI-TOF mass spectrometry. J. Am. Soc. Mass Spectrom. 18: 517–524.

Guo, S. and S. Dong. 2011. Graphene nanosheet: synthesis, molecular engineering, thin film, hybrids, and energy and analytical applications. Chem. Soc. Rev. 40: 2644–2672.

Hambrock, J., S. Rabe, K. Merz, A. Birkner, A. Wohlfart, R.A. Fischer et al. 2003. Low-temperature approach to high surface ZnO nanopowders and a non-aqueous synthesis of ZnO colloids using the single-source precursor $[MeZnOSiMe_3]_4$ and related zinc siloxides. J. Mater. Chem. 13: 1731–1736.

Harris, P.J.F. 2009. Carbon Nanotube Science Cambridge Univ. Press, Cambridge.

Hou, S., M.L. Kasner, S. Su and K. Patel. 2010. Highly sensitive and selective dopamine biosensor fabricated with silanized graphene. J. Phys. Chem. C 114: 14915–14921.

Hu, Y.H., H. Wang and B. Hu. 2010. Thinnest two-dimensional nanomaterial-graphene for solar energy. Chemsuschem. 3: 782–796.

Huang, K., L. Wang, T. Liu, M. Gan, L.L. Wang and Y. Fan. 2013. Synthesis and electrochemical performances of layered tungsten sulfide-graphene nanocomposite as a sensing platform for catechol, resorcinol and hydroquinone. Electrochim. Acta 107: 379–387.

Huang, M., S. Mao, H. Feick, H. Yan, Y. Wu, H. Kind et al. 2001. Room-temperature ultraviolet nanowire nanolasers. Science 292: 1897–1899.

Huang, X., Z.Y. Yin, S.X. Wu, X.Y. Qi, Q.Y. He, Q.C. Zhang et al. 2011. Graphene-based materials: Synthesis, characterization, properties, and applications. Small 7: 1876–1902.

Iijima, S. 1991. Helical microtubules of graphitic carbon. Nat. 354: 56–58.

Jo, S., H. Jeong, S. Bae and S.R. Jeon. 2008. Modified platinum electrode with phytic acid and single-walled carbon nanotube: application to the selective determination of dopamine in the presence of ascorbic and uric acids. Microchem. J. 88: 1–6.

Johnson, J.C., H.J. Choi, K.R. Knutsen, D. Schaller, P. Yang and R.J. Saykally. 2002. Single gallium nitride nanowire lasers. Nat. Mater. 1: 106–110.

Jones, M.P., J.B. Dilley, D. Drossman and M.D. Crowell. 2006. Brain-gut connections in functional GI disorders: anatomic and physiologic relationships. Neurogastroenterol. 18: 91–103.

Kaniyoor, A. and S. Ramaprablu. 2012. A Raman spectroscopic investigation of graphite oxide derived graphene. AIP Adv. 2: 2183.
Karkare, M. 2008. Nanotechnology: Fundamentals and applications. IK International publication, House Pvt. Ltd.
Kennedy, G.C. and J. Mitra. 1963. The effect of D-amphetimine on energy balance in hypothalamic obese rats. J. Nutr. 17: 569–573.
Kim, S.J. and D.W. Park. 2007. Synthesis of ZnO nanopowder by thermal plasma and characterization of photocatalytic property. Appl. Chem. 11: 377–380.
Knappertz, V.A., C.H. Tegeler, S.J. Hardin and W.M. McKinney. 1998. Vagus nerve imaging with ultrasound: anatomic and *in vivo* validation. Head Neck Surg. 118: 82–85.
Kumar, A., P. Vemula, P.M. Ajayan and G. John. 2008. Silver-nanoparticle-embedded antimicrobial paints based on vegetable oil. Nat. Mater. 7: 236–241.
Kwon, Y.J., K.H. Kim, C.S. Lim and K.B. Shim. 2002. Characterization of ZnO nanopowders synthesized by the polymerized complex method via an organochemical route. J. Ceram. Process Res. 3: 146–149.
Lane, R.F. and C.D. Blaha. 1990. Detection of catecholamines in brain tissue: surface-modified electrodes enabling *in vivo* investigations of dopamine function. Longmuir 6: 56–65.
Lee, M.S., G.D. Lee, C.S. Ju and S.S. Hong. 2005. Preparations of nanosized TiO_2 in reverse microemulsion and their photocatalytic activity. Solar Cells 88: 389–401.
Leibowitz, S.F. 1975. Catecholaminergic mechanisms of the lateral hypothalamus: Their role in the mediation of amphetamine anorexia. Brain Res. 98: 529–545.
Lerf, A., H. He, M. Forster and J. Klinowski. 1998. Structure of graphite oxide revisited. J. Phys. Chem. B 102: 4477–4482.
Li, D., M.B. Müller, S. Gilje, R.B. Kaner and G.G. Wallace. 2008. Processable aqueous dispersions of graphene nanosheets. Nat. Nanotechnol. 3: 101–105.
Lieverse, A.G., A.R.J. Girbes, D.J. VanVeldhuisen, A.J. Smit, J.G. Zijlstra, S. Meijer et al. 1995. The effects of ibopamine on glomerular filtration rate and plasma norepinephrine remain preserved during prolonged treatment in patients with congestive heart failure. Eur. Heart J. 16: 937–942.
Lima, M.D., N. Li, M.J. de Andrade, S. Fang, J. Oh, G.M. Spinks et al. 2012. Electrically, chemically, and photonically powered torsional and tensile actuation of hybrid carbon nanotube yarn muscles. Science 338: 928–932.
Liqiang, J., S. Xiaojun, X. Baifu, W. Baiqi, C. Weimin and F. Honggang. 2004. The preparation and characterization of La doped TiO_2 nanoparticles and their photocatalytic activity. J. Solid State Chem. 177: 3375–3382.
Liu, G., X. Wang, Z. Chen, H.M. Cheng and G. Lu. 2009. The role of crystal phase in determining photocatalytic activity of nitrogen doped TiO_2. J. Colloid. Inter. Sci. 329: 331–338.
Liu, Z., J.T. Robinson, X. Sun and H. Dai. 2008. PEGylated Nano-graphene oxide for delivery of water insoluble cancer drugs. J. Amer. Chem. Soc. 130: 10876–10877.
Macak, J.M., H. Hildebrand, U.M. Jahns and P. Schmuki. 2008. Mechanistic aspects and growth of large diameter self-organized TiO_2 nanotubes. J. Electroanal. Chem. 621: 254–266.
Macwan, D.P., S. Chaturvedi and P.N. Dave. 2011. A review on nano-TiO_2 sol–gel type syntheses and its applications. J. Mater. Sci. 46: 3669–3686.
Manjunatha, J.G., B.E. Kumara, S.G.P. Mamatha, E. Chandra, E. Niranjana and B.S. Sherigara. 2009. Cyclic voltammetric studies of dopamine at lamotrigine and TX-100 modified carbon paste electrode. Int. J. Electrochem. Sci. 4: 187–196.
Marcano, D.C., D.V. Kosynkin, J.M. Berlin, A. Sinitskii, Z. Sun, A. Slesarev et al. 2010. Improved synthesis of graphene oxide. ACS Nano 4: 4806–4814.
Marchand, R., L. Broham and M. Tournoux. 1980. TiO_2 (B) a new form of titanium dioxide and the potassium octatitanate $K_2Ti_8O_{17}$. Mater. Res. Bull. 15: 1129–1133.
Mazloum-Ardakani, M., H. Beitollahi, B. Ganjipour, H. Naeimi and M. Nejati. 2009. Electrochemical and catalytic investigations of dopamine and uric acid by modified carbon nanotube paste electrode. Bioelectrochemistry 75: 1–8.
Metraux, G.S., Y.C. Cao, R. Jin and C.A. Mirkin. 2003. Triangular nanoframes made of gold and silver. Nano Lett. 3: 519–522.
Michael, D.J. and R.M. Wightman. 1999. Electrochemical monitoring of biogenic amine neurotransmission in real time. J. Pharm. Biomed. Anal. 19: 33–46.
Mizuno, K., J. Ishii, H. Kishida, Y. Hayamizu, S. Yasuda, D.N. Futaba et al. 2009. A black body absorber from vertically aligned single-walled carbon nanotubes. Proc. Natl. Acad. Sci. 106: 6044–6047.
Mo, J.W. and B. Ogorevc. 2001. Simultaneous measurement of dopamine and ascorbate at their physiological levels using voltammetric microprobe based on overoxidized poly (1,2-phenylenediamine)-coated carbon fiber. Anal. Chem. 73: 1196–1202.
Moens, L., P. Ruiz, B. Delmon and M. Devillers. 1997. Enhancement of total oxidation of isobutene on bismuth-promoted tin oxide catalysts. Catal. Lett. 46: 93–99.
Moghaddam, A.B., T. Nazari, J. Badraghi and M. Ka-zemzad. 2009. Synthesis of ZnO nanoparticles and electrodeposition of polypyrrole/ZnO nanocomposite film. Int. J. Electrochem. Sci. 4: 247–257.
Musameh, M., N.S. Lawrence and J. Wang. 2005. Electrochemical activation of carbon nanotubes. Electrochem. Commun. 7: 14–18.
Naumenko, D., V. Snitka, B. Snopok, S. Arpiainen and H. Lipsanen. 2012. Graphene-enhanced Raman imaging of TiO_2 nanoparticles. Nanotechnology 23: 465703.

Nethravathi, C. and M. Rajamathi. 2008. Chemically modified graphene sheets produced by the solvothermal reduction of colloidal dispersions of graphite oxide. Carbon 46: 1994–1998.
Nichols, A.J., J.M. Smith, R.J. Shebuski and R.R. Ruffolo. 1987. Comparison of the effects of the novel inotropic agent, ibopamine, with epinine, dopamine and fenoldopam on renal vascular dopamine receptors in the anesthetized dog. J. Pharmacol. Exp. Ther. 242: 573–578.
Ookubo, A., E. Kanezaki and K. Ooi. 1990. ESR, XRD, and DRS studies of paramagnetic Ti^{3+} ions in a colloidal solid of titanium oxide prepared by the hydrolysis of $TiCl_3$. Langmuir 6: 206–209.
Owen, A.M., J. Doyon, A. Dagher, A. Sadikot and A.C. Evans. 1998. Abnormal basal ganglia outflow in Parkinson's disease identified with PET. Implications for higher cortical functions. Brain 121(Pt 5). Brain 121: 949–965.
Pan, Z.W., Z.R. Dai and L. Wang. 2001. Nanobelts of semiconducting oxides. Science 291: 1947–1949.
Park, S. and R.S. Ruoff. 2009. Chemical methods for the production of graphenes. Nat. Nanotechnol. 4: 217–224.
Peng, B., M. Locascio, P. Zapol, S. Li, S.L. Mielke, G.C. Schatz et al. 2008. Measurements of near-ultimate strength for multiwalled carbon nanotubes and irradiation-induced crosslinking improvements. Nat. Nanotechnol. 3: 626–631.
Perera, S.D., R.G. Mariano, K. Vu, N. Nour, O. Seitz, Y. Chabal et al. 2012. Hydrothermal synthesis of graphene-TiO_2 nanotube composites with enhanced photocatalytic activity. ACS Cat. 2: 949–956.
Phillips, L.G. and D.M. Barbeno. 1997. The influence of fat substitutes based on protein and titanium dioxide on the sensory properties of low fat milks. J. Dairy Sci. 80: 2726–2731.
Pop, E., D. Mann, Q. Wang, K. Goodson and H.J. Dai. 2006. Thermal conductance of an individual single-wall carbon nanotube above room temperature. Nano Lett. 6: 96–100.
Pumera, M., A. Ambrosi, A. Bonanni, E.L. Chng and H.L. Poh. 2010. Graphene for electrochemical sensing and biosensing. TrAC, Trends Anal. Chem. 29: 954–965.
Qi, X.Y., K.Y. Pu, H. Li, X.Z. Zhou, S.X. Wu, Q.L. Fan et al. 2010. Angew. Chem. Int. Ed. 49: 9426–9429.
Qu, L., L. Dai, M. Stone, Z. Xia and Z.L. Wang. 2008. Carbon nanotube arrays with strong shear binding-on and easy normal lifting-off. Science 322: 238–242.
Raj, C.R., K. Tokuda and T. Ohsaka. 2001. Electroanalytical applications of cationic self-assembled monolayers: square-wave voltammetric determination of dopamine and ascorbate. Bioelectrochemistry 53: 183–191.
Rao, C.N.R., A.K. Sood, K.S. Subrahmanyam and A. Govindaraj. 2009. Graphene: the new two-dimensional nanomaterial. Chem. Int. Ed. 48: 7752–7777.
Rao, C.N.R., K.S. Subrahmanyam, H.S.S. Ramakrishna Matte, U. Maitra, K. Moses and A. Govindaraj. 2011. Graphene: synthesis, functionalization and properties. Int. J. Mod. Phys. B 25: 4107–4143.
Razali, M.H., M.N. Ahmad-Fauzi, A.R. Mohamed and S. Sreekantan. 2013. Physical properties study of TiO_2 nanoparticle synthesis via hydrothermal method using TiO_2 microparticles as precursor. Adv. Mater. Res. 772: 365–370.
Ren, L., C.L. Pint, L.G. Booshehri, W.D. Rice, X. Wang, D.J. Hilton et al. 2009. J. Carbon nanotube terahertz polarizer. Nano Lett. 9: 2610–2613.
Sammes, N.M., G.A. Tompsett, H. Näfe and F. Aldinger. 1999. Bismuth based oxide electrolytes-structure and ionic conductivity. J. Eur. Ceram. Soc. 19: 1801–1826.
Samsonov, V.M., N. Yu. Sdobnyakov and A.N. Bazulev. 2003. On thermodynamic stability conditions for nanosized particles. Surf. Sci. 526: 532–535.
Sawa, A. and S.H. Snyder. 2002. Schizophrenia: diverse approaches to a complex disease. Science 296: 692–695.
Schniepp, H.C., J.L. Li, M.J. McAllister, H. Sai, M. Herrera-Alonso, D.H. Adamson et al. 2006. Functionalized single graphene sheets derived from splitting graphite oxide. J. Phys. Chem. 110: 8535–8539.
Selhofer, H. 1999. Comparison of pure and mixed coating materials for AR coatings for use by reactive evaporation on glass and plastic lenses. Vacuum Thin Films 315: 180–183.
Selvaraj, U., A.V. Prasadrao, S. Komerneni and R. Roy. 1992. Sol-gel fabrication of epitaxial and oriented TiO_2 thin films. J. Am. Ceram. Soc. 75: 1167–1170.
Sengupta, R., M. Bhattacharya, S. Bandyopadhyay and A.K. Bhowmick. 2011. Prog. Polym. Sci. 36: 638–670.
Shankar, S.S., A. Rai, B. Ankamwar, A. Singh, A. Ahmad and M. Sastry. 2004. Biological synthesis of triangular gold nanoprisms. Nat. Mater. 3: 482–488.
Sharma, P., S. Raghunathan, A. Giri and W. Miao. 2004. Nanomaterials: Manufacturing, processing and applications. pp. 2. *In:* J.A. Schwarz and C.I. Contescu (eds.). Dekker Encyclopedia of Nano Science. Dekker, New York.
Shi, W., H. Peng, N. Wang, C.P. Li, L. Xu, C.S. Lee et al. 2001. Free-standing single crystal silicon nanoribbons. J. Am. Chem. Soc. 123: 11095–11096.
Shokuhfar, T., M.R. Vaezi, S.K. Sadrnezhad and A. Sho-kuhfar. 2008. Synthesis of zinc oxide nanopowder and nanolayer via chemical processing. Int. J. Nanomanuf. 2: 149–162.
Sinha, N. and J. Yeow. 2005. Carbon nanotubes for biomedical applications. Nano Science 4: 180–195.
Sol, I., M.D. Rajfer, D. James, M.D. Rossen, W. John, M.D. Nemanich et al. 1987. Sustained hemodynamic improvement during long-term therapy with levodopa in heart failure: Role of plasma catecholamines. J. Am. Coll. Cardiol. 10: 1286–1293.
Song, J.H., X.D. Wang, E. Riedo and Z.L. Wang. 2005. Elastic property of vertically aligned nanowires. Nano Lett. 5: 1954–1958.
Song, Z., T. Cai, Z. Chang, G. Liu, J.A. Rodriguez and J. Hrbek. 2003. Molecular level study of the formation and the spread of MoO_3 on Au(111) by scanning tunneling microscopy and X-ray photoelectron spectroscopy. J. Am. Chem. Soc. 125: 8059–8066.

Stankovich, S., D.A. Dikin, R.D. Piner, K.A. Kohlhaas, A. Kleinhammes, Y. Jia et al. 2007. Synthesis of graphene-based nanosheets via chemical reduction of exfoliated graphite oxide. Carbon 45: 1558–1565.
Stathato, E., P. Lianos, F.D. Monte, D. Levy, D. Tsiourvas and D. Tsiourvas. 1997. Formation of TiO_2 nanoparticles in reverse micelles and their deposition as thin films on glass substrates. Langmuir 13: 4295.
Staudenmaier, L. 1898. Verfahren zur darstellung der graphitsaure, berichte der deutschen chemischen gesellschaft 31: 1481–1487.
Sugimoto, T., K. Okada and H. Itoh. 1997. Synthesis of uniform spindle-type titania particles by the gel-sol method. J. Colloid. Inter. Sci. 193: 140–143.
Sugimoto, T., X. Zhou and A. Muramatsu. 2003. Synthesis of uniform anatase TiO_2 nanoparticles by gel-sol method. 4. Shape control. J. Colloid. Interface Sci. 259: 53–61.
Suhr, J., N. Koratkar, P. Keblinski and P. Ajayan. 2005. Viscoelasticity in carbon nanotube composites. Nat. Mater. 4: 134–137.
Sun, X., Z. Liu, K. Welsher, J.T. Robinson, A. Goodwin, S. Zaric et al. 2008. Nano-Graphene oxide for cellular imaging and drug delivery. Nano Res. 1: 203–212.
Sun, Y.Q., Q.O. Wu and G.Q. Shi. 2011. Graphene based new energy materials. Energy Environ. Sci. 4: 1113–1132.
Taniguchi, N. 1974. On the basic concept of 'nano-technology' Jpn. Soc. Precis. Eng. 10: 5–10.
Thien, G.S.H., A. Pandikumar, N.M. Huang and H.N. Lim. 2014. Highly exposed {001} facets of titanium dioxide modified with reduced graphene oxide for dopamine sensing. Sci. Reports 4: 5044–5052.
Thompson, B.T. and B.A. Cockrill. 1994. Renal-dose dopamine: a siren song. Lancet 3: 44, 7–8.
Topoglidis, E., A.E.G. Cass, G. Gilardi, S. Sadeghi, N. Beaumont and J.R. Durrant. 1998. Protein adsorption on nanocrystalline TiO_2 films: An immobilization strategy for bioanalytical devices. Anal. Chem. 70: 5111–5113.
Vaezi, M.R. and S. Sadrnezhaad. 2007. Nanopowder synthesis of zinc oxide via solochemical processing. Mater. Des. 28: 515–519.
Valden, M., X. Lai and D.W. Goodman. 1998. Onset of catalytic activity of gold clusters on titania with the appearance of nonmetallic properties. Science 281: 1647–1650.
Van Veldhuisen, D.J., J. Brouwer, A.J. Man-in-'t-Veld, P.H. Dunselman, F. Boomsma and K.I. Lie. 1995. Progression of mild untreated heart failure during 6 months follow-up and clinical and neurohumoral effects of ibopamine and digoxin as monotherapy. Am. J. Cardiol. 75: 796–800.
Velasco, M. and A. Luchsinger. 1998. Dopamine: pharmacologic and therapeutic aspects. Am. J. Ther. 5: 37–43.
Wang, D.H., D.W. Choi, J. Li, Z.G. Yang, Z.M. Nie, R. Kou et al. 2009a. Self-assembled TiO_2–graphene hybrid nanostructures for enhanced Li-Ion insertion. ACS Nano 3: 907–914.
Wang, H., M. Liu, C. Yan and J. Bell. 2012a. Reduced electron recombination of dye-sensitized solar cells based on TiO_2 spheres consisting of ultrathin nanosheets with [001] facet exposed. Beilstein J. Nanotechnol. 3: 378–387.
Wang, J., G. Zhao and Z.H. Zhang. 2007a. Investigation on degradation of azo fuchsine using visible light in the presence of heat-treated anatase TiO_2 powder. Dyes and Pigments 75: 335–343.
Wang, L., Y. Zhang, Y. Du, D. Lu, Y. Zhang and C. Wang. 2012b. Simultaneous determination of catechol and hydroquinone based on poly (diallyldimethylammonium chloride) functionalized graphene-modified glassy carbon electrode. Solid State Electrochem. 16: 1323–1331.
Wang, L., Y. Meng, Q. Chen, J. Deng, Y. Zhang, H. Li et al. 2013a. Simultaneous electrochemical determination of dihydroxybenzene isomers based on the hydrophilic carbon nanoparticles and ferrocene-derivative mediator dual sensitized graphene composite. Electrochim. Acta 92: 216–225.
Wang, P., Z. Mai, Z. Dai, Y. Li and X. Zou. 2009b. Construction of Au nanoparticles on choline chloride modified glassy carbon electrode for sensitive detection of nitrite. Biosen. Bioelectron. 24: 3242–3247.
Wang, Q., D. Dong and N.Q. Li. 2001. Electrochemical response of dopamine at a penicillamine self-assembled gold electrode. Bioelectrochemistry 54: 169–175.
Wang, W., C. Lu, Y. Ni and Z. Xu. 2013b. Crystal facet growth behavior and thermal stability of {001} faceted anatase TiO_2: mechanistic role of gaseous HF and visible-light photocatalytic activity. Cryst. Eng. Comm. 15: 2537–2543.
Wang, X., N. Yang, Q. Wan and X. Wang. 2007b. Catalytic capability of poly (malachite green) films based electrochemical sensor for oxidation of dopamine B: Chemical 128: 83–90.
Wang, Z.J., X.Z. Zhou, J. Zhang, F. Boey and H. Zhang. 2009. Direct electrochemical reduction of single-layer graphene oxide and subsequent functionalization with glucose oxidase. J. Phys. Chem. 113: 14071–14075.
Wang, Z.L. 2004. Zinc oxide nanostructures: growth, properties and applications. J. Phys. Condens. Matter. 16: 829–858.
Wei, B.Q., R. Vajtai and P.M. Ajayan. 2001. Reliability and current carrying capacity of carbon nanotubes. Appl. Phys. Lett. 79: 1172–1174.
Wen, Z., S. Ci, S. Mao, S. Cui, G. Lu, K. Yu et al. 2013. TiO_2 nanoparticles-decorated carbon nanotubes for significantly improved bioelectricity generation in microbial fuel cells. J. Power Sources 234: 100–106.
Williams, G., B. Seger and P.V. Kamat. 2008. TiO_2-graphene nanocomposites. UV-assisted photocatalytic reduction of graphene oxide. ACS Nano 2: 1487–1491.
Wu, L., L.Y. Feng, J.S. Ren and X.G. Qu. 2012. Electrochemical detection of dopamine using porphyrin-functionalized graphene. Biosens. Bioelectron. 34: 57–62.
Wu, S.X., Z.Y. Yin, Q.Y. He, X.A. Huang, X.Z. Zhou and H. Zhang. 2010. Electrochemical deposition of semiconductor oxides on reduced graphene oxide-based flexible, transperant and conductive electrodes. J. Phys. Chem. 114: 11816–11821.

Xiao, L., Z. Chen, C. Feng, L. Liu, Z. Bai, Y. Wang et al. 2008. Flexible, stretchable, transparent carbon nanotube thin film loudspeakers. Nano Lett. 8: 4539–4545.
Xu, H.T., F. Kitamura, T. Ohsaka and K. Tokuda. 1992. Effect of electrochemical pretreatments on carbon fiber electrodes for detection of dopamine in the presence of ascorbic acid and/or metabolites. Electrochemistry 60: 1068–1074.
Xu, M., D.N. Futaba, T. Yamada, M. Yumura and K. Hata. 2010. Carbon nanotubes with temperature-invariant viscoelasticity from –196 degrees to 1000 degrees. C. Sci. 330: 1364–1368.
Xu, S., L. Yong and P. Wu. 2013. One-pot, green, rapid synthesis of flowerlike gold nanoparticles/reduced graphene oxide composite with regenerated silk fibroin as efficient oxygen reduction electrocatalysts. Mater. Inter. 5: 654–662.
Xu, Z., N. Gao, H. Chen and S. Dong. 2005. Biopolymer and carbon nanotubes interface prepared by self-assembly for studying the electrochemistry of microperoxidase-11. Langmuir 21: 10808–10813.
Xue, K.H., F.F. Tao, W. Xu, S.Y. Yin and J.M. Liu. 2005. Selective determination of dopamine in the presence of ascorbic acid at the carbon atom wire modified electrode. J. Electroanal. Chem. 578: 323–329.
Yang, N.L., J. Zhai, D. Wang, Y.S. Chen and L. Jiang. 2010. Two-dimensional graphene bridges enhanced photoinduced charge transport in dye-sensitized solar cells. ACS Nano 4: 887–894.
Yang, X., Y. Wang, X. Huang, Y. Ma, Y. Huang, R. Yang et al. 2011. Multi-functionalized graphene oxide based anticancer drug-carrier with dual-targeting function and pH-sensitivity. J. Mater. Chem. 21: 3448–3454.
Yavich, L. and J. Tiihonen. 2000. *In vivo* voltammetry with removable carbon fibre electrodes in freely-moving mice: dopamine release during intracranial self-stimulation. J. Neurosci. Method 104: 55–63.
Zhang, H. and J.B. Bandfield. 1998. Thermodynamic analysis of phase stability of nanocrystalline titania. J. Mater. Chem. 8: 2073–2076.
Zhang, H., X.J. Lv, Y.M. Li, Y. Wang and J.H. Li. 2010. P25-graphene composite as a high performance photocatalyst. ACS Nano 4: 380–386.
Zhang, Q., Y. Liu, Y. Duan, N. Fu, Q. Liu, Y. Fang et al. 2004. Mn_3O_4/graphene composite as counter electrode in dye-sensitized solar. RSC Adv. 4: 15091–15097.
Zhao, K., H. Song, S. Zhuang, L. Dai, P. He and Y. Fang. 2007. Determination of nitrite with the electrocatalytic property to the oxidation of nitrite on thionine modified aligned carbon nanotubes. Electrochem. Comm. 9: 65–70.
Zhu, C., S. Guo, Y. Fang and S. Dong. 2010a. Reducing sugar: new functional molecules for the green synthesis of graphene nanosheets. ACS Nano 4: 2429–2437.
Zhu, Y., S. Murali, W. Cai, X. Li, J.W. Suk, J.R. Potts et al. 2010b. Graphene and graphene oxide: synthesis, properties, and applications. Adv. Mater. 22: 390–3924.
Zou, G. 1999. Basic Nerve Pharmacology, Science Press, Beijing.

2

Fabrication and Application of Graphene Oxide-based Metal and Metal Oxide Nanocomposites

Babak Jaleh,[1,*] *Samira Naghdi,*[1] *Nima Shahbazi*[1] and *Mahmoud Nasrollahzadeh*[2]

Introduction

In the last two decades, the development of easy preparation methods of graphene like materials (Figure 1), such as graphene oxide (GO) and reduced graphene oxide (rGO) via reduction of graphite oxide, offers a wide range of possibilities for the preparation of GO based inorganic nanocomposites by the incorporation of various functional nanomaterials for a variety of applications. In this chapter, we discuss the current development of graphene and GO based metal and metal oxide nanocomposites, with a detailed account of their synthesis and properties (Figure 2). Specifically, much attention has been given to their wide range of applications in various fields and new technology. In this work, we explain all subjects in two parts. In the first part, we discuss about preparation of graphene, GO, reduced GO and GO based metal and metal oxide nanocomposites. In the other part, we explain about properties and application of what are mentioned.

[1] Department of Physics, Faculty of Science, Bu-Ali Sina University, Mahdieh street, Hamedan, 65174, Iran.
[2] Department of Chemistry, Faculty of Science, University of Qom, Qom 37185-359, Iran.
Emails: samira.naghdi@gmail.com; n.shahbazi86@yahoo.com; mahmoudnasr81@gmail.com
* Corresponding author: jaleh@basu.ac.ir, bkjaleh@yahoo.com

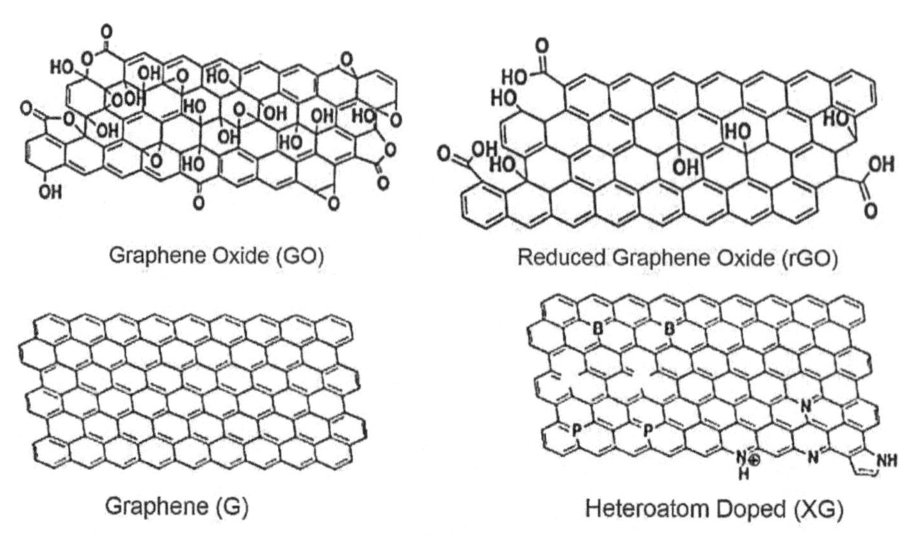

Figure 1. Graphene and related materials. Reproduced with permission from ref. (see Navalon et al. 2014). Copyright (2014) American Chemical Society.

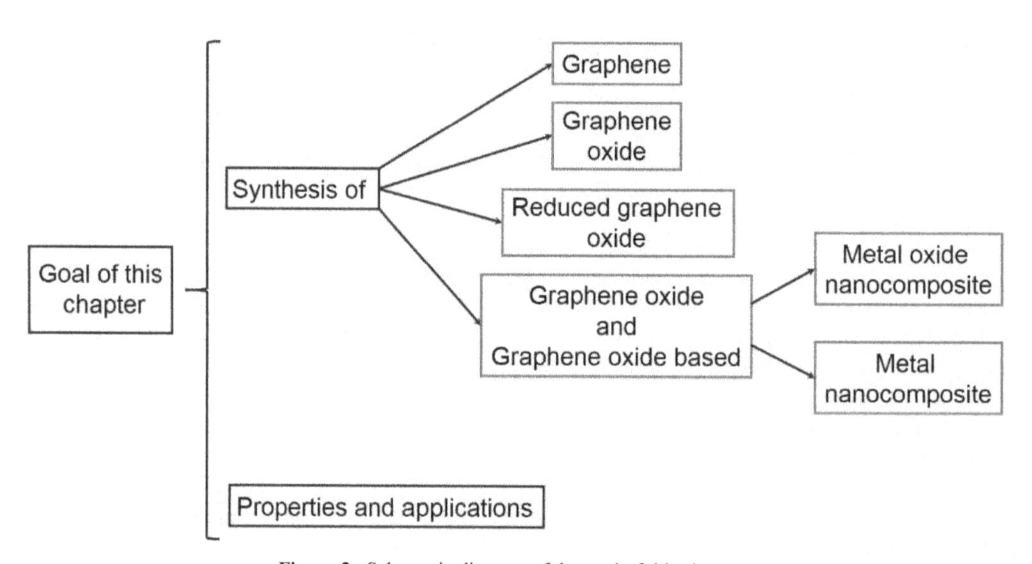

Figure 2. Schematic diagram of the goal of this chapter.

Introduction to graphene and different methods for the synthesis of graphene

Graphene was chosen as a name for a monolayer of two dimensional (2D) carbon atoms that were arranged in a honeycomb lattice structure (see Geim and Novoselov 2007). Since the finding of graphene in 2004, researchers started to develop innovative methods for synthesizing graphene. There are two basic branches for graphene production: the first branch is related to the methods that detach graphene from graphite and the second branch is related to the methods that grow graphene directly on a substrate by using carbon sources. Therefore, these methods are included in the graphene production method: mechanical exfoliation (see Geim and Novoselov 2007) and liquid phase exfoliation of graphite (see Yi and Shen 2015), epitaxial growth of graphene (see Tetlow et al. 2014), chemical vapor deposition of graphene (see Saner et al. 2010), and unzipping of carbon nanotubes (see Tiwary et al. 2015, Mohammadi et al. 2013, Xiao et al. 2014). Graphene can be produced from these methods with a relatively good quality and outstanding properties. The more important aspect in graphene production is the required properties. It means what you exactly expect from your product. Different techniques are focused on different requirements like good quality, big grain size or the quantity of production, and the last aspect is finding a method in order to reproduce the product with same quality and properties.

Exfoliation of graphite

Exfoliation of graphite is a method to produce graphene and it is based on graphite expansion by up to hundreds of times along the c axis (see Chung 1987) and also the simplest way of obtaining graphene sheets (see Fukada et al. 2012). In this process, usually graphene is peeled off from the graphite layer by layer. Van der Waals attraction between two graphene sheets is the obstacle in this method. Overcoming this force and separating the graphene sheets is a mechanical concern (see Yi and Shen 2015). In some methods, first graphite oxide is produced which makes exfoliated graphite oxide and at the end the separation between the layers happens (see Yi and Shen 2015). Here we discuss different methods to overcome the van der Waals force and producing graphene directly from graphite.

Micromechanical cleavage

In this method, a piece of a scotch tape is applied to overcome the van der Waals force between two layers of graphene in a Highly Ordered Pyrolytic Graphite (HOPG). If this method is repeated numerous times, the graphitic layer becomes thinner and finally graphene will be produced. This method, for the first time, was utilized in 2004 by the scientists who won the 2010 Nobel Physics prize. Graphene layers that are produced by this method are big in size and have good quality. The negative point of this method is the limitation of using this method for producing graphene in a big quantity because this method is very time consuming and tiring (see Wakeland et al. 2010).

Sonication

The next low-cost and fast method of graphene production is the sonication assisted liquid-phase exfoliation of graphite. A positive point of this method is the easy way of producing graphene, while the concentration of produced graphene is extremely low (0.01 mg/mL), which is a negative point for using this method for practical application. In this regard, many researchers have investigated to achieve high-concentration graphene by increasing the sonication time and the initial graphite concentration, mixing solvents, solvent exchange methods, adding surfactants and polymers, etc. (see Yi and Shen 2015, Tetlow et al. 2014).

Ball milling

Ball milling is a common method in the powder production industry. This method can be utilized for graphene production by exfoliating graphite by shear force (see Yi and Shen 2015, Tetlow et al. 2014, Saner et al. 2010). This method is one of the methods to produce high amount of graphene while the problem is the amount of defects in graphene flakes in this method that are induced by the high-energy collision of milling device. The amount of defects can't be controlled in this method because collisions among the grinding media in the milling device are unavoidable. Therefore, on the one hand, this method is a good method to produce large amount of graphene. On the other hand, the produced graphene has small size and the amount of defect is also high. Ball milling can be introduced in two different ways: wet ball milling and dry ball milling. The advantage of the wet ball milling is the production of graphene with small thickness of 10 nm, but it is a time consuming process (several ten hours). In dry ball milling, first graphite powder mix with chemically inert water-soluble inorganic salts and at the end of the process water washing and sonication of the milling products is utilized to achieve the final graphene product.

Other methods

There are three more methods that are less common among the mechanical methods of graphene production: first one is fluid dynamics where graphite flakes can move with the liquid and can thus be exfoliated repeatedly at different positions (see Yi and Shen 2015, Tetlow et al. 2014, Saner et al. 2010, Tiwary et al. 2015). Second one is detonation technique that relies on using shockwave and thermal energy for exfoliation of graphite. The final product of this method is GO rather than graphene. The third one is exfoliation assisted by a supercritical fluid. It depends on the high diffusivity, expansibility, and solvating power of the supercritical fluid. In this method, based on the expansibility and fluid power, supercritical fluid penetrates into the graphite layers and lead to graphene sheets (see Yi and Shen 2015).

Reduction of graphite oxide for graphene production

One of the safe and mild methods for exfoliating graphite and producing graphene sheets is oxidation of graphite. This method creates expanded structures of graphite oxide layers. Graphene sheets are produced from graphite oxide using a simple two-step process. First, graphite oxide is produced by some oxidation methods and later the graphite oxide reduces to graphene by utilizing reduction methods. Usually, some expansion-reduction agents are used to reduce graphite oxide. This method is a rapid, easy and low cast method for graphene production (see Wakeland et al. 2010). Graphene obtained by this method has limitation in the size of the in-plane sp^2 domains after the reduction of graphite oxide (Saner et al. 2010). As was mentioned before, graphene can be produced by micro-mechanical exfoliation of graphite, epitaxial growth, CVD, and the reduction of graphite oxide (see Fukada et al. 2012). As was mentioned in the previous section, the first three methods can produce graphene with a relatively perfect structure, while in comparison, graphite oxide has two important characteristics that makes this method more feasible for graphene production (see Pei and Chang 2012): this method is a cost-effective chemical method because of using graphite as raw material, and the resultant product is highly hydrophilic that can form stable aqueous solution to enable the assembly of macroscopic structures by simple and cheap solution processes. These characteristics are both effective in graphene production. The most important issue regarding graphite oxide is finding the best way for reduction, which somewhat restores the structure and properties of graphene. Different reduction processes result in different properties of reduced graphite oxide, which affect the final product.

Introduction to graphite oxide

Graphite oxide, first reported over 150 years ago (see Pei and Chang 2012), has recently attracted researcher's interest due to its function as a precursor for the low-cost and big quantity production of graphene. By means of ultrasonic, the oxidized layers of graphite oxide can be exfoliated in water. The exfoliated sheets with one or few layers of carbon atoms like graphene are named GO (see Wakeland et al. 2010). The synthesis methods and chemical structure of GO have been reviewed by scientists (see Kwon et al. 2009, Pei and Chang 2012). The most common method to produce GO is based on the method proposed by Hummers and Offeman in 1958. In this method, graphite was oxidized by water-free mixture of concentrated sulfuric acid, sodium nitrate and potassium permanganate. During recent years, this method has been developed and modified by scientists and named modified Hummers methods. There are different models to introduce GO, while the best GO model is proposed by Lerf and Klinowski, where the carbon plane is decorated with epoxy, carbonyl and hydroxylfunctional groups. Carbonyl groups present along the sheet edge and also organic carbonyl defects could be present within the sheet. Usually, an ideal graphene sheet is flat and consists of only trigonally bonded sp^2 carbon atoms (see Kim et al. 2015). While GO sheets consist partly of sp^2 and sp^3 carbon clusters of few nanometers in its structure. Therefore, GO is a random distribution of oxidized areas with oxygen functional groups, combined with non-oxidized regions with sp^2 carbon atoms (see Pei and Chang 2012).

Different methods of GO reduction

GO can be partially reduced to graphene by using reduction methods for removing the oxygen-functional groups. The reduced GO (rGO) sheets are usually considered as graphene because of their similar properties (see Kim et al. 2015). Reduction of GO has some positive points that makes scientists eager to put so much efforts in producing rGO, while some factors still alter the carbon structure. Because of defects and residual functional groups on rGO sheets, it is not correct to consider rGO as graphene. There are different physical and chemical methods to reduce GO and sometimes scientists even mix some methods to produce rGO with less defects and less oxygen functional groups.

Thermal reduction

Thermal reduction is one of the first and feasible methods to reduce GO. Based on the energy source, it can be divided in three sections, using heat, microwave and visible light for reduction. One of the first method to exfoliate graphite oxide to make graphene was rapid heating. The exfoliation mechanism is based on expansion of CO or CO_2 gases that exist in the spaces between graphene layers because of heat treatment which create enormous pressure between the stacked layers (Li et al. 2009). In this method, annealing temperature and atmosphere are very effective on rGO quality. To achieve high quality products, vacuum is needed; otherwise, the produced rGO contains so many defects (see Wang et al. 2011). Wu et al. (2012) reported the arc-discharge treatment as a method for exfoliating graphite oxide to prepare graphene (Li et al. 2009). They mentioned that annealing temperature and atmosphere are very important for the thermal reduction of GO. Next method to produce graphene by means of heat is using microwave irradiation, which is a very rapid and uniform method to reduce GO (see Coraux et al. 2008). Finally, photo irradiation is the last method for reducing GO by a thermal method. This method can be done with a single flash lamp (such as a xenon lamp that exists on a camera) or with advanced laser. The GO films usually undergo a huge expansion after photo reduction because of rapid degassing (see Coraux et al. 2008).

Chemical reduction

Chemical reduction is based on the chemical reactions of chemical reagents with GO. For the mass production of graphene, chemical reduction is more convenient compared with thermal reduction regarding requirement for environment and equipment. One of the common reagents to reduce graphene is hydrazine. This method was reported by Pei and Wang 2012. The chemical reduction by hydrazine and its products, e.g., dimethyl-hydrazine and hydrazine hydrate (Li et al. 2009), can be accomplished by adding these chemicals to a GO aqueous dispersion. The final product is agglomerated rGO nanosheets because of the increase in hydrophobicity. To solve the agglomeration problem, surfactant was added to the rGO to change the charge state of rGO layers. There are different chemicals which are well-known recently as reducing reagents, e.g., sodium borohydride ($NaBH_4$), lithium aluminum hydride, and sodium hydride, but the difficulty of using these reagents is their strong reaction with water. Water is the main solvent which is used for the exfoliation and dispersion of GO. Therefore, finding a reagent that is non-toxic and has a higher chemical stability in water, and produces rGO with less aggregation is the main goal in current researches on chemical reduction of GO (see Coraux et al. 2008).

Photocatalyst reduction

Photocatalyst reduction of GO is another method that can be achieved by using photo catalyst materials like ZnO, TiO_2, and $BiVO_4$ under ultraviolet (UV) irradiation (see Pei and Cheng 2012). A change in color from light brown to black or dark brown in GO color is one proof for reduction and achieving rGO because of rebuilding of the conjugated network in the graphene sheets.

Electrochemical reduction

Another process that displays good potential for the reduction of GO is based on electrochemical removal of oxygen functional groups (see Kwon et al. 2009). This method can simply be carried out in a normal electrochemical cell in a buffer solution at room temperature. This method doesn't need any kind of chemical reagent and the main reason for reduction by this method is the electron exchange between electrodes and GO. The positive point of this method is that this method doesn't need to use any toxic chemicals. Usually in this method, a thin layer of GO deposits on the surface of a piece of glass, ITO, or plastic and places in an electrochemical cell opposite an inert electrode. During the charging of the cell, reduction happens for GO. The researchers mention that this method for reduction can be controlled by pH of the buffer solution and lower pH is more promising for GO reduction.

Solvothermal reduction

Solvothermal method for GO reduction is a thermal process that performs in a vacuum-packed container; therefore, after heating, because of the high pressure in the container, the temperature can increase to a temperature above the solvent boiling point (see Coraux et al. 2008). In this method, water is used instead of reducing agent, so it offers a green reduction method. Zhou et al. (2013) reported the solvothermal method based on hydrothermal treatment of GO in water. They showed that water removes the oxygen functional groups of GO and also improves the aromatic structures of the carbon atoms in the rGO sheets (see Kwon et al. 2009). In this method, for achieving the rGO with better quality and without aggregation, the solution must have basic pH which helps in better dispersion of rGO in the solution.

Multi-step reduction

Usually by using one reduction method, rGO can be produced, but sometimes the quality or structure is not convenient for a specific purpose, for example in chemical reduction, usually the reagent can't eliminate all oxygen functional groups and most of them are still on the rGO sheets. Therefore, to improve the quality of the produced rGO, some reduction methods have been used and this is called multi-step reduction. Gao et al. (2011) used three different methods to reduce GO (see Wakeland et al. 2010). In their proposed method, first they used $NaBH_4$ for the deoxygenation, and in the next step, they used sulfuric acid for dehydration. Finally, they utilized thermal annealing in the Ar/H_2 atmosphere at 1100°. The rGO produced by their multi-step method was very good in quality and they could eliminate most of the oxygen functional groups from the carbon layers in rGO structure.

Chemical vapor deposition (CVD)

Since 2004, which was the first time it was introduced, graphene has attracted vast interest because of its unique physical and chemical properties. Among the different methods that were introduced for graphene production, CVD has been known as an effective method for production of graphene since 2008 (see Zhang et al. 2013). CVD is a reproducible method and can produce graphene with big grain size and with high growth rate, while this method needs high temperature. Furthermore, CVD is a complex method. During the CVD process, different gases (usually Ar, H_2 and carbon source gas) release into the chamber and pass through the furnace. At higher temperature, carbon precursors decompose into carbon radicals at the metal catalyst substrate surface and change to graphene. The quality of graphene that is produced by this method strongly depends on different aspects like the type of catalyst, growth temperature, pre-treatment, pressure, etc. Here we present some of the important factors in graphene growth.

Catalyst

In some works, researchers focused on the effect of catalyst on graphene deposition (see Naghdi et al. 2016a). Metal substrate, which also works as catalyst, determines the mechanism of graphene deposition, which definitely has great effect on the quality of graphene. Zou et al. investigated different transition metals, especially groups IVB-VIB metals that have good catalyst activities. They reported the role of these catalyst in single and multi-layer graphene growth at atmospheric pressure CVD (APCVD). They mentioned the perfect control of these catalysts in thickness of graphene sheets and uniformity of the products (see Zou et al. 2014). Wang et al. reported the synthesis of graphene with different thickness (1–5 layers) on a polycrystalline Co by using radio-frequency plasma-enhanced CVD at a temperature of 800°C. They showed the importance of Co crystallites structure on the formation of graphene (see Wang et al. 2010a).

Growth temperature

The growth temperature is also an important factor regarding graphene deposition. Depending on the catalyst, carbon source and pressure, growth temperature can be different. Usually, CVD from gaseous hydrocarbon sources needs high growth temperature and usually 1000°C is considered as a typical temperature for graphene growth. Recently, researchers have focused on finding temperature lower than 1000°C for graphene growth. Li et al. demonstrated CVD of graphene on Cu foils at 400°C by utilizing solid carbon precursors like PMMA and polystyrene, and they also used benzene as carbon precursor at 300°C. They could produce monolayer graphene with outstanding quality (see Li et al. 2011b). Kalita et al. demonstrated the deposition of graphene by surface wave plasma (SWP) CVD technique at 450°C on Cu foil. A gas mixture of C_2H_2, H_2 and Ar was used for the graphene coating experiments (see Kalita et al. 2014). Dathbun and Chaisitsak reported growth of high quality graphene film on a Cu foil by CVD and ethanol as a carbon source at different growth temperatures (650–850°C). They found that the quality of graphene and number of graphene layers are strongly dependent on the growth temperature (see Dathbun and Chaisitsak 2013).

Cooling rate

This parameter is one of the most important factor that can affect graphene structure (thickness and number of layers) and quality (domain size). Wu et al. reported deposition of large-area graphene on molybdenum foils by CVD. They showed that the cooling rate was the most important factor to control the thickness of graphene film (see Wu et al. 2012). Dathbun and Chaisitsak investigated graphene growth on a Cu foil by a CVD and using ethanol as a carbon source. They reported the effect of slow cooling rate and fast-cooling rate on the graphene structure. Their findings showed that the amount of defects in graphene sheets is sensitive to cooling rate but the number of layers is less sensitive to the cooling rate (see Dathbun and Chaisitsak 2013).

Pre-treatment

The surface morphology and structure of the substrate, depending on the pollution, metal oxides, or other factors, can be different. The growth behavior and quality of graphene both are dependent on the morphology of the substrate. Pre-treatment of the substrate, including pre-cleaning, pre-annealing, etc., can be useful to increase the crystallite structure of the substrate, decrease the defects of the surface and remove the pollution or native metal oxide from the surface. Kim et al. reported pre-cleaning of the Cu substrate by using nitric acid etchant to improve the graphene quality. They found nitric acid to be the most effective etchant agent among various acidic or basic solutions. Their results showed that the graphene, which was grown on the treated Cu surfaces, has better quality than the graphene grown on untreated Cu foil (see Kim et al. 2013). Kim et al. reported the CVD of high quality monolayer graphene on the pre-annealed Cu foil in a hydrogen atmosphere, followed by pre-washing the Cu foil in nitric acid solution (see Kim et al. 2014). They showed that these pre-treatment methods remove the rolling lines and native copper oxide from the Cu surface. The native Cu oxide and rolling lines on the surface can act as nucleation centers for graphene and result in multilayer graphene.

Hydrogen (H_2) flow rate

H_2 plays an important role in removing oxidize impurities from the substrate surface during CVD process. Hussain et al. investigated the effect of H_2 flow rate during low pressure CVD on the physical and electrical properties of graphene. They reported the effect of higher H_2 flow rate on the increasing graphene grain size. Furthermore, they observed more defects at lower H_2 flow rate (see Hussain et al. 2014).

Epitaxial growth of graphene

The epitaxial growth of graphene is one of the attractive methods to grow graphene on a hexagonal substrate. Both silicon carbide (SiC) and close-packed metals can be utilized as the substrate. In this method, by heating SiC in ultra-high vacuum (UHV) at temperatures between 1000 and 1500°C, Si sublimate, and carbon-rich remains on the surface (thermal decomposition). In this technique, the size of graphene can be selected by choosing the wafer scale which is of interest to the micro-electronics industry. The difference between CVD and epitaxial growth of graphene is the carbon source in the process. In thermal decomposition of SiC, carbon already exists in the substrate, while in CVD usually a source of carbon is used to grow the graphene sheet (see Tetlow et al. 2014). This method has some negative points: first, the SiC substrate is expensive and another problem is the difficulty in transferring the graphene sheet to other substrates. In comparison with the exfoliation of graphite and reduction of graphite oxide-defects which the produced graphene has the graphene layer in this method has good quality and also graphene product is single-domain (see Yang et al. 2013). However, as one atom thick, graphene is susceptible to defects from its substrate. For example, to have graphene with excellent electronic properties, substrate can be a very important factor. Other properties can also be affected by substrate; therefore, finding a suitable substrate is a big obstacle for the epitaxial growth of graphene (see Yang et al. 2013).

Yang et al. introduced hexagonal boron nitride (h-BN) as an excellent substrate for epitaxial growth of graphene, thanks to its character to improve graphene's electronic structure and its atomically flat surface. They were successful in growing single-domain graphene by epitaxial method on h-BN (see Yang et al. 2013).

Carbon ion implantation

Carbon ion implantation is another method that can produce graphene with very good quality and fewer amounts of defects. This method is suitable for different types of substrate, especially with Si microelectronics. Carbon ion implantation can damage the substrate, which can be controlled by controlling the kinetic energy of the incident ions. Usually in this method, high-temperature is vital for activation and graphene synthesis after carbon ion implantation. So finding the solution for using lower temperature in this method is one of the challenges for scientists. Kim et al. reported the synthesis of multi-layer graphene on thin Ni films on SiO_2/Si substrate by carbon ion implantation. Carbon ions bombarded the surface of the substrate at 500°C and this was followed by high-temperature activation annealing (600–900°C) to form a high quality multi-layer graphene on the substrate (see Kim et al. 2015).

Unzipping CNTs

Carbon nanotubes (CNTs) are allotropes of carbon with a cylindrical nanostructure that have been studied over the last two decades. As was mentioned before, there are different methods for graphene production, while the most significant problem related to graphene production is the quantity of products that is not enough for using graphene in different research areas. One of the recent methods that emerged as an alternative approach to produce industrial-scale graphene is unzipping the CNTs (see Tiwary et al. 2015, Xiao et al. 2014). Tiwary et al. reported the synthesis of graphene by unzipping CNTs using cryomill device (see Tiwary et al. 2015). In some works, using a chemical reaction to tear apart CNTs were investigated and one of the disadvantages of using this method for graphene production is using chemical oxidation agent for unzipping CNT which led to over-oxidation of graphene layer edges and creating defect sites. Mohammadi et al. reported the production of graphene by using chemicals for unzipping CNTs which grow on Si substrates. CNTs were placed horizontally on a Si substrate and unzipped in an etching process in a mixture of SF_6, H_2, and oxygen gases (see Mohammadi et al. 2013).

Some works involving graphene preparation are summarized in Table 1.

Table 1. Summary of the different methods for graphene production.

References	**Method**		**Advantages**	**Disadvantages**
Wakeland et al. 2010	Exfoliation of graphite	Micromechanical cleavage	Big graphene domain size with good quality	Low quantity of graphene production
Yi and Shen 2015 Tetlow et al. 2014		Sonication	Easy method	Low quantity of graphene production
Yi and Shen 2015 Tetlow et al. 2014 Saner et al. 2010		Ball milling	Production of high amount of graphene	High amount of defects in graphene flakes
Fukada et al. 2012 Pei and Chang 2012 Saner et al. 2010	Reduction of graphite oxide		Rapid, easy and low cast method	Small graphene domain size
Zhang et al. 2013 Zou et al. 2014 Wang et al. 2010a	Chemical vapor deposition (CVD)		Big graphene domain size and high growth rate	Needs high temperature and is a complex method
Tetlow et al. 2014 Yang et al. 2013	Epitaxial growth of graphene		Graphene product is single-domain with good quality	Using expensive substrate and difficulty for transferring the graphene sheet to other substrates
Kim et al. 2015	Carbon ion implantation		High quality products and suitable for different types of substrate	Needs high temperature
Tiwary et al. 2015 Xiao et al. 2014 Mohammadi et al. 2013	Unzipping CNTs		High quantity of graphene production	Low quality of products

Synthesis of graphene-based metal oxide composites

The increasing attachment in graphene-based metal oxides nanocomposites for several applications has led to a diversity of new processes being proposed for the preparation of nanocomposite materials (Omidvar et al. 2017a). Some lately established ways are considered in this chapter.

Solution mixing method

Solution mixing is an impressive and direct method. It has been widely used to manufacture graphene and GO-based metal oxide nanocomposites. At first, sol or solution of matter was prepared. Then graphene was gradually added to the solution and the mixture was sonicated to facilitate thorough mixing. Naghdi et al. prepared GO-TiO_2 composites by solution mixing (see Naghdi et al. 2016a). At first, TiO_2 powder dispersed deionized (DI) water to form a TiO_2 suspension. Then GO was dispersed in the TiO_2 suspension and it was followed by ultrasonication. The mixture was then refluxed and finally filtered. Figure 3 shows TEM image of GO/TiO_2 nanocomposite.

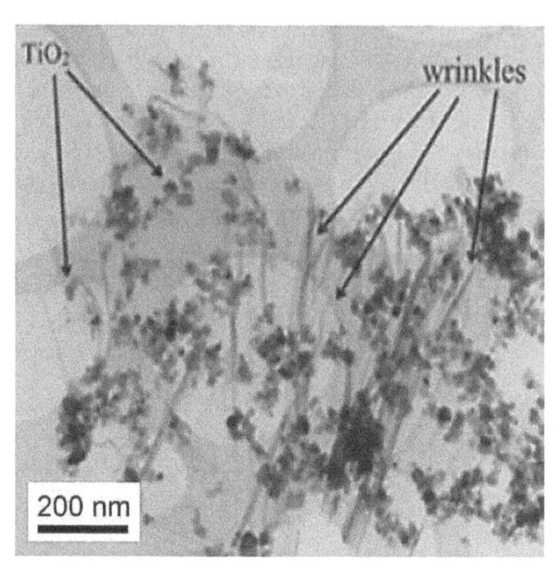

Figure 3. TEM image of GO/TiO_2 nanocomposite. Reproduced with permission from ref. (see Naghdi et al. 2016a). Copyright (2016) Sciencedirect.

Sol-gel method

The sol-gel method is a common approach for the fabrication of metal oxide structures and film coatings. It has been used for the *in situ* preparation of nanoparticles on graphene sheets. For instance, the direct increase of TiO_2 nanocrystals on GO sheets was attained by a two-step process. In the first method, amorphous TiO_2 was coated on GO sheets by hydrolysis, which was further crystallized into anatase nanocrystals by hydrothermal treatment in the second way. The method presents easy simple achievement to the GO/TiO_2 nanocrystals hybrids with a monotonous coating and powerful interplay between TiO_2 and the GO sheets. The key benefit of the *in situ* sol-gel method lies in the fact that the functional groups on GO/rGO prepare reactive and anchoring sites for nucleation and increase of nanoparticles, therefore the eventuate metal oxide nanoparticles are chemically bonded to the GO/rGO surfaces (see Hu et al. 2013).

Hydrothermal/solvothermal method

Solvothermal and hydrothermal synthetic approaches are one of the most popular synthetic strategies for the expansion of different graphene-based nanocomposite which works at a high temperature in a confined volume to produce numerous pressures. Hydrothermal and solvothermal process for the synthesis of graphene-based materials can be carried out at several temperature ranges, up to 190°C. By directing the experimental parameters like concentrations of forerunner solutions and the reaction time, graphene-TiO_2 nanocomposites with qualified crystal forms can be easily created using the hydrothermal method (see Wang et al. 2012).

Self-assembly

Self-assembly is an effective and usual antecedent method to congregate nano materials into ordered macroscopic structures. This has been operated to create functional materials like photonic crystals and composites' structures. To obtain an intermittent layered structure of the final composites, a novel process has been expanded to create the ordered graphene-based metal oxide hybrids though the surfactant assisted ternary self-assembly method (see Wang et al. 2010b). They used an anionic surfactant improved rGO as the starting material. This assembly method is significant in constructing layered composites materials.

Chemical reduction method

Chemical reduction is the next process among the most popular methods for the fabrication of several graphene-based nanocomposites. Several study groups have used a variety of reducing factors for the particular synthesis goals. For instance, in 2013, Jeong et al. successfully reported a simple and efficient reduction method for the synthesis of Au@Pd/graphene nanostructures (Figure 4) as a recyclable catalyst for ethanol oxidation (see Jeong et al. 2013). Also, Liu and co-workers reported the preparation of the Pt/Ni-G hybrids via a reduction method (Figure 5) (see Liu et al. 2013). The heterogeneous catalyst was successfully applied in the reduction of aromatic nitro compounds. In addition, Teymourian et al. used ammonia solution as a reducing agent for the synthesis of Fe_3O_4 magnetic nanoparticles/rGO hybrid nanosheets (see Teymourian et al. 2013). The main benefit of this method is that one can tune the degree of reduction and other attributes by using particular reducing factors. Furthermore, for most of the types, the responses are very energy-effective due to low temperatures and slow time. One of the obstacles in this regard is purifying the final products from different reducing agents, which in some cases is quite challenging.

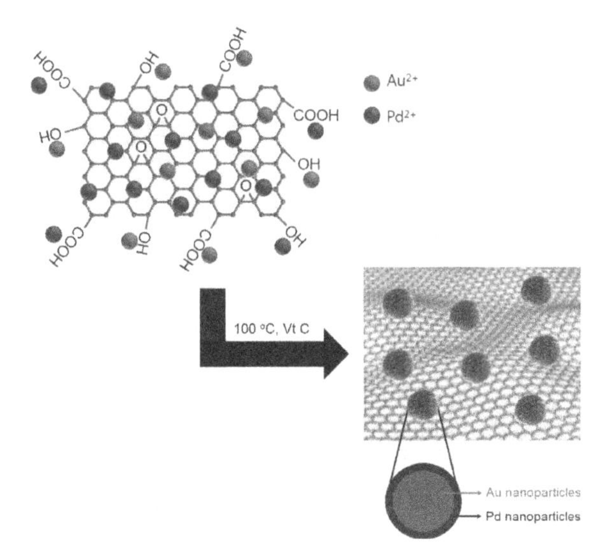

Figure 4. The synthetic scheme of Au@Pd/graphene nanocomposites. Reproduced with permission from ref. (see Jeong et al. 2013). Copyright (2013) The Royal Society of Chemistry.

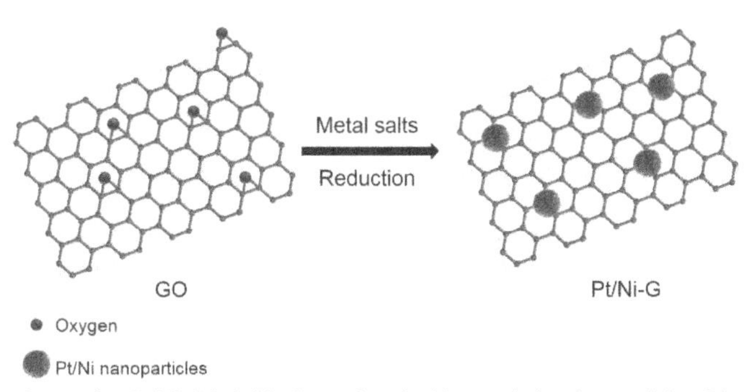

Figure 5. Illustration of preparing Pt/Ni-G hybrids. Reproduced with permission from ref. (see Liu et al. 2013). Copyright (2013) Hindawi Publishing Corporation.

Microwave-assisted method

Microwave (MW)-assisted synthesis is a beneficial and common method for the quick manufacture of nanoparticles and uniform particle size distribution. It is an identical heating way compared to the other formal heating systems. In addition, microwave can help the nucleation of nanoparticles and shorten the preparation time. There are several good examples of using this method. The microwave irradiation has been used to fabricate graphene-based metal oxide hybrids. For example dispersed titanium dioxide nanoclusters on rGO in a toluene-water system could be produced by a MW-irradiation-assisted process (see Luo et al. 2013). The principal benefit of the MW-assisted synthetic method includes quick reaction time, facility for scale-up manufacture and contamination-free final nanohybrid product. Naghdi et al. prepared GO-Fe_2O_3 composites by this method (see Naghdi et al. 2016b). At first, $FeCl_3$ was dissolved in deionized water, urea (as hydrolyzing agent) was separately dissolved in DI water and then the urea solution was slowly added to the $FeCl_3$ solution. Later, the mixture was gradually added to a solution of GO prepared by dispersing GO in DI water. After sonication, the mixture was heated with continuous stirring. Then the mixture was irradiated in a microwave oven. Finally, the mixture was washed with DI water to remove any excess ions and the GO/Fe_2O_3 composite was obtained by drying. Figure 6 shows the TEM image of GO-Fe_2O_3 nanocomposite.

Figure 6. TEM image of the GO-Fe_2O_3 nanocomposite. Reproduced with permission from ref. (see Naghdi et al. 2016b). Copyright (2016) Sciencedirect.

Electrochemical synthetic method

The electrochemical synthetic process is a popular method for transforming electronic states by controlling the outside power source to alter the Fermi energy level of the electrode material surface. Ye et al. successfully synthesized CuO nano needle/graphene/carbon nanofiber improved electrodes by electrochemical method (see Ye et al. 2013). The electrochemical synthetic method is quick and controllable. By this method, one can arrive at impurity-free nanomaterials. In addition, it would be difficult for large-scale production.

Plant mediated biogenic synthesis

In recent years, green synthesis of graphene-based nanocomposites has received much attention due to the growing need to develop environmentally benign technology in nanoparticles synthesis using plant extracts as biological materials under mild conditions (see Atarod et al. 2016, 2015, Maham et al. 2017, Maryami et al. 2016, Nasrollahzadeh et al. 2017, 2016a, 2016b, 2015). The present hydroxyl groups of phenolics in the plant extracts are directly responsible for the reduction of metal ions to metal nanoparticles (MNPs). The advantages of this environmentally benign and safe protocol include a simple reaction setup, very mild reaction conditions, and use of nontoxic solvents such as water, elimination of toxic and dangerous materials and cost effectiveness as well as compatibility for biomedical and pharmaceutical applications. Also, in this method there is no need to use high pressure, energy, temperature and toxic chemicals. However, the MNPs' agglomeration and their difficulty of separating from the reaction mixture restrict their application in organic reactions due to environmental problems. The dispersion of MNPs onto the surface of graphene and/or its subtypes such as graphite oxide, GO and rGO materials potentially provides a new way to develop catalytic materials. This behavior of graphene and its high specific surface area makes it an effective support for catalytic applications preventing, for example, the aggregation and hence deactivation of the catalytic species. For example, in 2016, for the first time, Nasrollahzadeh and co-workers reported the preparation of Ag NPs on GO/TiO_2 without applying any additional reductants just by using *Euphorbia helioscopia* L. leaf extract and showed its application as an effective catalyst for reduction of 4-nitrophenol (4-NP), Congo red (CR) and Methylene blue (MB) in aqueous media at an ambient temperature (see Nasrollahzadeh et al. 2016a).

Other methods

There are several methods that have also been studied for the preparation of various nanocomposites, like electrospinning, template-based synthesis, light- or radiation induced methods, etc. Moreover, each method has its particular benefits for special systems.

Properties and application

In the last years, numerous works have been done to prepare composites of graphene and GO with nano particles, often based on transition metal and metal oxide nanoparticles. A deficiency of knowledge in graphene chemistry barricaded a considerable development of this field for a long time. However, the appearance of methods for the fabrication of stable and homogeneous dispersions of graphene has changed the scenario. In specific, enormous interest has been generated in manufacturing graphene and GO based nanocomposites and develop them for various applications. The graphene-based nanocomposites include only two ingredients, though multicomponent composites have been fabricated for particular applications as well. The goal for extension of composite materials is to magnify their potential for pragmatic applications. Graphene has displayed some strange electrical, optical, mechanical and thermal properties, like very high carrier mobility, long range ballistic transport at room temperature, quantum confinement in nanoscale ribbons and single-molecule gas detection sensitivity. In this section, the potential applications of the aforesaid graphene and GO-based metal composites are reviewed.

Surface properties

Graphene, a nanometer-thick two-dimensional analog of fullerenes and carbon nanotubes, has lately sparked excellent excitement in the scientific community given its great mechanical, thermal and electronic properties. Also, as a pliable material with large particular surface area, graphene has proven to be a prominent building-block. But in its short history, graphene has revealed different potential applications in the manufacturing of devices, sensors, transparent conductive films, and composites. However, applied application of graphene primarily needs large scale manufacture. Surface properties of graphene are intimately relevant to graphite and can be showed in terms of the graphite structure. Similar to graphene, GO is fundamentally one-atom thick but can be as wide as tens of micrometers, resulting in an inimitable kind of material building block, characterized by various two length scales. GO sheets are very oxygenated having hydroxyl and epoxy functional groups on their basal plane; also, carbonyl and carboxyl groups are located at the sheet edges. These functional groups provide reactive sites for different surface modifications to make functionalized GO. The presence of these functional groups creates GO sheets to be forcefully hydrophilic which allows GO to be readily dispersed in water to form stable colloids (see Naghdi et al. 2015). The most attractive property of GO is that it can be reduced to graphene like sheets by removing the oxygen-containing groups with the recovery of a conjugated structure. The reduced GO sheets are usually considered as one kind of chemically derived graphene. Figure 7 illustrates the XPS spectra of carbon characteristic peak for GO and reduced GO.

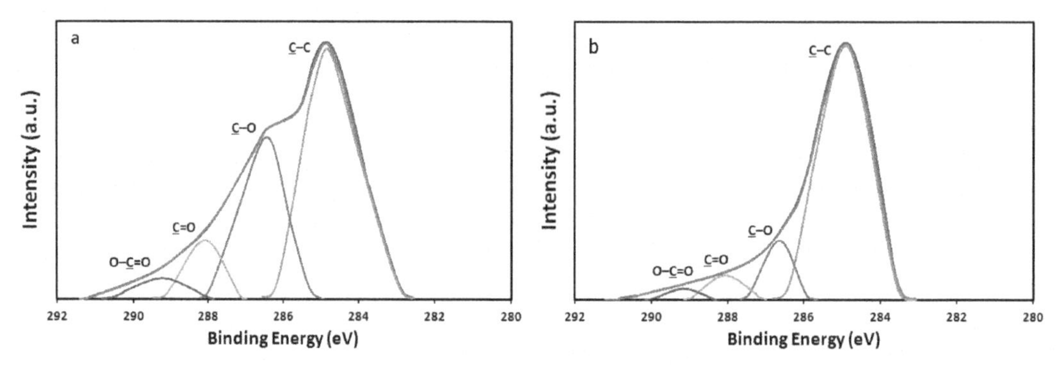

Figure 7. XPS spectra of C1s regions of the GO (a) and rGO (b). Reproduced with permission from ref. (see Naghdi et al. 2015). Bull. Chem. Soc. Jpn.

Electrical properties

Graphene has some prominent properties that makes it very attractive for applications in electronics. Also, the extremely high carrier mobility, μ, has obtained most consideration. The beneficial property of graphene is that it is a zero-overlap semimetal (with both holes and electrons as charge carriers) with extremely high electrical conductivity. Carbon atoms have a total of 6 electrons: 2 in the inner shell and 4 in the outer shell. The 4 electrons in an individual carbon atom exist for chemical bonding, though in graphene, each atom is connected to 3 other carbon atoms on the two dimensional plane, leaving 1 electron freely existing in the third dimension for electronic conduction. These highly-mobile electrons are showed as pi (π) electrons and are defined above and below the graphene sheet. These pi orbitals overlap and help to increase the carbon to carbon bonds in graphene. So, the electronic properties are dictated by the bonding and anti-bonding of these pi orbitals. Experiments have shown that the electronic mobility is extremely high, with previously reported consequences above 15,000 $cm^2 \cdot V^{-1} \cdot s^{-1}$ and theoretically potential limits of 200,000 $cm^2 \cdot V^{-1} \cdot s^{-1}$. It is showed that graphene electrons practice very much like photons in their mobility due to their absence of mass. These charge carriers are able to move sub-micrometer distances without scattering. The most significant specification about the early study on graphene transistors was the ability to continuously tune the charge carriers from holes to electrons. This influence is most pronounced in the tenuous samples whereas samples from multiple layers display much weaker gate dependence due to screening of the electric field by the other layers (see Goumri et al. 2016). At low temperatures and high magnetic fields, the exceptional mobility of graphene allows for the perception of the quantum hall effect for both electrons and holes. For more applied applications, one would like to utilize the powerful gate related to graphene for either sensing or transistor applications. Graphene has no band gap and, correspondingly, resistivity changes are small. So, a graphene transistor by its lot nature is plagued by a low on/off ratio. However, one way around this limitation is to carve graphene into thin ribbons. By shrinking the ribbon, the momentum of charge carriers in the transverse direction becomes quantized which concludes in the opening of a band gap. This band gap is commensurate to the width of the ribbon. This influence is pronounced in carbon nanotubes where a nanotube has a band gap commensurate to its diameter. The opening of a band gap in graphene ribbons has lately been perceived in wide ribbon devices lithographically patterned from large graphene flakes. The great resistance of GO is due to the entity of oxygen-containing groups, which could introduce defects to graphene. On the other side, deoxygenation could recover the GO conductivity to some extent. Lately, it has been illustrated that deoxygenation happens in GO when it is heated above 100°C, resulting in a thermal reduction. So, the sheet electrical resistance of the fabricated samples was investigated. GO functions like an electrical insulator because of the disturbance of its sp^2 bonding network which is significant to reduce the GO, so as to recover the honeycomb hexagonal lattice of graphene, in order to restore electrical conductivity (see Eda et al. 2009).

Optical properties

Tuning the optical properties of different materials has been of major interest due to their potential applications in optoelectronic system. Among several optical materials, GO has gained strong interest due to its versatility in different devices like flexible electronics, solar cells and chemical sensors. Recently, intense study has been carried out to understand the properties of GO and transform it as reduced GO in order to utilize in mentioned applications (see Khan et al. 2015). Graphene's inimitable optical properties generate an unbelievable high opacity for an atomic monolayer suspended in vacuum, with an evaluation white light absorbance of 2.3% and a negligible reflectance (< 0.1%). Absorbance develops linearly with the layers numbered from 1 to 5 (see Falkovsky et al. 2008). Also, the optical transition can be improved by changing the Fermi energy extremely through electrical gating. Another property of graphene is its photoluminescence (PL) that is conceivable to produce graphene luminescent by inducing a suitable band gap. Two routes have been proposed. The first method includes cutting graphene into nanoribbons and quantum dots. Another is a physical or chemical treatment with different gases to decrease the connectivity of the π electron network. For instance, it was shown that PL can be induced by oxygen plasma treatment of graphene single layer on silicon substrate covered with 100 nm SiO_2. This increase the chance of manufacturing hybrid structures by etching just the top layer, while keeping the underlying layer unaffected. Omidvar et al. prepared GO/Pd nanocomposite by solution mixing method and studied fluorescence of nanocomposite (see Omidvar et al. 2017a). Broad PL from solid GO and liquid GO suspension was also apperceived (see Kou et al. 2013). The advance chemical reduction of GO into rGO quenched the PL, whereas oxidation enhanced PL by making a disruption of the π network and opening a direct electronic band gap. Fluorescent organic mixture is very significant for the increase of low cost optoelectronic devices. Individually, blue fluorescence from aromatic or olefin molecules and their derivatives are significant for show and lighting applications. Blue PL was showed for GO thin films deposited from thoroughly exfoliated suspensions. The PL characteristic and its relation to the reduction of GO originate from the recombination of electron–hole pairs localized within small sp^2 carbon clusters embedded within the GO sp^3 matrix. The special electrical properties in conjunction with optical properties have fueled lot of profits in novel photonic and optoelectronics devices. Multiple promising applications using graphene have been offered, involving photodetectors, touch screens, light emitting devices, photovoltaics, transparent conductors, terahertz devices and optical limiters.

These carbon allotropes have been widely surveyed for their mechanical, electrical, thermal, optical and nonlinear optical (NLO) properties. About their NLO properties, it is shown that fullerenes have reverse saturable absorption at certain wavelengths, carbon nanotubes show ultrafast third-order nonlinearities and saturable absorption, and graphene shows ultra-broadband resonant NLO response (see Fakhri et al. 2016a). In addition, it is shown that the lately found GO has reverse saturable absorption and optical limiting properties. A variety of techniques are being used for evaluating the NLO properties of materials. Among them, the z-scan technique is considered one of the simplest ways for evaluating the real and imaginary parts of the complex nonlinear refractive index of materials. Fakhri et al. prepared GO/Au composite by solution mixing method and NLO properties have been measured using the z-scan technique (see Fakhri et al. 2016a).

Mechanical properties

Pure graphene structures indicate 2D plane sheets of covalently bonded carbon atoms that shape ideal hexagonal crystal lattices. Graphene specimens usually exist as either monolayers related to substrates made of another material. Graphene's special mechanical properties are as effective as its electrical and optical properties. The latter amount is about 200 times bigger than the one measured in steel. The evaluation of the Young's modulus yielded almost 0.5–1.0 TPa. These amounts, mixed with the relative low value of thin graphite, produce this material that is an ideal candidate for mechanical reinforcement as proposed in literatures. Moreover, due to its special elastic properties, graphene will be performed soon in flexible electronic devices such as transparent and stretchable screens, displays, sensors and antenna. We can observe that graphene is very strong and hard, and is a promising candidate for new applications such as pressure sensors and resonators. The experimental data on the Young modulus (E = 1 TPa) and the intrinsic strength (σ = 130 GPa) is displayed. These amounts of E and σ are very large and produce graphene to be extremely attractive for structural and other applications. This amount is one of the highest ever measured for real materials such as diamond. By the way, graphene can be easily bent, and this particular behavioral specification can be used in practice (see Ovidko 2013). These special mechanical properties of graphene are of superlative importance for its applications because they are highly essential (i) to be used as a super strong structural material; (ii) to realize and rein persistence of graphene used in electronics and energy storage; (iii) to plastically form curved graphene samples for electronics, optics and structural applications; (iv) to use composites with graphene inclusions as structural and/or functional materials. The melting temperature of thin samples reduces with the layer thickness and for considerably decreased thicknesses, the samples tend to become precarious. It is the basic reason why 2D materials like graphene were thought inconceivable to exist. Graphene was made to be not just thermodynamically stable up to high temperatures, but is also the strongest material to ever exist. The source of graphene's robustness lies in the σ bonds which link the carbon atoms in a solid honeycomb packed structure. Also, graphene flakes obtained by mechanical exfoliation of graphite have an extremely good crystalline quality. The strength of the atomic bonds ensures the lack of dislocation or other crystal defects. It is also confirmed by the high efficiency shown by the charge carriers which can move thousands of interatomic distances without scattering. GO composites are carbon-based materials with great efficiency and low cost. They have great Young's modulus and tensile strength. They are practical in a diversity of civil, mechanical and aerospace applications. They have also found widespread use since their creation. They have extremely advantageous mechanical properties, and at the same time, they are light and easy to prepare.

Thermal properties

Thermal conductivity shows the capability of a material to contact heat. Graphene is mostly thought to hold benefits over other materials because of its higher thermal conductivity. So, high thermal conductivity could propose very good heat sinking and low temperature rise during device operation. But, under high-field and high-temperature operating conditions, considerable dissipation and temperature rise can nevertheless happen in graphene devices. The thermal conductivity (κ) of graphene is dominated by phonon transport, namely diffusive conduction at high temperature and ballistic conduction (transport without scattering) at enough low temperature. Thermal stability is one of the key agents for better efficiency and dependability of electronic devices (see Shahil et al. 2012). Significant amount of heat generated during the device operation needs to be dissipated. Carbon allotropes like graphite and diamond have shown a higher thermal conductivity due to strong C-C covalent bonds. Graphene based nanoparticles are expected to display high thermal conductivity because graphene is inherently an excellent thermal conductor. The number of graphene layers and interlayer spacing are known to have a considerable effect on thermal conductivity. By increasing the number of layers and the spacing between them, the thermal conductivity can be considerably decreased. So, reduction of GO is significant to increment the thermal conductivity. GO is universally reduced by chemical and thermal behavior to remove the oxygen-containing functional groups (see Naghdi et al. 2016c). Efficient low temperature ways and environment friendly ways for reduction of GO are highly demanded.

Photocatalytic properties

In the last decades, substantial consideration has been given to the application of graphene and GO-based nanocomposites with semiconducting properties in photoelectrochemistry areas like electrochemical solar cells, photocatalytic degradation of organic pollutants, water splitting for hydrogen evolution, photocatalytic conversion of fuels, etc. An electron transition from valence band to conduction band in semiconductors by a photon generates an electron-hole pair. The photoelectrochemical efficiency and the visible light absorption by the semiconductor materials have been modified by various metal-ions doping, anion doping, adding co-catalyst, loading noble metal particles, dye sensitization and forming composite semiconductors. The great absorptivity, transparency, conductivity, diverse functionalities and controllability of graphene is due to its super photoelectrochemical efficiency by manufacturing its hybrids with semiconductors. The band gap and band potentials of the semiconductors are significant for the special kind of photocatalytic method. Graphene is considered to be a prominent candidate for hybrid photocatalytic materials with a work function around –4.4 eV, whereas significant semiconductor has their conduction band position at around –3.5 eV. It helps in the photogenerated electron transfer procedure from conduction band of the semiconductor to graphene surface. This delocalization of the photogenerated electron on graphene surface prevents the recombination of electron hole pairs within the semiconductor, leading to an enhancement of the photocatalytic efficiency. Graphene in nanocomposite photocatalytic materials also has a significant role by enhancing the solar light absorption efficiency. Graphene also reduces the band gap of the semiconductors leading to an effective solar light absorber. Also, graphene with its high surface area makes enough space for the adsorption of organic pollutants, which is one of the key factors for heterogeneous photocatalysis. In totality, TiO_2 is usually known as the most available and commercially cheapest photocatalyst. It has been well documented that GO is a strongly oxygenated graphene that is readily exfoliated in water to yield stable dispersions including mostly of single-layer sheets (see Stengl et al. 2013). The use of GO like the nano-size substrates for the formation of nanocomposites with metal and metal oxides is highly explored due to an idea to take a hybrid which could be combined with both the properties of GO like fascinating paper-shape material and the features of single nano-sized metal and metal oxide particles. Last research concentrates on the fabrication and applications of GO modified photocatalyst because of their particularly large surface area and great activity in most catalytic processes.

Catalytic properties

Carbon materials are usually used like catalytic supports due to their high surface area, great electrical conductivity, resistance to corrosion and structural stability. More information about the surface area of carbon supported heterogeneous catalysts is given in Table 2. GO, due to its inimitable structure and great properties, is considered a promising material for catalytic support in fuel cell electrodes. It suggests high conductivity, high surface area, exceptionally high mechanical strength and one of the fastest available electron transfer capabilities. More information about the general properties of graphene is given in Table 3 (see Julkapli et al. 2015). Nanotechnology suggests new insights in the development of advanced materials and composite for alternative energy sources. Study on materials and composites for catalytic applications focuses on making more active support surfaces and achieving lower precious metal content, optimizing the catalytic metal nanoparticles. Tuning the size and the morphology of the sample and the properties of the support material is essential to increase the catalyst efficiency. During recent years, graphene-based nanocatalysts have attracted extensive attention in catalyzed organic reactions due to their unique structures, big surface areas, easy recovery and recyclability, high thermal stability, high loading capacity of MNPs and other excellent properties (Figure 8). For example, Nasrollahzadeh et al. reported the preparation of the GO-ZnO nanocomposite and its application for the synthesis of

Table 2. Summary of surface area of carbon supported heterogeneous catalysts.

Catalyst	Total Surface Area ($m^2\ g^{-1}$)	Reference
PtRu supported carbon black support	29	Takasu et al. 2003
Pt catalyst supported graphitized carbon black	136	Bett et al. 1976
Multiwalled carbon nanotube supported platinum	42	Wenzhen et al. 2003
Pt supported carbon	250	Xingwen et al. 2003
Thiophene hydrodesulfurization of molybdenum supported activated carbon	1200	Calafata et al. 1996
Iron acetate supported macroporous carbon	40	Frédéric et al. 2003
Iron acetate supported mesoporous carbon	40	Frédéric et al. 2003
Iron acetate supported microporous carbon	40	Frédéric et al. 2003

Table 3. The properties of graphene.

Properties	Values
Tensile strength	130 GPa
Electron mobility	$15 \times 10^3\ cm^2\ V^{-1}$
Thermal conductivity	$4.84–5.30 \times 10^3\ WmK^{-1}$
Surface area to mass ratio	$2600\ m^2\ g^{-2}$
Superior charge carrier mobility	$2 \times 105\ cm^2\ V^{-1}\ s^{-1}$

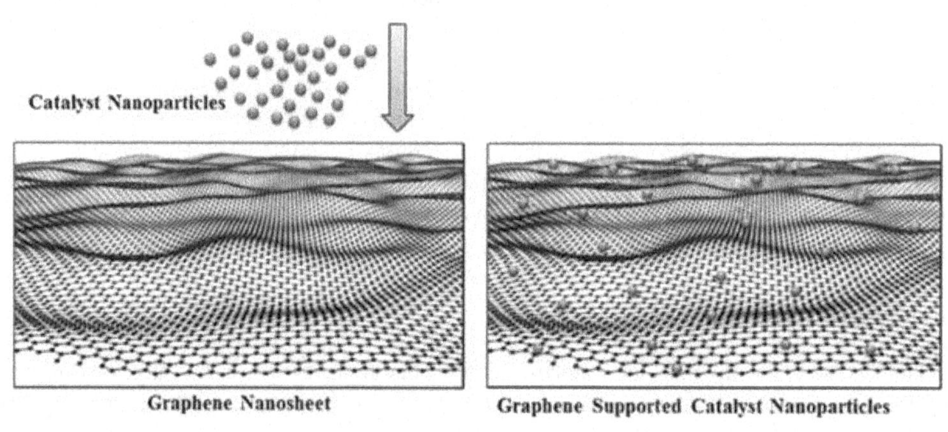

Figure 8. Penetration mechanism of heterogeneous nano catalyst towards graphene sheets. Reproduced with permission from ref. (Julkapli et al. 2015). Copyright (2015) Sciencedirect.

5-substituted 1*H*-tetrazoles in good to excellent yields (see Nasrollahzadeh et al. 2014). The results display great catalytic property of GO/ZnO nanocomposite.

Lithium ion batteries

The efficiency of Lithium ion batteries (LIBs) is greatly related to the physical and chemical properties of cathode and anode materials. Due to its great theoretical lithium storage capacity, graphite is commercially exploited as an anode material, while the storage capacity is not sufficient for high energy capacity, which sparked the study for other alternative anode materials. However, the very theoretical lithium storage capacity of a single graphene layer is higher than that of graphite and its applied applications in LIBs face severe limitations due its natural propensity for stacking. The concatenation of metal and metal oxide nanoparticles with high particular capacity into graphene not only prevents the aggregation of graphene layers but it also increases its lithium storage capacity. For instance, TiO_2 has long been targeted in high-performance lithium ion batteries (LIBs). Particularly, mesoporous anatase nanoparticles and carbon-coated TiO_2 nanoparticles comfort rapid ion diffusion and enhance the conductivity of bulk materials. Several rGO/TiO_2 nanocomposites with increased lithium storage capabilities compared to pure TiO_2 have been reported (see Khan et al. 2015).

Fuel cells

Fuel cell is one kind of electrical energy conversion device that transforms chemical fuels to electrical power by oxidizing fuels (anode) and decreasing oxygen from air (cathode) in the presence of catalysts. The challenge for fuel cells is to increase highly active and low-cost catalysts. Lately, graphene-based metal nanocomposite used as electrocatalysts have demonstrated great efficiency in fuel cells due to the large particular surface area and great conductivity of graphene and good catalytic activities of metals.

Solar cells

Solar photovoltaics (PV) capture the sun's energy and convert it into electricity. PV cells are prepared from layers of semiconducting material, and make an electric field across the layers when exposed to sunlight. When light is trapped into the cell, it is absorbed into the semiconducting material and causes electrons to break loose and flow. These flows of electrons are an electric current. Graphene is a 2D material with amazing characteristics. It is very strong and also astonishingly conductive and flexible. Solar cells need materials that are conductive and allow light to get through, so profiting from graphenes super conductivity and transparency. Graphene is indeed an excellent conductor; however, it is not very good at collecting the electrical current made inside the solar cell (see Kusumawati et al. 2016). Hence, researchers are looking for appropriate ways to modify graphene for this purpose. For instance, GO is less conductive but more transparent and a better charge collector which can be beneficial for solar panels.

Chemical sensors

Sensors are good devices that detect events that happen in the physical environment and usually respond with an electrical, mechanical or optical signal. The large surface area of graphene is able to increase the surface loading of desired biomolecules, and great conductivity and small band gap can be useful for conducting electrons between biomolecules and the electrode surface. Graphene is thought to become especially widespread in biosensors and diagnostics. Graphene also has considerable potential for enabling the growth of electrochemical biosensors. Graphene-based nanoelectronic devices have also been studied for application in DNA sensors, gas sensors, PH sensors, environmental contamination sensors, strain and pressure sensors (see Hazra et al. 2016).

Graphene-based supercapacitors

In recent years, global efforts have been made to find resources of energy that are clean and less limited. The decreasing sources of fossil fuels have forced the scientists to explore better substitutes for them such as wind and solar energy (see He et al. 2016, Li et al. 2016). For using the electrical energy which is provided by these types of energy resources, for a longer time and at any place we need, energy storage devices can help to store the electrical energy and release it at the time required. Supercapacitors are one the best devices for this aim that have attracted attention in recent years because of their ability in long cycle life, high power density, and rapid charge-discharge rate, which the great function of supercapacitors is highly dependent on the materials that have been chosen as their electrode (see He et al. 2016). Supercapacitors are of two basic categories (base on mechanism of energy storage): pseudo capacitors and electrochemical double-layer capacitors (EDLCs) that use carbon materials as their electrode because it's inexpensive and corrosion-resistant. Basically, EDLCs store energy by adsorption of anions and cations at the electrode-electrolyte interface, which is a reversible process. The capacitance of these devices is dependent on the specific surface area of the used electrode materials which are made of carbon (see Li et al. 2016, Chee et al. 2016). Recently, activated carbon and mesoporous carbon has been studied as the electrodes for EDLCs, while graphene is becoming more interesting for researchers as a basic material in the electrodes of supercapacitors (see Mohanapriya et al. 2016).

Advantages of using graphene in supercapacitors

As was mentioned in the previous part, among the different carbon materials, graphene has attracted the researchers' attention as electrode material for supercapacitors. Graphene is one of the most interesting materials and it is an excellent candidate for electrode in supercapacitors because of its high carrier mobility, corrosion resistance in aqueous electrolytes, large specific surface area, good flexibility, and strong mechanical strength (see Li et al. 2016, Mohanapriya et al. 2016, Balasubramaniam and Balakumar 2016). Kannappan et al. investigated a high energy density, power density and specific capacitance of highly porous graphene based supercapacitors. They reported that the retentivity of the capacitance after several tens of thousands of cycling is stable (see Kannappan et al. 2013). Liu et al. reported a supercapacitor with graphene-based electrodes with a perfect specific energy density. They mentioned that the energy density values are comparable to the Ni metal hydride battery, but in graphene based supercapacitor charge or discharge can be performed in seconds or minutes, which is because of specific surface area of single-layer (see Liu et al. 2010).

What is the difficulty in using graphene as an electrode in supercapacitors?

Though graphene-based materials emerged as perfect materials for electrode of supercapacitors, the application of graphene directly as material for electrode is not effective for energy storage. The basic reason is the strong van der Waals force between two graphene layers which causes the graphene sheets to aggregate easily (see Li et al. 2016), and as a result, electrode loses the specific surface area. Therefore, some ideas were developed to solve this problem, such as the synthesizing graphene in foams shape, three-dimensional porous graphene, and the design of sandwich-like structures with metal oxides or carbon nanomaterials as spacers between the sheets (see He et al. 2016).

Supercapacitors of nanocomposite

Supercapacitors are a significant class of energy storage devices that display high power density, long cycle life and excellent charge-discharge rates compared to common batteries. A capacitor works as an energy storage medium similar to an electrochemical battery. In the last years, graphene has attracted considerable attention as an electrode material in supercapacitors for its high surface area, chemical inertness, great flexibility, and superb electrical conductivity. Also, the applied use of the whole surface area of graphene is difficult; therefore, graphene has been widely used in combination with other electrochemically active metal oxides and hydroxides as electrode materials in supercapacitors. One of the example is using graphene/MnO_2 and Mn_3O_4 nanocomposite as electrode materials in supercapacitors. (see Yan et al. 2010).

Biological applications

Graphene and graphene-based nanomaterials have excellent potential indifferent biomedical applications. Various studies have been carried out on biomedical applications of graphene like drug/gene delivery, imaging, antibacterial and anti-cancer activities. However, due the low solubility in water, graphene or rGO has hardly been studied as an anti-bacterial agent. GO with oxygen containing functional group is water soluble and thus more biocompatible than graphene. Table 1 summarizes the major applications of most studied metal oxide/graphene nanocomposites (see Khan et al. 2015).

Antibacterial properties

All carbon nanomaterials are able to show antibacterial properties. The antibacterial activity of carbon nanomaterials is strongly dependent on surface chemistry, which surface chemistry determins critical factors like hydrophobicity or oxidation power. The antibacterial activity of graphene and GO is extensively documented and has already been utilized to create antibacterial materials for special applications and modern technology. Liu et al. evaluated the antibacterial activity of graphite, graphite oxide, GO and reduced GO. The results show that GO and rGO are both strongly antibacterial (see Liu et al. 2011). In these studies, GO showed stronger antibacterial activity than rGO. Most studies with antibacterial materials based on graphene explain nanocomposites of graphene with antibacterial nanoparticles. This study is governed by the principle that the antibacterial activity of the single component is lower than that of the composite materials. For instance, numerous researches explain the formation of antibacterial composites of graphene and silver nanoparticles, improving the antibacterial activity of silver (see Maas 2016).

Polymer nanocomposite

In the past decades, polymer nanocomposite has attracted great interest due to their excellent properties and possibility for different applications in new modern technology. The properties of nanocomposites are often a simple combination of the properties of various nanoparticles and polymeric matrix. Although sometimes novel specifications appear due to synergistic effects, graphene with fantastic high elastic modulus and great electrical conductivity has excellent prospects for use like the filler material for manufacture novel polymer nanocomposites designed for various applications in new technology. For example, Jaleh et al. prepared PVDF/rGO/ZnO by solution mixing method (see Jaleh et al. 2014). In their study, rGO and ZnO was added into PVDF to produce a composite film with two purposes. In the first, the results showed that the interface value and the β-phase content increased with increasing rGO/ZnO in composite. In the second, nanocomposite films have higher resistance to thermal degradation compared to pure PVDF. Fakhri et al. prepared PVDF/Au/Cu/GO composite (see Fakhri et al. 2016). Electroactive phase content and dielectric properties of PVDF were improved with Au and Cu doped GO.

Performance of graphene and GO as corrosion protection barrier

For some time, protection of the surfaces of reactive metals has developed into a significant industry that employs many different approaches, including the use of different types of coatings (see Kalita et al. 2014, Kirkland et al. 2012, Parasai et al. 2012, Chen et al. 2011, Naghdi et al. 2015). New approaches for manufacturing protective coatings introduce a high potential protective layer on the metal surface to protect against attack by a corrosive environment. While it protects against corrosion, this layer often modifies the physical properties of the metal surfaces and has a negative effect on the appearance, dimensions, optical properties, and even the thermal and electrical conductivity of the protected metals (see Kalita et al. 2014, Kirkland et al. 2012, Parasai et al. 2012, Chen et al. 2011, Naghdi et al. 2015, Montemor 2014). Therefore, a new challenge is to develop a coating with the ability to protect the metal surface from destructive factors while imparting few negative effects on the physical properties of the coated-metal. In the last decade, graphene and GO were introduced as a one-atom-thick layer that could provide a protective barrier between the metal surface and destructive environment. Graphene family became heavily favored because of their outstanding physical and chemical properties, including high electrical and thermal conductivity, minimal coating thickness, and thermal and chemical stability (see Kalita et al. 2014, Kirkland et al. 2012, Parasai et al. 2012, Chen et al. 2011, Naghdi et al. 2015, Berry 2013, Marimuthu et al. 2014, Saud et al. 2016). Application of graphene, GO and graphene-based material as corrosion and oxidation barrier was examined by several researchers. In this aspect, we discuss the application of graphene and GO as a barrier coating. The ability of graphene and GO to protect metal surfaces from destructive environment can be divided in three different aspects: protection from corrosion, from oxidation, and from microbial corrosion.

As corrosion resistance

First we start with the ability of graphene and GO as an anti-corrosion barrier. Using graphene as a coating on metals can provide the extremely lightweight coating that can't significantly alter optical properties, thermal and electrical conductivity, and size of the substrate. Furthermore, it can prevent charge transfer at the metal-electrolyte interface (see Kirkland et al. 2012). The corrosion-resistant alloys with the ability to develop a corrosion-resistant oxide film can also be disrupted by chloride ions which are present in seawater or other corrosive environment (see Raman and Tiwari 2014). Kirkland et al. have shown that graphene coatings upon Ni and Cu can protect them from being electrochemically corroded in aqueous media. They mentioned that both the thickness and structures of the graphene coating are effective in the ability of graphene as corrosion barrier (see Kirkland et al. 2012). Kyhl et al. deposited a single layer of graphene on Pt surface. They found that this graphene layer can protect the Pt surface from oxidation in air for the 6 months experiment, water for the 14 hour test period, and in NaCl solution for the 75 min test period at room temperature (see Kyhl et al. 2015). Dumee et al. demonstrated the growth of 3D networks of graphene across porous stainless steel substrates, and they investigated the anti-corrosion properties of the coating. Their results showed that the presence of the graphene coating effectively increased the specific surface area of the material and also increased the corrosion resistance and electrical conductivity of the material without altering the structure or properties of the native stainless steel (see Dumee et al. 2015).

As oxidation resistance

The impermeability of the gas that originates from the unique atomic structure of graphene introduced graphene as an excellent candidate for oxidation barrier coating of metallic substrates. The potential of graphene as a protection layer is based on its unique stability at extremely high temperatures (higher than 1500°C) and also its perfect physical and chemical properties that makes it a natural diffusion barrier which provides a physical separation between the destructive environment and protected metal (see Chen et al. 2011). Chen et al. demonstrated the ability of graphene coating to protect the surface of Cu and Cu/Ni alloy from air oxidation. They showed that the metal surface was very well protected from oxidation even after heating at 200°C in air for up to 4 hours (see Chen et al. 2011). Kang et al. investigated the oxidation resistance of rGO-coated Fe and Cu foils after heat treatment at 200°C in air for 2 h. These coatings were prepared by transferring multilayers rGO from a SiO_2 substrate onto the Fe and Cu substrate. The results showed that the bare metal surfaces were oxidized under heating process, but the rGO-coated metal surfaces were well protected from oxidation (see Kang et al. 2012).

As microbial corrosion resistance

Typically, corrosion occurs under acidic conditions, but in some environments it can happen by microbes even under neutral pH conditions and ambient temperatures, which is usually more complicated than the corrosion in acidic condition. In microbial corrosion, microbes develop biofilm coatings on the metal surface in an aqueous environment, and these biofilm layers may alter the metal-solution interface and lead to increase in metallic corrosion rate (see Sanar et al. 2010). This kind of corrosion limits the use of metallic structures in a variety of applications. A typical method to prevent microbially-induced corrosion (MIC) is developing passive physical or chemical layers on the metallic surfaces (see Sanar et al. 2010). Krishnamurthy et al. reported the use of graphene as a passive layer for MIC of metals for more than 2700 h. They investigated the effectiveness of the graphene coating from MIC by using Ni foam as an electrode in a microbial fuel cell. They showed that the MIC rates of Ni foam with graphene coating are effectively lower than the Ni foam without graphene coating (see Krishnamurthy et al. 2013). They also reported the better protection ability of an ultra-thin graphene layer as an anti-MIC coating in comparison to the two commercial polymeric coatings, Parylene-C (PA) and Polyurethane (PU). They showed that the dissolution rate of graphene-coated Ni in a corrosion cell was an order of magnitude lower than that of PA and PU coated Ni electrodes. Furthermore, they investigated the better performance of as-grown graphene vs. transferred graphene films for anti-MIC applications. They reported that the as-grown graphene coating was devoid of major defects, while wet transfer of graphene introduced large scale defects in graphene structure that made it less suitable for microbial corrosion prevention (see Krishnamurthy et al. 2015).

Some works involving graphene-based nanocatalysts applications are summarized in Table 4.

Table 4. Summary of the major applications of metal oxide-graphene nanocomposites.

Materials	Applications	References
Au/rG	Biosensor Electrocatalyst	Zhang et al. 2011 Hu et al. 2012
ZnO/rG	Photocatalyst	Chen et al. 2013
ZnO/G	Supercapacitors and electrochemical sensors	Dong et al. 2012a
ZnO/GO	Catalyst	Nasrollahzadeh et al. 2014
Ag/rGO	Biosensor for glucose	Maas 2016
Ag/TiO_2/GO	Catalyst	Nasrollahzadeh et al. 2016a
Cu-graphene	Biosensor for glucose	Luo et al. 2012
Cu/rGO	Catalyst	Fakhri et al. 2014
CuNiO/G	glucose sensors	Zhang et al. 2015
Pt-Pd/Nafion-graphene	Electrocatalyst	Yang et al. 2012
Pt/TiO_2/rGO	Catalyst	Zhao et al. 2012
Pt-Au/graphene	Electrocatalyst	Hu et al. 2011
Pd/GO	Catalyst	Omidvar et al. 2017a
CuO-N/rGO	Direct methanol fuel cells	Zhou et al. 2013
Ni-AL LDH/RG	Supercapacitor	Gao et al. 2011
Fe_2O_3/GO	Catalyst	Naghdi et al. 2016b
Fe_3O_4/rGO	Supercapacitors Sensors	Shi et al. 2011 Teymourian et al. 2013
V_2O_5/GO	Li-ion batteries	Liu et al. 2015
NA-NiONF/rGO	Nonenzymatic glucose sensor	Zhang et al. 2012
Co_3O/G	Supercapacitors	Dong et al. 2012b
Co_3O_4/G	Anode materials for Li-ion batteries	Yang et al. 2010
TiO_2/GO	Photocatalyst	Naghdi et al. 2016a
TiO_2/G	Adenine and Guanine sensors	Fan et al. 2011
TiO_2/graphene	Photocatalytic activity Lithium ion batteries and photocatalytic activity	Stengl et al. 2011 Li et al. 2011a
PtAu-MnO_2/G	Nonenzymatic glucose sensors	Xiao et al. 2013
MnO_2/GO	Sensor	Gan et al. 2013
MnO_2/GO	Fuel cell	Sheng et al. 2010
PdNi/GO	Electrocatalyst	Campesia et al. 2009
Pt-Ru/GO	Catalyst	Eve et al. 2002
Au@Pd/graphene	Direct ethanol fuel cells	Jeong et al. 2013
GO/Fe_3O_4/Pd	Catalyst	Omidvar et al. 2017b

References

Atarod, M., M. Nasrollahzadeh and S.M. Sajadi. 2015. Green synthesis of Cu/reduced graphene oxide/Fe_3O_4 nanocomposite using *Euphorbia Wallichii* leaf extract and its application as a recyclable and heterogeneous catalyst for the reduction of 4-nitrophenol and Rhodamine B. RSC Adv. 5: 91532–91543.

Atarod, M., M. Nasrollahzadeh and S.M. Sajadi. 2016. Green synthesis of Pd/RGO/Fe_3O_4 nanocomposite using Withania coagulans leaf extract and its application as magnetically separable and reusable catalyst for the reduction of 4-nitrophenol. J. Colloid Interf. Sci. 465: 249–258.

Balasubramaniam, M. and S. Balakumar. 2016. Tri-solvent mediated probing of ultrasonic energy towards exfoliation of graphene nanosheets for supercapacitor application. Mater. Lett. 182: 63–67.

Berry, V. 2013. Impermeability of graphene and its applications. Carbon 62: 1–10.

Bett, J.A.S., K. Kinoshita and P. Stonehart. 1976. Crystallite growth of platinum dispersed on graphitized carbon black: II. Effect of liquid environment. J. Catal. 41(1): 124–133.

Buasri, A., B. Ksapabutr, M. Panapoy and N. Chaiyut. 2013. Synthesis of biofuel from palm stearin using an activated carbon supported catalyst in packed column reactor. Adv. Sci. Lett. 19(12): 3473–3476.

Calafata, A., J. Lainea, A. López-Agudob and J.M. Palacios. 1996. Effect of surface oxidation of the support on the Thiophene hydrodesulfurization activity of Mo, Ni, and NiMo catalysts supported on activated carbon. J. Catal. 162(1): 20–30.

Campesia, R., F. Cuevasa, E. Leroya, M. Hirscherc, R. Gadioub, C. Vix-Guterlb et al. 2009. *In situ* synthesis and hydrogen storage properties of PdNi alloy nanoparticles in an ordered mesoporous carbon template. Microporous Mesoporous Mater. 117(1): 511–514.

Chee, W.K., H.N. Lim, Z. Zainal, N.M. Huang, I. Harrison and Y. Andou. 2016. Flexible graphene-based supercapacitors: A review. J. Phys. Chem. C 120: 4153–4172.

Chen, S., L. Brown, M. Levendorf, W. Cai, S.Y. Ju, J. Edgeworth et al. 2011. Oxidation resistance of graphene-coated Cu and Cu/Ni alloy. ACS Nano 5: 1321–1327.

Chen, Z., N. Zhang and Y.J. Xu. 2013. Synthesis of graphene-ZnO nanorod nanocomposites with improved photoactivity and anti-photocorrosion. Cryst. Eng. Commun. 15: 3022–3030.

Chung, D.D.L. 1987. Exfoliation of graphite. J. Mater. Sci. 22: 4190–4198.

Coraux, J., A.T. N'Diaye, C. Busse and T. Michely. 2008. Structural coherency of graphene on Ir(111). Nano Lett. 8(2): 565–570.

Dathbun, A. and S. Chaisitsak. 2013. Effects of three parameters on graphene synthesis by chemical vapor deposition. IEEE International Conference on Nano/Micro Engineered and Molecular Systems (NEMS).

Dong, X., Y. Cao, J. Wang, M.B. Chan-Park, L. Wang, W. Huang et al. 2012a. Hybrid structure of zinc oxide nanorods and three dimensional graphene foam for supercapacitor and electrochemical sensor applications. RSC Adv. 2: 4364–4369.

Dong, X.-C., H. Xu, X.-W. Wang, Y.-X. Huang, M.B. Chan-Park, H. Zhang et al. 2012b. 3D graphene-cobalt oxide electrode for high-performance supercapacitor and enzymeless glucose detection. ACS Nano 6: 3206–3213.

Dumee, L.F., L. He, Z. Wang, P. Sheath, J. Xiong, C. Feng et al. 2015. Growth of nano-textured graphene coatings across highly porous stainless steel supports towards corrosion resistant coatings. Carbon 87: 395–408.

Eda, G., C. Mattevi, H. Yamaguchi, H. Kim and M. Chhowalla. 2009. Insulator to semimetal transition in GO. J. Phys. Chem. C 113: 15768–15771.

Eve, S.S., A.D. Gregg and C.M. Lukehart. 2002. PteRu/carbon fiber nanocomposites: synthesis, characterization, and performance as anode catalysts of direct methanol fuel cells. A search for exceptional performance. J. Phys. Chem. B 106(4): 60–766.

Fakhri, P., B. Jaleh and M. Nasrollahzadeh. 2014. Synthesis and characterization of copper nanoparticles supported on reduced graphene oxide as a highly active and recyclable catalyst for the synthesis of formamides and primary amines. J. Mol. Catal. A Chem. 384: 17–22.

Fakhri, P., M.R. Rashidian Vaziri, B. Jaleh and N. Partovi Shabestari. 2016a. Nonlocal nonlinear optical response of graphene oxide-Au nanoparticles dispersed in different solvents. J. Opt. 18: 015502.

Fakhri, P., H. Mahmood, B. Jaleh and A. Pegoretti. 2016b. Improved electroactive phase content and dielectric properties of flexible PVDF nanocomposite films filled with Au- and Cu-doped graphene oxide hybrid nanofiller. Synthetic Metals 220: 653–660.

Falkovsky, L.A. 2008. Optical properties of graphene. Journal of Physics: Conference Series 129: 012004.

Fan, Y., K.-J. Huang, D.-J. Niu, C.-P. Yang and Q.-S. Jing. 2011. TiO_2-graphene nanocomposite for electrochemical sensing of adenine and guanine. Electrochim. Acta 56: 4685–4690.

Frédéric, J., M. Sébastien, D. Jean-Pol and L. Goran. 2003. Oxygen reduction catalysts for polymer electrolyte fuel cells from the pyrolysis of iron acetate adsorbed on various carbon supports. J. Phys. Chem. B 107: 1376–1386.

Fukada, S., Y. Shintani, M. Shimomura, F. Tahara and R. Yagi. 2012. Graphene made by mechanical exfoliation of graphite intercalation compound. Jpn. J. Appl. Phys. 51: 085101.

Gan, T., J. Sun, K. Huang, L. Song and Y. Li. 2013. A graphene oxide-mesoporous MnO_2 nanocomposite modified glassy carbon electrode as a novel and efficient voltammetric sensor for simultaneous determination of hydroquinone and catechol. Sens. Actuators B Chem. 177: 412–418.

Gao, Z., J. Wang, Z. Li, W. Yang, B. Wang, M. Hou et al. 2011. Graphene nanosheet/Ni2+/Al3+ layered double-hydroxide composite as a novel electrode for a supercapacitor. Chem. Mater. 23: 3509–3516.

Geim, A.K. and K.S. Novoselov. 2007. The rise of graphene. Nat. Mater. 6: 183–191.
Goumri, M., B. Lucas, B. Ratier and M. Baitoul. 2016. Electrical and optical properties of reduced graphene oxide and multi-walled carbon nanotube based nanocomposite: A comparative study. Opt. Mater. 60: 105–113.
He, X., N. Zhang, X. Shao, M. Wu, M. Yu and J. Qiu. 2016. A layered-template-nanospace-confinement strategy for production of corrugated graphene nanosheets from petroleum pitch for supercapacitors. Chem. Eng. J. 297: 121–127.
Hu, B., H. Ago, Y. Ito, K. Kawahara, M. Tsuji, E. Magome et al. 2012. Epitaxial growth of large-area single-layer graphene over Cu(111)/sapphire by atmospheric pressure CVD. Carbon 50: 57–65.
Hu, C., T. Lu, F. Chen and R. Zhang. 2013. A brief review of graphene-metal oxide composites synthesis and applications in photocatalysis. J. Chin. Adv. Mater. Soc. 1: 21–39.
Hu, J., F. Li, K. Wang, D. Han, Q. Zhang, J. Yuan et al. 2012. One-step synthesis of graphene-Au NPs by HMTA and the electrocatalytical application for O_2 and H_2O_2. Talanta 93: 345–349.
Hu, Y., H. Zhang, P. Wu, H. Zhang, B. Zhou and C. Cai. 2011. Bimetallic Pt-Au nanocatalysts electrochemically deposited on graphene and their electrocatalytic characteristics towards oxygen reduction and methanol oxidation. Phys. Chem. Chem. Phys. 13: 4083–4094.
Hussain, S., M.W. Iqbal, J. Park, M. Ahmad, J. Singh, J. Eom et al. 2014. Physical and electrical properties of graphene grown under different hydrogen flow in low pressure chemical vapor deposition. Nanoscale Res. Lett. 9: 546.
Jaleh, B. and A. Jabbari. 2014. Evaluation of reduced graphene/ZnO effect on properties of PVDF nanocomposite films. App. Sur. Sci. 320: 339–347.
Jeong, G.H., D. Choi, M. Kang, J. Shin, J.-G. Kang and S.-W. Kim. 2013. One-pot synthesis of Au@Pd/graphene nanostructures: electrocatalytic ethanol oxidation for direct alcohol fuel cells (DAFCs). RSC Adv. 3: 8864–8870.
Julkapli, N.M. and S. Bagheri. 2015. Graphene supported heterogeneous catalysts: An overview. Int. J. Hydrogen Energy 40: 948–979.
Kalita, G., M.E. Ayhan, S. Sharma, S.M. Shinde, D. Ghimire, K. Wakita et al. 2014. Low temperature deposited graphene by surface wave plasma CVD as effective oxidation resistive barrier. Corros. Sci. 78: 183–187.
Kang, D., J.Y. Kwon, H. Cho, J.H. Sim, H.S. Hwang, C.S. Kim et al. 2012. Oxidation resistance of iron and copper foils coated with reduced graphene oxide multilayers. ACS Nano 6: 7763–7769.
Kannappan, S., K. Kaliyappan, R.K. Manian, A.S. Pandian, H. Yange, Y.S. Lee et al. 2013. Graphene based supercapacitors with improved specific capacitance and fast charging time at high current density.
Kim, J., G. Lee and J. Kim. 2015. Wafer-scale synthesis of multi-layer graphene by high-temperature carbon ion implantation. Appl. Phys. Lett. 107: 033104.
Kim, S.M., A. Hsu, Y.H. Lee, M. Dresselhaus, T. Palacios, K.K. Kim et al. 2013. The effect of copper pre-cleaning on graphene synthesis. Nanotechnology 24: 365602–355609.
Kim, M.S., J.M. Woo, D.M. Geum, J.R. Rani and J.H. Jang. 2014. Effect of copper surface pre-treatment on the properties of CVD grown graphene. AIP Adv. 4: 127107.
Kirkland, N.T., T. Schiller, N. Medhekar and N. Birbilis. 2012. Exploring graphene as a corrosion protection barrier. Corros. Sci. 56: 1–4.
Khan, M., M. Nawaz Tahir, S. Farooq Adil, H. Ullah Khan, M. Rafiq, H. Siddiqui et al. 2015. Graphene based metal and metal oxide nanocomposites: synthesis, properties and their applications. J. Mater. Chem. A 3: 18753.
Kou, R., Sh. Tanabe, T. Tsuchizawa, K. Warabi, S. Suzuki, H. Hibino et al. 2013. Characterization of optical absorption and polarization dependence of single-layer graphene integrated on a silicon wire waveguide. Jpn. J. Appl. Phys. 52: 060203.
Krishnamurthy, A., V. Gadhamshetty, R. Mukherjee, Z. Chen, W. Ren, H.M. Cheng et al. 2013. Passivation of microbial corrosion using a graphene coating. Carbon 56: 45–49.
Krishnamurthy, A., V. Gadhamshetty, R. Mukherjee, B. Natarajan, O. Eksik, S.A. Shojaee et al. 2015. Superiority of graphene over polymer coatings for prevention of microbially induced corrosion. Sci. Rep. 5: 13858.
Kumar Hazra, S. and S. Basu. 2016. Graphene-oxide nano composites for chemical sensor applications. J. Carbon Res. 2: 12.
Kusumawati, Y., S.K. Daoud and T. Pauporte. 2016. TiO_2/graphene nanocomposite layers improving the performances of dye-sensitized solar cells using a cobalt redoxshuttle. J. Photochem. Photobiol. A Chem. 329: 54–60.
Kwon, S.Y., C.V. Ciobanu, V. Petrova, V.B. Shenoy, J. Bareno, V. Gambin et al. 2009. Growth of semiconducting graphene on palladium. Nano Lett. 9(12): 3985–3990.
Kyhl, L., S.F. Nielsen, A.G. Cabo, A. Cassidy, J.A. Miwa and L. Hornekær. 2015. Graphene as an anti-corrosion coating layer. Faraday Discuss 180: 495–509.
Li, X., W. Cai, J. An, S. Kim, J. Nah, D. Yang et al. 2009. Large-area synthesis of high-quality and uniform graphene films on copper foils. Science 324: 1312–1314.
Li, J., X. Huang, L. Cui, N. Chen and L. Qu. 2016. Preparation and supercapacitor performance of assembled graphene fiber and foam. Prog. Nat. Sci. Mater. Int. 26: 212–220.
Li, N., G. Liu, C. Zhen, F. Li, L. Zhang and H.M. Cheng. 2011a. Battery performance and photocatalytic activity of mesoporous anatase TiO_2 nanospheres/graphene composites by template-free self-assembly. Adv. Funct. Mater. 21: 1717–1722.
Li, Z., P. Wu, C. Wang, X. Fan, W. Zhang, X. Zhai et al. 2011b. Low-temperature growth of graphene by chemical vapor deposition using solid and liquid carbon sources. ACS Nano 5: 3385–3390.
Liu, Q., Z. Li, Y. Liu, H. Zhang, Y. Ren, C. Sun et al. 2015. Graphene-modified nanostructured vanadium pentoxide hybrids with extraordinary electrochemical performance for Li-ion batteries. Nat. Commun. 6: 1–10.

Liu, C., Z. Yu, D. Neff, A. Zhamu and B.Z. Jang. 2010. Graphene-based supercapacitor with an ultrahigh energy density. Nano Lett. 10: 4863–4868.
Liu, R., Q. Zhao, Y. Li, G. Zhang, F. Zhang and X. Fan. 2013. Graphene supported Pt/Ni nanoparticles as magnetically separable nanocatalysts. J. Nanomater. 2013: Article ID 602602, 1–7.
Liu, S., T.H. Zeng, M. Hofmann, E. Burcombe, J. Wei, R. Jiang et al. 2011. Antibacterial activity of graphite, graphite oxide, graphene oxide, and reduced graphene oxide: Membrane and oxidative stress. ACS Nano 5: 6971–6980.
Luo, J., S. Jiang, H. Zhang, J. Jiang and X. Liu. 2012. A novel non-enzymatic glucose sensor based on Cu nanoparticle modified graphene sheets electrode. Anal. Chim. Acta 709: 47–53.
Luo, Z., X. Ma, D. Yang, L. Yuwen, X. Zhu, L. Weng et al. 2013. Synthesis of highly dispersed titanium dioxide nanoclusters on reduced graphene oxide for increased glucose sensing. Carbon 57: 470–476.
Maas, M. 2016. Carbon nanomaterials as antibacterial colloids. Materials 9: 617.
Maham, M., M. Nasrollahzadeh, S.M. Sajadi and M. Nekoei. 2017. Biosynthesis of Ag/reduced graphene oxide/Fe_3O_4 using *Lotus garcinii* leaf extract and its application as a recyclable nanocatalyst for the reduction of 4-nitrophenol and organic dyes. J. Colloid Interf. Sci. 497: 33–42.
Marimuthu, M., M. Veerapandian, S. Ramasundaram, S.W. Hong, P. Sudhagar and S. Nagarajan. 2014. Sodium functionalized graphene oxide coated titanium plates for improved corrosion resistance and cell viability. Appl. Surf. Sci. 293: 124–131.
Maryami, M., M. Nasrollahzadeh, E. Mehdipour and S.M. Sajadi. 2016. Green synthesis of silver nanoparticles supported on reduced graphene oxide using Abutilon hirtum leaf extract and its application as a recyclable heterogeneous catalyst for the reduction of organic dyes in aqueous medium. Int. J. Hydrogen Energy 41: 21236–21245.
Mohammadi, S., Z. Kolahdouz, S. Darbari, S. Mohajerzadeh and N. Masoumi. 2013. Graphene formation by unzipping carbon nanotubes using a sequential plasma assisted processing. Carbon 52: 451–463.
Mohanapriya, K., G. Ghosh and N. Jha. 2016. Solar light reduced graphene as high energy density supercapacitor and capacitive deionization electrode. Electrochimica Acta 209: 719–729.
Montemor, M.F. 2014. Functional and smart coatings for corrosion protection: A review of recent advances. Surf. Coating Tech. 258: 17–37.
Naghdi, S., B. Jaleh and A. Ehsani. 2015. Electrophoretic deposition of graphene oxide on aluminum: characterization, low thermal annealing, surface and anticorrosive properties. Bull. Chem. Soc. Jpn. 88: 722–728.
Naghdi, S., B. Jaleh and N. Shahbazi. 2016a. Reversible wettability conversion of electrodeposition graphene oxide/titania nanocomposite coating: investigation of surface structure. App. Sur. Sci. 368: 409–416.
Naghdi, S., K.Y. Rhee, B. Jaleh and S.J. Park. 2016b. Altering the structure and properties of iron oxide nanoparticles and graphene oxide/iron oxide composite by urea. App. Sur. Sci. 364: 686–693.
Naghdi, S., K.Y. Rhee, M.T. Kim, B. Jaleh and S.J. Park. 2016c. Atmospheric chemical vapor deposition of graphene on molybdenum foil at different growth temperatures. Carbon Lett. 18: 37–42.
Nasrollahzadeh, M., B. Jaleh and A. Jabbari. 2014. Synthesis, characterization and catalytic activity of graphene oxide/ZnO nanocomposites. RSC Adv. 4: 36713–36720.
Nasrollahzadeh, M., M. Maham, A. Rostami-Vartooni, M. Bagherzadeh and S.M. Sajadi. 2015. Barberry fruit extract assisted *in situ* green synthesis of Cu nanoparticles supported on reduced graphene oxide-Fe_3O_4 nanocomposite as magnetically separable and reusable catalyst for the *O*-arylation of phenols with aryl halides under ligand-free conditions. RSC Adv. 5: 64769–64780.
Nasrollahzadeha, M., M. Atarod, B. Jaleh and M. Gandomi. 2016a. *In situ* green synthesis of Ag nanoparticles on graphene oxide/TiO_2 nanocomposite and their catalytic activity for the reduction of 4-nitrophenol, congo red and methylene blue. Ceram. Int. 42: 8587–8596.
Nasrollahzadeh, M., S.M. Sajadi, A. Rostami-Vartooni, M. Alizadeh and M. Bagherzadeh. 2016b. Green synthesis of Pd nanoparticles supported on reduced graphene oxide using barberry fruit extract and its application as a recyclable and heterogeneous catalyst for the reduction of nitroarenes. J. Colloid Interf. Sci. 466: 360–368.
Nasrollahzadeh, M., M. Atarod and S.M. Sajadi. 2017. Biosynthesis, characterization and catalytic activity of Cu/RGO/Fe_3O_4 for direct cyanation of aldehydes with $K_4[Fe(CN)_6]$. J. Colloid Interf. Sci. 486: 153–162.
Navalon, S., A. Dhakshinamoorthy, M. Alvaro and H. Garcia. 2014. Carbocatalysis by graphene-based materials. Chem. Rev. 114: 6179–6212.
Ovidko, I.A. 2013. Mechanical properties of graphene. Rev. Adv. Mater. Sci. 34: 1–11.
Omidvar, A., M.R. Rashidian Vaziri, B. Jaleh, N. Partovi Shabestari and M. Noroozi. 2016. Metal-enhanced fluorescence of graphene oxide by palladium nanoparticles in the blue-green part of the spectrum. Chin. Phys. B 25(11): 118102.
Omidvar, A., B. Jaleh and M. Nasrollahzadeh. 2017a. Preparation of the GO/Pd nanocomposite and its application for the degradation of organic dyes in water. J. Colloid Interf. Sci. 496: 44–50.
Omidvar, A., B. Jaleh, M. Nasrollahzadeh and H.R. Dasmeh. 2017b. Fabrication, characterization and application of GO/Fe_3O_4/Pd nanocomposite as a magnetically separable and reusable catalyst for the reduction of organic dyes. Chem. Eng. Res. Des. 121: 339–347.
Pei, S. and H.M. Cheng. 2012. The reduction of graphene oxide. Carbon 50: 3210–3228.
Prasai, D., J.C. Tuberquia, R.R. Harl, G.K. Jennings and K.I. Bolotin. 2012. Graphene: corrosion-inhibiting coating. Amer. Chem. Soc. 6: 1102–1108.

Raman, R.K.S. and A. Tiwari. 2014. Graphene: The thinnest known coating for corrosion protection. J. Miner. Met. Mater. Soc. 66: 637–642.
Saner, B., F. Okyay and Y. Yürüm. 2010. Utilization of multiple graphene layers in fuel cells. 1. An improved technique for the exfoliation of graphene-based nanosheets from graphite. Fuel 89: 1903–1910.
Saud, S.N., R. Hosseinian, H.R. Bakhsheshi-Rad, F. Yaghoubidoust, N. Iqbal, E. Hamzah et al. 2016. Corrosion and bioactivity performance of graphene oxide coating on Ti-Nb shape memory alloys in simulated body fluid. Mater. Sci. Eng. C 68: 687–694.
Shahil, K.M.F. and A.A. Balandin. 2012. Thermal properties of graphene and multilayer graphene: Application in thermal interface materials. Solid State Commun. 152: 1331–1340.
Sheng, C., Z. Junwu, W. Xiaodong, H. Qiaofeng and W. Xin. 2010. Graphene oxide-MnO_2 nanocomposites for supercapacitors. ACS Nano 4(5): 2822–2830.
Shi, W., J. Zhu, D.H. Sim, Y.Y. Tay, Z. Lu, X. Zhang et al. 2011. Achieving high specific charge capacitances in Fe_3O_4/reduced graphene oxide nanocomposites. J. Mater. Chem. 21: 3422–3427.
Stengl, V., D. Popelkova and P. Vlacil. 2011. TiO_2-graphene nanocomposite as high performace photocatalysts. J. Phys. Chem. C 115: 25209–25218.
Stengl, V., S. Bakardjieva, T.M. Grygar, J. Bludska and M. Kormunda. 2013. TiO_2-graphene oxide nanocomposite as advanced photocatalytic materials. Chem. Cent. J. 7: 41.
Takasu, Y., K. Kawaguchi, W. Sugimoto and Y. Murakami. 2003. Effects of the surface area of carbon support on the characteristics of highly-dispersed Pt-Ru particles as catalysts for methanol oxidation. Sci. Rep. 48(25): 3861–3868.
Teymourian, H., A. Salimi and S. Khezrian. 2013. Fe_3O_4 magnetic nanoparticles/reduced graphene oxide nanosheets as a novel electrochemical and bioeletrochemical sensing platform. Biosens. Bioelectron. 49: 1–8.
Tetlow, H., J. Posthuma de Boer, I.J. Ford, D.D. Vvedensky, J. Coraux and L. Kantorovich. 2014. Growth of epitaxial graphene: Theory and experiment. Phys. Rep. 542: 195–295.
Tiwary, C.S., B. Javvaji, C. Kumar, D.R. Mahapatra, S. Ozden, P.M. Ajayan et al. 2015. Chemical-free graphene by unzipping carbon nanotubes using cryo-milling. Carbon 89: 217–224.
Wakeland, S., R. Martinez, J.K. Grey and C.C. Luhrs. 2010. Production of graphene from graphite oxide using urea as expansion-reduction agent. Carbon 48: 3463–3470.
Wang, J., H. Yin, X. Meng, J. Zhu and S. Ai. 2011. Preparation of the mixture of graphene nanosheets and carbon nanospheres with high adsorptivity by electrolyzing graphite rod and its application in hydroquinone detection. J. Electroanal. Chem. 662: 317–321.
Wang, S.M., Y.H. Pei, X. Wang, H. Wang, Q.N. Meng, H.W. Tian et al. 2010a. Synthesis of graphene on a polycrystalline Co film by radio-frequency plasma-enhanced chemical vapor deposition. J. Phys. D Appl. Phys. 43: 455402.
Wang, Z.Y., B.B. Huang, Y. Dai, Y.Y. Liu, X.Y. Zhang, X.Y. Qin et al. 2012. Crystal facets controlled synthesis of graphene@ TiO_2 nanocomposites by a one-pot hydrothermal process. Cryst. Eng. Commun. 14: 1687–1692.
Wang, D.H., R. Kou, D.W. Choi, Z.G. Yang, Z.M. Nie, J. Li et al. 2010b. Ternary self-assembly of ordered metal oxide-graphene nanocomposites for electrochemical energy storage. ACS Nano 4: 1587–1595.
Wenzhen, L., L. Changhai, Z. Weijiang, Q. Jieshan, Z. Zhenhua, S. Gongquanand et al. 2003. Preparation and characterization of multiwalled carbon nanotube-supported platinum for cathode catalysts of direct methanol fuel cells. J. Phys. Chem. B 107: 6292–6299.
Wu, Y., G. Yu, H. Wang, B. Wang, Z. Chen, Y. Zhang et al. 2012. Synthesis of large-area graphene on molybdenum foils by chemical vapor deposition. Carbon 50: 5226–5231.
Xiao, B., X. Li, X. Li, B. Wang, C. Langford, R. Li et al. 2014. Graphene nanoribbons derived from the unzipping of carbon nanotubes: controlled synthesis and superior lithium storage performance. J. Phys. Chem. C 118: 881–890.
Xiao, F., Y. Li, H. Gao, S. Ge and H. Duan. 2013. Growth of coral-like PtAu-MnO_2 binary nanocomposites on free-standing graphene paper for flexible nonenzymatic glucose sensors. Biosens. Bioelectron. 41: 417–423.
Xingwen, Y. and Y. Siyu. 2007. Recent advances in activity and durability enhancement of Pt/C catalytic cathode in PEMFC: Part I. Physico-chemical and electronic interaction between Pt and carbon support, and activity enhancement of Pt/C catalyst. J. Power Sources 172(1): 133–144.
Yan, J., Z. Fan, T. Wei, W. Qian, M. Zhang and F. Wei. 2010. Fast and reversible surface redox reaction of graphene MnO_2 composites as supercapacitor electrodes. Carbon 48: 3825–3833.
Yang, W., G. Chen, Z. Shi, C.C. Liu, L. Zhang, G. Xie et al. 2013. Epitaxial growth of single-domain graphene on hexagonal boron nitride. Nature Mater. 12: 792–797.
Yang, X., Q. Yang, J. Xu and C.S. Lee. 2012. Bimetallic Pt Pd nanoparticles on Nafion-graphene film as catalyst for ethanol electro-oxidation. J. Mater. Chem. 22: 8057–8062.
Yang, S., G. Cui, S. Pang, Q. Cao, U. Kolb, X. Feng et al. 2010. Fabrication of cobalt and cobalt oxide/graphene composites: Towards high-performance anode materials for lithium ion batteries. ChemSusChem. 3: 236–239.
Ye, D., G. Liang, H. Li, J. Luo, S. Zhang, H. Chen et al. 2013. A novel nonenzymatic sensor based on CuO nanoneedle/ graphene/carbon nanofiber modified electrode for probing glucose in saliva. Talanta 116: 223–230.
Yi, M. and Z. Shen. 2015. A review on mechanical exfoliation for the scalable production of graphene. J. Mater. Chem. A 3: 11700–11715.

Zhang, X., Q. Liao, S. Liu, W. Xu, Y. Liu and Y. Zhang. 2015. CuNiO nanoparticles assembled on graphene as an effective platform for enzyme-free glucose sensing. Anal. Chim. Acta 858: 49–54.

Zhang, Y., Y. Wang, J. Jia and J. Wang. 2012. Nonenzymatic glucose sensor based on graphene oxide and electrospun NiO nanofibers. Sensors Actuators B Chem. 171-172: 580–587.

Zhang, Y., L. Zhang and C. Zhou. 2013. Review of chemical vapor deposition of graphene and related applications. Acc. Chem. Res. 46: 2329–2339.

Zhang, B., Y. Cui, H. Chen, B. Liu, G. Chen and D. Tang. 2011. New electrochemical biosensor for determination of hydrogen peroxide in food based on well-dispersive gold nanoparticles on graphene oxide. Electroanal. 23: 1821–1829.

Zhao, Y.H., Zhang, C. Huang, S. Chen and Z. Liu. 2012. Pt/titania/reduced graphite oxide nanocomposite: An efficient catalyst for nitrobenzene hydrogenation. J. Colloid Interf. Sci. 374: 83–88.

Zhou, R., Y. Zheng, D. Hulicova-Jurcakova and S.Z. Qiao. 2013. Enhanced electrochemical catalytic activity by copper oxide grown on nitrogen-doped reduced graphene oxide. J. Mater. Chem. A 1: 13179–13184.

Zou, Z., L. Fu, X. Song, Y. Zhang and Z. Liu. 2014. Carbide-forming groups IVB-IVB metals: a new territory in the periodic table for CVD growth of graphene. Nano Lett. 14: 3832–3839.

3

Tungsten Disulfide Polythiophene Nanocomposites

Nicole Arsenault,[1] *Rabin Bissessur*[1,*] and *Douglas C. Dahn*[2]

Introduction

This chapter introduces a new series of inorganic-polymer materials synthesized from graphene analogous materials and conducting polymers to produce conductive nanocomposites for possible application in lithium ion batteries. Recently, there has been a considerable amount of interest in combining inorganic compounds with organic polymers in hopes of taking advantage of the interesting properties that inorganic-polymer nanocomposites can exhibit. In general, organic polymers are advantageous as new materials because of their light weight, relative ease of fabrication and low cost, and ease of processability, and hence have been researched extensively for new electronic devices such as in energy storage devices, solar cells, and organic light emitting diodes (Kalyani and Dhoble 2012, Kumar et al. 2014, Li et al. 2012). However, organic polymers can have poor mechanical and thermal properties, as well as low electrical conductivities compared to inorganic compounds, which has inhibited them from overtaking inorganic compounds as new energy-saving materials (Jeon and Baek 2010). Inorganic compounds by comparison are more expensive to process on the industrial scale; however, they possess favorable mechanical and magnetic properties, as well as high electrical conductivity at the nanoscale level. With this in mind, research on incorporating inorganic nanomaterials within organic polymers has led to nanocomposites which exhibit properties found in both the polymers and inorganic nanomaterials (Jeon and Baek 2010, Haldorai et al. 2012).

In general, inorganic-organic polymer nanocomposites can be synthesized in the form of an intercalated nanocomposite or an exfoliated nanocomposite. Intercalated nanocomposites are those characterized by interchanging layers of polymer and inorganic materials to make highly ordered nanomaterials

[1] Department of Chemistry, University of Prince Edward Island, Charlottetown, Prince Edward Island, Canada, C1A 4P3.
[2] Department of Physics, University of Prince Edward Island, 550 University Avenue, Charlottetown, Prince Edward Island, Canada, C1A 4P3.
Emails: niarsenault@upei.ca; dahn@upei.ca
* Corresponding author: rabissessur@upei.ca

which are commonly characterized using powder X-ray diffraction techniques. In contrast, exfoliated nanocomposites have single or two-three layered sheets of inorganic compounds combined with organic polymers that are usually amorphous to form highly disordered systems. Schematic representations of the two are shown below (Figure 1). The synthesis of an intercalated nanocomposite generally involves the exfoliation of a layered inorganic material, followed by addition of the organic polymer, and then restacking of the inorganic substance to form the intercalated nanocomposite. On the other hand, an exfoliated nanocomposite is generally made using an inorganic layered compound that is synthesized in an exfoliated state, and then adding the polymer, or monomer followed by *in situ* polymerization. Because of the different degrees of disorder in the two materials, their chemical and physical properties such as mechanical strength, conductivity, hardness, and glass transition temperature, can vary substantially, affecting their applications. Therefore, it is imperative when making new nanocomposites that a variety of characterization techniques are used to determine the overall structure of the materials. The techniques used in this work include Fourier transform infrared spectroscopy (FT-IR), powder X-ray diffraction (PXRD), scanning electron microscopy (SEM), and total reflection X-ray fluorescence (TXRF).

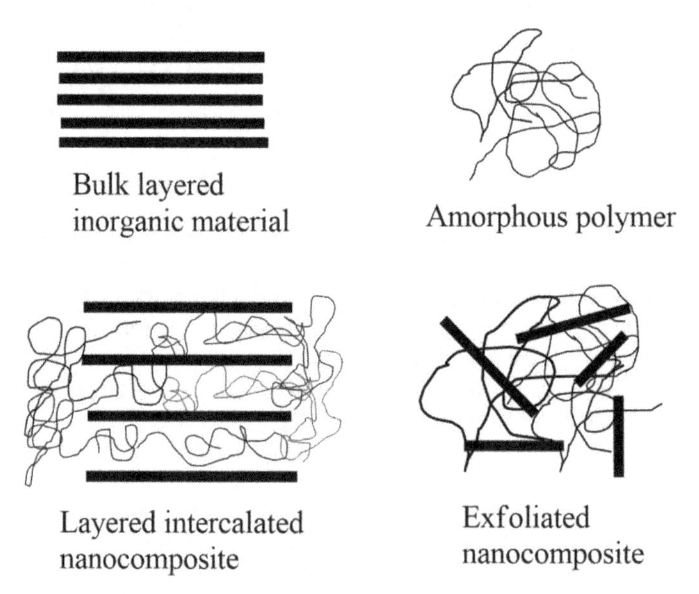

Figure 1. Schematic representation of layered versus exfoliated nanocomposites.

Graphene-analogous materials

Graphene and graphene-analogous materials have seen increasing interest in the materials' science community because of the discovery of graphene's interesting properties, including its high specific surface area, high thermoelectric behavior, favorable mechanical properties, and high electrochemical properties including its high electron mobility (Singh et al. 2011, Tang and Zhou 2013). Graphene is composed of carbon atoms arranged in a two dimensional honeycomb structure. By stacking graphene sheets on top of one another through weak van der Waals interactions, the three dimensional structure, graphite, is obtained. Compounds that fall under the category of graphene-analogous materials get their namesake because they also have similar honeycomb single layer structure, with weak interlayer interactions between the layers (Figure 2).

Common graphene-analogous materials that have been studied include BN nanosheets, silicene, SiC_2, and transition metal dichalcogenides (Tang and Zhou 2013). In particular, transition metal dichalcogenides (TMDs) are an interesting class of graphene-analogous materials due to their non-planar structure and intrinsic band gaps, which allow them to be tuned for specific properties in their semiconductor state, such as by reducing the number of layers in the TMD sample, or by applying external electric fields (Ramasubramaniam et al. 2011, Wang et al. 2015). By comparison, graphene has been shown to have its valence band and conduction band joined together, therefore having a zero-gap band gap in its unaltered two dimensional state, making it unable to have properties capable of controlling the current, such as those needed for transistors (Cooper et al. 2012, Tang and Zhou 2013).

TMDs have the general formula MX_2, where M is a transition metal sandwiched between two chalcogens, X, which is either selenium, sulfur, or tellurium (see Figure 3; Tang and Zhou 2013).

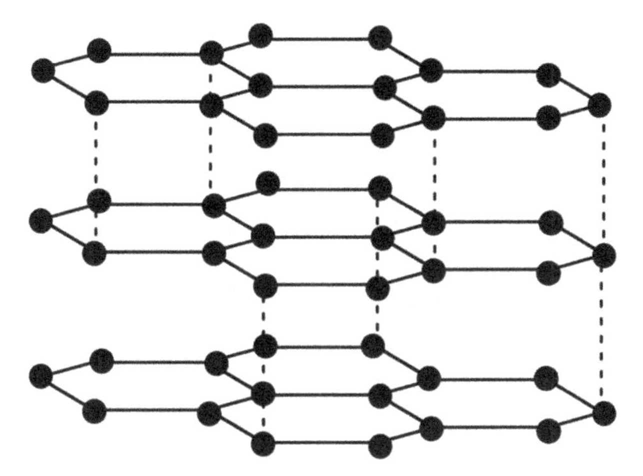

Figure 2. Honeycomb structure of graphene sheets stacked to form graphite.

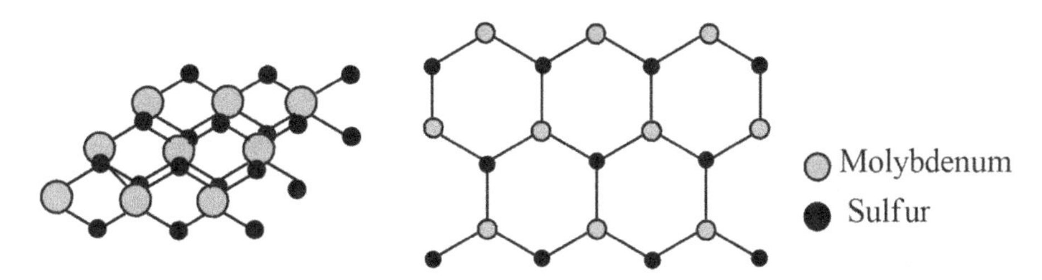

Figure 3. Hexagonal structure of MoS_2. Non-planar layers of Mo metal sandwiched between S atoms are shown.

Transition metal dichalcogenides as graphene-analogous materials

The most commonly studied TMDs are molybdenum disulfide, MoS_2, and tungsten disulfide, WS_2, both of which have the same hexagonal-layered structure, where strong covalent bonds bind the metal atoms and chalcogen sulfurs together, while weak interlayer interactions hold the layers together to form the bulk layered material (Tang and Zhou 2013). The weak interlayer interactions allow easy exfoliation of the MoS_2 and WS_2 into non-planar hexagonal nanosheets. There are a number of procedures that can be used to produce exfoliated MoS_2 and WS_2, the most common being lithium intercalation followed by immersion in water. While MoS_2 affords sufficient lithium intercalation and subsequent exfoliation, WS_2 proves challenging to exfoliate due to the low degree of lithium intercalation by similar techniques (Xu et al. 2007). Therefore, there has been a substantial lack of study on exfoliated and intercalated WS_2 compounds in comparison to MoS_2, and currently new methods of synthesizing highly exfoliated single-layered WS_2 nanomaterials are being investigated. So far, efforts have shown that using highly acidic treatments on layered WS_2 or synthesizing WS_2 directly into an exfoliated state in the solid phase have proven effective in synthesizing single-layered WS_2 materials (Wu et al. 2010, Matte et al. 2010).

In one such study, WS_2 was synthesized from thiourea and tungstic acid directly into its exfoliated state, in a simple solid-state reaction (Matte et al. 2010). Thiourea and tungstic acid were heated to 500°C under nitrogen, resulting in exfoliated WS_2 product shown by powder X-ray diffraction, Raman spectroscopy, and transmission electron microscopy.

By synthesizing WS_2 directly in the exfoliated single-layer state, we can take advantage of its interesting graphene-like properties that are not exhibited when in the bulk state, and investigate its use as lubricants, photoconductors, and battery materials (Wu et al. 2010).

Polythiophene in conducting materials

Polythiophene (PT) is a conjugated conductive polymer that has been broadly investigated in the past due to its torsional backbone and facile synthesis, robust stability in atmospheric conditions, and optoelectronic properties (Patil et al. 2012, Lee et al. 2014, Gumus et al. 2011). Hence, there are many potential applications of PT and its derivatives in areas such as LEDs, sensors, solar cells, and electronic thin film technologies (Patil et al. 2012, Lee et al. 2014, Khalid et al. 2014).

Polythiophene's ability to conduct electricity is due to the overlapping π orbitals in its conjugated backbone, which enables charge carriers to travel along the polymer chain when doped with a dopant, such as the $FeCl_4^-$ anion. Derived from iron(III) chloride ($FeCl_3$), $FeCl_4^-$ anion has been used extensively in the past. This is because of the low cost and mild oxidizing power of iron(III) chloride which initiates the polymerization of the thiophene monomer. In addition, the use of iron(III) chloride reduces the number of reagents needed for the polymerization reaction (Lee et al. 2014, Gumus et al. 2011, Xu et al. 2002). $FeCl_3$ oxidizes thiophene to form PT and $FeCl_4^-$, as well as a polaron or bipolaron depending on the amount of $FeCl_3$ used. A polaron consists of a positive charge and a radical electron spread across a couple of base units in the polymer chain, while a bipolaron consists of two positively charged carbons spread across a couple of base units in the polymer chain. The polaron or bipolaron can then move down the chain through resonance, thus enabling PT to conduct electric charge (Figure 4).

In general, the synthesis of PT is performed either through chemical oxidation as mentioned previously, or through electrochemical synthesis (Roncali 1992). Though electrochemical synthesis generally creates polymer chains with higher conductivity due to the pristine formation of long conjugated chains (Lee and Lim 2000), chemical synthesis is most often used in industry and in research because of the ability to produce PT on larger scales, and its simple procedure when using $FeCl_3$ as an oxidizer (Roncali 1997). Other chemical polymerization methods often performed involve nickel or palladium catalysts in cross-coupling reactions, which can be used at the research scale and, in general, produce highly conductive polymers, but are too expensive on the industrial scale (Martinez et al. 1995). Therefore, research into the synthesis of PT is still ongoing to find low cost, large scale and highly conductive polymers for materials' applications.

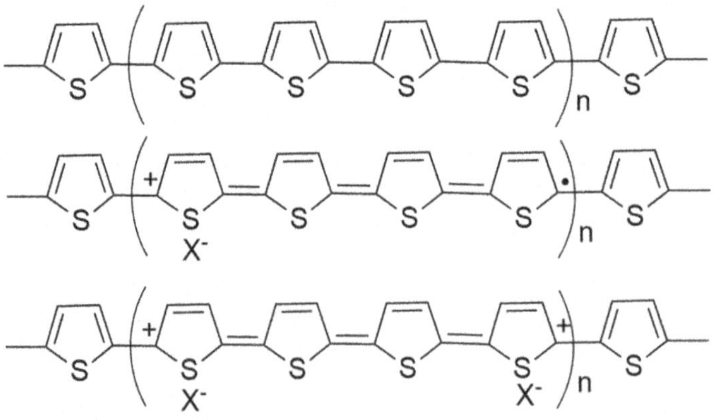

Figure 4. Redox forms of PT. Top: non-conductive reduced form. Middle: conducting half-oxidized polaronic form. Bottom: conducting fully oxidized bipolaronic form.

Polythiophene-transition metal dichalcogenide materials

As previously discussed, there are many applications of organic conjugated polymers, e.g., as battery materials for electrodes and electrolytes. Organic polymers in electronics are more attractive compared to inorganic or small molecules because of their light weight, flexibility, and ability to be fabricated in large-scale batches relatively easily and cheaply (Scheuble et al. 2015, Guo et al. 2013). However, the challenges associated with organic polymers have pushed researchers to combine organic polymers with inorganic materials in order to take advantage of both materials' properties. There have been a number of reports that combine inorganic substances with PT and PT derivatives to increase key traits, such as mechanical stability and conductivity (Advincula 2006, Zhai and McCullough 2004). One example used cerium(IV) phosphate with different ratios of conductive poly(3-methyl thiophene), and an increase in conductivity and isothermal stability was observed compared to the polymer alone (Khan and Baig 2014).

Though studies on different small inorganic salts and metal nanoparticles incorporated into PT materials are available, studies on combining PT with transition metal dichalcogenides are very few. One study by Lin et al. on intercalated PT/MoS_2 nanocomposites found that these new materials had increased conductivity compared to the layered MoS_2 and similar conductivity values compared to bulk PT (Lin et al. 2009). Another study on poly(3,4-ethylenedioxythiophene) (PEDOT), intercalated into VS_2 showed an increase in conductivity compared to the layered VS_2 alone, along with significantly increased discharge capacity when intercalated with lithium ions (Murugan et al. 2005).

Recently, exfoliated WS_2 combined with polyaniline (PANI) in different ratios were shown to exhibit significant enhancement in conductivity compared to bulk PANI (Lane et al. 2014). The high conductivity values of these nanocomposites are unprecedented; therefore, along similar lines, we choose to prepare new nanocomposites consisting of exfoliated WS_2 and conductive polythiophene in various mass percentages. The conductive properties of these nanocomposites were investigated in relation to pure polythiophene and pure exfoliated WS_2.

A variety of analytical techniques such as FTIR, PXRD, TXRF, and SEM were employed to determine the structural features of the PT, exfoliated WS_2, and the PT-WS_2 nanocomposites. The conductivity of PT-WS_2 nanocomposites, PT, and exfoliated WS_2 were characterized by the co-linear four-probe and Van der Pauw techniques. The co-linear four-probe technique was used for room temperature conductivity measurements in air in order to determine trends in conductivity between PT and the nanocomposites, while the Van der Pauw method was performed under vacuum at variable temperatures to determine conductivity trends as a function of temperature.

Electronic conductivity measurement on solid state samples

To measure the conductivity, σ, of a solid sample, we make use of the relationship between conductivity and resistivity, ρ, where conductivity is the reciprocal of resistivity. Resistivity is an inherent property of the material and is not measured directly. Instead, it is the resistance that can be measured directly for a specific sample by using Eq. 1, which relates the applied voltage (V) in volts to the current output (I) in amperes.

$$R = \frac{V}{I} \tag{1}$$

Resistivity is then determined from the resistance and the dimensions of the sample.

To determine the conductivity of a powdered sample, the material is usually pressed into a thin pellet and the electrodes are placed directly onto the pelletized sample. This method is relatively quick and easy; however, it can be difficult to get accurate conductivity measurements because of grain boundaries and air pockets that can be present in the pellet, restricting the flow of charge carriers. Additionally, different samples of the same material will have variations in the number of grain boundaries and air pockets, and thus slight differences in conductivity between samples will be observed. Therefore, conductivity is generally reported as an average value when the measurements are performed on solid pellets. Also, when using data from pellets to compare conductivities of different materials, it is preferable to investigate conductivity trends as a function of composition, rather than placing a lot of significance on the absolute conductivity values. Other aspects to consider are contact resistance from the probes and thermoelectric effects during conductivity measurements. Both of these effects can cause errors in conductivity values, but these can be reduced to insignificant levels when appropriate measurement techniques are used.

Co-linear four probe method

There are a number of different methods to set up a conductivity experiment, one of which is the co-linear four probe method. This uses four equally-spaced contact probes placed in a line on the surface of the sample as shown in Figure 5. A current I is passed through the sample via contacts one and four, and is determined by measuring the voltage V_1 across a standard resistor R_1, that is in series with the sample. The voltage V between contacts two and three is measured using a high impedance voltmeter. This ensures that the current through contacts two and three is extremely small, so that errors in the conductivity due to contact resistance are insignificant.

Resistivity is then calculated using Eq. 2, which gives resistivity as a function of resistance R = V/I, and thickness t (Smits 1958, Valdes 1954). This can then be used to determine the conductivity.

$$\rho = \frac{\pi}{ln2} Rt \tag{2}$$

For this equation to be valid, the sample thickness must be significantly smaller than the distance between the probes (Li et al. 2012).

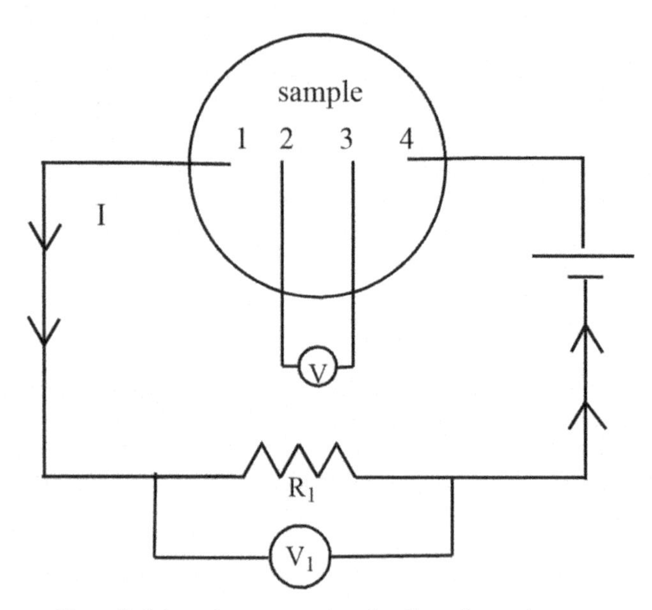

Figure 5. Schematic representation of co-linear four-probe setup.

van der Pauw method

Another technique used to determine the conductivity is the van der Pauw method (van der Pauw 1958). It uses four point contacts attached at the edge of the sample as shown in Figure 6. The sample can be any shape, but must be thin, homogeneous and of uniform thickness t. The contacts need not be evenly spaced. Two resistances R_a and R_b are determined by measuring the voltage between two adjacent contacts

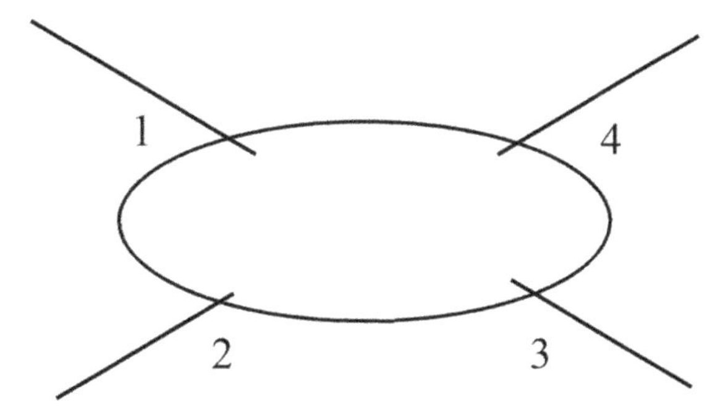

Figure 6. Placement of wires for van der Pauw conductivity measurements.

while current is passed through the sample via the other two contacts, as described in more detail below. The resistivity (ρ) of the sample is then found by solving the van der Pauw Eq. 3.

$$e^{-\frac{\pi R_a t}{\rho}} + e^{-\frac{\pi R_b t}{\rho}} = 1 \tag{3}$$

In order to improve accuracy and eliminate the effects of thermoelectric voltages, multiple measurements were used to get the resistances R_a and R_b. These sets of measurements use different combinations of contacts and different current directions. For example, to measure a resistance called $R_{12,43}$, we start with a current I (from a constant-current source) flowing into the sample at contact 1 and out at contact 2, and measure voltage V_+ between contact 4 (high) and contact 3 (low). We then reverse the current and measure the voltage again, calling it V_-. Reversing the current changes the sign of the voltage due to current flowing through the resistive sample, but does not alter thermal emfs in the circuit, so the effect of these can be eliminated by calculating

$$R_{12,43} = \frac{V_+ - V_-}{2I} \tag{4}$$

Current and voltage leads are then exchanged (4 and 3 are used for current, 1 and 2 for voltage) and the measurement process is repeated to determine $R_{43,12}$. A reciprocity theorem (van der Pauw 1958) implies $R_{12,43} = R_{43,12}$, so they are averaged to give R_a. Similarly, $R_{23,14}$ (2 and 3 used for current, 1 and 4 for voltage) and $R_{14,23}$ are averaged to give R_b Eq. 5.

$$R_a = \frac{R_{12,43} + R_{43,12}}{2} \quad R_b = \frac{R_{23,14} + R_{14,23}}{2} \tag{5}$$

Methods

Thiourea, tungstic acid, thiophene, and methanol were purchased from Sigma Aldrich and used as received. Anhydrous iron(III) chloride was bought from Anachemia and used as received, and spectrochemical grade chloroform was bought from Fisher and used as received. Layered tungsten(IV) sulfide was bought from Strem Chemicals. This was structurally characterized as a comparison material, but not used in the synthesis.

Synthesis of exfoliated tungsten disulfide

Exfoliated tungsten disulfide was synthesized using the procedure from Matte et al. (2010), where thiourea (3.66 g, 48 mmol) and tungstic acid (0.25 g, 1.0 mmol) were weighed in a 48:1 molar ratio, respectively, mechanically mixed together and placed in a ceramic vessel. The vessel was placed in a ceramic tube inserted into a split furnace, under constant nitrogen atmosphere. The split furnace was slowly ramped up to 500°C over a period of 3 hours, and the temperature was held at 500°C for 3.5 hours. The reaction mixture was then cooled to room temperature for 12 hours under a nitrogen atmosphere. The resulting product was black in color.

Synthesis of polythiophene, PT

PT was synthesized using an adapted method from Gumus et al. (2011). To a 250 mL Schlenk flask fitted with condenser and under nitrogen atmosphere, 15 mL of chloroform and 0.5 mL (6.2 mmol) of thiophene, along with approximately 2.0 g (12 mmol) of anhydrous $FeCl_3$, were added. The contents of the flask were then heated at 42°C for 3 hours with stirring, then stirred for an additional 1–2 days under nitrogen purge at room temperature until evaporation of the solvent. The dark grey precipitate was washed with 200 mL of methanol under vacuum filtration, and then dried under vacuum. Two products were collected, a red product and a black product.

Synthesis of exfoliated tungsten disulfide/PT nanocomposites

The following PT/WS_2 nanocomposites were synthesized in multiple trials based on the added mass percentage of WS_2 in the nanocomposite: 1%, 5%, 10%, 20%. The synthesis of the 10% nanocomposite is described below. The other nanocomposites were synthesized in a similar manner.

Exfoliated WS_2 (0.05 g) was dispersed in 15 mL of chloroform using a probe sonicator at 30% amplitude for a duration of 20 minutes. To a separate flask, 1.8 g of anhydrous $FeCl_3$ was added and stirred with 10 mL of chloroform under constant nitrogen flow. The WS_2 suspension was then added to the iron(III) chloride solution, and stirred for 30 min under nitrogen purge. 0.44 mL thiophene dissolved in 1 mL of chloroform was added dropwise by syringe to the reaction flask. The reaction mixture was then stirred under nitrogen at 42°C for 48 hours until all solvent evaporated. The product was washed with 200 mL of methanol and filtered by vacuum filtration. A red and a black product were collected.

Results and discussion

Structural characterization

FT-IR spectra of PT, exfoliated WS_2, and exfoliated PT-WS_2 nanocomposites were collected. The solid black PT showed similar IR peaks to the red PT product, and these FT-IR data are in very good agreement with the PT synthesized by Gumus et al. (2011). The spectra showed a strong peak between 780 and 789 cm^{-1}, assigned to the α-α' coupling of poly-2,5'-thiophene (Kelkar and Chourasia 2012). The C=C symmetric and asymmetric stretching were displayed at 1315–1324 cm^{-1} and 1194–1202 cm^{-1}, respectively. The C-H in plane bending was observed at 1103–1111 cm^{-1} and 1003–1035 cm^{-1} (Gumus et al. 2011). All of these IR peaks were found to be broad, suggesting that the samples were polymeric. The IR spectra of the exfoliated PT-WS_2 nanocomposites displayed similar peaks as the red and black PT products. This is attributed to the fact that a low percentage of the exfoliated WS_2 was used in the PT-WS_2 nanocomposites. As an illustration, the FT-IR spectrum of PT black-10% WS_2 is displayed in Figure 7.

Total X-ray fluorescence (TXRF) was also used to analyze the different synthesized nanomaterials to confirm the presence of WS_2. For both black and red samples of PT-WS_2 nanocomposites, TXRF confirms that tungsten was present in similar weight ratios. However, in the red sample there was significantly less retention of the iron dopant (6.4 KeV) required for conductivity (Figure 8). This finding is consistent with the conductivity data which demonstrate that the red PT-WS_2 nanocomposites are less conductive than the black PT-WS_2 nanocomposites (*vide infra*).

Scanning electron microscopy (SEM) and powder X-ray diffraction (PXRD) were also used to compare the bulk structure of the starting materials with the nanocomposites. First, exfoliated WS_2

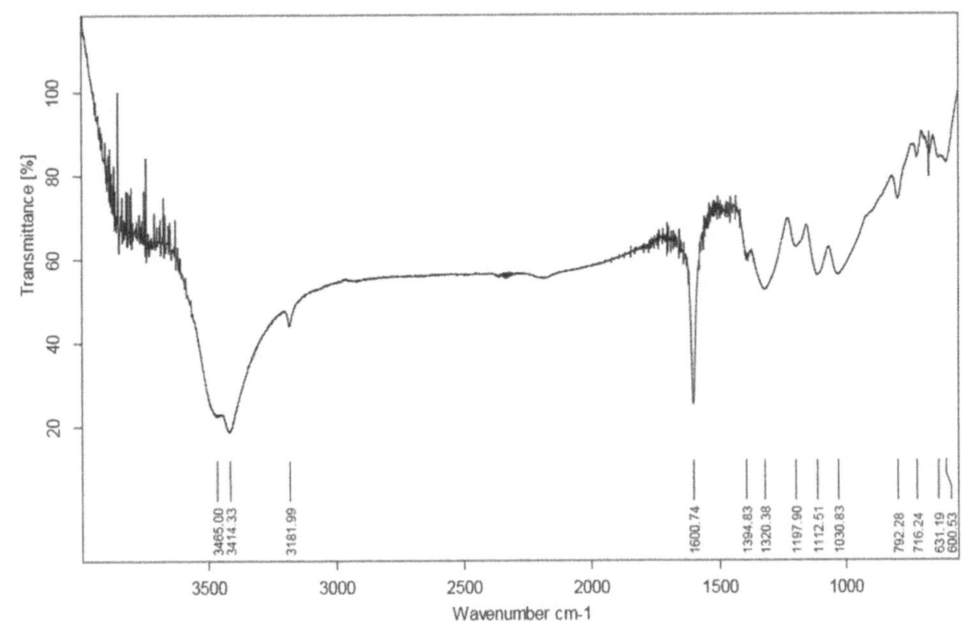

Figure 7. FT-IR spectrum of black PT-10% WS_2 nanocomposite.

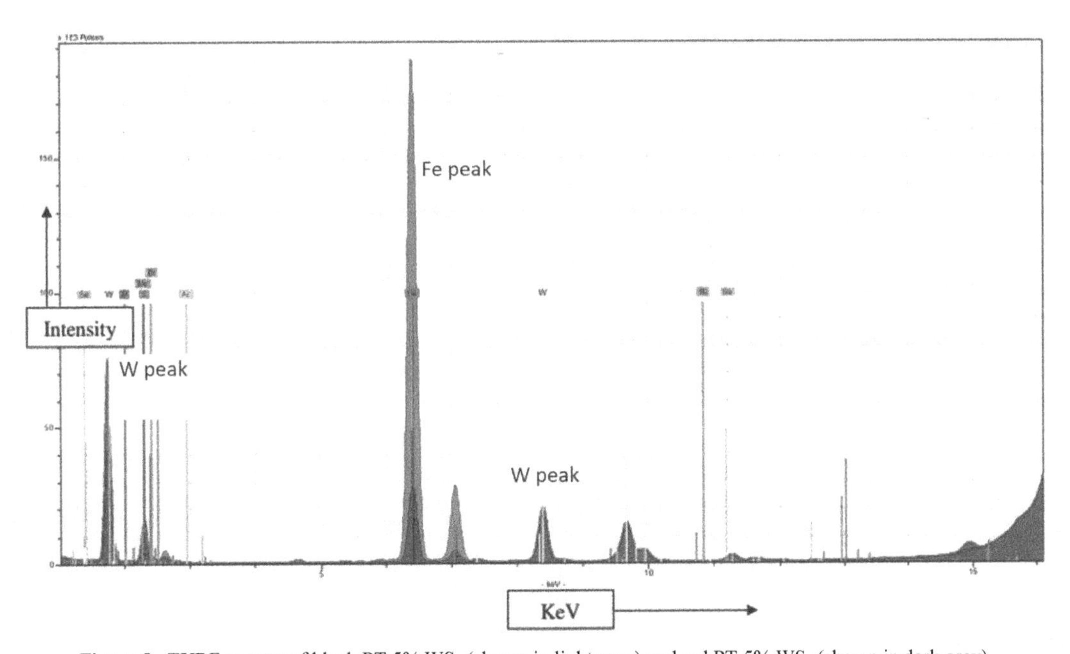

Figure 8. TXRF spectra of black PT-5% WS_2 (shown in light grey) and red PT-5% WS_2 (shown in dark grey).

was compared with commercially available layered WS_2. SEM micrographs showed distinct sheets in the layered WS_2 sample, while the exfoliated WS_2 displayed a highly disordered structure (Figure 9). PXRD data were in agreement with the SEM micrographs. PXRD revealed distinct peaks in the X-ray diffractogram of layered WS_2, with the most intense peak occurring at 6.13 Å ($2\theta = 14.437$). On the other hand, the powder pattern of exfoliated WS_2 showed no well-defined peaks, confirming its amorphous characteristic.

Figure 9. SEM micrographs of layered WS_2 (left, scale bar 5 μm) and exfoliated WS_2 (right, scale bar 10 μm).

SEM micrographs for the black PT product showed both crystalline rods and amorphous material, which agreed with the PXRD data that showed an amorphous curve along with diffraction peaks (Figure 10). The PXRD spectra of the red PT product also showed both disordered and ordered characteristics, with slightly more disordered characteristics compared to the black PT product. This suggests that the red PT product is most likely composed of short oligomeric chains. This combination of amorphous and crystalline rods is perhaps caused by the temperature used during the polymerization reaction. Previous studies on the synthesis of PT using $FeCl_3$ at different temperatures showed rod-like formation of PT when polymerized at 50°C (Kattimani et al. 2014). Thus, it appears that carrying out the polymerization reaction at 42°C results in a combination of rods and amorphous agglomerates.

This combination of crystalline and amorphous structure was also observed in the synthesized nanocomposites (Figure 11). No significant peak shifts in the PXRD patterns were seen in the nanocomposites compared to the pure PT, and no peaks from ordered WS_2 appeared as well, suggesting that the exfoliated WS_2 did not restack in the nanocomposites. For the black nanocomposites, SEM showed both crystalline and amorphous characteristics, while the red nanocomposites showed only amorphous agglomerates.

Electrical conductivity characterization

Electrical conductivity data at room temperature were obtained by using the co-linear four probe technique in air. Variable temperature conductivity measurements were obtained by the van der Pauw technique under vacuum. Samples of exfoliated WS_2, PT, and nanocomposites consisting of 1%, 5%, 10%, and 20% WS_2 were measured. Exfoliated WS_2 pellets demonstrated non-conductive behavior in agreement with the literature (Lane et al. 2016).

Polythiophene as synthesized was previously reported by the four probe technique to have a conductivity value of 7.2×10^{-3} S/cm (Gumus et al. 2011), while other reports of PT synthesized using $FeCl_3$ as the oxidizer and dopant have seen values as low as 10^{-5} S/cm (Gök et al. 2007). Because of this difference in conductivity values, it was important to optimize the conductivity value of our PT samples. Hence, the amount of $FeCl_3$ was increased. Using a 1:2 mole ratio of thiophene to $FeCl_3$ resulted in an increase in the conductivity of the PT samples compared to PT previously synthesized in the literature (Gumus et al. 2011).

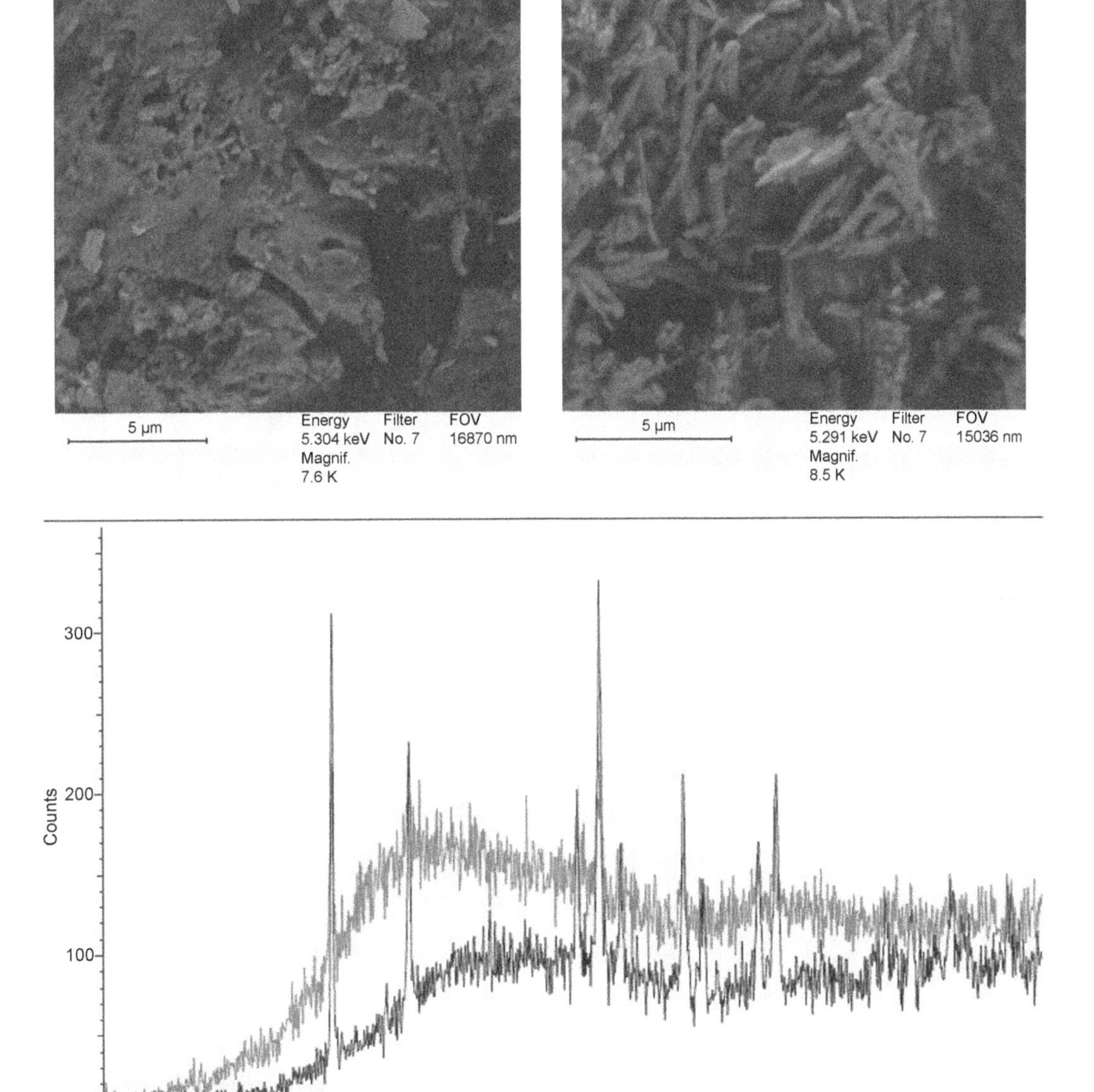

Figure 10. SEM micrographs (top) of black PT product showing both an amorphous (left, scale bar 5 μm) and crystalline (right, scale bar 5 μm) structure. PXRD diffractograms (bottom) of black PT product (shown in black color) and red PT product (shown in grey).

The average conductivity value of the black PT was $(1.4 \pm 0.3) \times 10^{-2}$ S/cm, while the conductivity of the red PT was $(9 \pm 4) \times 10^{-5}$ S/cm. This suggests that the red product is more likely to be short chains, as previously mentioned, and therefore more attention was focused on the black polymer and nanocomposite samples. The van der Pauw technique at room temperature under vacuum on the black PT agreed with the co-linear four probe technique, having a conductivity value of $(1.5 \pm 0.1) \times 10^{-2}$ S/cm. Conductivity measurements were also taken using the van der Pauw technique at variable temperatures

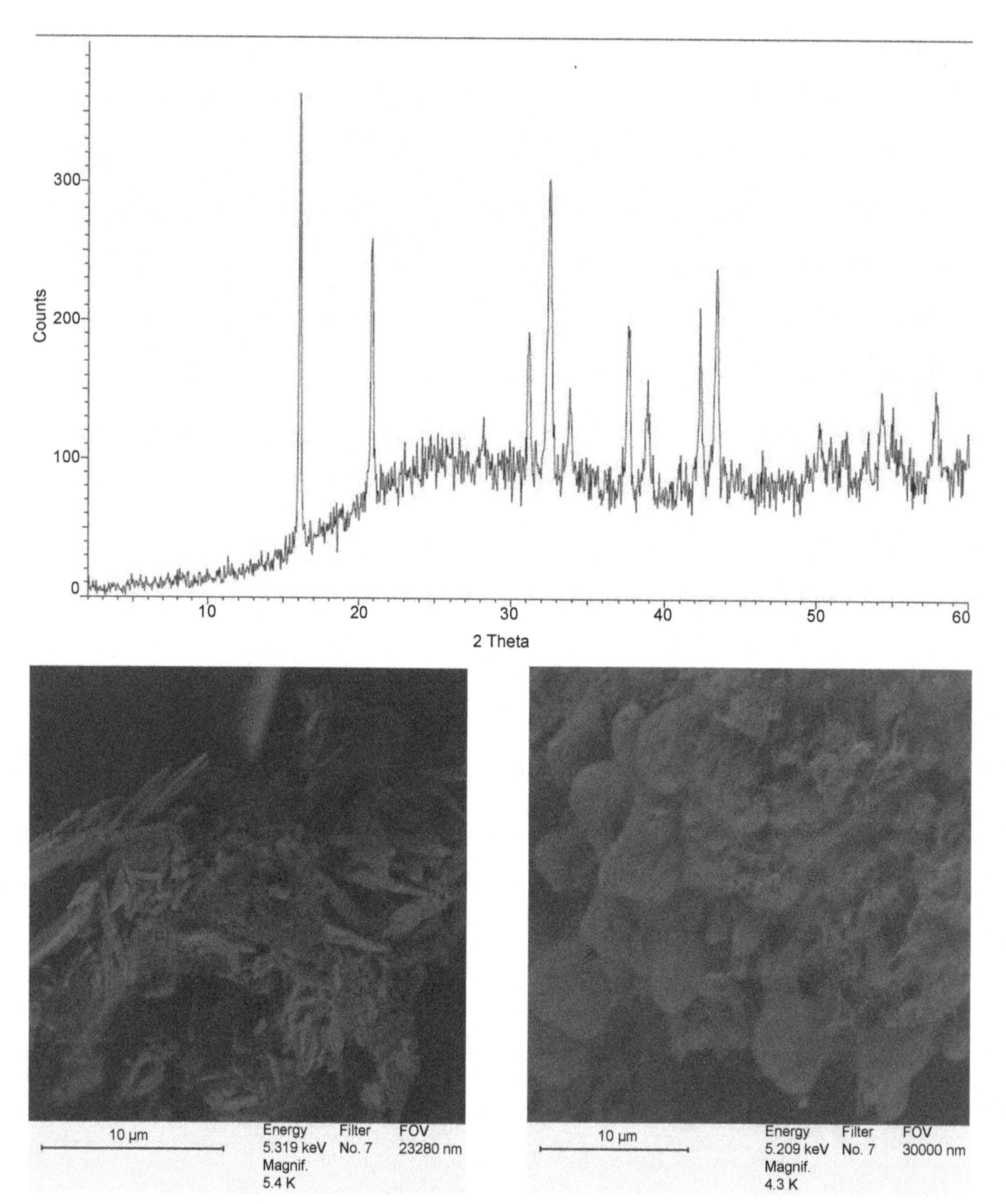

Figure 11. PXRD diffractogram (top) of black PT-20% WS_2 and SEM micrographs (bottom) of black PT-10% WS_2 (left, scale bar 10 µm) and red PT-10% WS_2 (right, scale bar 10 µm).

from 310 K to 200 K (Figure 12). The conductivity increased with increasing temperature, as discussed further below.

All of the black nanocomposites synthesized showed increased conductivity compared to the black PT product (Table 1). The 5% nanocomposite showed the highest conductivity, having over one order of magnitude higher conductivity than the black PT product. Variable-temperature van der Pauw measurements were made on some of the nanocomposites, and are shown in Figure 13. The 5%, 10%, and 20% nanocomposites all demonstrated higher conductivity values than PT, over the entire temperature range studied.

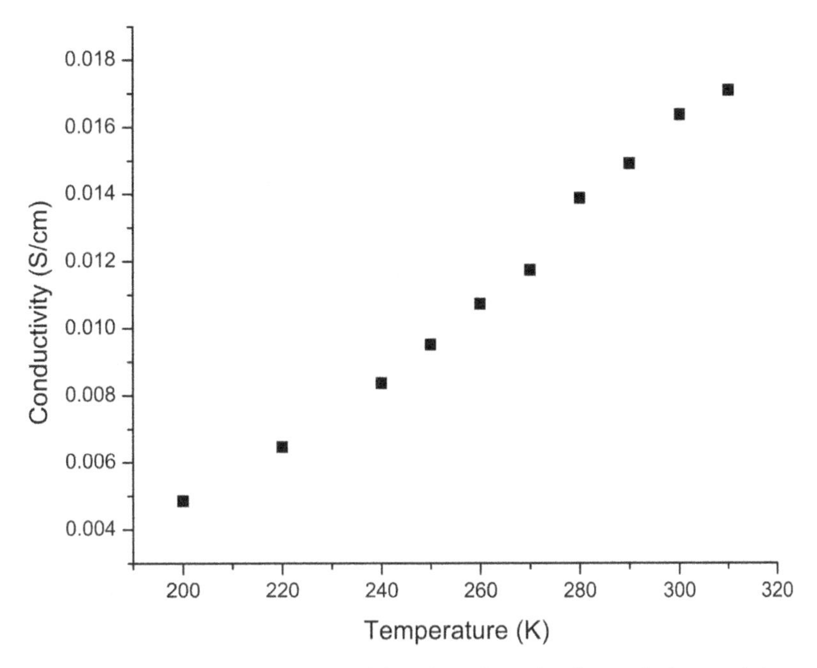

Figure 12. Variable temperature conductivity values of PT using the van der Pauw technique.

Table 1. Average room temperature conductivity values of nanocomposites, using the co-linear four probe technique.

% WS_2	Average Conductivity (S/cm)
0 (black PT)	$(1.4 \pm 0.3) \times 10^{-2}$
1	$(9 \pm 4) \times 10^{-2}$
5	$(3.7 \pm 0.7) \times 10^{-1}$
10	$(3.1 \pm 0.9) \times 10^{-1}$
20	$(2.1 \pm 0.4) \times 10^{-1}$

The conductivity of PT and all the nanocomposites increases with increasing temperature, as does the conductivity of semiconductors. However, a typical semiconductor has thermally-activated conductivity described approximately by an equation of the form

$$\sigma(T) = \sigma_0 e^{-E/kT}, \tag{6}$$

where σ_0 is a constant and E is an activation energy related to the semiconductor band gap. This equation does not fit the data in Figures 12 and 13.

The variable-temperature data for PT and most of the nanocomposites are consistent with the variable-range hopping (VRH or Mott law) model for conduction in disordered materials. In the VRH model, the resistivity ρ is given as a function of absolute temperature T by

$$\rho(T) = \rho_0 exp\left[\left(\frac{T_0}{T}\right)^{\frac{1}{d+1}}\right], \tag{7}$$

where d is the dimensionality of the material, and T_0 and ρ_0 are parameters that are nearly independent of temperature. VRH behavior has been reported in a number of conducting polymer and nanocomposite systems, including polythiophene and substituted polythiophenes (Kobayashi et al. 1984, Yamamoto et

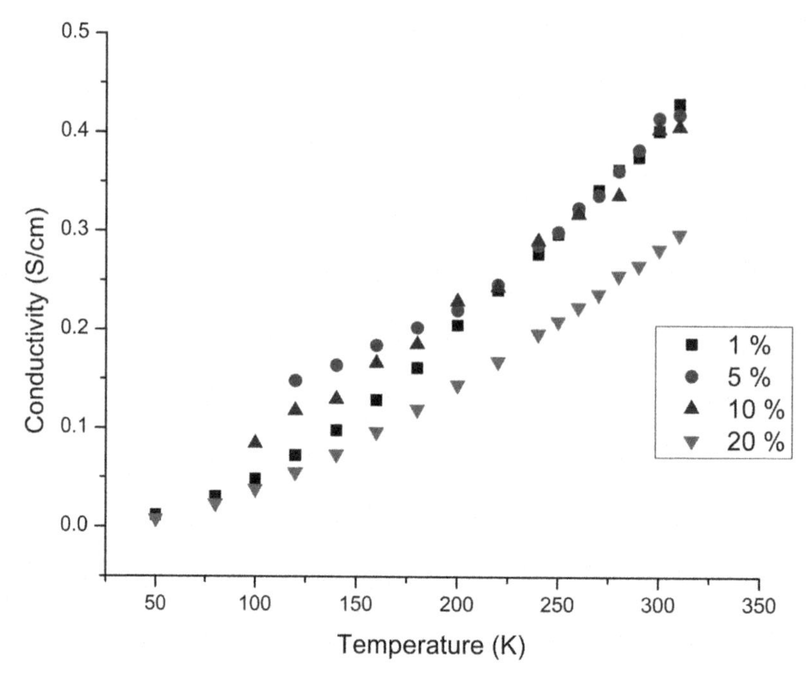

Figure 13. van der Pauw conductivity values of different PT-WS_2 nanocomposites under vacuum at different temperatures.

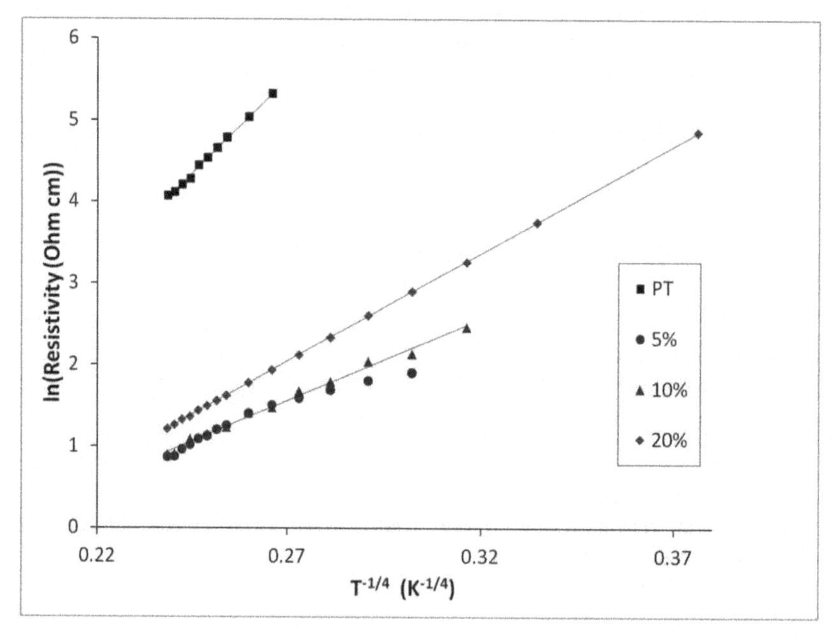

Figure 14. 3-dimensional VRH model for PT and nanocomposites.

al. 1999) and polyaniline/WS_2 nanocomposites (Lane et al. 2016). If Eq. 7 with $d = 3$ is valid, a plot of ln ρ as a function of $T^{-1/4}$ will be linear. Our data are plotted in this format in Figure 14. The straight-line behavior seen in the figure shows that a three-dimensional VRH model is a reasonable description of electrical conduction in all of the samples, except for 5% nanocomposite. The reason it had slightly different temperature dependence is not understood.

Summary

The first examples of nanocomposites composed of graphene analogous WS_2 and conductive polythiophene were synthesized and characterized by FTIR, TXRF, SEM, and PXRD. The electronic conductivity of these nanocomposites was measured by the co-linear four-probe technique and by the van der Pauw technique in air and under vacuum and at variable temperatures. The nanocomposites synthesized showed structural characteristics similar to PT, and all the black nanocomposites showed significantly higher conductivity values than the pure PT. The conductivity of the 5% nanocomposite was over one order of magnitude higher than pure PT, with the highest average conductivity at $(3.7 \pm 0.7) \times 10^{-1}$ S/cm. The variable-temperature conductivity showed variable-range hopping (Mott Law) behavior for most samples, with the conductivity decreasing with decreasing temperatures. The results shown here demonstrate the possibility for exfoliated graphene-analogous compounds to enhance the conductivity of conductive polymers. Similar conductivity enhancement has also been seen in polyaniline/WS_2 nanocomposites (Lane et al. 2016). Future studies on these types of nanocomposites are needed to determine how exfoliated graphene-analogous compounds play a role in the enhanced conductivity of the nanocomposites in order to better understand and design new conductive materials.

Acknowledgements

Financial support from the University of Prince Edward Island is gratefully acknowledged.

References

Advincula, R.C. 2006. Hybrid organic-inorganic nanomaterials based on polythiophene dendronized nanoparticles. Dalton Trans. 23: 2778–2784.

Cooper, D.R., B. D'Anjou, N. Ghattamaneni, B. Harack, M. Hilke, A. Horth et al. 2012. Experimental review of graphene. ISRN Condens. Matter Phys. 2012: 1–56.

Gök, A., M. Omastová and A.G. Yavuz. 2007. Synthesis and characterization of polythiophenes prepared in the presence of surfactants. Synth. Met. 157: 23–29.

Gumus, O.Y., H.I. Unal, O. Erol and B. Sari. 2011. Synthesis, characterization, and colloidal properties of polythiophene/borax conducting composite. Polym. Compos. 32: 418–426.

Guo, X., M. Baumgarten and K. Müllen. 2013. Designing π-conjugated polymers for organic electronics. Prog. Polym. Sci. 38: 1832–1908.

Haldorai, Y., J.J. Shim and K.T. Lim. 2012. Synthesis of polymer-inorganic filler nanocomposites in supercritical CO_2. J. Supercrit. Fluids 71: 45–63.

Jeon, I.Y. and J.B. Baek. 2010. Nanocomposites derived from polymers and inorganic nanoparticles. Materials 3: 3654–3674.

Kalyani, N.T. and S.J. Dhoble. 2012. Organic light emitting diodes: energy saving lighting technology—a review. Renewable Sustainable Energy Rev. 16: 2696–2723.

Kango, S., S. Kalia, A. Celli, J. Njuguna, Y. Habibi and R. Kumar. 2013. Surface modification of inorganic nanoparticles for development of organic-inorganic nanocomposites—a review. Prog. Polym. Sci. 38: 1232–1261.

Kattimani, J., T. Sankarappa, K. Praveenkumar, J.S. Ashwajeet, R. Ramanna, G.B. Chandraprabha et al. 2014. Structure and temperature dependence of electrical conductivity in polythiophene nanoparticles. Int. J. Adv. Res. Phys. Sci. 1: 17–21.

Kelkar, D. and A.B. Chourasia. 2012. Effect of dopant on thermal properties of polythiophene. Indian J. Phys. 86: 101–107.

Khalid, H., H. Yu, L. Wang, W.A. Amer, M. Akram, N.M. Abbasi et al. 2014. Synthesis of ferrocene-based polyhtiophene and their applications. Polym. Chem. 5: 6879–6892.

Khan, A.A. and U. Baig. 2014. Electrical and thermal studies on poly(3-methyl thiophene) and *in situ* polymerized poly(3-methyl thiophene)—cerium(IV)phosphate cation exchange nanocomposite. Composites: Part B 56: 862–868.

Kobayashi, M., J. Chen, T.-C. Chung, F. Moraes, A.J. Heeger and F. Wudl. 1984. Synthesis and properties of chemically coupled poly(thiophene). Synth. Met. 9: 77–86.

Kumar, B., B.K. Kaushik and Y.S. Negi. 2014. Organic thin film transistors: structures, models, materials, fabrication, and applications: a review. Polym. Rev. 54: 33–111.

Lane, B.C.S., R. Bissessur, A.S. Abd -El-Aziz, W.H. Alsaedi, D.C. Dahn, E. McDermott et al. 2016. Exfoliated nanocomposites based on polyaniline and tungsten disulfide. *In*: F. Yilmaz (ed.). Conductive Polymers. Publisher: Intech Open Access.

Lazo, R.M. 1950. Radiochemical studies on graphite ferric chloride. Masters' Thesis, The University of British Columbia, Vancouver, Canada.

Lee, J.M. and K.H. Lim. 2000. Electrochemical synthesis of conducting polythiophene in an ultrasonic field. J. Ind. Eng. Chem. 6: 157–162.

Lee, S.H., Y.S. Kim and J.H. Kim. 2014. Synthesis of polythiophene/poly(3,4-ethylenedioxythiophene) nanocomposites and their application in thermoelectric devices. J. Electron. Mater. 43: 3276–3282.
Li, G., R. Zhu and Y. Yang. 2012. Polymer solar cells. Nat. Photonics. 6: 153–161.
Li, J.C., Y. Wang and D.C. Ba. 2012. Characterization of semiconductor surface conductivity by using microscopic four-point probe technique. Phys. Procedia. 32: 347–355.
Lin, B.Z., C. Ding, B.-H. Xu, Z.-J. Chen and Y.-L. Chen. 2009. Preparation and characterization of polythiophene/molybdenum disulfide intercalation material. Mater. Res. Bull. 44: 719–723.
Lovinger, A.J., D.D. Davis, A. Dodabalapur and H.E. Katz. 1996. Comparative structures of thiophene oligomers. Chem. Mater. 8: 2836–2838.
Martinez, F., J. Rtuert, G. Neculqueo and H. Naarmann. 1995. Chemical and electrochemical polymerization of thiophene derivatives. Int. J. Polym. Mater. Polym. Biomater. 28: 51–59.
Mastragostino, M. and L. Soddu. 1990. Electrochemical characterization of "n" doped polyheterocyclic conducting polymers—I. polybithiophene. Electrochim. Acta. 35: 463–466.
Matte, H.S.S.R., A. Gomathi, A.K. Manna, D.J. Late, R. Datta, S.K. Patil et al. 2010. MoS_2 and WS_2 analogues of graphene. Angew. Chem. Int. Ed. 49: 4059–4062.
Murugan, A.V., C.S. Gopinath and K. Vijayamohanan. 2005. Electrochemical studies of poly(3,4-ethylenedioxythiophene) PEDOT/VS_2 nanocomposite as a cathode material for rechargeable lithium batteries. Electrochem. Commun. 7: 213–218.
Patil, B.H., A.D. Jagadale and C.D. Lokhande. 2012. Synthesis of polythiophene thin films by simple successive ionic layer adsorption and reaction (SILAR) method for supercapacitor application. Synth. Met. 162: 1400–1405.
Ramasubramaniam, A., D. Naveh and E. Towe. 2011. Tunable band gaps in bilayer transition-metal dichalcogenides. Phys. Rev. B 84: 205325–205335.
Roncali, J. 1992. Conjugated poly(thiophenes): synthesis, functionalization, and applications. Chem. Rev. 92: 711–738.
Roncali, J. 1997. Advances in the molecular design of functional conjugated polymers. pp. 311–342. *In*: T.A. Skothiem, R.L. Elsenbaumer and J.R. Reynolds (eds.). Handbook of Conductive Polymers. Marcel Dekker Inc. New York, USA.
Scheuble, M., M. Goll and S. Ludwigs. 2015. Branched tertiophenes in organic electronics: from small molecules to polymers. Macromol. Rapid Commun. 36: 115–137.
Singh, V., D. Joung, L. Zhai and S. Seal. 2011. Graphene based materials: past, present and future. Prog. Mater. Sci. 56: 1178–1271.
Smits, F.M. 1958. Measurement of sheet resistivities with the four-point probe. Bell Syst. Tech. J. 37: 711–718.
Tang, Q. and Z. Zhou. 2013. Graphene-analogous low-dimensional materials. Prog. Mater. Sci. 58: 1244–1315.
Valdes, L.B. 1954. Resistivity measurements on germanium for transistors. Proc. I.R.E. 42: 420–427.
Van der Pauw, I.J. 1958. A method of measuring the resistivity and hall coefficient on lamellae of arbitrary shape. Philips Res. Rep. 13: 1–9.
Wang, H., H. Yuan, S.S. Hong, Y. Li and Y. Cuo. 2015. Physical and chemical tuning of two-dimensional transition metal dichalcogenides. Chem. Soc. Rev. 44: 2664–2680.
Wu, Z., X. Zan and A. Sun. 2010. Synthesis of WS_2 nanosheets by a novel mechanical activation method. Mater. Lett. 64: 856–858.
Xu, B.H., B.-Z. Lin, D.-Y. Sun and C. Ding. 2007. Preparation and electrical conductivity of polyethers/WS_2 layered nanocomposites. Electrochim. Acta 52: 3028–3034.
Xu, J.M., H.S.O. Chan, S.-C. Ng and T.S. Chung. 2002. Polymers synthesized from (3-alkylthio)thiophenes by the $FeCl_3$ oxidation method. Synth. Met. 132: 63–69.
Yamamoto, T., M. Abla, T. Shimizu, D. Komarudin, B.-L. Lee and E. Kurokawa. 1999. Temperature-dependent electrical conductivity of p-doped poly(3,4-ethylenedioxythiophene) and poly(3-alkylthiophene)s. Polymer Bull. 42: 321–327.
Zhai, L. and R.D. McCullough. 2004. Regioregular polythiophene/gold nanoparticle hybrid materials. J. Mater. Chem. 14: 141–143.

Electrochemical Sensors/Biosensors Based on Carbon Aerogels/Xerogels

*Liana Maria Muresan** and *Aglaia Raluca Deac*

Introduction

Carbon aerogels and xerogels are microporous carbon materials that have received considerable attention in the literature over the past decade (Aragay 2012, Wang 2005, Rodrigues et al. 2011). These materials have exceptional properties such as high porosity, tunable surface area and pore volume, adjustable pore structure, controlled pore size distribution, low electrical resistivity and very good thermal and mechanical properties (Pekala et al. 1998). They can be produced in different forms (as powder, thin-film, cylinders, spheres, discs, or can be custom-shaped) depending on their applications (Rodrigues et al. 2011).

The particularly interesting properties of aerogels/xerogels recommend them for a wide range of applications such as electrode materials for capacitors or supercapacitors (Lee et al. 1999), adsorption materials for gas separation (Meena et al. 2005), catalyst supports (Kim et al. 2008), column packing materials for chromatography (Steinhart et al. 2007) biotechnological applications (Liang et al. 2003) and, last but not the least, for sensing devices (Chu and Lo 2008).

The electrochemical sensing devices based on carbon aerogels/xerogels are excellent sensing platforms with electrocatalytical activity and molecular recognition capabilities which exploit the outstanding properties of these materials mentioned above. In this context, recent advances in the preparation and characterization of electrochemical sensors using carbon aerogels/xerogels are reviewed with an emphasis on their application in heavy metals detection. The challenges for future research are also briefly discussed.

"Babes-Bolyai" University, Department of Chemical Engineering, 11, Arany Janos St., 400028 Cluj-Napoca, Romania.
Email: deac_raluca89@yahoo.com
* Corresponding author: limur@chem.ubbcluj.ro

Aerogels and xerogels

Aerogels are mesoporous nanomaterials in which there is a high volume of free space networked with nanometer solid domains (Rolison and Dunn 2001). They are "empty materials" which belong to a special class of synthetic porous ultralight materials (Aegerter et al. 1990). An aerogel is obtained by replacing the liquid phase of a gel by a gas in such a way that its solid network is retained, with only a slight or no shrinkage in the gel. It was first achieved in the 20th century by Kistler under supercritical conditions (Kistler 1931) but it is now possible to obtain it under ambient drying conditions as well. The shrinkage in the case of aerogels is less than 15%.

A **xerogel** is an open network formed by the removal of all swelling agents from a gel. It is obtained when the liquid phase of the gel is removed by evaporation. It may retain its original shape, but often cracks due to extreme shrinkage (more than 90%) are experienced while being dried. Therefore, it is important to underline that the method of drying will dictate whether an aerogel or xerogel will be formed (IUPAC 1997).

When the liquid solvent is removed by freeze-drying, a **cryogel** is obtained (Lozinski et al. 2003). Like aerogels, the cryogels have a meso- and macroporous structure depending on the experimental factors controlling the first preparation step. The resulting structure after low temperature pyrolysis of the gel is mainly meso- and microporous, while at higher temperatures, the structure is mostly macroporous.

Aerogels differ from xerogels mainly by the surface area and their porosity. Thus, aerogels have higher surface area and total pores volume than the xerogels prepared from the same precursor (Moreno-Castilla and Maldonado-Hodar 2005).

Only a limited number of materials can be obtained under the form of aerogel/xerogel. Most representatives are oxide-based aerogels (composed of silica-oxygen or metal-oxygen bonds), organic aerogels (consisting of resin-based and cellulose-based aerogel), carbon aerogels (carbonized plastic, CNT and graphene), chalcogenide aerogels and aerogels derived from natural materials, like gelatin, agar, egg albumin and rubber, etc. (Du et al. 2013). Among these, carbon aerogels deserve special attention due to their multiple applications, as will be illustrated in what follows.

Preparation of aerogels/xerogels

The first method used to prepare aerogels was **supercritical drying**. This method consists in drying assisted by the use of supercritical fluids, usually CO_2, leading to preservation of the high open porosity and superior textural properties of the wet gel in its dry form. Supercritical drying has major advantages in comparison with classical drying methods, avoiding the presence of vapor-liquid intermediates and surface tensions in the gel pores and thus preventing pore collapse phenomenon (García-González et al. 2012). Nevertheless, supercritical drying does not always preserve the wet gel structure (Job et al. 2005). The experimental conditions (pH, ratio of precursors, etc.) should be carefully chosen in order to avoid shrinkage and residual surface tensions.

In the 1960s, Teichner and Nicolaon prepared aerogels by using the **sol-gel method**, a much faster way to obtain these materials (Nicolaon and Teichner 1968). Their work opened the possibility to prepare a wide range of aerogels including metal oxides, composite and hybrid materials, etc. Nowadays, 90% of aerogels are obtained using sol-gel process.

The sol-gel method was first used to prepare inorganic materials such as ceramics or glasses but it was rapidly extended to other materials such as oxides, organic compounds, etc. It consists in progressive hydrolysis and condensation reactions of molecular precursors in a liquid medium in the presence of a catalyst (Brinker and Scherer 1990). Precursors used in sol-gel processing are metallic salts, alkoxides, organic precursors (formaldehyde, resorcinol, melamine, polyacrilonitrile, etc.) which can be solved in water or in different organic liquids. Gelation implies the transformation of a sol to a gel and can be strictly controlled by the factors affecting the process (precursor/catalyst molar ratios, solvent nature, working temperature, etc.). Gels are often aged in the mother liquor, then they are washed and dried to obtain xerogels (by simple evaporation), aerogels (by supercritical drying) or cryogels (by freeze drier) (Job et al. 2005, Pajonk 1995).

The advantages of sol-gel method are numerous: the processing temperature is low (often close to the ambient one), the method is waste-free and excludes the stage of washing (Wang and Bierwagen 2009). One of the reasons for using sol-gel technique is due to the hydrolysis and polycondensation reactions which enable a special relation between inorganic and organic compounds on molecular level. On the other hand, this method can ensure the obtaining of reproducible materials thanks to the possibility of adjusting the reaction parameters (Tomina et al. 2011).

The hydrolysis and condensation processes, which occur prior to the formation of sol, are taking place according to the following simplified reactions (Bach and Krause 2003):

$$M(OR)_n + H_2O \rightarrow M(OH)n + ROH \quad (1)$$

where M = Al, Si, Ti, V, Cr, Mo, W, etc.

$$2M(OH)_n \rightarrow MO_n + H_2O \quad (2)$$

It should be pointed out that in real systems, the equations are more complicated, especially because the hydrolysis is a multi-step process.

Carbon aerogels can be prepared starting from organic aerogels (e.g., resorcinol-formaldehyde, or melamine-formaldehyde aerogels) by pyrolysis at temperatures exceeding 500°C (Tamon and Ishizaka 2000). The carbonization temperature significantly influences the structural characteristics of carbon aerogels, affecting their microporosity, the surface area, the pores dimensions and the pores size distribution (Pierre and Pajonk 2002). At low temperatures, the mesopores and micropores volumes increase on behalf of macropores, whereas at high temperatures these volumes decrease.

By changing the precursor/catalyst molar ratio, the solvent nature or the catalyst, the microstructure of carbon aerogels can also be tailored. Thus, high precursor/catalyst ratios lead to colloidal aerogels consisting of spherical particles, while low ratios give rise to polymeric aerogels (Moreno-Castilla and Maldonado-Hodar 2005). The sol–gel synthesis is quite flexible and permits the preparation of gels, and finally of carbon gels, in different formats: monolith, powder, grain, pellets or coatings (Morales-Torres et al. 2011).

The most common method to prepare carbon aerogels is based on polymerization of resorcinol and formaldehyde, by using sodium carbonate as basic catalyst and deionized water as solvent (Halama and Szuzda 2010). Recently, by replacing water with acetone and sodium carbonate with perchloric acid, fractal carbon aerogels were prepared from resorcinol-formaldehyde precursors (Barbieri et al. 2001).

To prepare carbon xerogels, different drying methods can be used. For example, microwave drying was used to obtain porous carbon xerogels. Using this method, the procedure to obtain these materials is simplified, the time is reduced, no pretreatment is needed and textural properties are controlled. Although evaporation is the common drying method reported by literature, its' usage produces shrinkage in initial gel structure, while microwave drying method offers certain advantages, as was already mentioned (Zubizarreta et al. 2008).

Graphene based aerogels (Ji et al. 2013) can be prepared under strictly controlled conditions and offer distinctive opportunities for creating highly conductive composite materials. Their outstanding capacitive behavior is attributed mainly to the high degree of graphitization, large specific surface area and pore volume, and convenient pore size and pore size distribution. Cross linking and annealing are ways to significantly improve the mechanical properties and the surface area of these materials (Kohlmeyer et al. 2011). An interesting way to prepare graphene-based carbon aerogels is through the carbonization of chitosan aerogel and activation with KOH. The electrochemical behavior of graphene-based aerogels suggested that their 3D structure with well-developed and interconnected pore network provide ionic channels for electrochemical energy storage (Ji et al. 2013). 3D graphene aerogel with honeycomb-like porous structure and high C/O ratio was successfully prepared from graphene oxide dispersions in isopropanol/water solution by simple γ-ray irradiation and freeze-drying processes (He et al. 2016). Another simple method to prepare "pure" graphene aerogel is via reduction/self-crosslinking of graphene oxide dispersion induced by L-ascrobic acid and drying of the wet graphene gel (Zhang et al. 2011).

An interesting type of carbon aerogel is **carbon nanotube aerogel**. Carbon nanotubes (CNT) based aerogels (Kohlmeyer et al. 2011, Bryning et al. 2007) can be obtained from aqueous gel by critical-point drying or lyophilization of precursors. They can be reinforced, for example, with PVA polymer (Vigolo et al. 2000) which sensibly improves the strength and stability of the aerogel. Another possibility is to prepare CNT-based aerogels starting from a gel consisting of a 3D chemical assembly of CNTs in solution with a chemical cross-linker, followed by CO_2 supercritical drying and thermal annealing (Kohlmeyer et al. 2011). Thermally annealed CNT aerogels are mechanically stable, highly porous, exhibit excellent electrical conductivity and large specific surface area.

Sol-gel process can be used also to obtain template-oriented carbon aerogels by growing carbon on a lattice (template) of a zinc oxide crystals, and removing the zinc oxide in an oven, leaving just the carbon aerogel (Gao et al. 2007).

Doped carbon aerogels/xerogels

Aiming at modifying the structure, catalytic activity or conductivity of carbon aerogels, dopant elements are introduced into the carbon framework by different methods. Dopants can be metals (e.g., Fe (Fort et al. 2013, Gligor et al. 2013), Bi (Deac et al. 2015a, Deac et al. 2015b, Fort et al. 2015), Pt (Maldonado-Hodar et al. 2004), Ag (Sanchez-Polo et al. 2007), Co (Tian et al. 2010), Cr (Moreno-Castilla and Maldonado-Hodar 2005), etc.) but also non metals such as nitrogen (Bothelo-Barbosa et al. 2012).

The dopant can be introduced in the aerogel framework either in the initial precursor-solvent mixture, during the gelation step or after the carbon aerogel is formed (e.g., by surface deposition). The dopant can change the chemistry of the sol-gel process by catalyzing the polymerization or gelation process and, consequently, influences the surface morphology and pore texture of the aerogels, as well as their particles size (Moreno-Castilla and Maldonado-Hodar 2005). The initial pH of the precursor solution also influences the chemistry of the aerogel-producing process. Thus, the experimental conditions should be carefully controlled.

Metal doped carbon aerogels are better electrocatalysts than "pure" aerogels. However, an important issue is the necessity of a homogeneous distribution of the metals inside the carbon matrix in the form of nanoparticles. Moreover, during the preparation of the gel, a possible encapsulation of the metal particles in the carbon matrix should be avoided, which leads to a restricted access of gases and reactants to these particles, and consequently, to a diminished catalytic activity.

Nitrogen doped graphene aerogels exhibit improved electrochemical performance. For example, graphene-based nitrogen-doped porous carbon aerogels obtained by carbonization of a nitrogen-containing renewable biopolymer (Chitosan) (Hao et al. 2015) is a good electrode candidate for construction of a solid symmetric supercapacitor, which displays a high specific capacitance of about 197 Fg^{-1} at a current density of 0.2 A g^{-1}. A new class of nitrogen-doped carbonaceous nanofibers aerogels was prepared by hydrothermal carbonization with D (+)-glucosamine hydrochloride as the precursor (Song et al. 2016). N-doped carbon xerogels can also be prepared from a nitrogen-containing polymer precursor using melamine and urea as nitrogen sources that were incorporated into the polymer matrix via the sol–gel process (Barbosa et al. 2012).

The doping of carbon aerogels with N aim to increase the basicity of the surface and, consequently, the adsorption properties (Bothelo-Barbosa et al. 2012, Meng et al. 2014). Besides, N-doped carbon materials are efficient OER catalysts and promote H_2O_2 electroreduction (Cai et al. 2016).

Applications of carbon aerogels/xerogels

Carbon aerogels have a wide range of applications, varying from catalysis to environment protection and electroanalytical applications.

Most frequently, doped or undoped carbon aerogels are used as *catalysts* or *catalyst support* in water purification (Rasines et al. 2012) and for oxygen reduction reaction (Meng et al. 2014, Nagy et al. 2016, Seredych et al. 2016). Pt supported on carbon aerogels acted as catalysts in the toluene combustion reaction (Maldonado-Hodar et al. 2004) and in proton exchange membrane fuel cells (Smirnova et al. 2005). Carbon aerogels doped with transition metals (Co(II), Mn(II) and Ti(IV)) were used as catalysts in the photo-oxidation process of naphthalenesulphonic acids (Sanchez Polo and Rivera-Utrilla 2006) or in n-hexane conversion (Maldonado-Hódar 2011). In the latter case, the presence of a metal with different oxidation states improved the catalytic efficiency of the carbon aerogel through a synergetic effect, favoring alternately either cracking or the aromatization reaction.

Carbon aerogels doped with transition metals are very active in NO reduction, in absence of oxygen (Catalao et al. 2009) and those doped with rare-earth metals are good catalysts for Michael addition reaction (Kreek et al. 2014).

Carbon aerogels can be used as *adsorbents* for the removal of various chemical species. Thus, the adsorption capacity of carbon aerogel electrodes has been tested in NaCl solutions of several concentrations in order to evaluate the possibilities to use them for water desalination (Rasines et al. 2012). They were used also in the removal of organic dyes (Lin et al. 2015), and of other pollutants such as Bisphenol A from aqueous solutions (Hou et al. 2015). Three-dimensional nitrogen-doped graphene aerogels based on melamine were reported as excellent adsorbents for several metal ions such as Pb^{2+}, Cu^{2+} and Cd^{2+} and recycling performance for the removal of various oils and organic solvents (Xing et al. 2015).

Due to their 3D structure, high surface area, low electrical resistivity and excellent electrical conductivity, carbon aerogels can be used as electrode materials for batteries and supercapacitors (electrochemical double layer capacitors). In these devices, the storage of electricity is achieved trough the charging of electrical double layer existing at the electrode/electrolyte interface which is dependent on the porosity and surface area of the electrode material (Moreno-Castilla et al. 2012). Various metal (Co, Cu, Fe, and Mn)-doped carbon aerogels combine pseudo-capacitive property of metal oxide with electrochemical properties of carbon aerogel. The performance of carbon aerogels as supercapacitors depends strongly, but not exclusively, on the molar ratio of precursor to catalyst (Li et al. 2015). The relation between the electrochemical capacitance and the pore structure was widely investigated in the last decade, and the results reported in the literature show the proportionality between the electrochemical capacitance and specific surface area for the same kind of carbon materials. More than that, according to the literature data, the capacitance depends also on the pore volume, pore size distribution, particle size, electrical conductivity as well as the electrolyte composition and surface functional groups of the electrode materials (Halama et al. 2010).

One of the most widespread applications of carbon aerogels is in the field of electrochemical sensing devices, which deserve special attention due to their capabilities to detect various chemical species of practical interest. In order to build such devices, new electrode materials based on carbon aerogels were developed. The fabrication and application of these electrode materials can be achieved by several methods, shortly described in what follows.

Electrochemical sensors/biosensors based on C aerogels/xerogels

It is common knowledge that the electrode materials play a critical role in the construction of high-performance electrochemical sensing platforms for detecting target molecules through various analytical principles. Aerogels/xerogels-based materials have great potential for improving both selectivity and sensitivity of electrochemical sensors and biosensors. Besides their attractive characteristics described above, they have the ability to act as effective immobilization matrices and often generate a synergic effect among catalytic activity, conductivity, and biocompatibility to accelerate the signal transduction and to amplify bio-recognition. This is why they are used as materials to prepare electrodes used in electrochemical detection of molecular and ionic species.

Preparation of carbon aerogel/xerogel—modified electrodes

The carbon aerogel/xerogels can be used to obtain performing electrode materials by the modification of conventional electrode materials (especially carbon based), either by surface or bulk phase modification (Scheme 1).

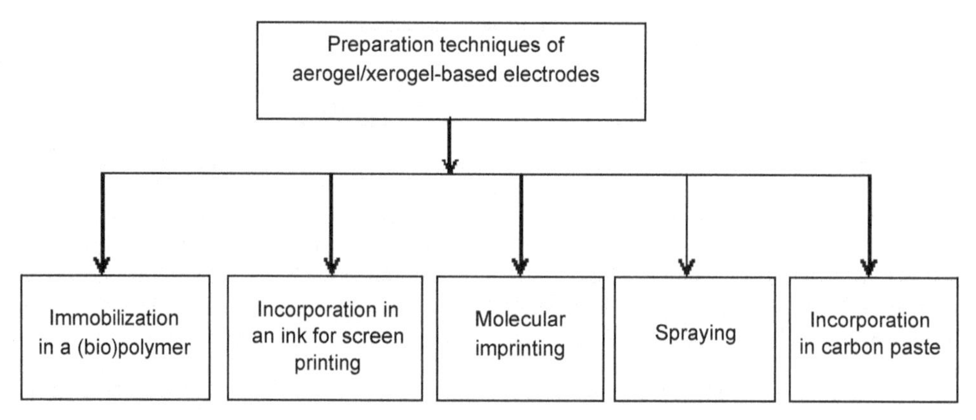

Scheme 1. Preparation techniques of modified electrodes based on aerogels/xerogels.

Surface modified electrodes

Surface modification of a conducting electrode material is one of the ways frequently used to prepare carbon aerogel-based modified electrodes. The substrate is generally a conventional electrode material, such as glassy carbon, boron-doped diamond, graphite, etc. and the immobilization methods include the use of a polymer (Nafion, Chitosan, polypyrrole, etc.), molecular imprinting, spraying etc.

The most used technique consists in the immobilization of the aerogel on the electrode surface by using a polymer, either preformed, or obtained by the *in situ* polymerization of a monomer. For example, a catalyst ink containing N-doped carbon nanofiber aerogel prepared by blending the catalyst powder with Nafion and ethanol was deposited on the surface of a polished glassy carbon electrode and tested for oxygen reduction (Meng et al. 2014). Composite aerogels incorporating layered MoS_2 nanoflowers and gold nanoparticles (AuNPs) were immobilized on the surface of glassy carbon by using a biopolymer (chitosan) in order to obtain a novel electrochemical aptamer-based biosensor (Fang et al. 2015). Chitosan was also used to immobilize a Bi doped carbon xerogel on the surface of glassy carbon for Pb^{2+} and Cd^{2+} traces voltammetric detection in aqueous solutions (Fort et al. 2015).

Molecular imprinting technique of a polymer onto a carbon aerogel surface by electropolymerization is another way to obtain novel electrochemical sensing platforms with electrocatalytical activity and molecular recognition capabilities. An important advantage in this case is that the film thickness on carbon aerogel can be strictly controlled by the electropolymerization time. As an example, molecularly imprinted polypyrrole on a carbon aerogel surface was successfully used for dopamine detection (Yang et al. 2015).

Another technique to prepare surface modified electrodes is based on the deposition of the active material containing aerogels on the surface of a conventional electrode material by spraying. Platinum/carbon aerogel composites were immobilized in this way on a boron-doped diamond (BDD) electrode by ink spreading followed by drying. The so-obtained modified electrode was used to prepare an acetyl cholinesterase (AChE) based biosensor in order to detect organophosphorous pesticides (Liu et al. 2014).

Bulk modified electrodes

The most common bulk modified electrodes are carbon paste electrodes (CPEs) due to their well-known advantages such as conductive entrapping matrix, low background current, wide potential window, versatility, etc. (Gorton 1995). Carbon paste is prepared by thoroughly mixing graphite powder with paraffin/mineral oils or silicone fluids (as binder) until homogenization. The mixture is put into a cavity of a Teflon holder and pyrolytic graphite in the bottom can be used for electric contact. Carbon paste electrodes (CPEs) are popular because they are easily obtainable at minimal costs and are especially suitable for preparing an electrode material modified with admixtures of other compounds, thus giving the electrode certain pre-determined desired properties.

A number of recent studies have been reported on the electrochemical performances of CPEs incorporating carbon aerogels/xerogels (Botelho Barbosa et al. 2012, Gligor et al. 2013). Fe doped mesoporous carbon aerogel obtained by sol-gel method coupled with supercritical drying with liquid CO_2, and followed by thermal pyrolysis, was incorporated in carbon paste and was successfully used for H_2O_2 amperometric detection (Gligor et al. 2013, Fort et al. 2013).

Various nanoparticles can be also used along with carbon aerogels/xerogels in order to improve the efficiency of the bulk modified electrode materials. CPEs containing carbon aerogels modified with gold nanoparticles (AuNPs) and decorated with Hemoglobin were used as transducers for two novel biosensors exploiting the synergetic effect with ionic liquids, and their catalytic ability to detect H_2O_2 and NO_2^- was studied (Peng et al. 2015). A new composite electrode consisting of carbon paste modified with carbon xerogel containing Bi nanoparticles (BiCXe) was reported as efficient for Pb^{2+} ions' determination at trace levels by using square wave anodic stripping voltammetry (Deac et al. 2015a).

An interesting development of the electrode materials based on carbon aerogels/xerogels is by combining them with different polymers. The high surface area of carbon aerogels can provide much more multidimensional spaces for electrochemical modification of conducting polymers, such as polypyrrole (Fang et al. 2015), polystyrene (Zhang et al. 2008), Nafion (Brigaudet et al. 2008), Chitosan, etc. (Zuo et al. 2015).

Applications of electrochemical sensor/biosensors based on carbon aerogels/xerogels

Electrochemical sensors are devices that transform chemical information, ranging from the concentration of a specific sample component to total composition analysis, into an electrical, analytically useful signal. In the case of electrochemical biosensors, a biological recognition element (biochemical receptor) is put in direct contact with an electrochemical transduction element to provide selective quantitative or semi-quantitative analytical information (Thevenot et al. 2001).

As already mentioned, carbon aerogels and xerogels are interesting materials for the construction of electrochemical sensors and biosensors, as they possess excellent conductivity, may promote the electrochemical oxidation of molecules on the electrode surface, and facilitate the charge transfer in the redox reactions (Wu et al. 2010). On the other hand, they have superior characteristics such as low mass density, controllable porosity, large specific area and good mechanical properties (Moreno-Castilla et al. 2012) that can be very useful in the construction of sensing devices.

Often, doped aerogels are preferred for the construction of electrochemical sensors/biosensors instead of undoped aerogels because they exhibit improved electrochemical performance when incorporated in electrode sensing devices. Thus, metal doped carbon aerogels combine the catalytic and conductive properties of metals with the large surface area and porosity of aerogels, offering an attractive sensing platform for detection of various analytes.

One example of successful electroanalytic application of metal-doped carbon xerogels is offered by Bi-modified carbon xerogels. As Bi is considered one of the less toxic metals, Bi-bulk and Bi-film electrodes are used as alternatives to replace toxic mercury electrode in detection of heavy metals' ions from aqueous solutions. The use of Bi electrodes is based on Bi ability to form low temperature alloys with heavy metals, (Kirk-Othmer 1978) favoring the accumulation of these ions during the preconcentration step of the stripping analysis.

Bi doped carbon xerogel (BiCXe) can be prepared by sol-gel method starting with polycondensation of resorcinol and formaldehyde in the presence of ammonium hydroxide and glycerol formal, followed by impregnation with a Bi salt, drying and pyrolysis (Deac et al. 2015a). The preparation of BiCXe involves several steps, as shown in Figure 1:

(i) polycondensation of precursors (resorcinol and formaldehyde in the presence of ammonium hydroxide and glycerol formal)

(ii) impregnation of the resulting resorcinol-formaldehyde gel (RF-gels) with Bi^{3+} salt

(iii) gel drying by evaporation at ambient temperature

(iv) pyrolysis of the Bi impregnated organic xerogel in Ar atmosphere

Further, the Bi-doped xerogel was used to prepare an appropriate electrode material by incorporating it into carbon paste. The BiCXe-CPEs successfully exploit the favorable electrochemical properties of carbon paste electrodes already mentioned (large potential window, versatility, etc.), and the unique electroanalytical characteristics of Bi-based xerogels (e.g., Bi-based electrodes are less susceptible to oxygen interference, exhibiting a lower background current for square wave anodic stripping voltammograms (Lee et al. 2007).

BiCXe-CPE were proven to be very efficient in environmental monitoring of toxic Pb^{2+} (Deac et al. 2015a) and Cd^{2+} ions (Deac et al. 2015b), in aqueous solutions and for the simultaneous detection of Pb^{2+} and Cd^{2+} ions (Deac et al. 2015a) by using anodic stripping analysis (SWASV) (Figure 2). In anodic stripping voltammetry, the analytes of interest (heavy metal ions, in this case) are electrodeposited on the working electrode during a reduction step, and oxidized from the electrode during the stripping step, the resulting current being measured.

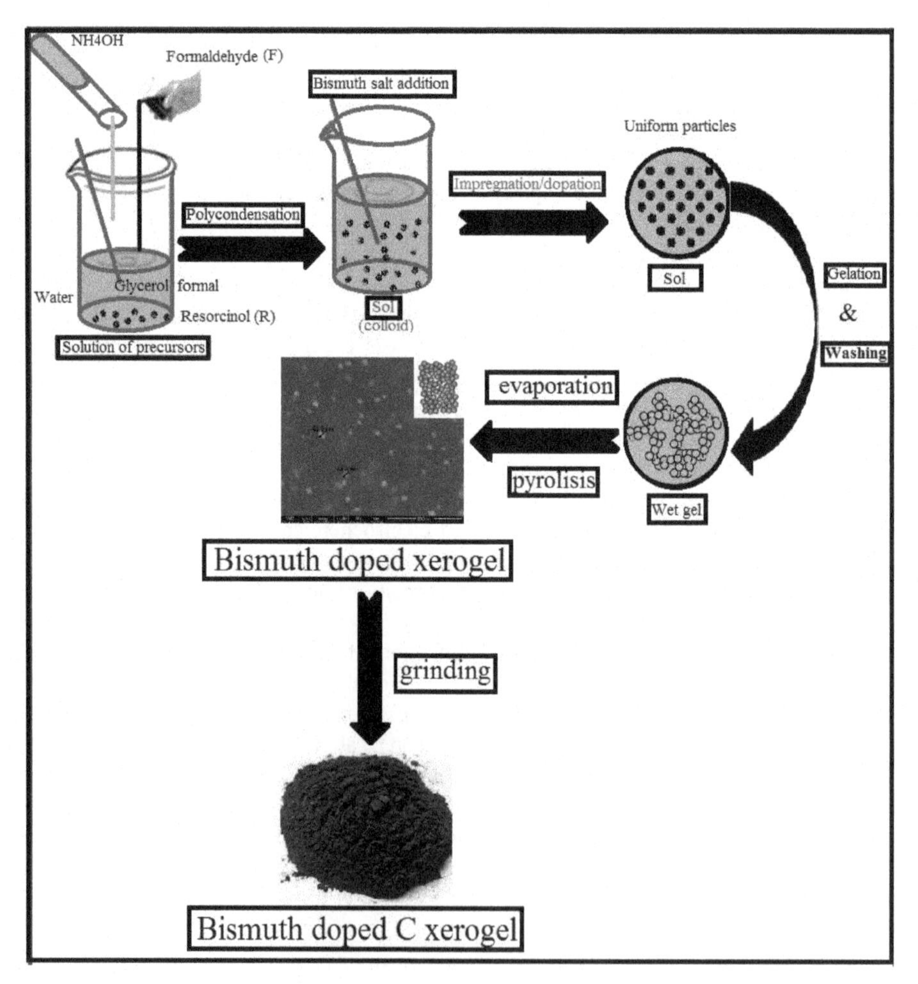

Figure 1. Steps involved in Bi-doped carbon xerogel preparation by sol-gel method.

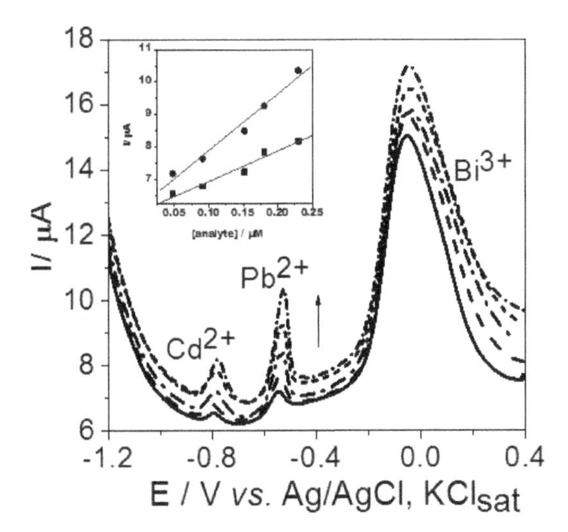

Figure 2. SWASVs responses for increasing concentrations of Cd^{2+} and Pb^{2+}, and the corresponding calibration curves at BiCXe-CPE electrode (Cd^{2+} circle, square for Pb^{2+}, respectively) (inset). Experimental conditions: electrolyte, 0.1 M acetate buffer (pH 4); starting potential, −1.3 V vs. Ag/AgCl, KClsat; deposition time 120 s under continuous stirring; scan rate, 0.050 V/s; frequency, 25 Hz; amplitude, 0.05 V; step potential, 0.004 V; equilibration time, 10 s.

BiCXe-CPEs are characterized by wide linear dynamic range and low detection limits (0.18 μg L^{-1} Pb^{2+}; 0.045 μM L^{-1} Cd^{2+}). Other advantages are easy handling and modification, low toxicity and low cost, which recommend them for the trace analysis of Pb^{2+} and Cd^{2+}. The analytical parameters of BiCXe-CPE were satisfactory when compared with the maximum admissible concentration of these ions required by the UE legislation for the drinking water (World Health Organization 2008).

Carbon aerogels can also be used for the obtaining of composite and hybrid electrode materials with good electroanalytical performances. Molecularly imprinted TiO_2/carbon aerogel was used as electrode material for the photoelectrochemical determination of atrazine (Zhang et al. 2011), while conductive aerogels composed of carbon nanotubes (CNTs) and cellulose were reported as sensitive vapor sensors (Qi et al. 2015). The development of reliable sensors for real-time tracing of hazardous volatile organic compounds and for monitoring of some specific pollutant gases in industries is of big importance. Carbon aerogel thin film composites with different sensitivity toward toluene, n-hexane, acetone, and water vapors at room temperature were prepared and characterized (Thubsuang et al. 2015). Activated carbon aerogel/polymer composites with appropriate polymer matrices exhibited excellent sensitivity and fast response towards organic vapors together with much enhanced mechanical properties.

In the case of biosensors, the main challenge consists in loading the biomolecules and keeping their bioactivity in the same time, which is often a difficult task. The large effective surface area and the porosity of aerogels allow the immobilization of biomolecules at or near the electrode surface, which facilitates the charge transfer between the electrode and the substrate (Fang et al. 2015). The effective reaction area is increased, while the electron transfer and ion diffusion paths are hindered, which lead to better electrochemical performance. Thus, some aerogel-based efficient biosensors were obtained, aiming to detect hemoglobin (Peng et al. 2015), H_2O_2 (Dong et al. 2013), organophosphorous pesticides (Liu et al. 2014), etc. In the construction of carbon-based hybrid materials, metal nanoparticles such as Ni, Pd and Au are often included for their high electrocatalytic behavior toward biomolecules in sensing application (Chen et al. 2012, Naruse et al. 2011). Hence, carbon aerogels incorporating gold nanoparticles and layered MoS_2 nanoflowers were prepared by using a facile hydrothermal route and were used to prepare a novel electrochemical aptamer-based biosensor (Fang et al. 2015).

Other examples illustrating different applications of carbon aerogels/xerogels-based electrochemical sensors/biosensors are presented in Table 1.

Table 1. Carbon aerogels/xerogels-based electrode materials for electrochemical detection of different chemical species.

No.	Electrode Material	Electrolyte	Analyte	Method	Ref.
1	Fe doped C aerogel—CPE	0.1 M phosphate buffer solution (pH 7.0)	H_2O_2	CV	(Fort et al. 2013, Gligor et al. 2013)
2	N-doped C xerogel	0.1 M KCl solution	$K_3Fe(CN)_6$	CV	(Barbosa et al. 2012)
3	N-doped C nanofiber aerogel	0.1 M KOH solution	O_2	CV; RDE	(Meng et al. 2014)
4	3D nitrogen-doped graphene aerogel	0.2 M PBS, pH 7	H_2O_2	cyclic voltammetry	(Cai et al. 2016)
5	Mesoporous carbon aerogels with 0–14% nitrogen content	H_2SO_4 5 M	oxygen reduction	Cyclic voltammetry	(Nagy et al. 2016)
6	C aerogel	phosphate buffer solution	Bisphenol A	electro-polymerization	(Hou et al. 2015)
7	C aerogel with molecularly imprinted pPy	0.1 M phosphate buffer solution (pH 7.0)	dopamine	DPV	(Li et al. 2008)
8	Dual aptamer biosensor using AuNPs/MoS_2/C aerogel	phosphate buffer solution; human serum samples	platelet-derived growth factor BB (PBGF-BB)	CV; DPV	(Fang et al. 2015)
9	Biosensor based on platinum-carbon aerogel composite	PBS pH 7.3	organophos-phorus pesticides	differential pulse voltammetry	(Liu et al. 2014)
10	AuNPs-C aerogel-Hemoglobin	0.1 M phosphate buffer solutions	H_2O_2; NO_2^-	EIS; CV; amperometry	(Peng et al. 2015)
11	Molecularly imprinted TiO_2/C aerogel	0.1 M KCl solution	atrazine	photo-electrochemical analysis	(Zhang et al. 2008)
12	Inorganic/organic doped carbon aerogel (CA) materials (Ni-CA, Pd-CA, and Ppy-CA)	0.1 M phosphate buffer solution (PBS, pH 7.0)	H_2O_2	EIS, amperometry	(Dong et al. 2015)
13	Carbon nanotube–cellulose composite aerogels	–	vapor sensing	monitoring the electrical resistance change	(Qi et al. 2015)
14	Polybenzoxazine-derived C aerogel composite	–	vapor sensor: toluene; n-hexane, acetone	electrical response measurements	(Thubsuang et al. 2015)
15	Glassy carbon electrode modified with chitosan/CNT aerogel-Cu_2O@CuO	NaOH solution	Glucose	cyclic voltammetry; amperometry	(Liu et al. 2015)
16	Nitrogen doped porous carbon aerogel	O_2 and N_2 saturated 0.1 M KOH	Oxygen reduction	Cyclic voltammetry	(Alatalo et al. 2016)
17	Carbon aerogels	A simulated waste water containing phenol	Removal of phenol	Electrochemical oxidation-cyclic voltammetry	(Guifen et al. 2009)

* CPE = carbon paste electrode; NPs = nanoparticles; CA = carbon aerogel; CNT = carbon nanotubes.

Conclusion

Carbon aerogels belong to a very promising class of materials with unique properties that have a wide range of applications varying from environmental clean-up, to energy storage devices, chemical sensors, catalysts, etc.

Doping of carbon aerogels/xerogels, either with metals or non-metals, is considered desirable, as it enhances the electrical conductivity, improves thermal/oxidation stability, advantageously modifies the surface morphology and pore texture of the aerogels, as well as their particles size and enhances their catalytic activity.

The electrochemical sensors and biosensors based on carbon aerogels represent viable solutions to detect various analytes of practical interest.

The challenges for future research concerning carbon aerogels/xerogels include:

- preparation of new single component or hybrid aerogels
- elaboration of novel methods that can be applied for the characterization of aerogels as well as their gel precursors
- tailoring of aerogels/xerogels properties in order to satisfy the requirements of different beneficiaries
- development of existing and potential applications in wide range of technological areas
- cost-effective mass production of aerogels/xerogels

To conclude, development of carbon aerogels/xerogel-based materials is foreseen as a promising and generous research field in the years to come.

References

Aegerter, M.A., L. Nicholas and M.M. Koebel. 1990. MRS Bulletin. Copyright © Materials Research Society 15: 30–36.

Ai, D., Z. Bin, Z. Zhihua and S. Jun. 2013. A special material or a new state of matter: A review and reconsideration of the aerogel. Materials 6: 941–968.

Alatalo, S.M., Q. Kaipei, H. Preuss, A. Marinovic, M. Sevilla, M. Sillanpaa et al. 2016. Soy protein directed hydrothermal synthesis of porous carbon aerogel for electrocatalitic oxygen reduction. Carbon 96: 622–630.

Aragay, G. 2012. Nanomaterials application in electrochemical detection of heavy metals. Electrochim. Acta 84: 49–61.

Bach, H.B. and D. Krause. 2003. Thin Films on Glass. Springer Verlag. Berlin-Heidelberg.

Barbieri, O., D.F. Ehrburger, T.P. Rieker, G.M. Pajonk, N. Pinto and R. Venkateswara. 2001. Small-angle X-ray scattering of a new series of organic aerogels. J. Non-Cryst. Solids 285: 109–115.

Botelho-Barbosa, M., J.P. Nascimento, P.B. Martelli, C.A. Furtado, N.D. Santina Mohallem and H.F. Gorgulho. 2012. Electrochemical properties of carbon xerogel containing nitrogen in a carbon matrix. Microporous Mesoporous Mater. 162: 24–30.

Brigaudet, M., S. Berthon-Fabry, Ch. Beauger, M. Chatenet and P. Achard. 2008. Influence of carbon aerogel texture on PEMFC performances. Fundamentals and Developments of Fuel Cells Conference 2008—FDFC 2008, Dec 2008, Nancy, France.

Brinker, C.J. and G.W. Scherer. 1990. Sol–Gel Science: The Physics and Chemistry of Sol–Gel Processing. Harcourt Brace Jovanovich (Academic Press, Inc.). Boston.

Bryning, M.B., D.E. Milkie, M.F. Islam, L.A. Hough, J.M. Kikkawa and A.G. Yodh. 2007. Carbon nanotube aerogels. Adv. Mater. 19: 661–664.

Cai, Z.X., X.H. Song, Y.Y. Chen, Y.R. Wang and X. Chen. 2016. 3D nitrogen-doped graphene aerogel: A low-cost, facile prepared direct electrode for H_2O_2 sensing. Sens. Actuators, B 22: 567–573.

Castilla, C.M., M.B. Dawidziuk, F. Carrasco-Marin and E. Morallon. 2012. Electrochemical performance of carbon gels with variable surface chemistry and physics. Carbon 50: 3324–3332.

Catalao, R.A., F.J. Maldonado-Hodar, A. Fernandes, C. Henriques and M.F. Ribeiro. 2009. Reduction of NO with metal-doped carbon aerogels. Appl. Catal. B. 88: 135–141.

Chen, Q.W., L.Y. Zhang and G. Chen. 2012. Facile preparation of graphene-copper nanoparticle composite by *in situ* chemical reduction for electrochemical sensing of carbohydrates. Anal. Chem. 84: 171–178.

Chu, C.S. and Y.L. Lo. 2008. Sens. Actuators, B 129: 120–125.

Deac, A.R., L.C. Cotet, G.L. Turdean and L.M. Muresan. 2015. Carbon paste electrode modified with Bi nanoparticles-carbon xerogel for Pb^{2+} determination by square wave anodic stripping voltammetry. Rev. Roum. Chim. 60: 697–705.

Deac, A.R., L.C. Cotet, G.L. Turdean and L.M. Muresan. 2015. Determination of Cd(II) using square wave anodic stripping voltammetry at a carbon paste electrode containing Bi-doped carbon xerogel. Studia UBB Chemia. LX 1: 203–212.

Dong, S., N. Li, G. Suo and T. Huang. 2013. Inorganic/organic doped carbon aerogels as biosensing materials for the detection of hydrogen peroxide. Anal. Chem. 85: 11739–11746.

Fang, L.X., K.J. Huang and Y. Liu. 2015. Novel electrochemical dual-aptamer-based sandwich biosensor using molybdenum disulfide/carbon aerogel composites and Au nanoparticles for signal amplification. Biosens. Bioelectron. 71: 171–178.

Fort, C.I., L.C. Cotet, V. Danciu, G.L. Turdean and I.C. Popescu. 2013. Iron doped carbon aerogel–New electrode material for electrocatalytic reduction of H_2O_2. Mat. Chem. Phys. 138: 893–898.

Fort, C.I., L.C. Cotet, A. Vulpoi, G.L. Turdean, V. Danciu, L. Baia et al. 2015. Bismuth doped carbon xerogel nanocomposite incorporated in chitosan matrix for ultrasensitive voltammetric detection of Pb (II) and Cd (II). Sens. Actuators, B 220: 712–719.

Gao, Y.P., C.N. Sisk and L.J. Hope-Weeks. 2007. A Sol–gel route to synthesize monolithic zinc oxide aerogels. Chem. Mater. 19: 6007–6011.

Gligor, D., L.C. Cotet and V. Danciu. 2013. Comparative study of two types of iron doped carbon aerogels for electrochemical applications. J. New Mat. Electrochem. Sensors 16: 97–101.

González, G.C.A., M.C. Camino-Rey, M. Alnaief, C. Zetz, I. Smirnova, C.A. García-González et al. 2012. J. Supercrit. Fluids 66: 297–306.

Gorton, L. 1995. Carbon paste electrodes modified with enzymes, tissues, and cells. Electroanal. 7: 23–45.

Guifen, L.V., W. Dingcai and F. Ruowen. 2009. Performance of carbon aerogels particle electrodes for the aqueous phase electro-catalytic oxidation of simulated phenol wastewater. J. Hazard. Mater. 165: 961–966.

Halama, A. and B. Szubzda. 2010. Carbon aerogels as electrode material for electrical double layer supercapacitors—Synthesis and properties. Electrochim. Acta 55: 7501–7505.

Hao, P., Z. Zhao, Y. Leng, J. Tian, Y. Sang, R.I. Boughton et al. 2015. Graphene-based nitrogen self-doped hierarchical porous carbon aerogels derived from chitosan for high performance supercapacitors. Nano Energy 15: 9–23.

Hou, Ch.H., Sh.Ch. Huang, P.H. Chou and W. Den. 2015. Removal of bisfenol A from aqueous solutions by electrochemical polymerization on a carbon aerogel electrode. J. Taiwan Inst. Chem. Eng. 51: 103–108.

IUPAC. Compendium of Chemical Terminology, 2nd ed. (the "Gold Book"). Compiled by A.D. McNaught and A. Wilkinson. Blackwell Scientific Publications, Oxford (1997). XML on-line corrected version: http://goldbook.iupac.org (2006-) created by M. Nic, J. Jirat, B. Kosata; updates compiled by A. Jenkins. ISBN 0-9678550-9-8. doi:10.1351/goldbook.

Ji, C-C., M.W. Xu, S.J. Bao, C.J. Cai, Z.J. Lu, H. Chai et al. 2013. Self-assembly of three-dimensional interconnected graphene-based aerogels and its application in supercapacitors. J. Coll. Interface Sci. 407: 416–424.

Job, N., A. Thery, R. Pirard, J. Marien, L. Kocon, J.N. Rouzaud et al. 2005. Carbon aerogels, cryogels and xerogels: influence of the drying method on the textural properties of porous carbon materials. Carbon 43: 2481–2494.

Kim, T.W., P.W. Chung, I.I. Slowing, M. Tsunoda, E.S. Yeung and V.S.Y. Lin. 2008. Structurally ordered mesoporous carbon nanoparticles as transmembrane delivery vehicle in human cancer cells. Nano Lett. 8: 3724–3727.

Kistler, S.S. 1931. Coherent expanded aerogels and jellies. Nature 127: 741–741.

Kohlmeyer, R.R., M. Lor, K. Deng, H. Liu and J. Chen. 2011. Preparation of stable carbon nanotube aerogels with high electrical conductivity and porosity. Carbon 49: 2352–2361.

Kreek, K., K. Kriis, B. Maaten, M. Uibu, A. Mere, T. Kanger et al. 2014. Organic and carbon aerogels containing rare-earth metals: Their properties and application as catalysts. J. Non-Crys. Solids 404: 43–48.

Lee, J., S. Yoon, T. Hyeon, S.M. Oh and K.B. Kim. 1999. Synthesis of a new mesoporous carbon and its application to electrochemical double-layer capacitors. Chem. Commun. 21: 2177–2178.

Lee, G.J., H.M. Lee and C.K. Rhee. 2007. Bismuth nano-powder electrode for trace analysis of heavy metals using anodic stripping voltammetry. Electrochem. Commun. 9: 2514–2518.

Li, J., X. Wang, Y. Wang, Q. Huang, C. Dai, S. Gamboa et al. 2008. Structure and electrochemical properties of carbon aerogels synthesized at ambient temperatures as supercapacitors. J. Non-Cryst. Solids 354: 19–24.

Liang, C., S. Dai and G.A. Guiochon. 2003. A graphitized-carbon monolithe column. Anal. Chem. 75: 4904–4912.

Lin, Y.F. and Ch.Y. Chang. 2015. Design of composite maghemite/hematite/carbon aerogel nanostructures with high performance for organic dye removal. Separation and Purification Technol. 149: 74–81.

Liu, Y. and M. Wei. 2014. Development of acetylcholinesterase biosensor based on platinum–carbon aerogels composite for determination of organophosphorus pesticides. Food Control 36: 49–54.

Liu, M., R. Liu and W. Chen. 2013. Graphene wrapped Cu_2O nanocubes: Non-enzymatic electrochemical sensors for the detection of glucose and hydrogen peroxide with enhanced stability. Biosens. Bioelectron. 45: 206–212.

Lozinski, V.I., I.Y. Galaev, F.M. Plieva, I.N. Savina, H. Jungvid and B. Matiasson. 2003. Polymeric cryogels as promising materials of biotechnological interest. Trends Biotechnol. 21: 445–51.

Maldonado, H.F.J., C. Moreno-Castilla and A.F. Perez-Cadenas. 2004. Catalytic combustion of toluene on platinum-containing monolithic carbon aerogels. Appl. Catal. B 54: 217–224.

Maldonado, H.F.J. 2011. Metal-doped carbon aerogels as catalysts for the aromatization of n-hexane. Appl. Catal. A 408: 156–162.

Meena, A.K., G.K. Mishra, P.K. Rai, C. Rajagopal and P.N. Nagar. 2005. Removal of heavy metal ions from aqueous solutions using carbon aerogel as an adsorbent. J. Hazard Mater. B 122: 161–170.

Meng, F., L. Li, Zh. Wu, H. Zhong, J. Li and J. Yan. 2014. Facile preparation of N-doped carbon nanofiber aerogels from bacterial cellulose as an efficient oxygen reduction reaction electrocatalyst. Ch. J. Catalysis 35: 877–883.
Moreno, C.C. and F.J. Maldonado-Hodar. 2005. Carbon aerogels for catalysis applications: An overview. Carbon 43: 455–465.
Nagy, B., S. Villar-Rodil, J.M.D. Tascon, I. Bakos and K. Laszlo. 2016. Nitrogen doped mesoporous carbon aerogels and implications for electrocatalytic oxygen reduction reactions. Microporous and Mesoporous Materials 230: 135–144.
Naruse, J., L. Hoa, Q. Sugano, Y. Ikeuchi, T. Yoshikawa, H.M. Saito et al. 2011. Development of biofuel cells based on gold nanoparticle decorated multi-walled carbon nanotubes. Biosens. Bioelectron. 30: 204–210.
Nicolaon, G.A. and S.J. Teichner. 1968. Préparation des aerogels de silice à partir d'orthosilicate de methyl en milieu alcoolique et leurs propriétés. Bull. Soc. Chim. Fr. 5: 1906–1911.
Othmer, K. 1978. Encyclopedia of Chemical Technology. John Wiley & Sons. New York. 3: 912–937.
Pajonk, G.M. 1995. Catalytic aerogels. Heterogen. Chem. Rev. 2: 129.
Pekala, R.W., J.C. Farmer, C.T. Alviso, T.D. Tran, S.T. Mayer, J.M. Miller and B. Dunn. 1998. Carbon aerogels for electrochemical applications. J. Non-Cryst. Solids 225: 74–80.
Peng, L., S. Dong, N. Li, G. Suo and T. Huang. 2015. Construction of a biocompatible system of hemoglobin based on AuNPs-carbon aerogel and ionic liquid for amperometric biosensor. Sens. Actuators, B 210: 418–424.
Pierre, A.C. and G.M. Pajonk. 2002. Chemistry of aerogels and their applications. Chem. Rev. 102: 4243–4265.
Qi, H., J. Liu, J. Pionteck, P. Pötschke and E. Mäder. 2015. Carbon nanotube–cellulose composite aerogels for vapour sensing. Sens. Actuators, B 213: 20–26.
Rasines, G., P. Lavela, C. Macias, M. Haro, C.O. Ania and J.L. Tirado. 2012. Mesoporous carbon black-aerogel composites with optimized properties for the electro-assisted removal of sodium chloride from brackish water. J. Electroanal. Chem. 671: 92–98.
Rodrigues, J.A., C.M. Rodrigues, P.J. Almeida, I.M. Goncalves, R.G. Compton and A.A. Barros. 2011. Increased sensitivity of anodic stripping voltammetry at the hanging mercury drop electrode by ultracathodic deposition. Anal. Chim. Acta 701: 152–156.
Rolison, D.R. and B. Dunn. 2001. Electrically conductive oxide aerogels: new materials in electrochemistry. J. Mater. Chem. 11: 963–980.
Sanchez-Polo, M., J. Rivera-Utrilla, E. Salhi and U. von Gunten. 2007. Ag-doped carbon aerogels for removing halide ions in water treatment. Water Research 41: 1031–1037.
Sanchez Polo, M. and J. Rivera-Utrilla. 2006. Photooxidation of naphthalenesulphonic acids in presence of transition metal-doped carbon aerogels. Appl. Catal. B 69: 93–100.
Seredych, M., K. László, E.R. Castellón and T.J. Bandosz. 2016. S-doped carbon aerogels/GO composites as oxygen reduction catalysts. J. Energy Chemistry 25: 236–245.
Smirnova, A., X. Dong, H. Hara, A. Vasiliev and N. Sammes. 2005. Evaluation of novel carbon aerogel-supported catalysts for PEM fuel cell application. J. Hydrogen Energy Sci. 30: 149–158.
Song, L.T., Z.Y. Wu, H.W. Liang, F. Zhou, Z.Y. Yu, L. Xu et al. 2016. Macroscopic-scale synthesis of nitrogen-doped carbon nanofiber aerogels by template-directed hydrothermal carbonization of nitrogen-containing carbohydrates. Nano Energy 19: 117–127.
Steinhart, M., C. Liang, G.W. Lynn, U. Gösele and S. Dai. 2007. Direct synthesis of mesoporous carbon microwires and nanowires chemistry of materials. Chem. Mater. 19: 2383–2385.
Tamon, H. and H. Ishizaka. 2000. Influence of gelation temperature and catalysts on the mesoporous structure of resorcinol-formaldehyde aerogels. J. Colloid Interface Sci. 223: 305–307.
Thevenot, D.R., K. Toth, R.A. Durst and G.S. Wilson. 2001. Electrochemical biosensors: recommended definitions and classification. Biosens. Bioelectron. 16: 121–131.
Thubsuang, S.U. D., S. Sahasithiwat, S. Wongkasemjit and T. Chaisuwan. 2015. Highly sensitive room temperature organic vapor sensor based on polybenzoxazine-derived carbon aerogel thin film composite. Mat. Science. Engineering, B 200: 67–77.
Tian, H.Y., C.E. Buckley, D.A. Shepard, M. Paskevicius, N. Hanna, H.Y. Tian et al. 2010. A synthesis method for cobalt doped carbon aerogels with high surface area and their hydrogen storage properties. Int. J. Hydrogen Energy 35: 13242–13246.
Tomina, Veronica V. et al. 2011. Synthesis of polysiloxane xerogels with fluorine-containing groups in the surface layer and their sorption properties. J. of Fluorine Chemistry 132: 1146–1151.
Torres, S.M., H.F.J. Maldonado, C.A.F. Pérez and Carrasco M. Francisco. 2011. Structural characterization of carbon xerogels: From film to monolith, Elsevier. (in press).
Vigolo, B., A. Pénicaud, C. Coulon, C. Sauder, R. Pailler, C. Journet et al. 2000. Macroscopic fibers and ribbons of oriented carbon nanotubes. Science 290: 1331–4.
Wang, J. 2005. Stripping analysis at bismuth electrodes: A review. Electroanalysis 17: 1341–1346.
Wang, D. and B.P. Gordon. 2009. Sol–gel coatings on metals for corrosion protection. Progress in Organic Coatings 64: 327–338.
World Health Organization, Guidelines for drinking water Quality, 3rd edition, World Health Organization (WHO) Geneva, 2008.
Wu, M.F., Y.N. Jin, G.H. Zhao, M.F. Li and D.M. Li. 2010. Electrosorption-promoted photodegradation of opaque wastewater on a novel TiO_2/carbon aerogel electrode. Environ. Sci. Technol. 44: 1780–1785.
Xing, L.B., S.F. Hou, J. Zhou, J.L. Zhang, W. Si, Y. Dong et al. 2015. Three dimensional nitrogen-doped graphene aerogels functionalized with melamine for multifunctional applications in supercapacitors and adsorption. J. of Solid State Chemistry 230: 224–232.

Yalei, H., J. Li, L. Li and J. Li. 2016. Gamma-ray irradiation-induced reduction and self-assembly of graphene oxide into three-dimensional graphene aerogel. Mat. Lett. 177: 76–79.

Yang, Zh., X. Liu, Y. Wu and Ch. Zhang. 2015. Modification of carbon aerogel electrode with molecularly imprinted polypyrrole for electrochemical determination of dopamine. Sens. Actuators, B 212: 457–463.

Zhang, B., X. Dong, W. Song, D. Wu, R. Fu, B. Zhao et al. 2008. Electrical response and adsorption of novel composites of polystyrene filled with carbon aerogel in organic vapors. Sens. Actuators, B 132: 60–66.

Zhang, X.T., Z.Y. Sui, B. Xu, S.F. Yue, Y.J. Luo, W.C. Zhan et al. 2011. Mechanically strong and highly conductive graphene aerogel and its use as electrodes for electrochemical power sources. J. Mater. Chem. 21: 6494–6497.

Zubizarreta, I., A. Arenillas, J.A. Menéndez, J.J. Pis, J.P. Pirard, N. Job et al. 2008. Microwave drying as an effective method to obtain porous carbon xerogels. J. of Non-Cryst. Solids 354: 4024–4026.

Zuo, L., Y. Zhang, L. Zhang, Y.E. Miao, W. Fan and T. Liu. 2015. Polymer/carbon-based hybrid aerogels: Preparation, properties and applications. Materials 8: 6806–6848.

From Nano- to Macro-engineering of Nanocomposites and Applications in Heterogeneous Catalysis

Guofeng Zhao, Ye Liu and *Yong Lu**

Introduction

Heterogeneous catalysts are involved in numerous amounts of chemical transformations by which raw materials-renewable, nonrenewable and intermediates derived therefrom-are converted to products such as fuels, basic and specialty chemicals, materials for construction and communications, fertilizers, pharmaceuticals, and a variety of consumer products (Figure 1; Dudukovic 2009). In these chemical transformations, it has been generally recognized that a heterogeneous catalyst can effectively and efficiently work only in combination with both its high catalytic performance (i.e., involving the catalytic side) and its well-organized structures from nano- to macro-scales (i.e., involving the engineering side) (Ertl et al. 2008). Therefore, the designing and preparation of high-performance heterogeneous catalyst from the perspectives of both sides is critical to allow us to address the emerging issues in the practical applications (Figure 1) and should deserve special attention.

As regards the catalytic side, the most general preparations of catalytic materials rely mainly on the classic methods such as hydrothermal transformations, precipitation, calcination and impregnation. More specialized methods have also being developed by using various precursors, including resinbound complexes, complexes bound by a supported ionic liquid phase and organometallic complex (Ertl et al. 2008, Serp and Kalck 2002, George 2010). However, the major challenge of these preparations is the difficulty of precisely controlling the average size and size distribution as well as the surface and local composition of the as-prepared nanoparticles. Additionally, these as-obtained catalysts usually exhibit a high deactivation rate due to the gradual migration and agglomeration of nanoparticles associated with the high-temperature and/or strongly exothermic reactions, such as catalytic methane combustion, oxidative

East China Normal University, School of Chemistry and Molecular Engineering, 3663 North Zhongshan Road., Shanghai 200062, China.

Email: gfzhao@chem.ecnu.edu.cn

* Corresponding author: ylu@chem.ecnu.edu.cn

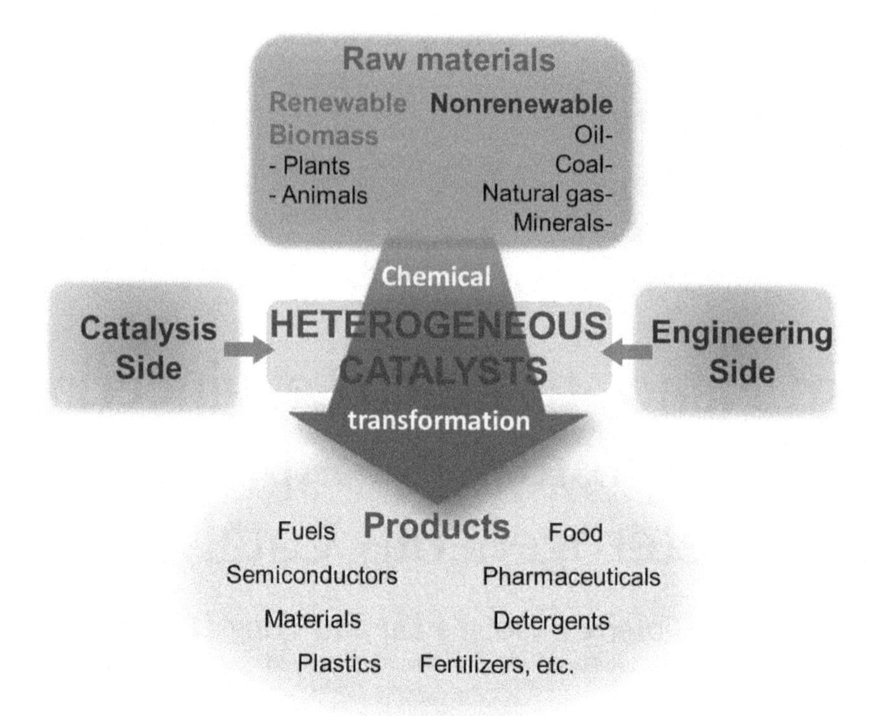

Figure 1. Inputs and outputs of chemical transformations over the heterogeneous catalysts as well as the design and preparation of catalysts from the perspectives of both catalysis and engineering sides.

dehydrogenation of alkanes (ethane, propane, and butane, for instance) to olefins, catalytic oxy-methane reforming, gas-phase selective oxidation of alcohols, and syngas methanation.

Nanocomposites not only simply mathematically add the respective functions of their building blocks, but also give birth to novel superior properties derived from the chemically intrinsic interactions, which is especially beneficial to the harsh chemical transformations. Particularly, the fine structural manipulation based on the modern nanotechnologies enables us to synthesize building blocks with precisely-controlled sizes and shapes and further tailor their positioning at nano-scale for fabricating new-generation of catalysts with high catalytic activity, selectivity and stability (Pernicone 2003, Somorjai et al. 2008). Various nanocomposites, which consist of metal-metal, metal-oxide and oxide-oxide with core@shell, heterostructure, and alloy structures, have been delicately designed and controllably synthesized, delivering improved activity and selectivity for the target reactions. For example, Yu and co-workers (Yu et al. 2005) pioneered the synthesis of nanostructured Au-Fe_3O_4 dumbbell with tuning Au from 2 to 8 nm and Fe_3O_4 from 4 to 20 nm; since then, many other dumbbells, Pt(Pd, Ag, ···)-Fe_3O_4(TiO_2, MnO, ···) (Zhang et al. 2014), have been further developed and exhibited excellent activity, especially in CO oxidation. Moreover, a series of Ni-based nanocrystals Ni_xM_y (M = Ga and Sn) was synthesized via a solution-based co-reduction strategy with uniform particle size and controlled composition (Liu et al. 2016). These as-prepared nonprecious-metal nanocrystals were taken as representative model catalysts for semi-hydrogenation of alkynes, and exhibited an excellent selectivity toward alkenes, which is much higher than the corresponding Pd-based catalysts. This result demonstrated that the Ni_xM_y are identified as an effective replacement for Pd-based catalysts. Also, promising synthetic approaches have been developed to protect the fine nanoparticles against sintering by tailoring core@shell nanostructures (Schärtl 2010, Alayoglu et al. 2008) and covering nanoparticles with an atomic layer (Lu et al. 2012). Thereinto, core@shell configuration with the encapsulation of nanoparticles by an outer oxide-shell is a promising protocol to produce catalysts with enhanced stability. One classic model of Pt@SiO_2 was prepared via colloidal synthesis of 14-nm Pt cores followed by 17-nm-thick SiO_2 shell polymerization around the Pt cores, exhibiting high-temperature stability (Joo et al. 2009). Recently, another core@shell system of Pd@CeO_2 was employed for CH_4 catalytic combustion and showed excellent thermal stability (Cargnello et al. 2012). Obviously, these nanocomposites could effectively enhance the catalyst activity, selectivity, and particularly, stability against nanoparticles sintering.

Despite these promising advances, following issues are industrially appealing yet challenging conundrums to solve: the preparations of these nanocomposites have become considerably complicated as the design is more precise, often requiring multiple steps and special techniques (Cargnello et al. 2010, Zhou et al. 2010, Lu et al. 2012); the lack of a general strategy for their large-scale production has limited the practical applications. Besides these limitations from the prospective of chemical synthesis side, another aspect from the perspective of engineering side is also particularly challenging. The practical use of these nanocomposites as catalysts in a fixed bed reactor requires the macroscopically shaped forms (such as microgranules or extruded pellets a few millimeters in size) rather than as-made powders. As a result, some frustrating problems emerged in these macroscopic forms: mass transfer limitation caused by the long pathway of reactants in the millimetric catalysts as well as heat transfer limitation caused by the weak thermal conductivity of catalysts, high pressure drop from the random packing of catalyst particles, non-regular flow pattern, and adverse effects of the binders used (always reducing the intrinsic catalyst activity and selectivity), which are harmful for the strongly exo-/endo-thermic and/or high-throughput reactions.

As regards the engineering side, structured catalysts and reactors (SCRs) is a promising avenue to solve these problems and achieve process intensification, thus being a hot topic in the heterogeneous catalysis (Cybulski and Moulijn 2006). SCRs are characteristic of the regular structures that are free of randomness at the reactor's level, opposite to the randomly packed beds of particles of various shapes. Several concepts for catalyst support structures have been established like pellets, honeycombs, and even wire meshes in the last decades. One of the most typical SCRs is the monolithic honeycomb ceramic catalyst extensively applied in the control of automotive emissions and the reduction of nitrogen oxides from power stations. Such honeycomb catalyst consists of thousands of opening parallel channels in millimetric diameter (to offer high void fraction and to allow low pressure drop at high flow rates through the catalyst bed) with catalytic washcoat in micrometric thickness on the channel walls (to improve mass transfer due to the short gas diffusion distance). Over the years, this type of SCRs has found other applications, such as catalytic combustion, partial oxidations, and liquid-phase hydrogenations (Cybulski and Moulijn 2006). However, using monolithic honeycomb still remains challenging because of their relatively low heat transfer and the lack of radial mixing, which should be improved for the endo-/exo-thermic and/or fast reactions (Heck et al. 2001).

Metal fiber/foam-based supports have attracted ever-increasing interest within the last decade (Reichelt et al. 2014). Besides the high surface to volume ratio, high void fraction, and internal diffusion as typically in honeycomb ceramic support, the other benefits of fiber/foam supports for catalytic applications could also be their unique three-dimensional (3D) network and open structure as well as high thermal conductivity and mechanical strength (Figure 2; Zhao et al. 2016a), which allow low pressure drop, high mass/heat transfer and especially high contacting efficiency of reactants resulting from the radial mixing (Dautzenberg 2004, Schaedler et al. 2011, Gascon et al. 2015). For the very fast and heat-transfer-controlled reactions, for instance, the high thermal conductivity of metal fiber/foam supports suppresses the formation of hot or cold spots within the catalyst bed, which thereby allows the full catalyst potential to be utilized (Sheng et al. 2011). Moreover, their metallic feature has unique form factors that provide a great flexibility in geometric appearance when filling up the structured reactors. However, the effective catalytic functionalizations and practical applications in heterogeneous catalysis of these fiber/foam supports remains problematic because the conventional washcoating techniques suffer from nonuniformity and exfoliation of coatings as well as binder contamination that will always reduce the intrinsic catalyst activity and selectivity.

Conceptually, one-step organization from "nano" to "macro" scales of different nano-scaled building blocks with tunable structures and functions into integrated-architectures based on chemical reactions represents a great promise to address above limited issues. In this chapter, we will revolve around the *in situ* chemical synthesis and engineering of the nanocomposites onto monolithic metal fiber/foam supports

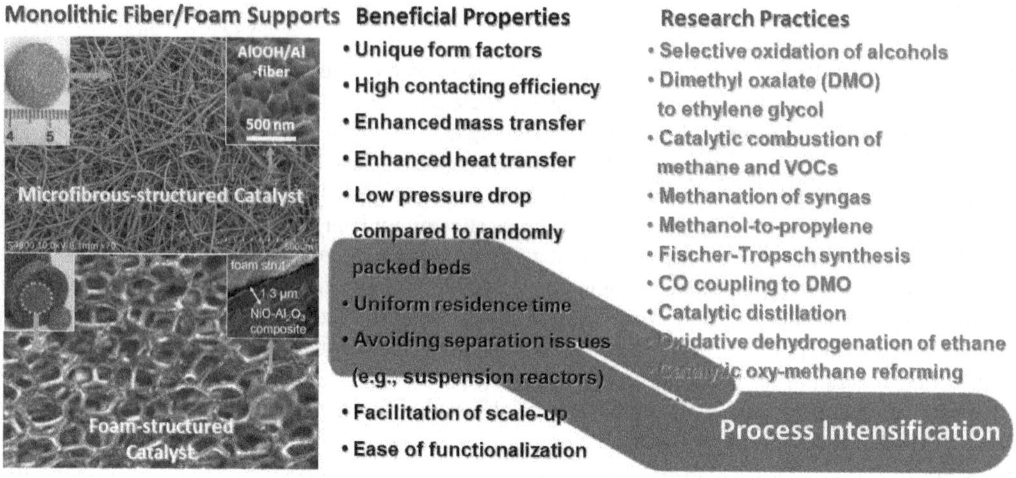

Figure 2. Fiber/foam-structured catalysts engineered from nano- to macro-scales: non-dip-coating preparation, beneficial properties and research practices in heterogeneous catalysis. Reprinted with permission from ref. Zhao, G., Y. Liu and Y. Lu. 2016a. Sci. Bull. 61: 745. Copyright@Elsevier.

in one step. A series of methods have been developed to achieve this goal, including galvanic deposition, wet chemical etching, *in situ* growth of zeolite and layered double hydroxides, endogenous growth via steam-only oxidation and hydrothermal treatment, one-step organization for structuring core@shell nanocomposites, and reaction-induced strategy for structuring oxide@oxide nanocomposites. The as-obtained fiber/foam structured catalysts are promising for applications in the chemical transformations involving energy/environment/chemical-production catalysis: gas-phase selective oxidation of alcohols to aldehydes/ketones and dimethyl oxalate hydrogenation to ethylene glycol (strong exothermicity), catalytic combustion of methane and methanation of CO and/or CO_2 (strong exothermicity and high-throughput operation), catalytic oxy-methane reforming (strong exothermicity and high reaction temperature), methanol-to-olefins process (strong exothermicity and reactant-diffusion dependence), and Fischer-Tropsch to lower olefins process (strong exothermicity and reactant-diffusion dependence).

Galvanic deposition technology

Galvanic replacement reaction is one of the most common chemical reactions, which can spontaneously proceed between the metal ions with higher potentials and the metallic substrates with lower potentials. For example, the gold nanoparticles can be automatically and strongly deposited onto Ni-microfiber surface by simply immersing the Ni-microfiber into a $HAuCl_4$ aqueous solution because of the great difference in potentials between Ni^0/Ni^{2+} (–0.23 V) and Au^0/Au^{3+} (1.69 V). Based on such galvanic replacement reaction, a series of fiber/foam structured catalysts have been fabricated from nano- to macro-scales in one step for several exothermic and/or high-throughput reactions, such as Au(Ag)/Ni(Cu)-microfiber for gas-phase selective oxidation of alcohols to aldehydes/ketones, Pd/Ni-foam for methane catalytic combustion, and AuPd/Cu-microfiber as well as AuPd-La_2O_3/Cu-microfiber for gas-phase hydrogenation of dimethyl oxalate to ethylene glycol.

Gas-phase selective oxidation of alcohols to aldehydes/ketones

The gas-phase selective oxidation of alcohols using oxygen (or air) as oxidant in the presence of heterogeneous catalysts is an attractive "green" process to produce "clean" aldehydes/ketones on an industrial scale. Given the strong exothermicity of this process, the major challenge is to develop an effective catalyst endowed with high activity/selectivity at low temperature (to avoid fast deactivation) coupled with high thermal conductivity (to rapidly dissipate reaction heat) to enhance catalyst production/energy efficiency and stability. Some silver-based catalysts including electrolytic silver and supported silver catalysts have been demonstrated for the gas-phase selective oxidation of alcohols in a packed bed reactor (Yamamoto et al. 2005, Shen et al. 2006, Magaev et al. 2008, Jia et al. 2012). The electrolytic silver catalyst has good thermal conductivity, but the catalytic activity is very low; supported silver catalysts such as those using Al_2O_3 and SiO_2 as supports have higher activity, but their weak thermal conductivity would induce hotspots in the reaction bed, which is not only a main cause of catalyst degradation, but also a hidden danger such as bed temperature runaway.

Hence, it is crucial to render a new kind of catalyst with both high low-temperature activity/selectivity and significantly intensified heat-transfer. To accomplish this goal, a novel silver catalyst supported on LTA zeolite film coated on a copper grid has been first reported for benzyl alcohol oxidation (Shen et al. 2006), delivering a 60% conversion with 90% benzaldehyde selectivity at 320°C. In order to further enhance the catalyst activity and heat-transfer, a series of microfibrous structured catalysts are developed. One early example is the sinter-locked Ni-microfibers (8 μm in diameter) structured nanosilver catalyst of Ag/Ni-microfiber (Deng et al. 2010, Mao et al. 2010), which is simply obtainable by immersing Ni-microfibers into an $AgNO_3$ aqueous solution to achieve the galvanic deposition of Ag nanoparticles on Ni-microfibers owing to the great potential difference between Ni^0/Ni^{2+} (–0.23 V) and Ag^0/Ag^+ (0.8 V). Some $Ni(NO_3)_2$ is generated from this galvanic reaction ($2AgNO_3 + Ni = 2Ag + Ni(NO_3)_2$), and after calcination, the nanocomposites of Ag-NiO are *in situ* formed and monolithically structured on Ni-microfibers. Originating from the high activity of "Ag-NiO" nanocomposites, the catalyst achieves a 90–95% benzyl alcohol conversion and a 95–97% benzaldehyde selectivity at a relatively low temperature of 300°C. Moreover, thanks to the high heat-transfer of Ni-microfiber, the catalyst is free of bulk-coke formation and is stable for at least 120 h.

Supported gold nanoparticles demonstrate wide versatility and excellent activity in many oxidation reactions (Chen and Goodman 2004, Bond et al. 2006, Min and Friend 2007). Several nanogold-based catalysts, such as Au/SiO_2 and $Au\text{-}Cu/SiO_2$ (Pina et al. 2008), have been developed for the gas-phase selective oxidation of benzyl alcohol. Considering the special property of the sinter-locked microfibers and the extraordinary activity of Au nanoparticles, a novel catalyst would be born with unique combination of high activity/selectivity with enhanced heat/mass transfer if Au nanoparticles could be easily and firmly embedded in a metallic microfibrous structure. A microfiber-supported nano-gold catalyst is successfully prepared by galvanically depositing gold nanoparticles on thin-sheet sinter-locked Cu-microfibers (8 μm in diameter) (Zhao et al. 2011a). The deposition proceeds automatically by exposing the Cu-microfibers to a $HAuCl_4$ aqueous solution because of the great difference in potentials between Cu^0/Cu^+ (0.19 V) and Au^0/Au^{3+} (1.69 V). The catalyst, with high heat-transfer, shows excellent low-temperature activity for the gas-phase selective oxidation of alcohols. For example, benzyl alcohol conversion of > 92% is obtained with a 97.2% benzaldehyde selectivity using the 3 wt% Au/Cu-microfiber catalyst at 250°C. Interestingly, the spectroscopic studies suggest that the low-temperature activity stems from the "$AuCu\text{-}Cu_2O$" nanocomposites with synergistic effect between AuCu-alloy and Cu_2O that are *in situ* formed during the reaction. These results indicate that the catalyst is endowed with high catalytic performance and good heat conductivity for rapidly dissipating reaction heat from catalyst bed, which is especially favorable for the selective oxidation of large alcohol molecules to suppress their thermal cracking and over oxidation. However, its reported lifetime is only 50 h at 250°C because the Cu microfibers are so slender that they are easily powdered during the long-term reaction. In order to improve the stability, another Au/Ni-microfiber catalyst of the same type is subsequently developed also with the aid of galvanic reaction between $HAuCl_4$ and Ni-microfiber (Zhao et al. 2011b). Via this galvanic reaction ($3Ni + 2HAuCl_4 = 2Au + 3NiCl_2 + 2HCl$), the composites of "$Au\text{-}NiCl_2$" are *in situ* formed and firmly anchored onto the Ni-microfiber surface; along with the activation in the reaction stream at 380°C, the "$Au\text{-}NiCl_2$" is then transformed into "Au@NiO" (a date-cake-like active nanocomposites with the small NiO segments, like dates, partially covering the large gold particle, like a cake, Figure 3) (Zhao et al. 2011b, 2013a). The resulting Au/Ni-microfiber catalyst is active, selective, and stable for the gas-phase selective oxidation of alcohols under mild conditions, while showing high heat-transfer ability. Taking the benzyl alcohol oxidation as an example, 94–96% conversion and 99% benzaldehyde selectivity can

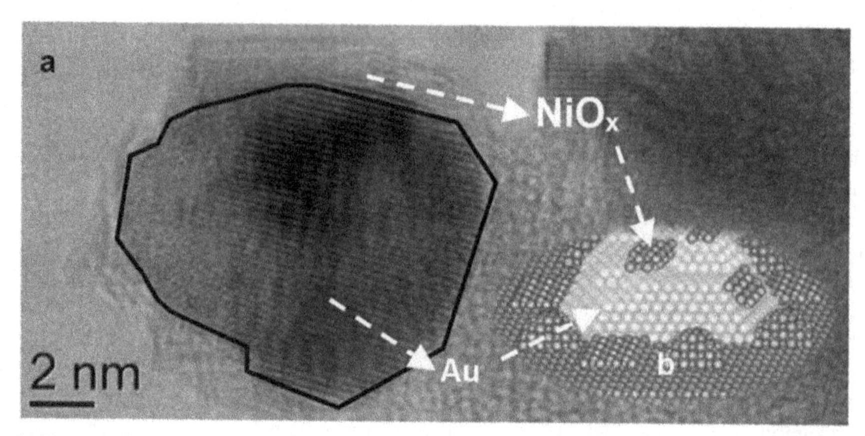

Figure 3. (a) Transmission electron microscope (TEM) image of the Au@NiO nanocomposites for the Au/Ni-microfiber catalyst. (b) The "date-cake" illustration of the partial coverage of Au particles (cake) with NiO segments (dates). Reprinted with permission from ref. Zhao, G., H. Hu, M. Deng and Y. Lu. 2011b. Chem. Commun. 47: 9642. Copyright@Royal Society of Chemistry.

be stably obtained within 700-h running at 250°C. The X-ray photoelectron spectroscopy combined with X-ray absorption near-edge structure reveals that a Ni_2O_3-Au^+ hybrid active site is defined to be the genesis of the high low-temperature activity and the Au@NiO nanocomposites facilitate the generation of high density Ni_2O_3-Au^+ sites during oxidation (Zhao et al. 2013b).

O_2-lean catalytic combustion of methane

Catalytic combustion of methane is an important technology for deoxygenation to upgrade coalbed methane (usually under oxygen-lean condition) as well as energy production and environment protection (usually under methane-lean condition) via burning methane with oxygen. Obviously, the strong exothermicity (ΔH_{298} of –802.7 kJ mol^{-1}) and high throughput operation for these processes require the catalyst to be not only active, selective and stable but also highly thermal conductive and highly permeable to maintain the intensified process efficiency and safety, which still remains greatly challenging. Towards this end, Precision Combustion Inc. has developed short channel length, high channel density supports in the form of honeycomb monolith (trademarked Microlith®) and high surface area ceramic coatings for them (Ersson and Järås 2006). These supports avoid substantial buildup of boundary layer and, further, greatly enhance the internal diffusion of reactants in reactors, but still face the following drawbacks: the catalytic functionalization via conventional dip-coating technique suffers from inhomogeneous coatings, relatively lower heat transfer than metallic supports, and the lack of radial mixing of the reactants (Heck et al. 2001).

Seeing the beneficial features of metal foams on the design of structured catalysts, including high void fraction, excellent thermal conductivity, and enhanced mechanical strength, a monolithic Ni-foam (typically 100 pores per inch) is employed to structure PdNi alloy to tailor the PdNi(alloy)/Ni-foam catalyst for the deoxygenation of coalbed methane (i.e., under oxygen-lean condition) (Zhang et al. 2015a). Via the galvanic replacement reaction between Pd^{2+} and Ni-foam, Pd nanoparticles are first deposited onto Ni-foam surface owing to the great potential difference between Ni^0/Ni^{2+} (–0.23 V) and Pd^0/Pd^{2+} (0.95 V); after an *in situ* activation in the reaction stream, more interestingly, Ni atoms from Ni-foam can alloy into Pd nanoparticles to form PdNi alloy. Clearly, the galvanic deposition combined with the *in situ* activation achieves the formation and monolithic engineering of PdNi alloy from nano- to macro-scales in one step. The as-prepared catalyst not only exhibits as-expected low pressure drop and high heat/mass transfer, but also achieves high activity, selectivity, and stability. The experimental and theoretical studies consistently reveal that Ni alloying into Pd nanoparticles modifies the electronic structure of surface Pd, which can activate surface O by decreasing the O adsorption energy, thereby leading to the increase in activity. In addition, the oscillatory behavior, commonly presented under oxygen-lean condition especially in Pd-based catalysts, is another challenge in CH_4 combustion and needs to be eliminated, since the reaction rate oscillation along with the reaction time and temperature may reduce the overall catalytic performance (Slinko et al. 2006, Figueroa and Newton 2014). Interestingly, the kinetic study of coalbed methane deoxygenation further reveals that PdNi alloy can suppress oscillatory behavior inherently by increasing the reaction order of O_2 from –0.6 to –0.3 (Zhang et al. 2016a). By combining the catalytic (high activity and oscillation elimination derived from PdNi alloy) and monolithic (high heat transfer and permeability) superiority, the PdNi(alloy)/Ni-foam catalyst exhibits good stability for at least 500 h with 95–100% O_2 conversion and 100% CO_2 selectivity at 320–350°C (Figure 4).

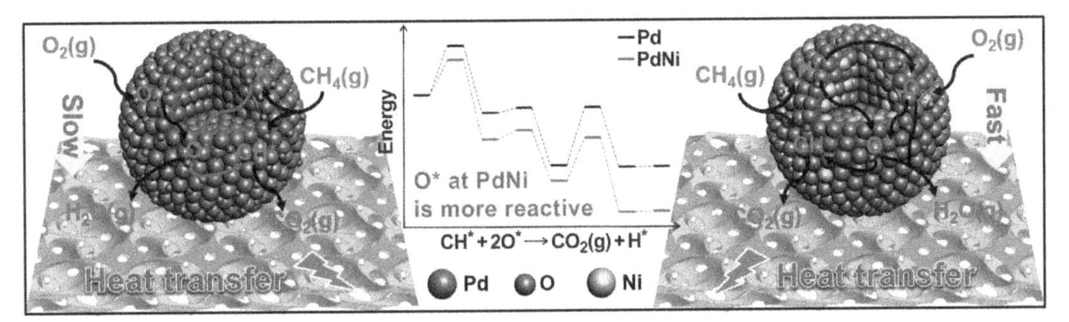

Figure 4. Combination of catalytic and monolithic superiority of the PdNi(alloy)/Ni-foam catalyst for deoxygenation of coalbed methane. Reprinted with permission from ref. Zhang, Q., X. Wu, G. Zhao, Y. Li, C. Wang, Y. Liu, X. Gong and Y. Lu. 2015a. Chem. Commun. 51: 12613. Copyright@Royal Society of Chemistry.

Dimethyl oxalate hydrogenation to ethylene glycol

Ethylene glycol is a versatile commodity chemical used in various applications such as antifreeze, solvents, and the manufacture of heat transfer agents, polymers, and many other chemical products. The hydrogenation of dimethyl oxalate to ethylene glycol is an advanced unit operation of chemical engineering in C1 chemistry, enabling the industrial level application in chemicals production from coal and natural gas conversion (Corma et al. 2007, Gong et al. 2012). Dimethyl oxalate hydrogenation has been considerably investigated in both liquid-phase and gas-phase. Ruthenium-based homogeneous catalysts are previously used for liquid-phase hydrogenation with a high ethylene glycol yield under milder conditions (95% at 7 MPa and 100°C), but greatly suffers from the problems of corrosion and separation (Teunissen and Elsevier 1997). From the standpoint of industry, the gas-phase process is more attractive because of the convenience of catalyst separation and higher production efficiency, especially for the ethylene glycol in a bulk form. The CuCr catalyst is reported to be highly active and selective in gas-phase dimethyl oxalate hydrogenation to ethylene glycol, but the toxicity of Cr greatly hinders its application (Zhu et al. 2010). Subsequently, Cr-free Cu-based catalysts have been intensively investigated, and Cu/SiO_2 is considered to be the most active for the gas-phase hydrogenation to ethylene glycol (Corma et al. 2007, Gong et al. 2012). However, the usage of silica support appears to be a fatal flaw due to its severe leaching caused by the reaction of SiO_2 with methanol in this reaction system (Wen et al. 2013). Moreover, dimethyl oxalate hydrogenation to ethylene glycol is strongly exothermic (ΔH of 126 kJ mol^{-1}), which normally generates hotspots that are the main cause for copper catalyst sintering deactivation. Therefore, from both the academic and industrial points of view, rendering a novel silica-free catalyst with a unique combination of high activity and selectivity, structural robustness, and excellent thermal conductivity is particularly desirable.

One novel microfibrous structured catalyst of Pd-Au/Cu-microfiber is first developed for the gas-phase dimethyl oxalate hydrogenation to ethylene glycol (Zhang et al. 2015b). The catalyst can be easily fabricated from nano- to macro-scales in one step by dipping Cu-microfibers into $HAuCl_4$ and $Pd(Ac)_2$ aqueous solution, and the Au and Pd nanoparticles are galvanically co-deposited on Cu-microfiber surface.

The as-prepared Au-Pd/Cu-microfiber catalyst is highly active, selective, and stable for the dimethyl oxalate hydrogenation to ethylene glycol, being capable of converting 97–99% dimethyl oxalate into ethylene glycol product at a selectivity of 90–93% with promising stability for at least 200 h. Notably, Cu-microfibers not only act as the support, but also offer catalytic species of Cu^+, which is enhanced by Au-Pd to selectively convert dimethyl oxalate into ethylene glycol. Accordingly, a ternary Pd-Au-Cu^+ nanocomposite is tentatively proposed, in which Cu^+ acts as an essential site for activating methoxy and carbonyl groups in dimethyl oxalate, and Au plays a key role in stabilizing Cu^+ sites to prevent their deep reduction to Cu^0, especially their Pd-catalyzed reduction, while Au-Pd alloy synergistically promotes H_2 activation as a result of strong Au-Pd electronic interaction. However, the Pd-Au/Cu-microfiber catalyst achieves an unacceptable low-temperature activity, and the optimal reaction temperature of 270°C is obviously higher than the common 180–200°C for other reported catalysts. Therefore, it is particularly desirable to improve the low-temperature activity of Pd-Au/Cu-microfiber, which makes significant sense for catalyst stability and industrial application.

Rare earth elements are often used as co-catalysts for different kinds of reactions because of their paramagnetism, mobility of the lattice oxygen, variable valence, and the ability to significantly improve the electron transfer property of catalysts (He et al. 2014). Therefore, the Pd-Au/Cu-microfiber catalyst is further modified by rare earth oxides to improve the low-temperature activity (Han et al. 2016a). Various rare earth oxides (2 wt%: La, Ce, Sm, Gd, Y, Er or Nd) are introduced via incipient wetness impregnation method with their corresponding nitrate aqueous solution followed by calcining at 300°C in air for 2 h. The pristine and modified catalysts are all tested in gas-phase dimethyl oxalate hydrogenation at 230°C. The pristine catalyst achieves a 97.3% dimethyl oxalate conversion but with only 43.9% ethylene glycol selectivity; in contrast, the rare earth oxides modification significantly improves the ethylene glycol selectivity to 82.0–93.4% with full dimethyl oxalate conversion under identical conditions. Particularly, the highest ethylene glycol yield of 93.4% is achieved over the 2La_2O_3-0.1Pd-0.5Au-CuO_x/Cu-microfiber at 230°C, which is reachable over the 0.1Pd-0.5Au-CuO_x/Cu-microfiber only at elevated temperature of 270°C. A long-term test of 2La_2O_3-0.1Pd-0.5Au-CuO_x/Cu-microfiber is carried out at 230°C, and this catalyst is stable for at least 500 h with nice maintenance of full dimethyl oxalate conversion and ~ 90% ethylene glycol selectivity, exhibiting promising industrial application prospect. The La_2O_3-assisted enhancement of electron deficiency of the PdAu alloy facilitates H_2 activation and the acidity of La_2O_3 promotes H-spillover while La_2O_3 additive shows the ability to stabilize the Cu^+ species (Figure 5); partially-reduced surface LaO_x species formed *in situ* during reaction is found to be active

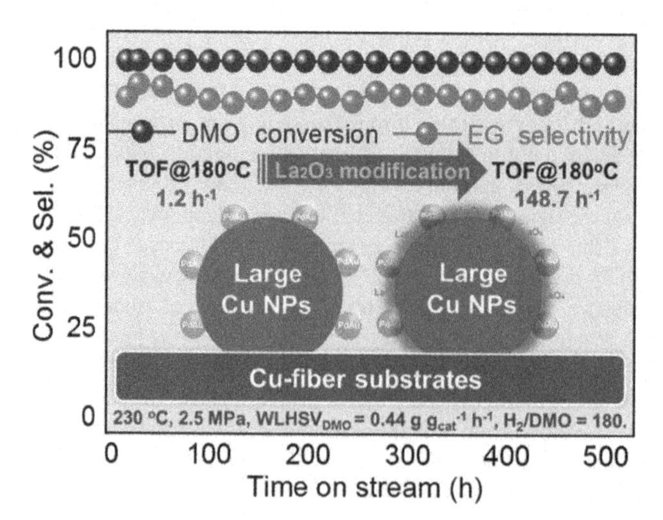

Figure 5. Schematic illustration of the Cu-microfiber-structured La_2O_3-PdAu(alloy)-Cu nanocomposites with low-temperature activity, selectivity and stability for the gas-phase dimethyl oxalate hydrogenation to ethylene glycol. Reprinted with permission from ref. Han, L., G. Zhao, Y. Chen, J. Zhu, P. Chen, Y. Liu and Y. Lu. 2016a. Catal. Sci. Technol. 6: 7024. Copyright@Royal Society of Chemistry.

for activating methoxy and carbonyl groups of dimethyl oxalate. As a result, remarkable improvement of low-temperature activity and stability is achieved over such a La_2O_3-modified microfibrous structured PdAu(alloy)-CuO_x nanocomposites catalyst.

Wet chemical etching technology

Even though the nanoparticles of precious metals (such as Ag, Au and Pd) can be effectively deposited on fiber/foam supports via galvanic replacement reaction, this strategy is invalid for the base metals with low potentials (such as Ni, Fe, Cu, Co, Mn, and Zn). Therefore, wet chemical etching method is developed to functionalize the metal fiber/foam supports. Taking the chemical etching of Ni-foam as an example (Li et al. 2015a), the Ni-foam (such as 100 pores per inch) tablets (2 mm thick, diameter of 16 mm, equal to the reactor inner diameter) are directly immersed into a chemical etching solution at desired temperature for some time. The chemical etching solution is composed of 1.0 mM sodium dodecyl sulfate ($C_{12}H_{25}OSO_3Na$), 0.2 M acetic acid (CH_3COOH), and 0.3 M aluminum nitrate ($Al(NO_3)_3$). The resulting product is washed with distilled water for several times, dried overnight, and calcined in air at 550°C for 2 h. Finally, the nanocomposites of NiO-Al_2O_3 can be *in situ* generated and firmly and homogeneously deposited on Ni-foam struts in one step. The monolithic structure and high heat/mass transfer of Ni-foam support are well retained after etching operation. Expectedly, the NiO-Al_2O_3/Ni-foam catalyst as well as its modified counterparts are qualified for the strongly exothermic and/or high-throughput reactions, such as methanation of CO and/or CO_2 and catalytic oxy-methane reforming.

Methanation of CO and/or CO_2

Production of substitute natural gas by methanation of CO and/or CO_2 generated from various carbon sources (coal, biomass, municipal solid waste, coke oven gas, etc.) is a promising way towards the clean utilization of coal and the future of sustainable energy, which has gained increasing attention in some parts of the world (e.g., China) (Zinoviev et al. 2010, Chen et al. 2014). The strong exothermicity ($CO + 3H_2 = CH_4 + H_2O$, ΔH of –206 kJ mol^{-1}; $CO_2 + 4H_2 = CH_4 + 2H_2O$, ΔH of –165 kJ mol^{-1}) and high throughput operation of this process require the catalyst to have excellent mass/heat transfer (towards effective temperature management) and high void fraction (to reduce the pressure drop). In a pioneering example of ruthenium catalyst both in pellet and in honeycomb form (Tucci and Thomson 1979), the latter demonstrates much lower pressure drop and also significantly higher selectivity likely resulting from high void fraction and internal diffusions. However, the heat transfer of honeycomb support is insufficient to rapidly dissipate the reaction heat released in the catalyst bed. Microchannel plate heat-exchanger technology offers the potential for solving temperature management problems by integrated reactor cooling (Kolb 2013), but the parallel and unconnected microchannels prohibit the mass mixing in radial direction.

Given the high void fraction, continuously interconnected open pores, high thermal conductivity, and enhanced mechanical strength of monolithic metal foam supports, a novel monolithic Ni-foam structured catalyst of Ni-Al_2O_3/Ni-foam is developed for the CO_2 methanation reaction (Li et al. 2015a). This catalyst is obtainable by facile modified wet chemical etching method, demonstrating a combination of high activity/selectivity, high stability, and significantly enhanced heat transfer. A uniform catalyst layer in about 3 μm thickness consisting of NiO-Al_2O_3 nanocomposites is efficiently *in situ* formed and firmly anchored onto the foam struts by directly immersing the Ni-foam support into a chemical etching solution followed by a calcination in air (mentioned in the previous paragraph). The parent NiO-Al_2O_3/Ni-foam catalyst is reduced in H_2 atmosphere to form the Ni-Al_2O_3/Ni-foam catalyst, which achieves a high CO_2 conversion of ~ 90% and very high methane selectivity of ~ 99.9% for a feed of H_2/CO_2 (4/1) at 320°C and 0.1 MPa with a gas hourly space velocity of 5000 h^{-1}, throughout the entire 1200 h test with 10.2 mL catalyst. Computational fluid dynamics calculation and experimental measurement consistently show a dramatic reduction of "hotspot" temperature due to the enhanced heat transfer. This approach allows the highly effective integration of greatly enhanced heat transfer stemming from continuously interconnected Ni-foam struts with high activity/selectivity of the NiO-Al_2O_3 nanocomposites.

The same kind of monolithic Ni-foam structured catalyst of Ni-CeO_2-Al_2O_3/Ni-foam is also developed by such modified wet chemical etching method for the methanation of syngas (a mixture of CO and H_2) (Li et al. 2015b). Just simply adding 0.02 M cerium nitrate ($Ce(NO_3)_3$) into the original chemical etching solution for NiO-Al_2O_3/Ni-foam preparation, a uniform layer in 3 μm thickness consisting of NiO-CeO_2-Al_2O_3 nanocomposites is also efficiently *in situ* formed and firmly anchored onto the Ni-foam struts in one step by directly immersing Ni-foam into the etching solution followed by calcination in air (Figure 6). The parent NiO-CeO_2-Al_2O_3/Ni-foam catalyst is reduced in H_2 atmosphere to form the Ni-CeO_2-Al_2O_3/Ni-foam catalyst, which delivers a catalytic performance (~ 100% CO conversion and 94% CH_4 selectivity) comparable to the traditional Ni-based particulate catalyst. Particularly, the catalyst can stably run throughout the entire 1500 h testing, stemming from the integration of enhanced heat/mass transfer and firmly anchoring of Ni-CeO_2-Al_2O_3 nanocomposites. Also, the computational fluid dynamics calculations and experimental measurements consistently show a high heat-transfer of this catalyst for the syngas methanation, which greatly reduces the "hotspot" temperature in the Ni-foam structured catalyst bed.

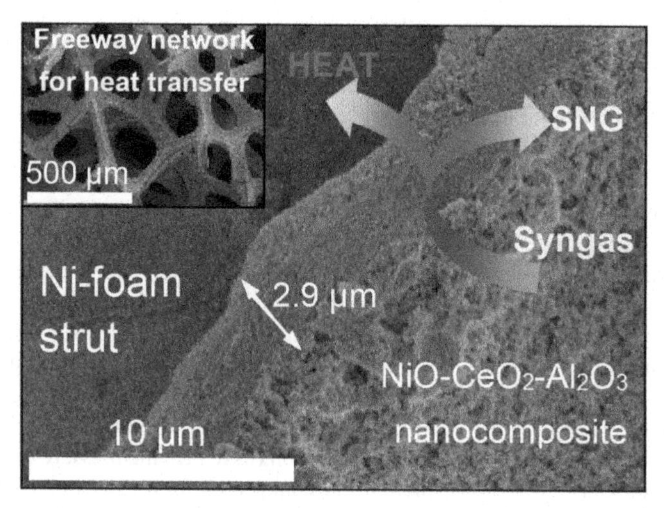

Figure 6. Monolithic Ni-CeO_2-Al_2O_3/Ni-foam catalyst developed by modified wet chemical etching of Ni-foam followed by calcination in air, providing unique combination of high activity/selectivity, excellent stability and enhanced heat transfer. Reprinted with permission from ref. Li, Y., Q. Zhang, R. Chai, G. Zhao, Y. Liu and Y. Lu. 2015b. ChemCatChem. 7: 1427. Copyright@Wiley and Sons.

Catalytic oxy-methane reforming

Methane reforming has been gaining considerable interest in the generation of syngas for subsequent production of energy, chemicals, and fuels (Caballero and Perez 2013). Steam reforming of methane is implemented commercially for the production of syngas, but this process is extremely endothermic and thus very energy- and capital-intensive. Thanks to the high production efficiency, and especially the beneficial H_2/CO ratio for the downstream synthesis (e.g., methanol synthesis and Fischer-Tropsch process), catalytic oxy-methane reforming (COMR) is attracting more and more attention as a compact and energy-efficient process (Choudhary and Choudhary 2008). Since the early 1990s, great efforts have been made on the development of COMR process. The noble metal-based catalysts (e.g., Pt and Rh) generally exhibit better performance and resistance to carbon deposition, but their wide-application is confined by the high cost; Ni-based catalysts are widely utilized due to their moderate cost and satisfactory performance (Lu et al. 1998, Chen et al. 2005, Li et al. 2011), but often suffer from sintering as well as coke formation. Most efforts have been made on improving the sintering- and coke-resistance of Ni-based catalysts, for instance, by adding promoters (alkali, alkali earth, and rare earth metal oxides) and/or utilizing different supports (Al_2O_3, ZrO_2, SiO_2, La_2O_3, and Y_2O_3) (Zhu et al. 2005, Tanaka et al. 2009, Makarshin et al. 2015). However, the formation of hotspots in the catalyst bed should be another important consideration because it is also one of the main causes for Ni-sintering and coke-formation (Liu and He 2012). Thus, to develop a novel catalyst system combining excellent catalytic performance with enhanced heat transfer is a promising strategy to overcome these major drawbacks.

A series of monolithic Ni-foam structured binary nanocomposites of Ni-MO_x (M = Al, Zr, or Y) to be used in the catalytic oxy-methane reforming are developed via one-step wet chemical etching method, of which the Ni-MO_x nanocomposites are formed *in situ* and embedded into the Ni-foam struts. The catalysts are obtained by directly immersing Ni-foam into the etching solution containing Al (or Zr, or Y) nitrate; after etching and calcining, the Ni-foam strut surface becomes coarse, in association with visible increase of the specific surface area. Among these catalysts, the Ni-Al_2O_3/Ni-foam possesses the largest specific surface area and the highest amount of NiO species (i.e., Ni active site precursors). More NiO is induced on Ni-Al_2O_3/Ni-foam and Ni-ZrO_2/Ni-foam, likely due to the lower pH value of $Al(NO_3)_3$ (pH of 2.9 to 3.5; start to end) and $Zr(NO_3)_4$ (1.1 to 3.9; start to end) than of $Y(NO_3)_3$ (3.6 to 4.9; start to end) in the whole etching process. Moreover, the quite different contents of Al_2O_3, ZrO_2, and Y_2O_3 might be caused by the different properties of nitrate salts. For example, serious self-hydrolysis of $Zr(NO_3)_4$ unavoidably makes Zr ions directly forming amorphous zirconiumhydroxide precipitate (Southon et al. 2002) rather than depositing on Ni-foam struts during etching process, leading to almost zero ZrO_2 content on the Ni-ZrO_2/Ni-foam. Overall, high NiO and Al_2O_3 contents lead to the highest specific surface area of the Ni-Al_2O_3/Ni-foam catalyst while low surface areas of the Ni-ZrO_2/Ni-foam and Ni-Y_2O_3/Ni-foam catalysts are due to their low MO_x or NiO content. As a result, Ni-Al_2O_3/Ni-foam exhibits the best catalytic performance: 86.4% CH_4 conversion with 96.6% H_2 selectivity and 91.2% CO selectivity for a fed of CH_4/O_2 (2/1) at 700°C using a high gas hourly space velocity of 100 L g^{-1} h^{-1}. It is anticipated that this assay will be a new point which may stimulate commercial exploitation of the new-generation structured catalyst technology for the high-throughput catalytic oxy-methane reforming reaction.

In situ growth of ZSM-5 zeolite

In recent years, the methanol-to-olefins process has been attracting particular attentions as an alternative route for the light olefins production from non-petroleum sources such as coal, natural gas, and biomass (Mokrani and Scurrell 2009). To date, large-scale implementation of this process has been successful in ethylene operation mode on a SAPO-34 zeolite catalyst in a fluidized bed reactor. However, the global demand for propylene is growing faster than for ethylene (Traa 2010). Hence, it is particularly desirable to develop catalysts that can selectively convert methanol to propylene (MTP). ZSM-5 zeolite-based catalysts for methanol-to-olefins process have been extensively studied to orient product selectivity toward light olefins and especially propylene and to further improve the catalyst stability, although Lurgi's MTP process based on a packed bed with a ZSM-5 particulate catalyst has been industrially demonstrated (Mokrani and Scurrell 2009). Most of the effort has been focused on ZSM-5 modification such as tuning the acidity, size- and/or morphology-controllable synthesis, and hierarchical design of the pore structure. In some cases, high selectivity toward propylene is obtainable with the increased propylene to ethylene ratio on ZSM-5 zeolite catalysts (Bleken et al. 2012, Liu et al. 2009). Despite these promising results, their practical use as catalysts in a fixed bed reactor is still particularly challenging, as microgranules or extruded pellets a few millimeters in size are required in real-world, macroscopically shaped forms rather than as-made powders. As a result, some frustrating problems are inevitable and unavoidable in these cases including mass/heat transfer limitations, high pressure drop, non-regular flow pattern, and adverse effects of the binders used, which will always reduce the intrinsic catalyst selectivity and activity.

The monolithic structured catalysts have been attracting growing interest in heterogeneous catalysis due to its improved hydrodynamics in combination with enhanced heat/mass transfer (Dautzenberg 2004, Renken and Kiwi-Minsker 2011, Zhao et al. 2011a). This recently becomes a source of inspiration for attempts to develop structured zeolite materials (Ivanova et al. 2007, Jiao et al. 2012) by growing zeolite coatings on monolithic porous supports such as SiC foam. However, only a few studies have been reported regarding the synthesis of ZSM-5 on SiC foam support for MTP applications because the uniform and extensive growth of ZSM-5 on SiC foam is still challenging due to the disadvantageous surface tension effect near the sharp edge of the cellular foam struts and ligaments. Simulation calculation shows that microstructured design of ZSM-5 can promote C2-C4 olefin selectivity to up to 71% with a high propylene selectivity of up to 49% (Guo et al. 2013); the benefits of microstructured ZSM-5 catalysts with regard to enhanced selectivity to light olefins, especially propylene, have not yet been demonstrated for ZSM-5/SiC-foam catalyst despite of visible lifetime improvement (Ivanova et al. 2007).

A new strategy of microfibrous structured design rather than the chemical and pore-tuning modifications of HZSM-5 zeolite is demonstrated to dramatically improve the selectivity and stability for MTP process (Wang et al. 2014, Wen et al. 2015, Wen et al. 2016). Recently, a microfibrous structured ZSM-5 zeolite catalyst of HZSM-5/SS-fiber is obtained in a macroscopic scale by the direct growth of ZSM-5 zeolite crystals onto a 3D microfibrous structure using 20-μm stainless steel microfibers (SS-microfibers) (Wang et al. 2014). This approach provides an efficient and effective combination of excellent thermal conductivity, large void volume (continuously adjustable), hierarchical porous structure from micro- to macro-size, binder-free *in situ* hydrothermal synthesis, and good rigidity/robustness. In addition, ZSM-5 zeolite layer can be facilely grown in a wide loading range of 5–30 wt%. Thanks to the above beneficial properties, such microfibrous structured design results in dramatic C2-C4 olefins (especially to propylene) selectivity and life-time improvements for MTP process compared with the particulate microporous zeolite. Using a feed of 30 vol% methanol in N_2, for example, at 480°C high propylene selectivity of ~ 46% is obtainable with a total C2-C4 olefin selectivity of ~ 70%, being much higher than that (~ 37%, C2-C4 olefin selectivity of ~ 64%) for the corresponding particulate ZSM-5 zeolite; the life-time of at least 210 h is achieved, almost 3-fold longer than the life-time of 60 h for the particulate HZSM-5 catalyst. In nature, such unprecedented performance is due to the propagation of olefin methylation/cracking cycle over the aromatic-based cycle in the methanol-to-hydrocarbon catalysis (Figure 7). Moreover, the kinetic and modeling studies also display the higher internal-diffusion efficiency and narrower residence time distribution of the reactants of such design, not only promoting the propylene formation but also improving the utilization efficiency of HZSM-5 (Wen et al. 2016).

Hierarchical tailoring is highly effective since the secondary mesopores network can shorten the internal-diffusion path length of reactants (Milina et al. 2014). Various methods have been intensively studied to create intracrystalline mesopores, such as desilication (involving controllably leaching silicon in an alkaline medium) and dual-templating (involving the usage of surfactants, block copolymers, organosilane agents, or hard templates) methods (Choi et al. 2006, Zhao et al. 2009, Milina et al. 2015, Wang et al. 2016a). In order to further prolong the lifetime to meet the standard of commercial run

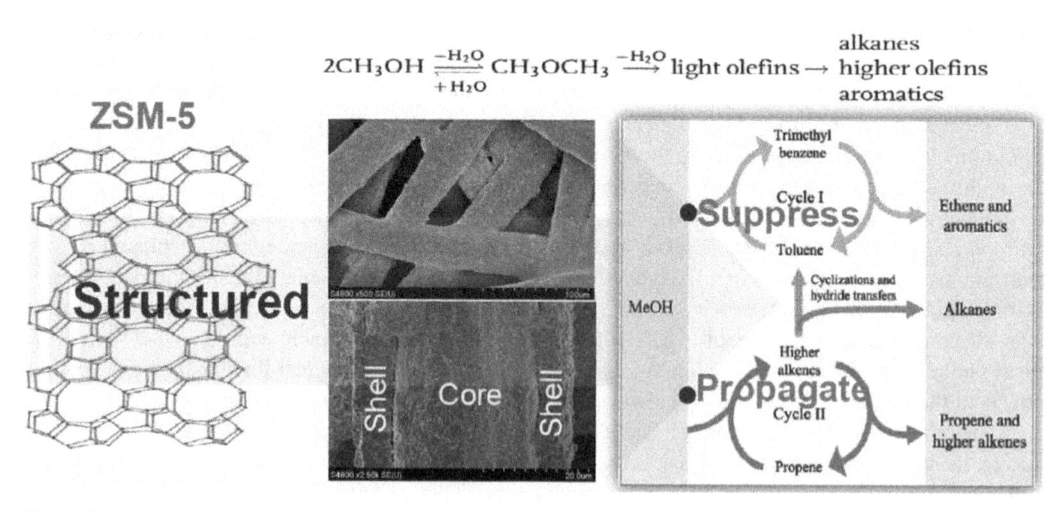

Figure 7. Suggested working principle for the MTP selectivity improvement over the microfibrous structured HZSM-5/SS-fiber catalyst. Reprinted with permission from ref. Wen, M., X. Wang, L. Han, J. Ding, Y. Sun, Y. Liu and Y. Lu. 2015. Micropor. Mesopor. Mater. 206: 8. Copyright@Elsevier.

(500–600 h, Lurgi's MTP process), the hierarchical tailoring of ZSM-5 zeolite is attempted to combine the microfibrous structured design. Very recently, further-developed MTP catalysts have been reported, enabling unique combination of the shortened internal-diffusion advantage of hierarchical zeolite design with the enhanced heat/mass transfer as well as improved hydrodynamics of microfibrous structured engineering (Ding et al. 2016). Given that caramel is economically affordable, eco-friendly and easily obtainable by simple hydrothermal pre-treatment of glucose, it is utilized as green templates to create intracrystalline mesopores. Hence, a one-pot coupling strategy is developed to synthesize the monolithic SS-microfiber structured catalyst of meso-HZSM-5/SS-microfiber from nano- to macro-scales by using caramel as a "green and cheap" template. Interestingly, caramel not only hierarchically tailors the ZSM-5 shell with micro-meso-macropore system, but also spontaneously tunes its acidic properties. Such ZSM-5 shell coated on SS-microfibers shows a dramatically prolonged lifetime of 845 h (> 90% methanol conversion) with high propylene selectivity of 48% and C2-C4 olefin selectivity of ~ 70%, mainly associated with both the shortened micropore diffusion path length as a result of a secondary mesopores network and the favorably-tuned acidic properties. More recently, the same authors reported another structured catalyst of this type (i.e., the free-standing stainless-steel (SS)-fiber@meso-HZSM-5 core-shell catalyst) engineered from micro- to macro-scales via cost-effective steam-assisted crystallization (SAC) method (Ding et al. 2017). Single-run lifetime of such catalysts for MTP process is strongly dependent on their preparation conditions but slightly on the product distribution. A volcano-shaped relationship for lifetime is observed against the SAC time, which is correlated well with the crystallization-time-dependent crystallinity, mesoporosity, and crystal size. The SS-fiber@meso-HZSM-5 catalyst achieves a single-run lifetime of 620 h with a high propylene selectivity of ~ 42% at 450°C using a methanol weight hourly space velocity of 1 h^{-1}, as the result of high crystallinity, well-developed mesoporosity, and small crystal size. Note that well-developed mesoporosity is paramount for the catalyst stability, because of the enhanced accommodation capacity of zeolite shell for receiving formed coke. It is anticipated that these findings will initiate attempts to develop the microfibrous structured catalysts or catalytic reactors, and more importantly, to inspire the research activities concerning the in-depth understanding of the promotion effect stemmed from the zeolites' modifications with combination of microfibrous structured design.

Endogenous growth via steam-only oxidation

Alumina (Al_2O_3) is the most widely used catalyst support for the heterogeneous catalysts due to its large surface area, excellent mechanical strength and brilliant hydrothermal stability. Similar to ZSM-5 zeolite, Al_2O_3 should be used in a structured way to avoid the questions emerging in its particulate applications. The manufacturing of Al_2O_3 immediate into honeycomb monoliths and open-cell foams is under development due to poor mechanical strength, while its deposition on ceramic and metallic materials by dip-coating techniques suffers from the nonuniformity and exfoliation of coatings (Cybulski and Moulijn 2006). Recently, a facile, green and generalized route is developed to monolithically structure AlOOH (convertible into γ-Al_2O_3 and/or α-Al_2O_3) 2D-nanosheets on Al-supports via steam-only oxidation (based on the reaction: $2Al + 4H_2O = 2AlOOH + 3H_2$) at desired temperature for some time (Wang et al. 2015). This *in situ* route permits the endogenous growth of 2D AlOOH and/or γ-Al_2O_3 nanosheets (ns-AlOOH or ns-Al_2O_3) from micro- to macro-scales in one step on various shaped monolithic Al substrates such as microfiber, foam, tube, and foil. Moreover, such approach paves a highly efficient and cost-effective way toward microfibrous structured catalyst and catalytic reactor technologies, being verified by several hot topic reactions such as Pd-catalyzed CO oxidative coupling to dimethyl oxalate, and FeMnK-catalyzed Fischer-Tropsch to lower olefins.

Monolithic Pd/ns-AlOOH/Al-fiber catalyst for dimethyl oxalate synthesis

Gas-phase CO coupling to dimethyl oxalate ($2CO + 2CH_3ONO = (COOCH_3)_2 + 2NO$, ΔH of –159 kJ mol^{-1}) is a crucial step from coal/biomass/natural gas to ethylene glycol as an attracting alternative non-oil route (Yue et al. 2012). One thin-sheet microfibrous structured Pd/ns-AlOOH/Al-fiber catalyst with low Pd-loading is monolithically engineered from micro- to macro-scales for this reaction, providing unique combination of high activity, selectivity, and good stability with high permeability and high thermal conductivity (Wang et al. 2015, 2016b). The support of ns-AlOOH/Al-fiber is initially prepared via endogenous growth of boehmite nanosheets (ns-AlOOH) on 3D network of 60-μm Al-microfiber; Pd nanoparticles are then placed onto the surface of ns-AlOOH rooted on the Al-microfiber via incipient wetness impregnation method with a toluene solution of palladium acetate (Figure 8; Wang et al. 2016b). Over a representative catalyst with very low Pd-loading of only 0.25 wt%, a high CO conversion of ~ 66% is achieved with a high dimethyl oxalate selectivity of ~ 94% for a feed of $CH_3ONO/CO/N_2$ (10/14/76, vol%) with a gas hourly space velocity of 3000 L kg^{-1} h^{-1}, and is stable for at least 200 h without deactivation. The Pd/ns-AlOOH/Al-fiber catalyst demonstrates two times higher intrinsic activity (expressed by turnover frequency) compared to a traditional Pd/α-Al_2O_3. The existence of Pd-hydroxyl synergistic interaction is paramount to the enhanced catalytic performance for the CO coupling reaction, in nature, as the result of hydroxyl-promoted adsorption of bridged CO on the Pd surface (Figure 8; Wang et al. 2016b).

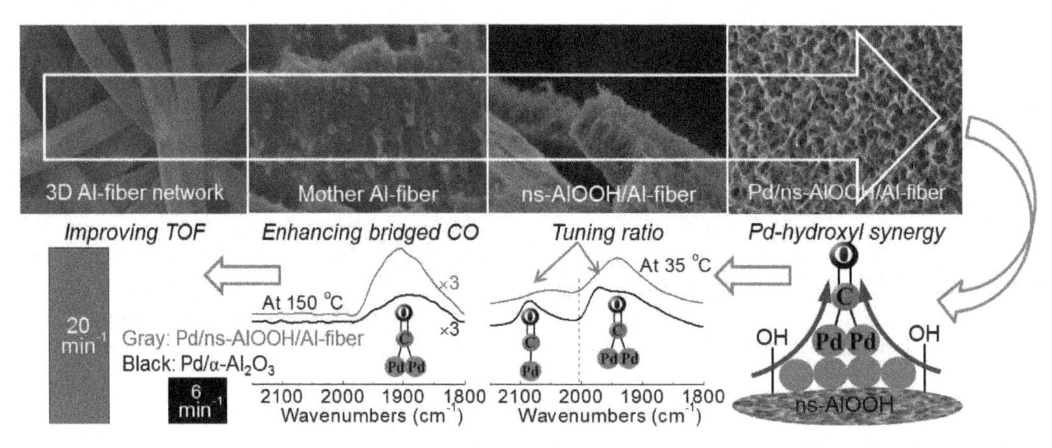

Figure 8. Upper: SEM images of 3D Al-fiber network, mother Al-fiber, ns-AlOOH/Al-fiber (prepared by steam-only oxidation), and Pd/ns-AlOOH/Al-fiber catalyst. Lower: schematic illustration of Pd-hydroxyl synergistic interaction for the enhancement of CO coupling reaction. Reprinted with permission from ref. Wang, C., L. Han, P. Chen, G. Zhao, Y. Liu and Y. Lu. 2016b. J. Catal. 337: 145. Copyright@Elsevier.

Monolithic FeMnK/ns-Al_2O_3/Al-fiber catalyst for Fischer-Tropsch to lower olefins

Lower olefins (C2-C4) are extensively used as key petrochemical feedstocks in the modern chemical industry, which are traditionally produced by steam cracking of naphtha or as byproducts of oil refining processes (Torres et al. 2012). With the ever-increasing shortfall of petroleum sources, the production of olefins via the non-petrochemical routes has attracted worldwide interest. The direct conversion of syngas (a mixture of H_2 and CO) into lower olefins based on Fischer-Tropsch process (named as FTO) has been known as an alternative process. To date, various catalytic components (such as Fe, Co, and Ru at bulk- and nano-scales), supports (ranging from oxides and molecular sieves to carbonaceous materials, such as Al_2O_3, MgO, carbon nanotubes, and carbon nanofibers) and promoters (Ag, Mn, and K) have been systematically studied to optimize the activity/selectivity associated with notably improved stability. The γ-Al_2O_3 supported Fe-Mn-K catalyst is considered to be a promising catalyst owing to its high CO conversion (> 60%) and olefins selectivity (> 60%) (Torres et al. 2012). In spite of these promising results, further industrial applications still remains particularly challenging because extruded pellets or microgranules are required in the real-world forms rather than as-made powders. Moreover, because of the strong exothermicity of FTO process, it is desirable to endow the catalyst with good heat conductivity for rapidly dissipating reaction heat from catalyst bed in a fixed-bed reactor. Therefore, the usage of microfibrous structured catalysts is a promising strategy to overcome these major drawbacks due to the enhanced heat/mass-transfer performance, high permeability (permitting high throughput operation), good mechanical stability, and improved hydrodynamics (to significantly improve the regular-flow and in-turn conversion and selectivity patterns) (Gascon et al. 2015).

Recently, a microfibrous structured FTO catalyst of Fe-Mn-K/ns-Al_2O_3/Al-fiber is developed by placing Fe-Mn-K composites on a monolithic ns-Al_2O_3/Al-fiber support (Han et al. 2016b). Firstly, the thin-sheet monolithic ns-Al_2O_3/Al-fiber support is synthesized according to the reported procedures (Wang et al. 2015): the pristine support of a microfibrous 3D network structure consisting of 10 vol% 60-μm Al-microfiber and 90 vol% voidage is functionalized by endogenously growing thin shell (~ 0.5 μm) of nanosheet AlOOH (ns-AlOOH) onto the Al-microfiber via steam-only oxidation treatment ($Al + H_2O \rightarrow$ ns-AlOOH + H_2) associated with ns-AlOOH transformation into ns-Al_2O_3 by calcination at 600°C. Active components of Fe and Mn as well as additive K are then placed onto the surface of γ-Al_2O_3 shell of the ns-Al_2O_3/Al-fiber support by incipient wetness impregnation method. By taking advantage of enhanced heat/mass transfer, large void volume and entirely opened network structure, this catalyst delivers a high iron time yield of 206.9 $\mu mol_{CO}\, g_{Fe}^{-1}\, s^{-1}$ at 90% CO conversion with 40% selectivity to C_2-C_4 olefins under optimal reaction conditions (350°C, 4.0 MPa, 10000 mL/(g·h)). In this case of microfibrous structured iron-based catalysts, K-modification, in nature, greatly facilitates the catalyst reduction/carbonization behavior and positively tunes the catalyst surface basicity.

Very recently, another microfibrous structured catalyst Fe-Mn-K/ns-Al_2O_3/Al-fiber of the same kind is obtained via surface impregnation combustion method (Han et al. 2016c). After being modified by K through impregnation method, the ns-Al_2O_3/Al-fiber support is functionalized with the active Fe-Mn nanocomposites via surface impregnation combustion method. The as-burnt Fe-Mn-K/ns-Al_2O_3/Al-fiber catalyst obtained under air atmosphere delivers a very high iron time yield of 202.3 $\mu mol_{CO}\, g_{Fe}^{-1}\, s^{-1}$ at a CO conversion of 89.6% with light olefins (C2-C4) selectivity of 42.1% C at 350°C and 4.0 MPa using a high gas hourly space velocity (GHSV) of 10000 mL/(g·h), which are very similar to the results for the impregnated catalyst. The hydrocarbon distribution of this catalyst remains constant throughout entire 225 h testing, with CH_4 selectivity of ~ 16%, lower olefins of ~ 38% and C_5^+ hydrocarbon of ~ 35%, despite the obvious volcano-shaped evolution of CO conversion. The effect of combustion atmospheres (air, N_2, and N_2 followed by air (N_2-air)) is investigated on the catalyst performance. It is found that combustion under air is helpful to form 6 nm Fe-Mn-K oxide particles with better reducibility and carburization properties, thereby leading to high-performance as-burnt catalyst for FTO. In contrast, under either N_2 or N_2-air atmosphere, smaller oxide particles (3–4 nm) are formed but suffer from deteriorated reducibility and carburization properties due to the strong support-metal interaction.

Notably, the resulting microfibrous structured ns-Al_2O_3-supported Fe-Mn-K catalysts possess unique physical properties in terms of void volume, pore structure, surface-to-volume ratio, permeability, thermal conductivity, particle size, and form factor. The use of a large void volume metallic microfibrous structure might facilitate interlayer heat and mass transfer, while the thin shell (0.5 μm) of Fe-Mn-K nanocomposites dispersed onto ns-Al_2O_3 might significantly reduce the intraparticle mass transfer. It is anticipated that this assay is a new point for iron-catalyzed FTO reaction, where the microfibrous catalysts and catalytic reactors considered in isolation can satisfy several of the most fundamental criteria needed for useful operation.

Endogenous growth via hydrothermal treatment

Similar to the chemical etching (see the section of "**Wet chemical etching technology**") and steam-only oxidation (see the section of "**Endogenous growth via steam-only oxidation**") technologies, the hydrothermal treatment is another technology to functionalize the monolithic metal fiber/foam supports. In a typical synthesis, ammonium chloride (NH_4Cl) and metal nitrates (such as manganese nitrate, $Mn(NO_3)_2$) with specified amount are simultaneously dissolved in distilled water under stirring at room temperature; subsequently, Ni-foam is immersed into the above mentioned homogenous solution and then the Ni-foam and solution are sealed inside a Teflon-lined stainless steel autoclave at 100–160°C for some time; then, the hydrothermally treated Ni-foam are washed with ethanol and deionized water for several times, dried overnight, and calcined in air at 400–600°C for some time; finally, the NiO-MnO_x nanosheets composites are uniformly *in situ* generated and firmly anchored onto the Ni-foam substrate. During hydrothermal reaction, Ni^{2+} ions are first dissolved from Ni-foam ($Ni + 2H^+ = Ni^{2+} + H_2$) and then co-precipitate with Mn^{2+} ions ($Ni^{2+} + Mn^{2+} + 4OH^- = Ni(OH)_2 + Mn(OH)_2$) to form Ni-Mn double hydroxides based on the pH-driven dissolution-precipitation mechanism (Tian et al. 2013). The Ni-foam not only serves as the monolith support, but also provides Ni^{2+} ions for the *in situ* fabrication of the NiO-MnO_x nanosheets in the hydrothermal reaction, which can lead to the strong adhesion between active species and support. Afterwards, the calcination procedure can lead to the generation of NiO-MnO_x nanosheet composites; simultaneously, pores are formed on the NiO-MnO_x nanosheet structure owing to the gas release during the decomposition of Ni-Mn hydroxide precursors.

Monolithic NiO-MnO_x nanosheet composites for de-NO_x process

Nitrogen oxides (NO_x) emitted from coal and fossil fuel combustion can cause great environmental problems including acid rain, photochemical smog, and greenhouse effect (Beirle et al. 2011, Felix et al. 2012). To date, the selective catalytic reduction of NO_x by ammonia (NH_3-SCR) has been considered as the most effective and economical technology for the removal of NO_x from the mobile or stationary sources (Mou et al. 2012, Wang et al. 2012). For decades, V_2O_5-MoO_3/TiO_2 and V_2O_5-WO_3/TiO_2 are used in many countries as industrially adopted NH_3-SCR catalysts (Kompio et al. 2012), but suffer from high operation temperature and low N_2 selectivity as well as toxicity (from V) to the environment and human health. Recently, many research efforts have been made on the manganese based catalysts due to their inherent environmentally benign character and the outstanding catalytic performance in the NH_3-SCR of NO_x (Thirupathi and Smirniotis 2012, Zhang et al. 2013a, Wan et al. 2014), and it has been demonstrated that the NiO-MnO_x could act as a promising candidate (Wan et al. 2014). However, for practical application, the active components are usually immobilized on the surface or adsorbed to the channel walls of ceramic monoliths or parallel passage reactors by wash coating, dip coating, impregnation, or extrusion (Pereda-Ayo et al. 2013). In the actual operation process, the random distribution of active components, the low inter-phase mass transfer ability, and the blockage of channels can lead to the decrease of catalytic activity (Li et al. 2012). Therefore, it is still a challenge to develop new synthetic strategies and catalyst supports in order to fabricate novel Ni-Mn monolith catalysts with high de-NO_x activity, mass transfer ability, and stability.

Owing to the high porosity, excellent mass/heat transfer ability, and enhanced mechanical strength, the Ni-foam is considered as a new prospective catalyst support. Combining the attractive properties of NiO-MnO_x binary nanocomposites and Ni-foam support, the Ni-foam structured 3D hierarchical monolith catalysts are developed by the following synthetic route as illustrated in Figure 9 (Cai et al. 2014). Porous NiO-MnO_x nanosheets successfully decorate Ni-foam as 3D hierarchical monolith de-NO_x catalysts via a simple hydrothermal reaction and calcination process. The catalysts exhibit excellent NH_3-SCR activity of NO and good stability: a wide temperature window for > 80% NO conversion ranging from 245°C to 360°C is exhibited and the maximum NO conversion of 91% is achieved at 270°C; the NO conversion rate is maintained at ca. 90% under the continuous running duration for at least 16 h without any signs of deactivation. The favorable properties arise from the 3D hierarchical structure, enriched active oxygen species and reducible species, and enhanced surface acidity as well as the synergetic effects between the well dispersed Mn and Ni species. In addition, the *in situ* formation of porous NiO-MnO_x nanosheets on Ni-foam brings about the strong adhesion between the active components and catalyst support, which leads to the expected stability during the catalytic process. Therefore, this eco-friendly and easily constructed 3D hierarchical catalysts with good catalytic performance might open up great opportunities for applications of monolith de-NO_x catalysts.

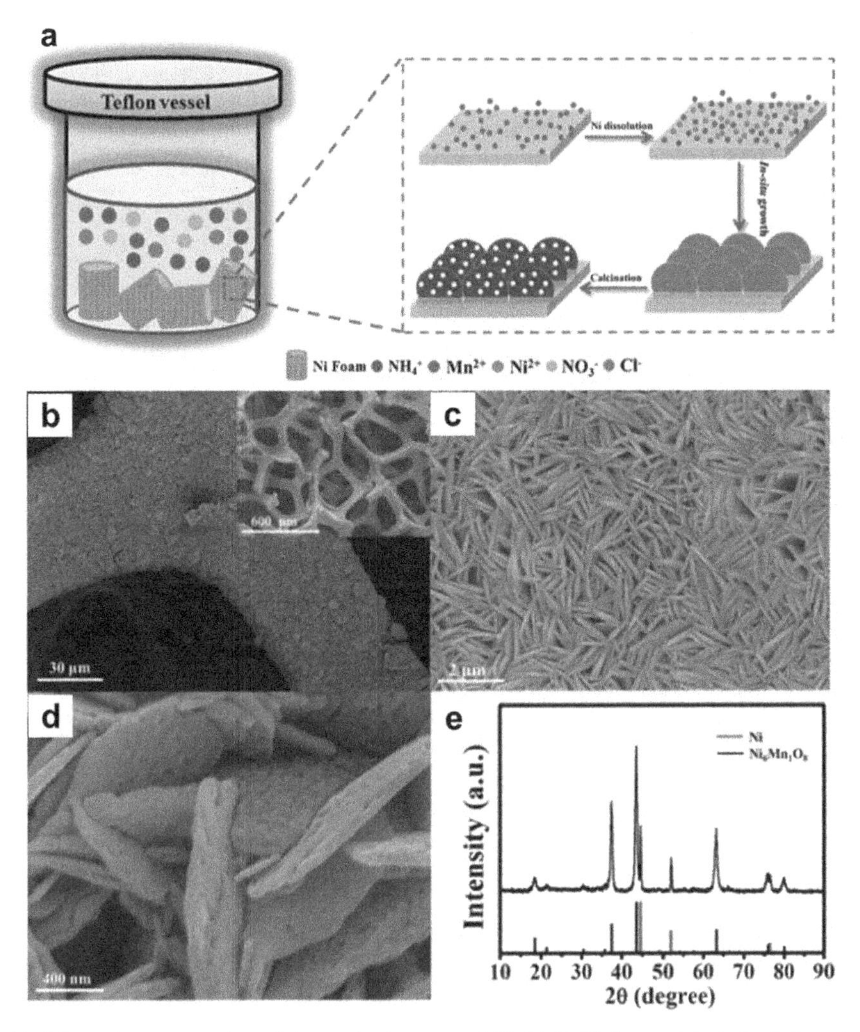

Figure 9. (a) Schematic representation of the synthesis route to NiO-MnO_x/Ni-foam catalyst. (b–d) Scanning electron microscope (SEM) images and (e) X-ray powder diffraction (XRD) pattern of the NiO-MnO_x/Ni-foam catalyst. Reprinted with permission from ref. Cai, S., D. Zhang, L. Shi, J. Xu, L. Zhang, L. Huang, H. Li and J. Zhang. 2014. Nanoscale 6: 7346. Copyright@Royal Society of Chemistry.

Monolithic NiO-MgO nanosheet composites for catalytic oxy-methane reforming

Similarly, the Ni-foam structured NiO-MgO nanosheets catalyst is also prepared by the hydrothermal treatment followed by a calcination treatment (Chai et al. 2016). Typically, magnesium nitrate ($Mg(NO_3)_2$) and ammonium chloride (NH_4Cl) with specified amount are simultaneously dissolved in distilled water under stirring at room temperature; subsequently, Ni-foam is immersed into the solution and then the Ni-foam and solution are sealed inside a Teflon-lined stainless steel autoclave at 100–160°C for some time; then, the hydrothermally oxidized Ni-foam are washed with distilled water for several times, dried overnight, and calcined in air at 400–600°C for some time. Finally, the NiO-MgO nanosheet composites are *in situ* generated and firmly anchored onto a Ni-foam substrate from nano- to macro-scales in one step. The representative NiO-MgO/Ni-foam catalyst consisting of 1.0% MgO and 18.7% NiO is the best structured catalyst for the catalytic oxy-methane reforming ($CH_4 + 0.5O_2 = CO + 2H_2$), being capable of converting 82.9% CH_4 into syngas at selectivity of 94.7% to H_2 and of 89.9% to CO for a feed of CH_4/

O_2 = 2/1 (vol/vol), at 700°C and a high GHSV of 100 L g^{-1} h^{-1}. Despite above interesting results, the improvement of their anti-sintering stability as well as carbon resistance is particularly desirable and remains challenging. Lighted by the preparation of structured nanosheet catalyst, a promising strategy is expected to be suggested by thermal decomposition of monolithic Ni-Mg-Al layered-double-hydroxides nanosheets that are hydrothermally grown onto Ni-foam.

In situ growth of layered double hydroxides

Layered double hydroxides (LDHs) belong to a family of naturally occurring and/or synthetic hydrotalcite-like materials presenting a $[Mg(OH)_2]$ brucite-like layered structure. These materials allow substitution of divalent metal cations by trivalent cations within their brucite-like layers to give a positively charged layer balanced by a wide variety of anions within their interlayer domain (Evans and Duan 2006). They are commonly represented by the formula $[M^{2+}_{1-x}M^{3+}_{x}(OH)_2](A^{n-})_{x/n}\cdot mH_2O$, where M^{2+} is a divalent cation such as Mg^{2+}, Ni^{2+}, Zn^{2+}, Cu^{2+} or Mn^{2+}; M^{3+} is a trivalent cation such as Al^{3+}, Fe^{3+} or Cr^{3+}; and A^{n-} is an interlayer anion such as CO_3^{2-}, SO_4^{2-}, NO_3^-, Cl^- or OH^-. Thus, the metal oxides can be obtained by simply designing an appropriate LDH precursor containing the desired catalytically active metal cations and by calcining the LDHs in a controlled manner. These oxides species have large surface areas, high thermal stability, and homogeneous distribution in the layers. Due to these special properties, the mixed oxides derived from LDHs are attracting growing interest in catalysis (He et al. 2012, Gardner et al. 2012, Sun et al. 2012). However, their application with nano-scale morphology still remains a challenge, and their weak thermal conductivity will induce hotspots and cause catalytic deactivation.

Zhang et al. report a facile strategy, by which the layered double hydroxides are fabricated on a porous anodic alumina/aluminum (PAO/Al) substrate via an *in situ* crystallization technique (Chen et al. 2009). Typically, metal nitrates (such as $Ni(NO_3)_2$) and NH_4NO_3 are dissolved in deionized water to form a mixed solution, and the pH of the solution is adjusted to 6.5 by adding dilute 1% aqueous ammonia solution. Then, the PAO/Al substrates are placed vertically in the above solution inside a Teflon-lined stainless steel autoclave, which is placed in a water bath at 75°C for 36 h. After completion of the LDH film growth, the substrates are taken out of the autoclave, rinsed with ethanol, and dried at room temperature, and the NiAl-LDH/PAO/Al is obtained. Subsequently, Wei et al. fabricate a monolithic binary Cu-Co catalyst via a facile two-step procedure involving the *in situ* growth of a ternary CuCoAl-LDH precursor onto a monolithic aluminum substrate followed by a calcination-reduction process (Figure 10), and demonstrate its application as an efficient catalyst for hydrogen generation from NH_3BH_3 decomposition (Li et al. 2013). The catalyst with a Cu/Co molar ratio of 1/1 yields a hydrolysis completion time less than 4.0 min at a rate of ~ 1000 mL (min^{-1} g_{cat}) under ambient conditions, comparable to the most reported noble metal catalysts (e.g., Ru and Pt). It is verified that the synergistic effect between highly dispersive metallic Cu and Co_3O_4 species plays a key role in the significantly enhanced activity of the Cu-Co catalyst. This work provides an effective strategy for the fabrication of excellent Cu-Co catalysts for NH_3BH_3 decomposition, which can be used as promising candidates in pursuit of practical implementation of NH_3BH_3 as a hydrogen storage material.

Recently, a Ni-foam-structured NiO-MgO-Al_2O_3 nanocomposite catalyst for catalytic oxy-methane reforming is developed by thermal decomposition of NiMgAl LDHs *in situ*, hydrothermally grown onto the Ni-foam (Chai et al. 2017a). Originating from the lattice orientation effect and topotactic decomposition of the LDH precursor, NiO, MgO, and Al_2O_3 are homogeneously distributed in the nanocomposite, and therefore, this catalyst shows enhanced coke- and sintering-resistance. At 700°C and a gas hourly space velocity of 100 L g^{-1} h^{-1}, 86.5% methane conversion and 91.8/88.0% H_2/CO selectivities are achieved with stability for at least 200 h. More recently, the authors present a FeCrAl-fiber-structured NiO-MgO-Al_2O_3 nanocomposite catalyst engineered from nano- to macro-scales in one-step, also via thermally decomposing NiMgAl LDHs (Chai et al. 2017b). Differently but interestingly, the NiMgAl LDHs are controllably grown onto the FeCrAl-fiber only through a γ-Al_2O_3/water interface-assisted method. By taking the advantage of homogeneous component-distribution in the LDHs-derived NiO-MgO-Al_2O_3 nanocomposites and enhanced heat transfer, this promising catalyst delivers satisfying performance with enhanced sintering/coke resistance in the dry reforming of methane ($CH_4 + CO_2 = 2CO + 2H_2$). At 800°C

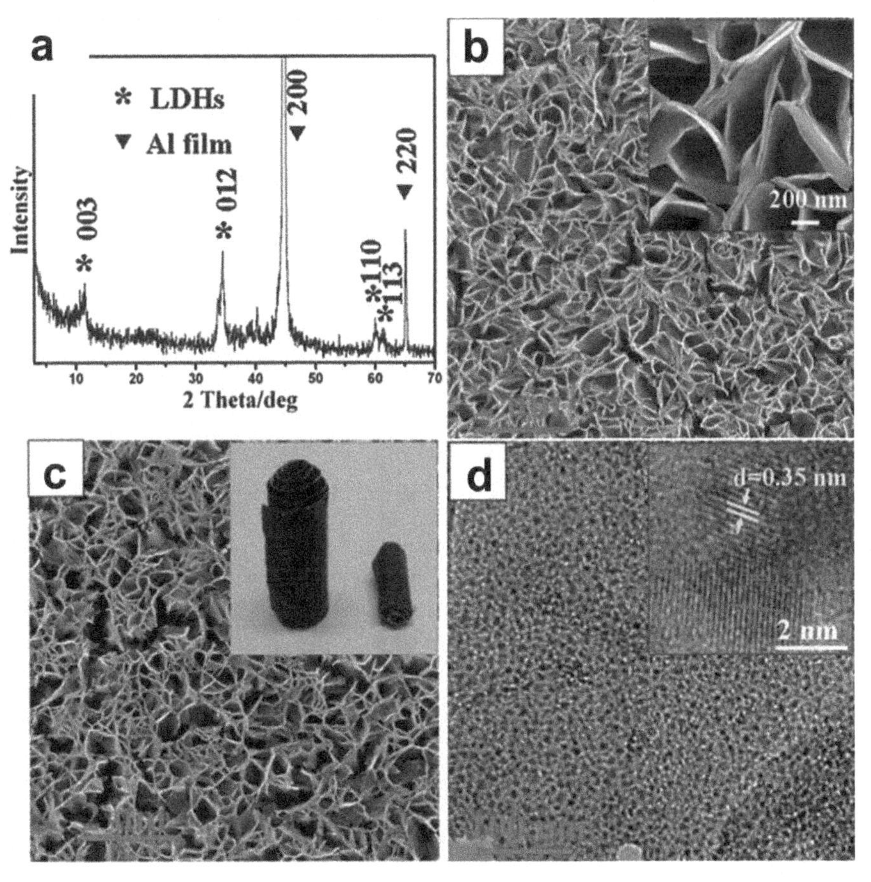

Figure 10. (a) XRD pattern of the CuCoAl-LDH film precursor on the Al substrate. SEM images of (b) the CuCoAl-LDH film precursor; the insets show its corresponding enlarged view, (c) the monolithic Cu-Co film catalyst; the insets show the photograph of the rolled Cu-Co film catalyst. (d) TEM images of the monolithic Cu-Co film catalyst; the inset shows the lattice fringe image assigned to metal Cu and spinel Co_3O_4, respectively. Reprinted with permission from ref. Li, C., J. Zhou, W. Gao, J. Zhao, J. Liu, Y. Zhao, M. Wei, D.G. Evans and X. Duan. 2013. J. Mater. Chem. A 1: 5370. Copyright@Royal Society of Chemistry.

and a gas hourly space velocity of 5000 mL g^{-1} h^{-1}, CH_4/CO_2 conversion is maintained almost constant at 91%/89% within the initial 90 h and then slides in a smooth downturn (to 80/85%) within another 180 h of reaction. We believe that this type of tailoring strategy and the as-obtained materials is opening up new opportunities for future applications in other high-throughput and high-temperature reactions.

One-step organization technology for structuring core@shell nanocomposites

As mentioned in the "Introduction", core@shell nanocomposites with fine nanoparticles encapsulated into the oxide-shell is a promising protocol to enhance catalyst stability (Schärtl 2010, Alayoglu et al. 2008). The outer shells isolate the catalytically active nanoparticles and prevent their sintering during catalytic reactions at high temperature and/or with strong exothermicity. One classic model of Pt@SiO_2 is prepared via colloidal synthesis of 14-nm Pt cores followed by the polymerization of 17-nm-thick SiO_2-shell around the Pt cores (Joo et al. 2009), exhibiting high stability during ethylene hydrogenation reaction (a typical strongly-exothermic reaction). Recently, a hierarchically structured catalyst is designed and prepared in two steps (Cargnello et al. 2012): the pre-organization of Pd@CeO_2 subunits composed of a 2 nm Pd core and a ceria (CeO_2) shell, and the homogeneous deposition onto a modified hydrophobic alumina. The Pd cores remain isolated even after heating the catalyst to 850°C, and the catalyst shows an excellent thermal stability for CH_4 oxidation. Very recently, a new Au@CeO_2 catalyst is facilely synthesized using a redox-coprecipitation strategy (Mitsudome et al. 2015), where a redox reaction between the core and shell precursors allows for the spontaneous formation of the core@shell nanocomposites in one step. Au@CeO_2 shows a high durability for the semihydrogenation of alkynes. However, several drawbacks greatly hamper their applications in heterogeneous catalysis (Cargnello et al. 2010, Zhou et al. 2010, Lu et al. 2012), such as their complicated synthesis, the lack of a strategy for large-scale production, and low mass/heat transfer and high pressure drop of randomly packed catalyst bed for strongly exo-/endo-thermic and/or high-throughput reactions.

On the basis of well-defined cross-linking molecules that possess functionally specified multi-groups, reported a chemically and economically affordable strategy is developed to organize analogous core@shell structures (i.e., fine nanoparticles encapsulated into oxide mesoporous-matrixes) and, meanwhile, to firmly embed them onto the monolithic structured supports in one step (Zhang et al. 2016b). The 3-aminopropyltriethoxysilane (APTES)-assistant organization of Pd@SiO_2/ns-Al_2O_3/Al-fiber is taken as an example to demonstrate this strategy (Figure 11). Typically, APTES is first added into the acetone solution of palladium acetate (Pd$(Ac)_2$), and the chelation between -NH_2 in APTES and single-metal-ion of Pd^{2+} takes place (Kim and Lee 2014). Meanwhile, silanisation occurs between -Si-O-CH_2-CH_3 and surface -OH to form Al-O-Si bonds on ns-AlOOH/Al-fiber which is obtained via endogenous growth of boehmite nanosheets (ns-AlOOH) on 3D network of 60-μm Al-microfiber (Wang et al. 2015). Subsequently, some water is added into the system and the rest -Si-O-CH_2-CH_3 polymerizes in the presence of water to produce Si-O-Si bonds (Lechevallier et al. 2012). After heating at 100°C for 2 h, a cross-linked organic-inorganic hybrid catalyst precursor of Pd^{2+}-APTES-AlOOH/Al-fiber with Pd^{2+} chelated in the network of APTES-AlOOH is afforded, which is finally transformed into the Pd@SiO_2/ns-Al_2O_3/Al-fiber catalyst after calcining at 450–600°C in air and then reducing at 150–400°C in H_2. Notably, calcination treatment also plays a pivotal role in the organization process because pyrolysis of the remaining alkyl chains in APTES can produce H_2 and CH_4, which creates mesopores inside the SiO_2 shell matrix (Kim and Lee 2014). The catalysts, with unique combination of both advantages of analogous core-shell structures and enhanced heat/mass transfer, are qualified for typical harsh reactions,

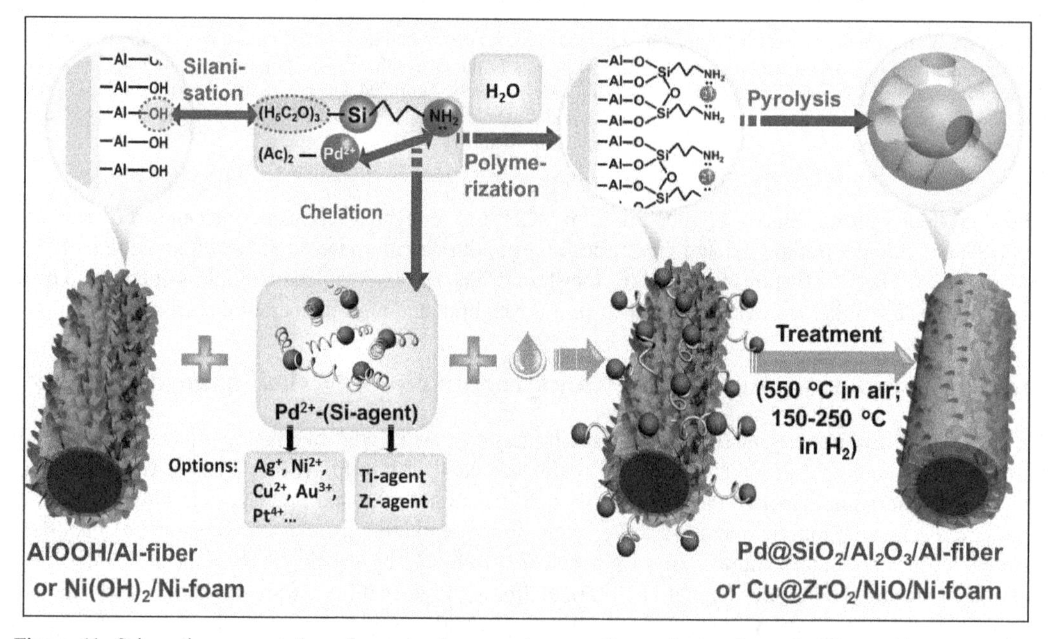

Figure 11. Schematic representation of one-step "macro-micro-nano" organization for embedding oxide-encapsulated-nanoparticles onto monolithic-substrates with the aid of well-defined cross-linking molecules. Reprinted with permission from ref. Zhang, Q., G. Zhao, Z. Zhang, L. Han, S. Fan, R. Chai, Y. Li, Y. Liu, J. Huang and Y. Lu. 2016b. Chem. Commun. 52: 11927. Copyright@Royal Society of Chemistry.

such as catalytic combustion of methane and volatile organic compounds, oxidative dehydrogenation of ethane, and gas-phase hydrogenation of dimethyl oxalate to methyl glycolate.

Catalytic combustion of low concentration methane and volatile organic compounds (VOCs)

The as-organised monolithic Pd@SiO_2/ns-Al_2O_3/Al-fiber catalyst is initially applied in catalytic combustion of methane (i.e., under methane-lean condition), a strongly exothermic process requiring a high throughput operation (Zhang et al. 2016b). Supported nano-Pd based catalysts deliver high activity and resistance to sulfur poison, but suffer from the sintering of Pd nanoparticles (Farrauto 2012). The Pd@SiO_2/ns-Al_2O_3/Al-fiber catalyst delivers a high activity owing to the highly dispersed Pd nanoparticles (2–3 nm): CH_4 conversion is gradually increased to 10% at 275°C, 50% at 325°C, and 90% at 365°C and complete CH_4 conversion is achieved at 400°C, using a high GHSV of 72,000 mL·$g_{cat.}^{-1}$·h^{-1} for a feed of 1.0 vol% CH_4 in air. Moreover, CH_4 conversion profile against temperature in the cooling operation almost overlaps the one obtained in heating operation mode, indicating that the reaction heat, which will make a higher CH_4 conversion, is not trapped due to the high thermal conductivity of this catalyst. The other advantage of such catalyst is the high permeability stemmed from its entirely open network structure with high void volume. As a result, this microfibrous structured catalyst bed, with complete CH_4 conversion at 445°C, generated a very low pressure drop of only 4,000 Pa m^{-1} even at a high GHSV of 100,000 mL·$g_{cat.}^{-1}$·h^{-1}.

Not surprisingly, the Pd@SiO_2/ns-Al_2O_3/Al-fiber assures activity maintenance throughout a 1,000-h test, and the surface morphology is well-preserved, revealing the thermal/hydrothermal stability of SiO_2 shell matrix. For comparison, a Pd/ns-Al_2O_3/Al-fiber catalyst is also prepared by highly dispersing Pd NPs (~ 2 nm) onto the ns-Al_2O_3/Al-fiber substrate via impregnation method. Under identical conditions, this contrastive catalyst exhibits a high initial activity comparable to the Pd@SiO_2/ns-Al_2O_3/Al-fiber, but CH_4 conversion is decreased drastically from 100% to 60% within only 140 h. TEM images clearly show serious sintering of Pd nanoparticles on the spent catalyst from 2 to 8 nm. These results in turn again confirm the pivotal role of the encapsulation by SiO_2 shell matrix in preventing Pd nanoparticles from sintering.

The Pd@SiO_2/ns-Al_2O_3/Al-fiber catalyst is also extended into the catalytic abatement of VOCs, including hydrocarbons, alcohols, aldehydes, and esters. Under the conditions of 1,000 ppm, single organic compound in air and GHSV of 72,000 mL·g_{cat}^{-1}·h^{-1}, T_{90} (temperature of 90% conversion) is 267°C for toluene, 200°C for propylene, 200°C for ethane, 138°C for methyl formate, 100°C for formaldehyde, and 60°C for methanol. Obviously, the catalyst shows promising, not even optimized, activity for various kinds of volatile organic compounds.

Other applications

Clearly, the APTES-assistant organization chemistry of Pd nanoparticles encapsulated into SiO_2 shell matrix and, meanwhile, anchored onto a monolithic support in one step, is straightforward. Interestingly, this strategy can potentially extend to other analogous core@shell nanocomposites (NiO, Cu, Ag, Pt and Au as cores; TiO_2 and ZrO_2 as shells) as well as monolithic $Ni(OH)_2$/Ni-foam support (Figure 11), giving to the present approach a wide applicability and versatility using cross-linking molecules with needed shell elements and nanoparticle precursors. The NiO@TiO_2/ns-Al_2O_3/Al-fiber catalyst is organised by the same method for the preparation of Pd@SiO_2/ns-Al_2O_3/Al-fiber and explored for the oxidative dehydrogenation of ethane, a pivotal process to produce ethene but with strong exothermicity. Clearly, the NiO@TiO_2 configuration effectively prevents NiO nanoparticles sintering, delivering desirable and more stable ethane conversion of 30% and ethene selectivity of 70% compared with the traditional NiO/TiO_2 catalyst. Additionally, another Cu@ZrO_2/ns-NiO/Ni-foam catalyst is tested in the gas-phase hydrogenation of dimethyl oxalate, and excitingly, delivers a full conversion and a high methyl glycolate (an important intermediate for synthesizing pharmaceutical products and perfumes) selectivity also with promising stability compared to the naked Cu supported on NiO/Ni-foam. The Au@SiO_2/ns-Al_2O_3/Al-fiber is tested in another strongly exothermic reaction, gas-phase oxidation of benzyl alcohol to benzaldehyde, which is unfortunately inert for this reaction. However, the Au cores can be facilely modified with other promoters like Cu to form Au-CuO@SiO_2/ns-Al_2O_3/Al-fiber, thereby leading to significant activity and stability enhancement compared to Au nanoparticles supported on CuO. These results clearly indicate that the organization from nano- to macro-scales of different building blocks with tunable structures and functions into monolithic structured architectures based on controllable chemical reactions represents a promising strategy to develop novel catalyst systems.

Reaction-induced technology for structuring oxide@oxide nanocomposites

The ever-increasing studies have demonstrated that the catalyst structures are sensitive to the ambient atmosphere and even can be induced into the optimum state in the reaction stream. One early example of this induced phenomenon is the bimetallic core@shell nanoparticles of $Rh_{0.5}Pd_{0.5}$ capable of undergoing segregation of the metals, driven by oxidizing and reducing environments (Tao et al. 2008): in NO and O_2 atmospheres, Rh migrates to the shell and almost completely oxidized; while in CO and H_2, the RhO_x is reduced, with the Rh atoms migrating to the core and the Pd atoms to the shell. Such observed changes in atomic distribution and chemical state are reversible. Moreover, the $PdZn_x$ homogeneous nanoalloy is *in situ* induced into $PdZn_y$@(x-y)ZnO heterostructure in the real methanol reforming process ($CH_3OH + H_2O = 3H_2 + CO_2$) (Friedrich et al. 2013), and Co_3O_4-nanorod supported single-atom Pt catalyst Pt_1/Co_3O_4 is also *in situ* induced into Pt_nCo_m/CoO_{1-x} for water-gas shift reaction ($CO + H_2O = H_2 + CO_2$) (Zhang et al. 2013b). Extrapolatedly, it is wondered whether a more optimum structure could be induced directly from the initial precursors, nitrates for instance, by reaction itself. So, one catalyst is prepared by supporting nitrates on monolithic support via impregnation method followed by directly undergoing benzyl alcohol oxidation at 300°C for 1 h for the gas-phase oxidation to benzaldehyde.

Recently, the thin-sheet sinter-locked Ti-microfiber-supported binary-oxide nanocomposites engineered on the micro- to macro-scales are developed for the gas-phase aerobic oxidation of benzyl alcohol to benzaldehyde (Zhao et al. 2016b). The catalyst is obtained by placing transient metal (e.g., Ni, Co, Cu, and Mn) nitrates onto a Ti-microfiber surface by impregnation method, and the supported nitrates are subsequently *in situ* transformed into the binary-oxide nanocomposites in the real reaction stream at 300°C. Among them, CoO-2.5-Cu_2O-2.5/Ti-microfiber is found to be the best catalyst, delivering 93.5% conversion of benzyl alcohol (boiling point of 210°C) with 99.2% selectivity to benzaldehyde at 230°C. The *in situ* induced formation of "CoO@Cu_2O" ensembles (i.e., larger CoO nanoparticles partially covered with smaller Cu_2O clusters and/or nanoparticles) is identified, which in nature results in a large Cu_2O-CoO interface and leads to a significant improvement in the low-temperature activity. Moreover, also via such reaction-induced method, the CoO@Cu_2O nanocomposites are structured on monolithic SiC-foam for gas-phase aerobic oxidation of ethanol to acetaldehyde (Zhao et al. 2017). By comparison, such CoO@Cu_2O nanostructure is exclusively induced in the reaction stream and demonstrates much higher activity than the counterparts by other preparation methods. For example, CoO-2.5@Cu_2O-2.5/SiC-foam delivers 92.2% ethanol conversion with 95.8% acetaldehyde selectivity at 260°C. In nature, the resulted abundant Cu_2O-CoO interface in the reaction-induced CoO@Cu_2O nanocomposites is responsible for the high catalytic activity for ethanol selective oxidation.

Final thoughts

The study of heterogeneous catalysis based on the modern nanoscience/nanotechnology as well as the monolithic engineering of nanocatalysts has a rich history extending for decades of years. This chapter is merely a snapshot of the current state of affairs. Obviously, the progress to date of the monolithically structuring of nanocomposites has been proven to be superior to the single-component catalysts that are randomly packed in reactor for the energy/environment/chemical-production catalysis. However, there still is enough room for improvement in these applications. The first is to develop more advanced strategies to effectively and efficiently engineer the nanocomposites from nano- to macro-scales in one step. Although some strategies such as galvanic deposition and chemical etching have been demonstrated to monolithically structure the nanocomposites, these methods are still junior because the size, shape and distribution of building blocks are hard to be controlled. Therefore, the more advanced methods that are based on the modern nanosynthesis should be urgently developed to resolve these problems. The second is to structure the nanocomposites with sintering- and coke-resistance properties for the ultrahigh temperature reactions such as catalytic oxy-methane reforming (usually at 700–800°C) and steam methane reforming (800–1000°C). To achieve this goal, the monolithic supports should be modified with the aim of high temperature stability and coke-resistance properties and further be combined with the opening of high-performance catalyst compositions. Therefore, new strategies should be developed to more effectively functionalize the supports via *in situ* growth of other catalytic materials such as hydrotalcites on them. Thirdly, the structured catalysts will find their place in other applications such as catalytic distillation for some chemicals production such as ethyl acetate from ethanol and acetic acid over the monolithic structured solid acid catalysts. Last but not least, the design of shapes and other features of the structured catalysts should be further developed on the basis of calculations. Unfortunately, most often there have no correlation to pressure drop and mass transfer between different structures for catalyst designs, and there should be a more intense focus on the description of hydrodynamics and transport properties. Anyhow, a close collaboration among the chemistry, materials, engineering, mathematics, and computer science, is highly required to develop a more advanced monolithically structured nanocomposite catalysts, which are attractive for industrial applications in heterogeneous catalysis.

Acknowledgements

We gratefully acknowledge the support of the Basic Key Project (18JC1412100) from the Shanghai Municipal Science and Technology Commission, the NSF of China (21773069, 21703069, 21473057, U1462129, 21273075), and the "973 program" (2011CB201403) from the MOST of China.

References

Alayoglu, S., A.U. Nilekar, M. Mavrikakis and B. Eichhorn. 2008. Ru-Pt core-shell nanoparticles for preferential oxidation of carbon monoxide in hydrogen. Nat. Mater. 7: 333–338.

Beirle, S., K.F. Boersma, U. Platt, M.G. Lawrence and T. Wagner. 2011. Megacity emissions and lifetimes of nitrogen oxides probed from space. Science 333: 1737–1739.

Bleken, F.L., S. Chavan, U. Olsbye, M. Boltz, F. Ocampo and B. Louis. 2012. Conversion of methanol into light olefins over ZSM-5 zeolite: Strategy to enhance propene selectivity. Appl. Catal. A 447-448: 178–185.

Bond, J.C., C. Louis and D.T. Thompson. 2006. Catalysis by Gold. Imperial College Press, London.

Caballero, A. and P.J. Perez. 2013. Methane as raw material in synthetic chemistry: the final frontier. Chem. Soc. Rev. 42: 8809–8820.

Cai, S., D. Zhang, L. Shi, J. Xu, L. Zhang, L. Huang et al. 2014. Porous Ni-Mn oxide nanosheets *in situ* formed on nickel foam as 3D hierarchical monolith de-NO_x catalysts. Nanoscale 6: 7346–7353.

Cargnello, M., N.L. Wieder, T. Montini, R.J. Gorte and P. Fornasiero. 2010. Synthesis of dispersible Pd@CeO_2 core-shell nanostructures by self-assembly. J. Am. Chem. Soc. 132: 1402–1409.

Cargnello, M., J.J.D. Jaen, J.C.H. Garrido, K. Bakhmutsky, T. Montini, J.J.C. Gamez et al. 2012. Exceptional activity for methane combustion over modular Pd@CeO_2 subunits on functionalized Al_2O_3. Science 337: 713–717.

Chai, R., Y. Li, Q. Zhang, G. Zhao, Y. Liu and Y. Lu. 2016. Free-standing NiO-MgO nanosheets *in-situ* controllably composited on Ni-foam as monolithic catalyst for catalytic oxy-methane reforming. Mater. Lett. 171: 248–251.

Chai, R., Y. Li, Q. Zhang, S. Fan, Z. Zhang, P. Chen et al. 2017a. Foam-structured NiO-MgO-Al_2O_3 nanocomposites derived from NiMgAl layered double hydroxides *in situ* grown onto nickel foam: A promising catalyst for high-throughput catalytic oxymethane reforming. ChemCatChem. 9: 268–272.

Chai, R., S. Fan, Z. Zhang, P. Chen, G. Zhao, Y. Liu et al. 2017b. Free-standing NiO-MgO-Al_2O_3 nanosheets derived from layered double hydroxides grown onto FeCrAl-fiber as structured catalysts for dry reforming of methane. ACS Sustainable Chem. Eng. 5: 4517–4522.

Chen, H., F. Zhang, T. Chen, S. Xu, D.G. Evans and X. Duan. 2009. Comparison of the evolution and growth processes of films of M/Al-layered double hydroxides with M = Ni or Zn. Chem. Eng. Sci. 64: 2617–2622.

Chen, L., Y. Lu, Q. Hong, J. Lin and F.M. Dautzenberg. 2005. Catalytic partial oxidation of methane to syngas over Ca-decorated-Al_2O_3-supported Ni and NiB catalysts. Appl. Catal. A 292: 295–304.

Chen, M.S. and D.G. Goodman. 2004. The structure of catalytically active Au on titania. Science 306: 252–255.
Chen, X., J.H. Jin, G.Y. Sha, C. Li, B.S. Zhang, D.S. Su et al. 2014. Silicon-nickel intermetallic compounds supported on silica as a highly efficient catalyst for CO methanation. Catal. Sci. Technol. 4: 53–61.
Choi, M., H.S. Cho, R. Srivastava, C. Venkatesan, D.H. Choi and R. Ryoo. 2006. Amphiphilic organosilane-directed synthesis of crystalline zeolite with tunable mesoporosity. Nat. Mater. 5: 718–723.
Choudhary, T.V. and V.R. Choudhary. 2008. Energy-efficient syngas production through catalytic oxy-methane reforming reactions. Angew. Chem. Int. Ed. 47: 1828–1847.
Corma, A., S. Iborra and A. Velty. 2007. Chemical routes for the transformation of biomass into chemicals. Chem. Rev. 107: 2411–2502.
Cybulski, A. and J.A. Moulijn (eds.). 2006. Structured Catalysts and Reactors. Marcel Dekker, New York.
Dautzenberg, F.M. 2004. New catalyst synthesis and multifunctional reactor concepts for emerging technologies in the process industry. Catal. Rev. 46: 335–368.
Deng, M., G. Zhao, Q. Xue, L. Chen and Y. Lu. 2010. Microfibrous-structured silver catalyst for low-temperature gas-phase selective oxidation of benzyl alcohol. Appl. Catal. B 99: 222–228.
Ding, J., Z. Zhang, L. Han, C. Wang, P. Chen, G. Zhao et al. 2016. A self-supported SS-fiber@meso-HZSM-5 core-shell catalyst via caramel-assistant synthesis toward prolonged lifetime for the methanol-topropylene reaction. RSC Adv. 6: 48387–48395.
Ding, J., P. Chen, S. Fan, Z. Zhang, L. Han, G. Zhao et al. 2017. Microfibrous-structured SS-fiber@meso-HZSM-5 catalyst for methanol-to-propylene: Steam-assisted crystallization synthesis and insight into the stability enhancement. ACS Sustainable Chem. Eng. 5: 1840–1853.
Dudukovic, M.P. 2009. Frontiers in reactor engineering. Science 325: 698–701.
Ersson, A.G. and S.G. Järås. 2006. Catalytic fuel combustion in honeycomb monolith reactors. pp. 215–241. *In*: A. Cybulski and J.A. Moulijn (eds.). Structured Catalysts and Reactors. Marcel Dekker, New York.
Ertl, G., H. Knözinger, F. Schüth and J. Weitkamp (eds.). 2008. Handbook of Heterogeneous Catalysis. Vol. 4. Wiley-VCH Verlag GmbH & Co. KGaA, Weinheim.
Evans, D.G. and X. Duan. 2006. Preparation of layered double hydroxides and their applications as additives in polymers, as precursors to magnetic materials and in biology and medicine. Chem. Commun. 6: 485–496.
Farrauto, R.J. 2012. Low-temperature oxidation of methane. Science 337: 659–660.
Felix, J.D., E.M. Elliott and S.L. Shaw. 2012. Nitrogen isotopic composition of coal-fired power plant NO_x: Influence of emission controls and implications for global emission inventories. Environ. Sci. Technol. 46: 3528–3535.
Figueroa, S.J.A. and M.A. Newton. 2014. What drives spontaneous oscillations during CO oxidation using O_2 over supported Rh/Al_2O_3 catalysts? J. Catal. 312: 69–77.
Friedrich, M., S. Penner, M. Heggen and M. Armbrüster. 2013. High CO_2 selectivity in methanol steam reforming through ZnPd/ZnO teamwork. Angew. Chem. Int. Ed. 52: 4389–4392.
Gardner, G.P., Y.B. Go, D.M. Robinson, P.F. Smith, J. Hadermann, A. Abakumov et al. 2012. Structural requirements in lithium cobalt oxides for the catalytic oxidation of water. Angew. Chem. Int. Ed. 51: 1616–1619.
Gascon, J., J.R. van Ommen, J.A. Moulijn and F. Kapteijn. 2015. Structuring catalyst and reactor-an inviting avenue to process intensification. Catal. Sci. Technol. 5: 807–817.
George, S.M. 2010. Atomic layer deposition: an overview. Chem. Rev. 110: 111–131.
Gong, J., H. Yue, Y. Zhao, S. Zhao, L. Zhao, J. Lv et al. 2012. Synthesis of ethanol via syngas on Cu/SiO_2 catalysts with balanced Cu^0-Cu^+ sites. J. Am. Chem. Soc. 134: 13922–13925.
Guo, W.Y., W.Z. Wu, M. Luo and W.D. Xiao. 2013. Modeling of diffusion and reaction in monolithic catalysts for the methanol-to-propylene process. Fuel Process. Technol. 108: 133–138.
Han, L., G. Zhao, Y. Chen, J. Zhu, P. Chen, Y. Liu et al. 2016a. Cu-fiber-structured La_2O_3-PdAu(alloy)-Cu nanocomposite catalyst for gas-phase dimethyl oxalate hydrogenation to ethylene glycol. Catal. Sci. Technol. 6: 7024–7028.
Han, L., C. Wang, G. Zhao, Y. Liu and Y. Lu. 2016b. Microstructured Al-Fiber@meso-Al_2O_3@Fe-Mn-K Fischer-Tropsch catalyst for lower olefins. AIChE J. 62: 742–752.
Han, L., C. Wang, J. Ding, G. Zhao, Y. Liu and Y. Lu. 2016c. Microfibrous-structured Al-fiber@ns-Al_2O_3 core-shell composite functionalized by Fe-Mn-K via surface impregnation combustion: as-burnt catalysts for synthesis of light olefins from syngas. RSC Adv. 6: 9743–9752.
He, L., Y. Huang, A. Wang, X. Wang, X. Chen, J.J. Delgado et al. 2012. A noble-metal-free catalyst derived from Ni-Al hydrotalcite for hydrogen generation from $N_2H_4{\cdot}H_2O$ decomposition. Angew. Chem. Int. Ed. 51: 6191–6194.
He, R., H. Jiang, F. Wu, K. Zhi, N. Wang, C. Zhou et al. 2014. Effect of doping rare earth oxide on performance of copper-manganese catalysts for water-gas shift reaction. J. Rare Earths 32: 298–305.
Heck, R.M., S. Gulati and R.J. Farrauto. 2001. The application of monoliths for gas phase catalytic reactions. Chem. Eng. J. 82: 149–156.
Ivanova, S., B. Louis, B. Madani, J.P. Tessonnier, M.J. Ledoux and C. Pham-Huu. 2007. ZSM-5 coatings on β-SiC monoliths: Possible new structured catalyst for the methanol-to-olefins process. J. Phys. Chem. C 111: 4368–4374.
Jia, L., S. Zhang, F. Gu, Y. Ping, X. Guo, Z. Zhong et al. 2012. Highly selective gas-phase oxidation of benzyl alcohol to benzaldehyde over silver-containing hexagonal mesoporous silica. Micropor. Mesopor. Mater. 149: 158–165.
Jiao, Y., C. Jiang, Z. Yang and J. Zhang. 2012. Controllable synthesis of ZSM-5 coatings on SiC foam support for MTP application. Micropor. Mesopor. Mater. 162: 152–158.

Joo, S.H., J.Y. Park, C.K. Tsung, Y. Yamada, P. Yang and G.A. Somorjai. 2009. Thermally stable Pt/mesoporous silica core-shell nanocatalysts for high-temperature reactions. Nat. Mater. 8: 126–131.

Kim, J. and D. Lee. 2014. Synthesis and properties of core-shell metal-ceramic microstructures and their application as heterogeneous catalysts. ChemCatChem. 6: 2642–2647.

Kolb, G. 2013. Microstructured reactors for distributed and renewable production of fuels and electrical energy. Chem. Eng. Process 65: 1–44.

Kompio, P.G.W.A., A. Brückner, F. Hipler, G. Auer, E. Löffler and W. Grünert. 2012. Liquid-phase glycerol hydrogenolysis to 1,2-propanediol under nitrogen pressure using 2-propanol as hydrogen source. J. Catal. 286: 237–247.

Lechevallier, S., P. Hammer, J.M. Caiut, S. Mazeres, R. Mauricot, M. Verelst et al. 2012. APTES-modified RE_2O_3: Eu^{3+} luminescent beads: structure and properties. Langmuir 28: 3962–3971.

Li, C., J. Zhou, W. Gao, J. Zhao, J. Liu, Y. Zhao et al. 2013. Binary Cu-Co catalysts derived from hydrotalcites with excellent activity and recyclability towards NH_3BH_3 dehydrogenation. J. Mater. Chem. A 1: 5370–5376.

Li, D., Y. Nakagawa and K. Tomishige. 2011. Methane reforming to synthesis gas over Ni catalysts modified with noble metals. Appl. Catal. A 408: 1–24.

Li, H., D. Zhang, P. Maitarad, L. Shi, R. Gao, J. Zhang et al. 2012. *In situ* synthesis of 3D flower-like NiMnFe mixed oxides as monolith catalysts for selective catalytic reduction of NO with NH_3. Chem. Commun. 48: 10645–10647.

Li, Y., Q. Zhang, R. Chai, G. Zhao, Y. Liu, F. Cao et al. 2015a. Ni-Al_2O_3/Ni-Foam catalyst with enhanced heat transfer for hydrogenation of CO_2 to methane. AIChE J. 61: 4323–4331.

Li, Y., Q. Zhang, R. Chai, G. Zhao, Y. Liu and Y. Lu. 2015b. Structured Ni-CeO_2-Al_2O_3/Ni-Foam catalyst with enhanced heat transfer for substitute natural gas production by syngas methanation. ChemCatChem. 7: 1427–1431.

Liu, H.M. and D.H. He. 2012. Recent progress on Ni-based catalysts in partial oxidation of methane to syngas. Catal. Surv. Asia 16: 53–61.

Liu, J., C. Zhang, Z. Shen, W.M. Hua, Y. Tang, W. Shen et al. 2009. Methanol to propylene: Effect of phosphorus on a high silica HZSM-5 catalyst. Catal. Commun. 10: 1506–1509.

Liu, Y., X. Liu, Q. Feng, D. He, L. Zhang, C. Lian et al. 2016. Intermetallic Ni_xM_y (M = Ga and Sn) nanocrystals: A non-precious metal catalyst for semi-hydrogenation of alkynes. Adv. Mater. 28: 4747–4754.

Lu, J., B. Fu, M.C. Kung, G. Xiao, J.W. Elam, H.H. Kung et al. 2012. Coking- and sintering-resistant palladium catalysts achieved through atomic layer deposition. Science 335: 1205–1208.

Lu, Y., Y. Liu and S.K. Shen. 1998. Design of stable Ni catalysts for partial oxidation of methane to synthesis gas. J. Catal. 177: 386–388.

Magaev, O.V., A.S. Knyazev, O.V. Vodyankina, N.V. Dorofeeva, A.N. Salanov and A.I. Boronin. 2008. Active surface formation and catalytic activity of phosphorous-promoted electrolytic silver in the selective oxidation of ethylene glycol to glyoxal. Appl. Catal. A 344: 142–149.

Makarshin, L.L., V.A. Sadykov, D.V. Andreev, A.G. Gribovskii, V.V. Privezentsev and V.N. Parmon. 2015. Syngas production by partial oxidation of methane in a microchannel reactor over a Ni-Pt/$La_{0.2}Zr_{0.4}Ce_{0.4}O_x$ catalyst. Fuel Process. Technol. 131: 21–28.

Mao, J.P., M.M. Deng, L. Chen, Y. Liu and Y. Lu. 2010. Novel microfibrous-structured silver catalyst for high efficiency gas-phase oxidation of alcohols. AIChE J. 56: 1545–1556.

Milina, M., S. Mitchell, P. Crivelli, D. Cooke and J. Pérez-Ramírez. 2014. Mesopore quality determines the lifetime of hierarchically structured zeolite catalysts. Nat. Commun. 5: 3922.

Milina, M., S. Mitchell, D. Cooke, P. Crivelli and J. Pérez-Ramírez. 2015. Impact of pore connectivity on the design of long-lived zeolite catalysts. Angew. Chem. Int. Ed. 54: 1591–1594.

Min, B.K. and C.M. Friend. 2007. Heterogeneous gold-based catalysis for green chemistry: Low-temperature CO oxidation and propene oxidation. Chem. Rev. 107: 2709–2724.

Mitsudome, T., M. Yamamoto, Z. Maeno, T. Mizugaki, K. Jitsukawa and K. Kaneda. 2015. One-step synthesis of core-gold/shell-ceria nanomaterial and its catalysis for highly selective semihydrogenation of alkynes. J. Am. Chem. Soc. 137: 13452–13455.

Mokrani, T. and M. Scurrell. 2009. Gas conversion to liquid fuels and chemicals: the methanol route—catalysis and processes development. Catal. Rev. Sci. Eng. 51: 1–145.

Mou, X., B. Zhang, Y. Li, L. Yao, X. Wei, D.S. Su et al. 2012. Rod-shaped Fe_2O_3 as an efficient catalyst for the selective reduction of nitrogen oxide by ammonia. Angew. Chem. Int. Ed. 51: 2989–2993.

Pereda-Ayo, B., U. De La Torre, M. Romero-Sáez, A. Aranzabal, J.A. González-Marcos and J.R. González-Velasco. 2013. Influence of the washcoat characteristics on NH_3-SCR behavior of Cu-zeolite monoliths. Catal. Today 216: 82–89.

Pernicone, N. 2003. Catalysis at the nanoscale level. Cattech 7: 196–204.

Pina, C.D., E. Falletta and M. Rossi. 2008. Highly selective oxidation of benzyl alcohol to benzaldehyde catalyzed by bimetallic gold-copper catalyst. J. Catal. 260: 384–386.

Reichelt, E., M.P. Heddrich, M. Jahn and A. Michaelis. 2014. Fiber based structured materials for catalytic applications. Appl. Catal. A 476: 78–90.

Renken, A. and L. Kiwi-Minsker. 2010. Microstructured catalytic reactors. Adv. Catal. 53: 47–122.

Schaedler, T.A., A.J. Jacobsen, A. Torrents, A.E. Sorensen, J. Lian, J.R. Greer et al. 2011. Ultralight metallic microlattices. Science 334: 962–965.

Schärtl, W. 2010. Current directions in core-shell nanoparticle design. Nanoscale 2: 829–843.

Serp, P. and P. Kalck. 2002. Chemical vapor deposition methods for the controlled preparation of supported catalytic materials. Chem. Rev. 102: 3085–3128.

Shen, J., W. Shan, Y. Zhang, J. Du, H. Xu, K. Fan et al. 2006. Gas-phase selective oxidation of alcohols: *In situ* electrolytic nano-silver/zeolite film/copper grid catalyst. J. Catal. 237: 94–101.

Sheng, M., H.Y. Yang, D.R. Cahela and B.J. Tatarchuk. 2011. Novel catalyst structures with enhanced heat transfer characteristics. J. Catal. 281: 254–262.

Slinko, M.M., V.N. Korchak and N.V. Peskov. 2006. Mathematical modeling of oscillatory behaviour during methane oxidation over Ni catalysts. Appl. Catal. A 303: 258–267.

Somorjai, G.A., F. Tao and J.Y. Park. 2008. The nanoscience revolution: Merging of colloid science, catalysis and nanoelectronics. Top. Catal. 47: 1–14.

Southon, P.D., J.R. Bartlett, J.L. Woolfrey and B. Ben-Nissan. 2002. Formation and characterization of an aqueous zirconium hydroxide colloid. Chem. Mater. 14: 4313–4319.

Sun, J., Y. Li, X. Liu, Q. Yang, J. Liu, X. Sun et al. 2012. Hierarchical cobalt iron oxide nanoarrays as structured catalysts. Chem. Commun. 48: 3379–3381.

Tanaka, H., R. Kaino, K. Okumura, T. Kizuka and K. Tomishige. 2009. Catalytic performance and characterization of Rh-CeO_2/MgO catalysts for the catalytic partial oxidation of methane at short contact time. J. Catal. 268: 1–8.

Tao, F., M.E. Grass, Y. Zhang, D.R. Butcher, J.R. Renzas, Z. Liu et al. 2008. Reaction-driven restructuring of Rh-Pd and Pt-Pd core-shell nanoparticles. Science 322: 932–934.

Teunissen, H.T. and C.J. Elsevier. 1997. Ruthenium catalysed hydrogenation of dimethyl oxalate to ethylene glycol. Chem. Commun. 667–668.

Thirupathi, B. and P.G. Smirniotis. 2012. Nickel-doped Mn/TiO_2 as an efficient catalyst for the low-temperature SCR of NO with NH_3: Catalytic evaluation and characterizations. J. Catal. 288: 74–83.

Tian, J., Z. Xing, Q. Chu, Q. Liu, A.M. Asiri, A.H. Qusti et al. 2013. PH-driven dissolution-precipitation: a novel route toward ultrathin $Ni(OH)_2$ nanosheets array on nickel foam as binder-free anode for Li-ion batteries with ultrahigh capacity. CrystEngComm. 15: 8300–8305.

Torres, H.M., J.H. Bitter, C.B. Khare, M. Ruitenbeek, A.I. Dugulan and K.P. de Jong. 2012. Supported iron nanoparticles as catalysts for sustainable production of lower olefins. Science 335: 835–838.

Tucci, E.R. and W.J. Thomson. 1979. Monolith catalyst favored for methanation. Hydroc. Proc. 58: 123–126.

Wan, Y., W. Zhao, Y. Tang, L. Li, H. Wang, Y. Cui et al. 2014. Ni-Mn bi-metal oxide catalysts for the low temperature SCR removal of NO with NH_3. Appl. Catal. B 148-149: 114–122.

Wang, C., L. Han, Q. Zhang, Y. Li, G. Zhao, Y. Liu et al. 2015. Endogenous growth of 2D AlOOH nanosheets on a 3D Al-fiber network via steam-only oxidation in application for forming structured catalysts. Green Chem. 17: 3762–3765.

Wang, C., L. Han, P. Chen, G. Zhao, Y. Liu and Y. Lu. 2016b. High-performance, low Pd-loading microfibrous-structured Al-fiber@ns-AlOOH@Pd catalyst for CO coupling to dimethyl oxalate. J. Catal. 337: 145–156.

Wang, N., W. Qian, K. Shen, C. Su and F. Wei. 2016a. Bayberry-like ZnO/MFI zeolite as high performance methanol-to-aromatics catalyst. Chem. Commun. 52: 2011–2014.

Wang, W., G. McCool, N. Kapur, G. Yuan, B. Shan, M. Nguyen et al. 2012. Mixed-phase oxide catalyst based on Mn-Mullite (Sm, Gd) Mn_2O_5 for NO oxidation in diesel exhaust. Science 337: 832–835.

Wang, X., M. Wen, C. Wang, J. Ding, Y. Sun, Y. Liu et al. 2014. Microstructured fiber@HZSM-5 core-shell catalysts with dramatic selectivity and stability improvement for the methanol-to-propylene process. Chem. Commun. 50: 6343–6345.

Wen, C., Y. Cui, W.L. Dai, S. Xie and K. Fan. 2013. Solvent feedstock effect: the insights into the deactivation mechanism of Cu/SiO_2 catalysts for hydrogenation of dimethyl oxalate to ethylene glycol. Chem. Commun. 47: 5195–5197.

Wen, M., X. Wang, L. Han, J. Ding, Y. Sun, Y. Liu et al. 2015. Monolithic metal-fiber@HZSM-5 core-shell catalysts for methanol-to-propylene. Micropor. Mesopor. Mater. 206: 8–16.

Wen, M., J. Ding, C. Wang, Y. Li, G. Zhao, Y. Liu et al. 2016. High-performance SS-fiber@HZSM-5 core-shell catalyst for methanol-to-propylene: A kinetic and modeling study. Micropor. Mesopor. Mater. 221: 187–196.

Yamamoto, R., Y. Sawayama, H. Shibahara, Y. Ichihashi, S. Nishiyama and S. Tsuruya. 2005. Promoted partial oxidation activity of supported Ag catalysts in the gas-phase catalytic oxidation of benzyl alcohol. J. Catal. 234: 308–317.

Yu, H., M. Chen, P.M. Rice, S.X. Wang, R.L. White and S. Sun. 2005. Dumbbell-like bifunctional Au-Fe_3O_4 nanoparticles. Nano Lett. 5: 379–382.

Yue, H., Y. Zhao, X. Ma and J. Gong. 2012. Ethylene glycol: properties, synthesis, and applications. Chem. Soc. Rev. 41: 4218–4244.

Zhang, L., D. Zhang, J. Zhang, S. Cai, C. Fang, L. Huang et al. 2013a. Design of meso-TiO_2@MnO_x-CeO_x/CNTs with a core-shell structure as $DeNO_x$ catalysts: promotion of activity, stability and SO_2-tolerance. Nanoscale 5: 9821–9829.

Zhang, L., L. Han, G. Zhao, R. Chai, Q. Zhang, Y. Liu et al. 2015b. Structured Pd-Au/Cu-fiber catalyst for gas-phase hydrogenolysis of dimethyl oxalate to ethylene glycol. Chem. Commun. 51: 10547–10550.

Zhang, Q., X. Wu, G. Zhao, Y. Li, C. Wang, Y. Liu et al. 2015a. High-performance PdNi alloy structured *in situ* on monolithic metal foam for coalbed methane deoxygenation via catalytic combustion. Chem. Commun. 51: 12613–12616.

Zhang, Q., X. Wu, Y. Li, R. Chai, G. Zhao, C. Wang et al. 2016a. High-performance PdNi nano-alloy catalyst *in-situ* structured on Ni-foam for catalytic deoxygenation of coalbed methane: Experimental and DFT studies. ACS Catal. 6: 6236–6245.

Zhang, Q., G. Zhao, Z. Zhang, L. Han, S. Fan, R. Chai et al. 2016b. From nano- to macro-engineering of oxide-encapsulated-nanoparticles for harsh reactions: One-step organization via cross-linking molecules. Chem. Commun. 52: 11927–11930.

Zhang, S., J. Shan, Y. Zhu, A.I. Frenkel, A. Patlolla, W. Huang et al. 2013b. WGS catalysis and *in situ* studies of CoO_{1-x}, $PtCo_n/Co_3O_4$, and Pt_mCo_m/CoO_{1-x} nanorod catalysts. J. Am. Chem. Soc. 135: 8283−8293.
Zhang, Z., X. Biao and X. Wang. 2014. Engineering nanointerfaces for nanocatalysis. Chem. Soc. Rev. 43: 7870–7886.
Zhao, J.J., Z.L. Hua, Z.C. Liu, Y.S. Li, L.M. Guo, W.B. Wu et al. 2009. Direct fabrication of mesoporous zeolite with a hollow capsular structure. Chem. Commun. 7578–7580.
Zhao, G., H. Hu, M. Deng, M. Ling and Y. Lu. 2011a. Au/Cu-fiber catalyst with enhanced low-temperature activity and heat transfer for the gas-phase oxidation of alcohols. Green Chem. 13: 55–58.
Zhao, G., H. Hu, M. Deng and Y. Lu. 2011b. Microstructured Au/Ni-fiber catalyst for low-temperature gas-phase selective oxidation of alcohols. Chem. Commun. 47: 9642–9644.
Zhao, G., J. Huang, Z. Jiang, S. Zhang, L. Chen and Y. Lu. 2013a. Microstructured Au/Ni-fiber catalyst for low-temperature gas-phase alcohol oxidation: Evidence of Ni_2O_3-Au^+ hybrid active sites. Appl. Catal. B 140-141: 249–257.
Zhao, G., H. Hu, W. Chen, Z. Jiang, S. Zhang, J. Huang et al. 2013b. Ni_2O_3-Au^+ hybrid active sites on NiO_x@Au ensembles for low-temperature gas-phase oxidation of alcohols. Catal. Sci. Technol. 3: 404–408.
Zhao, G., Y. Liu and Y. Lu. 2016a. Foam/fiber-structured catalysts: non-dip-coating fabrication strategy and applications in heterogeneous catalysis. Sci. Bull. 61: 745–748.
Zhao, G., S. Fan, L. Tao, R. Chai, Q. Zhang, Y. Liu et al. 2016b. Titanium-microfiber-supported binary-oxide nanocomposite with a large highly active interface for the gas-phase selective oxidation of benzyl alcohol. ChemCatChem. 8: 313–317.
Zhao, G., S. Fan, X. Pan, P. Chen, Y. Liu and Y. Lu. 2017. Reaction-induced self-assembly of CoO@Cu_2O nanocomposites *in situ* onto SiC-foam for gas-phase oxidation of bioethanol to acetaldehyde. ChemSusChem. 10: 1380–1384.
Zhou, H., H. Wu, J. Shen, A. Yin, L. Sun and C. Yan. 2010. Thermally stable Pt/CeO_2 hetero-nanocomposites with high catalytic activity. J. Am. Chem. Soc. 132: 4998–4999.
Zhu, J.J., J.G. van Ommen, A. Knoester and L. Lefferts. 2005. Effect of surface composition of yttrium-stabilized zirconia on partial oxidation of methane to synthesis gas. J. Catal. 230: 291–300.
Zhu, Y., S. Wang, L. Zhu, X. Ge, X. Li and Z. Luo. 2010. The influence of copper particle dispersion in Cu/SiO_2 catalysts on the hydrogenation synthesis of ethylene glycol. Catal. Lett. 135: 275–281.
Zinoviev, S., F. Miller-Langer, P. Das, N. Bertero, P. Fornasiero, M. Kaltschmitt et al. 2010. Next-generation biofuels: Survey of emerging technologies and sustainability issues. ChemSusChem. 3: 1106–1133.

6

Pilot Scale Treatment of Natural Rubber Processing Wastewater Using Organoclay

Rathanawan Magaraphan,[1] *Wasuthep Luecha*[1] and *Tarinee Nampitch*[2,*]

Introduction

Nowadays many industries use rubber products; however, the rubber production process generates large quantities of wastewater. Production of concentrated latex requires a centrifuge to remove skim natural rubber (NR) which contains 60% dry rubber content (DRC). The natural rubber latex industry is faced with the disposal of large amounts of wastewater containing small quantities of fine natural rubber latex particles spun out during the production process. Through acid coagulation, this light latex can be turned into skim rubber. The production process of concentrated natural rubber is shown in Figure 1.

The skim NR removed by this process can either be processed into skim rubber or removed as acidic wastewater with high total solid content, biological oxygen demand (BOD), and chemical oxygen demand (COD). Wastewater quality could be improved to meet industrial standards using prepared organoclay in the treatment process. A pilot plant to remove small rubber particles using modified clays, Dehyquart A-CA, Quaternium-18, Varisoft-PATC and Dehyquart F-75 was developed to improve wastewater quality. The pilot plant process consists of a mixer, sedimentation tank, main pump, and continuous adsorption operation as show in Figure 2. A new coagulating agent as 'nanomaterial' was

[1] Petroleum and Petrochemical College, Chulalongkorn University, Soi Chulalongkorn 12, Phayathai road, Pathumwan, Bangkok 10330, Thailand.
Emails: rathanawan.k@chula.ac.th; Lwasuthep@gmail.com

[2] Packaging and Materials Technology, Faculty of Agro-Industry, Kasetsart University, 50 Ngam Wong Wan Rd., Ladyaow Chatuchak Bangkok 10900.

* Corresponding author: fagitnn@ku.ac.th

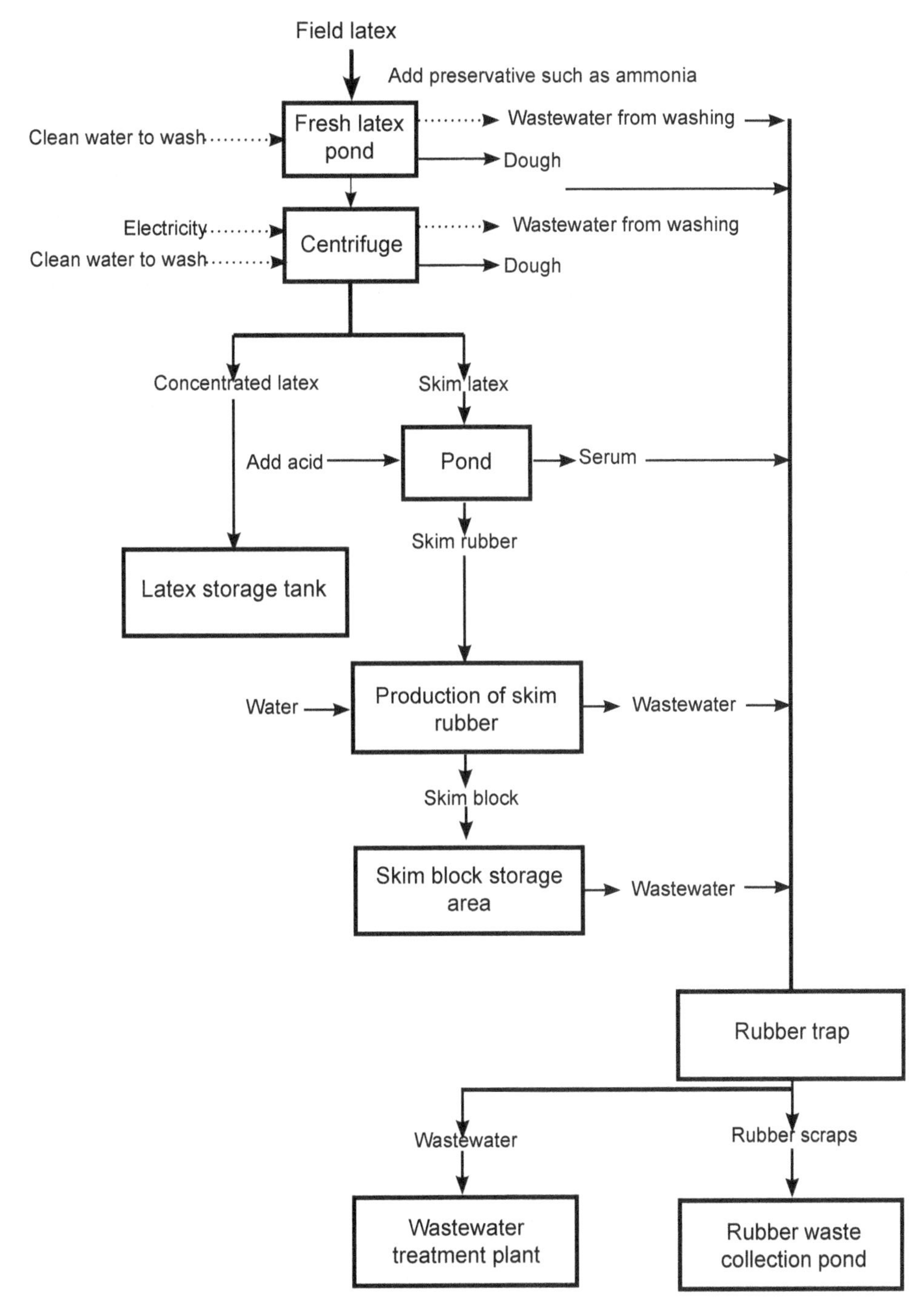

Figure 1. Latex production process.

used to capture the fine rubber particles and then coagulate the skim rubber nanocomposite into 'rubber nanocomposite masterbatch' as a valuable by-product for the rubber and plastic industries. The quality of wastewater improved because of the large reduction in BOD and COD from more than one hundred thousand to few tens of thousands or less. Thus, wastewater from rubber latex industries becomes less

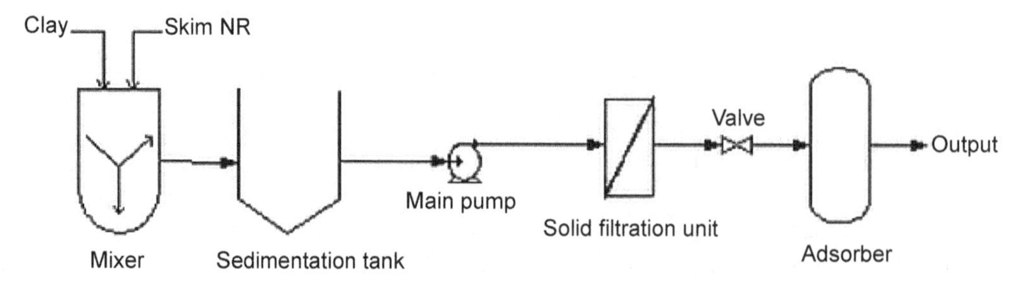

Figure 2. Waste water treatment process by modified clay.

of a burden, easier to treat and more environment friendly, leading to a reduction in both time and cost of wastewater treatment. In 2006, Nampitch and her co-worker reported the usage of organoclay to coagulate skim NR particles as a wastewater treatment process in the rubber industry. The effectiveness of organoclay to coagulate and separate skim NR particles from the serum by the separation process was demonstrated. Wastewater treated by this process is of better quality (Table 1). The characteristics of skim NR latex and wastewater after treatment with various organoclays prepared in our laboratory at 10% clay loading through pilot scale treatment are shown in Figure 1. The quality of skim rubber latex treated by commercial organoclays Cloisite 15A, Cloisite 20A, Cloisite 25A, and DDA-MMT are exhibited in Tables 2 and 3 (Nampitch and Magaraphan 2011).

Table 1. The quality of treated skim rubber latex.

Water Quality Index	Characterization Method	Standard (mg/L)	Skim Rubber Latex (mg/L)	Treated Skim Rubber Latex (mg/L)
pH value	pH Meter	5.5–9.0	10.1	8.48
Biochemical Oxygen Demand: BOD_5	5-Days BOD Test 5210 B.	< 20 mg/L	24,900	16,740 (–8,160 mg/L, 32.77% removal)
Chemical Oxygen Demand: COD	Open Reflux Method 5220 B.	< 120 mg/L	131,868	15,015 (–116,853 mg/L, 88.61% removal)
Total Dissolved Solids: TDS	Total Dissolved Solids Dried at 180°C 2540 C.	< 3,000 mg/L	52,500	24,800 (–27,700 mg/L, 52.76% removal)
Total Suspended Solids: SS	Total Suspended Solids Dried at 103–105°C 2540 D.	< 50 mg/L	3,413	874 (–2,539, 74.39% removal)
Sulfide: H_2S	Iodometric Method 4500-S2-F.	< 1.0 mg/L	115	29.3 (–85.7, 74.52% removal)
Total Kjeldahl Nitrogen: TKN	Macro-Kjeldahl Method 4500-Norg B.	< 100 mg/L	12,606	3,525 (–9,081, 72.04% removal)
Ammonia	Ammonia-Selective Electrode Method 4500-NH_3 D.	–	5,200	1,665 (–3,535 mg/L, 67.98% removal)
Fat, Oil and Grease	Partition-Gravimetric Method 5520 B.	< 5.0 mg/L	270	142 (–128 mg/L, 47.41% removal)
Heavy Metal				
Zn	Atomic Absorption Spectrometric Method 3500-Fe B. (3111B.)	< 5.0 mg/L	250	7.072
Cu		< 2.0 mg/L	0.25	< 0.10
Mn		< 5.0 mg/L	< 0.1	< 0.10
Fe		–	6.85	1.306

Table 2. Characteristic of skim NR latex and wastewater after treatment with various organoclay at 10% clay loading (Nampitch and Magaraphan 2011).

Before treatment	After treatment with various organoclay	Quality indicator							
		BOD (mg/L)	COD (mg/L)	%BOD removal	%COD removal	BOD/COD	% Transmittance at 650 nm (dilute factor 0.1)	Magnesium content (ppm)	%Ash
Skim NR latex		19250	132496				3.062	2.533	0.387
	Closite 15A	10600	45472	44.97	65.68	0.15/1	0.883	11.855	0.4532
	Closite 20A	9800	55664	49.09	57.99	0.23/1	0.805	4.995	0.4261
	Closite 25A	9800	65856	49.09	50.3	0.15/1	62.35	6.988	0.901
	DDA-MMT	12800	87808	33.51	33.73	0.15/1	79.33	3.424	0.5699

Table 3. Soluble proteins' contents and % protein loss of organoclay nanocomposites at various percent clay loading (Nampitch and Magaraphan 2011).

Organoclay composite	soluble proteins' contents (μg/g)						% protein loss					
	0% clay	1% clay	3% clay	5% clay	10% clay	20% clay	0% clay	1% clay	3% clay	5% clay	10% clay	20% clay
Skim NR	13,289											
Skim NR/Closite 15A		940	647	811	4,244	3,178		92.9265	95.1313	93.8972	68.0638	76.0855
Skim NR/Closite 20A		1,344	308	394	340	357		89.8864	97.6823	97.0427	97.4415	97.3136
Skim NR/Closite 25A		1,206	476	351	343	442		90.9248	96.4181	97.3587	97.4189	96.6739
Skim NR/DDA-MMT		105	251	187	112	113		99.2099	98.1112	98.5928	99.1572	99.1497

Quality indicators used in this research were BOD, COD, %BOD, %COD, BOD/COD, %transmittance, magnesium content, %ash, soluble protein content and %protein loss. The quality of wastewater skim NR latex before treatment improved after treatment with various organoclays as Cloisite 15A, Cloisite 20A, Cloisite 25A, and DDA-MMT. Some organoclays including Cloisite 15A, Cloisite 20A, and Cloisite 25A were sourced from Southern Clay (USA), whereas others were produced from the intercalation of alkylammonium ions into the clay interlayers as shown in Figure 3.

The %transmittance and %ash of wastewater improved dramatically when employing DDA-MMT, with reduction of BOD and COD at 33–49% and 34–66%, respectively. Nowadays, companies such as M^2 Polymer Technologies, Inc. use organoclay to treat various types of wastewater. The innovation and development of wastewater treatment processes in industry have greatly reduced environmental pollution. Commercial organoclays such as Cloisite 15A, Cloisite 20A, and Cloisite 25A are obtained originally from montmorillonite, with bentonite (7 baht/kg) as a cheaper source. Bentonite produced in Lopburi Province, Thailand can be modified with various surfactants to obtain new organoclays which can be effectively used as the water treatment agents Dehyquart A-CA, Varisoft PATC, Quaternium 18, and Dehyquart F-75. Skim NR latex was first stirred in a mixer with various organoclays (Figure 2). After that, a sedimentation tank was used to separate the serum and sludge. The sludge was separated from this process to prepare nanocomposites as masterbatches for blending with various types of polymer to improve the properties of various products. The serum obtained consisted of a variety of compositions as shown in Table 4.

The serum composition obtained from skim NR is shown in Table 4. This serum can be used as nitrogen fertilizer, antioxidant, and nutrient agar, while it also contains the effective bacteria that can be developed to produce the biodegradable plastic PHA. After separation in the sedimentation tank, the serum was passed through the adsorber containing packing media obtained in the laboratory (Figure 4). The rubber nanocomposite as a sludge in the sedimentation tank can be used to blend with

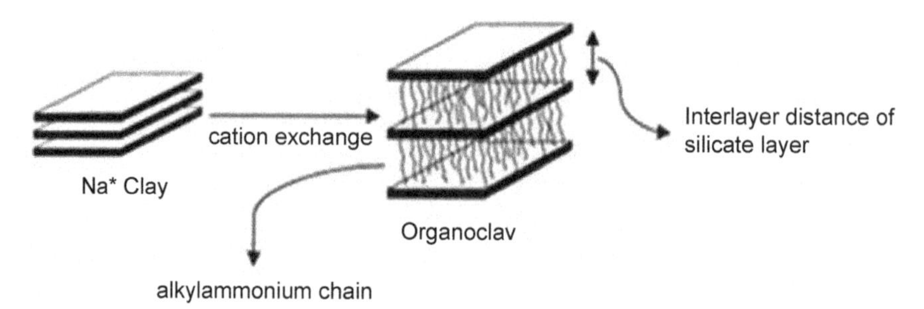

Figure 3. Schematic representation of intercalated alkylammonium ions into interlayer of clay (Nampitch and Magaraphan 2011).

Table 4. The composition of serum obtained from skim NR.

Parameter	Unit
pH	9.45
Total sugars	6.87 g/L
Reducing sugars	0.26 g/L
Protein	0.12 g/L

Figure 4. Packing media.

various chemicals and polymers to obtain heat-resistant products such as rubber gaskets, rubber paving blocks, and engine mounts. Many case studies have demonstrated polymer-clay nanocomposites as materials with improved thermal properties. Rubber nanocomposites can also be blended with polystyrene to produce high impact polystyrene products.

Organoclay

Clay minerals consist of hydrous aluminum silicates that possess high plasticity. Clays are plastic when mixed with water and become brittle when dry. Clay mineral structures are mostly sheet silicates grouped into phyllosilicates consisting of Al-octahedral and Si-tetrahedral sites. Thailand contains large clay resources (Limpanart et al. 2006). Clay minerals are used to produce ceramic products such as dishes, plates, memorial gifts, and sanitary ware. Nowadays, new technology in the development of these clay minerals has resulted in organoclay which is widely used in diverse industries. Dr. John W. Jordan has researched organoclay engineering materials since 1941 as substances to increase viscosity. Clay minerals are hydrous aluminum phyllosilicates with various amounts of iron, alkali metals, alkaline earths, and other cations as Na^+, Mg^+ and Ca^+ existing in the interlayer of clays. The cations are used to replace ion-exchange reactions with cationic surfactant and modification of these clay minerals leads to organoclay which possesses hydrophilic or organophilic surface properties resulting in good dispersion of organoclay in the organo-matrix. The cation exchange process is a specific characteristic of each clay mineral which depends on the capability of ion-adsorption and still remains during the occurring cationic exchange process. The cations or molecules most commonly used are Na^+, Ca^{2+}, Mg^{2+}, H^+, K^+ and NH_4^+, and are called exchangeable cations. The cationic exchange process reaction is shown below (Limchareonvanich et al. 2003).

$$X^{\bullet}\ \text{clay} + Y^+ \rightarrow Y^{\bullet}\ \text{clay} + X^+$$

Where X^+ is the cation contained inside the space between layers of clay

Y^+ is the ion of surfactant molecules

The reaction occurs from left to right and depends on the nature of X^+ and Y^+. Figure 5 shows an ion-exchange reaction resulting in the replacement of cations contained inside the space of the clay layer with cations of the surfactant molecules, which leads to increase of the interlayer space.

Some authors have studied the arrangement of surfactant molecules inside the galleries of layered silicates as shown in Figure 6 (Lagaly 1982, Bonczek et al. 2002, Lee and Kim 2002). They revealed that

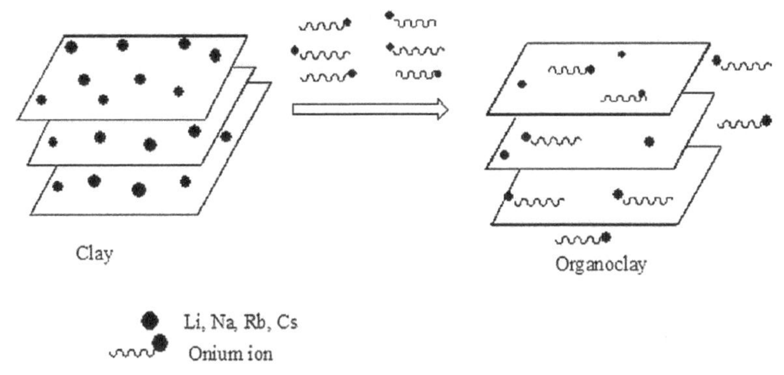

Figure 5. The model of ion-exchange reaction between surfactant molecules and cations.

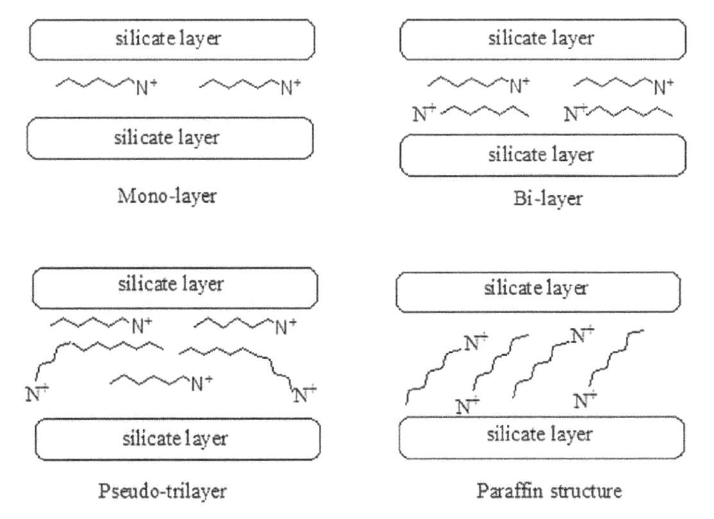

Figure 6. The arrangement of surfactant molecules in the interlayer of clay.

molecules of hexadecyltrimethylammonium bromide (HDTMA) have four arrangements as monolayer, bilayer, pseudotrimolecular layer, and paraffin complex with d-spacing lattice parameters as 1.37, 1.77, 2.17, and 2.20 nm, respectively.

Preparation of organoclay

Organoclay was prepared by a cationic exchange process in an aqueous solution with vigorous stirring of 1 g of sodium montmorillonite which was dispersed in 30 ml of distilled water for 24 hours with three surfactants which were prepared with excess concentrated 1 N HCl. The precipitate was filtered and washed with hot distilled water until no chloride was detected with 1 N $AgNO_3$ solution. It was then dried at 60°C for 24 hours. The organophilic clay was crushed with a mortar and particles were collected.

Characterization of prepared organoclays

Determination of cation exchange capacity (CEC)

The CEC is the number of positive charges (cations) which a representative sample can hold. CEC was determined according to ASTM C837-81 (Standard test method for methylene blue index of clay). The clay was first dried at 105 ± 5°C for 24 hours to remove humidity, and then 2 g of clay was blended with 30 ml of distilled water in a 600 ml beaker at room temperature. The solution was adjusted to pH 2.5–3.8 with 0.1 N HCl. Methylene blue solution in a burette was then added by titration. The solution was dropped on filter paper to determine the endpoint and the methylene blue index (MBI). The equation to calculate the MBI is shown below.

$$\mathrm{MBI} = \frac{\mathrm{E} \times V}{W} \times 100 \tag{1}$$

MBI = methylene blue index (meq/100 g)
E = concentration methylene blue (meq/ml)
V = Volume of methylene blue solution (ml)
W = Weight of sodium bentonite (g)

The CEC of the original unmodified Na^+ clay with different experiments is shown in Table 5. The CEC of unmodified Na^+ clay was equal to 49.74 meq/100 g clay, and the chemical compounds of the original unmodified Na^+ clay are shown in Table 6. The original unmodified Na^+ clay contained various

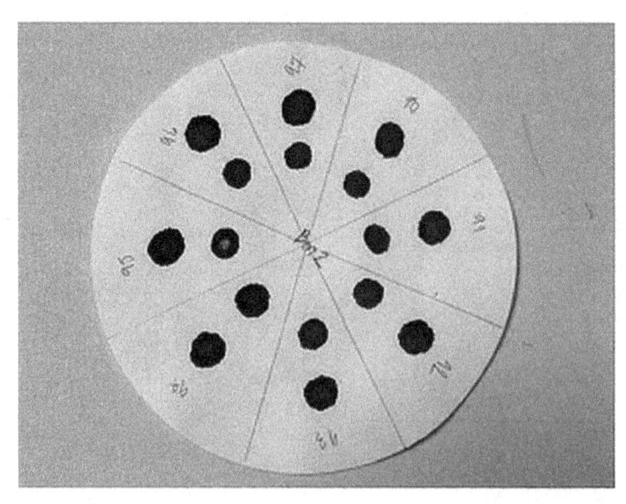

Figure 7. Representative of the drop of methylene blue in filter paper (Nampitch and Magaraphan 2013).

Table 5. Cation exchange capacity (CEC) of the original unmodified Na^+ clay.

The Number of Experiments	The Weight of Original Unmodified Na^+ clay (g)	The Amount of Methylene Blue (ml)	Cation Exchange Capacity (CEC) (meq/100 g)
1	2.0044	100	49.89
2	2.0883	103	49.32
3	2.0266	101	49.84
4	2.0242	101	49.90
5	2.0105	100	49.74
Average			49.74

Table 6. The chemical compound of the original unmodified Na^+-clay.

Chemical compound (on dry basis)	%
Silicon dioxide (SiO_2)	46 – 60
Aluminum oxide (Al_2O_3)	14 – 17
Ferric oxide (Fe_2O_3)	6 – 8
Sodium oxide (Na_2O)	0.5 – 1.5
Magnesium oxide (MgO)	1.5 – 3.0
Calcium oxide (CaO)	1.0 – 2.5
Potassium oxide (K_2O)	0.1 – 1.0
Titanium dioxide (TiO_2)	0.2 – 1.5
Loss on ignition (LOI)	7 – 12
Moisture content	8 – 12 %
Dry particle size	Min. 75% pass through 200 mesh

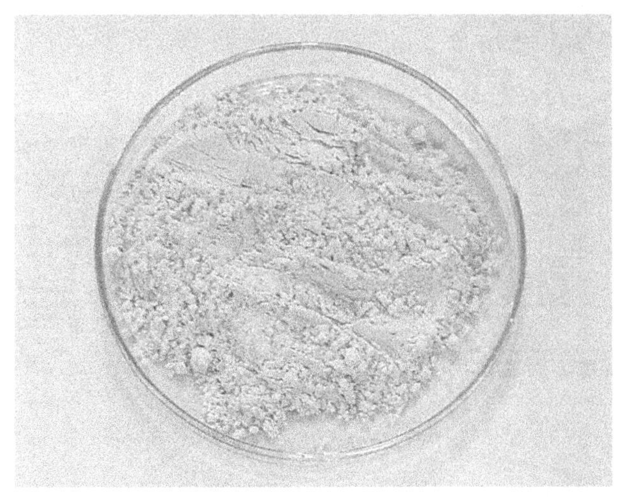

Figure 8. The original unmodified Na^+ clay (Nampitch and Magaraphan 2013).

components such as silicon dioxide (SiO_2), aluminum oxide (Al_2O_3), ferric oxide (Fe_2O_3), sodium oxide (Na_2O), magnesium oxide (MgO), calcium oxide (CaO), potassium oxide (K_2O), and titanium dioxide (TiO_2) with maximum content of silicon dioxide.

X-ray diffraction measurements

The dried rubber sheet was characterized by X-ray Diffraction (XRD) to determine the structures of the prepared organoclays and bentonite. Samples were tested by a diffractometer (Bruker AXS Model D8) with 2 theta ranging from 1.5 to 30° by K-α radiation (λ = 1.542 Å). The scanning rate was 0.02°/sec. Bragg's law was employed to determine the d-spacing of the silicate layers as shown in Eq. 2:

$$\lambda = 2d\sin\theta \tag{2}$$

where λ corresponds to the wavelength of the X-ray radiation used, d corresponds to d_{001}, and θ is the measured diffraction angle. The interlayer distance of the silicate layers (i.e., d-spacing or basal spacing, d_{001}) defined as the width between two adjacent silicate layers was determined from XRD evaluation.

Nanocomposite materials are prepared from the combination of two or more components at the nano scale (10^{-9} m). The added substances may be organic, inorganic, amorphous, crystalline, or semi-crystalline. The objective of the incorporation of these components is to obtain improved material properties and many materials are used to reinforce the matrices. Nowadays, there is great interest regarding the usage of organoclay nano-materials as reinforcement. The various dispersion characteristics of organoclay in the matrix lead to different properties of the nanocomposite. Incorporation of organoclay and polymer was studied to obtain the nanocomposite materials. Dispersion of organoclay in a polymer matrix is classified into three types as intercalation of the polymer chains between the layered clays (Figure 10). The absence of intercalation of the polymer chains into the layered clays leads to conventional composites (Figure 9).

Table 7 shows the interlayer distances of bentonite and the organoclays Dehyquart A-CA, Quaternium-18, Varisoft PATC, and Dehyquart F-75. The basal spacing of bentonite expanded from 12.85 Å to 40.87 Å, 35.40 Å, 41.64 Å and 35.55 Å for Dehyquart A-CA, Quaternium-18, Varisoft PATC, and Dehyquart F-75, respectively, because of the ion exchange reaction between clay and alkylammonium substance, which led to the change in surface characteristics from hydrophilic to hydrophobic. The surfactant molecules were intercalated as a paraffin complex inside the clay layers. There was an attractive force between the anionic surface of the clays and the positively charged head of surfactant molecules.

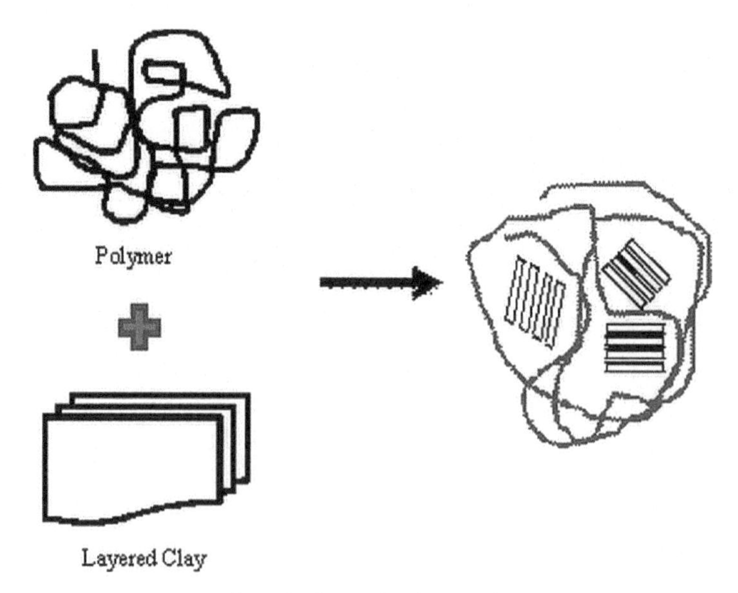

Figure 9. Conventional composite.

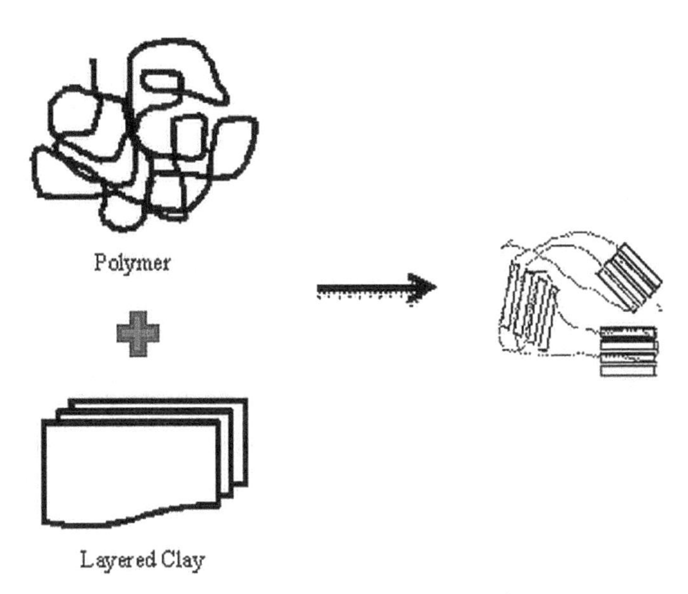

Figure 10. Intercalated clay nanocomposite.

Table 7. The gallery height of the layered structure of various types of organoclays.

Type	Nominal d(001) (A°)	Δd(001) (A°)
Bentonite	12.85	0
Dehyquart A-CA	40.87	+ 28.02
Quaternium-18	35.40	+ 22.55
Varisoft PATC	41.64	+ 28.79
Dehyquart F-75	35.55	+ 22.70

These results corresponded with Lagaly (Lagaly 1982) and Bonczek (Bonczek et al. 2002), for the value of d_{001} obtained from the XRD technique corresponded to the conformation of surfactant molecules inside the clay layers. A value of d_{001} higher than 22 Å corresponded to the conformation of surfactant molecules inside the clay layers as paraffin complexes.

Determination of particle size and particle size distribution

Determination of particle size and particle size distribution of prepared organoclay with sizes ranging from 0.05 to 900 microns was analyzed by Mastersizer 2000 (Malvern).

Table 8 shows the results of particle size and particle size distribution of the modified organoclays. Less than 10, 50, and 90% by volume of Dehyquart A-CA showed particle sizes equal to 3.18, 1.93, and

Table 8. Particle size and particle size distribution of various organoclays modified by surfactants.

Type of Organoclay	Particle Size (μm)		
	10 percent by volume	50 percent by volume	90 percent by volume
Dehyquart A-CA	3.18	10.93	54.36
Quaternium-18	3.52	20.07	84.01
Varisoft PATC	2.66	9.67	27.69
Dehyquart F-75	3.91	20.68	84.63

54.36 μm, respectively. Particle size distribution of Dehyquart A-CA is shown in Figure 11. Less than 10, 50, and 90% by volume of Quaternium-18 particles showed particle size equal to 3.52, 20.07, and 84.01 μm, respectively. Particle size distribution of Quaternium-18 is shown in Figure 12. Particle size distributions of Varisoft-PATC and Dehyquart F-75 are shown in Figures 13 and 14.

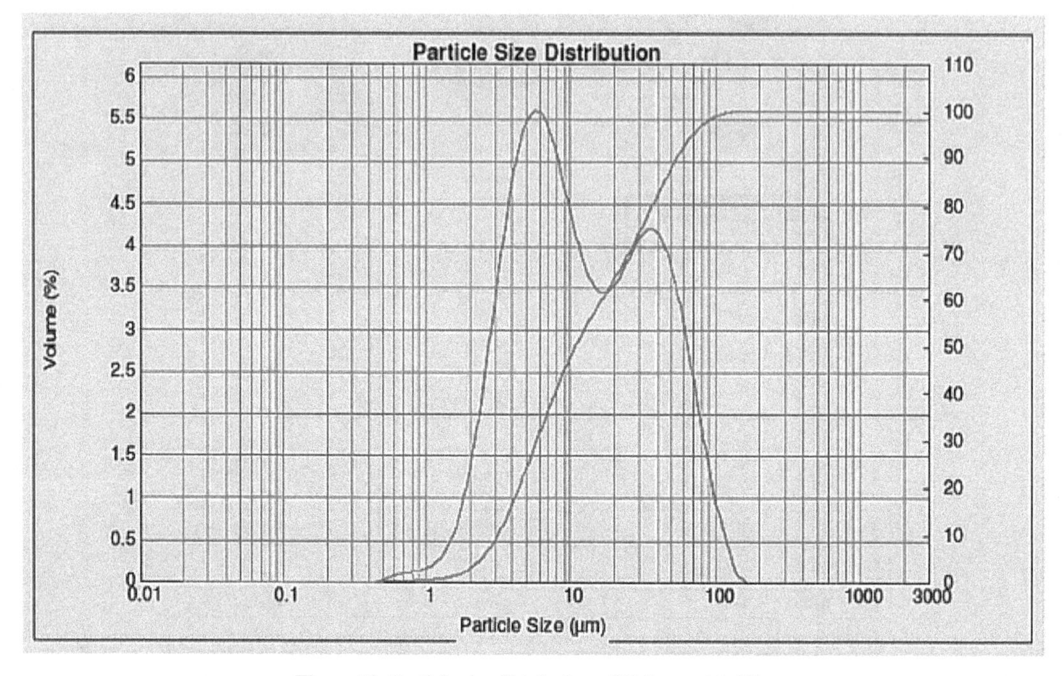

Figure 11. Particle size distribution of Dehyquart A-CA.

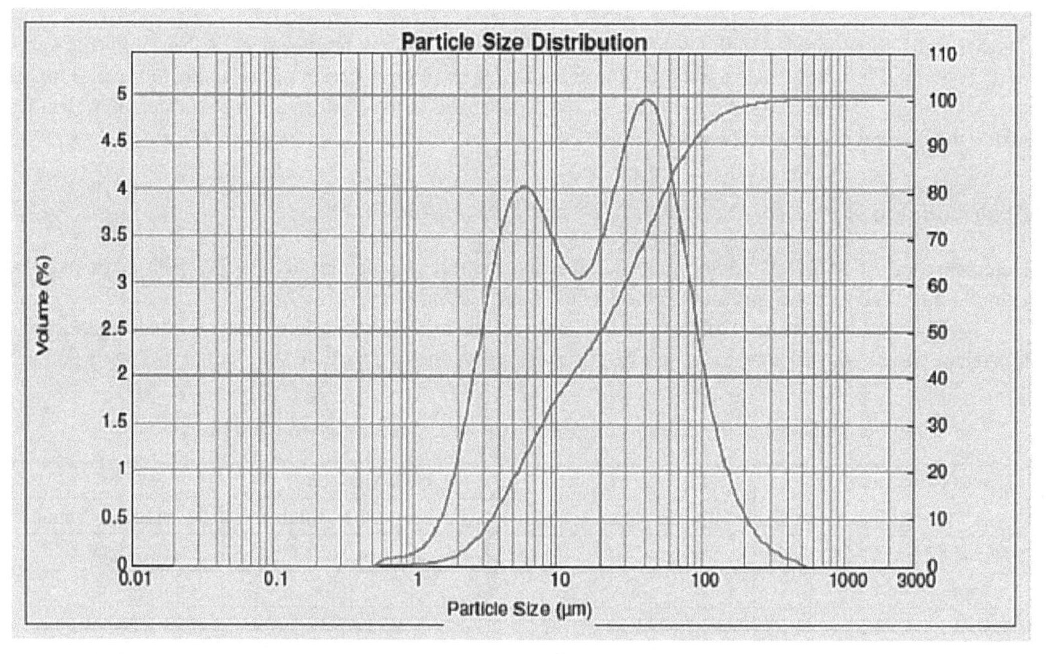

Figure 12. Particle size distribution of Quaternium-18.

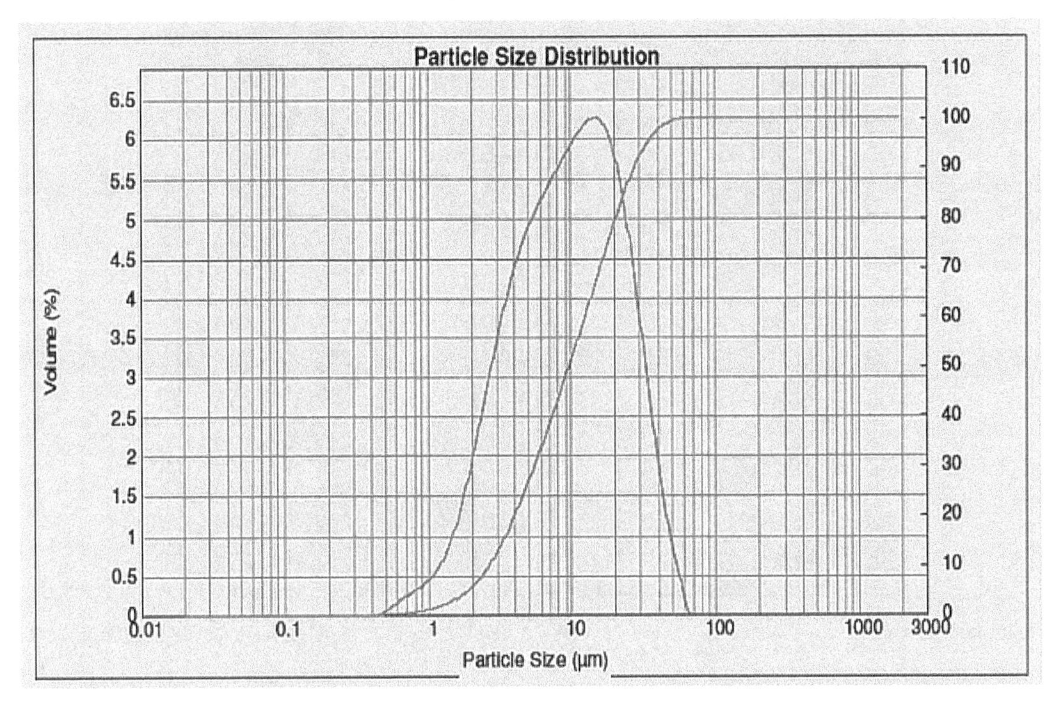

Figure 13. Particle size distribution of Varisoft-PATC.

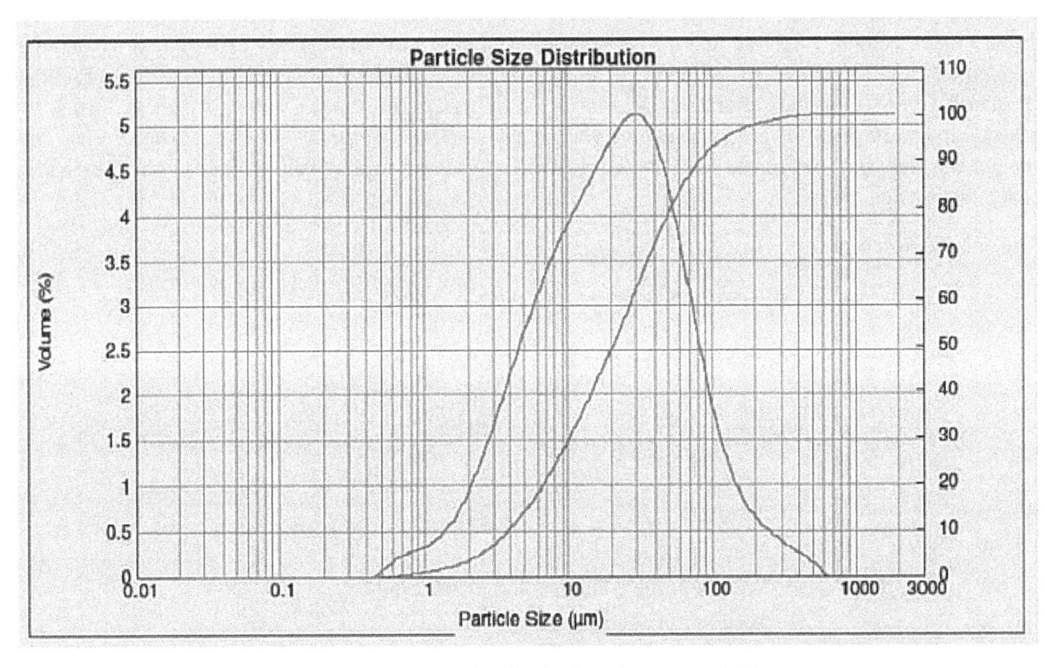

Figure 14. Particle size distribution of Dehyquart F-75.

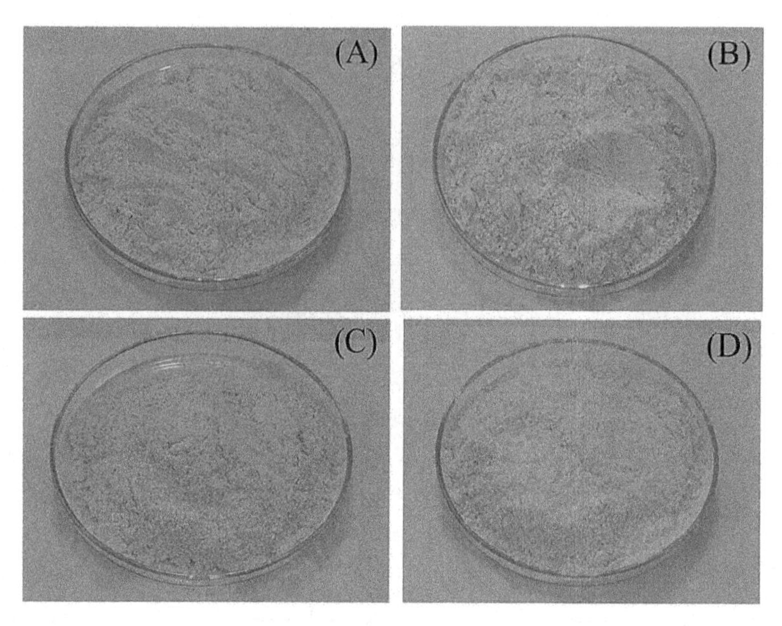

Figure 15. Representative of organoclay, (A) Dehyquart A-CA, (B) Dehyquart F-75, (C) Quaternium 18 and (D) Varisoft-PATC (Nampitch and Magaraphan 2013).

Determination of density and water content

The density of organoclay was analyzed by an Ultrapycnometer 1000 (Quantachrome) using the gas displacement technique. Helium gas (99.99%) at 22.5–22.6°C was employed to determine dried sample in microcell. Water content of the prepared organoclays was determined according to ASTM D-2216 (Standard Test Method for Laboratory Determination of Water (Moisture) Content of Soil and Rock by Mass). Firstly, 20 g of prepared organoclay was dried at 110 ± 5°C for 12–16 hours. After drying, the sample was kept in a desiccator until cooled. Finally, the sample was reweighed to calculate the water content using Eqs. 3–5.

The calculation of water content

$$W_{water} = W_{container+wet\ sample} - W_{container+dry\ sample} \tag{3}$$

$$W_{dry\ sample} = W_{container+dry\ sample} - W_{container} \tag{4}$$

$$m = W_{water}/W_{dry\ sample} \times 100 \tag{5}$$

Where

W_{water}	=	Water weight
$W_{container+wet\ sample}$	=	Weight of the container and a wet sample
$W_{container}$	=	Weight of the container
$W_{dry\ sample}$	=	Weight of a dry sample
m	=	Moisture content

Table 9 shows the density and water content (%) of various organoclays. The densities of Dehyquart A-CA, Quaternium-18, Varisoft PATC, and Dehyquart F-75 were 1.92, 1.71, 1.99, and 1.61 g/cm^3,

Table 9. Density and water content (%) of various organoclays.

Type of Organoclay	Density (g/cm^3)	Water Content (%)
Dehyquart A-CA	1.92	3.1%
Quaternium-18	1.71	2.8%
Varisoft PATC	1.99	2.2%
Dehyquart F-75	1.61	2.5%

respectively. Experimental results according to ASTM D-2216 showed water content (%) of the various clays as 3.1%, 2.8%, 2.2%, and 2.5% for Dehyquart A-CA, Quaternium-18, Varisoft PATC, and Dehyquart F-75, respectively.

Thermal properties of organoclay

Thermal properties of prepared organoclays were characterized by thermogravimetric analysis (TGA). Prepared organoclay (5 mg) was used to determine the thermal property. Thermogravimetric analysis was conducted using a Thermal Analyzer in the temperature range of 50–1,000°C at a heating rate of 10°C/min. Figure 16 shows the TGA results, leading to the determination of composition, as shown in Table 10.

Composition and content of skim NR and organoclay in skim NR-organoclay nanocomposites obtained from sludge in the sedimentation process are shown in Table 10. Skim NR-Dehyquart A-CA, skim NR-Dehyquart F-75, skim NR-Varisoft-PATC, and skim NR-Quaternium 18 nanocomposites contained skim NR content at 45.5, 51.3, 47.4, and 59.2% by weight, respectively.

These skim NR-organoclay nanocomposites could be used as a masterbatch to blend with other polymers such as polystyrene to improve the brittleness of this polymer. Figure 17 exhibits product examples such as plates, containers, rulers, cups, decorations, and floor blocks obtained from polystyrene-skim rubber nanocomposites. These product examples were tested according to various ASTM standards and their properties are shown in Table 11.

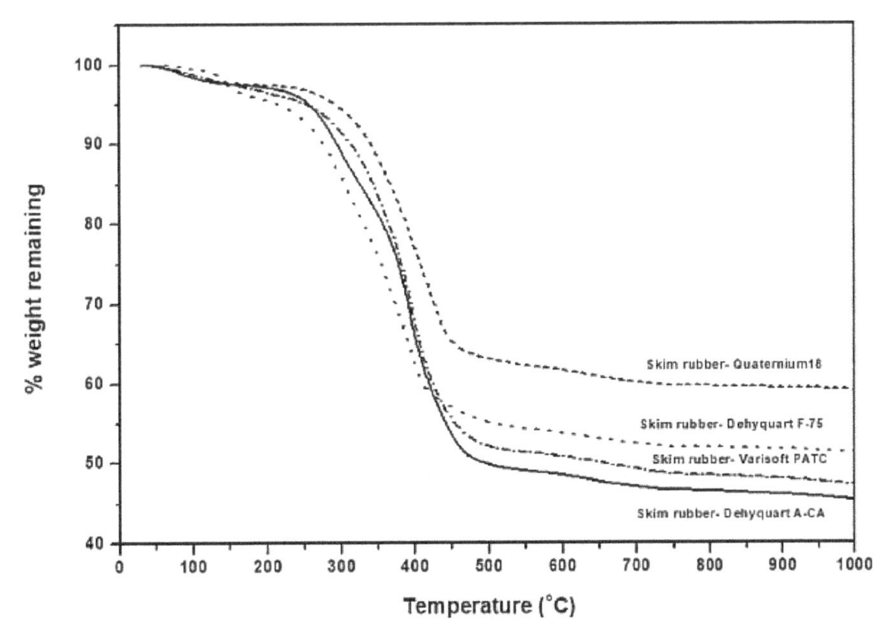

Figure 16. TGA results of skim NR-organoclay nanocomposites.

Table 10. The composition of skim NR-organoclay nanocomposites obtained from sludge in sedimentation process.

Type of Skim NR/Clay	Composition	Percent by Weight
Skim NR/Dehyquart A-CA	Skim NR content (decompose at 300–600°C)	35.6
	Dehyquart A-CA content (composition cannot decompose)	45.5
Skim NR/Dehyquart F-75	Skim NR content (decompose at 300–600°C)	41.9
	Dehyquart F-75 content (composition cannot decompose)	51.3
Skim NR/Varisoft-PATC	Skim NR content (decompose at 300–600°C)	44.6
	Varisoft-PATC content (composition cannot decompose)	47.4
Skim NR/Quaternium 18	Skim NR content (decompose at 300–600°C)	35.4
	Quaternium 18 content (composition cannot decompose)	59.2

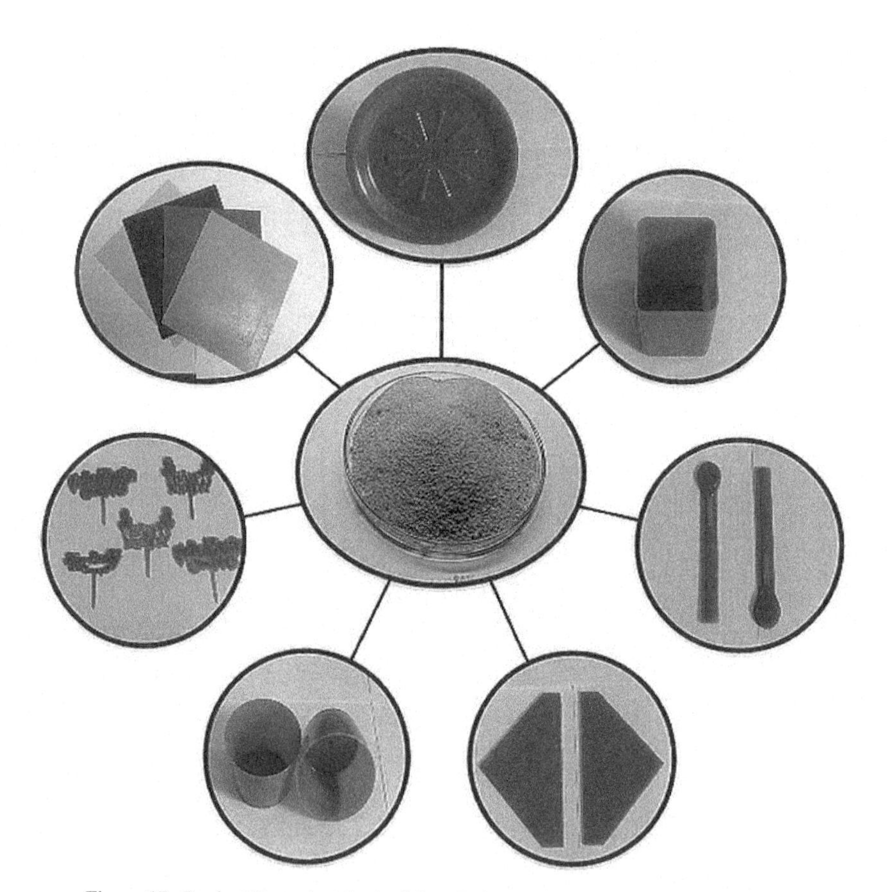

Figure 17. Product Examples obtained from Polystyrene-skim rubber nanocomposite.

Table 11. The properties of product examples obtained from polystyrene-skim rubber nanocomposite.

Properties	
Specific Gravity: ASTM D-792	1.03
Tensile Strength @ Yield: ASTM D-638	6,073 psi
Tensile Modulus: ASTM D-638	86,080 psi
Flexural Strength @ Yield: ASTM D-790	13,448 psi
Flexural Modulus: ASTM D-790	485,483 psi
Izod Impact: ASTM D-256	1.52 ft-lbs/in
Hardness (Rockwell "M")	66.68
Heat Deflection Temperature (264 psi, unannealed): ASTM D-648	191 °F

Determination of kinetic model

Determination of kinetic models of adsorption show three models as first-order Lagergren equation, pseudo-second-order equation, and second-order rate equation (Paune et al. 1998, Ake et al. 2003, Tsai et al. 2009) as shown in Equations 6, 7, and 8.

$$Log\frac{(q_e - q_t)}{q_e} = Log\, q_e - \frac{K_L}{2.303}t \quad \text{first order Lagergren equation} \quad (6)$$

$$\frac{t}{q_t} = \frac{1}{K/q_e^2} + \frac{t}{q_e} \quad \text{pseudo-second-order equation} \quad (7)$$

$$\frac{1}{(q_e - q_t)} = \frac{1}{q_e} + K_2 t \quad \text{The second-order rate equation} \quad (8)$$

Where K_L = Lagergren rate equation of adsorption (min^{-1})

k' = The pseudo-second-order rate constant of adsorption ($g\ mg^{-1}\ min^{-1}$)

q_e and q_t = The amount of metal ion sorbed (mg/g) at equilibrium and at time t, respectively

k_2 = The second-order rate constant ($g\ mg^{-1}\ min^{-1}$)

A kinetic study of the removal of rubber particles from skim rubber latex by batch adsorption was conducted as shown in Table 12. The kinetic models, equilibrium time, amount of adsorption at equilibrium (q_e), and the Lagergren rate equation of adsorption (K_L) as a first-order equation of various organoclays are shown in Table 12.

After the sedimentation process, this research obtained sludge containing skim NR-organoclay nanocomposites that could be blended with polystyrene to develop high impact polystyrene which can be

Table 12. Kinetic models of adsorption in various types of organoclays.

Type of Organoclay	**Kinetic Model**	**R^2**	**Equilibrium Time (hours)**	**q_e (mg/g)**	**Lagergren Rate Equation of Adsorption (K_L)**
Dehyquart A-CA	First order Lagergren kinetic	0.8800	40	980.86	9.212×10^{-4}
Dehyquart F-75	First order Lagergren kinetic	0.9017	40	80.27	9.212×10^{-4}
Quaternium 18	First order Lagergren kinetic	0.989	40	259.48	1.1515×10^{-3}
VARISOFT-PATC	First order Lagergren kinetic	0.9884	40	919.88	1.1515×10^{-3}

injected to obtain various products. Figure 18 illustrates the process for skim rubber removal as a rubber nanocomposite masterbatch. The coagulation agents employed were Dehyquart A-CA, Dehyquart F-75, Quaternium 18, and Varisoft-PATC. After coagulation of the skim rubber particles with the prepared organoclays, a higher quality of wastewater was obtained. Skim NR-organoclay nanocomposites, as a by-product in the pilot process, were ground to blend with other polymers to improve the brittleness of materials (Figure 19).

Figure 18. The illustration of process for skim rubber removal as rubber nanocomposite masterbatch.

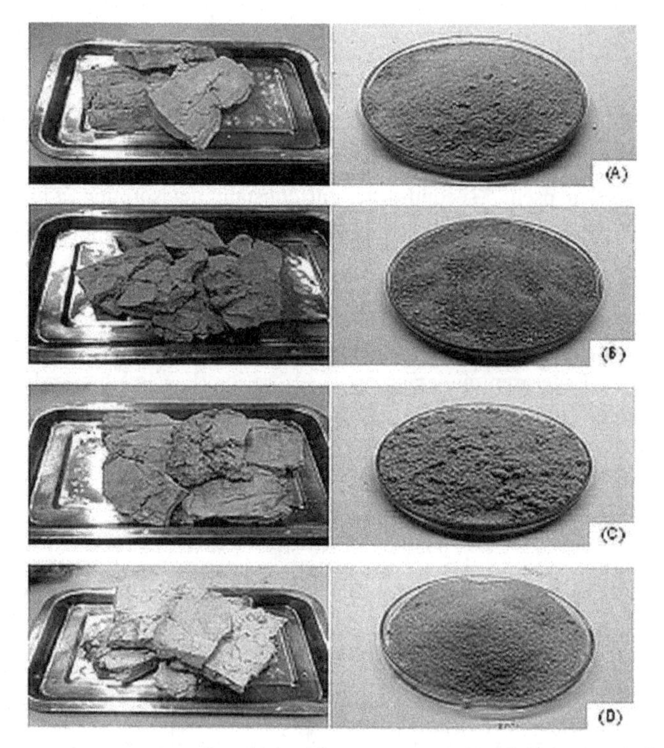

Figure 19. Skim NR-organoclay nanocomposites obtained from treatment with (A) Dehyquart A-CA (B) Dehyquart F-75 (C) Quaternium 18 (D) Varisoft-PATC.

Composition of the skim NR-organoclay nanocomposites was determined by thermogravimetric analysis at a temperature range of 50–1,000°C with a heating rate of 10°C/min under nitrogen gas at 25 ml/min.

Acknowledgements

The authors would like acknowledge the financial support of the Agricultural Research Development Agency (ARDA) and we also thank the Rubber Research Institute of Thailand for generously permitting the authors to use testing machines.

References

Ake, C.L., M.C. Wiles, H.J. Huebner, T.J. McDonald, D. Cosgriff, M.B. Richardson et al. 2003. Porous organoclay composite for the sorption of polycyclic aromatic hydrocarbons and pentachlorophenol from groundwater. Chemosphere 51: 835–844.

Bonczek, J.L., W.G. Harris and P. Nkedi-Kizza. 2002. Monolayer to bilayer transitional arrangements of hexadecyltrimethylammonium cations on Na-montmorillonite. Clays and Clay Minerals 50: 11.

Lagaly, G. 1982. Layer charge heterogeneity in vermiculites. Clays and Clay Minerals 30: 215–222.

Lee, S.Y. and S.J. Kim. 2002. Expansion characteristics of organoclay as a precursor to nanocomposites. Colloids Surf A Physicochem Eng Asp 211: 19–26.

Limchareonvanich, S., K. nopichetwattana and W. Fujitniran. 2003. The Study of the Concentration of Surfactant on the Preparation of Organoclay. 5–8. Department of Materials Science, Chulalongkorn University.

Limpanart, S., S. Khunthon and T. Srikhirin. 2006. Surface modification of bentonites. 4–10. The Metallurgy and Materials Science Research Institute, Chulalongkorn University: Chulalongkorn University.

Nampitch, T. and R. Magaraphan. 2011. Effect of coagulating skim Nr particles as Nr–clay nanocomposite: Properties and structure. Rubber Chem Technol 84: 114–135.

Nampitch, T. and R. Magaraphan. 2013. Adsorption of organic pollutants by organoclay for wastewater treatment in rubber industry, Bangkok, The agricultural Research Development Agency.

Paune, F., J. Caixach, I. Espadaler, J. Om and J. Rivera. 1998. Assessment on the removal of organic chemicals from raw and drinking water at a Llobregat river water works plant using GAC. Water Res 32: 3313–3324.

Tsai, W.-T., K.-J. Hsien and H.-C. Hsu. 2009. Adsorption of organic compounds from aqueous solution onto the synthesized zeolite. J Hazard Mater 166: 635–641.

7

Physicochemical Properties and Biological Activity of Polymethylmethacrylate/ Fullerene Composites

Olga V. Alekseeva, Nadezhda A. Bagrovskaya and *Andrew V. Noskov**

Introduction

The promising direction in polymer materials science is the fabrication of hybrid materials based on organic-inorganic systems. A number of publications deal with various types of fillers for polymers and various methods of composites production (Dittrich et al. 2013, Ayatollahi et al. 2011, Fonseca et al. 2013, Chen et al. 2007, Burnside and Giannelis 2000, Lazzeri et al. 2005, Meenakshi and Sudhan 2016, Hattab and Benharrats 2015, Alekseeva et al. 2014).

Currently, fullerene-containing polymers and composites have attracted much attention of researchers because of the prospective usage of these materials in electronics, biotechnology, optics, medicine, pharmaceutics, etc. Insertion of these fillers results in the modification of the original polymer matrix, which can lead to the creation of materials with improved physical and chemical properties, biological activity and the main service characteristics (mechanical, electrical).

There are two acknowledged procedures for the preparation of two different types of fullerene-containing polymers and related products. It is shown that fullerene doped polymeric materials can be produced by covalent bonding of fullerene molecules with polymeric circuits or as a result of formation of polymer-fullerene complexes due to donor-acceptor interactions (Sibileva et al. 2004). Meanwhile, noncovalent interactions of polymer with fullerene, in opinion of authors of ref. (Krakovyak et al.

G.A. Krestov Institute of Solution Chemistry, Russian Academy of Sciences, Akademicheskaya str., 1, Ivanovo, Russia.
Emails: ova@isc-ras.ru; sal@isc-ras.ru
* Corresponding author: avn@isc-ras.ru

2002), can provide uniform distribution of nano-carbonic particles in a polymeric matrix. It is suggested that significant importance in the non-covalent linkage of fullerenes in a composite is connected with aromatic rings present in the structure of a macromolecule (Krakovyak et al. 2002) which increases fullerene-connecting ability of polymers. Moreover, owing to their structural features, fullerenes are capable of multipoint noncovalent interactions. This phenomenon is of special significance for interaction of fullerene molecules with numerous electron-donor macromolecular fragments (Badamshina and Gafurova 2008).

Scientific interest in the doping of polymers with fullerene is likely related to the simplicity of fullerene incorporation, either in its native form or as solutions in organic solvents, and to the use of minor amounts of modifying agents. Furthermore, less significant changes in the electron structure and, hence, less dramatic changes in the characteristics of fullerene molecules in the absence of any covalent interaction between fullerene molecules and polymer chain fragments were expected. These modifiers, when introduced in small amounts (up to several percent), can be nucleating agents and can affect the degree of crystallinity of the polymer (Potalitsin et al. 2006).

Among the high-molecular compounds, which are widely used as matrices for fullerene composites, one can mark out polymethylmethacrylate (PMMA). The main advantages of this polymer are its accessibility, i.e., environmental resistance, ease, and low cost. PMMA is a transparent thermoplastic material that allows it to be successfully used in many industries: aviation, instrumentation, electronics, building sector, food industry, as well as in the production of medical supplies (Arshad et al. 2009, Harper 2000, Andrade et al. 2010, Thangamani et al. 2010).

Polymethylmethacrylate is a well-known film-forming polymer often used for different modifications with low molecular compounds of special properties, including fullerenes. PMMA is well dissolved in benzene, toluene, o-xylene which are also solvents for fullerenes. It is a widespread procedure of polymer-fullerene composite formation that consists of preparation of the base-polymer solution and fullerene solution in the same organic solvent and mixing them with following evaporation of the solvent.

In the present paper, we report on study of the thermal, electrical properties, and antioxidant effect of both polymethylmethacrylate films and PMMA/fullerene composite films, as well as on intermolecular interactions of macromolecules of polymer and filler in composite.

Polymethylmethacrylate ("Aldrich", US; $M = 1.2 \cdot 10^5$) and fullerenes C_{60} ("NeoTechProduct", Russia) were used. A solvent casting of perspective components from solutions was employed for preparing the mixtures of C_{60} with polymer. Preliminary purification of organic solvent (toluene) was made by standard techniques (Coetzee 1982). PMMA/fullerene composite films were being produced as follows. Fullerene batches were dissolved in toluene in the required concentrations. Then PMMA batches were dissolved in all obtained solutions, and the mixed solutions were stirred for about 1 day before being cast into thin films. After casting, the solvent was slowly evaporated over several days to produce the composite films. By this technique, samples of PMMA/fullerene composites with various filler percentage (up to 3 wt. %) were prepared. Unmodified PMMA film was made by the solution cast method as well.

DSC investigation of the polymethylmethacrylate films filled with fullerene

A research on the phase transition from the glassy state to elastic one was performed using DSC 204 F 1 apparatus (Netzsch, Germany) in argon atmosphere (15 ml/min). To remove solvent residues, the films were dried under vacuum (80 mm Hg) at 90°C for 4 hours. A stack of films with a diameter of 4 mm was placed in a press-fitted aluminum crucible covering the pierced lid. Most of the samples were in the range of 4–5 mg. Heating of the samples was carried out at up to 160°C with a scan rate of 10°C/min. An empty aluminium crucible was used as reference.

Phase transition from the glassy state to elastic one was characterized by the following parameters:

T_1 is the extrapolated temperature of the phase transition onset;

T_2 is the extrapolated temperature of the phase transition end;

T_g is the temperature of DSC curve inflection taken as the glass transition point.

Typical DSC curves for the films of pure polymer and composite filled with 0.05 and 0.1 wt. % of fullerene are shown in Figure 1(a). It is seen that for all examined materials there is a reversible phase transition from the glassy state to elastic one, which manifests itself as a step of heat flow in endothermic direction. This is the second-order phase transition.

However, the thermal behavior of composites containing 0.5, 1 or 3 wt. % of C_{60} is more complex. Namely, two steps are observed in DSC curves (Figure 1(b)). According to Tugov and Kostrykina (1989), the thermograms of this type are characteristic of filled polymers with two glass transition temperatures, which correspond to transitions in the "soft phase" and "hard phase". Glass transition temperature of "solid phase", T_g^{solid}, is usually above the glass transition of temperature "soft phase", T_g^{soft}, which is associated with lower mobility segments. For PMMA/C_{60} composites examined in this study, the difference is 20–30°C (Figure 1(b)).

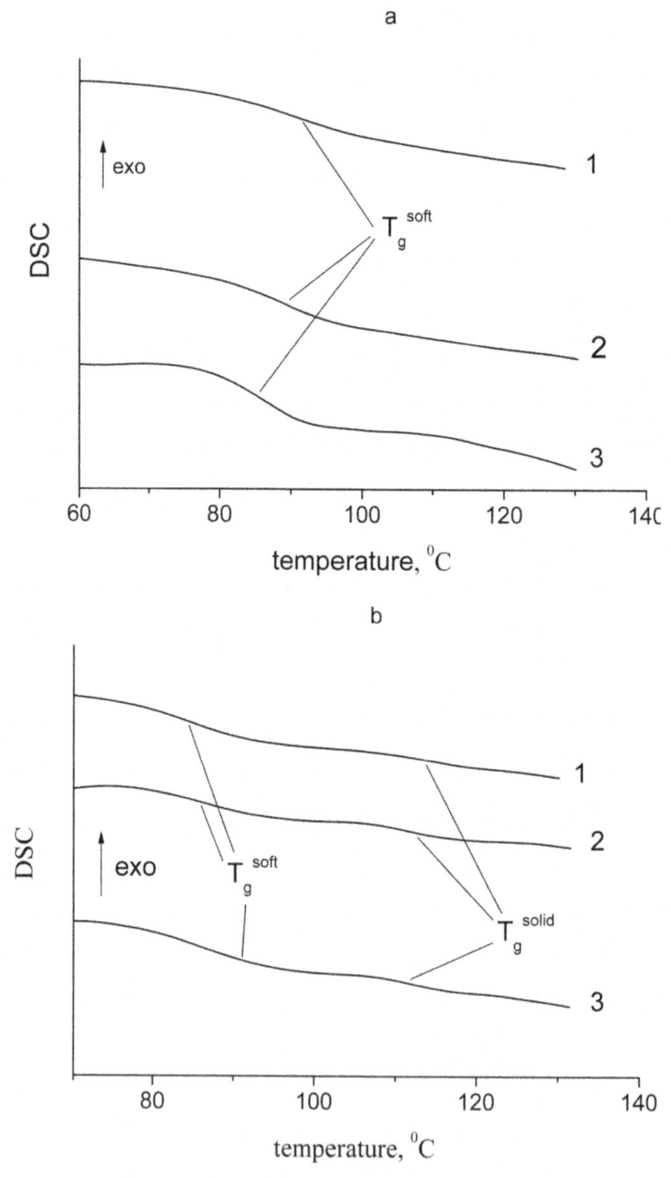

Figure 1. DSC curves of the PMMA/C_{60} composite films with various filler concentration, wt. %: (a) 0 (*1*); 0.05 (*2*); 0.1 (*3*); (b) 0.5 (*1*); 1 (*2*); 3 (*3*).

Table 1 shows the average values of the characteristic temperatures of phase transition for the "soft phase" obtained from the analysis of thermograms of the materials studied. It can be seen, for composites with a filler concentration in the range of 0.05 to 1 wt. %, value of T_g^{soft} is less than the corresponding value for the unmodified PMMA. This means that the fullerene particles have a plasticizing effect on the polymer structure. Moreover, the greatest effect is achieved at 0.5 wt. % of C_{60}: for such films, the glass transition temperature, T_g^{soft}, is minimal (Figure 2).

With further increase in the filler concentration, the T_g^{soft} value increases. For composite containing 3 wt. % of C_{60}, the glass transition temperature of "soft phase" exceeds the value of T_g^{soft} for pure PMMA indicating deceleration of polymer chains' mobility due to their interaction with carbon nanoparticles (Zanotto et al. 2012).

It can be seen in Table 1 that the rest of characteristic temperatures of phase transition for the "soft phase" (T_1, T_2), as well as T_g^{soft}, are non-monotonically dependent on the film composition in the investigated range of C_{60} concentrations (up to 3 wt. %). In this case, the minimum values of these parameters, as a minimum of T_g^{soft}, are at 0.5 wt. % of filler.

As noted above, for composites containing 0.5, 1 or 3 wt. % of fullerenes, two steps are observed in the DSC curve (Figure 1(b)) which corresponds to phase transitions in "soft phase" and "solid phase". Table 2 shows the values of the characteristic parameters of the transition for "solid phase" in PMMA/C_{60} composites researched. The analysis of these values shows that for films with fullerene concentration of 0.5–3 wt. %, this phase transition occurs in the temperature range of 107–118°C.

Table 1. Parameters of phase transition for "soft phase" in pure polymethylmethacrylate and composite films with various contents of fullerene.

C_{60} Content, wt. %	T_1, °C	T_g^{soft}, °C	T_2, °C
0.0	77.1	89.4	101.1
0.05	77.9	88.8	95.7
0.1	77.9	86.4	92.3
0.5	76.0	84.4	91.6
1	79.3	87.0	93.5
3	81.5	90.3	97.2

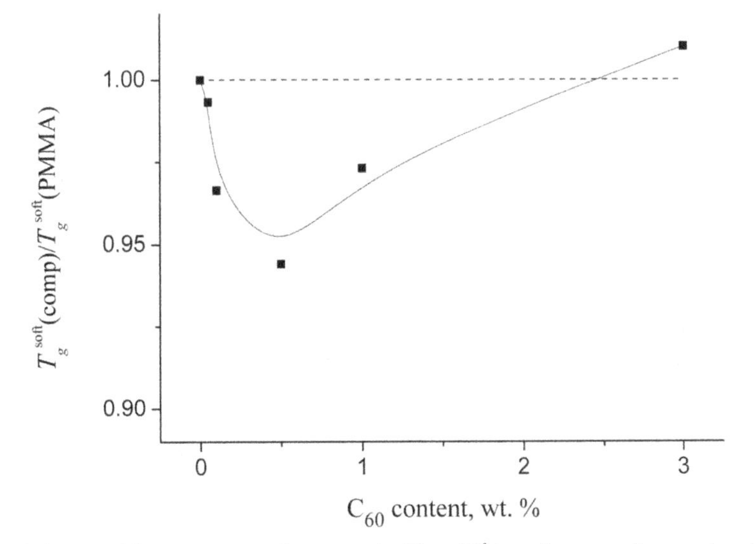

Figure 2. Ratio of glass transition temperatures for composite films, T_g^{soft}(comp), to one of pure polymethylmethacrylate film, T_g^{soft}(PMMA), vs. fullerene content.

Table 2. Parameters of phase transition for "solid phase" in pure polymethylmethacrylate and composite films with various contents of fullerene.

C_{60} Content, wt. %	T_1, °C	T_g^{solid}, °C	T_2, °C
0.5	108.9	114.7	118.2
1	107.0	111.3	116.3
3	107.6	111.5	116.1

Occurrence of the second glass transition temperature (T_g^{solid}) was also found in refs. (Tsagaropoulos and Eisenberg 1995a, b), dedicated to the study of the thermomechanical properties of some polymer composites. To explain this effect, an idea concerning an interfacial layer on the surface modifier particles was developed. In this layer, the mobility of the polymer chains is reduced. Furthermore, it was shown in refs. (Tsagaropoulos and Eisenberg 1995a, Robertson and Rackaitis 2011) that such features of thermal behavior are characteristic only of composites with nanometer-sized filler particles (7 nm) and are not observed in the case of micron particles (< 44 microns), as well as unmodified polymers. Thus, occurrence of the second glass transition temperature for the composites containing C_{60} nanoparticles is consistent with the findings of other researchers.

Dielectric parameters of the PMMA/fullerene composite films

In this section, we determine the dielectric parameters of PMMA films filled with small (up to 3 wt %) additions of fullerenes C_{60}. To determine these characteristics of composites (capacitance, *C*, and dielectric loss tangent, *tanδ*), we used Solartron 1255 frequency response analyzer (UK). The numerical values of the parameters were obtained at room temperature using a two-electrode cell with round clamp electrodes 19.8 mm in diameter.

The dielectric constant (ε') of the film substance was calculated using the flat-plate capacitor formula:

$$C = \frac{\varepsilon' \varepsilon_0 S}{d}, \quad (1)$$

where *S* is the electrode area, *d* is the film thickness, and $\varepsilon_o = 8.854 \cdot 10^{-12}$ F·m^{-1} is the electric constant. The ac resistivity was determined by the relationship

$$\rho = \left(2\pi f \varepsilon' \varepsilon_0 \tan\delta\right)^{-1}, \quad (2)$$

where *f* is the frequency of alternating current.

The frequency dependences of the capacitance of both polymethylmethacrylate and composite films with various fullerene contents are shown in Figure 3. It can be seen for all the samples studied that capacity decreases monotonically with increasing frequency up to 10^1–10^3 Hz. In case of modified polymers, *C*(*f*) curves are above than the curve for pure PMMA.

With further increase in the frequency, curves practically coincide and the capacity of samples does not change. Such course of the frequency dependence of the capacitance is characteristic of the polar dielectrics, which includes PMMA.

Table 3 shows the dielectric constant values computed from the capacitance at a frequency of 1000 Hz by Eq. (1). For unmodified polymer, the value of ε is equal to 2.5 that is close to the literature data. As can be seen in the Table 3, doping of fullerene into PMMA increases the dielectric constant. It is a striking example of fullerene homeopathy effect that a small addition of C_{60} leads to significant changes in the physico-chemical properties of the polymer.

Also, Table 3 shows the value of resistivity, calculated using Eq. (2). It can be seen that the specific resistance decreases with growth in the fullerene concentration which also indicates that the conductivity increases. In materials with higher conductivity, reduction of static charge is more probable. Consequently, doping PMMA with fullerenes prevents the accumulation of static electricity in the polymer.

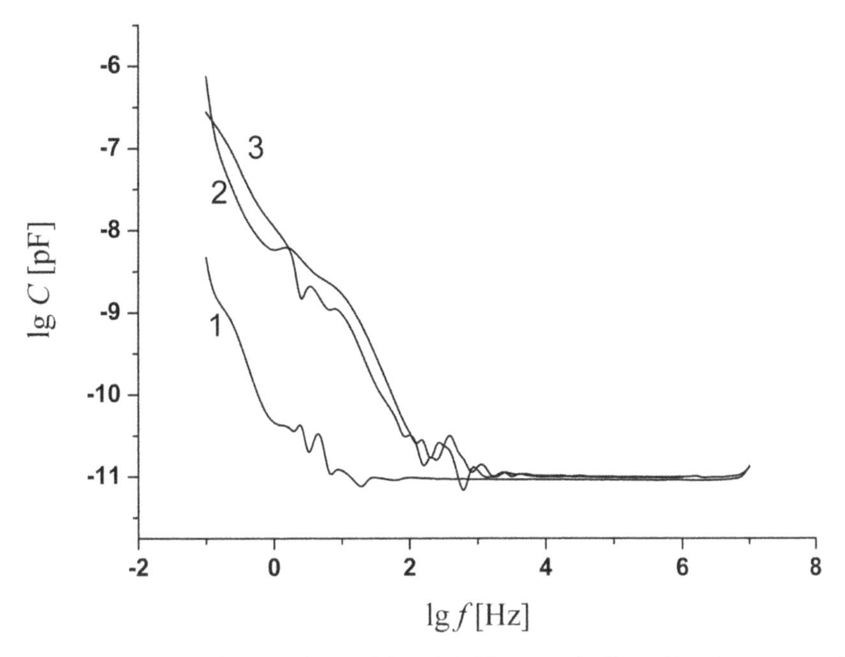

Figure 3. Frequency dependences of the capacitance of the PMMA/C_{60} composite films with various contents of C_{60}, wt %: 0 (*1*), 0.05 (*2*), (*3*).

Table 3. Dielectric parameters of the PMMA/C_{60} composite films as functions of C_{60} content.

C_{60} Content, wt. %	*d*, mm	*tanδ*	ε	$\sigma \cdot 10^9$, $Sm \cdot m^{-1}$
0	0.74	0.02	2.50	3.18
0.05	0.77	0.04	3.37	8.26
0.1	0.79	0.11	3.37	20.15
0.5	0.78	0.28	3.15	49.49
1	0.83	0.24	2.95	39.91
3	0.77	0.28	3.97	62.66

It is known that introduction of dielectric into an electric field causes polarization of material. A decrease in the polarization in time (dielectric relaxation) can be quantitatively characterized by the position of the peak in the frequency dependence of the dielectric loss tangent. For the PMMA/fullerene films examined in this study, these peaks were not observed at room temperature. Apparently, the induced dipoles arise very rapidly under these conditions, and the dielectric loss is low. This is confirmed, in particular, by the results of measuring the dielectric loss tangent: in the frequency range of $10^3 \div 10^6$ Hz, tanδ values are equal to 10^{-2} (in case of 0.05 wt. % of C_{60}) and 10^{-1} (in case of 0.1–3 wt. % of C_{60}).

Study of intermolecular interactions between macromolecules of polymethylmethacrylate and fullerene in the PMMA/C_{60} composite films

The nature of the interaction of PMMA with fullerene has been studied in the analysis of electronic and vibrational spectra of the pure and fullerene-containing films.

Optical absorption spectra for fullerene in the toluene solution were recorded in the wavelength range 280–800 nm in quartz cell (size is equal to 1.0 cm) by the Spectrophotometer U-2001 (HITACHI, Japan) with a working range of 190–1100 nm. A deuterium lamp was a light source for UV zone, and photometric accuracy is ±0,002 Abs. Optical absorption spectra for polymethylmethacrylate films and fullerene-polymethylmethacrylate film were recorded by the Spectrophotometer U-2001 (HITACHI, Japan) with a reflection attachment SMART MULTI-BOUNCE HATR (ZnSe 45° crystal).

Infrared spectra of films were recorded by Avatar 360 FT-IR ESP spectrometer (Termo Nicolet, US).

Figure 4(a, b) shows UV spectra of fullerene in toluene solution (a) and PMMA films and composites with C_{60} concentrations in range of 0.05–1 wt. %. (b).

In the spectrum of the fullerene in toluene solution (Figure 4), an intense absorption band at $\lambda = 334$ nm is observed in the ultraviolet region (200–400 nm), corresponding to allowed π-π* electronic transitions. In the visible part of the spectrum, there are weak bands in the region of 400–500 nm.

Study of electronic absorption spectra of fullerene in the polymeric matrix shows hypsochromic shift of the band at $\lambda = 334$ nm by 3 nm (Table 4). Intensity of this band increases with the C_{60} concentration. It should be noted that the PMMA has no absorption bands in this region of spectrum (Figure 5, spectrum 1).

Based on the analysis of the electronic spectra, it can be assumed that in composites, an intermolecular complex forms. It contains the electron-donor functional groups of PMMA (C=O, C-O-C) and π-electron system of the fullerene molecule.

Additional information about interaction nature of fullerene with PMMA was obtained from the analysis of the vibrational spectra of the original and modified films. Figure 6 shows the IR spectra of

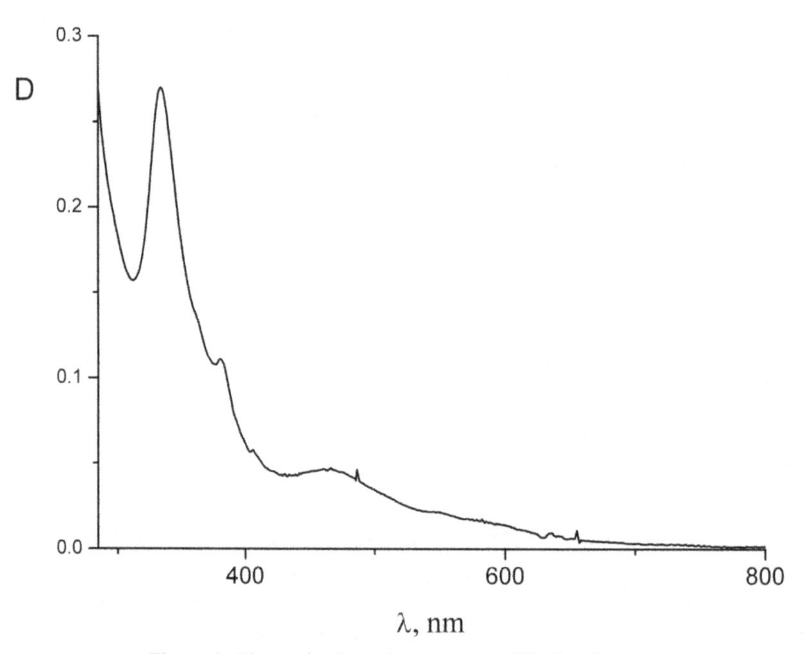

Figure 4. Electronic absorption spectrum of C_{60} in toluene.

Table 4. Optical properties of fullerene in various systems.

C_{60} in Toluene Solution		PMMA/C_{60} Composites							
λ, nm	D	0.05 wt. % of C_{60}		0.1 wt. % of C_{60}		0.5 wt. % of C_{60}		1 wt. % of C_{60}	
		λ, nm	D	λ, nm	D	λ, nm	D	λ, nm	D
334	0.270	331	0.511	331	1.056	331	3.377	331	9.999
380	0.111	379	0.248	379	0.488	379	1.116	379	2.761
406	0.058	405	0.146	405	0.282	405	0.592	405	1.179
466	0.047	469	0.139	469	0.261	469	0.559	469	1.094

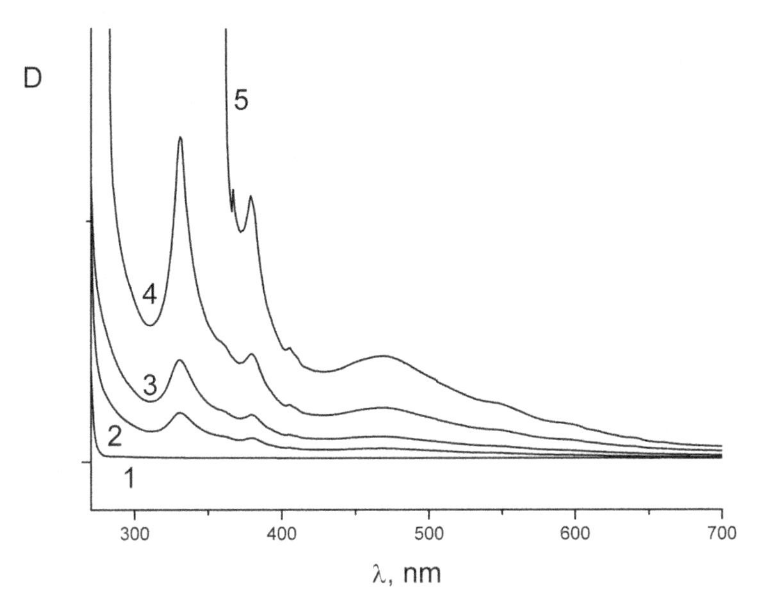

Figure 5. Electronic absorption spectra of pure polymethylmethacrylate (*1*) and PMMA/C_{60} composite films with various contents of fullerene, wt. %: 0.05 (*2*), 0.1 (*3*), 0.5 (*4*), 1 (*5*).

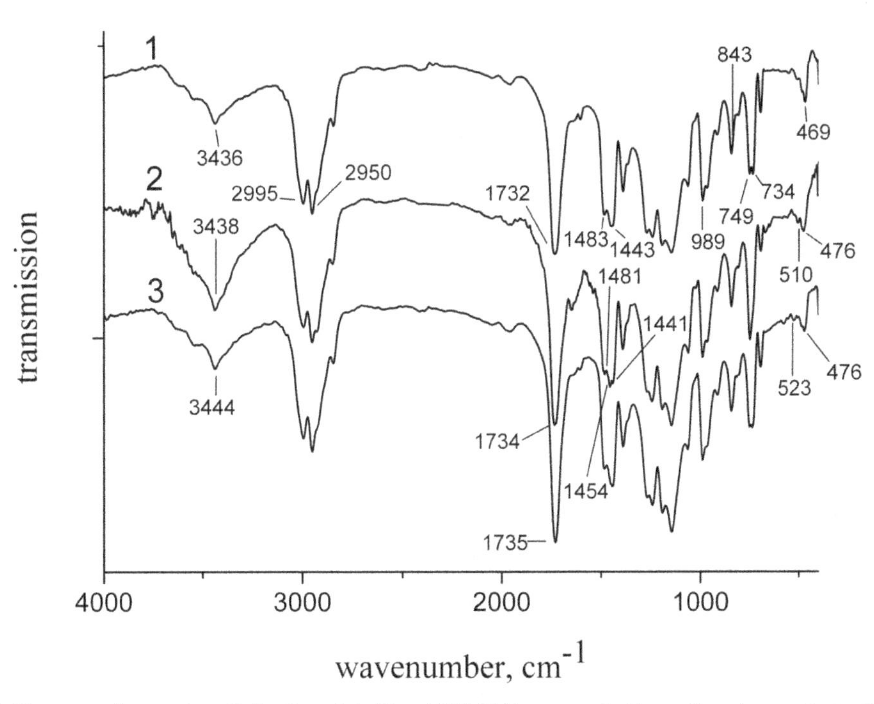

Figure 6. IR spectra of pure polymethylmethacrylate (*1*) and PMMA/C_{60} composite films with various contents of fullerene, wt. %: 1 (*2*), 3 (*3*).

original PMMA and PMMA/fullerene composites containing 1 and 3 wt. % of filler. The IR spectrum of PMMA (Figure 6, spectrum *1*) indicates the details of functional groups. The broad peak at 3436 cm^{-1} and the sharp intense peak at 1732 cm^{-1} appear due to the presence of ester carbonyl group (C=O) stretching vibration.

The band with two maxima at 2995 cm^{-1} and 2950 cm^{-1} refers to the stretching vibrations of C-H bond in the O-CH_3, CH_3 and CH groups. The deformation vibrations of CH_2 and CH_3 manifest band with two peaks (1443 and 1483 cm^{-1}). The broad peak ranging from 1260–1000 cm^{-1} can be explained owing to the C-O (ester bond) stretching vibration. The broad band from 950–650 cm^{-1} is due to the bending of C-H. The band at 989 cm^{-1} belongs to the ν (C-O-C), mixed with the γ (CH_3-O). The band at 843 cm^{-1} corresponds to CH_2 rocking vibration bands. Doublet at 734 cm^{-1} and 749 cm^{-1} corresponds to (CH2) vibrations mixed with (C-C). The band at 469 cm^{-1} belongs to the vibration ν (C-C-O) group (Dechant et al. 1972).

Figure 7 shows the IR spectrum of the fullerene molecule. Active four vibrations with absorption bands at 527, 578, 1180 and 1427 cm^{-1} can be seen (Konarev and Lyubovskaya 1999).

In the spectra of PMMA/C_{60} composite containing 1 and 3 wt. % C_{60}, there are significant changes in comparison with the spectrum of the original PMMA (Figure 6, spectra *2* and *3*). In the spectra of the composites, the absorption bands at 3436 cm^{-1} and 1732 cm^{-1} are shifted by 6 and 2 cm^{-1}, respectively, to higher frequencies region. There is a splitting of the band in region of 1540–1410 cm^{-1}, with the advent of peak at 1454 cm^{-1}, which may be due to the presence of fullerene in the spectrum of the modified polymer. The appearance of new bands in the spectra at 510 cm^{-1} (for composites with 1 wt. % of C_{60}) and 523 cm^{-1} (for composites with 3 wt. % of C_{60}) is noted. This indicates the presence of C_{60} in the composite in unbound form. We can assume a non-covalent interaction of polymethylmethacrylate donor macromolecules with fullerene acceptor molecule.

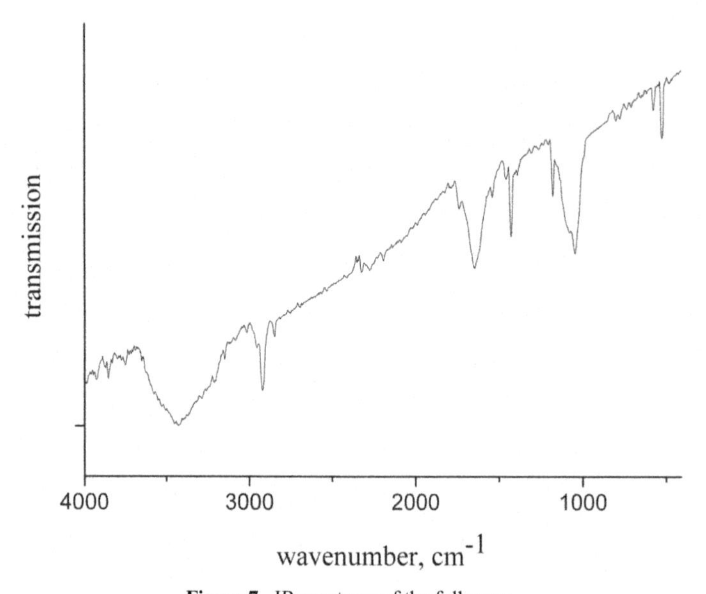

Figure 7. IR spectrum of the fullerene.

Antioxidant activity of polymethylmethacrylate and polymethylmethacrylate/fullerene nanocomposites

It is considered that integration of fullerenes into polymer matrix can produce biocomposites which have medical potential as drug transporters, antiseptics and antioxidants. Today, regulation of free-radical processes is adjusted by both natural and synthetic pharmaceutical compositions (Okovitiy 2003). As any other medicine, some antioxidants may produce adverse events. So finding of safe preparations with high antioxidant activity is still crucial.

Influence of PMMA/C_{60} film composites on free-radical oxidation of lipids in biologic fluid (blood serum) was researched *in vitro*. Subject of research was native blood serum of 10 patients managed in V.N. Gorodkov Research Institute of Maternity and Childhood (Ivanovo, Russia). Specimen of pure PMMA or composite film containing 0.1, 0.5, 1 or 3 wt. % of fullerene was put into test tube with blood serum (1 ml). System was incubated for 1 hour at 4°C. Then, the film specimen was removed from tube.

The parameters of lipid peroxidation in serum after exposure of the film nanomaterials were determined by chemiluminescent analysis. The induced chemiluminescence (ChL) tests were performed by BChL-07 luminometer (Medozons, Russia). Hydrogen peroxide and ferric sulfate have been used as inductors of ChL. 0.1 ml of serum, 0.4 ml of phosphate buffer (pH 7.5), 0.4 ml of 0.01 M ferric sulfate and 0.2 ml of 2% hydrogen peroxide were put into cuvette. Luminescence was registered for 40 s.

To estimate the intensity of lipid peroxidation, following parameters have been used:

J_{max} is maximum intensity of ChL during the experiment. Value of J_{max} quantifies the level of free radicals, i.e., it gives an idea of the potential ability of the blood serum to free radical lipid peroxidation;

$tan\alpha$ is tangent of maximum slope angle of ChL curve towards time axis. This value characterizes the decay rate of free radical oxidation, i.e., it quantifies an effectiveness of the antioxidant system;

A is an area covered by intensity curve or total light sum. Value of A is inversely proportional to the antioxidant activity of the sample;

$Z = AJ_{max}^{-1}$ is normalized light sum.

Free radical processes in serum have been studied after exposure of pure PMMA films and fullerene-containing composite films. The mean values of ChL parameters in native serum without film adding were used as controls 6–8 measurements required for each film were carried out on the same day. The results have been expressed as percentages relative to controls and were given as mean values ± standard deviations. A p-value of 0.05 was chosen as the significance limit.

Kinetics of chemiluminescence in serum were researched after exposure of pure PMMA and nanocomposite films. Peak of chemiluminescence due to free radical production was in 2 s of reaction. This can be explained by production of active oxygen species (HO_2^*, O_2^*, O_2^-, OH^-).

Table 5 shows the main parameters obtained from analysis of ChL curves. It can be seen that for pure PMMA film, the ChL parameters were approximate to controls. In case of PMMA/C_{60} composites, there is the extreme dependence of the luminescence intensity, J_{max}, on the C_{60} concentration. An analogous dependence is observed for the light sum, A. Note that in all cases for values J_{max} and A, there are significant differences compared to control ($p < 0.05$). This indicates a change in the concentration of free radicals and reactive oxygen species in the system. However, value of tanα was significantly changed only for films containing 3 wt % of fullerenes ($p < 0.05$).

Thus, changing the content of nanocarbon particles in the polymer matrix allows the specific regulation of processes of free radical oxidation of lipids in biological fluids.

Table 5. Chemiluminescence parameters in blood serum after exposure of pure PMMA film and fullerene-containing composites.

Fullerene Concentration, wt. %	**A, %**	**J_{max}, %**	**tanα, %**	**Z, %**
Controls	100.0	100.0	100.0	100.0
0	106.0 ± 9.0	106.0 ± 7.0	99.0 ± 10.5	101.0 ± 4.0
0.1	114.0 ± 8.0*	113.5 ± 5.5*	107.0 ± 8.0	101.0 ± 3.0
0.5	108.0 ± 7.0*	108.5 ± 11.0*	104.0 ± 9.0	102.0 ± 4.5
1.0	108.0 ± 8.0*	107.0 ± 11.0*	106.5 ± 11.5	102.0 ± 3.5
3.0	80.0 ± 6.0*	82.0 ± 6.0*	91.0 ± 4.0*	97.0 ± 2.0*

* significant differences compared to control ($p < 0.05$).

Conclusion

The experimental results presented in current study have shown fullerene to be quite an effective modifier for polymethylmethacrylate. However, to successfully obtain PMMA/C_{60} composite materials with high performance, there is need for further systematic study of all factors affecting the formation of the composite structure and its properties. It is hoped that a better understanding of the principles for such composites formation will stimulate the development of modern technologies for creation of medical supplies on the basis of these materials.

Acknowledgments

DSC measurements were performed in the centre for joint use of scientific equipment "The upper Volga region centre of physico-chemical research".

The work was supported by Russia Foundation for Basic Research (Grant N 15-43-03034-a).

References

Alekseeva, O.V., N.A. Bagrovskaya and A.V. Noskov. 2014. Effect of C_{60} filling on structure and properties of composite films based on polystyrene. Arab. J. Chem. http://dx.doi.org/10.1016/j.arabjc.2014.09.008 (in press).

Andrade, C.K.Z., R.A.F. Matos, V.B. Oliveira, J.A. Duraes and M.J.A. Sales. 2010. Thermal study and evaluation of new menthol-based ionic liquids as polymeric additives. J. Therm. Anal. Calorim. 99: 539–543.

Arshad, M., K. Masud, M. Arif, S. Rehman, M. Arif, J.H. Zaidi et al. 2009. The effect of $AlBr_3$ additive on the thermal degradation of PMMA: a study using TG-DTA-DTG, IR and PY-GC-MS techniques. J. Therm. Anal. Calorim. 96: 873–881.

Ayatollahi, M.R., S. Shadlou and M.M. Shokrieh. 2011. Multiscale modeling for mechanical properties of carbon nanotube reinforced nanocomposites subjected to different types of loading. Compos. Struct. 93: 2250–2259.

Badamshina, E.R. and M.P. Gafurova. 2008. Characteristics of fullerene C_{60}-doped polymers. Polym. Sci. B. 50: 215–225.

Burnside, S.D. and E.P. Giannelis. 2000. Nanostructure and properties of polysiloxane-layered silicate nanocomposites. J. Polym. Sci. Polym. Phys. 38: 1595–1604.

Chen, K., C.A. Wilkie and S. Vyazovkin. 2007. Nanoconfinement revealed in degradation and relaxation studies of two structurally different polystyrene–clay systems. J. Phys. Chem. 111: 12685–12692.

Coetzee, J.F. (ed.). 1982. Recommended Methods for Purification of Solvents and Tests for Impurities. Pergamon Press, Oxford.

Dechant, J., R. Danz, W. Kimmer and R. Schmolke. 1972. Ultrarotspektroskopische untersuchungen an polymeren. Akademie-Verlag, Berlin (in German).

Dittrich, B., K.-A. Wartig, D. Hofmann, R. Mülhaupt and B. Schartel. 2013. Flame retardancy through carbon nanomaterials: Carbon black, multiwall nanotubes, expanded graphite, multi-layer graphene and graphene in polypropylene. Polym. Degrad. Stab. 98: 1495–1505.

Fonseca, M.A., B. Abreu, F.A.M.M. Gonçalves, A.G.M. Ferreira, R.A.S. Moreira and M.S.A. Oliveira. 2013. Shape memory polyurethanes reinforced with carbon nanotubes. Compos. Struct. 99: 105–111.

Harper, C.A. 2000. Modern Plastics Handbook. McGraw-Hill, New York.

Hattab, Y. and N. Benharrats. 2015. Thermal stability and structural characteristics of PTHF–Mmt organophile nanocomposite. Arab. J. Chem. 8: 285–292.

Konarev, D.V. and G.N. Lyubovskaya. 1999. Donor–acceptor complexes and radical ionic salts based on fullerenes. Russ. Chem. Rev. 68: 19–38.

Krakovyak, M.G., T.N. Nekrasova, T.D. Arsenyeva and E.V. Anufriyeva. 2002. Noncovalent interactions between polymers and fullerene C_{60} in organic solvents. Polym. Sci. B. 44: 271–276.

Lazzeri, A., S.M. Zebarjad, M. Pracella, K. Cavalier and R. Rosa. 2005. Filler toughening of plastics. Part 1—The effect of surface interactions on physico-mechanical properties and rheological behaviour of ultrafine $CaCO_3$/HDPE nanocomposites. Polymer 46: 827–844.

Meenakshi, K.S. and E.P.J. Sudhan. 2016. Development of novel TGDDM epoxy nanocomposites for aerospace and high performance applications—Study of their thermal and electrical behaviour. Arab. J. Chem. 9: 79–85.

Okovitiy, S.V. 2003. Clinical pharmacology of antioxidants. FARMindex: Praktik. 5: 85–111 (in Russian).

Potalitsin, M.G., A.A. Babenko, O.S. Alekhin, N.I. Alekseev, O.V. Arapov, N.A. Charykov et al. 2006. Caprolons modified with fullerenes and fulleroid materials. Rus. J. Appl. Chem. 79: 306–309.

Robertson, C.G. and M. Rackaitis. 2011 Further consideration of viscoelastic two glass transition behavior of nanoparticle-filled polymers. Macromolecules 44: 1177–1181.

Sibileva, M.A., E.V. Tarasova and N.I. Matveyeva. 2004. Interaction of fullerenes with polyethylene oxide, poly-N-vinylpyrrolidone, and DNA in water–chloroform–polymer systems. Rus. J. Phys. Chem. A 78: 526–532.

Thangamani, R., T.V. Chinnaswamy, S. Palanichamy, S. Bojja and A.W. Charles. 2010. Thermal degradation studies on PMMA–HET acid based oligoesters blends. J. Therm. Anal. Calorim. 100: 651–660.

Tsagaropoulos, G. and A. Eisenberg. 1995a. Direct observation of two glass transitions in silica-filled polymers. Implications to the morphology of random ionomers. Macromolecules 28: 396–398.

Tsagaropoulos, G. and A. Eisenberg. 1995b. Dynamic mechanical study of the factors affecting the two glass transition behavior of filled polymers. Similarities and differences with random ionomers. Macromolecules 28: 6067–6077.

Tugov, L.I. and G.L. Kostrykina. 1989. Khimiya i fizika polimerov (Polymer chemistry and physics). Khimiya, Moscow (in Russian).

Zanotto, A., A. Spinella, G. Nasillo, E. Caponetti and A.S. Luyt. 2012. Macro-micro relationship in nanostructured functional composites. Express Polym. Lett. 6: 410–416.

Advances in Nanobiocatalysis

Strategies for Lipase Immobilization and Stabilization

Patel Vrutika,[1] *Ashok Pandey,*[2] *Christian Larroche*[3] and *Datta Madamwar*[1,*]

Introduction

Enzymes are naturally tailored protein catalysts synthesized to perform under physiological conditions. However, biotransformations imply the use of enzymes under conditions that may depart significantly from physiological states. The challenge consists in building catalysts that preserve functional properties of enzymes but are robust enough to withstand harsh process conditions. Enzymes are catalysts of exquisite specificity, being well appreciated for the synthesis of pharmaceuticals and fine chemicals (Woodley 2008). On the other hand, many enzymes are rather promiscuous catalysts since they are capable of catalyzing several reactions and/or transforming many substrates, in addition to the ones for which they are physiologically specialized, or evolved (Khersonsky and Tawfik 2010).

According to the International Union of Biochemistry (IUB), enzymes are divided into six classes: oxidoreductases, transferases, hydrolases, lyases, isomerases, and ligases. Hundreds of enzymes are used industrially, over half are from fungi, over one-third are from bacteria, with the remainder originating from animal (8%) and plant (4%) sources. Over 500 commercial products are made using enzymes. The industrial enzyme market reached U$ 1.6 billion in 1998 and in 2009, the market was U$ 5.1 billion. In the 1980s and 1990s, microbial enzymes replaced many plant and animal enzymes and they have found use in many industries including food, detergents, textiles, leather, pulp and paper, diagnostics, and therapy (Sánchez and Demain 2011).

[1] Post Graduate Department of Biosciences, Vadtal Road, Satellite Campus, Post Box # 39, Sardar Patel University, Vallabh Vidyanagar, Bakrol - 388 120, Gujarat, India.
Email: vrutikaptl_19@yahoo.com

[2] Centre of Innovative and Applied Bioprocessing (a national institute under Dept. of Biotechnology, Ministry of S&T, Govt. of India) C-127, 2nd Floor, Phase 8 Industrial Area, SAS Nagar Mohali-160 071, Punjab, India.

[3] Université Clermont Auvergne, Université Blaise Pascal, Institut Pascal, BP 20206, F-63174 Aubière cedex, France.

* Corresponding author: datta_madamwar@yahoo.com

On the other hand, enzyme immobilization widened the scope of application allowing less stable, intracellular and non-hydrolytic enzymes to be developed as process catalysts and biocatalysts in non-aqueous media. This approach allowed the opening up of a vast field of enzyme applications in reactions of organic synthesis, with an exquisite selectivity, especially for the synthesis of pharmaceuticals and bioactive compounds (Illanes et al. 2009). Enzyme activity under mild conditions is also a valuable attribute for the production of labile compounds, having profound technological implications, significantly reducing the costs of equipment, energy and downstream operations. However, the use of enzyme catalysts in organic synthesis has been difficult to adopt by an industry not sufficiently acquainted to deal with biological materials. The bottlenecks of enzyme technology are their high cost, instability and poor performance under reactor conditions, narrow substrate specificity and requirements of complex cofactors.

Biomolecules such as proteins and enzymes exhibit comparable dimension to that of nanoparticles (NPs). Thus, by integrating biomolecules and NPs into hybrid conjugates, new functional chemical entities that combine the unique electronic, optical, and catalytic properties of metallic or semiconductor NPs with the unique recognition and catalytic properties of biomolecules might be envisaged. Indeed, substantial progress has been accomplished in recent years in the use of biomolecule–NP hybrid systems as functional units for nanobiotechnology, and several detailed review articles have summarized the different nanobiomolecular constructs and their potential applications.

This review addresses recent advances in the development of lipase–NP conjugates. Its aim is to introduce some facets of nanobiotechnology and, together with the other articles, to highlight the broadness and perspectives of the topic.

Lipase

According to the Nomenclature Committee of the International Union of Biochemistry and Molecular Biology, enzymes are classified into six main classes (Table 1). One of the most important classes is hydrolases (E.C.3.-.-.-), which catalyzes the hydrolytic cleavage of different types of chemical bonds. Many commercially-critical enzymes belong to this class, e.g., proteases, amylases, acylases, lipases, and esterases (Zhang and Kim 2010).

Lipases are simply hydrolytic enzymes that catalyze hydrolysis reactions by breaking down triacylglycerides into free fatty acids and glycerols, which act under aqueous conditions on the carboxyl ester bonds present in triacylglycerols to liberate fatty acids and glycerol (Woolley and Petersen 1994, Jaeger et al. 1999). Hydrolysis of glycerol esters carrying an acyl chain, which comprises less than 10 carbon atoms in length, with tributyrylglycerol (tributyrin) as the standard substrate, usually indicates the presence of an esterase. Most lipases are able to hydrolyze esterase substrates. These reactions usually continue with high regio- and/or enantio-selectivity, making lipases a valuable group of biocatalysts in organic chemistry.

Structure of lipases

Based upon X-ray crystallography studies, the following structural features of lipases are identified: (i) All the lipases are members of "α/β-hydrolase fold" family, i.e., have a structure which is composed of a core of predominantly parallel β strands surrounded by α helices (Ollis et al. 1992, Nardini and Dijkstra 1999, Carrasco-Lopez et al. 2009). (ii) The active nucleophilic serine residue rests at a hairpin turn between a β strand and α helix in a highly conserved pentapeptide sequence Gly-X-Ser-X-Gly, forming a characteristic β-turn-α motif named the 'nucleophilic elbow' (Nardini and Dijkstra 1999, Carrasco-Lopez et al. 2009). Exceptionally, lipase B from *Candida antarctica* (CALB) does not have a conserved pentapeptide sequence Gly-X-Ser-X-Gly around the active site which is present in most of the other lipases (Uppenberg et al. 1994). (iii) The active site of lipases is formed by a catalytic triad consisting of amino acids serine, histidine and aspartic acid/glutamic acid (Brady et al. 1990, Winkler et al. 1990, Carrasco-Lopez et al. 2009). Presence of a lid or flap is composed of an amphiphilic α helix peptide

Table 1. Lipase and enzyme classification according to EC number.

EC number	Enzyme
EC 1.-.-.-	Oxidoreductases
EC 2.-.-.-	Transferases
EC 3.-.-.-	Hydrolases
EC 3.1.-.-	Acting on ester bonds
EC 3.1.1.-	Carboxylic ester hydrolases
EC 3.1.1.3	**Triacylglycerol lipase (= lipase, in general)**
EC 3.2.-.-	Glycosylases
EC 3.3.-.-	Acting on ester bonds
EC 3.4.-.-	Acting on peptide bonds
EC 3.5.-.-	Acting on carbon-nitrogen bond, other than peptide bonds
EC 3.6.-.-	Acting on acid anhydrides
EC 3.7.-.-	Acting on carbon-carbon bonds
EC 3.8.-.-	Acting on halide bonds
EC 3.9.-.-	Acting on phosphorous-nitrogen bonds
EC 3.10.-.-	Acting on sulfur-nitrogen bonds
EC 3.11.-.-	Acting on carbon-phosphorous bonds
EC 3.12.-.-	Acting on sulfur-sulfur bonds
EC 3.13.-.-	Acting on carbon-sulfur bonds
EC 4.-.-.-	Lyases
EC 5.-.-.-	Isomerases
EC 6.-.-.-	Ligases

Note: EC numbers and their descriptions are adapted from Nomenclature Committee of the International Union of Biochemistry and Molecular Biology. Lipase is highlighted in bold to show its position among the EC classification.

sequence that covers the active site (Schmid and Verger 1998, Kapoor and Gupta 2012). Pleiss et al. (1998) subdivided lipases into three subgroups on the basis of the geometry of the binding site (i) lipases with a hydrophobic, crevice-like binding site located near the protein surface (lipases from *Rhizomucor* and *Rhizopus*); (ii) lipases with a funnel-like binding site (lipases from *C. antarctica, Pseudomonas* and mammalian pancreas); and (iii) lipases with a tunnel-like binding site (lipase from *Candida rugosa*) (Kapoor and Gupta 2012).

Mechanism of lipase action

The catalytic mechanism for lipase-catalyzed hydrolysis is similar to that for serine proteases, i.e., it involves formation of two tetrahedral intermediates (Gandhi et al. 2000, Kapoor and Gupta 2012). The mechanism involves the nucleophilic attack of hydroxyl group of serine residue (present in the active site) on carbon from the ester bond of susceptible substrate. This results in the formation of tetrahedral intermediate which then loses an alcohol molecule to give an acyl-enzyme intermediate. A water molecule then attacks the complex (nucleophilic attack) to give tetrahedral intermediate, which finally, loses an acid molecule (Ribeiro et al. 2011, Petkar et al. 2006). Similarly, in a different case when alcohol acts a nucleophile (transesterification reaction), ester is the expected product and when amine acts a nucleophile (amidation reaction), amide is the desired product. In most of the cases, it is observed that formation of the acyl enzyme is fast; hence, deacylation is the rate-determining step.

Interfacial activation

The activity of lipases is low on monomeric substrates but as soon as an aggregate supersubstrate (such as an emulsion or a micellar solution) is formed, the lipase's activity increases dramatically; this phenomenon is called interfacial activation (Schmid and Verger 1998, Sarda and Desnuelle 1958). Desnuelle et al. (1960) gave the hypothesis that this activation was due to a conformational change resulting from the adsorption of the lipase onto a hydrophobic interface. In 1990, X-ray crystallographic structures of several lipases showed the presence of α-helical fragment (termed the "lid") covering the active center which suggested that displacement (opening) of the lid might occur during interfacial activation (Figure 1), hence allowing access to the otherwise inaccessible active center (Brzozowski et al. 1991, Groshulski et al. 1993, Derevenda and Sharp 1993). In the presence of a hydrophobic surface (such as drop of oil) (Miled et al. 2001, Reis et al. 2009), the lid of the lipase moves away and turns the 'closed' form of the lipase into an 'open' form and allows the interaction between its hydrophobic internal face (Figure 1) and the hydrophobic residues that usually surround the lipase active center with the substrate (Carrasco-Lopez et al. 2009, Uppenberg et al. 1994).

Van Tilbeurgh et al. (1993) crystallized human pancreatic lipase in two conformational states: a closed (inactive) one, in which the catalytic triad in the active site is covered by a helical "lid" and an "open" (active) one, obtained by crystallization in the presence of micelles, in which the "lid" has been displaced. Wilson et al. (2006) reported that when lipase from *Alcaligenes* sp. was immobilized via interfacial adsorption on a very hydrophobic support (octadecyl-sephabeads), it exhibited 135% of catalytic activity for the hydrolysis of p-nitrophenyl propionate as compared to the soluble enzyme. Hydrophilic moieties like arginine and chitosan have also been found to cause interfacial activation in lipases from *R. miehei* and *C. rugosa*, respectively (Chiou et al. 2004). There are lipases which have a lid but do not exhibit interfacial activation, e.g., lipase from *Pseudomonas glumae*, lipase from *Pseudomonas aeruginosa*, pancreatic lipases from coypu, and lipase from *Staphylococcus hyicus* showed interfacial activation with only some substrates. Lipase B from *C. antarctica* and lipase from guinea-pig also do not undergo interfacial activation.

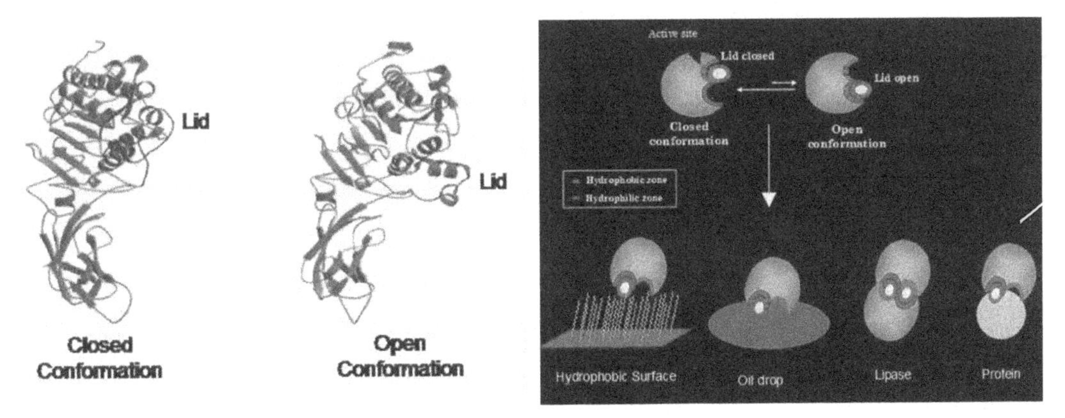

Figure 1. (a) Closed and open confirmation of lipase (b) Mechanism for interfacial activation of lipase. Van Tilbeurgh et al. (1993).

Specificity of lipases

(a) Positional specificity (Regiospecificity)

Lipases can be divided into three major groups on the basis of their ability to hydrolyze glycerides (Ribeiro et al. 2011, Krishna and Karanth 2002). Matori et al. (1991) determined the positional specificity index (PSI) value of many microbial preparations of lipases and divided them into three groups: PSI values of the first group were 100, i.e., completely 1,3 specific, those of the second group were between 70 and 80 and those of the third group were between −20 and +30, i.e., relatively non-specific.

Non-specific lipases: These are the ones which catalyze reactions on all hydroxyl groups of triglyceride and thus can remove fatty acid from any position of the triglyceride (Ribeiro et al. 2011, Krishna and Karanth 2002, Uhlig 1998, Saxena et al. 1999). These catalyze the complete breakdown of triglyceride to glycerol and free fatty acid. Diglycerides and monoglycerides are formed as intermediates in the reaction mixture (Saxena et al. 1999). These intermediates are hydrolyzed more rapidly than the triglyceride and so, these do not accumulate in the reaction (Saxena et al. 1999).

1,3-Specific lipases: These are the ones which can catalyze reactions only on primary hydroxyl groups of triglyceride (Ribeiro et al. 2011, Krishna and Karanth 2002, Uhlig 1998, Saxena et al. 1999). Thus, these preferentially release fatty acids from positions 1 and 3 to give free fatty acid and di- and or monoglycerides.

Fatty acid specific lipases: Some lipases prefer hydrolysis of those esters which are formed from long-chain fatty acids with double bonds in between C-9 and C-10 (Krishna and Karanth 2002, Jensen 1974).

(b) Stereo-specificity

Stereospecificity is defined as the ability of lipases to distinguish between sn-1 and sn-3 position on the triglyceride. Lipases may show insignificant stereoselectivity or be very stereoselective. The stereoselectivity of the same enzyme may vary according to the structure of the substrate (Sonnet 1988). Lipases from *Pseudomonas* sp. and *P. aeruginosa* exhibit sn-1 preference with trioctanoin as a substrate while *C. antarctica* lipase B shows sn-3 preference, with high stereospecificity. All other microbial lipases show medium or low sn-1(3) stereospecificity toward trioctanoin (lipases from *Rhizopus oryzae, R. miehei* and *C. rugosa* hydrolyze trioctanoin preferably at sn-1 with low stereoselectivity) (Rogalska et al. 1993).

Lipase mediated biocatalysis in non-aqueous media

Traditionally, enzymatic catalysis was carried out primarily in aqueous systems as water forms the natural milieu for these biomolecules. Developments in application of enzymes in nearly anhydrous medium containing traces of water have stimulated research in achieving various kinds of enzymatic transformations (Klibanov 2001). These reactions include chiral synthesis or resolution, production of high-value pharmaceutical substances, modification of fats and oils, synthesis of flavor esters and food additives, and production of biodegradable polymers, peptides, proteins and sugar-based polymers (Dandavate et al. 2008, Ozyilmaz et al. 2010, Raghavendra et al. 2010, Patel and Madamwar 2015). One of the most influential parameters affecting enzymatic activity in aqueous solution is pH. But it has no meaning in organic solvents. Instead, it has been found that enzymes in such media have a 'pH memory': their catalytic activity reflects the pH of the last aqueous solution to which they were exposed. Alternatively, the enzymatic activity is maximized, by adding appropriate buffer pairs of acids and their conjugated compounds.

The main advantages that biocatalysis in organic solvent offer compared to traditional aqueous enzymology can be summarized as: (i) increased solubility of non-polar substrates and products, which markedly speeds up overall reaction rates; (ii) reversal of thermodynamic equilibrium in favor of synthesis over hydrolysis, allowing reactions usually not favored in aqueous solutions to occur (e.g., transesterification, thioesterification, aminolysis); (iii) drastic changes in the enantioselectivity of the reaction when one organic solvent is changed to another; (iv) suppression of unwanted water-dependent side reactions, which often degrade common organic reagents; and (v) elimination of microbial contamination in the reaction mixture (Halling 2000, Ogino and Ishikava 2001, Ru et al. 1999, Castro and Knubovets 2003, Torres and Castro 2004). Also, it has been reported that low-water environments can be used to stabilize enzyme conformations that exhibit unpredictable catalytic properties.

Stabilization of enzyme in organic solvents

Following three different strategies are employed to enhance enzyme activity and stability in non-conventional media:

(i) Isolation of intrinsically stable enzymes

Lipases are most commonly employed in non-aqueous solvent systems. Thermostable lipases have been isolated from many sources including *P. flurescens, Bacillus* sp., *B. coagulans, B. cereus, B. stearothermophilus, Geotrichum* sp., *Aeromonas sobria* and *P. aeruginosa*. Organic solvent stability is required in addition to thermostability. Recently, *P. aeruginosa* PST-01 (Ogino et al. 1995), *Pseudomonas* sp. strain S5 (Baharam et al. 2003), *Pseudomonas stutzeri* LC2-8 (Cao et al. 2012), *Burkholderia multivorans* V2 (Dandavate et al. 2009) and *Pseudomonas* sp. DMVR46 (Patel et al. 2015) have been isolated and reported to demonstrate greater stability towards solvents.

(ii) Environmental engineering approaches

Biotransformation engineering approaches involve the tuning of all physicochemical parameters of reaction media. Changes in the microenvironment of the biocatalyst not only are able to modulate the enzyme activity and stability but also to shape the enzyme selectivity. One of the main variables in biotransformations is the water content in the enzyme microenvironment. The water activity is relevant since it involves the effect of water mass action on the chemical equilibrium (Castro and Knubovets 2003). Different types of reaction media can be classified in two main groups from a physicochemical point of view: continuous (without boundaries inside the medium) or discontinuous (with border inside the medium or heterogeneous or multiphasic), as previously described by Davidson et al. (1997). Many substances and solvents, as organic solvents (with different physiochemical properties), gases (supercritical fluids) and ionic liquids can be mixed together in different proportions to fit in one of above-mentioned groups (Illanes et al. 2012).

(i) Solvents

Selection of solvent is a key task to obtain best results for a particular enzymatic reaction. Apart from these, the activity of enzyme always depends on the essential water which is tightly bound to the active surface of enzyme and plays an important role in catalysis. Hence, selected solvent should be such that it should not disrupt the essential water of enzyme, which in turn would increase the catalytic activity, thermal and operation stability of enzyme (Ma et al. 2012). The best parameter used for selection of suitable solvent for biocatalysis is the log P value of the solvent. A hydrophobicity parameter, Log P, was first proposed for microbial epoxidation of propene and 1-butene (Laane et al. 1987). The Log P value is the logarithm of P, where P is defined as the partition coefficient of a given compound in standard n-octanol/water two phase. In general, it is inspected that when biocatalytic reaction is conducted in polar organic solvents having a Log P value < 2, the reaction rate/product yield is low, whereas in solvents bearing Log P between 2–4, it is moderate to good. Considerable reaction rate/product yield is observed in non-polar solvents having a Log P > 4.27. This is because polar solvents have a tendency to strip out the essential water present on enzyme active site. The non-polar solvents do not strip out these water molecules and hence remarkable enzyme activity is observed in non-polar solvents like n-hexane, iso-octane, etc.

In non-polar solvents, enzymes are insoluble but the solubility of hydrophobic substrates and products is improved. The enhanced solubility of substrates and products in non-polar solvents implies the stabilization of ground states of the molecules, and consequently, the decrease of the biocatalytic reaction rate. Additionally, non-polar solvents tend to rigidify the protein structure, but a delicate balance in the dynamics of the protein structure between protein flexibility and stiffness of the protein shell structure is required (Illanes et al. 2012).

(ii) Semisolid systems

A prominent advantage of enzyme catalysis can be obtained in an aqueous medium if working at very high substrates concentration, i.e., the use of substrates' concentrations beyond the limit of solubility (Youshko et al. 2004), or even in solid-state (Basso et al. 2006). Thus, semisolid systems stem as a very promising technology that avoids the use of obnoxious chemicals and allows obtaining very high product concentrations. Enzyme catalysis in nearly solid or semi-solid systems has been studied, in which the reaction mixture consists of solid reactants suspended in a comparatively small volume of liquid phase (Ulijn et al. 2003). The advantages of using solid systems include environmental innocuousness, high conversion yields in reversal of hydrolytic reactions and high enzyme stability. Nevertheless, mass transfer limitations and mixing problems may represent an important drawback, especially when scaling up to production level is required (Erbeldinger et al. 1998).

(iii) Physical stabilization of enzyme

(i) Immobilization

For technical and economical reasons, most chemical processes catalyzed by enzymes require the stabilization, re-use or continuous use of the biocatalyst for a very long time. From an industrial perspective, simplicity and cost-effectiveness are key properties of immobilization techniques, but the long term industrial re-use of immobilized enzymes also requires the preparation of very stable derivatives having the right functional properties for a given reaction (Cao 2005). The immobilization methods can be broadly divided into two categories: carrier bound or carrier free, depending on the inclusion of an inert matrix.

For carrier-bound, catalyst support binding can be physical or chemical, involving weak or covalent bonds. In general, physical bonding is comparatively weak and is hardly able to keep the enzyme fixed to the carrier under industrial conditions. The support can be a synthetic resin, an inorganic polymer such as zeolite or silica, or a biopolymer. Entrapment involves inclusion of an enzyme in a polymer network (gel lattice) such as an organic polymer or a silica sol–gel, or a membrane device such as a hollow fiber or a microcapsule. Entrapment requires the synthesis of the polymeric network in the presence of the enzyme (Cao et al. 2003).

Carrier-free biocatalysts are novel type enzyme catalysts bearing the advantages of the high concentration of active enzyme within the biocatalyst particle and reduced cost (Roessl et al. 2010). In this case, the enzyme protein constitutes its own support so that concentrations close to the theoretical packing limit are obtained (Cao 2005). Therefore, carrier-free immobilized enzymes are advantageous as catalysts in processes where high productivity and yield is required or in the case of labile enzymes that cannot be properly stabilized by conventional immobilization to solid supports (Illanes et al. 2009). Carrier-free immobilized enzymes are prepared by direct chemical crosslinking of the protein containing the enzyme, using mainly glutaraldehye as crosslinking agent. This strategy has been applied to cross linked enzymes in solution (CLEs), to cross linked enzyme crystals (CLECs) and more recently, to cross linked enzyme aggregates (CLEAs). CLEAs have advantages over CLEs of better mechanical properties and higher yields of activity and are simpler and much cheaper to produce than CLECs, which require a purified crystal protein as starting material.

(ii) Nanobiocatalysis

Recent development in nanotechnology has provided a wealth of diverse nanoscale carriers that could be applied to enzyme immobilization. Immobilization of the enzymes on nanostructured materials has been recognized as a promising approach to enhance enzyme performances (Verma et al. 2013, Kim et al. 2008, Ansari and Hussain 2012). The nanobiocatalyst (NBC) is an emerging innovation that synergistically fuses nanotechnology and biotechnology breakthroughs. NBC is a specifically functionalized enzyme–nanocarrier assembly, which promises exciting advantages in improving enzyme stability, capability and engineering performances, and allowing the creation of a microenvironment surrounding the enzyme catalysts for maximal reaction efficiencies (Kim et al. 2008, Verma et al. 2013, Gupta et al. 2011). Enzyme immobilization using nanostructured carriers can significantly increase life cycles of the biocatalyst, hence reducing the cost of the biocatalytic process. Betancor and Luckarift (2008) described enzyme immobilization onto the nanoscale material as a versatile new technology, which offers advantages including low cost, rapid immobilization and reaction, similarity of nano size, mild conversion conditions, robust activities, mobility, high loading, minimum diffusional limitations, self-assembly and stability. A revolutionary class of biocatalysts can be developed by introducing the unique properties of the nanoscale material such as mobility, confining effects, solution behaviors and interfacial properties into NBCs. To date, functional nanomaterials have been used for the development of NBCs such as nanofibre (NF) scaffolds (Plessis et al. 2012), nanotubes (NTs) (Wang and Jiang 2011), nanoparticles (NPs) (Johnson et al. 2011), nanocomposites (NCs) (Tran et al. 2012) and nanosheets (NSs) (Ma et al. 2012, Patel et al. 2015). The increasing interest of NBCs has become a driving force for the fabrication of nanocarriers with unique properties and structures. Advanced nanocarriers such as NFs nanopores and nanocontainers can significantly enhance the engineering performances of enzymes. An integrated process for industrial bioprocesses using the NBCs is proposed as sketched in Figure 2, which illustrates the research approach from the synthesis of nanocarriers and enzyme immobilization strategy for execution of a specific function and evaluation of bioprocessing performance of the NBCs.

Functionalities of nanobiocatalysis

Engineering of the nanomaterials, called nanomaterial functionalization, is a process designed to improve suitability for their application in nanobiocatalysis (Johnson et al. 2011). The properties of nanomaterials are easily changed, and the functionalized materials are used as nanoscaffolds. The chemical functionalization of nanomaterial is a well established technique for grafting desirable functional groups onto their surface to obtain nanomaterials with desired properties (Shim et al. 2002, Bourlinos et al. 2003). The development of novel nanocarriers with unique functions and characteristics comprises (i) the introduction of functional groups on the surface of the nanocarriers for immobilizing various enzymes or responding to external stimuli, (ii) construction of special structures for increasing the surface area,

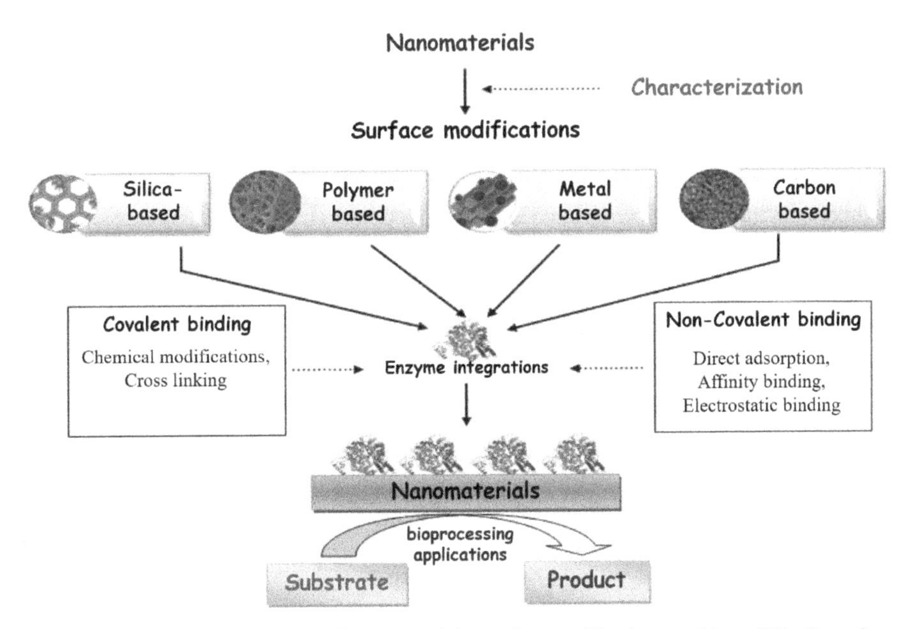

Figure 2. Integrated process for utilization of nanomaterials, surface modifications and immobilization of enzyme for industrial bioprocessing applications.

facilitating substrate diffusion, recycling nanocarriers or confining enzymes inside nanocages, and (iii) improving process ability of nanocarriers such as mechanical and thermal stability (Pavlidis et al. 2010).

Investigating the organization and role of enzymes immobilized on nanomaterials is essential for understanding enzyme-nanomaterial interactions and designing of functional protein-nanomaterial conjugates (Cruz et al. 2010). Numerous analytical techniques are used for characterization of enzymes immobilized on support/particles (Ganesan et al. 2009). The conformational changes of enzymes upon immobilization have been investigated by many researchers using sophisticated analytical techniques such as Brunauer-Emmett-Teller (BET) analysis, Atomic Force Microscopy (AFM), Scanning Electron Microscope (SEM), Transmission Electron Microscopy (TEM), Circular Dichroism (CD) spectroscopy, Fourier Transform Infrared Spectroscopy (FTIR), XPS (X-ray Photoelectron spectroscopy) and X-Ray diffraction (XRD) pattern. The size and morphology for enzyme immobilized nanomaterials is characterized by TEM and SEM. The binding of lipase by either covalent or adsorption method on the nanomaterials was confirmed by FTIR, XPS and TEM observations. The influence of physical/covalent attachments on the secondary structure of enzyme can be analyzed by CD spectroscopy (Ji et al. 2010). AFM has been quite useful to confirm enzyme immobilization *in vitro* on the graphene oxide nanosheet (Zhang et al. 2010). The lipase-bound nanomaterials exhibited the same XRD patterns as the native nanomaterials (Huang et al. 2003).

Advancement of nanosized carriers as support for enzyme immobilization

The advantages of enzymes immobilized on micron-sized particles are inherited when nanomaterials are used as solid supports for enzyme immobilization (Kim et al. 2006a, 2006b, Gupta et al. 2011, Verma et al. 2013, Ansari and Husain 2012). The following factors, presented in the context of nanomaterials, are important considerations in evaluating the suitability of a material for enzyme immobilization.

(a) Surface area to volume ratio: Nanomaterials have large surface area to volume ratios. Nanofibers offer two-thirds of the surface to volume ratio of (nearly) spherical particles of the same diameter (Wang et al. 2006, Gupta et al. 2011). These are easier to prepare, easier to handle, and offer larger flexibility in reactor design.

(b) Enzyme loading: Nanometer-scale materials offer high enzyme loading due to their large relative surface area. High enzyme loading leads to better biocatalytic activity and stability. This makes the nanoscaffold an ideal support for enzyme immobilization as compared to conventional materials (Verma et al. 2013).

(c) Mass transfer: Apparent enzyme activity could be improved due to the relieved mass transfer limitation of substrates in nanostructures when compared to macro-scale matrices in conventional enzyme immobilization. Enzymes immobilized on nanomaterials have low mass transfer resistance and thus have high activity and stability (Kim et al. 2006b, Verma et al. 2013).

(d) Flow rate/Mobility: When enzymes are immobilized on micron-sized carriers, shaking or stirring is required in a batch reactor. However, some very novel findings from the research group of Wang suggest (Wang 2006) that well-dispersed nanoparticles show Brownian motion. According to the Stokes-Einstein equation, the mobility and diffusivity of the nanoparticles have to be smaller than those of free enzymes, owing to their relatively larger size. "This mobility difference may point to a transitional region between the homogeneous catalysis with free enzyme and heterogeneous catalysis with immobilized enzyme" (Wang 2006).

(e) Bioreactor design: Enzyme immobilization is fundamental for the development of reactors, biosensors or micro total analysis systems (Song et al. 2012). Nanomaterials, especially nanofibers, offer larger flexibility in reactor design as they are easier to prepare and easier to handle (Nair et al. 2007). The small pressure drop and high flow rate of the nanofiber membrane represents most important advantages of enzyme-immobilized nanofiber membrane bioreactors over the traditional enzyme-immobilized membranes and fixed bed bioreactors. An enzyme immobilized fibrous membrane bioreactor was established with a continuous steady hydrolysis conversion at a constant flow rate under optimum condition (Huang et al. 2008).

(f) Recovery and reusability: Recovery after use may be distinguished from reusability and refers to the ease with which the immobilized enzymes can be removed/separated from the reaction components for reuse. Use of smart carriers allows one to develop stimuli-sensitive immobilized enzymes, which can function as homogeneous catalysts but separate out conveniently from the reaction mixture just like heterogeneous catalysts (Safarik and Safarikova 2009, Ren et al. 2011). Nanoparticles allow reusability and stability of the attached enzyme as compared to conventional matrices, where centrifugation/filtration is the only option to separate the enzyme from the product. Such operation leads to enzyme leaching/ instability due to mechanical shear while mixing the pellet with the appropriate buffer to begin a new reaction (Yiu and Keane 2012).

Strategies for development of lipase mediated nanobiocatalysis

Recent development of lipase-mediated nanobiocatalysis using different novel nanomaterials variants is discussed herewith in the following section. All the various methods are also represented in Table 2 stating strategies for lipase immobilization on nanostructured materials.

Table 2. Immobilization of lipase on various nanostructured materials, immobilization methods and properties.

Lipase Sources	Conjugation Method	Nanomaterial Type	Feature	References
Nanoparticles				
Candida rugosa	Covalent	γ-Fe_2O_3	Long-term stability	Dyal et al. 2003
C. rugosa	Covalent	Silica	High synthetic activity, good reusability	Dandavate et al. 2009
P. cepecia	Covalent	Zirconia	High stable lipase	Chen et al. 2009
Buekholderia cepacia	Adsorption	Magnetic	High thermal stability, high product yield	Andrade et al. 2010
C. rugosa	Covalent	Chitosan	High lipase loading, high activity retention	Wu et al. 2010
C. rugosa	Adsorption	Polylactic acid	Enhancement of enzyme activity and stability	Chronopoulou et al. 2011
Nanofibers				
C. antarctica	Adsorption	Polysulfone	Enhanced thermal stability	Wang et al. 2006
C. antarctica	Covalent	PAN	High operational stability	Li et al. 2008
C. rugosa	Covalent	Poly(acrylonitrile-co-2-hydroxyethyl methacrylate)	Improved stability	Huang et al. 2008
Nanotubes				
C. antarctica	Adsorption	CNT	Protect lipase inactivation	Pavildis et al. 2010
C. rugosa	covalent	CNT	Thermal stability and enhanced catalytic activity	Pavildis et al. 2012
C. antarctica	Covalent	CNT	Higher operational stability	Raghavendra et al. 2013
Pseudomonas sp. DMVR46	Covalent	CNT	Higher stability and reusability	Patel and Madamwar 2015
Nanosheets				
Lipozyme CalB, C. rugosa	Adsorption, covalent	Graphene oxide	High activity and stability	Pavildis et al. 2012a
C. rugosa	Covalent, Adsorption	Exfoliated graphene oxide	Highly stable in organic solvent	Patel et al. 2015

Nanoparticles

Nanoparticles (NP) present useful platform for nanobiocatalysis, demonstrating distinctive properties with wide-range of industrial applications (Johnson et al. 2011). The surface and core properties of versatile nanoparticles can be engineered as per the need of the applications (De et al. 2008). Nanoparticles scaffolds improve the biocatalytic efficiency of a diversity of lipases, in terms of activity, stability and reusability. Different techniques and methodologies are explored and developed for retaining the enzymes on or in the nanomaterials viz: physical adsorption via electrostatic interactions, hydrophobic interactions, hydrogen bonding or van der Waals forces, covalent binding, cross-linking of enzymes or physical entrapment or encapsulation.

Magnetic iron oxide shows higher significance among nanoparticles because of its biocompatibility, stability, large surface area, super-paramagnetic property and low cost synthesis. Lipase was bound to γ-Fe_2O_3 nanoparticles via covalent bond resulting in higher enzyme loading (5.6 wt%) on γ-Fe_2O_3 nanoparticles (20 nm). However, operational stability of the immobilized enzyme was also greatly enhanced. *Candida rugosa* lipase (CRL) was covalently bound onto Fe_3O_4 magnetic nanoparticles via carbodiimide activation (Dyal et al. 2003).

Huang et al. (2003) reported immobilization of lipase on Fe_3O_4 magnetic nanoparticle (12.7 nm) exhibiting 1.41-fold enhanced activity, 31-fold improved stability, and enhanced tolerance to variation in pH as compared to the free enzyme. Cui et al. (2010) used TEOS modified Fe_3O_4 nanoparticles treated with 3-aminopropyltriethoxysilane (APTES), an organosilane, which is a bifunctional molecule containing a trialkoxy group and an organic head-group functionality that determines the final chemical character of the modified surface (NH_2). Nano-sized magnetite (NSM) particles (10 nm) were coated by an alkyl benzenesulfonate (ABS) to provide a hydrophobic surface. Furthermore, immobilized PPL showed enhanced durability in the reuse for hydrolysis of olive oil (Ponvel et al. 2009). Jiang et al. (2009) studied CRL immobilized magnetic nanoparticles supported with ionic liquids resulting in higher synthesis of butyl oleate and retained 60% of its initial activity after eight repeated reactions. *Burkholderia cepacia* lipase (BCL) was immobilized on superparamagnetic nanoparticles (10 nm) using adsorption and chemisorption methodologies applied as a recyclable biocatalyst in the enzymatic kinetic resolution of (RS)-1-(phenyl) ethanol via transesterification reactions (Rebelo et al. 2010). Nano-sized Fe_3O_4 particles were used for immobilization of *Mucor javanicus* lipase. This lipase was covalently bounded to surface of activated nano-sized magnetite (NSM) particles and later cross-linked to form a CLEA (cross-linked enzyme aggregate). The immobilized lipase was reused as catalyst for preparing 1,3-DAGs for 10 cycles at 55°C with only a 10% loss of activity (Meng et al. 2014).

Chitosan is widely used as support for enzyme immobilization due to its different geometric configurations, such as powder, flakes, hydrogels, membranes, fibers and presence of hydroxyl and amino groups. Jenjob et al. (2012) studied chitosan-functionalized poly(methyl methacrylate) (PMMACH) particles prepared by negatively charged PMMA particles and positively charged chitosan via spinning disk processing. The bioconjugation of prepared PMMA-CH with *C. rugosa* lipase AY leads to easy access of converting substrate into product. Huang et al. (2007), in his study, immobilized *C. rugosa* lipase on nanofibrous membrane using glutaraldehyde as cross-linker. Immobilized lipase retained 56.2% of initial activity after 30 days compared to free lipase that retained 36.6% activity after 10 days. Kuo et al. (2012) prepared magnetic Fe_3O_4-chitosan nanoparticles by coagulation of aqueous solution of chitosan. The prepared nanoparticles were utilized for covalent immobilization of lipase from *C. rugosa* using N-(3-dimethylaminopropyl)-Nethylcarbodiimide (EDC), and N-hydroxysuccinimide (NHS) as coupling agents. In another study, Wu et al. (2010) prepared two different sizes (7 and 10 nm) of chitosan nanoparticles in a-water-in-oil (W/O) microemulsion using two different solvents and precipitants. The lipase activity retention for small particles was as high as 66.7% in comparison to that of large nanoparticles (62.8%). This study displayed the importance of nanoparticles size on enzyme loading potential and biocatalytic efficiency.

Silica nanostructure has gained higher attention in biomedical field due to its well-defined chemical structure, which allows modification of its structure by amine, carboxyl and thiol groups, methacrylates, enzymes, proteins and DNA (Souza et al. 2011). Zhang et al. (2011) reported functionalization of silica NPs and immobilization of lipase by aldehyde, cyanogen, epoxy or carbodiimide groups. Kim et al. (2006) studied the immobilization of *M. javanicus* lipase on silica nanoparticles using glycidyl methacrylate (GMA) which was attached onto the surface of the nanoparticles. Glutaraldehyde (GA) or 1,4-phenylene diisothiocyanate (NCS) were employed as a coupling agent. Gustafsson et al. (2012) synthesized three types of mesoporous silica particles' varying particles morphology as well as particles' size (1000 nm, 300 nm and 40 nm), using lipase from *Mucor miehei* and *Rhizopus oryzae*. As a result, both lipases were highly active when 300 nm particles were used for immobilization. Dandavate et al. (2009) studied the immobilized CRL by covalent attachment to glutaraldehyde-activated silica nanoparticles (100 nm). The activation energy of immobilized lipase was lower in comparison to free lipase. Tran et al. (2012) implemented affinity binding of lipase with ferric silica NPs by grafting long alkyl groups. This nanobioconjugation resulted in an excellent lipase binding efficiency of up to 97%, leading to enhanced stability without major loss of transesterification activity.

Zirconia nanoparticles are a polymorphic material occurring in three temperature-dependent forms: monoclinic or baddeleyte (room temperature to 1170°C), tetragonal (1170–2370°C) and cubic (2370–2700°C). The high chemical, thermal and pH stability of zirconia makes it more feasible for fabrication as supports to immobilize proteins and enzymes (Treccani et al. 2013), biosensors (Tong et al. 2007, Zong et al. 2007) and enzyme-based bioreactors (Reshmi et al. 2007). Chen et al. (2008) reported immobilization of lipases from *C. rugosa* and *Pseudomonas cepacia* on the modified zirconia nanoparticles by adsorption in aqueous solution. Chen et al. (2009) studied *Pseudomonas cepacia* lipase (PCL) immobilization on modified zirconia nanoparticles used for asymmetric synthesis in organic media. Various carboxylic acids with different alkyl chain lengths (valeric acid, caprylic acid, stearic acids, oleic acids, linoleic acid and 1,10-decanedicarboxylic acid) were grafted to zirconia nanoparticles. PCL immobilized on modified nanoparticles with stearic acid gave the best activity/enantioselectivity for the resolution of (R,S)-1-phenylethanol (E-value > 91.6) through acylation in isooctane.

Nanofibers

Nanofibers offer a number of attractive features compared to other nanostructures which relieve mass transfer limitation of substrates/products due to their reduced thickness and make it easier to recover and reuse (Lee et al. 2008). Conducting nanofibers as immobilization matrix show superior properties over other carrier materials, such as their environmental stability (high stability to extremes of temperature, pH and resistance toward microorganisms), direct electron transfer capability between an enzyme and a polymer, ease of preparation, higher enzyme loading per unit mass with reduced diffusion resistance with a shortened path for substrate diffusion, higher conductivity and more facile fabrication (Ghosh et al. 2012). Electrospinning is a frequently used technique to synthesize NFs. Fabrication through the electrospinning produce various sizes of fibrous mats (Li et al. 2012, Park et al. 2013).

Huang et al. (2011) studied immobilization of CRL on a cellulose nanofiber membrane. Cellulose acetate (CA) was electrospun into a non-woven nanofiber membrane and then oxidized by $NaIO_4$ to create aldehyde groups. Effects of pH, temperature, and additive concentration were scrutinized on the adsorption capacity of polysulfone nanofibres. K_M and V_{max} for the immobilized lipases were higher and lower in comparison to those for free lipase, respectively. Furthermore, the immobilization resulted into enhanced thermal stability for immobilized enzyme (Wang et al. 2006). Nakane et al. (2007) studied a polyvinyl alcohol (PVA) nanofibers-immobilized lipase. The specific surface area of the nanofiber (5.96 m^2/g) was ca. 250 times larger than that of PVA-film-immobilized lipase (0.024 m^2/g). The esterification activity of the nanofiber bound lipase was equivalent to that of commercially available immobilized lipase (Novozym-435). The nanofiber-immobilized lipase showed higher activity for the esterification than lipase powder and film attributed to high specific surface area and high dispersion state of lipase molecules in PVA matrix. Sakai et al. (2010) utilized polycaprolactone (PCL) for immobilization onto electrospun PAN fibers used for the conversion of (S)-glycidol with vinyl n-butyrate to glycidyl n-butyrate in isooctane. The rate of reaction was 23-fold higher than the initial material. After ten reaction cycles, the initial reaction rate remained at 80% of the original rate. An et al. (2011) used polystyrene-poly(styrene-co-maleic anhydride) nanofiber for immobilization of CRL using coatings onto the nanofibers, yielding high activity and stability, and creating an economically viable nanobiocatalytic system for efficient use of expensive enzymes. The enzymatic activity of the encapsulated nanofiber was higher in the hydrolysis reaction than transesterification reaction (Song et al. 2012b).

Carbon-based nanomaterials

Carbon-based nanomaterials have been extensively employed for enzyme immobilization owing to their inertness, biocompatibility and thermal stability (Yu et al. 2009, Guo et al. 2009). Among them, carbon NTs (CNTs), nanodiamond and graphene derivatives appeared to be the most striking nanocarriers for producing the NBC assembly. Carbon nanotubes (CNTs) acquire exceptional structural, mechanical, thermal and biocompatibility properties (Asuri et al. 2007). CNTs, including both single-walled carbon nanotubes (SWNTs) and multi-walled carbon nanotubes (MWNTs), characteristically consist wide range of diameter from an order of one to tens of nanometers and a length of up to several hundred micrometers (Sotiropoulou et al. 2003, Shi et al. 2007, Lee et al. 2010). Numerous methods for covalent binding and physical adsorption have been studied using pristine or functional CNTs. CNTs possess a hydrophobic nature which is crucial for driving enzyme binding. Enzyme attachment onto single walled CNTs can be attained via simple physical adsorption without chemical modifications through sequential steps (Goh et al. 2012). To fully explore potential applications of CNTs, it is often required to increase CNTs' solubility in common organic solvents through appropriate functionalization. Efforts have been made for biological functionalization of CNTs applied in biosensor and enzyme immobilization (Shi et al. 2007, Yang et al. 2007, Wang and Jiang 2011). Covalent binding has been reported for modifying CNT carriers via introducing carboxylic acid groups by oxidation of CNTs and activated using carbodiimide (Azamian et al. 2002, Pedrosa et al. 2010).

Yu et al. (2005) used the peptide nanotube for the immobilization of lipase through hydrogen bonding between nanotube amidic groups and complementary groups on the protein surface. Immobilized CRL exhibited a high thermal stability as well as reusability. Shah et al. (2007) reported immobilization of CRL on MWCTs through physical adsorption and observed elevated retention of their catalytic activity (97%). Furthermore, the immobilized biocatalyst showed 2.2- and 14-fold increase in initial rates of transesterification using hexane and water immiscible ionic liquid [Bmim] [PF_6] as compared to lyophilized powdered enzyme. Lipase was covalently attached to MWNTs and used for the resolution of the model compound (R, S)-1-phenyl ethanol in n-heptane as reaction medium. Reusability and high resolution efficiency were superior with MWNT-lipase without reduction of the selectivity in native lipase (Ji et al. 2010). Pavlidis et al. (2010) investigated the immobilization of CALB on functionalized MWCNTs through physical adsorption, where MWCNTs were functionalized with carboxyl-, amine- and ester-terminal groups on their surface. Lipase immobilized MWCNT exhibited high catalytic activity and increased storage and operational stability. To overcome the enzyme leaching 3D structure change and diffusion resistance, Wang and Jiang (2011) introduced a detailed site-specific enzyme immobilization method. Briefly, the method was based on specific interaction between Histagged enzyme and single-walled carbon nanotubes modified with Nα, Nα-bis(carboxymethyl)-L-lysine hydrate. The developed nanoscale biocatalyst can maintain high enzyme activity and stability without any need for enzyme purification. The resulting interaction of enzyme-carbon nanotube significantly improves the thermal stability and catalytic activity of enzymes. The rigid structure for enzyme after immobilization was confirmed by fluorescence and circular dichroism studies (Pavlidis et al. 2012a, 2012b). Raghavendra et al. (2013) immobilized CRL onto MWCNTs using two different methods. One method of enzyme immobilization involved carbodiimide chemistry while in the other approach, the cross linker (3-Aminopropyl) triethoxysilane (APTES) followed by succinic acid anhydride (SAA) were employed prior to carbodiimide activation. The lipase-MWCNTs conjugates were applied for synthesis of the flavor ester 'pentyl valerate' in cyclohexane and effects of solvent, temperature and agitation on ester synthesis were studied. Upon subject to reusability studies for 50 cycles, the bionanoconjugates were found to be highly sturdy and exhibited ~ 79% activity (immobilization using carbodiimide) whereas the nanoconjugate, prepared using APTES and SAA, retained only up to ~ 30% activity. In another study, the surface of multiwalled carbon nanotubes (MWCNTs) was functionalized with a mixture of concentrated acids to create an interface for purified lipase immobilization and exploited for the synthesis of ethyl butyrate (Patel and Madamwar 2015).

Graphene has attracted considerable interest among nanostructured materials for their unique structural, physiochemical properties as well as supports for biomacromolecules immobilization (Cang-Rong and Pastorin 2009, Pavlidis et al. 2010). Graphene oxide (GO) is a monolayer of carbon atoms packed into a dense hexagonal network similar to standard of honeycomb crystal structure. In addition to the incredibly large specific surface area (two accessible sides), the abundant oxygen containing surface functionalities, such as epoxide, hydroxyl and carboxylic groups, and high water solubility render GO as a promising application (Li et al. 2008, Park and Ruoff 2009). There are ample reports for using GO based carriers for immobilization of enzymes, either by non-specific physical adsorption or by covalent coupling (Zhou et al. 2012, Wang et al. 2011, Patel et al. 2015). Also, its high surface area and low production cost make graphene an ideal candidate for enzyme immobilization.

Nanodiamond, a carbon derivative nanomaterial, has great potential for biological applications, due to its chemical stability, biological compatibility, non-toxicity and the unbreakable fluorescence from nitrogen vacancy centers. The surface of nanodiamond provides a unique platform for conjugation of biological molecules. It has been reported that the modified surface of nanodiamond can be conjugated with bio-molecules such as DNA, antigen, cytochrome c and hormone (Krueger 2008, Liu et al. 2008). Wei et al. (2010) reported the covalent immobilization of enzyme on nanodiamond prepared by detonation (dND), which was performed by the diimide-activated amidation of dND-bound carboxylic acids under the experimental conditions designed to avoid denaturing of bonded enzyme. Nguyen et al. (2007) demonstrated that carboxylated/oxidized diamond crystallites can be utilized as a solid support for non-covalent protein immobilization.

Nanoporous materials

Nanoporous materials generally consist of nanometer-scale sized pore spaces and large interior surfaces making novel nanoporous material for enzyme immobilization (Kim et al. 2006b). The stability of adsorbed enzymes is dependent on the pore size of nanosaffold and charge interaction. When the pore size of the enzyme carrier is smaller than the enzymes, enzyme molecules are incapable of entering the pore, resulting in a low enzyme loading. On the other hand, nanocarriers with pore sizes larger than the enzyme may reduce the enzyme's activity, which may be attributed to enzyme leaching or reduction in the stability of the enzyme (Ikemoto et al. 2010, Sang and Coppens 2011). The porous structures of the nanocariers can also be beneficial for hybridizing other components to form multiple functional NCs or assemblies. Pore-size is probably the most important parameter in the enzyme immobilization process (Serra et al. 2008), while other textural properties, such as the nature of the pores (channel-like or cage-like), the connectivity of the porous network, total pore volume as well as surface area, do not have obvious effects. The most suitable pore diameter for PPL immobilization is reported as 13 nm (Kang et al. 2007).

The influences of pore diameter and cross-linking method on the immobilization efficiency of CRL (molecular weight between 45 and 60 kDa) in a mesoporous material (15.6 nm) have been investigated amongst five kinds of SBA-15 with pore-sizes ranging from 6.8–22.4 nm. Gao et al. (2010) showed that the activities of immobilized CRL are much higher than that of free lipase retaining 80.5% of the initial activity after six cycles in 48 h. Mesoporous silica particles (15.8 nm) were used as carrier for immobilization of CRL (Nikolic et al. 2009). The average pore size of the material was 15.8 nm, which allowed enzyme adsorption inside the pores and high enzyme loading. PPL was successfully incorporated into the ionic liquid modified SBA-15 material by physical adsorption (PPL-ILSBA). The pore structure and surface properties of SBA-15 materials have changed due to IL modification, resulting in enriching the lipase catalysis environment (Zou et al. 2010). Matsura et al. (2011) developed a microreactor containing lipase-nanoporous material composite and employed it in the hydrolysis reaction of a triglyceride. Lipase used as a model enzyme was encapsulated in two types of folded-sheet mesoporous silicas, FSM4 (4 nm) and FSM7 (7 nm). The lipase-FSM composites contained in the microreactor displayed higher enzymatic activity than those in a batch experiment.

Nanocomposites

Metal-based nanomaterials have been widely used for fabricating NBCs. Among them, magnetic NPs are one exceptional example due to their high recyclability for biocatalysts. However, magnetic NP causes damage due to erosion of reacting agents (Lu et al. 2007). In order to avoid environmental reaction, the deposition of a layer of silica/gold/polymer like poly ethylene glycol (PEG) on the magnetic nanocores, chitosan, polyvinyl alcohol and polyethyleneimine (PEI) is coated onto the surface of metal NPs creating nanocomposites (NC) or hybrid nanomaterials (Long et al. 2004, Drechsler et al. 2004). NC not only adds high versatility to nanoparticle for surface modification but also improves biocompatibility for broader biological application due to their hydrophilic properties. This makes NC an ideal nanosupport for enzyme immobilization (Chaubey et al. 2009).

Dyal et al. (2003) first reported magnetic (maghemite) nanoparticles for the immobilization of CRL and its use in the hydrolysis of long chain synthetic ester. Fabrication of novel hierarchically ordered porous magnetic nanocomposites with interconnecting macroporous windows and meso-microporous walls containing well-dispersed magnetic nanoparticles were used as a support to immobilize lipase for the efficient hydrolysis of esters (Sen et al. 2010). Functionalized superparamagnetic particles were prepared by graft polymerization of glycidyl methacrylates and methacryloxyethyl trimethyl ammonium chloride onto the surface of modified-Fe_3O_4 nanoparticles. *Candida rugosa* (CRL) was immobilized onto the polymer-grafted magnetic nanoparticles. Immobilized CRL had better thermal stability compared to free lipase (Yang et al. 2008). In a recent study, the core-shell nanocomposite Fe_3O_4-SiO_2 was employed for *B. cepacia* lipase C20 immobilization (Tran et al. 2012). The immobilized lipase exhibited high catalytic activity as well as high reusability displaying potent commercial applications. Similar encouraging results like high thermal stability and good reusability have also been obtained using *Pseudomonas cepecia* lipase (PCL) on magnetic silica nanocomposite.

Processability and reusability of nanobiocatalytic system

Reutilization of the biocatalyst is an exceptionally significant necessity in transforming NBCs for commercialization. Recovery of enzymes along with nanoparticles is a significant issue and downstream processing after the reaction is very intricate. However, microcarriers or microbeads can be straightforwardly separated from reaction medium due to the density variation (Misson et al. 2015). Nanostructured materials offer feasible environment for enzymatic catalysis and reusability for easy separation as well as stability for enzymes in organic solvents (Patel et al. 2010). Enzyme immobilization using magnetic technology on magnetic nanocarriers is reported to elevate recovery and reusability of the NBCs (Fuertes and Tartaj 2007). Ngo et al. (2012) reported improved technique for controlling the size of magnetic silica NPs. The developed technique was also applied for polymer-based nanocarriers' synthesis. As a result, excellent magnetic response was obtained by coating the magnetic particles with polyaniline (Neri et al. 2011).

Nanobiocatalytic systems yield much higher enzyme loading capacity, enzyme activity and stability, as well as mass transfer efficiency, their processability due to costs for manufacturing nanocarrier. Among all the nanostructure materials, polymerbased nanoparticles comprise of incredible potential for meeting the requirement of industrial applications in terms of cost, reproducibility and functionality (Persona et al. 2013). Yet another approach to diminish enzyme purification-associated cost is merging of immobilization and purification in the equivalent step. This method allows selectively binding of target enzymes using functional groups for immobilization and eliminating impurities (Solanki and Gupta 2011). A high transesterification activity was obtained through simultaneous enzyme purification and immobilization. Briefly, nanomaterials were first coated with PEI, i.e., rich in amino groups, leading to greater binding selectivity and two fold purification of the enzyme. As a result of purification, this enormously increased the initial activity of purified enzyme about 110 times that of the unpurified free enzyme.

Final thoughts

Nanobiocatalytic systems, in which enzymes are incorporated into nanostructured materials, lately have emerged as a rapidly growing R&D field for biotechnological industries. Recent development in nanobiotechnology has provided richness of various nanoscale scaffolds that could potentially be applied to the flourishing of nanobiocatalytic system driven industrial bioprocesses. In this review, numerous examples for enzyme mediated nanobiocatalysis were discussed with exhilarating benefits for improving enzyme stability and activity by creating unique environment surrounding the enzyme catalysts for utmost reaction efficiencies. As a consequence, enzyme immobilization using nanostructure material drastically amplifies life cycles of biocatalyst for reusability, reducing overall cost of the biocatalytic process. However, immobilization of multi-enzymes on the functionalized nanocarriers to form nanobiocatalytic assembly still remains a challenge. With efforts from material scientists, bioprocessing engineers and biochemists, advanced multifunctional nanobiocatalytic systems could be effectively commercialized in the very near future.

Acknowledgements

The authors would like to acknowledge University Grants Commission (UGC) grant no. F. 42 167/2013 (SR), New Delhi for financial support. DM and CL acknowledge the support received from the French government research program "Investissements d'avenir" through the IMobS3 Laboratory of Excellence (ANR-10-LABX-16-01).

References

An, H.J., H.J. Lee, S.H. Jun, S.Y. Hwang, B.C. Kim, K. Kim et al. 2011. Enzyme precipitate coatings of lipase on polymer nanofibers. Bioprocess Biosyst. Eng. 34: 841–847.

Ansari, S.A. and Q. Husain. 2012. Potential applications of enzymes immobilized on/in nanomaterials: a review. Biotechnol. Adv. 30: 512–523.

Asuri, P., S.S. Bale, R.C. Pangule, D.A. Shah, R.S. Kane and J.S. Dordick. 2007. Structure, function, and stability of enzymes covalently attached to single-walled carbon nanotubes. Langmuir 23: 12318–12321.

Azamian, B.R., J.J. Davis, K.S. Coleman, C.B. Bagshaw and M.L.H. Green. 2002. Bioelectrochemical single-walled carbon nanotubes. J. Am. Chem. Soc. 124: 664–665.

Baharum, S.N., A.B. Salleh, C.N.A. Razak, M. Basri, M.B.A. Rahman and R.N.Z.R.A. Rahman. 2003. Organic solvent tolerant lipase by *Pseudomonas* sp. strain S5: stability of enzyme in organic solvent and physical factors affecting its production. Anal. Microbiol. 53: 75–8.

Basso, A., P. Spizzo, M. Toniutti, C. Ebert, P. Linda and L. Gardossi. 2006. Kinetically controlled synthesis of ampicillin and cephalexin in highly condensed systems in the absence of a liquid aqueous phase. J. Mol. Catal. B: Enzym. 39: 105–111.

Betancor, L. and H.R. Luckarift. 2008. Bioinspired enzyme encapsulation for biocatalysis. Trends Biotechnol. 26: 566–572.

Bourlinos, A.B., D. Gournis, D. Petridis, T. Szabo, A. Szeri and I. Dekany. 2003. Graphite oxide: chemical reduction to graphite and surface modification with primary aliphatic amines and amino acids. Langmuir 19: 6050–6055.

Brady, L., A.M. Brzozowski, Z.S. Derewenda, E. Dodson, G. Dodson, S. Tolley et al. 1990. A serine protease triad forms the catalytic centre of a triacylglycerol lipase. Nature 343: 767–70.

Brzozowski, A.M., U. Derewenda, Z.S. Derewenda, G.G. Dodson, D.M. Lawson, J.P. Turkenburg et al. 1991. A model for interfacial activation in lipases from the structure of a fungal lipase-inhibitor complex. Nature 351: 491–4.

Cang-Rong, J.T. and G. Pastorin. 2009. The influence of carbon Nanotubes on enzyme activity and structure: investigation of different immobilization procedures through enzyme kinetics and circular dichroism studies. Nanotechnology 20: 255102.

Cao, L., L. van Langen and R.A. Sheldon. 2003. Immobilized enzymes: carrier-bound of carrier-free? Curr. Opin. Biotechnol. 14: 387–394.

Cao, L. 2005. Carrier-bound Immobilized Enzymes. Principles, Application and Design. Wiley-VCH, Weinheim.

Cao, Y., Y. Zhuang, C. Yao, B. Wu and B. He. 2012. Purification and characterization of an organic solvent-stable lipase from *Pseudomonas stutzeri* LC2-8 and its application for efficient resolution of (R, S)-1-phenylethanol. Biochem. Eng. J. 64: 55–60.

Carrasco-Lopez, C., C. Godoy, B. de las Rivas, G. Fernandez-Lorente, J.M. Palomo, J.M. Guisan et al. 2009. Activation of bacterial thermo alkalophilic lipases is spurred by dramatic structural rearrangements. J. Biol. Chem. 284: 4365–72.
Castro, G.R. and T. Knubovets. 2003. Homogeneous biocatalysis in organic solvents and water-organic mixtures. Crit. Rev. Biotechnol. 23: 195–231.
Chaubey, A., R. Parshad, S.C. Taneja and G.N. Qazi. 2009. *Arthrobacter* sp. lipase immobilization on magnetic sol-gel composite supports for enatioselectivity improvement. Process Biochem. 44: 154–160.
Chen, Y.Z., C.T. Yang, C.B. Ching and R. Xu. 2008. Immobilization of lipases on hydrophobilized zirconia nanoparticles: highly enantioselective and reusable biocatalysts. Langmuir 24: 8877–8884.
Chen, Y.Z., C.B. Ching and R. Xu. 2009. Lipase immobilization on modified zirconia nanoparticles: studies on the effects of modifiers. Process Biochem. 44: 1245–1251.
Chiou, S.H. and W.T. Wu. 2004. Immobilization of *Candida rugosa* lipase on chitosan with activation of the hydroxyl groups. Biomaterials 25: 197–204.
Chronopoulou, L., G. Kamel, C. Sparago, F. Bordi, S. Lupi, M. Diociaiutic et al. 2011. Structure-activity relationships of *Candida rugosa* lipase immobilized on polylactic acid nanoparticles. Soft Matter.
Cruz, J.C., P.H. Pfromm, J.M. Tomich and M.E. Rezac. 2010. Conformational changes and catalytic competency of hydrolases adsorbing on fumed silica nanoparticles: I. Tertiary structure. Colloids Surf. 79: 97–104.
Cui, Y., Y. Li, Y. Yang, X. Liu, L. Lei, L. Zhou et al. 2010. Facile synthesis of amino-silane modified superparamagnetic Fe_3O_4 nanoparticles and application for lipase immobilization. J. Biotechnol. 150: 171–174.
Dandavate, V., H. Keharia and D. Madamwar. 2008. Ethyl isovalerate synthesis using *Candida rugosa* lipase immobilized on silica nanoparticles prepared in non-ionic micelles. Process Biochem. 44: 349–352.
Dandvate, V., J. Jinjala, H. Keharia and D. Madamwar. 2009. Production, partial purification and characterization of organic solvent tolerant lipase from *Burkholderia multivorans* V2 and its application for ester synthesis. Bioresour Technol. 100: 3374–3381.
Davidson, B.H., J.W. Barton and G.R. Petersen. 1997. Nomenclature and methodology for classification of nontraditional biocatalysis. Biotechnol. Prog. 13: 512–518.
De, M., P.S. Ghosh and V.M. Rotello. 2008. Applications of nanoparticles in biology. Adv. Mater. 20: 4225–4241.
Derevenda, Z.S. and A.M. Sharp. 1993. News from the interface: the molecular structures of triacyglyceride lipases. Trends Biochem. Sci. 18: 20–5.
Desnuelle, P., L. Sarda and G. Ailhaud. 1960. Inhibnition de la lipase pancréatique par le diéthyl-p-nitrophényl phosphate en emulsion. Biochem. Biophys. Acta 37: 570–95.
Drechsler, U., N.O. Fischer, B.L. Frankamp and V.M. Rotello. 2004. Highly efficient biocatalysts via covalent immobilization of *Candida rugosa* lipase on ethylene glycol-modified gold–silica nanocomposites. Adv. Mater. 16: 271–274.
Dyal, A., K. Loos, M. Noto, S.W. Chang, C. Spagnoli, K.V.P.M. Shafi et al. 2003. Activity of *Candida rugosa* lipase immobilized on gamma-Fe_2O_3 magnetic nanoparticles. J. Am. Chem. Soc. 125: 1684–1685.
Erbeldinger, M., X. Ni and P.J. Halling. 1998. Effect of water and enzyme concentration on thermolysin-catalyzed solid-to-solid peptide synthesis. Biotechnol. Bioeng. 59: 68–72.
Fuertes, A. and P. Tartaj. 2007. Monodisperse carbon–polymer mesoporous spheres with magnetic functionality and adjustable pore-size distribution. Small 3: 275–279.
Gandhi, N.N., N.S. Patil, S.B. Sawant, J.B. Joshi, P.P. Wangikar and D. Mukesh. 2000. Lipase catalyzed esterification. Catal. Rev. 42: 439–80.
Ganesan, A., B.D. Moore, S.M. Kelly, N.C. Price, O.J. Rolinski, D.J.S. Birch et al. 2009. Optical spectroscopic methods for probing the conformational stability of immobilized enzymes. Chem. Phys. Chem. 10: 1492–1499.
Gao, S., Y.J. Wang, X. Diao, G.S. Luo and Y.Y. Dai. 2010. Effect of pore diameter and cross-linking method on the immobilization efficiency of *Candida rugosa* lipase in SBA-15. Bioresour. Technol. 101: 3830–3837.
Ghosh, S., S.R. Chaganti and R.S. Prakasham. 2012. Polyaniline nanofiber as a novel immobilization matrix for the anti-leukemia enzyme L-asparaginase. J. Mol. Catal. B: Enzym. 74: 132–137.
Goh, W.J., V.S. Makam, J. Hu, L. Kang, M. Zheng, S.L. Yoong et al. 2012. Iron oxide filled magnetic carbon nanotube-enzyme conjugates for recycling of amyloglucosidase: toward useful applications in biofuel production process. Langmuir. 28: 864–873.
Groshulski, P., Y. Li, J.D. Schrag, F. Bouthillier, P. Smith, D. Harrison et al. 1993. Insights into interfacial activation from an open structure of *Candida rugosa* lipase. J. Biol. Chem. 268: 12843–7.
Guo, L., S. Zeng, J. Li, F. Cui, X. Cui, W. Bu et al. 2009. An easy co-casting method to synthesize mesostructured carbon composites with high magnetic separability and acid resistance. New J. Chem. 33: 1926–1931.
Gupta, M.N., M. Kaloti, M. Kapoor and K. Solanki. 2011. Nanomaterials as matrices for enzyme immobilization. Artif. Cells Blood Substit. Biotechno. 39: 98–109.
Gustafsson, H., E.M. Johansson, A. Barrabinoa, M. Oden and K. Holmberg. 2012. Immobilization of lipase from *Mucor miehei* and *Rhizopus oryzae* into mesoporous silica-The effect of varied particle size and morphology. Colloids Surf. B 100: 22–30.
Halling, P.J. 2003. Biocatalysis in low-water media: understanding effects of reaction conditions.
Huang, S.H., M.H. Liao and D.H. Chen. 2003. Direct binding and characterization of lipase onto magnetic nanoparticles. Biotechnol. Prog. 19: 1095–1100.
Huang, X.J., D. Ge and Z.K. Xu. 2007. Preparation and characterization of stable chitosan nanofibrous membrane for lipase immobilization. Eur. Polym. J. 43: 3710–3718.

Huang, X.J., A.G. Yu and Z.K. Xu. 2008. Covalent immobilization of lipase from *Candida rugosa* onto poly(acrylonitrile-co-2-hydroxyethyl methacrylate) electrospun fibrous membranes for potential bioreactor application. Bioresour. Technol. 99: 5459–5465.
Huang, X.J., P.C. Chen, F. Huang, Y. Ou, M.R. Chen and Z.K.J. Xu. 2011. Immobilization of *Candida rugosa* lipase on electrospun cellulose nanofiber membrane. J. Mol. Catal. B: Enzym. 70: 95–100.
Ikemoto, H., Q. Chi and J. Ulstrup. 2010. Stability and catalytic kinetics of horseradish peroxidase confined in nanoporous SBA-15. J. Phys. Chem. 114: 174–180.
Illanes, A., L. Wilson and C. Aguirre. 2009. Synthesis of cephalexin in aqueous medium with carrier-bound and carrier-free penicillin acylase biocatalysts. Appl. Biochem. Biotechnol. 157: 98–110.
Illanes, A., C. Ana, W. Lorena and G.R. Castro. 2012. Recent trends in biocatalysis engineering. Bioresour. Technol. 115: 48–57.
Jensen, R.G. 1974. Characteristics of the lipase from the mold *Geotrichum candidum*: a review. Lipids 9: 149–57.
Ji, P., H.S. Tan, X. Xu and W. Feng. 2010. Lipase covalently attached to multiwalled carbon nanotubes as an efficient catalyst in organic solvent. Aiche J. 56: 3005–3011.
Jiang, Y., C. Guo, H. Xia, I. Mahmood, C. Liu and H. Liu. 2009. Magnetic nanoparticles supported ionic liquids for lipase immobilization: enzyme activity in catalyzing esterification. J. Mol. Catal. B: Enzym. 58: 103–109.
Johnson, P.A., H.J. Park and A.J. Driscoll. 2011. Enzyme nanoparticle fabrication: magnetic nanoparticle synthesis and enzyme immobilization. Methods Mol. Biol. 679: 183–191.
Kang, Y., J. He, X.D. Guo, X. Guo and Z.H. Song. 2007. Influence of pore diameters on the immobilization of lipase in SBA-15. Ind. Eng. Chem. Res. 46: 4474–4479.
Kapoor, M. and M.N. Gupta. 2012. Lipase promiscuity and its biochemical applications. Process Biochem. 47: 555–569.
Khersonsky, O. and D.S. Tawfik. 2010. Enzyme promiscuity: a mechanistic and evolutionary perspective. Annu. Rev. Biochem. 79: 471–505.
Kim, J., H. Jia and P. Wang. 2006b. Challenges in biocatalysis for enzyme based biofuel cells. Biotechnol. Adv. 24: 296–308.
Kim, J., J.W. Grate and P. Wang. 2008. Nanobiocatalysis and its potential applications. Trends Biotechnol. 26: 639–646.
Kim, M.I., H.O. Ham, S.D. Oh, H.G. Park, H.N. Chang and S.H. Choi. 2006a. Immobilization of *Mucor javanicus* lipase on effectively functionalized silica nanoparticles. J. Mol. Catal. B: Enzym. 39: 62–68.
Klibanov, A.M. 2001. Improving enzymes by using them in organic solvents. Nature 409: 241–246.
Krishna, S.H. and N.G. Karanth. 2002. Lipases and lipase-catalyzed esterification reactions in nonaqueous media. Catal. Rev. 44: 499–591.
Krueger, A. 2008. The structure and reactivity of nanoscale diamond. J. Mater Chem. 18: 1485–1492.
Kuo, C.H., Y.H. Liu, C.M.J. Chang, J.H. Chen, J.H.C. Chang and C.J. Shieh. 2012. Optimum conditions for lipase immobilization on chitosan-coated Fe_3O_4 nanoparticles. Carbohydrate Polym. 87: 2538–2545.
Laane, C., S. Boeren, V. Kees and C. Veeger. 1987. Rules for optimization of biocatalysis in organic solvents. Biotechnol. Bioeng. 30: 81–87.
Lee, G., H. Joo and J. Lee. 2008. The use of polyaniline nanofibre as a support for lipase mediated reaction. J. Mol. Catal. B: Enzym. 54: 116–121.
Lee, S.H., T.T.N. Doan, K. Won, S.H. Ha and Y.M. Koo. 2010. Immobilization of lipase within carbon nanotube-silica composites for non-aqueous reaction systems. J. Mol. Catal. B: Enzym. 62: 169–172.
Li, D., M.B. Muller, S. Gilje, R.B. Kaner and G.G. Wallance. 2008. Processable aqueous dispersions of graphene nanosheets. Nat. Nanotechnol. 3: 101–105.
Li, Y., J. Quan, C.B. White, G.R. Williams, J.X. Wu and L.M. Zhu. 2012. Electrospun polyacrylonitrileglycopolymer nanofibrous membranes for enzyme immobilization. J. Mol. Catal. B: Enzym. 76: 15–22.
Liu, K.K., M.F. Chen, P.Y. Chen, T.F.J. Lee, C.C. Cheng, C.C. Chang et al. 2008. Alpha-bungarotoxin binding to target cell in a developing visual system by carboxylated nanodiamond. Nanotechnol. 19: 1–10.
Long, J.W., M.S. Logan, C.P. Rhodes, E.E. Carpenter, R.M. Stroud and D.R. Rolison. 2004. Nanocrystalline iron oxide aerogels as mesoporous magnetic architectures. J. Am. Chem. Soc. 126: 16879–16889.
Lu, A.H., E.L. Salabas and F. Schuth. 2007. Magnetic nanoparticles: synthesis, protection, functionalization, and application. Angew Chem. Int. Ed Engl. 46: 1222–1244.
Ma, Y.X., Y.F. Li, G.H. Zhao, L.Q. Yang, J.Z. Wang, X. Shan et al. 2012. Preparation and characterization of graphite nanosheets decorated with Fe_3O_4 nanoparticles used in the immobilization of glucoamylase. Carbon 50: 2976–2986.
Matori, M., T. Asahara and Y. Ota. 1991. Positional specificity of microbial lipases. J. Ferment. Bioeng. 72: 397–8.
Matsura, S.I., R. Ishii, T. Itoh, S. Hamakawa, T. Tsunoda, T. Hanaoka et al. 2011. Immobilization of enzyme-encapsulated nanoporous material in a microreactor and reaction analysis. Chem. Eng. J. 167: 744–749.
Meng, X., G. Xu, Q. Zhou, J. Wu and L. Yang. 2014. Highly efficient solvent-free synthesis of 1, 3-diacylglycerols by lipase immobilized on nano-sized magnetite particles. Food Chem. 143: 319–324.
Miled, N., F. Beisson, J. De Caro, A. De Caro, V. Arondel and R. Verger. 2001. Interfacial catalysis by lipases. J. Mol. Catal. B: Enzym. 11: 165–71.
Nair, S., J. Kim, B. Crawford and S.H. Kim. 2007. Improving biocatalytic activity of enzyme-loaded nanofibers by dispersing entangled nanofiber structure. Biomacromolecules 8: 1266–1270.
Nardini, M. and B.W. Dijkstra. 1999. α/β Hydrolase fold enzymes: the family keeps growing. Curr. Opin. Struct. Biol. 9: 732–7.

Neri, D.F.M., V.M. Balcao, F.O.Q. Dourado, J.M.B. Oliveira and J.A. Teixeira. 2011. Immobilized b-galactosidase onto magnetic particles coated with polyaniline: support characterization and galactooligosaccharides production. J. Mol. Catal. B Enzym. 70: 74–80.
Ngo, T.P.N., W. Zhang, W. Wang and Z. Li. 2012. Reversible clustering of magnetic nanobiocatalysts for high performance biocatalysis and easy catalyst recycling. Chem. Commun. 48: 4585.
Nguyen, T.T.B., H.C. Chang and V.W.K. Wu. 2007. Adsorption and hydrolytic activity of lysozyme on diamond nanocrystallites. Diamond Relat. Mater. 16: 872–876.
Nikolic, M., V. Srdic and M. Antov. 2009. Immobilization of lipase into mesoporous silica particles by physical adsorption. Biocatal. Biotransform. 27: 254–262.
Ogino, H., K. Yusui, T. Shiotani, T. Ishihara and H. Ishikawa. 1995. Organic solvent tolerant bacterium which secretes an organic solvent-stable proteolytic enzyme. Appl. Environ. Microbiol. 61: 4258–62.
Ogino, H. and H. Ishikawa. 2001. Enzymes which are stable in the presence of organic solvents. J. Biosci. Bioeng. 91: 109–116.
Ollis, D.L., E. Cheah, M. Cygler, B. Dijkstra, F. Frolow, S.M. Franken et al. 1992. The α/β hydrolase fold. Protein Eng. 5: 197–211.
Ozyilmaz, G. and E. Gezer. 2010. Production of aroma esters by immobilized *Candida rugosa* and porcine pancreatic lipase into calcium alginate gel. J. Mol. Catal. B: Enzym. 64: 140–145.
Park, J.M., M. Kim, H.S. Park, A. Jang, J. Min and Y.H. Kim. 2013. Immobilization of lysozyme-CLEA onto electrospun chitosan nanofiber for effective antibacterial applications. Int. J. Biol. Macromol. 54: 37–43.
Park, S. and R.S. Ruoff. 2009. Chemical methods for the production of graphenes. Nat. Nanotechnol. 4: 217–223.
Patel, V., H. Gajera, A. Gupta, L. Manocha and D. Madamwar. 2015. Synthesis of ethyl caprylate in organic media using Candida rugosa lipase immobilized on exfoliated graphene oxide: Process parameters and reusability studies. Biochem. Eng. J. 95: 62–70.
Patel, V. and D. Madamwar. 2015. Lipase from solvent-tolerant *Pseudomonas* sp. DMVR46 strain adsorb on multiwalled carbon nanotubes: application for enzymatic biotransformation in organic solvents. Appl. Biochem. Biotechnol. 117: 1313–1326.
Pavlidis, I.V., T. Tsoufis, A. Enotiadis, D. Gournis and H. Stamatis. 2010. Functionalized multi-wall carbon nanotubes for lipase immobilization. Adv. Eng. Mater. 12: 179–183.
Pavlidis, I.V., T. Vorhaben, D. Gournis, G.K. Papadopoulos, U.T. Bornscheuer and H. Stamatis. 2012a. Regulation of catalytic behaviour of hydrolases through interactions with functionalized carbon-based nanomaterials. J. Nanoparticle. Res. 14: 842.
Pavlidis, I.V., T. Vorhaben, T. Tsoufis, P. Rudolf, U.T. Bornscheuer, D. Gournis et al. 2012b. Development of effective nanobiocatalytic systems through the immobilization of hydrolases on functionalized carbon-based nanomaterials. Bioresour. Technol. 115: 164–171.
Pedrosa, V.A., S. Paliwal, S. Balasubramanian, D. Nepal, V. Davis, J. Wild et al. 2010. Enhanced stability of enzyme organophosphate hydrolase interfaced on the carbon nanotubes. Colloids Surf. B Biointerfaces 77: 69–74.
Persano, L., A. Camposeo, C. Tekmen and D. Pisignano. 2013. Industrial upscaling of electrospinning and applications of polymer nanofibers: a review. Macromol. Mater. Eng. 298: 504–520.
Petkar, M., A. Lali, P. Caimi and M. Daminati. 2006. Immobilization of lipases for non-aqueous synthesis. J. Mol. Catal. B: Enzym. 83–90.
Pleiss, J., M. Fischer and R.D. Schmid. 1998. Anatomy of lipase binding sites: the scissile fatty acid binding site. Chem. Phys. Lipids. 93: 67–80.
Plessis, D.M., M. Botes, L.M.T. Dicks and T.E. Cloete. 2012. Immobilization of commercial hydrolytic enzymes on poly (acrylonitrile) nanofibers for anti-biofilm activity. J. Chem. Technol. Biotechnol. 88: 585–593.
Ponvel, K.M., D.G. Lee, E.J. Woo, I.S. Ahn and C.H. Lee. 2009. Immobilization of lipase on surface modified magnetic nanoparticles using alkyl benzenesulfonate. Korean J. Chem. Eng. 26: 127–130.
Raghavendra, T., A. Basak, L. Manicha, A. Shah and D. Madamwar. 2013. Robust nanobioconjugates of *Candida antarctica* lipase B- multiwalled carbon nanotubes: Characterization and application for multiple usages in non-aqueous biocatalysis. Bioresous. Technol. 140: 103–110.
Rebelo, L.P., C.G.C.M. Netto, H.E. Toma and L.H. Andrade. 2010. Enzymatic kinetic resolution of (RS)-1-(Phenyl)ethanols by *Burkholderia cepacia* lipase immobilized on magnetic nanoparticles. J. Braz. Chem. Soc. 21: 1537–1542.
Reis, P., K. Holmberg, H. Watzke, M.E. Leser and R. Miller. 2009. Lipases at interfaces: a review. Adv. Colloid Interface Sci. 147: 237–50.
Ren, Y., J.G. Rivera, L. He, H. Kulkarni, D.K. Lee and P.B. Messersmith. 2011. Facile, high efficiency immobilisation of lipase enzyme on magnetic iron oxide nanoparticle via a biomimetic coating. BMC Biotechnol. 11: 63.
Reshmi, R., G. Sanjay and S. Sugunan. 2007. Immobilization of α-amylase on zirconia: a heterogeneous biocatalyst for starch hydrolysis. Catal. Commun. 8: 393–399.
Ribeiro, B.D., A.M. de Castro, M.A.Z. Coelho and D.M.G. Freire. 2011. Production and use of lipases in bioenergy: a review from the feedstocks to biodiesel production. Enzyme Res.
Roessl, U., J. Nahalka and B. Nidetzky. 2010. Carrier-free immobilized enzymes for biocatalysis. Biotechnol. Lett. 32: 341–350.
Rogalska, E., C. Cudrey, F. Ferrato and R. Verger. 1993. Steroselective hydrolysis of triglycerides by animal and microbial lipases. Chirality 5: 24–30.
Safarik, I. and M. Safarikova. 2009. Magnetic nano and microparticles in biotechnology. Chem. Pap. 63: 497–505.

Sakai, S., Y.P. Liu, T. Yamaguchi, R. Watanabe, M. Kawabe and K. Kawakami. 2010. Production of butyl-biodiesel using lipase physically adsorbed onto electrospun polyacrylonitrile fibers. Bioresour. Technol. 101: 7344–7349.
Sang, L.C. and M.O. Coppens. 2011. Effects of surface curvature and surface chemistry on the structure and activity of proteins adsorbed in nanopores. Phys. Chem. Chem. Phys. 13: 6689.
Sarda, L. and P. Desnuelle. 1958. Action de la lipase pancreatique sur les esters en Cmulsion. Biochem. Biophys. Acta 30: 513–21.
Saxena, R.K., P.K. Ghosh, R. Gupta, W.S. Davidson, S. Bradoo and R. Gulati. 1999. Microbial lipases: potential biocatalysts for the future. Curr. Sci. 77: 101–15.
Schmid, R.D. and R. Verger. 1998. Lipases: interfacial enzymes with attractive applications. Angew Chem. Int. Ed. 37: 1608–33.
Sen, T., I.J. Bruce and T. Mercer. 2010. Fabrication of novel hierarchically ordered porous magnetic nanocomposites for biocatalysis. Chem. Commun. 46: 6807–6809.
Serra, E., A. Mayoral, Y. Sakamoto, R.M. Blanco and I. Diaz. 2008. Immobilization of lipase in ordered mesoporous materials: effect of textural and structural parameters. Microporous Mesoporous Mater. 114: 201–213.
Shah, S., K. Solanki and M.N. Gupta. 2007. Enhancement of lipase activity in non-aqueous media upon immobilization on multi-walled carbon nanotubes. Chem. Cent. J. 1: 30.
Shi, Q., D. Yang, Y. Su, J. Li, Z. Jiang, Y. Jiang et al. 2007. Covalent functionalization of multi-walled carbon nanotubes by lipase. J. Nanopart. Res. 9: 1205–1210.
Shim, M., N.W.S. Kam, R.J. Chen, Y. Li and H. Dai. 2002. Functionalization of carbon nanotubes for biocompatibility and biomolecular recognition. Nano. Lett. 2: 285–288.
Solanki, K. and M.N. Gupta. 2011. Simultaneous purification and immobilization of *Candida rugosa* lipase on superparamagnetic Fe_3O_4 nanoparticles for catalyzing transesterification reactions. New J. Chem. 35: 2551.
Song, Y.S., H.Y. Shin, J.Y. Lee, C. Park and S.W. Kim. 2012. β-galactosidase immobilized microreactor fabricated using a novel technique for enzyme immobilization and its application for continuous synthesis of lactulose. Food Chem. 133: 611–617.
Sonnet, P.E. 1998. Lipase selectivities. J. Am. Oil Chem. Soc. 65: 900–5.
Sotiropoulou, S., V. Gavalas, V. Vamvakaki and N.A. Chaniotakis. 2003. Novel carbon materials in biosensor systems. Biosens. Bioelectron. 18: 211–215.
Souza, K.C., N.D. Mohallem, de EMB and S. Sousa. 2011. Nanocompositos magneticos: potencialidades de aplicacoes em biomedicine. Quim Nova. 34: 1692–1703.
Tong, Z., R. Yuan, Y. Chai, Y. Xie and S. Chen. 2007. A novel and simple biomolecules immobilization method: electrodeposition ZrO_2 doped with HRP for fabrication of hydrogen peroxide biosensor. J. Biotechnol. 128: 567–575.
Torres, S. and G.R. Castro. 2004. Non-aqueous biocatalysis in homogeneous solvent systems. Food Technol. Biotechnol. 42: 271–277.
Tran, D.T., C.L. Chen and J.S. Chang. 2012. Immobilization of *Burkholderia* sp. lipase on a ferric silica nanocomposite for biodiesel production. J. Biotechnol. 158: 112–119.
Treccani, L., T.Y. Klein, F. Meder, K. Pardun and K. Rezwan. 2013. Functionalized ceramics for biomedical, biotechnological and environmental applications. Acta Biomater. 9: 7115–7150.
Uhlig, H. 1998. Industrial enzymes and their applications. New York: John Wiley & Sons.
Ulijn, R.V., L.D. Martin, L. Gardossi and P.J. Halling. 2003. Biocatalysis in reaction mixtures with undissolved solid substrates and products. Curr. Org. Chem. 7: 1333–1346.
Uppenberg, J., H.M. Trier, S. Patkar and T.A. Jones. 1994. The sequence, crystal structure determination and refinement of two crystal forms of lipase B from *Candida antarctica.* Structure 2: 293–308.
van Tilbeurgh, H., M.P. Egloff, C. Martinez, N. Rugani, R. Verger and C. Cambillau. 1993. Interfacial activation of the lipase-procolipase complex by mixed micelles revealed by X-ray crystallography. Nature 362: 814–20.
Verma, M., C. Barrow and M. Puri. 2013. Nanobiotechnology as a novel paradigm for enzyme immobilization and stabilization with potential applications in biodiesel production. Appl. Microbiol. Biotechnol. 97: 23–39.
Wang, L. and R. Jiang. 2011. Reversible His-tagged enzyme immobilization on functionalized carbon nanotubes as nanoscale biocatalyst. Methods Mol. Biol. 743: 95–106.
Wang, Y., Z.H. Li, J. Wang, J.H. Li and Y.H. Lin. 2011. Graphene and graphene oxide: biofunctionalization and applications in biotechnology. Trends Biotechnol. 29: 205–212.
Wang, Z.G., J.Q. Wang and Z.K. Xu. 2006. Immobilization of lipase from *Candida rugosa* on electrospun polysulfone nanofibrous membranes by adsorption. J. Mol. Catal. B: Enzym. 42: 45–51.
Wei, L., W. Zhang, H. Lu and P. Yang. 2010. Immobilization of enzyme on detonation nanodiamond for highly efficient proteolysis. Talanta 80: 1298–1304.
Wilson, L., J.M. Palomo, G. Fernandez-Lorente, A. Illanes, J.M. Guisan and R. Fernandez- Lafuente. 2006. Improvement of the functional properties of a thermostable lipase from alcaligenes sp. via strong adsorption on hydrophobic supports. Enzyme. Microb. Technol. 38: 975–80.
Winkler, F.K., A. D'Arcy and W. Hunziker. 1990. Structure of human pancreatic lipase. Nature 343: 771–4.
Woodley, J.M. 2008. New opportunities for biocatalysis: making pharmaceutical processes greener. Trends Biotechnol. 26: 321–327.
Woolley, P. and S.B. Petersen. 1994. Lipases: Their Structure, Biochemistry and Application. UK: Cambridge University Press.
Wu, Y., Y. Wang, G. Luo and Y. Dai. 2010. Effect of solvents and precipitant on the properties of chitosan nanoparticles in a water-in-oil microemulsion and its lipase immobilization performance. Bioresour. Technol. 101: 841–844.

Yang, W., P. Thordarson, J.J. Gooding, S.P. Ringer and F. Braet. 2007. Carbon nanotubes for biological and biomedical applications. Nanotechnol. 18: 412001.
Yang, Y., Y. Bai, Y. Li, L. Lin, Y. Cui and C. Xia. 2008. Preparation and application of polymer-grafted nanoparticles for lipase immobilization. J. Magn. Magn. Mater. 320: 2350–2355.
Yiu, H.H.P. and M.A. Keane. 2012. Enzyme-magnetic nanoparticle hybrids: new effective catalysts for the production of high value chemicals. J. Chem. Technol. Biotechnol. 87: 583–594.
Youshko, M.I., H. Moody, A. Bukhanov, V.H.J. Boosten and V.K. Svedas. 2004. Penicillin acylase-catalyzed synthesis of β-lactam antibiotics in highly condensed aqueous systems: beneficial impact of kinetic substrate supersaturation. Biotechnol. Bioeng. 85: 323–329.
Yu, C.H., A. Al-Saadi, S.J. Shih, L. Qiu, K.Y. Tam and S.C.T. Tsang. 2009. Immobilization of BSA on silica-coated magnetic iron oxide nanoparticle. J. Phys. Chem. 113: 537–543.
Yu, L., I.A. Banerjee, X.Y. Gao, N. Nuraje and H. Matsui. 2005. Fabrication and application of enzyme-incorporated peptide nanotubes. Bioconj. Chem. 16: 1484–1487.
Zhang, C. and S.K. Kim. 2010. Research and application of marine microbial enzymes: status and prospects. Mar Drugs 8: 1920–34.
Zhang, J., F. Zhang, H. Yang, X. Huang, H. Liu, J. Zhang et al. 2010. Graphene oxide as a matrix for enzyme immobilization. Langmuir 26: 6083–6085.
Zhang, Y.W., M.K. Tiwari, M. Jeya and J.K. Lee. 2011. Covalent immobilization of recombinant *Rhizobium etli* CFN42 xylitol dehydrogenase onto modified silica nanoparticles. Appl. Microbiol. Biotechnol. 90: 499–507.
Zhou, L., Y. Jiang, J. Gao, X. Zhao and Q. Zhou. 2012. Oriented immobilization of glucose oxidase on graphene oxide. Biochem. Eng. J. 69: 28–31.
Zong, S., Y. Cao, Y. Zhou and H. Ju. 2007. Reagentless biosensor for hydrogen peroxide based on immobilization of protein in zirconia nanoparticles enhanced grafted collagen matrix. Biosens. Bioelectron. 22: 1776–1782.
Zou, B., Y. Hu, D. Yu, J. Xia, S. Tang, W. Liu et al. 2010. Immobilization of porcine pancreatic lipase onto ionic liquid modified mesoporous silica SBA-15. Biochem. Eng. J. 53: 150–153.

9

Metal Oxide Based Heterojunction Nanoscale Materials for Chemiresistive Gas Sensors

Keerthi G. Nair, V.P. Dinesh and *P. Biji**

Introduction

Conductometric (chemiresistive) sensors are comprised of a significant part of the gas sensor component market. While a large variety of materials such as, metal oxides, carbon based materials, polymeric materials have been extensively used for gas sensor applications (Korotcenkov and Cho 2013), metal oxide sensors remain a widely used choice for detection of a range of gas species (Ramgir et al. 2010). These devices offer low-cost portable device possessing, elevated sensitivity, quick response, virtual simplicity and compatibility. The working principle of a classic metal oxide based resistive gas sensor is based on the change of the state of equilibrium of the surface adsorbed oxygen in the presence of target analyte molecules. The resulting variation in concentration of chemisorbed oxygen is documented as change in resistance of the gas-sensing materials (Cuenya and Kolmakov 2008). In general, reducing gases (CO, H_2, CH_4, etc.) lead to an increase in the conductivity in *n*-type semiconductors and a decrease in conductivity in *p*-type materials, respectively, whereas the effect of oxidizing gases (O_3, etc.) acts vice versa. The sensor response (sensitivity) of such devices, to be exact, the capability of a sensor to detect a given concentration of an analyte gas is usually estimated as the ratio of their electrical resistance ($S = R_{gas}/R_{air}$, or R_{air}/R_{gas}) calculated in air and in the atmosphere containing the target gas. The velocity of sensor response is explained as the response or recovery time, which characterizes the time taken for the sensor output to achieve 90% of its saturation value after exposure to the analyte gas (Tiemann 2007). Numerous materials have been reported to be active for metal oxide sensor design including both single and multi-component oxides (Korotcenkov et al. 2013). In practice, nanoscale materials have

Nanosensor Laboratory, Department of Chemistry, PSG College of Technology, Avinashi Road, Peelamedu, Coimbatore-641 004, Tamil nadu, India.
Emails: keerthiak89@gmail.com; vp.dinesh@gmail.com
* Corresponding author: bijujal23@yahoo.co.in

found promising applications in solid-state gas sensors (Barsan and Weimar 2001). Nanocrystalline and poly-crystalline materials have the most favorable blend of vital properties for sensor applications as well as high surface area as a result of small crystallite size, inexpensive design technology, and permanence of both structural and electro-physical properties.

Metal-oxide based gas sensors

Metal Oxides (MOS) possess unique electronic, chemical, and physical properties that are often highly sensitive to changes in the chemical environment. In reality, majority of the commercial solid state chemical sensors are based on appropriately structured metal oxides (SnO_2, ZnO, TiO_2, etc.) that have proven to be capable of detecting variety of gases with good stability, high sensitivity and low production cost (Barsan et al. 2007). The gas sensing mechanism in these materials is governed by the reactions which take place at the sensor surface between the sensitive layer and the target gas molecules leading to conductivity changes. It involves chemisorption of oxygen on the metal oxide surface pursued by charge transfer throughout the reactions of oxygen with target gas molecules (Tiemann 2007). The adsorbed gas molecules introduce electrons into or remove electrons from the semiconducting material, depending on their reducing or oxidizing nature, respectively (Yamazoe et al. 2003). This mechanism results in a change of the film conductivity corresponding to the gas concentration. Deep research and improvement have been conducted to design highly selective, sensitive and stable gas sensors, as Seiyama first observed gas sensing effects in Zinc Oxide (ZnO) (Seiyama et al. 1962). Later, the range of sensitive materials was extended to SnO_2, TiO_2, WO_3, In_2O_3 and other oxides (Morrison 1981). Semiconductor metal oxide based gas sensors are chosen for automotive, domestic, environment and emission monitoring, industrial and medical applications (Yamazoe 1991). Although semiconductor metal oxide gas sensors are promising, high power consumption, low selectivity and lack of long term stability have prohibited their use in more demanding applications (Gopel and Schierbaum 1995). In the literature, several approaches are reported to reduce these limitations, such as use of catalysts and promoters, multi-sensor array systems, temperature modulation and using materials in nanostructured forms (Korotcenkov et al. 2013).

The performance of MOS gas sensors improves with a reduction in the size of the oxide particles, as sensing phenomenon is a surface reaction during the interaction process. As a result, the performance of a gas sensor is directly correlated to porosity, granularity and ratio of exposed surface area to volume (Barsan and Weimer 2001). Current progress in the synthesis, structural characterization and investigation of physical properties of nanostructured metal oxides offer the prospect to significantly progress the response of gas sensors based on these materials. There are plenty of reports and reviews existing in literature on several MOS-based gas sensors (Tiemann 2007, Yamazoe 1991, Korotcenkov and Cho 2014). Enormous efforts have been performed in recent years by various groups to improve the performance parameters by suitable materials engineering. Lao et al. (2002) fabricated and analyzed the performance of each SnO_2 single crystal nanoribbons and found the detection limit for NO_2 as 3 ppm with fast response/recovery times. The variation in the electrical conductivity was observable even close to RT and was altered by molecular adsorption on surface states aided by ultraviolet (UV) light with an energy near the SnO_2 bandgap (Lee et al. 2008). Comini et al. (2009) deposited SnO_2 nanobelts on Platinum interdigitated electrodes and investigated their behavior in the range 300°C–400°C. The device showed excellent sensitivity towards CO, ethanol, and NO_2 which could be detected down to few ppb concentrations. While CO and ethanol adsorption resulted in an enhanced conductivity, the electrical resistivity of the nanobelts was increased by NO_2. Kolmakov et al. (2005) used nanoporous alumina as a template for synthesizing arrays of parallel tin nanowires, which can be converted to polycrystalline SnO_2 nanowires of restricted composition and dimension. Conductance measurements of individual nanowires were performed in inert, oxidizing, and reducing environments in the temperature range 25°C–300°C. Configured as a CO sensor, a detection limit of hundreds of ppm in dry air at 300°C was measured with a sensor response time of 30 s. McCue and Ying (2007) observed that SnO_2–In_2O_3 nanocomposite exhibited high response and selectivity towards NO_x and CO, with a sensitivity based on the composition and calcination temperature. Performance of the sensor was further improved through the introduction of small quantity of metals or further oxides as dopants and surface coatings (Kolmakov et al. 2005).

Vertically aligned ZnO nanowire arrays were grown alongside substrate by Cheng et al. (2011), where the gas sensor device showed good sensing properties towards NO_2. Similarly, Comini et al. (2002) investigated zinc oxide nanowires' networks using an evaporation method with standardized copper addition. Sensor response was found to be increased towards ethanol, acetone, NO_2, and CO gases at 400°C by addition of Cu. Transistors based on both single and multiple In_2O_3 nanowires operating at room temperature was found to detect NO_2 down to ppb levels (Comini et al. 2002). Table 1 gives an overview of parameters of metal-oxide based nanosensor used for detection of various gases including NO_2 at varied operating temperatures.

Table 1. Gas sensing properties of chemo-resistors based on various metal oxides.

Materials		**Target Gas**	**Lowest Detection Concentration (ppm)**	**Working Temperature (°C)**	**Reference**
SnO_2	Nanorods	Ethanol	50	300	Wang et al. 2008
	Nanoparticles	H_2	10	300	Wang et al. 2008
	Nanowire	Humidity, H_2	100	300	Huang et al. 2009
In_2O_3	Nanowires	Ethanol	100	370	Xiang et al. 2004
	Nanowhiskers	H_2S	0.2	RT	Kaur et al. 2008
	Nanowires	H_2S	0.5	268.5	Xu et al. 2006a
ZnO	Nanostructures	H_2S	25	RT	Kalyamwar 2013
	Nanoparticles	H_2	10	RT	Wang et al. 2005
	Nanorods	Ethanol	100	325	Ge et al. 2008
TeO_2	Nanowires	NO_2	10	26	Liu et al. 2007
	Nanorods	H_2S	10	26	Liu et al. 2007
	Nanorods	NH_3	200	26	Liu et al. 2007
CuO	Nanowires	NO_2	2	300	Kim et al. 2008
	Nanoparticles	Ethanol	5	200	Gou et al. 2008
CdO	Nanowires	NO_2	1	100	Guo et al. 2008

Metal-oxide sensors often display remarkable changes in their electrical properties, e.g., their work function and electrical conductivity, upon exposure to O_2, CO, NO_2 and other reactive gases (Teimann 2007). Gas sensors normally function in air at temperatures between 100°C and 400°C. Under these conditions, the surface of the sensing layer, or the particles present in the sensing film, in the case of porous materials, is covered with adsorbed oxide species. Specifically, chemisorbed oxygen or superoxide adions ($O^-_{(ads)}$ or $O_2^-{}_{(ads)}$) play vital roles in the sensing mechanism by involving in the gas/material interaction (Datskos et al. 2001). Therefore, the work function and surface conductivity are receptive to these gases and gets altered according to the nature of gas. This effect was first observed by Brattain and Bardeen (1953) for Ge based sensors and later supported by Heiland (1954) for ZnO, followed by an important work done by Seiyama et al. (1962) and Taguchi (1971) which had taken the metal oxide based sensors to the next level. Since then, much advancement had come to improve the performance of sensor in terms of selectivity, sensitivity, stability, recovery and response time.

Working principle of metal-oxide sensors

In general, the sensing of *n*-type semiconductor is based on the depletion layer formation in the presence of oxygen environment. The ionosorbed oxide species act as electron acceptors due to the relative energetic position with respect to the Fermi level (E_f) as shown in the Figure 1 (Barsan and Weimer 2001). As mentioned earlier, the surface predominant oxygen species greatly depend on the temperature.

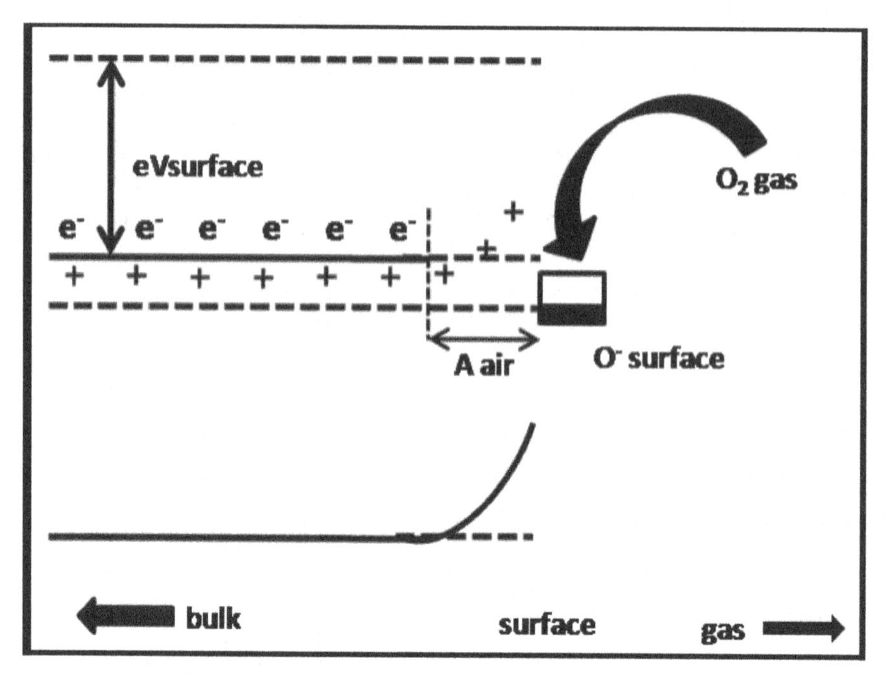

Figure 1. Simplified model illustrating band bending in a wide bandgap semiconductor after chemisorption of charged species (here the ionosorption of oxygen) on surface sites. E_C, E_V, and E_F denote the energy of the conduction band, valence band, and the Fermi level, respectively, while L_{air} denotes the thickness of the space-charge layer, and $eV_{surface}$, the potential barrier. The conducting electrons are represented by e– and + represents the donor sites (Barsan and Weimar 2001).

It is established in the literature that the reactive oxygen species on the surface of ZnO differs according to the temperature, viz., < 200°C O_2^- is dominant, 200–400°C O^- is prominent and > 400°C O^{2-} prevails (Kim and Lee 2014). The required electrons for this process originate from donor sites, i.e., intrinsic oxygen vacancies and are extracted from the conduction band (E_c) and trapped at the surface, leading to an electron-depleted surface region, known as depletion layer or space-charge layer Λ_{air} (Samson and Fonstad 1973, Jarzebski and Marton 1976, Maier and Gopel 1988, Gopel and Schierbaum 1995). The maximum surface coverage of about 10^{-3} to 10^{-2} cm^{-1} ions is dictated by the Weisz limitation, which describes the equilibrium between the Fermi level and energy of surface-adsorbed species (Weisz 1953, Lampe et al. 1997).

Band bending of semiconducting metal-oxides occurs due to the difference in their surface charges, in general, the surface potential barrier height, $eV_{surface}$ in the range of 0.5 to 1.0 eV. The height ($eV_{surface}$) and depth (Λ_{air}) of the band bending can be determined by the surface charge, which is determined by the amount and type of adsorbed oxygen. At the same time, L_{air} depends on the Debye length L_D, which is a characteristic of the semiconductor material for a particular donor concentration and can be given as,

$$L_D = \sqrt{\frac{\varepsilon_0 \varepsilon K_B T}{e^2 n_d}} \tag{1}$$

where, ε_o is the permittivity of free space, ε is the dielectric constant, K_B is Boltzmann's constant, T is the operating temperature, e is the electron charge and n_d is the carrier concentration (donor concentration (Barsan and Weimar 2001)).

In polycrystalline sensing materials, electronic conductivity occurs along percolation paths via grain-to-grain contacts and therefore depends on the value of $eV_{surface}$ of the adjacent grains. $eV_{surface}$ represents the Schottky barrier. The conductance, G of the sensing material in this case can be written as (Madou and Morrison 1989).

$$G \approx \exp\left(\frac{-eV_{surface}}{K_B T}\right) \tag{2}$$

As per the above equation, the height of the Schottky barrier is directly related to the conductance of the material (n or p type). As a consequence, the height of Schottky barrier is reduced, which results in the increased conductance of nanomaterials as a whole. In nanomaterials, the size effect comes into picture due to their large surface to volume ratio (Lenaerts et al. 1995). Hence, minimizing the particle diameter, D will lead to converging Schottky barriers, if the radius, $r = D/2$ is in the range of the space-charge layer. This means that with further decrease in the radius, i.e., smaller than Λ_{air}, the depleted zones start to overlap and, consequently, the electrical properties are primarily resolute by surface states as shown in Figure 2. Therefore, a pronounced dependence of the sensitivity on the particle size is estimated with enhanced sensitivity towards smaller dimensions. If the barrier height of adjacent Schottky barriers fall below the thermal energy, that is, $eV_{surface} \leq k_B T$, the so called "flat-band condition" is realized and the energetic difference between surface and bulk vanishes, that is, the conductance is proportional to the difference of the Fermi level, E_F and the bottom of the conduction band, E_C, thus leading to the formation of a derived Eq. 2 relating conductance, conduction band and Fermi level (Schierbaum et al. 1991).

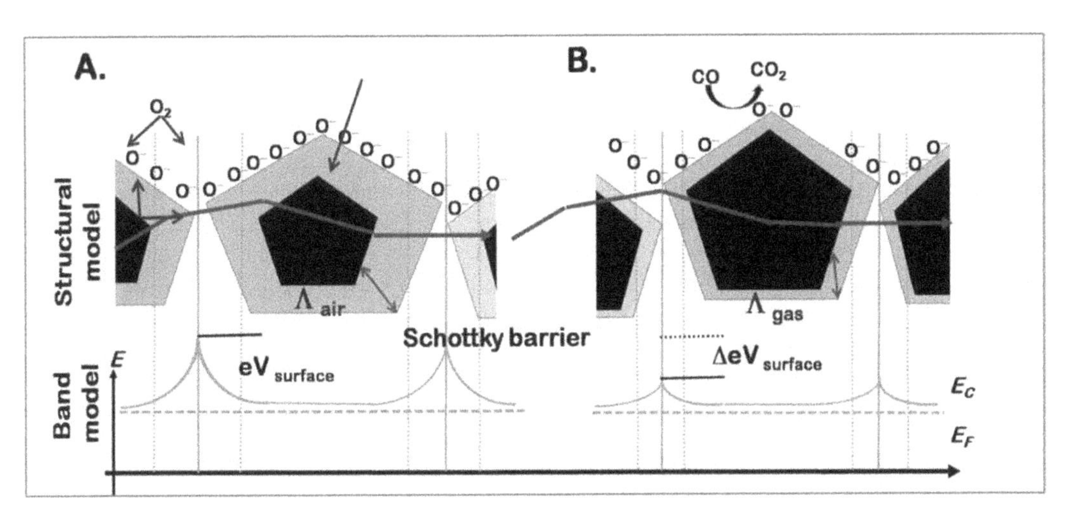

Figure 2. Structural and band model showing the role of intergranular contact regions in determining the conductance over a polycrystalline metal oxide semiconductor: (a) initial state, and (b) effect of CO on Λ_{air} and $eV_{surface}$ for large grains (Schierbaum et al. 1991).

Structural parameters of metal oxides controlling gas-sensing characteristics

As mentioned earlier, the fundamentals of resistive type sensor operation are based on the changes in resistance (or conductance) of the material as induced by the surrounding gas. The changes are caused by various processes, which may be able to happen both at the surface and in the bulk of gas-sensing material (Kohl 2001). Possible processes, which can manage gas-sensing properties, are illustrated in Figure 3.

Investigation has established that all processes indicated in Figure 3 including catalysis, reduction/re-oxidation, adsorption/desorption and diffusion are significant in gas sensors and prejudiced by structural parameters of the sensor material. This affirms that gas-sensing properties are structurally susceptible as well. It has been shown in the previous reports (Samson and Fonstad 1973, Jarzebski and Marton

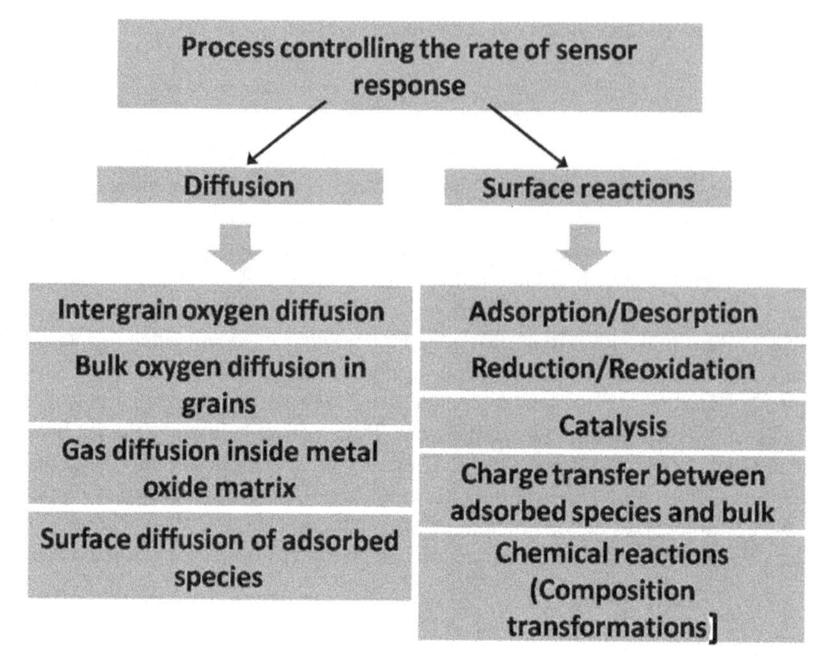

Figure 3. Possible processes controlling the sensor response of metal oxides (Kohl 2001).

1976, Maier and Gopel 1988, Gopel and Schierbaum 1995) that the influence of the above-mentioned factors on gas-sensing distinctiveness takes place throughout the transformation in the efficient area of inter-grain and inter-agglomerate contacts, active parameters of adsorption/desorption, band bending, number of surface sites, concentration of charge carriers, primary surface band bending, coordination number of metal atoms on the surface, etc. The major parameters influencing the gas sensing properties are discussed in the following sections in detail.

The effect of surface doping

Pure metal-oxides usually are not able to comply with all the requirements to act as a perfect chemiresistor, due to certain constraints, such as high operating temperature, formation of reactive oxide species on the surface, and poisoning effect, thereby lowering the sensitivity. To overcome the inherent limitations of the pure base material, doping with metals or oxides has established a profound impact on their sensor performance (Yamazoe et al. 2003). In this case, doping is an addition of catalytically active sites on the surface of the base material. Preferably, the doping process develops the sensor performance by increasing the sensitivity, favoring the selective interaction with the target analyte, thus increasing the selectivity and decreasing the response and recovery time, respectively, which results in the reduction of the working temperature. SnO_2-based sensors can also be promoted by the addition of little quantity of different metals, for example, Pt, Pd, and Ag (Kolmakov and Moskovits 2004). The metal additives change the sensor properties in a slightly difficult manner. To understand the promoting effects, it is necessary to consider both the catalytic activity of the metals or metal oxides and the surface properties of the oxide semiconductors. Two types of interactions among metal additive and oxide semiconductor have been envisaged, as shown in Figure 4 (Yamazoe et al. 2003), such as chemical sensitization and electronic sensitization effects. In chemical sensitization, the metal additive and the target gas are spilled-over to the semiconductor surface to react with the adsorbed oxygen. The metal additive thus catalytically assists the chemical reaction of the gas on the semiconductor. As a result, the surface coverage with oxygen, and therefore $eV_{surface}$, is reduced and accompanied by a change in conductance, while the catalytic cluster

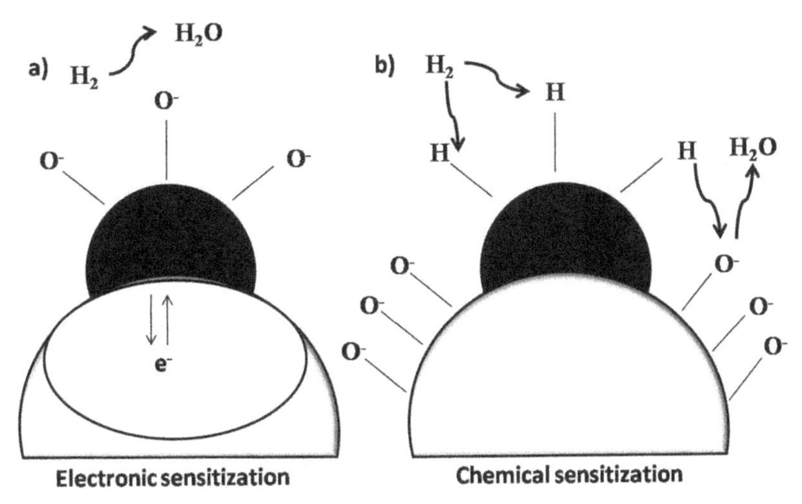

Figure 4. Mechanism of sensitization by metal or metal oxide additives: (a) electronic sensitization, where the additive is an acceptor for electrons and the redox state/chemical potential is changed by reaction with the analyte; (b) chemical sensitization by activation of the analyte (H_2) followed by spill over and change of the surface oxygen concentration (Yamazoe et al. 2003).

itself remains unchanged. During electronic sensitization, there is no such mass transfer between the additive and the oxide happens. Here, the additive in the oxidized state operates as a tough acceptor of electrons from the oxide, suggesting a surface space charge layer which is robustly depletive of electrons in the oxide close to the interface (Epifani 2001, Pagnier et al. 1999). When the additive is reduced in contact with the target gas, it relaxes the space charge layer by replacing the electrons to the oxide. The difference of work function of metal-oxides between the oxidized and reduced state is often large, which brings a huge enhancement in the response (R_a/R_g) to the gas. This type of sensitization has thus far been observed for SnO_2 sensors impregnated with Ag, Pd, or Cu (Chowdhuri et al. 2002, Chowdhuri et al. 2003, Korotcenkov et al. 2005). Such a modification in the oxidation state of the additive is liable for the endorsement of the gas response.

Grain size influence

It is well-known that the electro-physical properties of polycrystalline materials strongly depend on their microstructure (Brinzari et al. 1999, Barsan and Weimar 2001, Xu et al. 1991, Rothschild and Komem 2004). It has been established that the grain size and the width of the necks are the main parameters that control gas-sensing properties in metal oxide sensors as well. Moreover, in the frame of modern gas sensor models the role of grains size and necks size on sensor response may be attributed to the fundamentals of gas sensor operation (Xu et al. 1991, Rothschild and Tuller 2006, Schierbaurn et al. 1991). Usually it is displayed through the so-called "dimension effect", e.g., a comparison of the grain size (d) or necks width (X) the Debye length (L_d) can be given as (Rothschild and Tuller 2006);

$$L_D = \frac{1}{4}\frac{eKT}{2pe^2N} \tag{3}$$

where, k is the Boltzmann constant, T is the absolute temperature, e is the dielectric constant of the material and N is the concentration of charge carries.

Figure 5 illustrates the potential distribution across the neck and the role of necks in the conductivity of polycrystalline metal oxide environment. It is obvious that the width of the necks resolves the height of the potential barrier for current carriers, while the length of the necks decides the depletion-layer

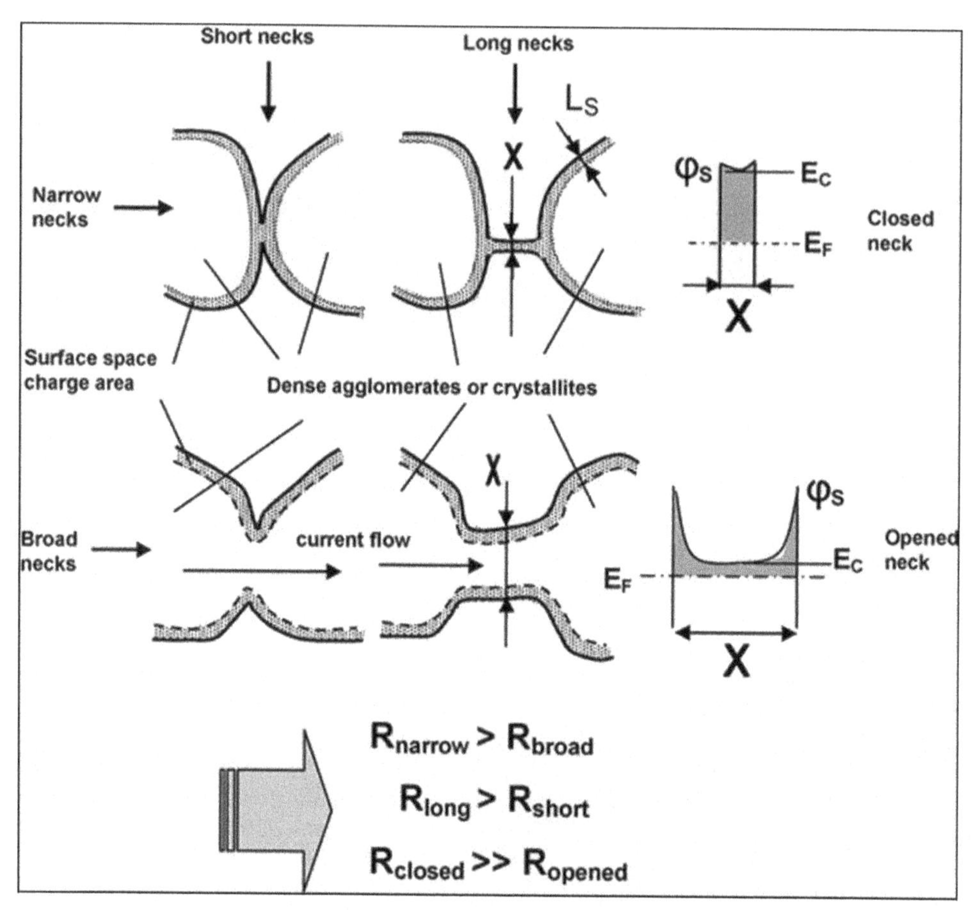

Figure 5. Diagram illustrating the role of necks in the conductivity of polycrystalline metal oxide matrix and the potential distribution across the neck. (Reprinted with permission from Korotcenkov 2007.)

width of the potential barrier. It is necessary to note that the enlargement of the necks length amplifies the role of necks in the constraint of metal oxide conductivity and likewise in gas sensing effects. The grain size would find out the depth of valley on the potential distribution within grains. The distribution of the potential appears alike to those observed in individual grains except that at the boundary of a 1-D structure with ohmic contact, the potential barrier would be noticeably lower than the potential barrier between grains.

Challenges in metal-oxide based gas sensors

Though metal-oxide based gas sensor paves the key for the futuristic real time devices, there are certain challenges which still need to be addressed to make it a big success. Lots of challenges still prevail in the MOS based gas sensors, as the field requires knowledge in solid-state physics, defect thermodynamics, kinetics and defect engineering (Kim et al. 2014). As the chemisorption of analyte molecules on the surface of semiconductors relies on charge transfer processes that mainly focuses on semiconductor junction physics and catalysis, it is still a wonder that a detailed understanding pertaining to the operation of chemical sensors remains unclear. Apart from that, problems in the 3 "*S*" have to be dealt with say, Stability, Selectivity, and Sensitivity. MOS sensors are promising candidates because of their attractive features like low-cost, simplicity and compatibility with modern electronic devices.

However, chemiresistive type MOS sensors have the main disadvantages of poor selectivity, where many intervening gases should be properly eliminated to choose the target analyte molecules. Secondly, long term stability is required for an active sensor to be in progress. The MOS based sensors are kept at elevated temperatures to attain their equilibrium as their working temperatures are really higher. Due to the long term exposure, these MOS sensors generally loses their sensitivity and stability. Hence, to address these issues, researchers have focused their attention to hybrid nanoscale materials having heterojunctions. The following sections provide a detailed account of heterojunction based MOS gas sensors.

Heterojunctions at metal oxide semiconductors

Heterojunctions are contacts between two different materials with interesting electrical or electro optical properties. Although a heterojunction, in general, is defined as the physical interface between two dissimilar materials, its usage in semiconductor research is normally restricted to a junction between two different monocrystalline semiconductor materials. Such heterojunctions can be classified as abrupt or graded according to the distance during which the transition from one material to the other is ended up across the interface. In the former case, the transition occurs within few atomic distances (< 1 μm), while in the latter, it takes place over distances of the order of several diffusion lengths. Another classification, which is often used in literature, involves naming the heterojunction by the type of conductivity present on both regions of the junction. If both semiconductors concerned have related types of conductivity, the junction is described as isotype heterojunction, otherwise it is termed as anisotype heterojunction.

It was Gubanov (Gubanov 1951, Schewchun and Wei 1964) who first theoretically analyzed these combinations, but heterojunction research was taken up only after Kroemer (Kroemer 1957, Oldham et al. 1963) suggested that anisotype heterojunctions might exhibit enormously elevated booster efficiencies in contrast to homojunctions. The first isotype and anisotype heterojunctions were fabricated by Anderson (Anderson 1960). He also published his landmark article in 1962 (Anderson 1962), which proposed a coherent model for heterojunctions. Such a model was essentially an extension of the Schottky model for metal-semiconductor diodes and similar to the Schottky model and described the critical parameters of the junction in terms of the parameters of the two component materials. The Schottky model predicts that

$$\varphi_n = \varphi_m - \chi \quad (4)$$

where, φ_n is the Schottky barrier for the interface between a given metal and an *n*-type semiconductor, φ_m is the metal work function and χ is the electron affinity of the semiconductor. The work function, φ_m, of a semiconductor is defined as the energy required for capturing an electron from the Fermi level through the bulk surface to the energy level of free space (or vacuum) outside the material. The work function therefore, being dependent on the Fermi level, varies as a function of the doping level. Electron affinity, χ defined as the energy to take an electron from the conduction band edge to the vacuum level, is more convenient to use since it is a material property that is invariant with normal doping. Some features of the heterojunction energy diagram of Figure 6 are quite similar to the corresponding features of other classes of semiconductor interfaces.

Anderson (1962) recognized the band discontinuities and the "built-in potential" as the basic parameters of a heterojunction. The latter is given by:

$$V_D = V_D^1 + V_D^2 \quad (5)$$

where, V_D^1 and V_D^2 as shown in Figure 6, are the band-bending potentials of both sides of the junction. The band bending is required to keep the Fermi energy constant everywhere in the system, whereas extreme from the junction its distance as of the valence (or conduction) band edge is entirely determined by doping. In the specific case illustrated in Figure 6, the band bending corresponds to an *n-p* heterojunction.

Recently, Jayaseelan and Biji (2016) explored the charge transport and adsorption kinetics of wet-chemically synthesized CuO nanocuboids towards H_2S sensing. Adsorption of oxygen (O_2) on the CuO

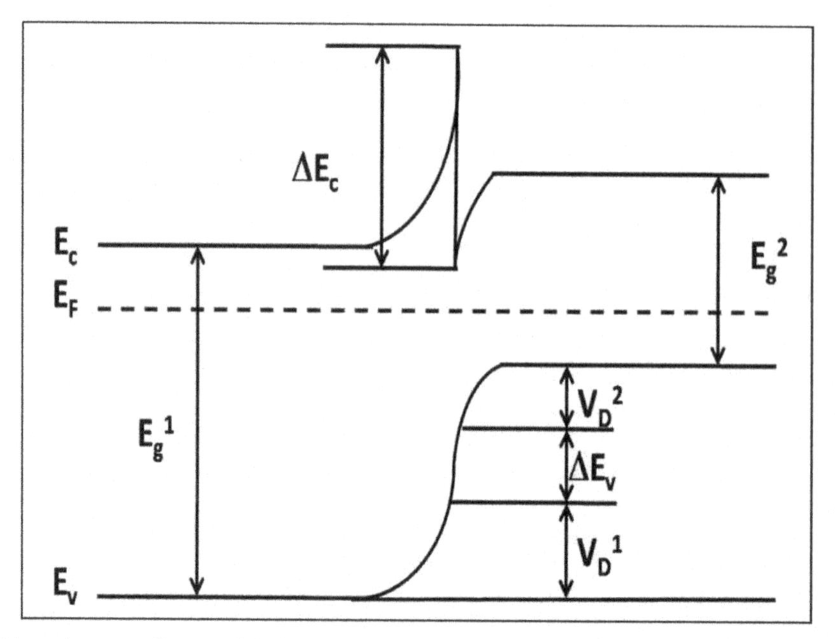

Figure 6. Schematic energy diagram of the interface between two different semiconductors with forbidden gaps E_g^1 E_g^2 E_c and E_v are the conduction and valence band edges, and E_F is the Fermi level. V_D^1 and V_D^2 measure the band bendings of the two sides of the junction. The difference between the two gaps is accommodated by the valence and conduction band discontinuities, ΔE_v and ΔE_e. The exact values of such discontinuities depend on the lineup of the two band structures (Margaritondo 1988).

(111) surface resulted in the formation of ionosorbed O_2^- species, which amplified the hole density and improved the surface conductivity of CuO nanocuboids. The H_2S sensing mechanism was found to be associated with local suppression/expansion of the hole-accumulation layer of *p*-CuO nanocuboids rather than the thermally activated carriers. Exposure to H_2S gas molecules was found to decrease the band bending energy as a function of concentration. The gas response of the semiconducting metal oxide and the concentration of analyte gas can be correlated as,

$$S_{Gas} = \frac{Rg}{R0} = 1 + (P_{Gas}A)^{\beta} \tag{6}$$

$$\log(S_{Gas-} - 1) = \beta\log(P_{Gas}) + \log A \tag{7}$$

where, A is the prefactor and β the characteristic exponent for the concentration dependence and P_{Gas} is the target gas partial pressure, which is directly related with concentration. The value of β depends on the surface species and stoichiometry of the elementary reactions on the surface.

$$q\Delta V = -2K_B T.\ln\{\frac{R_{gas}}{R_{air}}\} \tag{8}$$

where, K_B is Boltzmann constant, T is the temperature, R_{gas} sensor resistance in gas environment and R_{air} sensor resistance in air environment. The value of $q\Delta V$ was calculated with respect to the sensor response towards various concentration of H_2S gas. The work function (φ) of the material has two components that can be prejudiced by the surface reactions

$$\phi = qV + \chi + (E_C - E_F)_{Bulk} \tag{9}$$

$$\phi_{air} = qV_{air} + \chi_{air} + (E_C - E_F)_{Bulk} \tag{10}$$

$$\phi_{gas} = qV_{gas} + \chi_{gas} + (E_C - E_F)_{Bulk} \tag{11}$$

$$\Delta\chi = \Delta\phi - q\Delta V \tag{12}$$

where, qV_{air} and qV_{Gas} is surface band bending in air and gas environment. Similarly, G_{air}, G_{gas}, φ_{air} and φ_{gas} are electron affinity and work function in air and gas environment, respectively. Since the gas interaction are restricted to the surface, the material's bulk won't be affected; therefore, the value of $(E_C - E_F)_{bulk}$ does not change. G is the electron affinity that describes the measure of surface dipole moment. The contribution of electrons by reducing gas at the CuO surface result in the shift of Fermi level (E_F) towards conduction band edge and lower the work function. The negative sign of '$q\Delta V$' indicates the decrease in band bending after H_2S gas exposure (Jayaseelan et al. 2016).

Band bending at both surfaces of the interface is, moreover, present for *p-n* and metal-semiconductor interfaces. The band discontinuities can be present as opposing or, irregular, to heterojunction interfaces. On the one hand, they add to the flexibility in designing devices modified to distinctive tasks. Alternatively, they also add to the complexity of the interfaces and of the devices. These details clarify the whole evolution of heterojunction research. The design flexibility, owing to the occurrence of two dissimilar semiconductors with two different sets of parameters is a powerful incentive for the development of heterojunction technology. Potentially, heterojunction devices could revolutionize solid-state electronics and set up an exceptional degree of autonomy in tailoring devices to their applications. However, the complexity of the heterojunction interfaces has made it impractical to apply the similar empirical approach that has been so successful for other kinds of interfaces (Margaritondo 1988).

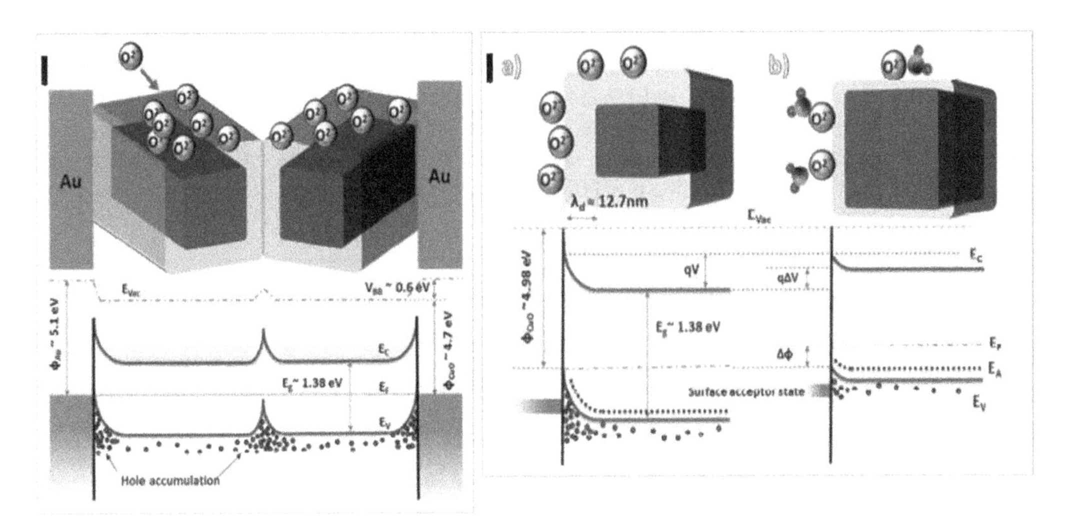

Figure 7. (I) Ohmic contact formation at p-type CuO and Au electrode junction under equilibrium condition (b) Energy band's representation of a p-type CuO nanocuboid surface exposed in (II) (a) air and (b) H_2S environment at 200°C. (Reprinted with permission from Jayaseelan et al. 2016.)

Heterojunction metal oxides based gas sensors

Heterojunction nanomaterials, with their promising physical and chemical properties, can be highly utilized in gas sensor technology to address the issues while using MOS sensors. Fermi levels across such interfacial regions of heterojunctions equilibrate to the same energy levels, resulting in the charge transfer process and formation of depletion layer. Heterojunction interface is a complex system, as two dissimilar materials are involved during the formation process. The counter parts can be either *n*-type or *p*-type or *n-p* or *p-n* junction types. In general, *n*-type is the most widely used category and hence combination of *n-p* types can lead to an alteration in their Fermi level which in-turn affects the sensing phenomenon. The improvements in sensing performance of these heterojunction nanocomposites have been attributed to many factors, including electronic effects such as: band bending due to Fermi level equilibration (Kusior et al. 2012, Chen et al. 2008), charge carrier separation (Yu et al. 2012), depletion layer manipulation (Choi et al. 2009, Wang et al. 2012, Liu et al. 2013) and increased interfacial potential barrier energy (Wang et al. 2010a), chemical effects, decrease in activation energy (Gu et al. 2012), targeted catalytic activity (Rumyantesva et al. 2006), synergistic surface reactions (Costello et al. 2003), geometrical effects, such as grain refinement (Chen et al. 2006a), surface area enhancement (Liangyuan et al. 2008), and increased gas accessibility (Zeng et al. 2012). Understanding the mechanisms so as to manage the sensing performance in these heterostructures will be necessary for upcoming progresses in this field.

Types of heterojunction materials

Researchers started to develop heterojunctions by integrating single nanoparticles with the metal oxides because such particles were showing much better properties than the bulk materials (Kar et al. 2011, Iwamoto et al. 1978, Liu et al. 2013). Later, in the late 1980s, researchers found that heterogeneous, composite colloidal semiconductor particles have better efficiency and in some cases they even develop some new properties (Lyson-Sypien et al. 2013, Costello et al. 1999). More recently, during the early 2000s, concentric multilayer semiconductor nanoparticles were synthesized by researchers with the view to improve the property of such semiconductor materials. Hence, the terminology "Heterojunctions" was adopted (Costello et al. 1999, Zhang et al. 2012a, Lyson-Sypien et al. 2013). Moreover, there has been a slow enhancement in research activities because of the remarkable requirement for progressively advanced materials fueled by the demands of modern sensor technology. Simultaneously, various types of core-shell heterojunction nanostructures were developed by various groups as depicted in Figure 8.

Heterojunction materials exist in various forms, such as mixed composite structures, bi/multilayers, decorated with secondary phases, mixed 1D and 2D materials and core-shell structures. Among the above mentioned heterojunctions, inhomogeneous core-shell structures are the promising candidates for gas sensing due to their catalytic behaviors. Also, they provide maximum interfacial area by minimizing the bulk interfaces and thus aids in sensing properties.

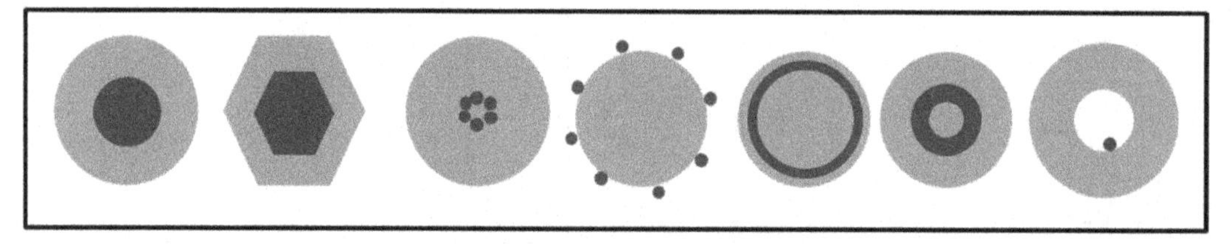

Figure 8. Schematics showing the types of core-shell heterojunctions.

Constitutions of core-shell heterojunction materials

A variety of constitutions of core/shell heterojunction nanomaterials have been developed by various groups so far for gas sensing applications (Rajib and Santanu 2011). The classification mainly depends on the application area, where they are used and the properties exerted by them while forming a core-shell heterojunction. Generally, core-shell heterojunctions are made of inorganic and organic materials. They exist in many forms such as, inorganic/inorganic, inorganic/organic or, organic/inorganic types of combinations as described below.

Inorganic/Inorganic

In this category, the core and shell are made up of metal, metal oxide or any other inorganic materials. These types of materials have a wide range of applications in the field of gas sensors, optoelectronics, information storage, semiconductor industry, optical bioimaging, biological labeling, etc. (Lu et al. 2005). Depending on the nature of the shell material, core/shell heterojunctions can be broadly classified into two categories: either silica-containing ones or those comprised of any other inorganic material.

Class I. Silica containing core-shell heterojunction materials

In this class of core-shell nanoparticles, silica is the base material with combination of other inorganic materials such as metals, metal oxides, metal chalcogenides, etc. The different types of silica based inorganic/inorganic core-shell nanoparticles have been prepared by various methods such as reduction, precipitation, sol-gel, Stober method, etc. (Wang et al. 2006).

Class II. Non-silica based core-shell heterojunction materials

In this class of core-shell nanoparticles, non-silica based materials are synthesized with combination of a different inorganic material such as metal, metal oxides, and metal chalcogenides, etc. The different types of non-silica based inorganic/inorganic core-shell nanoparticles have been prepared by various methods, such as chemical reduction, hydrothermal, electro-chemical deposition, etc. (Sun and Li 2005).

(a) Metal-metal oxide semiconductor heterojunctions

Metal/semiconductor contacts are primary component of any semiconductor device. At the same time, such contacts are not supposed to have a resistance as small as that of two connected metals. In particular, a huge mismatch between the Fermi energy of the metal and the semiconductor can result in high-resistance rectifying contact. Appropriate selection of materials can offer a low-resistance ohmic contact. Alternative method is to generate a tunnel contact. Such a contact consists of a thin barrier, achieved by heavily doping the semiconductor, through which charge carriers can readily tunnel. Thin interfacial layers too affect the formation, which is examined for metal/semiconductors contacts (Potje-Kamloth 2008).

The band bending model of metal–semiconductor contacts was first developed by Schottky and Mott to describe their rectifying effect (Schottky 1938, 1939, Mott 1938, 1939). Figure 9 depicts the ideal energy band diagrams of metal and *n*-type semiconductor contacts. When the semiconductor and metal are in contact, the free electrons will transfer between metal and semiconductor due to the work function difference. If the metal work function (Φ_m) is superior to that of the semiconductor (Φ_s), that is, $\Phi_m > \Phi_s$ as shown on the left of Figure 8, the electrons will flow from the semiconductor to the metal. The electron transfer will persist until the Fermi levels of metal ($E_{F,m}$) and semiconductor ($E_{F,s}$) are aligned. Under equilibrium, a Helmholtz double layer will be formed at the metal/semiconductor interface, where the metal is negatively charged and the semiconductor is charged positively in close proximity to its surface owing to electrostatic induction. Due to the low concentration of free charge carriers in the semiconductor, the electric field between metal and semiconductor interfaces cannot be effectively screened in the semiconductor. This causes the free charge carrier concentration near the semiconductor surface to be depleted compared to bulk. This region is called the space charge region. In the *n*-type semiconductor (majority charge carriers are electrons), when $\Phi_m > \Phi_s$, the electrons are depleted in the space charge region, and this region is therefore called as depletion layer and is characterized by excess positive charge. When $\Phi_m < \Phi_s$, as shown on the right side of Figure 9, the electrons are accumulated in the space charge region due to the electron transfer from the metal to the semiconductor, and this region is called as accumulation layer. In general, when the Fermi level of the metal is lower than that of the semiconductor, charge will flow to the metal causing the semiconductor Fermi level to decrease, and vice versa (Zhang et al. 2012).

In the space charge region, the energy band edges in the semiconductor also shift constantly owing to the electric field between the semiconductor and the metal in accordance with charge transfer, which

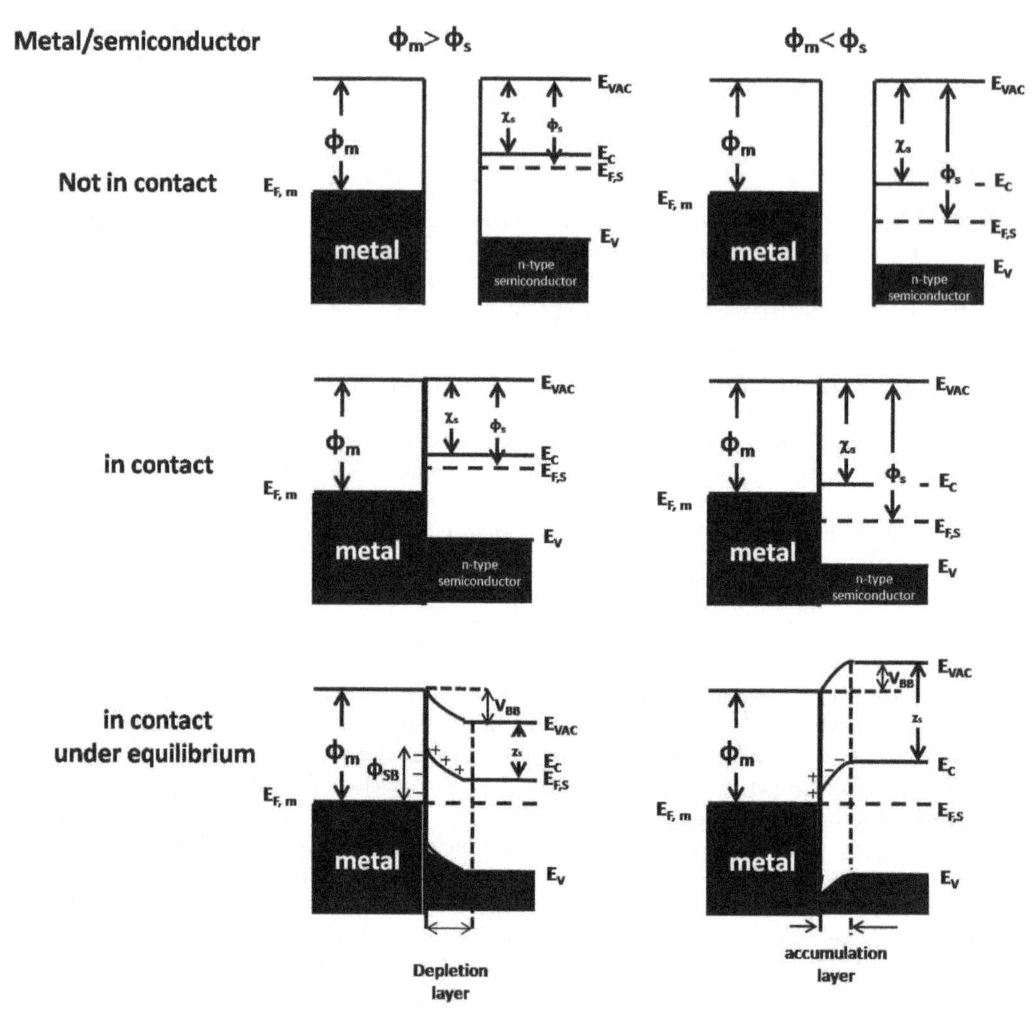

Figure 9. Energy band diagrams of metal and n-type semiconductor contacts. E_{vac}, vacuum energy; E_c, energy of conduction band minimum; E_v, energy of valence band maximum; Φ_m, metal work function; Φ_s, semiconductor work function; χ_s, electron affinity of the semiconductor (Zhang et al. 2012).

is called band bending. The energy bands bend upward towards the interface, when $\Phi m > \Phi s$, while the edges bend downward toward the interface when $\Phi m < \Phi s$ (Zhang et al. 2012). Also, for an n-type semiconductor when $\Phi m > \Phi s$, a Schottky barrier (Φ_{SB}) will be formed at the metal–semiconductor interface which can be represented as

$$\Phi_{SB} = (\Phi_m - \chi_s) \tag{13}$$

or

$$\Phi_{SB} = eVd + \xi = V_{BB} + \xi \tag{14}$$

where, χ_s is the electron affinity of the semiconductor, defined as the difference in energy between an electron at the bottom of the conduction band just inside the surface and an electron at rest outside the surface, e is the electronic charge, ξ is the difference between the bottom of the conduction band E_C

and the Fermi level of the semiconductor and eV_d is the band bending at zero bias voltage. The built-in potential V_{BB} or the contact potential formed between the metal and the semiconductor is given by

$$V_{BB} = eV_d = \frac{eN_D W_s^2}{2\varepsilon_s} \tag{15}$$

where N_D is the donor concentration, ε_s is the permittivity of the semiconductor, and W_S is the depletion layer width. Equation 15 has been referred to as the Mott-Schottky limit. While obtaining this, a number of key hypothesis have to be made, namely that (1) the surface dipole contributions to Φ_m and χ_s do not alter when the metal and semiconductor are brought into contact, (2) there are no localized states on the surface of the semiconductor and (3) there is a perfect contact between the semiconductor and the metal (i.e., there is no interfacial layer). For an *n*-type semiconductor, when $\Phi_m < \Phi_s$, the contact is biased so that electrons flow from the semiconductor to the metal. They encounter no barrier. If a bias voltage is applied such that electrons flow in the reverse direction, the comparatively high concentration of electrons in the region (where the semiconductor bands are bent downward; usually referred to as the accumulation region) behaves like a cathode, which is capable of providing copious supply of electrons. The current is then determined by the bulk resistance of the semiconductor and the applied voltage. Such a contact is termed as ohmic contact. A *p*-type semiconductor/metal junction, for which Φ_m exceeds Φ_s, represents an ohmic contact. Since, holes have trouble in going beneath the barrier potential (Potje-Kamloth 2008). The barrier height Φ_b for an ideal contact between a metal and a *p*-type semiconductor is given by

$$\Phi_b = E_g - (\Phi_m - \chi_s) \tag{16}$$

Metal loading leads to the development of clusters. Depending on the noble metal deposited, the loading, and nature of interacting gas, these clusters will be in metallic or oxidized form. Anyhow, the contact of the additive with the semiconducting oxide built a barrier that is wholly characterized by the electron affinity of the semiconductor, the density of surface states of the semiconductor that are placed inside the energy band gap and the work function of the metal. All of these three contributions produce a Schottky barrier through the formation of a depletion region in the semiconductor surface in contact with the cluster. Ultimately, the surface states formed by the presence of the additive can hold the Fermi level of the semiconductor to that of the additive.

Kolmakov et al. 2005, compared the sensing performance of an individual nanostructure (nanowire or nanobelt) before and after it was sensitized with catalytically active Pd nanoparticles. Also, they observed dramatic enhancement in sensitivity in terms of the catalytic action of Pd nanoparticles, which pre-dissociate the adsorbing species delivering atomic (rather than molecular) species to the surface of the nanostructure where they become chemisorbed.

The dramatic progress in sensing performance analyzed upon sensitization with Pd was ascribed to the collective consequence of spillover of atomic oxygen created catalytically on the Pd particles then migrating onto the tin oxide, and the reverse spillover effect in which faintly bound molecular oxygens migrate to the Pd and are catalytically dissociated (Figure 10). Therefore, both the liberation of activated species to, and the capture of precursors from the SnO_2 nanostructure surface, are endorsed by means of catalytically active Pd nanoparticles (Kolmakov et al. 2005).

Semiconductor metal oxides are themselves immensely vigorous for this task; however, this task can be enhanced to bring a large transformation in the sensitivity by doping noble metals, acidic or basic oxides to the metal oxide surface (Yamazoe et al. 2003). Thus, the sensitivity of MOS gas sensors increases enormously by the deposition of catalyst on metal oxide semiconductors. The catalyst too influences the selectivity of gas sensor (Wöllenstein et al. 2000). Hence, noble metal additives are frequently used as sensitizers to tune the gas selectivity, sensitivity and to lesser the operating temperature (Jhang and Colbow 1997, Montmeat et al. 2002). Mainly, there are two ways wherein the catalysts can influence the inter-granular contact region and so change the film resistance. One is the spillover mechanism or chemical sensitization, the other being the Fermi energy control or electronic sensitization (Hübner et al. 2012). In chemical sensitization, the dissociation of molecular oxygen is catalytically activated by

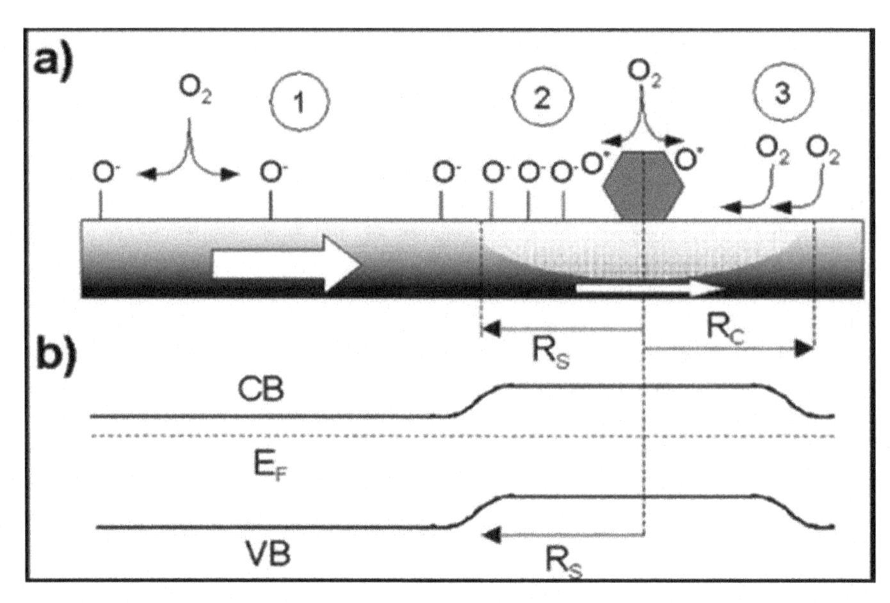

Figure 10. (a) Schematic depiction of the three major processes taking place at a SnO_2 nanowire/nanobelt surface: (1) ionosorption of oxygen at defect sites of the pristine surface; (2) molecular oxygen dissociation on Pd nanoparticles followed by spillover of the atomic species onto the oxide surface; (3) capture by a Pd nanoparticle of weakly adsorbed molecular oxygen that has been diffused along the tin oxide surface to the Pd nanoparticle's vicinity (followed by process 2). RS is the effective radius of the spillover zone, and RC is the radius of the collection zone. (b) Band diagram of the pristine SnO_2 nanostructure and in the vicinity (and beneath) a Pd nanoparticle. The radius of the depletion region is determined by the radius of the spillover zone. (Reprinted with permission from Kolmakov et al. 2005.)

the noble metal additives. Later, the molecular oxygen is captured by the conductance band electrons and gets adsorbed on the metal oxide surface. This results in an enhanced electron withdrawal from the metal oxide than for the bare metal oxide. In electronic sensitization, noble metal behaves as an electron acceptor on the surface of semiconductor oxide, which results in an enhancement of the depletion layer (Matsushima et al. 1988). Therefore, the change in resistance in noble metals loaded metal oxide will be more as compared with the bare metal oxides, thereby gas senor response will also improved. Thus, noble metal aids in upgrading the sensing performance. Conventionally, surface modification of metal oxides entails the deposition of noble metals on the surface of metal oxides. However, this exercise passivates the effective surface area of metal oxides occupied in gas sensing. Also, at higher temperatures the mobility of the metal nanoparticles increases which disables the catalytic activity, hence may end up with stability issues upon heating the noble metal nanoparticles (Arnal et al. 2006). The mobility of metal nanoparticles on oxide support results in the formation of either a shunting layer or an active membrane filter, which efficiently hinders the penetration of the targeting gas into the surface of the gas sensing matrix. The polluting of noble metal nanoparticles by various chemicals containing sulfur (H_2S, SO_2, and thiols) or phosphorus is another problem of their application (Subramanian et al. 2003). Therefore, it was challenging, until the emergence of core@shell structure, to surmount above drawbacks along with the advancement in performance and stability of sensor.

Zhang et al. (2010) have fabricated electrospun Pd–SnO_2 composite nanofibers towards H_2 sensing. The highest response value of 8.2 and enormously fast response–recovery performance are obtained for 100 ppm H_2 detection at 280°C, which demonstrates improved responses and lower operating temperature than that of the pure SnO_2 nanofibers. Step 1 is the dissociation of molecular oxygen on the sensor surface (Figure 11 Ia).

In this step, the PdO nanoparticles play a role as catalyst for triggering the dissociation of molecular oxygen, the atomic products then diffuse to the metal oxide support. This stage really enhances both the amount of oxygen that can repopulate opportunities on the SnO_2 surface and the rate at which this repopulation takes place, resulting in a greater and faster degree of electron withdrawal from the SnO_2

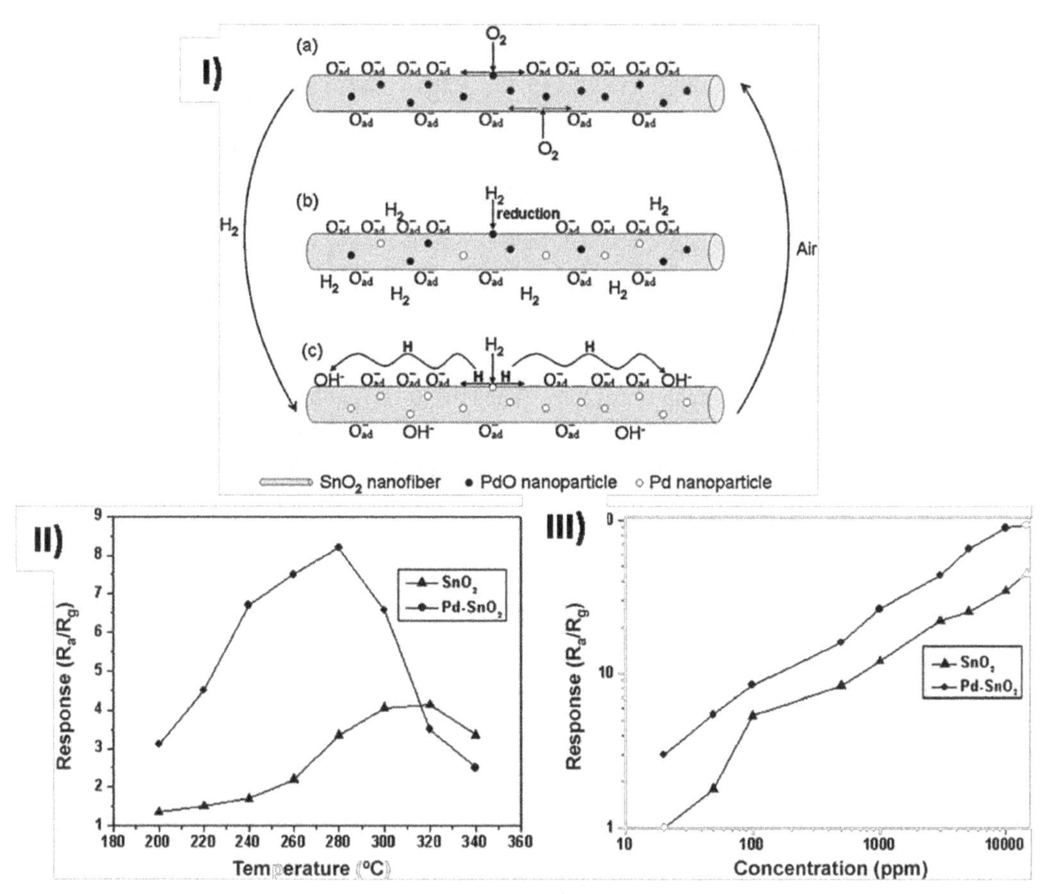

Figure 11. (I) Schematic model of the three steps for the sensitization mechanism of the as-prepared nanofiber sensor: (a) molecular oxygen dissociation on PdO nanoparticles followed by spill-over of the atomic species onto the SnO_2 surface; (b) reduction of PdO nanoparticles by H_2; (c) spill-over of H_2 on the SnO_2 surface and reactive with the oxygen species. (II) Responses of pure SnO_2 and Pd–SnO_2 nanofibers to 100 ppm H_2 as a function of operating temperatures. (III) Dependence of response on H_2 concentration for pure SnO_2 and Pd–SnO_2 nanofibers (Reprinted with permission from Zhang et al. 2010).

than for the pure SnO_2 nanofibers (Pd is a far better oxygen dissociation catalyst than SnO_2). This step is well-established in the catalysis articles and identified as the "spill-over" effect (Zhang et al. 2010).

When hydrogen is introduced, PdO is reduced to metallic palladium, and electrons are given back to SnO_2, which is defined as Step 2 (Figure 11 Ib). In Step 3 (Figure 11 Ic), hydrogen molecules adsorbed on palladium simultaneously spillover the surface of SnO_2, activating the reaction between hydrogen and the adsorbed oxygen. In this step, the dissociation of H_2 at the Pd surface forms atomic H that leads to additional reaction schemes. Reviewing the basic reaction scheme between adsorbed oxygen species on the SnO_2 surface and molecular H_2 is a single step process presented below

$$H_2(g) + O_{ad}^{-}(SnO_2) \rightarrow H_2O(g) + (SnO_2) + e^- \quad (17)$$

With the dissociation of H_2 at the Pd surface, additional reaction schemes are supplied

$$H_2(g) \rightarrow H + H(\text{dissociation}) \quad (18)$$

$$H + SnO_2 \rightarrow H_{ad}(SnO_2) \quad (19)$$

$$H_{ad}(SnO_2) + O_{ad}^-(SnO_2) \rightarrow OH_{ad}^-(SnO_2) \quad (20)$$

$$OH_{ad}^-(SnO_2) + H_{ad}(SnO_2) \rightarrow H_2O(g) + (SnO_2) + e^- \quad (21)$$

In these reaction paths, $H_2O(g)$ is liberated as the final reaction product, resulting in accumulation of electrons at the surface which is in charge for the conductance increase, hence a better sensitivity to H_2 of the Pd–SnO_2 nanofiber sensor than pure SnO_2 nanofiber.

Liu et al. synthesized Pure and Co-doped SnO_2 nanofibers using electrospinning method for H_2 sensing (Liu et al. 2010). The most widely accepted model is that the change in resistance of the SnO_2 gas sensors is mainly caused by the adsorption and desorption of the gas molecules on the surface of the SnO_2 film. When SnO_2 nanofibers are exposed to air, oxygen adsorbs on the exposed surface of the SnO_2 and ionizes to O^- or O_2^- (O^- is believed to be dominant), resulting in a decrease of the carrier concentration and electron mobility (Windischmann et al. 1979). When the SnO_2 nanofibers are exposed to a reducing gas (such as H_2 in this case), the reducing gas reacts with the adsorbed oxygen molecules and liberates the trapped electrons back to the conduction band, thereby increasing the carrier concentration and carrier mobility of SnO_2. Thus the resistance change of the SnO_2 sensors can be found depicted in Figure 12 (Liu et al. 2010).

Xu et al. 2011 fabricated Pristine and Al-doped SnO_2 nanofibers via electrospinning towards H_2 sensing. Comparing with the pristine SnO_2 nanofibers, Al-doped SnO_2 composite nanofibers show better hydrogen sensing properties, with the Al–SnO_2 metastable solid solution having the greatest performance, such as higher sensitivity and rapid response (~ 3 s) and recovery (less than 2 s). There are several reasons which should be considered to explain the enhanced sensing performances based on Al-doped SnO_2 nanofibers. It is well known that *n*-type semiconducting metal oxides are exactly stoichiometric which cannot chemisorb oxygen. The oxygen vacancies play a critical role in determining the sensing performances. To reinstate the stoichiometry in n-type semiconducting metal oxide, oxygen molecules will absorb on their surfaces and generate chemisorbed oxygen species (O_2^-, O^{2-}, O^-), resulting in high resistance. When reductive gas is introduced at reasonable temperature, the reductive target gas will respond with oxygen species on the outer surface of *n*-type semiconducting metal oxide and increase the electron concentrations. In this case, the partial substitution of Sn^{4+} cations with lower valence Al^{3+} cations at low concentrations of Al proved that difference in ionic radius of Al^{3+} cations and Sn^{4+} cations will generate more oxygen vacancies through the SnO_2 crystals (Comini 2006, Ishihara et al. 1994, Huang et al. 1998) resulting in higher sensing performances. In these composite fibers, the Al_2O_3 nanoclusters act as the catalytic sites ("spill over" effect) for redox processes and oxygen dissociation, resulting in improved sensing performance in contrast with the pristine SnO_2 nanofibers. Zhang et al. 2010, reported

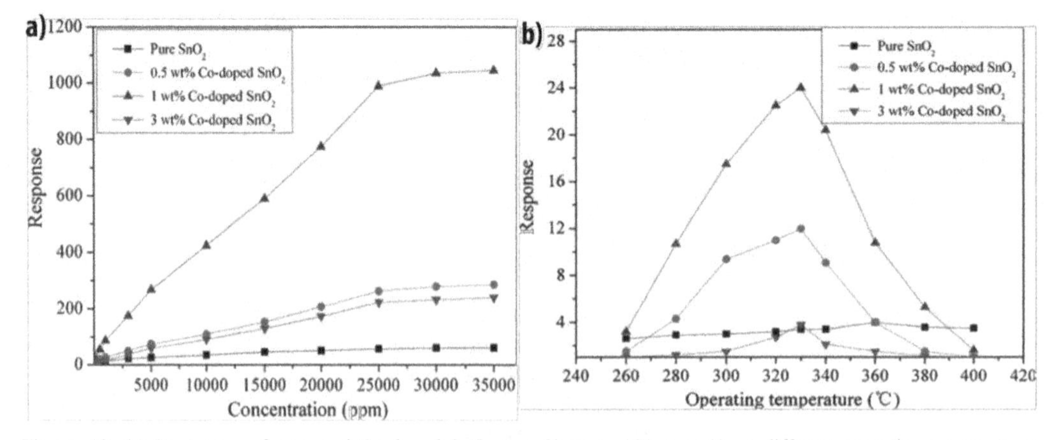

Figure 12. (a) Responses of pure and Co-doped SnO_2 nanofibers to 100 ppm H_2 at different operating temperatures, (b) Responses of pure and Co-doped SnO_2 nanofibers to different concentrations of H_2 at 330°C (Reprinted with permission from Liu et al. 2010).

that remarkably sensitive hydrogen sensors were produced using Pd-nanoparticle-decorated, single VO nanowires. The high sensitivity arose from the large downward shift in the insulator to metal transition temperature following the adsorption on and integration of atomic hydrogen, produced by dissociative chemisorption on Pd, in the VO_2, producing ~ 1000-fold increase in current.

Although noble metal nanoparticles are widely used in the functionalization of metal oxides for enhanced gas sensing properties, these noble metals will suffer from undesired aggregation arising from its low melting point and increased migration at high operating temperatures which results in a loss of catalytic activity. Therefore, in order to diminish this effect, by encapsulating noble metal nanoparticles in a protective shell, the resulting hybrid core-shell structure could effectively increase the stability of catalyst against undesirable aggregation during practical operation. Recently, Dinesh et al. (2015, 2017a, b, c) developed ZnO based core-shell heterojunction nanostructures comprising ZnO as core material (nanospheres, nanorods, ultralong nanorods, and nanofibers) with Au nanoclusters as catalytic shell layer and were chosen as the model systems to understand the mechanism of gas-material interactions towards NO_2 gas by analyzing their structure-property relationship. Detailed NO_2 gas sensing property analysis of these ZnO@Au core-shell heterojunction nanostructures have been correlated with their structural and morphological analysis. These sensors based on ZnO@Au nanorods could detect a trace-level concentration of NO_2 gas (as low as 500 ppb) using conventional electrodes compared to lithography techniques at a lower operating temperature of 150°C with excellent sensitivity and selectivity. The presence of Au nanoclusters on the surface of ZnO nanorods was found to enhance the sensor performance due to catalytic sensitization effects during NO_2 adsorption and formation of Schottky contacts at the interface as shown in Figures 13 and 14. The comparative analysis of structure-property relationship of the heterojunction materials proved that the ultra-long ZnO@Au core-shell heterojunction nanorods are excellent low-temperature NO_2 gas sensor material, owing to their favored direct electron transport, combined spill-over and back spill-over effect, and higher activation energy due to the presence of catalytic Au nanoclusters on the surface.

Chava et al. (2016) developed a new Au@In_2O_3 metal core@semiconductor shell nanoparticles for gas sensing devices. Here, Au metal nanoparticles act as a good catalyst to modify the surface reactions of metal oxide semiconductors toward better sensing performance. Au@In_2O_3 core-shell nanoparticles showed a greater sensitivity and selectivity towards H_2 gas with the highest response of 34.38 at operating temperature of 300°C to 100 ppm gas level, whereas In_2O_3 nanoparticles showed a response of 9.26 only. This remarkably enhanced hydrogen gas sensing performance of Au@In_2O_3 core shell nanoparticles over In_2O_3 nanoparticles was due to the electronic sensitization effects induced by changing Schottky barrier at the interface between the Au nanoparticles and the In_2O_3 shell and also due to the chemical sensitization of the catalytic metal Au nanoparticles, i.e., they activate the target gas by dissociation and subsequent spillover of dissociation fragments onto the gas sensing material.

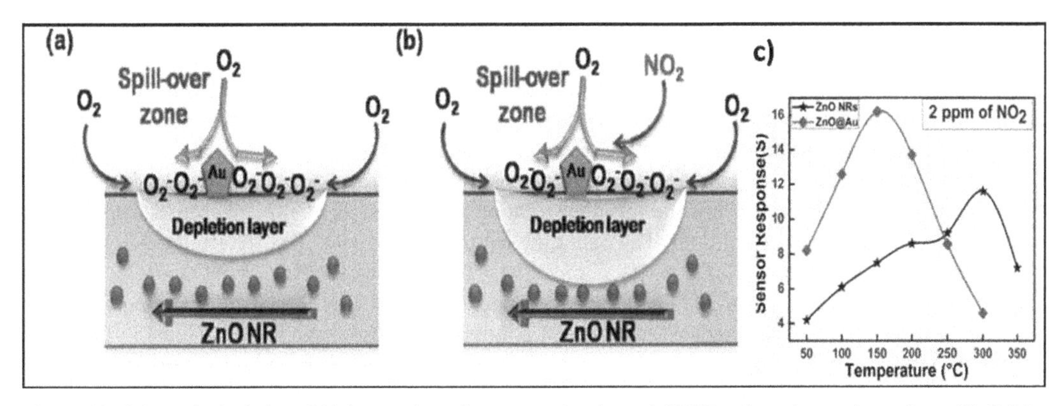

Figure 13. Schematic depiction of (a) ionsorption of oxygen molecules and (b) NO_2 adsorption on the surface of ZnO@Au core–shell nanorods. (c) Sensor response graph for 2 ppm of NO_2 gas as a function of temperature for ZnO nanorods and ZnO@Au core–shell nanorods sensors. (Reprinted with permission from Dinesh et al. 2015.)

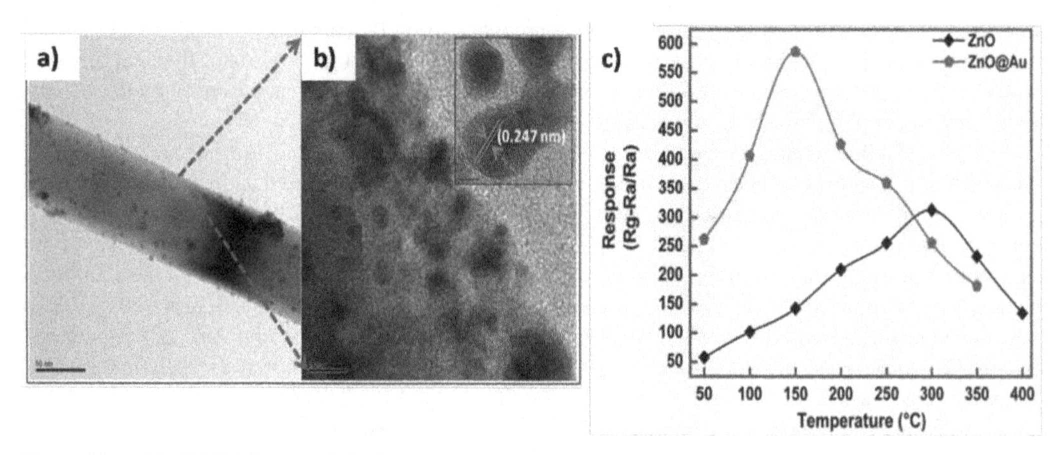

Figure 14. (a–b) HRTEM images of ultralong ZnO@Au heterojunction nanorods (inset showing the HRTEM image of Au). (c) Sensor responses to 2 ppm of NO_2 versus operating temperature dynamic sensor response graph for ultralong ZnO nanorods and ultralong ZnO@Au heterojunction nanorods sensors. (Reprinted with permission from Dinesh et al. 2017b.)

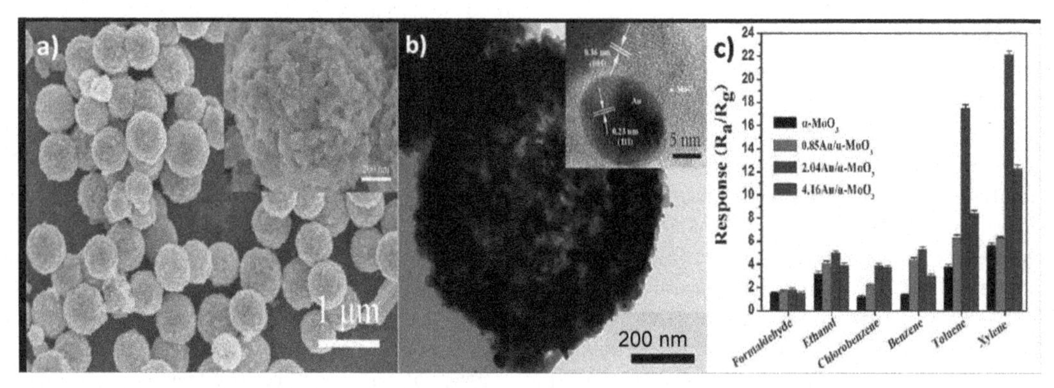

Figure 15. SEM (a) and TEM (b) images of Au decorated α-MoO_3 hollow spheres and (c) the responses of Au decorated α-MoO_3 sensors towards various gases. (Reprinted with permission from Sui et al. 2016.)

Sui et al. (2016) prepared sensor based on Au decorated α-MoO_3 hollow spheres, which exhibits improved gas sensing performance to BTX, especially the enhanced selectivity to methyl benzene (toluene and xylene) with negligible cross-responses to interfering gases of ethanol, formaldehyde, benzene and chlorobenzene (Figure 15). The maximum response is increased; the optimum working temperature and response time to 100 ppm toluene and xylene are lowered by 2.04 wt% Au loading. The enhanced gas sensing properties might be ascribed to the "spillover effect" and catalytic effect of Au nanoparticles as well as the hierarchical hollow nanostructure of the nanomaterials.

Dinesh et al. (2017) elucidated a controlled synthesis strategy for the growth of TiO_2@Au heterojunction nanorods and its effectual implementation in ultrasensitive NO_2 gas sensor applications under atmospheric pressure conditions. The operating temperature of TiO_2@Au heterojunction nanorods based sensor was found to be 250°C which is much less as compared to alternate NO_2 sensors (400°C) with the lowest detection limit of 500 ppb (Dinesh et al. 2017a). The TiO_2@Au heterojunction nanorods, shown in Figure 16, exhibited higher sensitivity at atmospheric pressure conditions (sensitivity = 140) compared to vacuum conditions (sensitivity = 8) because of the changes in surface O_2 ionosorption properties of the hybrid material at different oxygen partial pressure and the existence of mixed phases in TiO_2 nanorods.

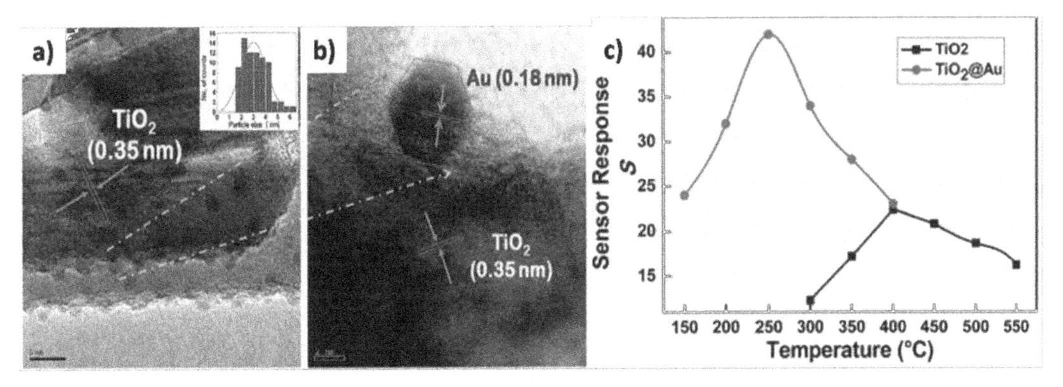

Figure 16. (a–b) HR-TEM images of TiO_2@Au heterojunction nanorods. Inset shows histogram of particle size distribution. (c) Sensor response graph for 2 ppm of NO_2 gas as a function of operating temperature for TiO_2 nanorods and TiO_2@Au heterojunction nanorods. (Reprinted with permission from Dinesh et al. 2017a.)

Metal oxide—metal oxide isotype heterojunctions

Band bending can also occur in *n–n* and *p–p* heterojunctions (Sen et al. 2010). Ga_2O_3-core/ZnO-shell nanorods were fabricated towards NO_2. The Ga_2O_3–ZnO heterojunction acts as a lever in electron transfer by which electron transfer is facilitated or restrained, resulting in superior sensing properties of the core–shell nanorod sensor (Figure 17). Besides, the recovery time of the core–shell nanorods was approximately 1/3 that of the bare-Ga_2O_3 nanorods at a NO_2 concentration of 10 ppm and almost half at other NO_2 concentrations, even if the response time of the former is longer than that of the latter (Jin et al. 2012).

Singh et al. synthesized In_2O_3–ZnO core–shell nanowires towards the CO, H_2 and ethanol while pristine In_2O_3 nanowires have revealed an advanced response towards the NO_2. The presence of ZnO shell layer creates heterojunctions (In_2O_3–ZnO) which provides additional energy barrier at the junction, augmenting the change in resistance of the nanowires towards gas exposure (Figure 18). Combinations of homo and heterointerfaces formed at the junctions in the In_2O_3–ZnO core–shell nanowires sensor improved the sensitivity towards reducing gases by lowering the potential barrier heights along the charge carrier path (Singh et al. 2011).

Choi et al. fabricated heterogeneous branched nanostructures in which SnO_2 nanowire branches were grown thermally on the Au-deposited ZnO nanofiber stems. The mean diameter of the SnO_2 branches grown on the ZnO stems was approximately 40–60 nm. The ZnO–SnO_2 stem-branch NNH was tested for its potential in chemical gas sensors in terms of NO_2 and CO (Figure 19). The ZnO–SnO_2 NNH showed excellent sensing responses to ppm levels of NO_2 and CO in a reproducible manner, highlighting its potential as a platform for highly sensitive chemical sensors (Choi et al. 2013). Modulation of the depletion width along the branch nanowires and potential barriers at nanograins in the stem nanofibers in addition to both complex homojunctions and heterojunctions amid the stems and branches are likely to be the reason for the good sensing ability. Further literature for gas sensing performance of Metal Oxide—Metal Oxide Isotrope heterojunction nanomaterials is summarized in Table 2.

Metal oxide—metal oxide anisotype heterojunction

Zeng et al. (2010) proposed that the electron migration between the MOS can help to facilitate additional oxygen adsorption at the surface due to the larger electron density (Zeng et al. 2010). Kusior et al. (2012) also proposed this mechanism in the same system. The interface at a *p–n* junction has fewer free electrons due to electron–hole recombination, and rising resistance, at the same time as the interface at an *n-n* junction merely transfers electrons into the lower-energy conduction band, forming an "accumulation layer" in the MOS rather than forming a depletion layer. The accumulation layer can be minimized

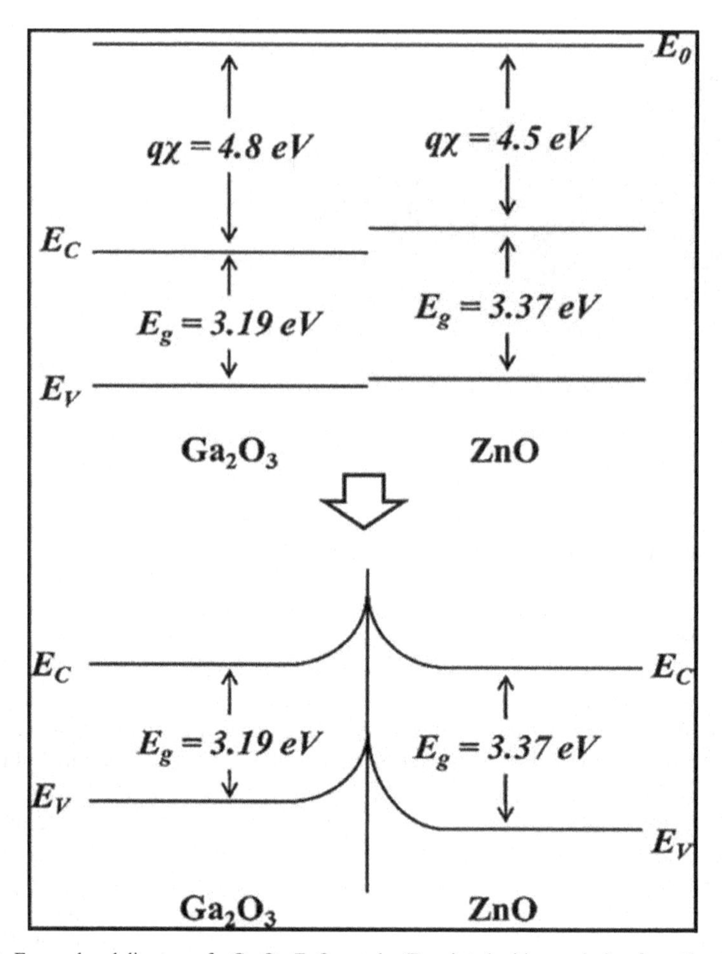

Figure 17. Energy band diagram of a Ga_2O_3–ZnO couple. (Reprinted with permission from Jin et al. 2012.)

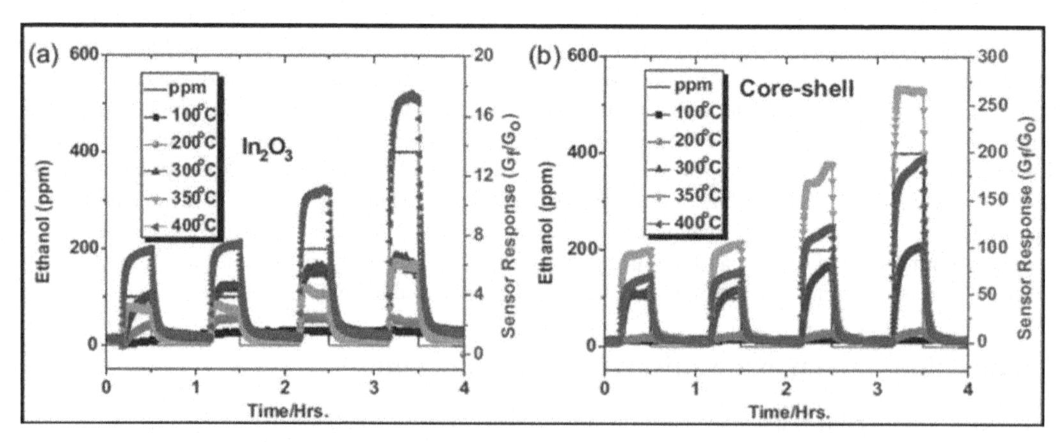

Figure 18. I. Illustration of the charge transport through (a) homo-structures nanowire junction followed by cross section contact and formation of an energy barrier at the junction, (b) hetero-structured core–shell nanowire junction followed by the cross sectional view with two depletion regions and formation of three energy barriers at the interfaces. II. A representative plot of dynamic sensor responses obtained from the devices based on (a) pure In_2O_3 nanowires and (b) In_2O_3–ZnO core–shell nanowires. (Reprinted with permission from Singh et al. 2011.)

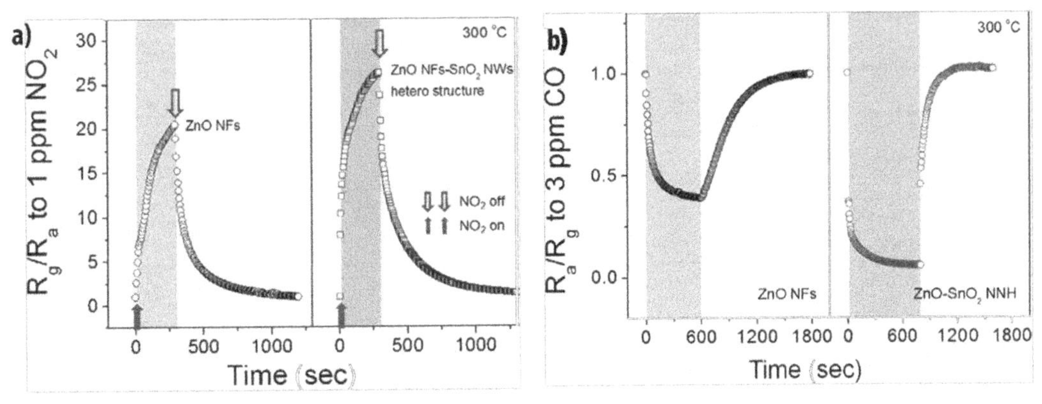

Figure 19. (a) Response curves of the sensors fabricated with bare ZnO nanofibers and with ZnO–SnO_2 NNH to 1 ppm NO_2. (b) Response curves of the sensors fabricated with bare ZnO nanofibers and ZnO–SnO_2 NNH to 3 ppm CO. (Reprinted with permission from Choi et al. 2013.)

by subsequent oxygen adsorption on the surface, further increasing the potential energy barrier at the interface, thereby enhancing the response. The mechanism pertaining to heterojunction materials is similar to that of metal oxides, where oxygen molecules in air dissociate and get adsorbed on the surface of oxides forming ionosorbed species. This ionosorption changes the base resistance of the heterojunction materials, resulting in formation of depletion layer within the Debye length of the material (λ), typically in the order of 2–100 nm. The Debye length determines the resistance of the material when exposed to air and in-turn exposed to analyte gas.

Sensing properties of *p–n* CuO–ZnO core–shell nanowires (C–S NWs) for reducing gases were explained by Kim and his group. The sensing mechanism in the p–n C–S NWs is the resistance discrepancy caused by the radial modulation of an electron-depletion region in the ZnO shell layer during the interaction among the reducing gas molecules and the adsorbed oxygen species, suggesting a universal sensing principle operating in *p–n* type C–S structures depicted in Figure 20 (Kim et al. 2016).

The majority of oxide composites and heterojunctions in recent literatures state about their dispersions of oxide constituents with a little focus towards the mechanism governing their sensing. Korotcenkoc et al. (2013) detailed the various dispersions resulting from the doping In_2O_2 with several other precursors through sol-gel method and resulted in obtaining n-p type variants. But, they failed to explain the mechanism response for their enhanced sensing behavior. Similarly Ivanovskaya et al. (2003) prepared a complex heterojunction bi-layered film of [γ-Fe_2O_3-In_2O_3(1:1)]/In_2O_3 for the detection of ethanol, but a comparatively simpler film comprising of γ-Fe_2O_3-In_2O_3 performed well than the previous bi-layered film because of the complexity involved in understanding the mechanism of the heterojunctions (Ivanovskaya et al. 2003). Chen et al. demonstrated that the growth direction and orientation of the secondary SnO_2 nanorods can have an interfacial orientation relationship with the α-Fe_2O_3 core nanorods resulting in the formation of mixed 1D-1D structures (Chen et al. 2008). This presents unique opportunities to control the crystallographic nature of the interface by using core nanorods of dissimilar crystallographic growth directions. It has been shown that the family of exposed planes in the nanostructure can affect the gas-solid inter-action (Chen et al. 2008). In some cases, the growth kinetics can be manipulated to let the power of the aspect ratio and morphology of the secondary features. A very open, porous structure allows simple gas accessibility to a well-defined heterojunction interface and still there is a controversy prevailing in relating the crystallographic planes which makes the interfacial layers and its influence on electronic properties. In a similar work carried by Na et al. (2011), an *n*-type ZnO nanowire sensor was decorated with *p*-type Co_3O_4 nanoislands. The normal ambient resistance of the nanowires in air (R_a) will be even higher than without the heterojunction due to the depletion region by the heterojunction interface expanding into the ZnO nanowire, decreasing the width of the charge conduction channel (Sonker et al. 2015). Alali et al. fabricated CeO_2/$ZnCo_2O$ composite which exhibited excellent response toward ethanol

Table 2. Gas sensing performance of Metal Oxide—Metal Oxide Isotype heterojunction nanomaterials.

Analyte Gas	Composition	Synthesis Route	Morphology	Temperature	Concentration	Reference
Ethanol	50 wt% SnO_2–ZnO	Electrospinning	SnO_2–ZnO Composite nanofibers	360°C	2500 ppm	Khorami et al. 2011
H_2S	5 wt% ZnO–SnO_2	Ball mill	ZnO–SnO_2 powder	250°C	50 ppm	Wagh et al. 2004
NO_2	40% In_2O_3–SnO_2	Co-precipitation	Particles	200°C	1000 ppm	Chen et al. 2006
Ethanol	Al_2O_3–ZnO	LDH solution	Particle	260°C	1000 ppm	Xu et al. 2013
Ethanol	In_2O_3/ZnO	Electrospun	Nanofiber bi-layer	210°C	100 ppm	Zhang et al. 2012
H_2	ZnO@SnO_2	MBE/PLD	SnO_2-coated ZnO Nanowires	400°C	500 ppm	Tien et al. 2007
H_2	20 wt% SnO_2–TiO_2	Co-precipitation	Mixed nanoparticles	400°C	20 ppm	Shaposhnik et al. 2011
Ethanol	MoO_3@SnO_2	Hydrothermal-Solution	SnO_2 nanoparticle-coated MoO_3 nanobelts	300°C	500 ppm	Xing et al. 2011
CO	20 wt% WO_3–MoO_3	RF sputtering	Mixed nanocrystals	200°C	15 ppm	Comini et al. 2002
H_2	(0.005 mol MoO_3)–SnO_2	Sol–gel method	Mixed	240°C	1000 ppm	Ansari et al. 2002
CH_4	25 wt% In_2O_3–SnO_2	Co-precipitation	Coated nanoparticles	300°C	850 ppm	Chen et al. 2008
Ethylene	0.3 wt% WO_3–SnO_2	Solution precipitation	Mixed nanoparticles	300°C	6 ppm	Pimtong-Ngam et al. 2007
Butanol	50 wt% ZnO–SnO_2	Mechanical mixing	Mixed particles	350°C	5 ppm	Costello et al. 2003
TMA	10 wt% ZnO–SnO_2	Precipitation hydrothermal	Mixed SnO_2 nanoparticles/ZnO nanorods	330°C	50 ppm	Zhang and Zhang 2008
NO_2	Ga_2O_3-ZnO	Thermal evaporation/ALD	Ga_2O_3-core/ZnO-shell nanorods	300°C	100 ppm	Jin et al. 2012
Ethanol	In_2O_3–ZnO	Two-step growth process	In_2O_3–ZnO core–shell nanowire	350°C	100 ppm	Singh et al. 2011
NO_2 **CO**	ZnO–SnO_2 stem-branch NNH	Thermal Evaporation	ZnO–SnO_2 stem-branch NNH	300°C	1 ppm 3 ppm	Choi et al. 2013

compared with pure CeO_2, in which the responses of CeO_2 and CeO_2/$ZnCo_2O_4$ to 100 ppm ethanol at an optimal temperature of 180°C were 5 and 12.4, respectively (Figure 21). Good selectivity of composite CeO nanotubes toward ethanol was observed after exposure to various gases with a rapid response (10s) and recovery time of 15s being measured (Alali et al. 2016).

Lin et al. synthesized a new type of acetylene gas sensor based on the hollow NiO/SnO_2 heterostructure. Compared with the pure SnO_2 gas sensor, the response of the hollow NiO/SnO_2 heterostructure gas sensor to 100 ppm acetylene (C_2H_2) was elevated to 13.8 from 5.4 at the finest operating temperature

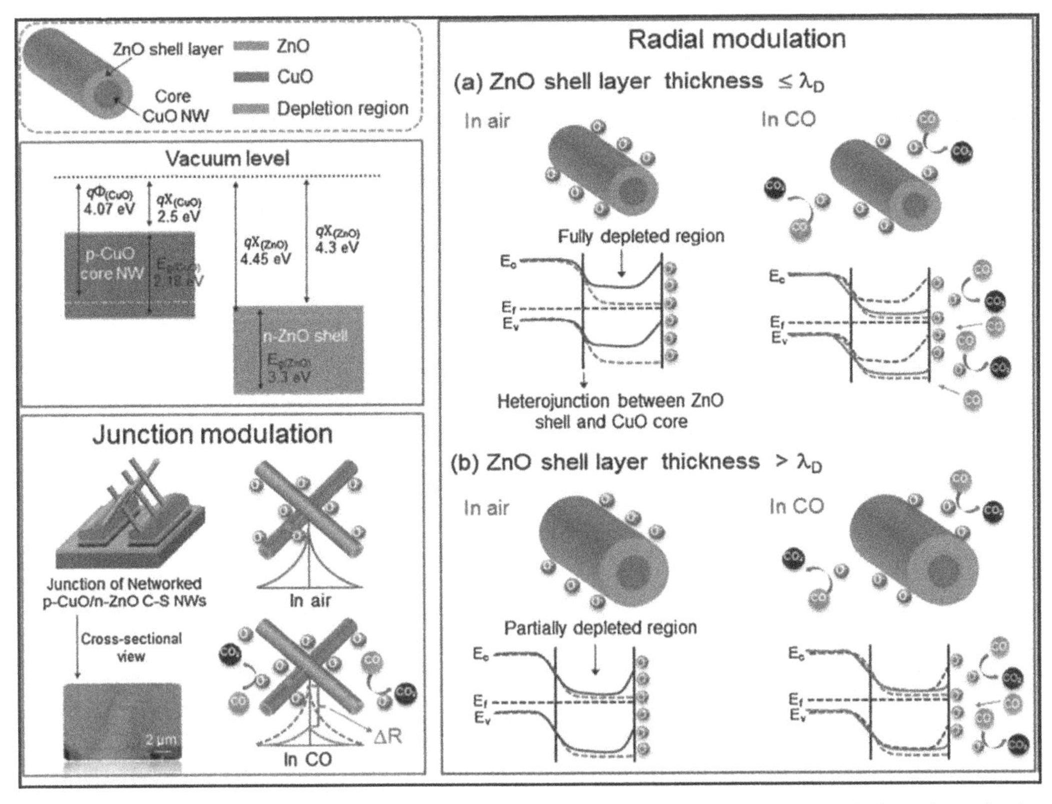

Figure 20. Schematic of the reducing gas sensing mechanism in the CuO–ZnO C–S NWs. E_c and E_F indicate the conduction band energy and Fermi energy level, respectively, in cases of ZnO shell layers (a) thinner and (b) thicker than ZnO's Debye length. (Reprinted with permission from Kim et al. 2016.)

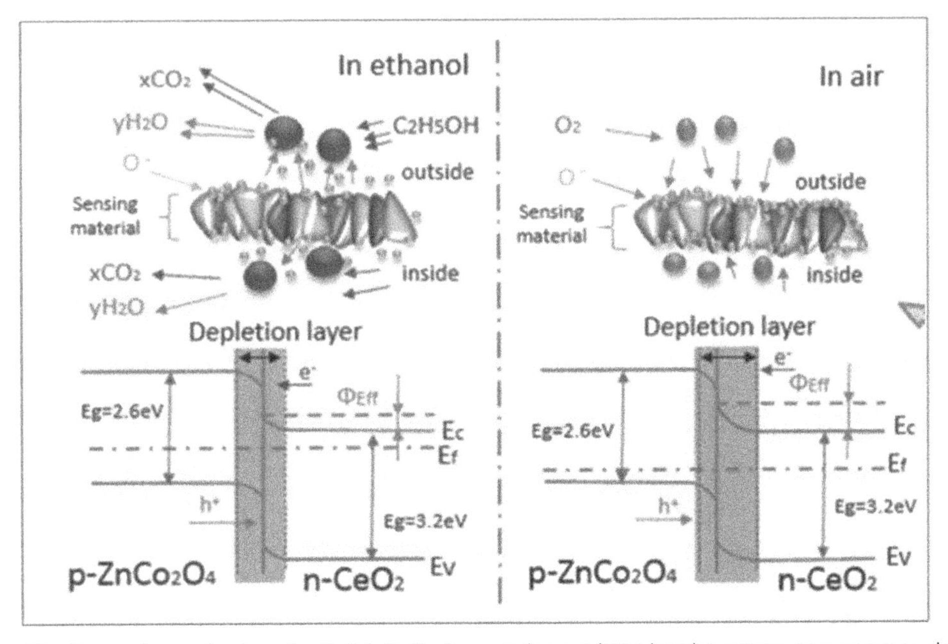

Figure 21. Gas sensing mechanism of n-CeO_2/p-$ZnCo_2O_4$ nanotubes n-p heterojunction sensors on exposure to air and ethanol gas with the energy band diagram. (Reprinted with permission from Alali et al. 2016.)

of 206°C. At the interface between NiO and SnO_2 nanoparticles, many p–n junctions are generated. The electrons transform from n-type SnO_2 to p–type NiO while the holes transform from p–type NiO to n-type SnO_2. Until the system obtains equalization at the Fermi level, the extensive depletion regions are created leading to a notable reduction in conductivity (Figure 22). When the sensor based on the NiO/

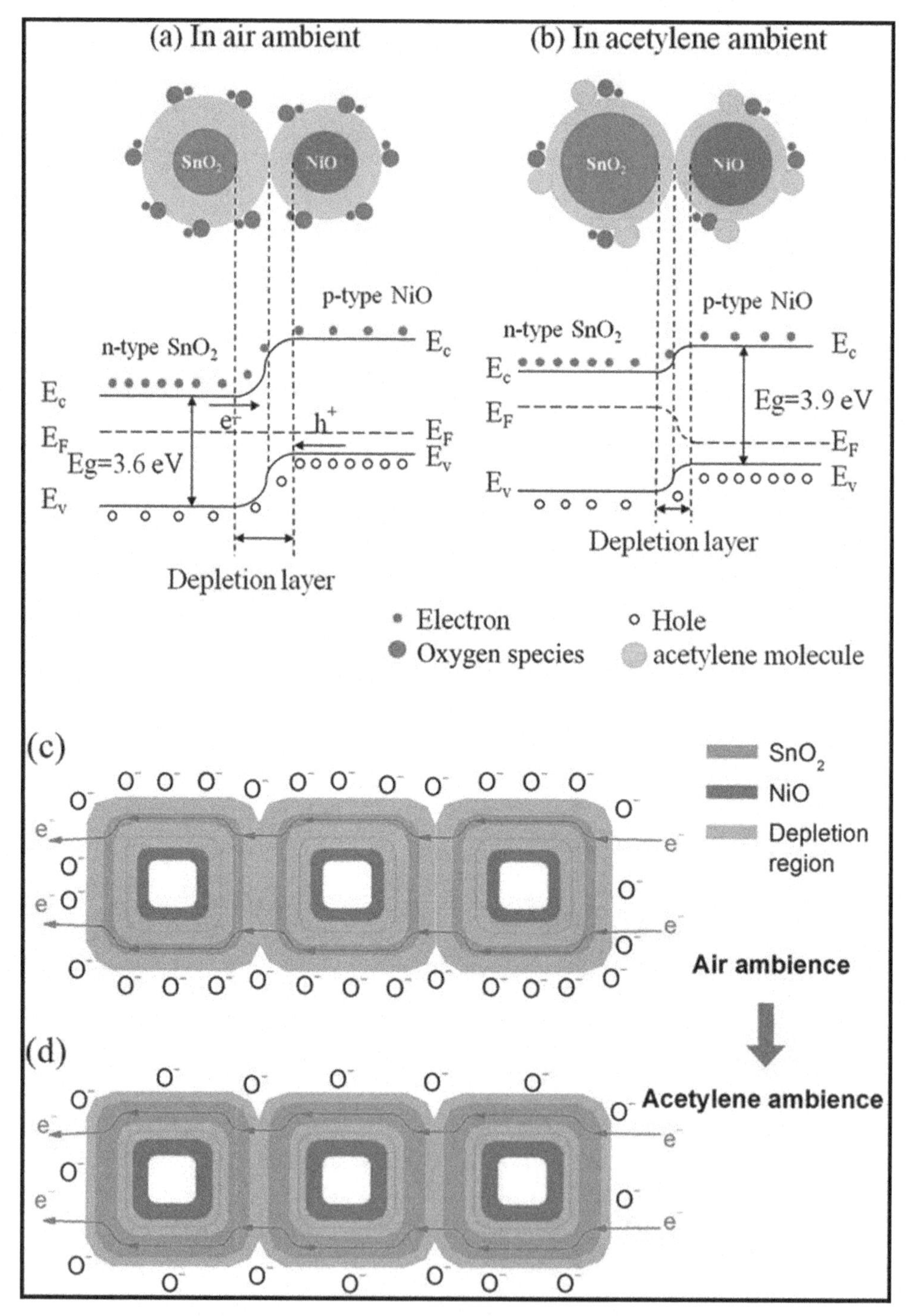

Figure 22. Diagram showing the mechanism of hollow NiO/SnO_2 heterostructure to acetylene. Reprinted with permission from Lin et al. (2015).

SnO_2 heterostructure is exposed to acetylene ambient, the electrons trapped by absorbed oxygen species and NiO nanoparticles are fed back to SnO_2 through surface interactions, which shrink p–n junction depletion regions and decrease the barrier height (Lin et al. 2015). As a result, the conducting channel will be widened and the conductivity might strengthen considerably. Therefore, the sensor response is remarkably improved.

A reduced cross-sectional area accessible for charge transmission in the nanowire will result in an increased resistance. Additionally, charge conduction across the *p–n* interface will further contribute to the increase of the resistance. When an oxidizing gas such as NO_2 is introduced, any additional increase in the resistance due to adsorption is now minimized. However, when a reducing gas is established, a huge drop in the resistance is possible due to the initial value of R_a being exceedingly high, as response is typically measured as R_a/R_g (Wlodarski et al. 2010). As shown in Figure 23, *p*-type nanoparticles on an n-type core compliment the alteration in the depletion region of the core induced by oxygen adsorption. Oxygen creates an accumulation region in the *p*-type nanoparticles which formulates it further heavily *p*-type and makes a stronger depletion effect at the interface, motivating the depletion region advanced into the core nanowire in conjunction with oxygen adsorbed on the exposed regions of the *n*-type core (Nguyen et al. 2014). Heterostructures using *n*-type material coated *n*-type nanowires can also be well thought-out in a related behavior. Both of these mechanisms should be considered in any study utilizing a randomly-oriented mixture of coated nanowires. Additionally, these resistance-based mechanisms will not always agree with experimental results and other factors, such as increased surface area, and increased defect sites, and catalytic activity towards the analyte gas should be considered.

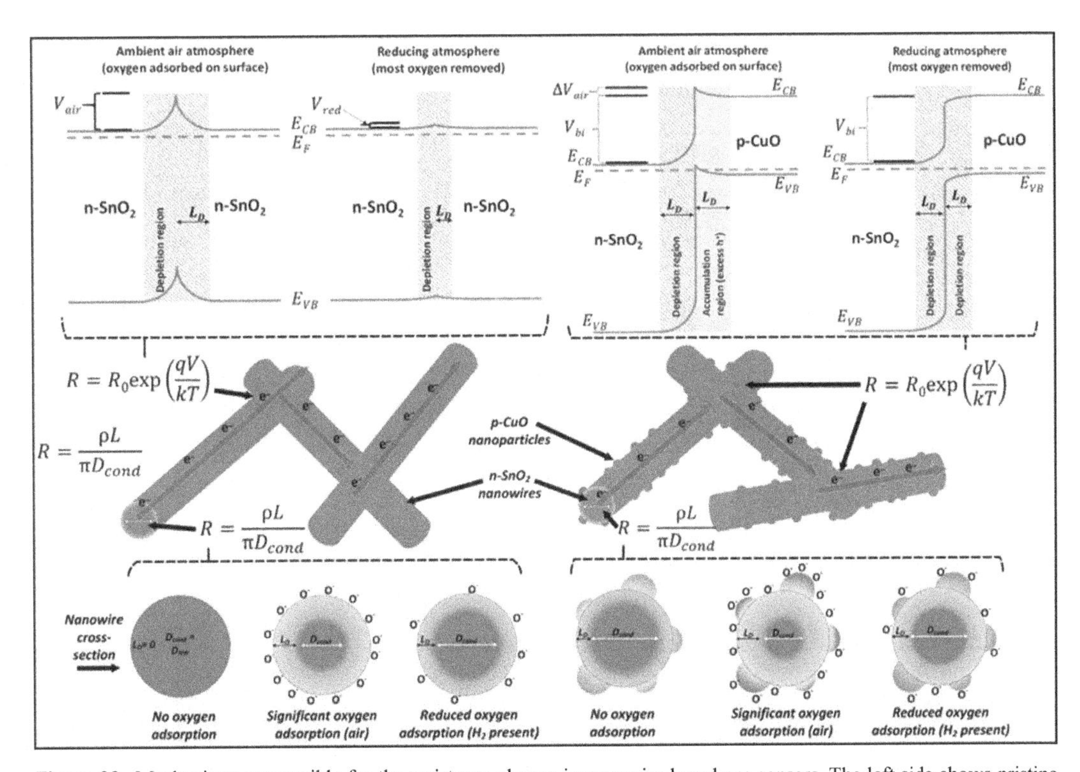

Figure 23. Mechanisms responsible for the resistance change in nanowire-based gas sensors. The left side shows pristine *n*-type SnO_2 nanowires with the interfacial mechanisms depicted at top and the conduction channel mechanism depicted at bottom. The right side similarly shows these mechanisms in n-type SnO_2 nanowires coated with discrete *p*-type CuO nanoparticles. (Reprinted with permission from Miller et al. 2014.)

Table 3. Gas sensing performance of Metal Oxide—Metal Oxide Anisotype heterojunction nanomaterials.

Analyte Gas	Composition	Synthesis Route	Morphology	Temperature	Concentration	Reference
Ethanol	5 wt% La_2O_3–SnO_2	Powder solution coating	La_2O_3-coated SnO_2 nanoparticles	300°C	1000 ppm	Jinkawa et al. 2000
H_2	1 wt% Co_3O_4–SnO_2	Precipitation	Mixed nanoparticles	250°C	1000 ppm	Choi et al. 2005
H_2S	CuO/SnO_2	Sputtering-oxidation	Single bi-layer	433°K	16.6 ppm	Vasiliev and Rumyantseva 1999
H_2S	1 mol% CeO_2–SnO_2	Solution precipitation	Mixed nanoparticles	300°K	5 ppm	Fang et al. 2000
SO_2	1 mol% NiO–SnO_2	Pechini method	NiO-coated SnO_2	25°C	18 ppm	Hidalgo and Castro 2005
Ethanol	4.5 wt% Cr_2O_3–ZnO	Electrospun	Nanofibers	300°C	100 ppm	Wang et al. 2010b
Ethanol	ZnO@Co_3O_4	VLS-thermal evaporation	Co_3O_4 nanoisland-coated ZnO nanowires	400°C	100 ppm	Na et al. 2011
H_2	TiO_2/NiO	DC Sputtering	100 nm TiO_2 over 10 nm NiO film	200°C	10,000 ppm	Kosc et al. 2013
H_2S	ZnO@3 wt% CuO	Traditional hydrothermal/ solution synthesis	ZnO porous nanorods@ CuO nanoparticles	100°C	100 ppm	Wang et al. 2012
NH_3	ZnO@Cr_2O_3	Powder film dip-coating	Cr_2O_3-coated ZnO powder	RT	300 ppm	Patil et al. 2007
LPG	ZnO@0.47 wt% Cr_2O_3	Powder film dip-coating	Cr_2O_3-coated ZnO powder	350°C	100 ppm	Patil and Patil 2009
O_2	90% CeO_2–TiO_2	Sol–gel method	Mixed	420°C	10^3 ppm	Zheng et al. 2010
Ethanol	ZnO–Co_3O_4	Solution	ZnO-coated Co_3O_4 nanoparticles	170°C	100 ppm	Liu et al. 2013
Acetylene	NiO/SnO_2	hydrothermal method	Hollow NiO/SnO_2	206°C	100 ppm	Lin et al. 2015
H_2	NiO/SnO_2	electrospinning technique	NiO/SnO_2 nanofibers	320°C	5 ppm	Wang et al. 2010c
CO **C_6H_6**	CuO–ZnO	thermal oxidation/ALD	CuO–ZnO core–shell nanowires	300°C 300°C	1 ppm 1 ppm	Kim et al. 2016

Inorganic/organic core-shell heterojunction materials

Inorganic/organic core/shell heterojunctions are made of metal, a metallic compound, metal oxide, or a silica core with a polymer shell or else a shell of some other high density organic materials. One of the major rewards of the organic coating on the inorganic material is the increase in oxidation stability of the metal cores under normal environment. In addition, they exhibit enhanced biocompatibility for bio applications. The polymer-coated inorganic core-shell materials have a broad spectrum of applications, ranging from catalysis to additives, pigments, paints, cosmetics and inks (Min et al. 2009). Depending on the material properties of the core particles, they can be broadly classified into two different groups: (i) magnetic/organic and (ii) non-magnetic/organic core/shell nanoparticles. Different inorganic/organic core/shell nanoparticles are listed in Table 4.

Table 4. Various combinations of inorganic core/organic shell heterojunction materials.

Types	Core		Shell		Ref.
Core-shell	**Method**	**Basic Reagent**	**Method**	**Basic Reagent**	
Fe/PIB	TD	$Fe(CO)_5$	–	PIB	Burke et al. 2002
Fe/PS	TD	$Fe(CO)_5$	–	PS	Burke et al. 2002
Fe/dextran	Wet chemical reaction	$FeCl_3$	–	Dextran	Molday and Mackenzie 1982, Molday and Molday 1984
Fe_3O_4/PEGMA	Co precipitation	$FeCl_3.6H_2O$, $FeCl_2.4H_2O$	RAFT	PEGMA	Qin et al. 2010
$CoFe_2O_4$/DTPA-CS	Low-temperature solid-state method	$CoCl_2.6H_2O$, $FeCl_3.6H_2O$, NaCl	Emulsion cross-linking polymerization	CS, DTPA	Qin et al. 2010
Au/aryl polyether	Reduction	$HAuCl_4$	Self-assembly	Disulfide dendrons	Gopidas et al. 2003
Au/PSS and PDADMAC	Reduction	$HAuCl_4$	Layer by layer assembly	PSS, PDADMAC	Gittins and Caruso 2001
Ag-Au/PEG-HA heterojunction	Reduction	$AgNO_3$	Precipitation	MEO_2MA, PEGDMA	Wu et al. 2010
SiO_2/PMMA	Stober method	TEOS	EP	MMA	Palkovits et al. 2005
SiO_2/PS	Sol-gel	TEOS	DP	Styrene	Shin et al. 2010
SiO_2/PS	Stober method	TEOS	Polymerization	Styrene	Haldorai et al. 2010
TiO_2/PS_2	Sol-gel	TIP, TMAO	Polymerization	Styrene	Maliakal et al. 2005
TiO_2-SiO_2/PMMA	Hydrolysis	Silicic acid, $TiCl_4$	Polymerization	MMA	Xu et al. 2006
SnO_2/graphene	High temperature	$SnCl_2.H_2O$, PEG	Reduction	Graphene	Du et al. 2010

Organic/inorganic core-shell heterojunction materials

Organic/inorganic core-shell heterojunctions are structurally just the reverse of the previous types described above. The core material of this particular class of core-shell heterojunctions is made of a polymer, such as polystyrene, poly(vinyl pyrrolidone), poly(ethylene oxide), poly(vinyl benzyl chloride), polyurethane, dextrose, surfactant, and different copolymers, such as acrylonitrile butadiene styrene, poly-(styrene acrylic acid), and styrene methyl methacrylate. The shell can also be made from different materials, such as metals, metal oxides, metal chalcogenides, or silica. A range of different organic/inorganic core/shell heterojunctions based on different core materials are listed in Table 5.

Heterojunction gas sensor materials based on 2D materials

Recently, 2D materials, with thicknesses of single to several atomic layers, such as graphene, transition metal dichalcogenides (TMDs), such as MoS_2 or WS_2, $MoSe_2$ as well as black phosphorus (also known as phosphorene) have been suggested as efficient sensing materials for next generation gas sensing systems. Two-dimensional (2D) nanomaterials such as graphene and metal sulfides have been found to be very sensitive to low concentration NO owing to their distinctive structure and electrical properties. However, the high charge carrier concentration of graphene limits its response to the electron-transfer-induced perturbations. In contrast, the transitional metal sulfides such as Mo/W/SnS_2 are a family of materials with a wide variety of electronic properties. Among these metal sulfides, tin disulfide (SnS_2) has a similar two dimensional nanostructure and has been attracting increasing attention with several desirable features such as a narrow band gap, large surface-to-volume ratio, and a larger electronegativity. However, the

Table 5. Inorganic core/organic shell heterojunctions.

Material	**Core**		**Shell**		**Reference**
Core-shell	**Method**	**Basic Reagent**	**Method**	**Basic Reagent**	
PS/Cu, PS/Cu_2O	EP	Styrene	Hydrolysis with calcinations		Kawahashi and Shiho 2000
PS/Ag	DP	Styrene	Reduction	$Ag(NH_3)_2OH$, $SnCl_2$	Yang et al. 2008
PS/TiO_2		PS beads	Stober method	TEOS	Dai et al. 2005
PS/Fe_3O_4	EP	Styrene	Co-precipitation	$FeCl_2$, $2H_2O$	Moore et al. 2000
PU/SiO_2	Polymerization	PA, adipic acid	Stober method	TEOS	Chen et al. 2006
PVP/SiO_2		PVP	Stober method	TEOS	Park et al. 2003
CS-PAA/SiO_2		Chitosan, PA	Ion-exchange process	Na_2SiO_3, cation exchange	Tsai and Li 2006
CTAB/SiO_2		CTAB	Stober method	TEOS, $Na_2S_2O_3$	Fowler et al. 2001a, 2001b
SDS and TPAB/SiO_2		SDS, TPAB	Stober method	TEOS	Wang et al. 2006
ABS/TiO_2	Radical EP	AN-butadine, styrene	Hydrolysis	$Ti(SO_4)_2$	Fan and Gao 2006
PSAA/TiO_2	Polycondensation	Melamine, HCHO	Hydrolysis	$Ti(SO_4)_2$	Sgraja et al. 2006
SMA/TiO_2	Radical co-polymerization	Styrene, MMA	Hydrolysis	TBOT	Song et al. 2009
MF/SiO_2	EP	Melanine, HCHO	Hydrolysis	Nano-SiO_2	Mahdavian et al. 2007
PS/ZnS	DP	Styrene, MAA	Precipitation	Zinc acetate, $SC(NH_2)_2$	Bi et al. 2010

sensing performances of pure SnS_2 is poor to low NO_2 concentrations at lower temperatures than 100°C (Gu et al. 2017).

Gu et al. (2017) reported SnS_2 decorated with SnO_2 nanoparticle by oxidizing pristine SnS_2 in air at 300°C. The SnS_2 decorated with SnO_2 nanoparticle was used as an effective sensing material for detecting low concentrations of NO_2 at low temperature (80°C).

It was found that the sensor with the decoration of SnO_2 nanoparticles exhibited higher response than that of the pristine SnS_2, while the SnO_2 exhibited no response to NO_2 at low temperature (80°C) (Figure 24).

There are many other members of the TMD family, including the noble metal TMDs, which have thus far garnered less attention. Studies have theoretically predicted the properties of these materials and touted them for use in assorted applications, but little work has been done to date on their synthesis. Platinum diselenide ($PtSe_2$) is one such material, which has not been previously assessed for use in device applications. Bulk $PtSe_2$ is a semimetal with zero band gap, which has been synthesized by chemical methods and is used as a photocatalytic material in nanocomposites with graphene. Theoretical studies have suggested the emergence of a band gap in monolayer $PtSe_2$. Monolayer $PtSe_2$ can provide an enhanced mobility and thus is of great interest for electronic applications. Sajjad et al. (2017) reported first-principles' calculations of the structural and electronic properties of monolayer 1T-$PtSe_2$ with adsorbed NO_2, NO, NH_3, H_2O, CO_2, and CO molecules. The favorable adsorption sites have been determined by analysis of the adsorption energy. All molecules draw charge from monolayer $PtSe_2$, except for NH_3, which acts as charge donor. The DOS reveals that the adsorption does not significantly alter the valence and conduction bands of monolayer $PtSe_2$, while partially occupied impurity states are introduced by NO_2

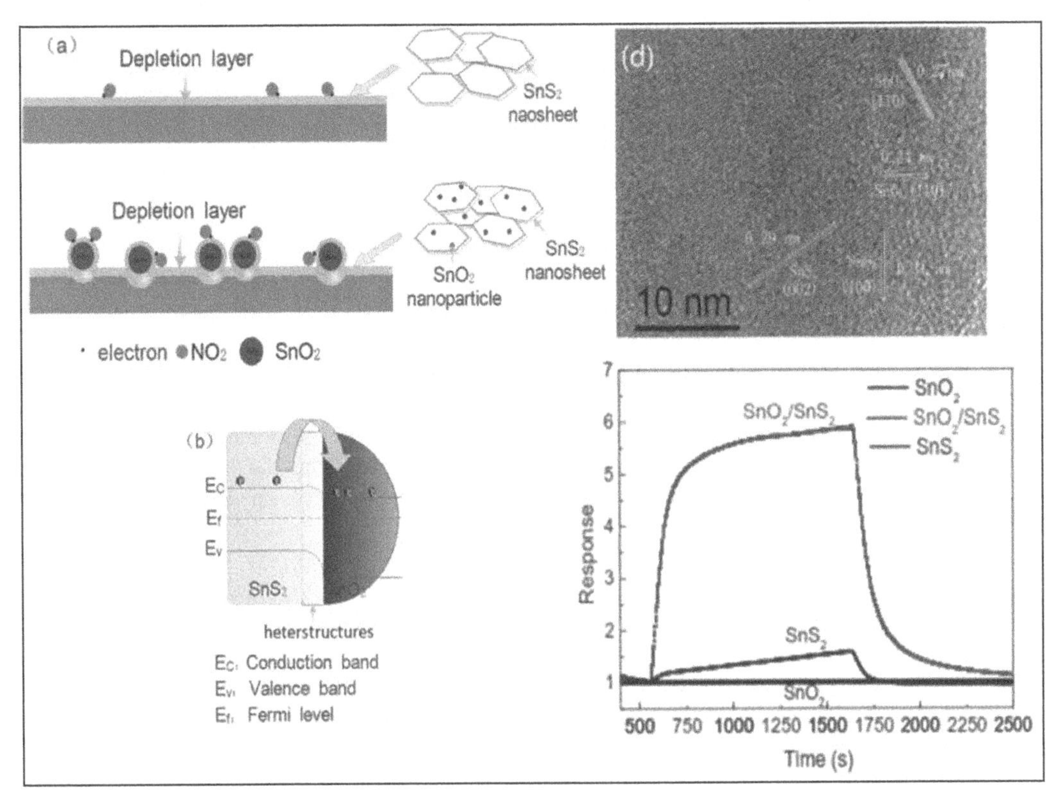

Figure 24. (a) The schematic diagram of possible gas sensing mechanism of SnS_2 and SnO_2/SnS_2 based sensors, (b) Electron transfer at the SnO_2/SnS_2 interface. (c) The response curves of the sensors of SnS_2, SnO_2/SnS_2 nanocomposites and SnO_2 to 8 ppm NO_2 at 80°C. (Reprinted with permission from Gu et al. 2017.)

and NO. High adsorption energies turn out to be characteristic of monolayer $PtSe_2$ and indicate potential in gas sensor applications.

Challenges in core-shell heterojunction materials as gas sensors

The discovery of novel materials, such as core-shell heterojunction materials opened up a new paradigm in sensor technology which can drastically enhance the gas sensor performances. But, though these special heterojunction materials improved the sensor performance, they open up the complexity in design and configurations of materials and finally understanding their physico-chemical mechanism behind their sensing properties. These aspects further challenged the researchers working in the field anticipating to fill the gap prevailing in this technology. Although several researchers have attempted to manage the thickness and consistent coating of the shell using these methods, these methods are still not well established, and proper control is very difficult. The main difficulties are (i) agglomeration of core particles in the reaction medium, (ii) favored structure of separate particles of shell material rather than coating on the surface of the core materials, (iii) incomplete coverage of the core surface and (iv) control of the reaction rate.

Challenges pertaining to the mechanism of gas sensing in heterojunctions

As observed from the above discussions, the gas sensing performances in the heterojunction nanomaterials is a complex phenomenon (Miller et al. 2014). The sophisticated analytical methods in the current generation made a pathway to study in depth the crystallographic, defect states, surface characteristics and morphologies of such heterojunction materials and correlating to their sensing behaviors. As far as heterojunctions are concerned, two important mechanisms control the sensing, first being the change in the potential barrier height as a result of formation of interfaces. These interfaces react in ultrafast manner, making them more sensitive to the analyte gas. The second mechanism pretending to the heterojunctions is the narrowing of the charge conduction channel along the materials, especially 1D nanomaterials, which favors directed electron transport (Kim et al. 2013). This 1D factor pretends to be the most influential factor in sensing; similarly, depletion layer formations in these kinds of heterostructures are still unclear making them difficult to understand exact mechanism.

Stability parameters

Though the heterojunction metaloxides are well known for their chemical stability relative to organic and hybrid sensors, small drifts in sensor resistance occur at the interfacial regions as particle is scaled down to nanodimensions. These small changes in the base resistance values of sensors occur due to the high operating temperature and long time exposure (Kim et al. 2013). Thus, either the thermal stability of the heterojunction nanomaterials should be studied in detail or bringing down the operating temperature can address the issue.

Reproducibility

Another important phenomenon to be considered when taking the material to device applications is reproducibility. Sensors should be repeatable and should have device-to-device reproducibility while considering them for large scale device applications. Thus, fabrication methods should effectively control the heterojunction formation with appropriate thickness (nm) of the interfaces. As the main parameter, Debye length directly affects the sensor response and hence change in the λ between sensor devices can affect the commercial value (Miller et al. 2014).

Integration of materials into micro-fabricated devices

The most important challenge in the current sensor technology is integration of the heterojunction nanomaterials with the silicon based microelectronic devices. Complex nano-heterostructures often have precise multi-step synthesis processes that can bind compatibility with architectures required in versatile microelectronic devices. In general, most of the MOS based heterojunctions works on higher temperature, where alumina substrates should be used, after forming a thin film which should be continuously heated from a temperature of 100°C to 500°C (Miller et al. 2014). Hence, it is not desirable to use silicon and alumina heaters together in device formation, which again leads to complexity in fabrication. Instead of using MEMS microplate platform, low temperature 1D core-shell heterojunction nanomaterials can be synthesized to address the issues. Hence, it's understood from the above mentioned challenges that pretending in the core-shell heterojunctions nanomaterials has opened certain issues and problems to be addressed by the upcoming researchers.

Operando studies

From the beginning of metal-oxide-based gas sensors, a lot of effort has been made to explain the mechanism responsible for gas sensing (Williams 1987, Madou et al. 1989, Sberveglieri 1992). Regardless of progress in recent years, a lot of key issues stay behind the matter of controversy, for instance, the discrepancy between electrical and spectroscopic investigations, over and above the lack of demonstrated mechanistic explanation of the surface reactions concerned in gas sensing. Herein, the "simultaneous measurement of the gas response and the purpose of molecular adsorption properties are necessary for a better understanding of gas sensing mechanisms" (Batzill and Diebold 2006).

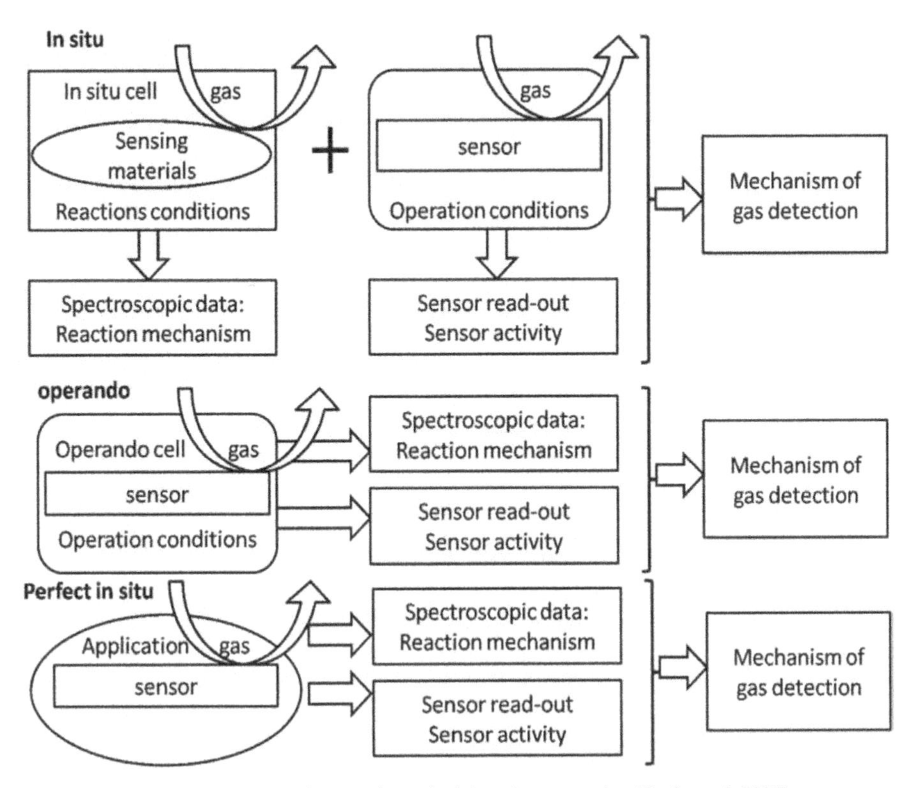

Figure 25. *In situ* and operando methodology in gas sensing (Gurlo et al. 2007).

This measurement can be performed either on clean and precise surfaces in ultrahigh vacuum (UHV) surroundings or at temperatures and pressures that imitate genuine sensor operating conditions (Bell et al. 2002). Continuous improvement has been made for the duration of the past few years for the final approach, that is, the use of *in situ* and operando spectroscopic techniques (Figure 25). The terms "*in situ*" and "operando" derive from the field of heterogeneous catalysis (Hunger and Weitkamp 2001, Gurlo and Riedel 2007) in which "*in situ* spectroscopy" represents spectroscopic techniques and measurements for learning catalysts *in situ*—"under reaction conditions or conditions relevant to reaction conditions". The term "operando spectroscopy" was established only in recent times to demonstrate techniques that are able to differentiate a "working" catalyst. This methodology merges *in situ* spectroscopic investigation with real-time monitoring of the catalytic performance. *In situ* studies based on XPS, Raman, DRIFT and TEM can be used to characterize the heterojunction nanomaterials in the presence of gas, thereby strengthening our understanding.

Discrepancies between a characteristic in situ and operando experiments

(a) *In situ spectroscopy*: Spectroscopic characterization of sensing material under conditions relevant to operation conditions. In this, the sensing performance of this material may be not characterized or may be characterized in a separate experiment.

(b) Operando spectroscopy: Spectroscopic characterization of an active sensing element in real time and under operating conditions with the synchronized read-out of the sensor activity and simultaneous monitoring of gas composition.

These definitions conclude the boundary conditions under which an "operando" experiment is performed:

- On a sensing element, which itself is a complicated device and consists of a number of parts: in solid-state devices with an electrical response, for example, the sensing film is deposited onto a substrate to which electrodes for an electrical read-out are attached ("transducer"); consequently, the evaluation of their interfaces is of supreme significance for understanding the general sensing mechanism;
- In real time: a sensor is developed to respond to the changes in the atmospheric gas as quick as possible; consequently, it demands a swift spectroscopic response;
- Under operating conditions: these can differ from ambient environments (RT and atmospheric pressure) to high temperatures and pressures;
- With simultaneous read-out of sensor activity: Concentration of the gas to be measured is transduced by the sensor into an electrical or other convenient output, depending upon the *modus operandi* of the sensor (optical, mechanical, thermal, magnetic, electronic, or electrochemical) and the transducer technology;
- With simultaneous monitoring of gas composition: on-line gas analysis in gas sensing shows a dual role: the output compositions and concentrations offer data regarding reaction products and possible reaction paths and the input concentration verifies the sensor input data (concentration of the component to be detected) (Gurlo et al. 2007).

Operando DRIFT spectroscopy and isotopic labeled gases were able to propose a surface vacancy based reception model for the gas sensing with undoped SnO_2. They also found that the difference in the calcination temperature determines different surface terminations for the SnO_2 materials (Degler et al. 2015). Degler and his group, for the first time, demonstrated the feasibility and benefit of operando UV/vis spectroscopy on SnO_2-based gas sensors. The obtained values for the optical band gaps for both materials are in line with other reports. The gas sensing experiments demonstrated that SnO_2 either gets reduced in the presence of reducing gases like CO or oxidized when increasing the oxygen concentration in a nitrogen atmosphere. However, in the absence of atmospheric oxygen the reduction remains surficial and does not impinge on the entire grain, even for relatively small grains (d = 12.5 nm). Consequently, the proportion of reduced and unaltered materials sturdily depends on the grain size. These findings reveal the usefulness of operando UV/vis-DRS for gas sensor research by analyzing other sensing material as well as doped systems (Degler et al. 2015). Pavelko et al. explained the usage of modulation excitation diffuse reflectance infrared Fourier transform spectroscopy (DRIFTS) and resistance measurements at 300°C. Their results reveal that the material synthesized from tin tetrachloride manifests higher affinity to chemisorbed water than the one made from tin hydroxide acetate (SnO_2 Ac) (Pavelko et al. 2013). The latter was shown to exhibit a strong correlation between the evolution of surface OH groups (linking type, implicated in hydrogen bonding) and electric resistance upon increasing concentration of CO. On the SnO_2 Cl surface, the amount of ITOD groups is only ca. 20% lower than that of BBODs, while on SnO_2 Ac the difference is 80%. Harbeck et al. performed DRIFT measurement on SnO_2 sensors at different temperatures between room temperature and 300°C. Presence of different surface OH groups as well as coordinated water on the SnO_2 sensor surface was studied using DRIFT. Their strength alters with temperature. Throughout the temperature cycles, the bands' peak positions are reversibly changed but their intensity remained the same. CO measurements were carried out at 300°C at different humidity levels (0 and 50% r.h.) on un-doped and Pd doped sensors. In the presence of CO in the observed spectra: a decrease of the OH groups on the SnO_2 surfaces, the appearance of gaseous CO_2 and CO in the pores of the sensitive layer and an increase of hydrated protons and of the free charge concentration. The effects are vividly influenced by the water vapor concentration, temperature, and dopands (Pd) and can be linked with concurrently performed sensor resistance measurements (Harbeck et al. 2003).

Conclusions and future perspectives

Metal-oxide based gas sensors often display remarkable changes in their electrical properties, for example, their work function and electrical conductivity, upon exposure to O_2, CO, NO_2, H_2S, NH_3, H_2 and other reactive gases. The fundamentals of resistive type sensor operation are based on the changes in resistance (or conductance) of the material as induced by the surrounding gas. The changes are caused by various processes, which can occur both at the surface and also in the bulk of gas-sensing material. This chapter explained in detail about the sensing mechanism of different types of metal oxides heterojunctions and its sensing property variations towards various gases (both reducing and oxidizing gases). Core–shell structures are very promising due to the maximization of interfacial heterojunction area, which formulates the electronic interactions of the interface are mainly dominant. The large surge in recent 1D nanostructure research has developed an excellent platform for future researchers to create 1D core–shell structures simply by pertaining deposition routes, for instance, spin-coating, ALD, sputtering and electrospinning to existing methods. These processes might still permit for multiple dissimilar shell layers of controlled thickness that may give enhanced sensing responses. Numerous distinctive morphologies known as hierarchical structures have also been synthesized through self-assembly and vapor phase growth processes. The properties of these core-shell heterojunction are not only size dependent but are also linked with the actual geometry. The concept of coupling a nanometer or subnanometer scale oxide-phase to a metal surface to form a metal–oxide heterojunction material with novel gas sensing properties is an attractive approach in the essential exploration for evolving phenomenon in low-dimensional gas sensors materials, and it may open new paradigm of materials' design with a view toward gas sensor applications. Though micro/nanostructure manipulation is often the claim for improved performance, also, there are likely electronic and chemical effects that take part, that were not carefully elucidated. Heterojunction based sensors have the potential to improve sensitivity and stability compared to conventional gas sensors which can work at very low operating temperature. The special physical and chemical properties of hybrid core-shell heterostructures make their electrical responses extremely sensitive to the species adsorbed on the surface, lowers the operating temperature by increasing the reaction rate and active sites at the interface favoring adsorption of analyte molecules. But, the catalytic sensitization effects in such 1D core-shell heterojunction nanostructures introduces additional complexity towards understanding the underlying mechanism of selective interactions of the sensing materials. In this scenario, development of dimension controlled core-shell heterojunctions and analyzing their structure-property relationship can significantly contribute towards understanding the underlying complex sensing mechanism and to predict the reason behind their anomalous behavior.

This chapter has greatly extended our perceptive of the role of nanostructured heterojunctions as gas sensing materials. Further, new materials and heterostructure designs will require advance study of mechanisms and influencing factors on gas sensing performance. Future information in nanostructured heterojunction gas sensors must have a double focus on (1) novel materials and novel heterojunction interfaces, and (2) the mechanisms that preside over the sensing performance.

Acknowledgements

The authors acknowledge the funding from DST-SERB and facilities and support provided by PSG Sons and Charities, Coimbatore.

References

Alali, K.T., T. Liu, J. Liu, Q. liu, Z. Li, H. Zhang et al. 2016. Fabrication of $CeO_2/ZnCo_2O_4$ N-P heterostructural porous nanotubes via electrospinning technology for enhanced ethanol gas sensing performance. RSC Adv. Copyright@Royal Society of Chemistry.

Anderson, R.L. 1960. Germanium-gallium arsenide heterojunctions. I. B. M. J. Res. and Develop. 4: 283.

Anderson, R.L. 1962. Experiments on Ge-GaAs heterojunctions. Solid-State Electronics Pergamon Press 5: 341–351.

Ansari, Z.A., S.G. Ansari, T. Ko and J. Oh. 2002. Effect of MoO_3 doping and grain size onSnO_2-enhancement of sensitivity and selectivity for CO and H_2 gas sensing. Sens. Actuators B: Chem. 87: 105–114.

Arnal, P.M., M. Comotti and F. Schüth. 2006. High-temperature-stable catalysts by hollow sphere encapsulation. Angew. Chem. Int. Ed. 45: 8224–8227.

Barsan, N. and U. Weimar. 2001. Conduction model of metal oxide gas sensors. Journal of Electroceramics 7: 143–167.

Barsan, N., D. Koziej and U. Weimar. 2007. Metal oxide-based gas sensor. Sensors and Actuators B: Chemical 121: 18.

Batzill, M. and U. Diebold. 2006. Characterizing solid state gas reponses using surface charging in photoemission: water adsorption of SnO_2. J. Phys. Condens. Matter. 18: L129–L134.
Bell, N.A., J.S. Brooks, S.D. Forder, J.K. Robinson and S.C. Thorpe. 2002. Backscatter Fe-57 Mössbauer studies of iron(II) phthalocyanine. Polyhedron. 21: 115–118.
Bi, C., L. Pan, Z. Guo, Y. Zhao, M. Huang, X. Ju et al. 2010. J. Q. Mater. Lett. 64: 1681.
Brattain, W.H. and J. Bardee. 1953. Surface Properties of Germanium. The Bell System Technical Journal, Volume XXXII.
Brinzari, V., G. Korotcenkov and J. Schwnk. 1999. Optimization of thin film gas sensors for environmental monitoring through theoretical modeling. SPIE Conference on Chemical Microsensors and Applications II.
Burke, N.A.D., H.D.H. Stover and F.P. Dawson. 2002. Magnetic nanocomposites: preparation and characterization of polymer-coated iron nanoparticles. Chem. Mater. 14: 4752–4761.
Chava, R.K., S.Y. Oh and Y.T. Yu. 2016. Enhanced H_2 gas sensing properties of Au@In_2O_3 core-shell hybrid metal semiconductor hetero nanostructures. CrystEngComm.
Chen, A., X. Huang, Z. Tong, S. Bai, R. Luo and C.C. Liu. 2006a. Preparation, characterization and gas-sensing properties of SnO_2–In_2O_3 nanocomposite oxides. Sens. Actuators B: Chem. 115: 316–321.
Chen, A., S. Bai, B. Shi, Z. Liu, D. Li and C.C. Liu. 2008. Methane gas-sensing and catalytic oxidation activity of SnO_2–In_2O_3 nanocomposites incorporating TiO_2. Sens. Actuators B: Chem. 135: 7–12.
Chen, Y.C., S.X. Zhou, H.H. Yang and L.M. Wu. 2006b. J. Sol-Gel Sci. Technol. 37: 39.
Choi, S.-W., J.Y. Park and S.S. Kim. 2009. Synthesis of SnO_2–ZnO core–shell nanofibers via a novel two-step process and their gas sensing properties. Nanotechnology 20: 465603.
Choi, S.-W., A. Katoch, G.J. Sun and S.S. Kim. 2013. Synthesis and gas sensing performance of ZnO–SnO_2 nanofiber–nanowire stem-branch heterostructure. Sensors and Actuators B 181: 787–794. Copyright@Elsevier.
Choi, U.-S., G. Sakai, K. Shimanoe and N. Yamazoe. 2005. Sensing properties of Au-loaded SnO_2-Co_3O_4 composites to CO and H_2. Sens. Actuators B: Chem. 107: 397–401.
Chowdhuri, A., P. Sharma and V. Gupta. 2002. H_2S gas sensing mechanism of SnO_2 films with ultrathin CuO dotted islands. J. Appl. Phys. 92: 2172.
Chowdhuri, A., V. Gupta and K. Sreenivas. 2003. Fast response H_2S gas sensing characteristics with ultra-thin CuO islands on sputtered SnO_2. Sensors and Actuators B 93: 572–579.
Comini, E., M. Ferroni, V. Guidi, G. Faglia, G. Martinelli and G. Sberveglieri. 2002. Nanostructured mixed oxides compounds for gas sensing applications. Sens. Actuators B: Chem. 84: 26–32.
Comini, E. 2006. Metal oxide nano-crystals for gas sensing. Analytica Chimica Acta 568: 28–40.
Comini, E., C. Baratto, G. Faglia, M. Ferroni, A. Vomiero and G. Sberveglieri. 2009. Quasi-one dimensional metal oxide semiconductors: Preparation, characterization and application as chemical sensors. Progress in Materials Science 54(1): 1–67.
Dai, Z., F. Meiser and H.J. Mohwald. 2005. Colloid Interface Sci. 288: 298.
Datskos, P.G., S. Rajic, M.J. Sepaniak, N. Lavrik, C.A. Tipple, L.R. Senesac and I. Datskou. 2001. Chemical detection based on adsorption-induced and photoinduced stresses in microelectromechanical systems devices. J. Vac. Sci. Technol. B 19(4).
de Lacy Costello, B.P., R. Ewen, P.R. Jones, N. Ratcliffe and R.K. Wat. 1999. A study of the catalytic and vapour-sensing properties of zinc oxide and tin dioxide in relation to 1-butanol and dimethyldisulphide. Sens. Actuators B: Chem. 61: 199–207.
de Lacy Costello, B.P.J., R.J. Ewen, N.M. Ratcliffe and P.S. Sivanand. 2003. Thick filmorganic vapour sensors based on binary mixtures of metal oxides. Sens. Actuators B: Chem. 92: 159–166.
Degler, D., S. Wicker, U. Weimar and N. Barsan. 2015. Identifying the active oxygen species in SnO_2 based gas sensing materials: an operando IR spectroscopy study. J. Phys. Chem. C 119(21): 11792–11799.
Dinesh, V.P., P. Biji, Arun K. Prasad, S. Dhara, A. Anuradha and M. Kamaruddin. 2015. Rapid synthesis and characterization of hybrid ZnO@Au core–shell nanorods for high performance, low temperature NO_2 gas sensor applications. Applied Surface Science 355: 726–735. Copyright@Elsevier.
Dinesh, V.P., P. Biji, Arun K. Prasad, S. Dhara, M. Kamaruddin and A.K. Tyagi. 2017a. Highly sensitive, atmospheric pressure operatable sensor based on Au nanoclusters decorated TiO_2@Au heterojunction nanorods for trace level NO_2 gas detection. J. Mater. Sci. Mater. Electron. Copyright@Springer.
Dinesh, V.P., S. Abdulla and P. Biji. 2017b. An emphatic study on role of spill-over sensitization and surface defects on NO_2 gas sensor properties of ultralong ZnO@Au heterojunction NRs. Journal of Alloys and Compounds 712: 811–821. Copyright@Elsevier.
Dinesh, V.P., S. Abdulla and P. Biji. 2017c. Highly monodispersed mesoporous, heterojunction ZnO@Au micro-spheres for tracelevel detection of NO_2 gas. Microporous and Mesoporous Materials.
Du, Z., X. Yin, M. Zhang, Q. Hao, Y. Wang and T. Wang. 2010. *In situ* synthesis of SnO_2/graphene nanocomposite and their application as anode material for lithium ion battery. Mater. Lett. 64: 2076.
Epifani, M. 2001. Sol–Gel processing and characterization of pure and metal-doped, SnO_2 thin films. J. Am. Ceram. Soc. 84(1): 48–54.
Fan, W. and L. Gao. 2006. J. Am. Chem. Soc. 297: 157.
Fang, G., Z. Liu, C. Liu and K. Yao. 2000. Room temperature H_2S sensing properties and mechanism of CeO_2–SnO_2 sol–gel thin films. Sens. Actuators B: Chem. 66: 46–48.
Fowler, C.E., D. Khushalani and S. Mann. 2001a. Facile synthesis of hallow silica microspheres. J. Mater. Chem. 11: 1968–1971.

Fowler, C.E., D. Khushalani and S. Mann. 2001b. Chem. Commun. 2028.
Ge, C., D. Zhang, A. Wang, H. Yin, M. Ren, Y. Liu, Jiang.
Ge, C., Z. Bai, M. Hu, D. Zeng and S. Cai. 2008. Preparation and gas-sensing property of ZnO nanorod-bundle thin films. Materials Letters 62(15-31): 2307–2310.
Ghosh Chaudhuri, R. and S. Paria. 2011. Core/shell nanoparticles: Classes, properties, synthesis mechanisms, characterization, and applications. Chem. Rev. 112: 2373–2433.
Gittins, D.I. and F. Caruso. 2001. Tailoring the polyelectrolyte coating of metal nanoparticles. J. Phys. Chem. B 105: 6846.
Gopel, W. and K.D. Schierbaum. 1995. SnO_2 sensors: current status and future prospects. Sensors and Actuators B: Chemical 26: 1.
Gopidas, K.R., J.K. Whitesell and M.A. Fox. 2003. Metal-core–organic shell dendrimers as unimolecular micelles. J. Am. Chem. Soc. 125: 14168.
Gou, X., G. Wang, J. Yang and J. Park. 2008. Chemical synthesis, characterisation and gas sensing performance of copper oxide nanoribbons. J. Mater. Chem. 18: 965–969.
Gu, Ding, Xiaogan Li ,Yangyang Zhao and Jing Wang. 2017. Enhanced NO_2 sensing of SnO_2/SnS_2 heterojunction based sensor. Sensors and Actuators B 244: 67–76. Copryright@Elsevier.
Gu, H., Z. Wang and Y. Hu. 2012. Hydrogen gas sensors based on semiconductor oxide nanostructures. Sensors 12: 5517–5550.
Gubanov, A.I. 1951. The theory of the contact of two semiconductors of the same type of conductivity. Zh. Tekh. Fiz. 21: 304.
Guo, Z., M. Li and J. Liu. 2008. Highly porous CdO nanowires: preparation based on hydroxy- and carbonate-containing cadmium compound precursor nanowires, gas sensing and optical properties. Nanotechnology 19: 24.
Gurlo, A. and R. Riedel. 2007. *In situ* and operando spectroscopy for assessing mechanisms of gas sensing. Angew. Chem. Int. Ed. 46: 3826–3848.
Haldorai, Y., W.S. Lyoo, S.K. Noh and J. Shim. 2010. Polystyrene-coated alumina powder via dispersion polymerization for indirect selective laser sintering applications. J. React. Funct. Polym. 70: 393.
Harbeck, S., A. Szatvanyi, N. Barsan, U. Weimar and V. Hoffmann. 2003. Drift studies of thick film un-doped and Pd-doped SnO_2 sensors: temperature changes effect and CO detection mechanism in the presence of water vapour. Thin Solid Films 436: 76–83.
Heiland, G. 1954. Zum Einfluss von Wasserstoff auf die elektrische Leitf ahigkeit von ZnO Kristallen. Zeit. Phys. 138: 459–464.
Hernandez-Ramirez, F., J.D. Prades, R. Jimenez-Diaz, T. Fischer, A. Romano-Rodriguez, S. Mathur et al. 2007. On the role of individual metal oxide nanowires in the scaling down of chemical sensors. Phys. Chem. Chem. Phys. 11: 7105–7110.
Hidalgo, P. and R. Castro. 2005. Surface segregation and consequent SO_2 sensor response in SnO_2–NiO. Chem. Mater. 17: 4149–4153.
Huang, H., Y.C. Lee, O.K. Tan, W. Zhou, N. Peng and Q. Zhang. 2009. High sensitivity SnO_2 single-nanorod sensors for the detection of H_2 gas at low temperature. Nanotechnology 20: 115501.
Huang, K., R.S. Tichy and J.B. Goodenough. 1998. Superior Perovskite oxide-ion conductor; strontium- and magnesium-doped LaGaO3 I. phase relationships and electrical properties. Journal of the American Ceramic Society 81: 2565–2575.
Hübner, M., D. Koziej, J.D. Grunwaldt, U. Weimar and N. Barsan. 2012. An Au clusters related spill-over sensitization mechanism in SnO_2-based gas sensors identified by operando HERFD-XAS, work function changes, DC resistance and catalytic conversion studies. Phys. Chem. Chem. Phys. 14: 13249–13254.
Hunger, M. and J. Weitkamp. 2001. *In situ* IR, NMR, EPR, and UV/Vis spectroscopy: Tools for new insight into the mechanisms of heterogeneous catalysis. Angew. Chem. 113: 3040–3059.
Iwamoto, M., Y. Yoda, N. Yamazoe and T. Seiyama. 1978. Study of metal oxide catalysts by temperature programmed desorption. 4. Oxygen adsorption on various metal oxides. J. Phys. Chem. 82: 2564–2570.
Ishihara, T., H. Matsuda and Y. Takita. 1994. Doped $LaGaO_3$ perovskite type oxide as a new oxide ionic conductor. Journal of the American Chemical Society 116: 3801–3803.
Ivanovskaya, M., D. Kotsikau, G. Faglia and P. Nelli. 2003. Influence of chemical composition and structural factors of Fe_2O_3/In_2O_3 sensors on their selectivity and sensitivity to ethanol. Sensors Actuators B Cem. 96: 498–503.
Jarzebski, Z.M. and J.P. Marton. 1976. Physical properties of SnO_2 materials. J. Electvochem. Soc.: Reviews and News 123: 9.
Jayaseelan, D. and P. Biji. 2016. New insights towards electron transport mechanism of highly efficient p-Type CuO (111) nanocuboids-based H2S gas sensor. J. Phys. Chem. C 120(7): 4087–4096.
Jin, C., S. Park, H. Kim and C. Lee. 2012. Ultrasensitive multiple networked Ga_2O_3-core/ZnO-shell nanorod gas sensors. Sensors and Actuators B 161: 223–228. Copyright@Elsevier.
Jinkawa, G. Sakai, J. Tamaki, N. Miura and N. Yamazoe. 2000. Relationship between ethanol gas sensitivity and surface catalytic property of tin oxide sensors modified with acidic or basic oxides. J. Mol. Catal. A: Chem. 155: 193–200.
Jhang, Z. and K. Colbow. 1997. Surface silver clusters as oxidation catalysts on semiconductor gas sensors. Sens. Actuators, B 40: 47–52.
Kalyamwar, V.S., F.C. Raghuwanshi, N.L. Jadhao and A.J. Gadewar. 2013. Zinc oxide nanostructure thick films as H_2S gas sensors at room temperature. Journal of Sensor Technology 3: 31–35.
Kar, A., M.a. Stroscio, M. Meyyappan, D.J. Gosztola, G.P. Wiederrecht and M. Dutta. 2011. Tailoring the surface properties and carrier dynamics in SnO_2 nanowires. Nanotechnology 22: 285709.
Kaur, M., N. Jain, K. Sharma, S. Bhattacharya, M. Roy and A.K. Tyagi. 2008. Room-temperature H_2S gas sensing at ppb level by single crystal In_2O_3 whiskers. Sensors and Actuators B 133: 456–461.

Kawahashi, N. and H.J. Shiho. 2000. Copper and copper compounds as coatings on polystyrene particles and as hollow spheres. Mater. Chem. 10: 2294.

Khorami, H.A., M. Keyanpour-Rad and M.R. Vaezi. 2011. Synthesis of SnO_2/ZnO composite nanofibers by electrospinning method and study of its ethanol sensing properties. Appl. Surf. Sci. 257: 7988–7992.

Kim, H.J. and J.-H. Lee. 2014. Highly sensitive and selective gas sensors using p-type oxide semiconductors: overview. Sens. Actuators B: Chem. 192: 607–627.

Kim, I.D., A. Rothschild and H.L. Tuller. 2013. Acta Mater. 61: 974–1000.

Kim, J.H., A. Katoch and S.S. Kim. 2016. Optimum shell thickness and underlying sensing mechanism in p–n CuO–ZnO core–shell nanowires. Sensors and Actuators B 222: 249–256. Copyright@Elsevier.

Kim, Y.S., I.S. Hwang, S.J. Kim, C.Y. Lee and J.H. Lee. 2008. CuO nanowire gas sensors for air quality control in automotive cabin. Sensors and Actuators B: Chemical 135(1): 298–303.

Kohl, D. 2001. Function and applications of gas sensors. J. Phys. D: Appl. Phys. 34: R125–R149.

Kolmakov, A. and M. Moskovits. 2004. Chemical sensing and catalysis by one-dimensional metal-oxide nanostructures. Annu. Rev. Mater. Res. 34: 151–180.

Kolmakov, A., D.O. Klenov, Y. Lilach, S. Stemmer and M. Moskovits. 2005. Enhanced gas sensing by individual SnO_2 nanowires and nanobelts functionalized with Pd Catalyst particle. Nano. Lett. 5: 667–673.

Kolmakov, A., D.O. Klenov, Y. Lilach, S. Stemmer and M. Moskovits. 2005. Enhanced gas sensing by individual SnO_2 nanowires and nanobelts functionalized with Pd Catalyst particles. Nano Letters 5: 667. Copyright@American Chemical Society.

Korotcenkov, G., V. Golovanov, V. Brinzari, A. Cornet, J. Morante and M. Ivanov. 2005. Distinguishing feature of metal oxide films' structural engineering for gas sensor applications. Journal of Physics: Conference Series 15: 256–261.

Korotcenkov, G. 2007. Metal oxides for solid-state gas sensors: What determines our choice? Materials Science and Engineering B 139: 1–23. Copyright@Elsevier.

Korotcenkov, G. and B.K. Cho. 2013. Engineering approaches for the improvement of conductometric gas sensor parameters. Part 1: Improvement of sensor sensitivity and selectivity (Short survey). Sensors and Actuators B: Chemical 188: 709.

Korotcenkov, G. and B.K. Cho. 2014. Engineering approaches to improvement of conductometric gas sensor parameters. Part 2: Decrease of dissipated (consumable) power and improvement stability and reliability. Sensors and Actuators B: Chemical 198: 316.

Kosc, I., I. Hotovy, V. Rehacek, R. Griesseler, M. Predanocy, M. Wilke et al. 2013. Sputtered TiO_2 thin films with NiO additives for hydrogen detection. Appl. Surf. Sci. 269: 110–115.

Kroemer, H. 1957. Theory of a wide-gap emitter for transistors. Proc. IRE 45: 1535.

Kusior, A., M. Radecka, M. Rekas, M. Lubecka, K. Zakrzewska, A. Reszka et al. 2012. Sensitization of gas sensing properties in TiO_2/SnO_2 nanocomposites. Proc. Eng. 47: 1073–1076.

Lao, J.Y., J.G. Wen and Z.F. Ren. 2002. Hierarchical ZnO nanostructures. Nano Letter 2: 1287.

Lee, M.-J., H.-W. Cheong, L.D.N. Son and Y.-S. Yoon. 2008. Surface reaction mechanism of acetonitrile on doped SnO~2 sensor element and its response behavior. Japanese Journal of Applied Physics 47: 2119.

Lenaerts, S., J. Roggen and G. Maes. 1995. FT-IR characterisation of tin dioxide gas sensor materials under working conditions. Spectrochim. Acta A 51883–894.

Liangyuan, C., B. Shouli, Z. Guojun, L. Dianqing, C. Aifan and C.C. Liu. 2008. Synthesis of ZnO–SnO_2 nanocomposites by microemulsion and sensing properties for NO_2. Sens. Actuators B: Chem. 134: 360–366.

Lin, Ying, Chao Li, Wei Wei, Yujia Li, Shanpeng Wen, Dongming Sun et al. 2015. A new type of acetylene gas sensor based on a hollow heterostructure. RSC Adv. 5: 61521. Copyright@Royal Society of Chemistry.

Liu, L., C.C. Guo, S. Li, L. Wang, Q. Dong and W. Li. 2010. Improved H_2 sensing properties of Co-doped SnO_2 nanofibers. Sensors and Actuators B 150: 806–810. Copyright@Elsevier.

Liu, Y., G. Zhu, J. Chen, H. Xu, X. Shen and A. Yuan. 2013. Co_3O_4/ZnO nanocomposites for gas-sensing applications. Appl. Surf. Sci. 265: 379–384.

Liu, Z., T. Yamazaki and Y. Shen. 2007. Room temperature gas sensing of p-type TeO_2 nanowires. Appl. Phys. Lett. 90: 173119.

Lu, X., X. Liang, Z. Sun and W. Zhang. 2005. Ferromagnetic Co/SiO_2 core/shell structured nanoparticles prepared by a novel aqueous solution method. Materials Science and Engineering B 117: 147–152.

Lyson-Sypien, B., A. Czapla, M. Lubecka, E. Kusior, K. Zakrzewska, M. Radecka et al. 2013. Gas sensing properties of TiO_2–SnO_2 nanomaterials. Sens. Actuators B: Chem. 1–10.

Madou, M.J. and S.R. Morrison. 1989. Chemical Sensing With Solid State Devices. Academic Press, San Diego.

Mahdavian, A.R., M. Ashjari and A.B. Makoo. 2007. Eur. Polym. J. 43: 336.

Maier, J. and W. Gopel. 1988. Investigations of the bulk defect chemistry of polycrystalline Tin(IV) oxide. Journal of Solid State Chemistry 72(2): 293–302.

Maliakal, A., H. Katz, P.M. Cotts, S. Subramoney and P. Mirau. 2005. Inorganic oxide core, polymer shell nanocomposite as a high K gate dielectric for flexible electronics applications. J. Am. Chem. Soc. 127: 14655.

Margaritondo, G. 1988. Perspectives in condensed matter physics, electronic structure of semiconductor heterojunctions. A Critical Reprint Series: Volume 1.

Matsushima, S., Y. Teraoka, N. Miura and N. Yamazoe. 1988. 100 Jpn. J. Appl. Phys. 27: 1798–18002.

McCue, J.T. and J.Y. Ying. 2007. SnO_2–In_2O_3 nanocomposites as semiconductor gas sensors for CO and NOx detection. Chem. Mater. 19: 1009–1015.

Miller, Derek R., Sheikh A. Akbar and Patricia A. Morris. 2014. Nanoscale metal oxide-based heterojunctions for gas sensing: A review. Sensors and Actuators B 204: 250–272T. Copyright@Elsevier.
Molday, Robert S. and Mackenzie Donald. 1982. Immunospecific Ferromagnetic iron-dextran reagents for the labeling and magnetic separation of cells. Journal of Immunological Methods 52: 353–367.
Molday, R.S. and L. Molday. 1984. Separation of cells labeled with immunospecific iron dextran microspheres using high gradient magnetic chromatography L. FEBS Lett. 170: 232.
Montmeat, P., C. Pijolat, G. Tournier and J.P. Viricelle. 2002. The influence of a platinum membrane on the sensing properties of a tin dioxide thin film. Sens. Actuators, B 84: 148–159.
Moore, L.R., M. Zborowski, M. Nakamura, K. McCloskey, S. Gura, M. Zuberi et al. 2000. J. Biochem. Biophys. Methods 44: 115.
Morrison, S.R. 1981. Sensors and Actuators 2: 329.
Mott, N.F. 1938. Note on the contact between a metal and an insulator or semi-conductor. Proc. Cambridge Philos. Soc. 34: 568.
Mott, N.F. 1939. The theory of crystal rectifiers. Proc. R. Soc. Lond. A 171: 27–38.
Na, C.W., H.-S. Woo, I.-D. Kim and J.-H. Lee. 2011. Selective detection of NO_2 and C_2H_5OH using a Co_3O_4-decorated ZnO nanowire network sensor. Chem. Commun. 47,4: 5148–5150.
Nguyen, H., C.T. Quy, N.D. Hoa, N.T. Lam, N.V. Duy, V.V. Quang et al. 2014. Sensors and Actuators B: Chemical 193: 888.
Oldham, W.G., A.R. Riben, D.L. Feucht and A.G. Milnes. 1963. N-n semiconductor heterojunctions. Solid State Elec. 6: 121.
Pagnier, T., M. Boulova, A. Galerie, A. Gaskov and G. Lucazeau. 1999. *In situ* coupled raman and impedance measurements of the reactivity of nanocrystalline SnO_2. Journal of Solid State Chemistry 143: 86–94.
Palkovits, R., H. Althues, A. Rumplecker, B. Tesche, A. Dreier, U. Holle et al. 2005. Polymerization of w/o microemulsions for the preparation of transparent SiO_2/PMMA nanocomposites. Langmuir 21: 6048.
Park, J.H., C. Oh, S.-I. Shin, S.K. Moon and S.G. Oh. 2003. J. Colloid Interface Sci. 266: 107.
Patil, D.R., L.a. Patil and P.P. Patil. 2007. Cr_2O_3-activated ZnO thick film resistors for ammonia gas sensing operable at room temperature. Sens. Actuators B: Chem. 126: 368–374.
Patil, D.R. and L.a Patil. 2009. Cr_2O_3-modified ZnO thick film resistors as LPG sensors. Talanta 77: 1409–1414.
Pavelko, Roman G., Helen Daly, Michael Hubner, Christopher Hardacre and Eduard Llobet. 2013. Time-resolved DRIFT, MS, and resistance study of SnO_2 materials: The role of surface hydroxyl groups in formation of donor states. J. Phys. Chem. 117(8): 4158–4167.
Pimtong-Ngam, Y., S. Jiemsirilers and S. Supothina. 2007. Preparation of tungsten oxide–tin oxide nanocomposites and their ethylene sensing characteristics. Sens. Actuators A: Phys. 139: 7–11.
Potje-Kamloth, K. 2008. Semiconductor junction gas sensors. Chem. Rev. 108: 367–399.
Qin, R., F. Li and W. Jiang. 2010. Synthesis and characterization of diethylenetriaminepentaacetic acid-chitosan-coated cobalt ferrite core/shell nanostructures. Chen. M. Mater. Chem. Phys. 122: 498.
Qin, S., L. Wang, X. Zhang and G. Su. 2010. Grafting poly(ethylene glycol) monomethacrylate onto Fe_3O_4 nanoparticles to resist nonspecific protein adsorption. Appl. Surf. Sci. 257: 731.
Ramgir, N., S. Yang and Y. Zacharias. 2010. Nanowire-based sensors. Small 6(16): 1705–1722.
Roldan Cuenya, B. and A. Kolmakov. 2008. Nanostructures: Sensor and catalytic properties. *In*: S. Seal (ed.). Functional Nanostructures (Springer Verlag, New York). Chap. 6, p. 305.
Rothschild, A. and Y. Komem. 2004. The effect of grain size on the sensitivity of nanocrystalline metal-oxide gas sensors. J. Appl. Phys. 95: 6374.
Rothschild, A. and H. Tuller. 2006. Gas sensors: New materials and processing approaches. Journal of Electroceramics 17: 1005.
Sajjad, M., E. Montes, N. Singh and U. Schwingenschlögl. 2017. Superior gas sensing properties of monolayer PtSe2. Adv. Mater. Interfaces, 1600911.
Samson, S. and C.G. Fonstad. 1973. Defect structure and electronic donor levels in stannic oxide crystals. J. Appl. Phys. 44(4618): 10.
Sberveglieri, G. 1992. Gas sensors. Kluwer, Dordrecht, p. 409.
Schierbaum, K.D., R. Kowalkowski, U. Weimar and W. Gopel. 1991. Conductance, work function and catalytic activity on SnO_2-based gas sensors. Sens. Actuators B: Chem. 205–214.
Schottky, W.Z. 1938. Semiconductor theory of the barrier layer. The Natural Sciences 26(52): 843–843.
Schottky, W.Z. 1939. On the semiconductor theory of the barrier and peak rectifiers. Phys. 113: 367–414.
Seiyama, T., A. Kato, K. Fujiishi and M. Nagatani. 1962. A new detector for gaseous components using semiconductive thin films. Anal. Chem. 34: 1502–1503.
Sgraja, M., J. Bertling, R. Kummel and P.J.J. Jansens. 2006. Mater. Sci. 41: 5490.
Shaposhnik, D., R. Pavelko, E. Llobet, F. Gispert-Guirado and X. Vilanova. 2011. Hydrogen sensors on the basis of SnO_2–TiO_2 systems. Proc. Eng. 25: 1133–1136.
Shewchun, J. and L.Y. Wei. 1964. Germanium-silicon alloy heterojunction. J. Electrochemical Society 111: 1145.
Shin, K., J.-J. Kim and K.-D.J. Suh. 2010. A facile process for generating monolithic-structured nano-silica/polystyrene multi-core/shell microspheres by a seeded sol–gel process method. Colloid Interface Sci. 350: 581.
Singh, N., A. Ponzoni, R.K. Gupta, P.S. Lee and E. Comini. 2011. Synthesis of In_2O_3–ZnO core–shell nanowires and their application in gas sensing. Sensors and Actuators B 160: 1346–1351. Copyright@Elsevier.
Song, C., W. Yu, B. Zhao, H. Zhang, C. Tang, K. Sun et al. 2009. Catal. Commun. 10: 650.

Sonker, Rakesh K. and B.C. Yadav. 2015. Growth mechanism of hexagonal ZnO nanocrystals and their sensing application. Materials Letters 160(1): 581–584.
Subramanian, V., E.E. Wolf and P.V. Kamat. 2003. Green emission to probe photoinduced charging events in ZnO–Au nanoparticles. Charge distribution and fermi-level equilibration. Langmuir 19: 469–474.
Sui, L.L., X.F. Zhang, X. Cheng, P. Wang, Y.M. Xu and S. Gao et al. 2016. Au-loaded hierarchical MoO_3 hollow spheres with enhanced gas sensing performance for the detection of BTX (benzene, toluene and xylene) and the sensing mechanism. ACS Appl. Mater. Interfaces. Copyright@ American Chemical Society.
Sun, X. and Y. Li. 2005. Ag@C Core/shell structured nanoparticles: Controlled synthesis, characterization, and assembly. Langmuir 21: 6019–6024.
Taguchi, N. 1971. U.S. Patent 3,631,436.
Tiemann, M. 2007. Porous metal oxides as gas sensors. Chemistry—A European Journal 13: 8376.
Tien, L.C., D.P. Norton, B.P. Gila, S.J. Pearton, H.-T. Wang, B.S. Kang et al. 2007. Detection of hydrogen with SnO_2-coated ZnO nanorods. Appl. Surf. Sci. 253: 4748–4752.
Tsai, M.S. and M.J.J. Li. 2006. Non-Cryst. Solids 352: 2829.
Vasiliev, R. and M. Rumyantseva. 1999. Effect of interdiffusion on electrical and gas sensor properties of CuO/SnO_2 heterostructure. Mater. Sci. 57: 241–246.
Wagh, M., L. Patil, T. Seth and D. Amalnerkar. 2004. Surface cupricated SnO_2–ZnO thick films as a H_2S gas sensor. Mater. Chem. Phys. 84: 228–233.
Wang, B., L.F. Zhu, Y.H. Yang, N.S. Xu and G.W. Yang. 2008. Fabrication of a SnO_2 nanowire gas sensor and sensor performance for hydrogen. J. Phys. Chem. C 112(17): 6643–6647.
Wang, G., H. Wu, D. Wexler, H. Liu and O. Savadogo. 2010a. Ni@Pt core–shell nanoparticles with enhanced catalytic activity for oxygen reduction reaction. Journal of Alloys and Compounds 503: L1–L4.
Wang, H.T., B.S. Kang, F. Ren, L.C. Tien, P.W. Sadik and D.P. Norton et al. 2005. Hydrogen-selective sensing at room temperature with ZnO nanorods. Applied Physics Letters 86(24).
Wang, L., Y. Kang, Y. Wang, B. Zhu, S. Zhang, W. Huang et al. 2012. CuO nanoparticle decorated ZnO nanorod sensor for low-temperature H_2S detection. Mater. Sci. Eng. C 32: 2079–2085.
Wang, W., Z. Li, W. Zheng, H. Huang, C. Wang and J. Sun. 2010b. Cr_2O_3-sensitized ZnO electrospun nanofibers based ethanol detectors. Sens. Actuators B: Chem. 143: 754–758.
Wang, Y.D., K.Y. Zang and S.J. Chua. 2006. Fonstad, C. G. Applied Physics Letters 89.
Wang, Z., Z. Li, J. Jinghui Sun, H. Hongnan Zhang, W. Wei Wang and W.Wei Zheng. 2010c. Improved hydrogen monitoring properties based on p-NiO/n-SnO_2 heterojunction composite nanofibers. J. Phys. Chem. C. 114: 6100–6105.
Weisz, P.B. 1953. Effects of electronic charge transfer between adsorbate and solid on chemisorption and catalysis. The Journal of Chemical Physics 21: 1531.
Williams, D.E. 1987. Conduction and gas response of semiconductor gas sensors. *In*: P.T. Moseley and B.C. Totfield (eds.). Solid State Gas Sensors. Adam Hilger, Philadelphia, p. 71.
Windischmann, H. and P. Mark. 1979. A model for the operation of a thin films tin oxide conductance modulation carbon monoxide sensor. J. Electrochem. Soc. 126: 627–630.
Wlodarski, W., M. Shafiei, J. Yua, R. Arsata, K. Kalantar-zadeha, E. Comini et al. 2010. Reversed bias Pt/nanostructured ZnO Schottky diode with enhanced electric field for hydrogen sensing. Sensors and Actuators B: Chemical 146: 507.
Wöllenstein, J., H. Böttner, M. Jaegle, W.J. Becker and E.Wagner. 2000. Material properties and the influence of metallic catalysts at the surface of highly dense SnO_2 films. Sens. Actuators, B 70: 196–202.
Wu, W., J. Shen, P. Banerjee and S. Zhou. 2010. Biomaterials 31: 7555.
Xiang feng, C, W. Caihong, J. Dongli and Z. Chenmonu. 2004. Ethanol sensor based on indium oxide nanowires prepared by carbothermal reduction reaction. Chemical Physics Letters 399(4-6).
Xing, L., S. Yuan, Z. Chen, Y. Chen and X. Xue. 2011. Enhanced gas sensing performance of SnO_2/MoO_3 heterostructure nanobelts. Nanotechnology 22: 1–6.
Xu, C., J. Tamaki, N. Miura and N. Yamazoe. 1991. Grain size effects on gas sensitivity of porous SnO_2-based elements. Sensors and Actuators B: Chemical 3: 147.
Xu, P., Z. Cheng, Q. Pan, J. Xu, Q. Xiang, W. Yu et al. 2006a. Hydrothermal synthesis of In_2O_3 for detecting H_2S in air. Sensors and Actuators B 115: 642–646.
Xu, P., H. Wang, R. Lv, Q. Du, W. Zhong and Y.J. Yang. 2006b. Polym. Sci. Part A: Polym. Chem. 44: 3911.
Xu, Xiuru, Jinghui Sun, Hongnan Zhang, Zhaojie Wang, Bo Dong et al. 2011. Effects of Al doping on SnO_2 nanofibers in hydrogen sensor. Sensors and Actuators B 160: 858–863.
Xu, Q.-H., D.-M. Xu, M.-Y. Guan, Y. Guo, Q. Qi and G.-D. Li. 2013. $ZnO/Al_2O_3/CeO_2$ composite with enhanced gas sensing performance. Sens. Actuators B: Chem. 177: 1134–1141.
Yamazoe, N. 1991. New approaches for improving semiconductor gas sensors. Sensors and Actuators B: Chemical 5: 7.
Yamazoe, N., G. Sakai and K. Shimanoe. 2003. Oxide semiconductor gas sensors. Catal. Surv. Asia 7: 63–75.
Yang, Z., L. Yang, Z. Zhang, N. Wu, J. Xie and W. Cao. 2008. Colloids, Surf. A 312: 113.
Yu, X., G. Zhang, H. Cao, X. An, Y. Wang, Z. Shu et al. 2012. ZnO@ZnS hollow dumbbells–grapheme composites as high-performance photocatalysts and alcohol sensors. New J. Chem. 36: 2593.
Zeng, W., T. Liu and Z. Wang. 2010. Sensitivity improvement of TiO_2-doped SnO_2 to volatile organic compounds. Phys. E Low-Dimensional Syst. Nanostruct. 43: 633–638.

Zeng, Y., Y. Bing, C. Liu, W. Zheng and G. Zou. 2012. Self-assembly of hierarchical $ZnSnO_3$–SnO_2 nanoflakes and their gas sensing properties. Trans. Nonferrous Met. Soc. China 22: 2451–2458.

Zhang, H., Z. Lia, L. Liu, X. Xu., Z. Wang and W. Wang. 2010. Enhancement of hydrogen monitoring properties based on Pd–SnO_2 composite nanofibers. Sensors and Actuators B 147: 111–115. Copyright@Elsevier.

Zhang, W.-H. and W.-D. Zhang. 2008. Fabrication of SnO_2–ZnO nanocomposite sensor for selective sensing of trimethylamine and the freshness of fishes. Sens. Actuators B: Chem. 134: 403–408.

Zhang, X.-J. and G.-J. Qiao. 2012a. High performance ethanol sensing films fabricated from ZnO and In_2O_3 nanofibers with a double-layer structure. Appl. Surf. Sci. 258: 6643–6647.

Zhang, Z. and J.T. Yates. 2012b. Band bending in semiconductors: Chemical and physical consequences at surfaces and interfaces. Chem. Rev. 112(10): 5520–5551.

10

Dual-Carbon Enhanced Composites for Li/Na-Ion Batteries

*Xing-Long Wu,** *Chao-Ying Fan* and *Jing-Ping Zhang*

Introduction

Along with the rapid developments of society and economy, energy consumption has been increasing more and more. It can be imagined that the conventional fossil fuels (such as, coal and petroleum) will be exhausted in the near future, finally leading to the serious energy crisis as well as the increasingly worsened environmental pollution due to the heavy combustion usage of these fossil fuels. In order to stop the overuse of fossil fuels to solve the energy crisis and alleviate the environmental pollutions, exploiting "green" and renewable energies (abbreviated as "GR-En") has presently become one of the most urgent research missions for the scientists working in the energy areas. Among various GR-En, solar energy and wind energy are the two most abundant and readily accessible types, making them attractive extensively. However, both are intermittent and unstable energies (e.g., solar energy can be received only in daytime with variable power due to the constantly changing sunlight intensity, but not at night), meaning that we cannot input them into the Grid continuously. This is one major inherent shortcoming, limiting their practical applications. In order to utilize these intermittent and unstable GR-En to power and serve us well in our daily lives, employing rechargeable energy storage devices (RESDs) to stably store/release GR-En, large-scale RESDs stabilize the GR-En into Grid and portable RESDs facilitate our everyday power requirements, should be the most effective strategy. Hence, developing highly efficient RESDs becomes one of the most significant and urgent research objectives for the scientists majoring in energy storage.

Among all kinds of RESDs, rechargeable batteries are one of the most popular and widely used technologies concerned with electrochemical methods of energy conversion due to their highly efficient energy-storage ability. Rechargeable batteries commonly include lithium ion battery (LIB), sodium ion

Northeast Normal University, Department of Chemistry, No. 5268, Renmin Street, Changchun, Jilin 130024, Republic of China.
Emails: fancy242@nenu.edu.cn; jpzhang@nenu.edu.cn
* Corresponding author: xinglong@nenu.edu.cn

battery (SIB), lithium air/oxygen ($Li-O_2$) battery, lithium sulfur (Li-S) battery, other advanced batteries (such as magnesium ion batteries and aluminum ion batteries), and so on. Among them, LIB is the state-of-the-art technology and has gained versatile applications in the market of portable electronic devices (such as laptops, cellular phone), electric vehicles (EVs), plug-in hybrid EVs (PHEVs) and large-scale energy storage stations, which mainly benefits from their merits of high voltage, low self-discharge, long cycle life, low toxicity and high reliability. For the SIB, it stores electrical energy under the similar work mechanism as that of LIB, making them cutting-edge for energy storage because of cheaper and more abundant sodium resources. However, it is still a highly challenging issue to obtain feasible electrode materials for SIB with superior electrochemical properties in terms of high specific capacity, acceptable rate performance, and long cycle life, because larger sodium ions lead to the sluggish electrode kinetics including slower ionic diffusion and lower electronic conductivities in comparison with LIB. Nevertheless, LIB and SIB are two significant RESDs with similar energy-storage mechanism and are worthy of further studies to improve their performances to power us better.

In this chapter, we will first introduce the work mechanism of LIB and SIB briefly, and point out what are the key factors for them to achieve outstanding energy-storage performances. After that, a dual-carbon-enhanced (DCE) strategy, which is an effective approach to improve the electrochemical properties of electrode materials, will be presented and then exemplified in several advanced cathode and anode materials for LIB and SIB. The mainly discussed materials include lithium iron phosphate ($LiFePO_4$) cathode, silicon (Si) and oxide anodes for LIB, NASICON-type sodium vanadium phosphate ($Na_3V_2(PO_4)_3$) cathode and antimony (Sb) anode for SIB. After reading this chapter, readers can learn not only the basic charge/discharge mechanisms and research frontiers (mainly electrode materials) of LIB and SIB but also one effective carbon-incorporation-involving strategy to improve the electrochemical properties and practicability of electrode materials, which has been extensively verified in several material systems by our group and many others.

Work mechanisms of LIB and SIB

The commercially successful LIB was first marketed by Sony Corporation using the carbonaceous material (coke derived from coal product) and layered-structural lithium cobalt oxide ($LiCoO_2$) as anode and cathode, respectively, in 1991. Such LIB is completely different from previously commercialized but rapidly failed lithium batteries due to the safety problems. The main difference and advancement of LIB are free from metallic lithium anode in comparison to lithium batteries, which can greatly lower the level of usage dangers including firing, combustion and explosion while actual using. Hence, LIB rapidly became the most successful and popular energy storage devices to power our daily lives.

One common LIB mainly consists of two electrodes (anode and cathode with low and high potential, respectively), electrolyte to promise the Li^+ transferring between two electrodes, porous separator with the abilities of e^--insulating and Li^+-conducting in conjunction with electrolyte, and other accessories (including Cu/Al foils as current collectors for anode/cathode, can/casing, tab leaf, and so on). As shown in Figure 1, taking graphite anode and $LiFePO_4$ cathode as the example, the charge (electrical energy storage) processes are: (i) Li ions migrate very slowly towards the surface of $LiFePO_4$ particles in the crystal lattices, and then transfer into electrolyte across the interface between $LiFePO_4$ and electrolyte, which are the Li^+ deintercalation processes from $LiFePO_4$ particulates, corresponding to the valence increase of Fe atoms from +2 to +3 and hence rise of cathode potential. (ii) The deintercalated Li ions diffuse from the cathode area to the anode side driven by the electric field in electrolyte. At the same time, electrons escape from cathode and move to anode through the external circuits. (iii) The Li ions further transfer from the anode area of electrolyte into the crystal lattices of graphite anode across the interface between graphite and electrolyte, migrate from the surface of graphite particulates into the interiors in the interlayers between graphene layers, and finally couple with electrons to lower the anode potential. As a result, the potential difference between cathode and anode rises to achieve charge. For the discharge processes, the directions of Li^+ transfer and e^- move are reversed in comparison to charging processes, resulting in the decrease of cell voltage and powering the electrical equipment. The following

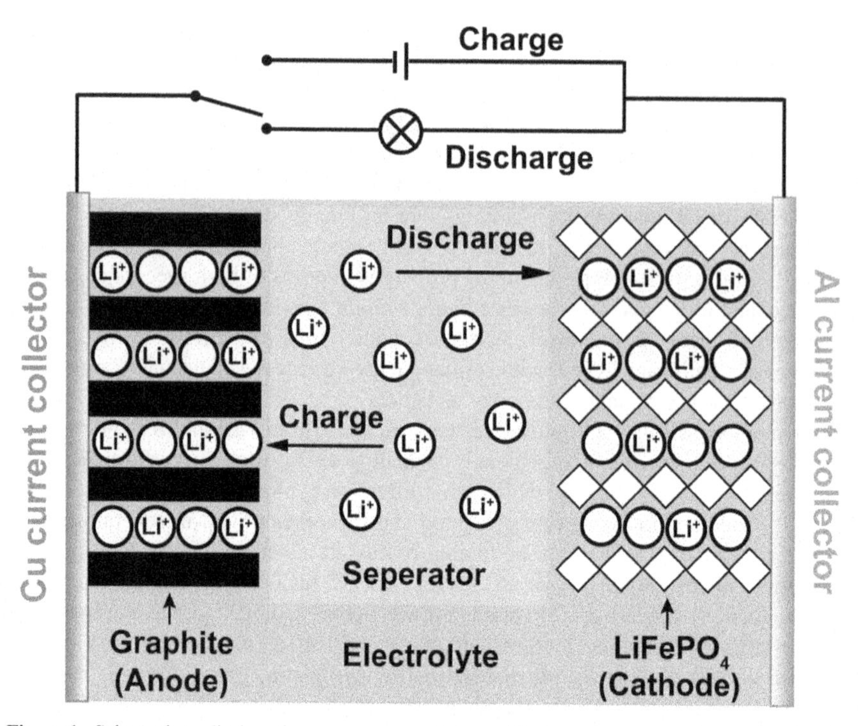

Figure 1. Schematic to disclose the components and charge/discharge processes of commercial LIB.

electrochemical Eqs. (1), (2) and (3), in which C_6 represents the graphite anode and x is in the range of 0–1, further show the changes of chemical compositions of cathode and anode materials upon charge/discharge. During the whole charge/discharge processes, the materials convert reversibly between $LiFe(II)PO_4$ and $Fe(III)PO_4$ on cathode, while between C_6 and LiC_6 on anode.

$$\textbf{Cathode:}\ LiFePO_4 \underset{\text{discharge}}{\overset{\text{charge}}{\rightleftharpoons}} Li_{1-x}FePO_4 + xLi^+ + xe^- \tag{1}$$

$$\textbf{Anode:}\ C_6 + xLi^+ + xe^- \underset{\text{discharge}}{\overset{\text{charge}}{\rightleftharpoons}} Li_xC_6 \tag{2}$$

$$\textbf{Overall:}\ LiFePO_4 + C_6 \underset{\text{discharge}}{\overset{\text{charge}}{\rightleftharpoons}} Li_xC_6 + Li_{1-x}FePO_4 \tag{3}$$

For the cutting-edge room temperature SIB, work mechanism as well as structures, components and systems are almost the same as LIB except that Li ions are replaced by Na ions. Briefly, a SIB is also constituted by two Na^+ insertable/extractable materials (i.e., cathode and anode), which are separated by the porous separator immersed in e^--insulating and Na^+-conducting electrolyte. During the energy storage/release processes, Na ions shuttle between cathode and anode, indicating that SIB is also the typical rocking-chair battery. In comparison to LIB, the much higher sodium abundance in the earth crust and cheaper price make Na ions very attractive as charge carriers for rechargeable batteries. However, due to the larger ionic radius (1.02 Å for Na^+ vs. 0.76 Å for Li^+), the kinetics of electrochemical processes (mainly the ionic diffusion in crystalline cathode and anode lattices) in SIB will be more sluggish in comparison to those in LIB, although these processes are already very slow for LIB (e.g., the apparent Li diffusion coefficient in pure $LiFePO_4$ is about 10^{-12}–10^{-16} cm^2/s). This implies that many electrode materials of LIB and SIB cannot exhibit their true electrochemical properties. For example, one electrode material usually delivers the lower specific capacity than the theoretical value, and exhibits poor rate and cycling properties in comparison to the expected data. Hence, one of the main research tasks for LIB and SIB is to enhance the electrochemical properties of electrode materials through diverse methods.

Significance of dual-carbon-enhanced (DCE) strategy

Carbonaceous materials and their application for LIB and SIB

Carbon is a chemical element with the symbol "C", which is in Row 2 Group 14 in the periodic table. The three best-known and traditional allotropes for carbon element are graphite, diamond, and amorphous carbon due to the different bonding configurations between carbon atoms. With the rapid development of materials, science and technology, lots of other carbon allotropes/structures have been found and prepared in recent decades; they include fullerenes (such as Bucky Balls C_{60} and C_{70}), carbon nanotubes, carbon nanobuds, carbon nanofoam, carbyne, and definitely the most famous graphene. Hence, the types of carbonaceous materials have been greatly developed and enriched, making them to be widely used in various fields including electrochemistry, batteries, catalysis, environmental protection, biomedicine, food engineering, and so on.

For the LIB and SIB applications, the carbonaceous materials have been extensively used as anode active material for ionic storage at low potential, and additive added into electrode to construct effective conductive networks for fast electron conduction. This is mainly due to their characteristics on structure and physicochemical properties including electrochemical stability in the cell systems, high conductivity, suitable particle size and specific surface area of carbon black, and so on. For example, the graphite anode microparticles can maintain their structural stability with low rate of volumetric changes during lithiation/delithiation processes, and Li-storage activity after even thousands of cycles; the acetylene black with nanoscale particle size (about 40–50 nm in diameter) can electrically connect all cathode/anode particulates to synergistically facilitate the ionic transport and storage processes during cycling. In addition to the direct use in the cell package procedures as active material or conductive additive, recent studies have extensively demonstrated that carbonaceous materials can also play a more important role in improving the electrochemical properties of electrode materials for LIB and SIB, if employing them to modify the electrode materials in the material preparation procedures, which will be further discussed in the following sections.

DCE strategy and its significance for the performance enhancement of electrode materials

In recent years of LIB and SIB studies, carbon-incorporation strategies have been believed to be the most effective methods to improve the electrochemical properties of electrode materials. The traditional carbon-incorporation strategies/structures include carbon coating on the active particulates (Zhang et al. 2008, Su et al. 2013, 2012), construction of three-dimensional carbon network (Xin et al. 2012b, Cao et al. 2010), addition of graphene (Jiang et al. 2012, Wang et al. 2010), carbon nanotubes (Xin et al. 2012a, Wang et al. 2016b) or amorphous porous carbon (Wu et al. 2009), hollow core-shell nanostructures with carbon layer as the shell (Xue et al. 2012), sandwich-like layer-by-layer structures (Ji et al. 2010), and so on. These studies mainly utilized one or more characteristics/merits (such as high conductivity, superior flexibility, and excellent electrochemical stability during cell cycling) of carbonaceous materials, which are highly helpful to improve the electrochemical properties of electrode materials. In addition, the low price and high abundance of carbon also makes it easy for practical applications. However, any kind of carbonaceous material cannot own all the required characteristics/merits, making the improvement effect of one carbon incorporation limited. For example, carbon nanotubes are very hard to provide high specific surface area and excellent dispersion to electrically connect all electrode material particulates, although they have high electrical conductivity; fullerenes are nonconductive and extremely expensive for battery applications, although they own excellent redox stability in the electrochemical processes. Therefore, we believe, two or more carbon incorporation must be more effective to improve the electrochemical properties of electrode materials to better levels.

DCE strategy is a preparation suggestion for advanced composites to enhance the electrochemical properties of electrode materials for LIB and SIB through carbon incorporation, in which two types of carbonaceous materials should be employed because different physicochemical merits derived from two different carbonaceous materials can complement their disadvantages. For example, carbon nanotubes can significantly increase the conductivity of electrode materials modified by amorphous porous carbon, while amorphous porous carbon has the ability to stabilize the particulate surfaces of electrode materials and serve as ion-buffering reservoirs, if such two carbonaceous materials are simultaneously incorporated into one electrode material to form the corresponding composite, although one dimensional carbon nanotubes cannot effectively coat the surfaces of electrode materials and amorphous carbon usually cannot exhibit the highly electrical conductivity. Hence, it can be rationally deduced that the DCE strategy provides us an effective research approach to enhance the electrochemical properties of electrode materials for LIB and SIB because different carbonaceous materials have different physicochemical characteristics and play different roles of property enhancements. Our group, as well as many other research groups, had employed this DCE strategy to enhance the electrochemical properties of some electrode materials to much higher levels in previous studies. Table 1 shows some recently positive results in LIBs and SIBs in

Table 1. The summary of recent achievements on LIBs and SIBs by the DCE strategy in other groups, mainly concentrated on phosphate cathode and alloyed anode.

	Composite		Current Rate	Cycle Number	Reversible Capacity	Capacity Decay Rate	Reference
Cathode materials for LIBs and SIBs	LFP NP@NC@RGO	LFP@RGO	10 C	700	42	0.054%	(Oh et al. 2017)
		LFP NP@NC			89	0.0033%	
		LFP NP@NC@RGO			112	0.0054%	
	LVP/C@NCF	LVP/C	10 C	1000	75.8	0.01%	(Zhang et al. 2017)
		LVP/C@NCF			102.5	0.0048%	
	(C@NVP)@pC		10 C	1000	83	0.017%	(Zhu et al. 2014)
	NVP@C@CMK-3	NVP@C	5 C	900	19	0.09%	(Jiang et al. 2015)
		NVP@CMK-3		1000	82	0.023%	
		NVP@C@CMK-3		2000	78	0.016%	
	NVP:RGO-CNT		10 C	2000	107	0.002%	(Zhu et al. 2016)
Anode materials for LIBs and SIBs	Si/graphite/CNTs@C for LIBs		100 mA g^{-1}	100	700	0.187%	(Yang et al. 2017)
	CoP@C polyhedrons-RGO-NF for SIBs		100 mA g^{-1}	100	473.1	0.21%	(Ge et al. 2017)
	Sb/NC+CNTs for SIBs	Sb/NC	100 mA g^{-1}	200	108	0.41%	(Liu et al. 2016)
		Sb/CNTs			408	0.13%	
		Sb/NC+CNTs for SIBs			476	0.06%	
	P/C@RGO for SIBs	P/C	100 mA g^{-1}	100	1200	0.45%	(Lee et al. 2017b)
		P/C@RGO			1900	0.05%	
	DCS-Si for LIBs		1 C	200	1100	0.05%	(Chen et al. 2017b)
	Sm-Si@C-Gr for LIBs		1000 mA g^{-1}	100	1200	0.16%	(Lee et al. 2017a)
	RGO/Ge/RGO for LIBs		1 C	500	1085	0.01%	(Wang et al. 2017)
	CNTS/SnO_2@C for LIBs		100 mA g^{-1}	100	854	0.2%	(Luo et al. 2015)
	GF-CNTs@SnO for SIBs		100 mA g^{-1}	600	540	0.012%	(Chen et al. 2017a)
	SnO_2@C embedded into CF for LIBs		100 mA g^{-1}	100	928.9	0.037%	(Ao et al. 2017)
	MSHSs@C/C	MSHSs	1000 mA g^{-1}	500	194	–	(An et al. 2017)
		MSHSs@C		500	520	–	
		MSHSs@C/C		1000	616	0.018%	
	CNTS@SnO_2@G	SnO_2@G	100 mA g^{-1}	60	559	–	(Zhou et al. 2017)
		CNTS@SnO_2		70	168	–	
		CNTS@SnO_2@G		100	947	–	

other groups, mainly concentrated on the phosphate cathode and alloyed anode (metal oxides are rarely used as the anode of SIBs, so the individual table about the metal oxides as the anode of LIBs is listed in the place where the metal oxides are discussed). Moreover, in the following section, we will, in detail, exemplify the effectiveness and significance of DCE strategy for the LIB and SIB applications via several specific examples mainly studied by our group.

Case studies of DCE strategy for LIB application

$LiFePO_4$ cathode for LIB

Olivine-structural $LiFePO_4$ (abbreviated as LFP) was first reported by Padhi et al. (1997), and rapidly attracted extensive interest as a promising cathode material for LIB due to its appealing merits including the intrinsic thermal stability (making it the safest cathode material for LIB), high theoretical capacity (170 mA h g^{-1}), environmental benignity and low cost. However, there is still one big challenge in using it for high-performance LIB, which is to tackle the sluggish transport kinetics of mass and charge. In order to overcome the Li^+/e^--transport limitations, tremendous efforts have been made which include lattice doping with foreign atoms, particle size decreasing into nanometer scale, surface coating or admixing with electronically conductive materials, especially carbon. For example, in the composite of embedding LFP nanoparticles into a nanoporous carbon matrix (LFP-NP@NPCM), the electrically conducting nanoporous carbon matrix with mesopores can be considered not only as a mixed conducting three-dimensional network that facilitates the Li^+ and e^- migrate to reach each $LiFePO_4$ nanoparticle, but also as an electrolyte reservoir for high-rate charging/discharging (Wu et al. 2009). This makes such LFP-NP@NPCM nanocomposite to exhibit superior electrochemical properties in terms of high-rate capabilities (e.g., 60% capacity retention as current density increases about 50 times) and excellent cycling performance with a very low capacity loss of less than 3% over 700 cycles at a rate of 1.5 C.

Although carbonaceous material had been demonstrated as an effective modifier to enhance the electrochemical properties of $LiFePO_4$ cathode material, the commonly incorporated carbon in most of the $LiFePO_4$/C composites is usually amorphous, which is mainly due to the poor graphitizing ability of precursors (such as the traditionally used glucose, sucrose, and poly(ethylene glycol)) at the sintering temperature of forming pure olivine phase. This makes it very difficult to increase the rate performance of LFP cathode to an ultrahigh level (> 60 C) which is a very essential factor for the applications of EVs and PHEVs and fast storage of wind and solar energy. In our one previous study (Wu et al. 2013), it was demonstrated that introducing the second carbonaceous phase (carbon nanotube) into amorphous-carbon-coated LFP nanocomposite (the finally formed material was named as LFP@C/CNT) exhibited very important and strong ability to improve the high-rate capabilities, low-temperature and cycling performances of $LiFePO_4$-based composite when used as cathode material for LIB.

As shown in Figure 2 for the LFP@C/CNT, $LiFePO_4$ nanoparticles were first coated by the uniform amorphous carbon layers of 2–3 nm in thickness, forming the primary carbon-coated LFP (LFP@C) nanoparticles. This is the first carbon modification for $LiFePO_4$ cathode. In order to achieve the dual-carbon enhancement, all LFP@C nanoparticles were further connected by the wire-like carbon nanotubes (Figure 2b). Due to the optimal three-dimensional networks composed of carbon nanotubes as well as the stabilization effect of amorphous coating carbon for $LiFePO_4$ nanoparticles, the as-designed DCE composite of LFP@C/CNT exhibits the best electrochemical properties in comparison to pure LFP nanoparticles and two single-carbon-enhanced composites, LFP@C and carbon nanotubes incorporated LFP (LFP/CNT). The enhanced electrochemical properties are shown in Figure 3, and will be analyzed in the following two paragraphs.

Figure 3a is the representative galvanostatic curves obtained at 0.2 C (completing the charge or discharge process in 1/0.2 = 5 hours), clearly disclosing that the DCE LFP@C/CNT nanocomposite exhibits the highest specific capacity of nearly 160 mA h g^{-1} (a value is very close to the theoretical capacity of 170 mA h g^{-1} for $LiFePO_4$ if subtracting the carbon content of ~ 5.7 wt% in the nanocomposite). More importantly, the LFP@C/CNT also has the smallest polarization value (about 45.6 mV), the difference

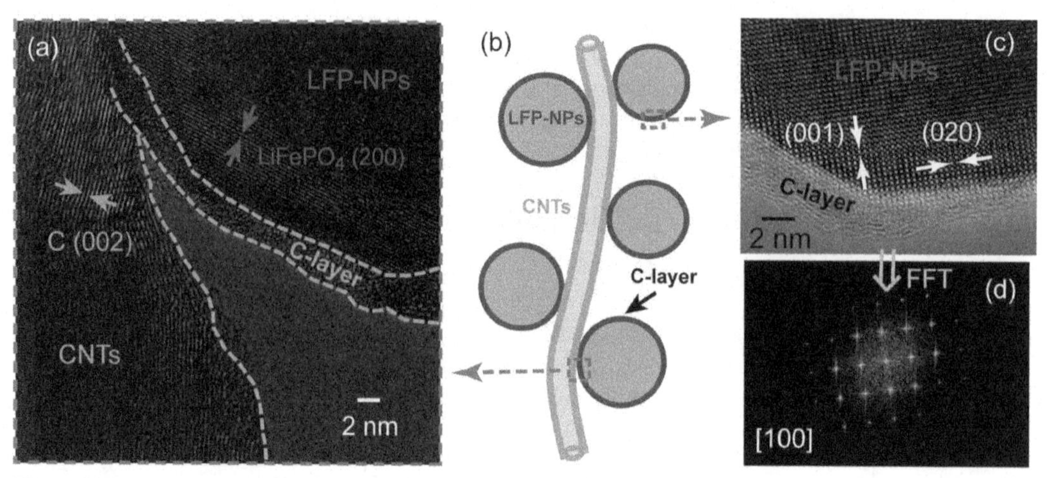

Figure 2. (a, c) HRTEM images and (b) schematic illustration of the DCE LFP@C/CNT nanocomposite. (d) The corresponding FFT of HRTEM image in (c). Reprinted with permission from Wu, X.L., Y.G. Guo, J. Su, J.W. Xiong, Y.L. Zhang and L J. Wan. 2013. Adv. Energy Mater. 3: 1155. Copyright@Whily.

between the charge and discharge plateaus is 63.3 mV for LFP/CNT, 87.4 mV for LFP@C, and 332.7 mV for pure LFP, respectively, suggesting that dual-carbon incorporation improves the kinetics of $LiFePO_4$ cathode more compared to the single-carbon incorporation and definitely the pure LFP sample. The comparisons of redox peak positions and current densities in cyclic voltammogram (CV) patterns (Figure 3b) for the four samples also disclose the similar results and rules. In addition to the charge/discharge processes, the DCE LFP@C/CNT nanocomposite also exhibits the best high-rate capabilities, the longest cycling stability, and the most outstanding low-temperature properties. As disclosed in Figures 3c–f, the orders of such electrochemical properties for the four samples are: (i) rate performance, LFP@C/CNT > LFP/CNT > LFP@C > LFP; and (ii) cycling and low-temperature performances, LFP@C/CNT > LFP@C > LFP/CNT > LFP.

From these electrochemical data as well as charge/discharge processes of galvanostatic curves and CV patterns, it can be rationally concluded that (Wu et al. 2013) (1) both samples of single-carbon incorporation have enhanced the electrode kinetics (including the Li diffusion and electron conduction) of the $LiFePO_4$ cathode to some extent; (2) the DCE LFP@C/CNT nanocomposite combines the merits of amorphous carbon coating and highly conductive carbon nanotubes for property enhancement, and shows a synergistic effect to improve all the electrochemical properties to the best levels; (3) two carbonaceous materials play different roles for property enhancement: the conducting carbon nanotubes construct an efficiently conductive network to optimize the e^- transport pathways and hence be the main contributor for the enhancement of high-rate capabilities, whereas the amorphous carbon coating is mainly responsible for the facilitation of the Li diffusion processes and stabilizing the surface of $LiFePO_4$ nanoparticles, which are beneficial for the improvement of low-temperature performance and cycling stability.

Si anode for LIB

Si is a chemical element located at Row 3 Group 14 in the periodic table, which most widely exists in the Earth's crust, dusts, sands, and planetoids as silica or silicates. In the Earth's crust, over 90% mass is composed of silicate minerals, making Si the second most abundant element (about 28% by mass). Recently, Si-based materials have been considered as one of the most promising anode candidates for the next-generation high-energy LIB due to its high theoretical capacity of 4200 mA h g^{-1}, which is above eleven times compared to the theoretical capacity of 372 mA h g^{-1} for the widely commercial graphite anode. Such high theoretical capacity is mainly due to its alloying Li-storage mechanism with 4.4 Li uptake per Si atom. Unfortunately, during the lithiation/delithiation processes, the Si particles will undergo very huge volumetric changes of up to 420%. This makes Si anode suffer from very high

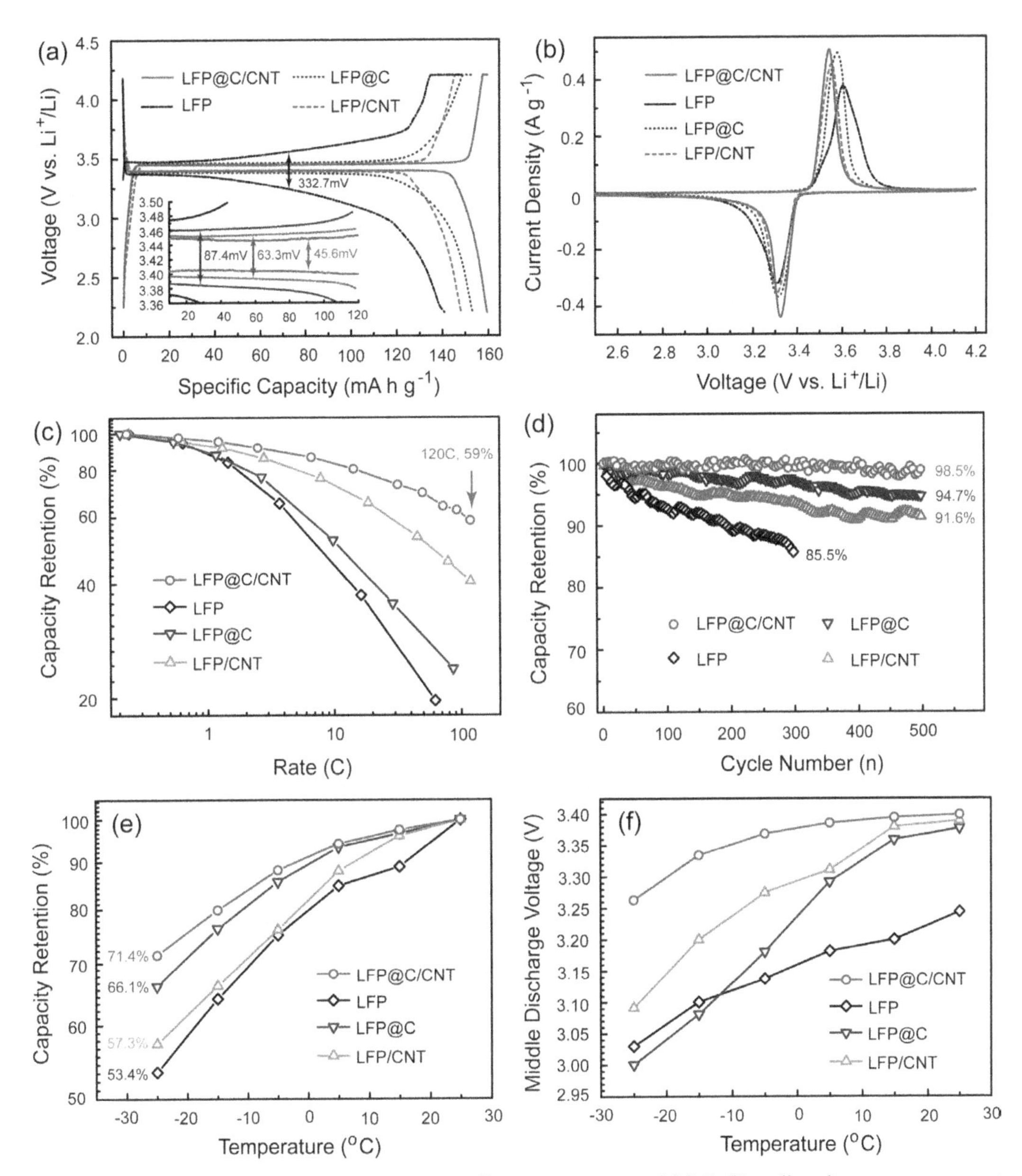

Figure 3. The comparisons of (a) charge/discharge profiles at a current rate of 0.2 C, (b) cyclic voltammogram curves at a sweep rate of 0.1 mV s^{-1}, (c) high-rate capabilities, (d) cycling performances, as well as the dependency of (e) capacity retention and (f) middle discharge voltage on temperature in the range of –25°C to 25°C for LFP@C/CNT, LFP@C, LFP/CNT, and pure LFP. Reprinted with permission from Wu, X.L., Y.G. Guo, J. Su, J.W. Xiong, Y.L. Zhang and L.J. Wan. 2013. Adv. Energy Mater. 3: 1155. Copyright@Whily.

irreversible capacity loss and poor capacity retention, thus impeding its practicability as anode material in LIB. For instance, the initial coulombic efficiency of 10 μm Si particles was as low as 25% and the reversible capacity rapidly dropped up to 70% after only five cycles (McDowell et al. 2012). Hence, in order to practically use the Si-based material as high-capacity anode for LIB, the biggest obstacles include the huge volume variation and the derivative problems (such as the invalidation of e^-/Li^+-transport pathways, pulverization of active particles, cracking of electrodes, and the formation of excessively thick solid electrolyte interphase (SEI) layers) during lithiation/delithiation processes.

Among all the methods to solve the above-mentioned difficulties of practical LIB applications, carbon-incorporated ones should be the most popular and effective strategies. Because the topic of this chapter is to discuss the significance of DCE strategy for the property improvement of electrode materials, herein we just exhibit the power of DCE strategy to enhance the Li-storage properties of Si-based anode according to one of our recent study (Wang et al. 2016a). In that study, Diatomite, a kind of biological sedimentary mineral with high content and wide distribution in Earth's crust as well as low price, was selected as the Si source to facilitate the practical applications of this preparation method. In addition, two carbonaceous precursors of commercially available flaky graphite and glucose were also good choices for practical applications. The highly conductive graphite and glucose-derived amorphous carbon were employed to simultaneously modify the porous Si microparticles obtained from the magnesiothermic reduction of Diatomite, preparing the DCE Si-based micro/nanocomposites (abbreviated as Si/C/G). In the Si/C/G, all primary components of porous Si, flaky graphite and amorphous coating carbon were merged together, forming the micrometer-sized secondary particulates, which is the necessary characteristic for practical applications. In addition, three contrasting samples were also prepared. They include two single-carbon modified composites (amorphous carbon coated Si/C and graphite incorporated Si/G) and one pure Si material.

From the electrochemical test results shown in Figure 4, the DCE Si/C/G exhibits the best Li-storage properties in terms of the most outstanding cycling stability and the highest rate capabilities at almost all of the current densities in comparison to the other three contrasts. Figure 4a compares the galvanostatic curves of the initial three cycles in the voltage window of 0.01–1.5 V vs. Li^+/Li at the current density of 0.1 A g^{-1}. For the pure Si anode material, although the reversible Li-storage capacity is up to about 1350 mA h g^{-1} at the first cycle, the value decays rapidly to only 580 mA h g^{-1} with an attenuation rate of about 28.5% per cycle. This may be ascribed to the problems derived from the huge volume variations during cycling, including pulverization of active Si particles, cracking of the whole electrode, excessively thick SEI layers, and invalidation of transport pathways for electrons and ions. After carbon coating

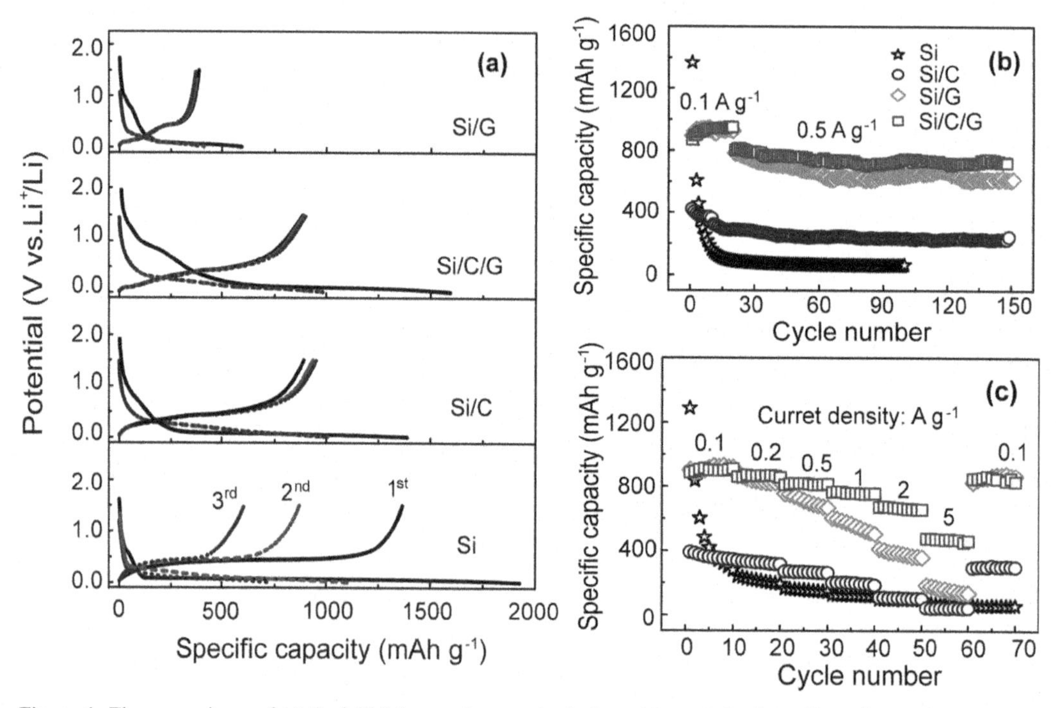

Figure 4. The comparisons of (a) the initial three galvanostatic discharge/charge behaviors, (b) cycling performance and (c) rate capabilities of all four Si-based anode materials (pristine Si, single-carbon-modified Si/C and Si/G, and DCE Si/C/G composite) for LIBs. Reprinted with permission from Wang, J., D.-H. Liu, Y.-Y. Wang, B.-H. Hou, J.-P. Zhang, R.-S. Wang et al. 2016a. Power Sources 307: 738. Copyright@Elsevier.

regardless of single or dual carbonaceous materials, the discharge/charge curves exhibit the much better overlapping, implying the much improved cycling stability. Figure 4b discloses the difference of cycling performance directly. All the carbon-incorporated Si-based composites present higher capacity retention compared to pure porous Si. The reversible capacities for Si/G, Si/C and Si/C/G composites are around 361, 927 and 938 mA h g^{-1}, respectively, after 10 cycles at 0.1 A g^{-1}, while this value for pristine porous Si is almost zero, demonstrating that the pristine Si has lost its Li-storage activity after only 10 cycles. More importantly, the DCE Si/C/G exhibits the best cycling stability in comparison to the two single-carbon modified contrasts. For example, the DCE Si/C/G composite can still deliver a specific capacity of 802.7 mA h g^{-1} with a retention of above 90% after 150 cycles at 0.5 A g^{-1}. Both the capacity value and retention are higher than those of the two contrasts. Moreover, the Si/C/G composite also exhibits the best rate performance among the four samples as clearly shown in Figure 4c. For example, at the high current density of 5 A g^{-1}, the Si/C/G composite can still deliver a specific capacity of about 470 mA h g^{-1}, while the other three contrasts have already been unable to store lithium. It can be also concluded that the best Li-storage properties from the DCE Si/C/G composite should be mainly attributed to the presence of two carbonaceous modifiers, amorphous coating carbon and conductive graphite. While the former can mainly buffer the volumetric expansion/extraction of active Si particles, the latter can provide more effective pathways for the electron transportation during the lithiation and delithiation processes.

Oxide anode for LIB

For the anode materials of LIB, three main Li-storage mechanisms include intercalation reaction, alloying reaction, and conversion reaction. The intercalation mechanism is mainly based on the Li insertion/extraction into/out the host crystal lattices with open frameworks. The most commonly used graphite anode in commercial LIB is the typical representative for this mechanism due to its layered crystal structures to facilitate the Li^+ intercalation and deintercalation. Another famous host with intercalation mechanism is the spinel $Li_4Ti_5O_{12}$ ($Li[Li_{1/3}Ti_{5/3}]O_4$) with three-dimensional Li transport pathway. Regarding the anode materials based on alloying reaction, the most famous and studied one should be the above-discussed Si material due to its highest theoretical Li-storage capacity. Similar to the alloying anode, materials based on conversion reaction can also deliver higher specific capacity than the intercalation anode materials. The conversion mechanism was first found in 3d transition metal oxides (Poizot et al. 2000) followed by fluorides, sulfides and nitrides with different redox potential. Taking oxides as the examples, the lithiation/delithiation processes are mainly related to the reversible redox reaction, as described in the following Eq. (4) in the half cells using metallic Li as the counter electrode:

$$M_xO_y + 2yLi^+ + 2ye^- \underset{\text{charge}}{\overset{\text{discharge}}{\rightleftharpoons}} xM + yLi_2O \tag{4}$$

where M represents the 3d transition metals including Cr, Mn, Fe, Co, Ni, Cu, Zn, and so on. Upon discharge (lithiation), metal oxides are reduced electrochemically to nanometer sized metallic clusters (about 2–8 nm), dispersed in the *in situ* formed amorphous Li_2O matrix. Owing to the nanometric nature of formed composite electrodes, the conversion reactions are almost reversible. Thus, the produced metallic clusters can react with fresh Li_2O matrix to return to metal oxide in the following charge (delithiation) process. Owing to the multiple Li^+-storage (e^--transfer) characteristics per transition metal oxide host, most of these oxides can deliver high theoretical reversible capacities (usually 600–1000 mA h g^{-1}). Hence, 3d transition metal oxides are one kind of promising anode materials for high-energy LIB. However, almost all of these pure oxides usually exhibit inapplicable cycling and rate performances when used due to the huge volumetric variations, interfacial instability, low conductivity, and so on. In this section, we will mainly exemplify the power of DCE strategy to improve the Li-storage properties of two such 3d transition metal oxides. First, the relevant works reported by other groups are listed in Table 2, and then, the recent results of our group are displayed in detail.

The first example is manganese oxide (MnO) anode (Liu et al. 2015), which may be the most promising oxide for LIB because of its lower voltage hysteresis (the potential difference between lithiation

Table 2. Summary of recent works about metal oxides as the anode of LIBs by other groups employing the DCE strategy.

<table>
<tr><th colspan="2">Composite</th><th>Cycle Rate</th><th>Cycle Number</th><th>Reversible Capacity</th><th>Capacity Decay Rate</th><th>Reference</th></tr>
<tr><td colspan="2">C/MnO/rGO</td><td>1000 mA g^{-1}</td><td>250</td><td>577.8</td><td>0.015%</td><td>(Liu et al. 2017)</td></tr>
<tr><td colspan="2">MnO@C-CNT</td><td>1000 mA g^{-1}</td><td>100</td><td>655</td><td>–</td><td>(Dai et al. 2016)</td></tr>
<tr><td colspan="2">MnO@N-GSC/GR</td><td>2000 mA g^{-1}</td><td>1000</td><td>812</td><td>–</td><td>(Zhang et al. 2016a)</td></tr>
<tr><td colspan="2">graphene@NiO@C</td><td>200 mA g^{-1}</td><td>50</td><td>754</td><td>–</td><td>(Wang et al. 2016c)</td></tr>
<tr><td colspan="2">C@NiCo-$NiCoO_2$/CX</td><td>100 mA g^{-1}</td><td>100</td><td>861</td><td>–</td><td>(Zhang et al. 2016b)</td></tr>
<tr><td colspan="2">$CoMoO_4$/PPy NWAs on CC</td><td>1200 mA g^{-1}</td><td>1000</td><td>764</td><td>0.01%</td><td>(Chen et al. 2015)</td></tr>
<tr><td colspan="2">graphene paper@Fe_3O_4 NAs@C</td><td>2000 mA g^{-1}</td><td>1000</td><td>852</td><td>0.006%</td><td>(Guo et al. 2017)</td></tr>
<tr><td rowspan="2">$ZnFeO_4$@C/N-GAs</td><td>$ZnFeO_4$@C/N-GAs</td><td rowspan="2">100 mA g^{-1}</td><td rowspan="2">100</td><td>952</td><td>–</td><td rowspan="2">(Yao et al. 2017)</td></tr>
<tr><td>$ZnFeO_4$/N-GAs</td><td>487</td><td>0.41%</td></tr>
</table>

and delithiation curves, below 0.8 V) compared to other transition metal oxides, suitable delithiation potential (around 1.0 V vs. Li^+/Li), high density (5.43 g cm^{-3}), as well as high theoretical capacity of 756 mA h g^{-1}, relatively low cost and environmental benignity. However, the intrinsically low conductivity of pure MnO anode makes it suffer from inferior rate capacity and poor cycling stability arising from kinetic limitations, the rapid capacity fading resulting from severe agglomeration and drastic volume changes during the lithiation/delithiation processes. Furthermore, the repeating volume changes during cycling could also give rise to the pulverization of the whole MnO electrodes, and hence the Mn shedding and dissolution into electrolyte (Liu et al. 2015). For example, as shown in Figure 5, the commercially available MnO material delivers very poor Li-storage properties in terms of low reversible capacity (an initial capacity of only 220 mA h g^{-1} at a low current density of 0.08 A g^{-1}), rapid capacity attenuation upon cycling, and very poor rate performance (e.g., Li-storage inactivity as current density increases to 1.5 A g^{-1}). After single carbon modification (carbon coating on the MnO nanoparticles, MnO@C NPs), all of these electrochemical properties improve significantly. For example, the delivered Li-storage capacity has been increased to about 500 mA h g^{-1} after carbon coating. More significantly, both DCE MnO-based composites exhibit the much improved electrochemical properties compared to the MnO@C NPs. For the carbon nanotubes enhanced MnO@C composite (MnO@C/CNTs), its reversible Li-storage capacity is further increase to about 670 mA h g^{-1}, and it can still deliver a specific capacity of 594 mA h g^{-1} at higher current density of 0.38 A g^{-1}. More interestingly, the reduced graphene oxide enhanced MnO@C composite (MnO@C/RGO) exhibits further improved Li-storage properties in comparison to MnO@C/CNTs. Not only the reversible Li-storage capacity is increased greatly to about 840 mA h g^{-1}, but more importantly the MnO@C/RGO can still deliver a high specific capacity of about 448.2 mA h g^{-1} at a very high current density of 7.6 A g^{-1}. It can also be readily concluded that the superiority sequence of the four samples is MnO@C/RGO > MnO@C/CNTs >> MnO@C NPs > commercial MnO when used as anode materials for LIBs. This suggests that different types of carbonaceous material in the akin DCE composite will enhance electrochemical properties to a different level. In this case, it is demonstrated that RGO plays a much greater role in the improvement of electrochemical properties in comparison to carbon nanotubes in the DCE systems due to its more superior flexibility, higher specific surface area and higher conductivity to construct better three-dimensional conductive networks. Nevertheless, the DCE strategy is further confirmed as a more effective method to improve the Li-storage properties in comparison to single carbon modification using the MnO-based anode material.

The second example is iron oxide (Fe_3O_4) based anode composite, in which the RGO, already proved to be the more excellent carbonaceous material to improve the Li-storage properties in comparison to others, was used as one of the carbonaceous additives to construct the conductive networks (Hou et al. 2015). In order to demonstrate the effectiveness and advancement of DCE strategy in this system, carbon coating on the primary RGO/Fe_3O_4 composite was further implemented as the preparation procedures shown in Figure 6(a–e), controllably forming the DCE RGO/Fe_3O_4@C composite. In the

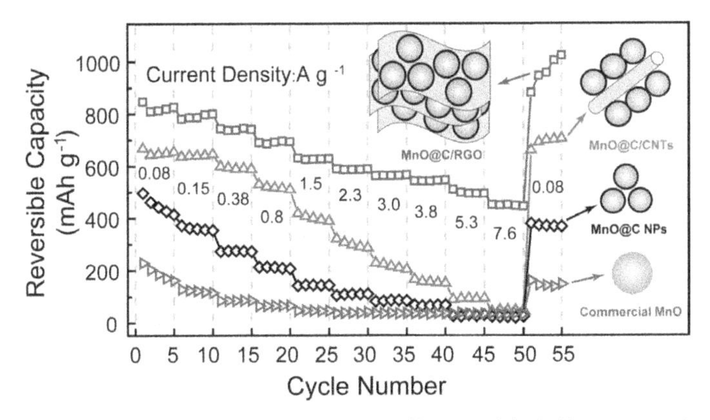

Figure 5. The comparison of rate performance, disclosing the significance of the DCE strategy to improve the Li-storage properties of MnO-based anode materials for LIB. The contrasts include the pure commercial MnO, carbon coated MnO nanoparticles (MnO@C NPs), and two DCE samples (carbon nanotubes incorporated MnO@C (MnO@C/CNTs) and reduced graphene oxide incorporated MnO@C (MnO@C/RGO)). Reprinted with permission from Liu, D.-H., H.-Y. Lü, X.-L. Wu, B.-H. Hou, F. Wan, S.-D. Bao et al. 2015. J. Mater. Chem. A 3: 19738. Copyright@Royal of Society Chemistry.

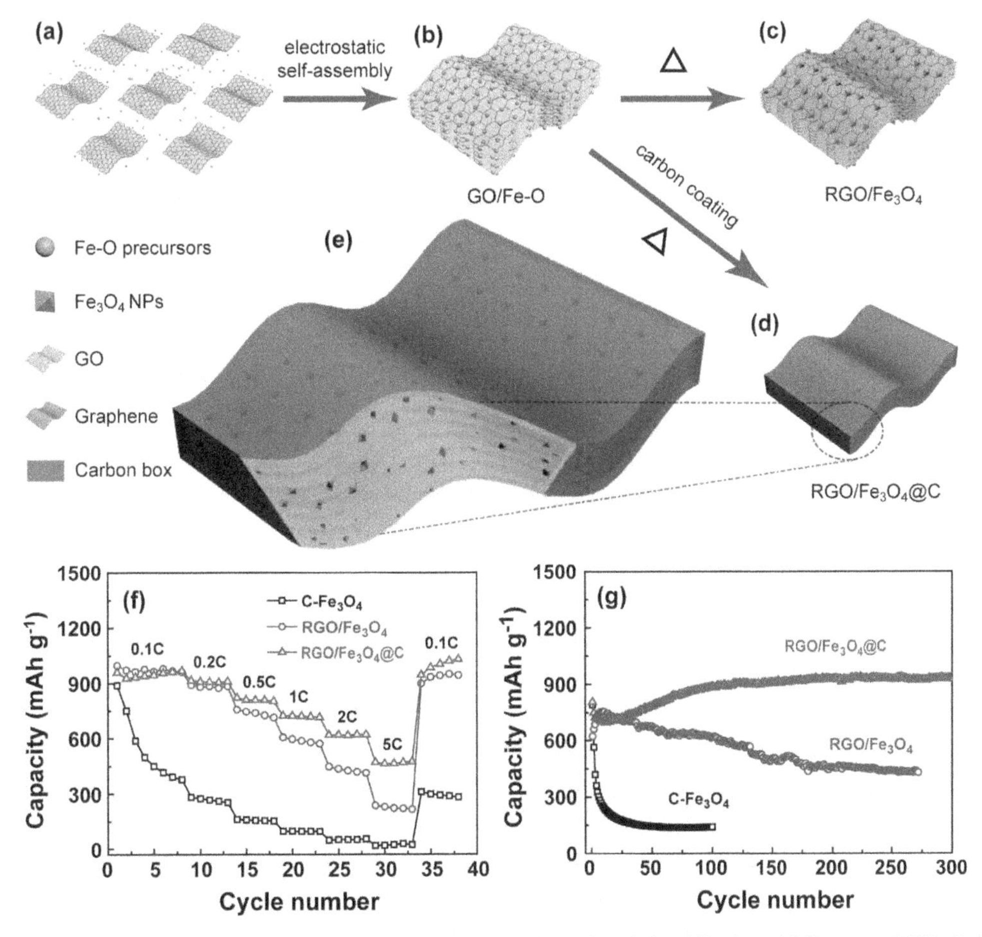

Figure 6. (a–e) Schematic representation of the preparation procedures for pristine G/Fe_3O_4 and full-protected G/Fe_3O_4@C nanocomposites. The comparisons of (d) rate and (g) cycling performances between pristine G/Fe_3O_4, full-protected G/Fe_3O_4@C nanocomposite, and C-Fe_3O_4. Reprinted with permission from Hou, B.-H., X.-L. Wu, Y.-Y. Wang, H.-Y. Lü, D.-H. Liu, H.-Z. Sun et al. 2015. Part. Part. Syst. Char. 32: 1020. Copyright@Whily.

preparation processes, the iron/oxygen-containing precursors (Fe-O) are uniformly adhered to on the three-dimensional networks composed of graphene oxide (GO) nanosheets, first forming the GO/Fe-O microparticles due to the electrostatic self-assembly. After coating the GO/Fe-O microparticles via the glucose-assisted hydrothermal treatment, the designed RGO/Fe_3O_4@C composite can be easily obtained via annealing at 400°C for 1 h. For preparing the contrast of single carbon modified RGO/Fe_3O_4, the hydrothermal carbon coating was not included in the preparation procedures. Because all Fe_3O_4 nanoparticles have been completely protected by the carbon coating in the RGO/Fe_3O_4@C composite to avoid their direct contact with electrolyte (as shown in Figures 6d and 6e), we can consider the RGO/Fe_3O_4@C is a fully protected composite, which has the ability to optimize the interfaces between G/Fe_3O_4@C microparticles and electrolyte by constructing more stable surface electrolyte interphase (SEI) when used as anode materials for LIB. In comparison, for the RGO/Fe_3O_4 contrast, many Fe_3O_4 nanoparticles are still exposed on the surface of composite microparticles (Figure 6c). Figures 6f and 6g compare the rate and cycling performances of RGO/Fe_3O_4@C, RGO/Fe_3O_4, and the commercial Fe_3O_4 (C-Fe_3O_4), respectively. As disclosed, the RGO/Fe_3O_4@C delivers not only the highest rate capabilities (Figure 6f) but also the best cycling stability (Figure 6g) in comparison to both RGO/Fe_3O_4 and C-Fe_3O_4. For instance, the RGO/Fe_3O_4@C composite can still deliver a reversible capacity of up to 473.7 mA h g^{-1} at 5 C, which is obviously higher than 225 and 23 mA h g^{-1} for RGO/Fe_3O_4 and C-Fe_3O_4 contrasts, respectively, although three samples deliver almost equal specific capacities of 900–1000 mA h g^{-1} at low rate of 0.1 C. Furthermore, as the cycling performance of 0.5 C shows in Figure 6g, the RGO/Fe_3O_4@C exhibits no capacity attenuation even after 300 cycles with a clear activation process, which may have originated from the gradual infiltration of electrolyte into the mesopores existing in the composite. For the two contrasts, RGO/Fe_3O_4 shows obvious capacity decay especially in the initial 200 cycles, while the C-Fe_3O_4 becomes Li-storage inactivity very rapidly. The above data, as well as more detailed results of structural characterizations and electrochemical tests in the published paper (Hou et al. 2015), demonstrate that the DCE (fully protected) strategy has strong power to improve the Li-storage properties of Fe_3O_4 anode to the best levels in comparison to the single carbon modified RGO/Fe_3O_4 and commercial Fe_3O_4 product.

Case studies of DCE strategy for SIB application

As discussed in the sections of "Work Mechanisms of LIB and SIB" and "Significance of Dual-Carbon-Enhanced (DCE) Strategy", the charge carriers of Na ions are much larger in size than Li ions, therefore, many of the SIB electrode materials exhibit more sluggish electrode kinetics for ionic and electronic transports, more serious structural and volume variations during sodiation/desodiation process, and hence poorer electrochemical properties in comparison to LIB electrode materials. Due to the superiority of DCE strategy to improve the electrochemical properties of electrode materials as demonstrated above in the LIB system, it is highly expected that such strategy can also play an important role in enhancing the properties of SIB electrode materials. In this section, we will further show the power of DCE strategy to improve the SIB performances using two key electrode materials ($Na_3V_2(PO_4)_3$ cathode and Sb anode).

$Na_3V_2(PO_4)_3$ cathode for SIB

Among the cathode materials, NASICON structural $Na_3V_2(PO_4)_3$ (NVP) should be one of the most promising candidates due to its high theoretical energy density (about 400 W h kg^{-1}), superior thermal stability (up to 450°C), and highly covalent three-dimensional framework with large interstitial channels for fast Na^+ migration. Nevertheless, for the phosphate cathode materials, low electrical conductivity is their inherent shortcoming, making pure NVP exhibit low coulombic efficiency, poor rate and cycling performances, and hence weak practicality in comparison to $Li_3V_2(PO_4)_3$ (LVP) with the same NASICON structure for LIB. In order to improve the electrochemical properties of NVP cathode for SIB by the DCE strategy, employing disodium ethylenediamintetraacetate (molecular formula: $Na_2(C_{10}H_{16}N_2O_8)$, abbreviated as: Na_2EDTA) as both sodium and nitrogen-doped (N-doped) carbon sources, a NVP-based composite with both N-doped coating carbon and RGO modifications (NVP/C/RGO) had been prepared (Guo et al. 2015). In the NVP/C/RGO composite, all NVP nanoparticles are effectively coated by the *in situ* formed N-doped carbon layers due to the thermal decomposition of nitrogen-containing Na_2EDTA precursor, and further electrically connected by the highly conductive RGO layers. For comparison, two contrasts, pure NVP and single-carbon modified NVP nanoparticles (NVP/C), were also prepared under akin preparation conditions, just with the replacement or no use of carbon sources.

Figures 7(a–c) are the representative TEM images of the three NVP-based materials. As shown in Figure 7a as well as the corresponding HRTEM image (not shown here but can be found in the original paper), in the NVP/C/RGO composite, all well-crystallized NVP nanoparticles of 100–200 nm are first coated by an amorphous N-doped carbon layer and then well dispersed in the RGO networks. The

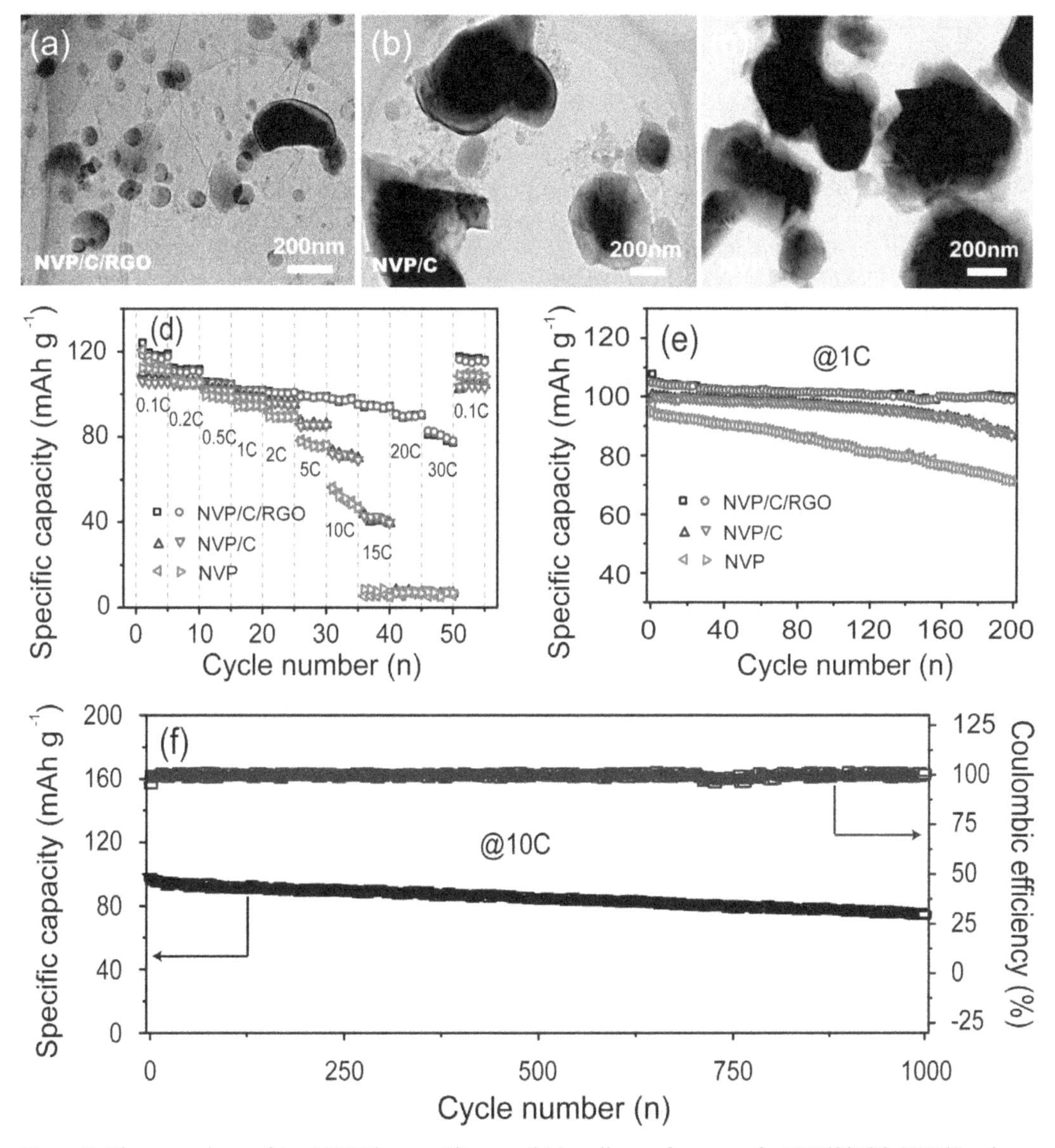

Figure 7. The comparisons of (a–c) TEM images, (d) rate and (e) cycling performances for NVP/C/RGO, NVP/C and pure NVP. (f) The long-term cycling performance of DCE NVP/C/RGO composite. Reprinted with permission from Guo, J.-Z., X.-L. Wu, F. Wan, J. Wang, X.-H. Zhang and R.-S. Wang. 2015. Chem. Eur. J. 21: 17371. Copyright@Whily.

highly conductive three-dimensional construction facilitates the transport of electrons upon charging/discharging. In the NVP/C contrast (Figure 7b), NVP particles are mainly in the size of 500–700 nm. The obviously larger (about 3–5 times) size of NVP particles in NVP/C compared to NVP/C/RGO demonstrates that the RGO incorporation in the preparation procedures can also restrict the growth of active NVP particles, hence achieving shorter Na diffusion time in NVP crystal lattice according to the diffusion equation ($t_d = L^2/D$, where t_d, L and D represent the diffusion time, diffusion distance and diffusion coefficient, respectively) during sodiation/desodiation. For the pure NVP material (Figure 7c), NVP particles are the largest one among all their samples, suggesting that carbon coating also has the ability to decrease particle size. Owing to the existence of dual carbon modification and the smallest NVP particles, the DCE NVP/C/RGO composite is expected to exhibit a better electrochemical properties as cathode material for SIB in comparison to the single carbon modified NVP/C and pure NVP.

Figures 7d first compares rate capabilities of three NVP-based contrasts from 0.1 to 30 C. As shown, the specific capacities delivered by all three samples are very close at the low rate of < 2 C, although NVP/C/rGO nanocomposite exhibits slightly higher values than NVP/C and definitely pristine NVP at various rates. For example, at 0.1 C rate, the specific capacity of the 1st cycle for NVP/C/rGO is up to 116.8 mA h g^{-1} (very close to the theoretical capacity of 117.6 mA h g^{-1} for NVP cathode), which is obviously higher than 106 and 112.3 mA h g^{-1} for NVP/C and NVP, respectively. More significantly, the NVP/C/rGO exhibits much improved high-rate performance. For example, it can deliver a specific capacity of 80 mA h g^{-1} even at a very high rate of up to 30 C, whereas other two contrasts have already been inactivated for Na insertion/desertion as rates are higher than 10 C. Figure 7e is the comparison of cycling performance of three contrasts at 1 C and 2.3–3.9 V vs. Na^+/Na, revealing that the superior NVP/C/rGO nanocomposite also owns the best cycle life. For example, after 200 cycles, the capacity retentions of NVP/C and pure NVP are only 87.7 and 75.2%, respectively, both of which are obviously lower than 94.4% for the NVP/C/rGO composite. More interestingly, the DCE NVP/C/rGO composite also exhibits outstanding cycling stability with 1000 cycles at 10 C (Figure 7f). One of the main reasons for the much improved high-rate and long-term cycling performances of NVP/C/rGO composite should be its dual carbon (N-doped coating carbon and highly conductive RGO nanosheets) incorporated structures, which cannot only enhance the electrical conductivity of as-prepared materials but also decrease the Na diffusion time in crystal lattice due to the reduced particle size and more active sites for fast Na^+ insertion/extraction.

Sb anode for SIB

Among the SIB anode materials, metallic Sb is the promising one because of its high theoretical capacity of 660 mA h g^{-1} and very suitable Na-storage potential (0.5–0.8 V vs Na^+/Na). However, it will suffer from tremendous volumetric expansion/contraction during Na-uptake/release and hence poor practical Na-storage performance due to the alloying Na-storage mechanism corresponding to the formation of Na_3Sb alloy in the highest Na content (Wan et al. 2016b, a). Hence, many reports had focused on its improvements of Na-storage properties via preparing diverse nanostructures or nanocomposites, such as porous Sb hollow nanospheres, Sb/graphene composite, Sb/C nanofibers, Sb/CNTs, and Sb/acetylene black. In this section, we will further employ the Sb anode material to illustrate the power of DCE strategy in SIB anode applications (Lü et al. 2016), in which the Na-storage properties of Sb-based anode have also been significantly enhanced. In addition to the improvement of electrochemical properties for Sb anode, this study also reconfirmed that addition of the second carbon modifier (RGO nanosheets) can inhibit the growth of active Sb particles to ensure them in nanometer size, as well as completely protect all Sb nanoparticles from being inactivated during cycling.

Figure 8a is the representative TEM image of Sb-based composite incorporated by both RGO and amorphous carbon (Sb/C/RGO). As disclosed, the Sb nanoparticles mainly concentrate on 20–50 nm, and are wrapped in the RGO-based networks. For the TEM images of single-carbon incorporated Sb-based composite (Sb/G), there is an obvious difference with Sb/C/RGO in the same scale, which can be learnt from the original publication (Lü et al. 2016). In addition to nanoscale morphology,

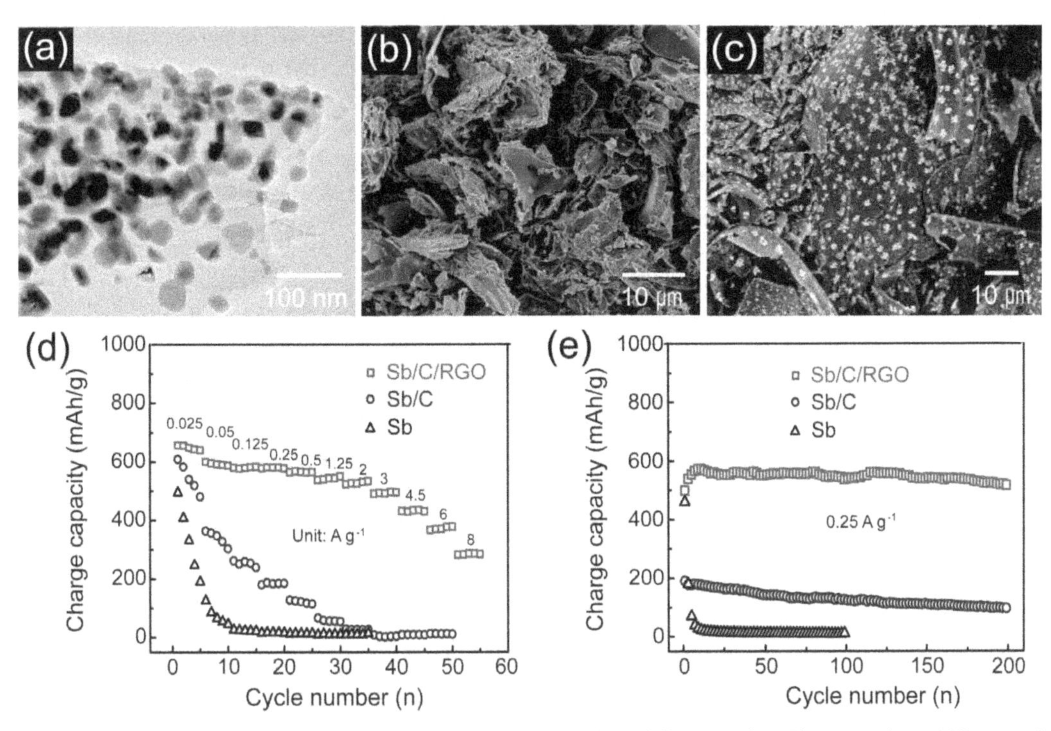

Figure 8. The (a) TEM and (b, c) SEM images of (b) Sb/C/RGO and (c) Sb/C composites. The comparison of (d) rate and (e) cycling performances of Sb/C/RGO, Sb/C and pure Sb anode materials for SIB. Reprinted with permission from from Lü, H.-Y., F. Wan, L.-H. Jiang, G. Wang and X.-L. Wu. 2015. Part. Part. Syst. Char. 33: 204. Copyright@Whily.

Figures 8b and 8c further compare the SEM images of both composites in the micrometer scale. It can be observed that both Sb/C/RGO and Sb/C nanocomposites are composed of irregular plate-like particles of > 10 μm. The difference is there are still lots of Sb microparticles (bright points of 1.5–2.5 μm in the SEM image of Figure 8c) on the surface of plate-like particles in Sb/C, while there is no smooth surface in Sb/C/RGO. This suggests that addition of RGO nanosheets effectively prevents the thermal fusion/aggregation of Sb nanoparticles and protects them completely. Hence, the Sb/C/RGO composite exhibits much improved Na-storage properties in comparison to Sb/C as well as pure Sb material, as demonstrated by the electrochemical data in Figures 8d and 8e. In the comparative patterns of rate capabilities (Figure 8d), the Sb/C/RGO can still deliver a high reversible capacity of 530 mA h g^{-1} as current density is up to 2 A g^{-1}, while both Sb/C and pure Sb have already become almost Na-storage inactive. More importantly and excitingly, the Sb/C/RGO exhibits the ability of achieving Na-storage capacity of about 290 mA h g^{-1} even at an ultrahigh current density of 8 A g^{-1}. In addition, Sb/C/RGO delivers the highest capacity (650 mA h g^{-1}) in comparison to 609 mA h g^{-1} and 494 mA h g^{-1} for Sb/C and pure Sb, respectively, at the low current density of 0.025 A g^{-1}. Furthermore, Sb/C/RGO also exhibits the best cycling performance compared to the other two contrasts (Figure 8e). For example, there is no obvious capacity decay for Sb/C/RGO electrode with an initial capacity of 499 mA h g^{-1} after 200 cycles at 0.25 A g^{-1}, while the capacity retention of Sb/C is only around 51% and pure Sb becomes inactive rapidly for Na storage within the initial several cycles under the same test conditions.

Conclusion and outlook

From the first commercialization of LIB in 1991 to now, about 25 years have passed. During this period, LIB has obtained substantial progress with more and more applications. In these processes, the incorporation of carbonaceous materials into electrodes and/or active materials has played the most significant role in improving the performances of LIB. In addition to the traditional approaches based on the single-carbon incorporation, the dual-carbon-enhanced (DCE) strategy was gradually realized and proposed in very recent years. The power of such DCE strategy to enhance the electrochemical properties has been demonstrated in several systems of electrode (including cathode and anode) materials as discussed above. With the development of SIB, this DCE strategy has also been employed to improve the electrochemical properties of SIB electrode materials. In the following researches for LIB and SIB, I believe that the DCE suggestion can also play an important role in the domains of electrode and full cell. For example, the addition of two conductive carbonaceous additives (e.g., carbon nanotubes and carbon black, graphene and carbon black) may have the capability of enhancing the energy-storage performances of cell packages to higher levels in comparison to the traditional processes with the only addition of carbon black. Furthermore, such DCE strategy can also be employed by other battery systems, such as Li-S batteries, Mg ion batteries, Al ion batteries and so on, to improve their performances. In addition, due to the diversity of carbonaceous materials with various physicochemical properties, it is suggested to develop triple-carbon or multi-carbon enhanced strategies under different carbon combinations to further improve the electrochemical properties of electrode materials. Therefore, there is still a lot of research to be done for the carbon-incorporation enhancement of electrochemical properties in the future studies.

Acknowledgements

We gratefully acknowledge the financial supports from the National Natural Science Foundation of China (51602048) and the Fundamental Research Funds for the Central Universities (2412017FZ013).

References

An, W., J. Fu, S. Mei, L. Xia, X. Li, H. Gu et al. 2017. Dual carbon layer hybridized mesoporous tin hollow spheres for fast-rechargeable and highly stable lithium-ion battery anodes. J. Mater. Chem. A 5: 14422–14429.

Ao, X., J. Jiang, Y. Ruan, Z. Li, Y. Zhang, J. Sun et al. 2017. Honeycomb-inspired design of ultrafine SnO_2@C nanospheres embedded in carbon film as anode materials for high performance lithium- and sodium-ion battery. J. Power Sources 359: 340–348.

Cao, F.-F., X.-L. Wu, S. Xin, Y.-G. Guo and L.-J. Wan. 2010. Facile Synthesis of mesoporous TiO_2-C nanosphere as an improved anode material for superior high rate 1.5 V rechargeable Li ion batteries containing $LiFePO_4$-C cathode. J. Phys. Chem. C 114: 10308–10313.

Chen, M., D. Chao, J. Liu, J. Yan, B. Zhang, Y. Huang et al. 2017a. Rapid pseudocapacitive sodium-ion response induced by 2D ultrathin tin monoxide nanoarrays. Adv. Funct. Mater. 27: 1606232.

Chen, S., L. Shen, P. A. Aken, J. Maier and Y. Yu. 2017b. Dual-functionalized double carbon shells coated silicon nanoparticles for high performance lithium-ion batteries. Adv. Mater. 29: 1605650.

Chen, Y., B. Liu, W. Jiang, Q. Liu, J. Liu, J. Wang et al. 2015. Coaxial three-dimensional $CoMoO_4$ nanowire arrays with conductive coating on carbon cloth for high-performance lithium ion battery anode. J. Power Sources 300: 132–138.

Dai, C., M. Wang, J. Yang, L. Hu and M. Xu. 2016. Fabrication of MnO@C-CNTs composite by CVD for enhanced performance of lithium ion batteries. Ceram. Int. 42: 18568–18572.

Ge, X., Z. Li and L. Yin. 2017. Metal-organic frameworks derived porous core/shell CoP@C polyhedrons anchored on 3D reduced graphene oxide networks as anode for sodium-ion battery. Nano Energy 32: 117–124.

Guo, J.-Z., X.-L. Wu, F. Wan, J. Wang, X.-H. Zhang and R.-S. Wang. 2015. A superior $Na_3V_2(PO_4)_3$-based nanocomposite enhanced by both N-doped coating carbon and graphene as the cathode for sodium-ion batteries. Chem. Eur. J. 21: 17371–17378.

Guo, J., H. Zhu, Y. Sun, L. Tang and X. Zhang. 2017. Pie-like free-standing paper of graphene paper@Fe_3O_4 nanorod array@ carbon as integrated anode for robust lithium storage. Chem. Eng. J. 309: 272–277.

Hou, B.-H., X.-L. Wu, Y.-Y. Wang, H.-Y. Lü, D.-H. Liu, H.-Z. Sun et al. 2015. Full protection for graphene-incorporated micro-/nanocomposites containing ultra-small active nanoparticles: the best Li-storage properties. Part. Part. Syst. Char. 32: 1020–1027.

Ji, H.X., X.L. Wu, L.Z. Fan, C. Krien, I. Fiering, Y.G. Guo et al. 2010. Self-wound composite nanomembranes as electrode materials for lithium ion batteries. Adv. Mater. 22: 4591–4595.

Jiang, K.-C., X.-L. Wu, Y.-X. Yin, J.-S. Lee, J. Kim and Y.-G. Guo. 2012. Superior hybrid cathode material containing lithium-excess layered material and graphene for lithium-ion batteries. ACS Appl. Mater. Interfaces 4: 4858–4863.

Jiang, Y., Z. Yang, W. Li, L. Zeng, F. Pan, M. Wang et al. 2015. Nanoconfined carbon-coated $Na_3V_2(PO_4)_3$ particles in mesoporous carbon enabling ultralong cycle life for sodium-ion batteries. Adv. Energy Mater. 5: 1402104.

Lü, H.-Y., F. Wan, L.-H. Jiang, G. Wang and X.-L. Wu. 2016. Graphene nanosheets suppress the growth of Sb nanoparticles in an Sb/C nanocomposite to achieve fast Na storage. Part. Part. Syst. Char. 33: 204–211.

Lee, B., T. Liu, S.K. Kim, H. Chang, K. Eom, L. Xie et al. 2017a. Submicron silicon encapsulated with graphene and carbon as a scalable anode for lithium-ion batteries. Carbon 119: 438–445.

Lee, G.-H., M.R. Jo, K. Zhang and Y.-M. Kang. 2017b. A reduced graphene oxide-encapsulated phosphorus/carbon composite as a promising anode material for high-performance sodium-ion batteries. J. Mater. Chem. A 5: 3683–3690.

Liu, B., D. Li, Z. Liu, W. Xie, Q. Li, P. Guo et al. 2017. Carbon-wrapped MnO nanodendrites interspersed on reduced graphene oxide sheets as anode materials for lithium-ion batteries. Appl. Surface Sci. 394: 1–8.

Liu, D.-H., H.-Y. Lü, X.-L. Wu, B.-H. Hou, F. Wan, S.-D. Bao et al. 2015. Constructing the optimal conductive network in MnO-based nanohybrids as high-rate and long-life anode materials for lithium-ion batteries. J. Mater. Chem. A 3: 19738–19746.

Liu, X., Y. Du, X. Xu, X. Zhou, Z. Dai and J. Bao. 2016. Enhancing the anode performance of antimony through nitrogen-doped carbon and carbon nanotubes. J. Phys. Chem. C 120: 3214–3220.

Luo, B., T. Qiu, B. Wang, L. Hao, X. Li, A. Cao et al. 2015. Freestanding carbon-coated CNT/SnO_2 coaxial sponges with enhanced lithium-ion storage capability. Nanoscale 7: 20380–20385.

McDowell, M.T., S. Woo Lee, C. Wang and Y. Cui. 2012. The effect of metallic coatings and crystallinity on the volume expansion of silicon during electrochemical lithiation/delithiation. Nano Energy 1: 401–410.

Oh, J., j. Lee, T. Hwang, J.M. Kim, K.-D. Seoung and Y. Piao. 2017. Dual layer coating strategy utilizing N-doped carbon and reduced graphene oxide for high-performance $LiFePO_4$ cathode material. Electrochim. Acta 231: 85–93.

Padhi, A.K., K.S. Nanjundaswamy and J.B. Goodenough. 1997. Phospho-olivines as positive electrode materials for rechargeable lithium batteries. J. Electrochem. Soc. 144: 1188–1194.

Poizot, P., S. Laruelle, S. Grugeon, L. Dupont and J.M. Tarascon. 2000. Nano-sized transition-metal oxides as negative-electrode materials for lithium-ion batteries. Nature 407: 496–499.

Su, J., X.-L. Wu, C.-P. Yang, J.-S. Lee, J. Kim and Y.-G. Guo. 2012. Self-assembled $LiFePO_4$/C nano/microspheres by using phytic acid as phosphorus source. J. Phys. Chem. C 116: 5019–5024.

Su, J., X.-L. Wu, J.-S. Lee, J. Kim and Y.-G. Guo. 2013. A carbon-coated $Li_3V_2(PO_4)_3$ cathode material with an enhanced high-rate capability and long lifespan for lithium-ion batteries. J. Mater. Chem. A 1: 2508–2514.

Wan, F., J.-Z. Guo, X.-H. Zhang, J.-P. Zhang, H.-Z. Sun, Q. Yan et al. 2016a. *In situ* binding Sb nanospheres on graphene via oxygen bonds as superior anode for ultrafast sodium-ion batteries. ACS Appl. Mater. Interfaces 8: 7790–7799.

Wan, F., H.-Y. Lü, X.-H. Zhang, D.-H. Liu, J.-P. Zhang, X. He et al. 2016b. The *in-situ*-prepared micro/nanocomposite composed of Sb and reduced graphene oxide as superior anode for sodium-ion batteries. J. Alloy Compd. 672: 72–78.

Wang, B., X.-L. Wu, C.-Y. Shu, Y.-G. Guo and C.-R. Wang. 2010. Synthesis of CuO/graphene nanocomposite as a high-performance anode material for lithium-ion batteries. J. Mater. Chem. 20: 10661–10664.

Wang, B., J. Jin, X. Hong, S. Gu, J. Guo and Z. Wen. 2017. Facile synthesis of the sandwich-structured germanium/reduced graphene oxide hybrid: an advanced anode material for high-performance lithium ion batteries. J. Mater. Chem. A 5: 13430–13438.

Wang, J., D.-H. Liu, Y.-Y. Wang, B.-H. Hou, J.-P. Zhang, R.-S. Wang et al. 2016a. Dual-carbon enhanced silicon-based composite as superior anode material for lithium ion batteries. J. Power Sources 307: 738–745.

Wang, X., D. Chen, Z. Yang, X. Zhang, C. Wang, J. Chen et al. 2016b. Novel metal chalcogenide SnSSe as a high-capacity anode for sodium-ion batteries. Adv. Mater. 28: 8645–8650.

Wang, X., L. Zhang, Z. Zhang, A. Yu and P. Wu. 2016c. Growth of 3D hierarchical porous NiO@carbon nanoflakes on graphene sheets for high-performance lithium-ion batteries. Phys. Chem. Chem. Phys. 18: 3893–3899.

Wu, X.-L., L.-Y. Jiang, F.-F. Cao, Y.-G. Guo and L.-J. Wan. 2009. $LiFePO_4$ nanoparticles embedded in a nanoporous carbon matrix: superior cathode material for electrochemical energy-storage devices. Adv. Mater. 21: 2710–2714.

Wu, X.L., Y.G. Guo, J. Su, J.W. Xiong, Y.L. Zhang and L.J. Wan. 2013. Carbon-nanotube-decorated nano-$LiFePO_4$ @C cathode material with superior high-rate and low-temperature performances for lithium-ion batteries. Adv. Energy Mater. 3: 1155–1160.

Xin, S., L. Gu, N.H. Zhao, Y.X. Yin, L.J. Zhou, Y.G. Guo et al. 2012a. Smaller sulfur molecules promise better lithium-sulfur batteries. J. Am. Chem. Soc. 134: 18510–18513.

Xin, S., Y.G. Guo and L.J. Wan. 2012b. Nanocarbon networks for advanced rechargeable lithium batteries. Acc. Chem. Res. 45: 1759–1769.

Xue, D.J., S. Xin, Y. Yan, K.C. Jiang, Y.X. Yin, Y.G. Guo et al. 2012. Improving the electrode performance of Ge through Ge@C core-shell nanoparticles and graphene networks. J. Am. Chem. Soc. 134: 2512–2515.

Yang, Y., Z. Wang, Y. Zhou, H. Guo and X. Li. 2017. Synthesis of porous Si/graphite/carbon nanotubes@C composites as a practical high-capacity anode for lithium-ion batteries. Mater. Lett. 199: 84–87.

Yao, L., H. Deng, Q.-A. Huang, Q. Su and G. Du. 2017. Three-dimensional carbon-coated $ZnFe_2O_4$ nanospheres/nitrogen-doped graphene aerogels as anode for lithium-ion batteries. Ceram. Int. 43: 1022–1028.

Zhang, L.-L., Z. Li, X.-L. Yang, X.-K. Ding, Y.-X. Zhou, H.-B. Sun et al. 2017. Binder-free $Li_3V_2(PO_4)_3$/C membrane electrode supported on 3D nitrogendoped carbon fibers for high-performance lithium-ion batteries. Nano Energy 34: 111–119.

Zhang, W.-M., X.-L. Wu, J.-S. Hu, Y.-G. Guo and L.-J. Wan. 2008. Carbon coated Fe_3O_4 nanospindles as a superior anode material for lithium-ion batteries. Adv. Funct. Mater. 18: 3941–3946.

Zhang, Y., P. Chen, X. Gao, B. Wang, H. Liu and S. Dou. 2016a. Nitrogen-doped graphene ribbon assembled core–sheath MnO@graphene scrolls as hierarchically ordered 3D porous electrodes for fast and durable lithium storage. Adv. Funct. Mater. 26: 7754–7765.
Zhang, Z., Q. Li, Z. Li, J. Ma, C. Li, L. Yin et al. 2016b. Partially reducing reaction tailored mesoporous 3D carbon coated NiCo-$NiCoO_2$/carbon xerogel hybrids as anode materials for lithium ion battery with enhanced electrochemical performance. Electrochim. Acta 203: 117–127.
Zhou, D., X. Li, L.-Z. Fan and Y. Deng. 2017. Three-dimensional porous graphene-encapsulated CNT@SnO_2 composite for high-performance lithium and sodium storage. Electrochim. Acta 230: 212–221.
Zhu, C., K. Song, P.A. Aken, J. Maier and Y. Yu. 2014. Carbon-coated $Na_3V_2(PO_4)_3$ embedded in porous carbon matrix: an ultrafast Na-storage cathode with the potential of outperforming Li cathodes. Nano Lett. 14: 2175–2180.
Zhu, C., P. Kopold, P.A. Aken, J. Maier and Y. Yu. 2016. High power–high energy sodium battery based on threefold interpenetrating network. Adv. Mater. 2016: 2409–2416.

11

Nanocomposites for IT SOFC Cathodes and Oxygen Separation Membranes

Vladislav A. Sadykov,[1,2,*] *Nikita F. Eremeev,*[1,2] *Yulia E. Fedorova,*[1,2,3] *Vasily A. Bolotov,*[1] *Yuri Yu. Tanashev,*[1] *Tamara A. Krieger,*[1,2] *Arkady V. Ischenko,*[1,2] *Anton I. Lukashevich,*[1] *Vitaliy S. Muzykantov,*[1] *Ekaterina M. Sadovskaya,*[1,2] *Vladimir V. Pelipenko,*[1] *Alexey S. Bobin,*[1,2] *Oleg Ph. Bobrenok,*[5] *Nikolai F. Uvarov,*[4,6] *Artem S. Ulikhin*[6] and *Robert Steinberger-Wilckens*[7]

Introduction

Design of materials for solid oxide fuel cells (SOFC) is a complex problem including preparation of materials and optimization of methods of their deposition on electrolyte and sintering. There is a necessity for lowering operating temperature down to intermediate temperature (IT) range to expand areas of SOFC application. Hence, this is an important task in materials science, electro- and solid state chemistry of searching appropriate materials for electrolytes and electrodes to provide good and stable performance at lower working temperatures. Another significant problem related to chemical engineering is the design of such fuel cells including deposition and sintering of functional layers (Huang et al. 2012, Irvine et al.

[1] Boreskov Institute of Catalysis SB RAS, 630090 Novosibirsk, Russian Federation.
[2] Novosibirsk State University, 630090 Novosibirsk, Russian Federation.
[3] Novosibirsk State Pedagogical University, 630126 Novosibirsk, Russian Federation.
[4] Novosibirsk State Technical University, 630073 Novosibirsk, Russian Federation.
[5] Institute of Thermal Physics SB RAS, Novosibirsk, Russian Federation.
[6] Institute of Solid State Chemistry and Mechanochemistry SB RAS, Novosibirsk, Russian Federation.
[7] University of Birmingham, Edgbaston, B15 2TT, United Kingdom.
Email: yeremeev21@gmail.com
* Corresponding author: sadykov@catalysis.ru

2016, Sadykov et al. 2015b). Moreover, such fuel cells allow the usage of hydrogen or syngas as fuel. Catalytic reactors with oxygen separation membranes allow the transformation of biogas or natural gas into syngas by reaction with oxygen transferred through membranes from the air side to the fuel side. Materials with ionic or mixed ionic-electronic conductivity (MIEC) are generally used for creating SOFC components and catalytic membrane functional layers (Sadykov et al. 2010a, 2011b). Microwave heating is a promising technique in SOFC and oxygen separation membranes' design due to cheaper, faster and stronger sintering of cathode layers at lower temperatures compared to conventional methods (Janney and Calhoun 1992). It was shown for Pr nickelates-cobaltites and their nanocomposites as well as for other systems in our previous works (Sadykov et al. 2012, 2016a).

Supported oxygen separation membrane typically consist of compressed porous metallic (generally Ni-Al foam) or cermet substrate, porous and dense MIEC layers (perovskite or perovskite—fluorite nanocomposite, etc.), dense buffer nanocomposite (spinel—GDC) and porous catalytic (mostly Pt-containing) layers (Shelepova et al. 2016, Sadykov et al. 2011b). One of the most important problems associated with such membranes is achieving necessary tightness and density of dense MIEC layer at not very high temperatures. As mentioned above, microwave sintering (MWS) technique followed by deposition of the MIEC opens ability to reach high density (~ 6 g/cm^3) and low residual porosity (below ~ 5%) of the nanocomposite at T ~ 870–1000°C, that is impossible for conventional sintering (CS), which was demonstrated for Pr nickelate-cobaltite – Y-doped ceria nanocomposites (Sadykov et al. 2016a).

Materials with high ionic (oxygen) and low electronic conductivity are used as SOFC electrolytes. Commonly, these are complex oxides with the fluorite-like (F) structure, generally yttria-stabilized zirconia ($Zr_{0.84}Y_{0.16}O_{2-\delta}$, YSZ) (Ormerod 2003). Though YSZ conductivity is not very high and a lot of other materials are used (Sc- and Yb-stabilized zirconia, etc.), YSZ provides the best combination of chemical stability and ionic conductivity amongst ZrO_2-based fluorites (Sammes and Du 2005). Yttria- and gadolinia-doped ceria ($Ce_{0.9}Y_{0.1}O_{2-\delta}$, YDC, $Ce_{0.9}Gd_{0.1}O_{2-\delta}$, GDC) are interesting materials as well. One of the ways to improve their ionic (oxygen) conductivity is co-doping with Pr (Borchert et al. 2005, Sinev et al. 1996, Sadykov et al. 2014b, 2015a). As anode material, Ni/YSZ cermet has a good catalytic activity and stability in the oxidation of hydrogen fuel at SOFC operation conditions (Atkinson et al. 2004, Sadykov et al. 2011a). Traditional and state-of-the-art cathode materials based upon Sr-doped lanthanum manganites (LSM), ferrites (LSF), ferrites-nickelates (LSFN) and ferrites-cobaltites (LSFC) have high electronic or mixed ionic-electronic conductivity (MIEC) sufficient for working in operating conditions; however, despite of this advantage, they have high thermal expansion coefficient compared to electrolyte, interact with YSZ electrolyte forming low-conducting zirconates and are unstable to carbonization (Murray et al. 1998, Minh 1993, Basu 2007, Sadykov et al. 2010a). This causes deterioration of SOFC performance and even failure of the cell. Figure 1 shows the most important problems of chemical stability associated with using traditional materials (e.g., LSM) such as formation of isolating layers of La and Sr zirconates (or cerates in the case of using doped CeO_2-based electrolyte) and interaction with CO_2 in the gas mixture to form metal carbonates and hydroxycarbonates (carbonization). To avoid these problems, the use of Sr-free perovskites with Pr cations occupying A-sites can be suggested.

Materials based upon praseodymium nicketales $Pr_{2-x}NiO_{4+\delta}$ (PN) with Ruddlesden—Popper (RP) structure and praseodymium nickelates-cobaltites $PrNi_{1-x}Co_xO_{3-\delta}$ (PNC) with perovskite structure are among the most promising SOFC cathodes due to stability to carbonization, compatibility with YSZ, high mixed ionic-electronic conductivity and oxygen bulk mobility/surface reactivity (Boehm et al. 2005, Hjalmarsson and Mogensen 2011, Nishimoto et al. 2011, Sadykov et al. 2014a, b). Layered RP phases $Ln_2NiO_{4+\delta}$ have a moderate thermal expansion coefficient compatible with a lot of solid electrolytes (Pikalova and Kolchugin 2016). Moreover, such oxides are known to have a high oxygen mobility due to cooperative mechanism of oxygen migration involving weakly bound interstitial oxygen of rock salt ($Ln_2O_{2+\delta}$) layers and oxygen vacancies in perovskite ($LnNiO_{3-\delta}$) layers (Sadykov et al. 2013b, Minervini et al. 2000, Li and Benedek 2015, Sadykov et al. (2018)) along with a high surface reactivity. One of the ways to enhance their performance is to provide a slight deficiency in A position, thus increasing concentration of oxygen vacancies and, hence, oxygen mobility. Incorporation of guest cations into rock salt layers (such as alkaline earth metal cations (Pikalova and Kolchugin 2016) in the case of doping or

Ce/Zr cations (Sadykov et al. 2013b) in the case of nanocomposites) can violate cooperative migration mechanism. Nanocomposites comprised of Pr nickelates-cobaltites with Y-doped ceria (PNC–YDC) are known to be promising materials primarily due to their good oxygen transport properties. Strong cation redistribution resulting in Pr incorporation into the fluorite domains' increasing oxygen mobility as mentioned above along with well-developed perovskite—fluorite interface are responsible of forming fast and broad oxygen diffusion channel involving up to 90% of total bulk oxygen anions (Sadykov et al. 2014a, b, 2015a). Working characteristics of various cathode materials developed in the last decade are compared in Table 1.

This work summarizes results of research aimed at design of stable to carbonation Sr-free materials for IT SOFCs cathodes and functional layers of oxygen separation membranes comprised of Pr nickelates/nickelates-cobaltites with perovskite (P) or Ruddlesden–Popper (R-P) type structures and their nanocomposites with Y/Gd-doped ceria with fluorite (F) structure. Nanocrystalline fluorites, Sm-doped praseodymium nickelates-cobaltites $Pr_{1-x}Sm_xNi_{1-y}Co_yO_{3-\delta}$ and $Pr_{2-x}NiO_{4+\delta}$ oxides were synthesized by modified Pechini route. Nanocomposites were prepared via ultrasonic dispersion of F and P (R-P) powders in isopropanol. Samples were sintered at temperatures up to 1300°C using both conventional and microwave heating. Phase composition, morphology, microstructure and elemental composition of domains have been characterized by XRD and HRTEM/SEM with EDX analysis. Bulk and surface oxygen mobility and reactivity have been studied by using O_2 temperature-programmed desorption,

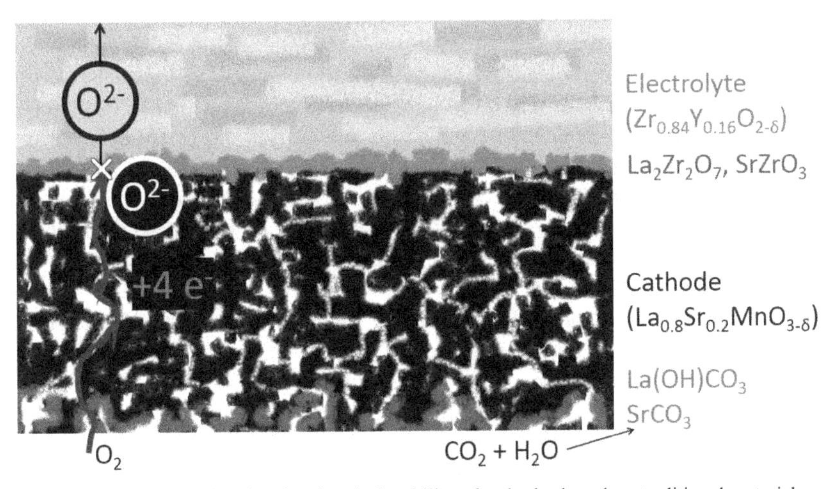

Figure 1. Problems related to the chemical stability of cathodes based on traditional materials.

Table 1. Maximum power density of single button SOFC with various cathodes.

Cathode	Anode Half-cell	*T*, [°C]	P_{max}, [mW/cm²]	Reference
LSM | LSM–ScCeSZ	YSZ | Ni/YSZ	600 700	260 580	(Sadykov et al. 2010c)
LSFN	YDC | YSZ | Ni/YSZ	700	330	(Duval et al. 2010)
LSFN–GDC	GDC | YSZ | Ni/YSZ	700	240	(Sadykov et al. 2010a)
LSFC	GDC | YSZ | Ni/YSZ	700	160	(Sadykov et al. 2010a)
$Pr_2NiO_{4+\delta}$	SDC | Ni/SDC YDC | GDC | Ni/YSZ	600 600	650 400	(Fan et al. 2012) (Ferchaud et al. 2011)
$Pr_2NiO_{4+\delta}$–Ag	LNSDC| Ni/LNSDC	600	700	(Fan et al. 2012)
$PrNi_{0.6}Co_{0.4}O_{3-\delta}$	SDC | Ni/SDC	700	700	(Huang et al. 2012)
$PrNi_{0.6}Co_{0.4}O_{3-\delta}$ – SDC	SDC | Ni/SDC	700	1090	(Huang et al. 2012)

oxygen isotope exchange techniques (including $C^{18}O_2$ SSITKA) and relaxation (weight loss, unit cell parameters and electric conductivity) methods. Due to redistribution of elements between phases, P/RP and F domains were nanosized and disordered even in dense nanocomposite ceramics. That provides fast oxygen diffusion in such domains and their interfaces with $D_O \sim 10^{-6}$ $cm^2 \cdot s^{-1}$ at 700°C that exceeds values for LSFC and LSFC–GDC by 2–3 orders of magnitude. Selected compositions tested as thin film SOFC cathodes and oxygen separation membranes functional layers demonstrated promising performance.

Experimental

Individual oxides with perovskite (P) $Pr_{1-x}Sm_xNi_{1-y}Co_yO_{3-\delta}$ (PSNC, $x = 0$–0.3, $y = 0$–0.6), Ruddlesden–Popper (RP) $Pr_{2-x}NiO_{4+\delta}$ (PN, $x = 0$–0.3) and fluorite (F) $Ce_{0.9}M_{0.1}O_{2-\delta}$ (M = Y, Gd) (YDC, GDC) structures were prepared via modified Pechini route. Metal nitrates hexahydrates were dissolved in water with citric acid monohydrate being dissolved in ethylene glycol, then the solutions obtained were mixed, thermally decomposed to xerogel, burned and sintered at 700°C. Nanocomposites were obtained from P/RP and F powders in 1:1 weight ratio by ultrasonic dispersion in isopropanol. Powders obtained were pressed into pellets and then sintered at 800–1300°C. For microwave sintering (MWS), nanocomposites powders were presintered at 900°C, mechanically activated in high power planetary ball mill AGO-2 (Novic ltd, Russia), pressed into pellets and then sintered at 870–1100°C in specially designed set-up.

All samples obtained were characterized by X-ray diffraction (XRD) and transmission electron microscopy (TEM) with energy-dispersive X-ray spectroscopy (EDX) analysis. XRD patterns were obtained for powdered or pelletized samples with a D8 Advance (Bruker, Germany) diffractometer using Cu-K_α monochromatic radiation (λ = 1.5418 Å) in 2θ range 5–90°. Plasticine was used to fix a pellet in the cell before measurements. The TEM images were obtained with a JEM-2010 (JEOL, Japan) instrument. Analysis of the local elemental composition was carried out by using an energy-dispersive EDX spectrometer equipped with Si(Li) detector.

In situ synchrotron XRD (SXRD) studies were performed at the "Precision Diffractometry" station at the synchrotron radiation facilities of the VEPP-3 storage ring (Siberian Synchrotron and Terahertz Radiation Center (SSTRC), Novosibirsk, Russia). Wavelength of 1.0211 Å was set by a single reflection from perfect flat crystal of Si(220).

Specific surface area of samples was determined by Brunauer–Emmett–Teller technique using Sorbi-M (Meta ltd, Russia) device. True density and residual porosity of densified samples were estimated by weighing the sample before and after putting into ethanol or propan-2-ol.

For conductivity measurements, four electrode *dc* current (Van der Pauw) technique was used. Voltage was measured by measuring-feeding device IPU-01 (ISSCM SB RAS, Russia).

Temperature programmed desorption of oxygen (TPD O_2) in He stream (flow rate 2.5 l/h) was carried out for ground samples with the temperature ramp 5°C/min from ambient to 880°C followed by 70 min plateau to characterize the bonding strength of oxygen with the surface along with the amount of highly mobile oxygen. Outlet gas concentration was measured by TEST-1 (Boner, Russia) gas analyzer. Desorption rate was calculated from the outlet oxygen volume fraction knowing flow rate and specific surface area of the sample. Effective activation energies were estimated by analyzing the ascending part of the peak using Wigner–Polanyi equation assuming pseudo-zero reaction order.

The oxygen mobility and surface reactivity of powdered samples were studied by the oxygen isotope exchange with $^{18}O_2$ and $C^{18}O_2$ using isothermal (IIE) and temperature programmed (TPIE) modes in closed and open reactors with mass spectra control of the gas phase isotope composition (Sadykov et al. 2010). The isotope gas composition was analyzed using SRS QMS200 (for closed reactor experiments) and UGA200 (for flow reactor experiments) (Stanford Research Systems, USA) mass spectrometers. Dependencies on time of $^{16}O^{18}O$ (or $C^{16}O^{18}O$) molecules fraction $f_{16-18}(t)$ and ^{18}O atoms fraction $\alpha(t)$ in the outlet stream (referred to as isotope kinetic curves) were analyzed by using isotope kinetic equations. To calculate oxygen tracer diffusion coefficients $D_O(T)$ and surface exchange constants $k_{ex}(T)$ depending on temperature, analysis of the oxygen isotope exchange data was carried out using approaches earlier described in detail (Sadovskaya et al. 2010, Sadykov et al. 2015c).

Oxygen chemical diffusion coefficients and chemical surface exchange constant of MIEC materials were estimated by analysis of their conductivity, weight or unit cell volume relaxation after step-wise change of O_2 content in the N_2 stream. Numeric modeling was performed using ECRPro 3.0.0.0 software (Ananyev 2011). Modeling equations of the relaxation process are not given for brevity and can be found elsewhere (Sadykov et al. 2014c, Steinberger-Wilckens et al. 2009).

For preparation of button-size cells, anode half-cells YDC/YSZ/NiO+YSZ provided by H.C. Starck (Germany; type AS4 with thin YSZ layer and YDC buffer layer) were used. Functionally graded cathodes were made using nanocomposite and P/RP layers. Cathode layers with thickness up to 20 microns were supported by screen printing from specially prepared inks using DEK 248 (DEK, UK) printer, dried and sintered at temperatures up to 1100°C (Sadykov et al. 2010a, 2014a). For small cells, Pt and Ni gauzes were used as current collectors. Polarization curves were obtained using wet hydrogen as fuel and dried compressed air as oxidant. Cell Test System 1400 (Solartron Analytical, UK) and P-150J (Elins LLC, Russia) were used for measurements.

For preparation of oxygen separation membranes, Ni/Al foam substrates were used. Mixture of powders sintered at 700 and 1300°C and powder sintered at 700°C were used for preparation of rough and thin fraction suspensions, respectively. A few layers were deposed on the substrate according to the following sequence: rough fraction, thin fraction and GDC+$MnFe_2O_4$ buffer layer. After deposition, each layer was sintered at 1100°C in Ar atmosphere. Then Pt (1.4 wt. %) – $Sm_{0.15}Pr_{0.15}Ce_{0.35}Zr_{0.3}O_{2-\delta}$ catalytic layer was deposited and sintered at 900°C in Ar atmosphere. For performance studies, the membranes obtained were put into the reactor and then tested in specially built setup. Air and methane (pure or with addition of CO_2) were supplied to the counterpart sides of the membrane.

Results and discussion

Structural and textural characteristics

For $Pr_{2-x}NiO_{4+\delta}$ (PN) samples, nanostructured RP phase formation from the mixture of Pr_6O_{11} and NiO simple oxides (produced by burning the polymeric precursor) was observed at high (> 1000°C) temperatures (Sadykov et al. 2013b). Main phase for these samples was *Fmmm* A_2BO_4-like RP oxide. Some Pr_6O_{11} *Fm3m* fluorite admixture was detected, its content decreasing with increasing x. The content of this admixture was also shown to be lower for samples sintered in N_2. PN samples sintered at 1250°C were single phases. Figure 2 shows diffraction patterns of PN samples sintered at 1100°C depending on sintering atmosphere and Pr deficiency. PN–GDC nanocomposites comprised of RP and F phases. Insignificant NiO segregation was observed. Cation redistribution between RP and F phases was revealed. XRD data were fully confirmed by TEM with EDX analysis. According to the SXRD data, the symmetry of PN oxides crystal lattices increases with heating in air with *Fmmm* structure, changing to the *I4/mmm* tetragonal one at 450°C and then to *Bmab* orthorhombic one with a lower symmetry at 600°C (Sadykov et al. 2014c).

Specific surface area obtained from BET data was found to be ~ 0.4–0.5 m^2/g for the PN samples, ~ 1 m^2/g for PN–GDC samples sintered at 1100°C and ~ 0.2 m^2/g for the $Pr_2NiO_{4+\delta}$ sample sintered at 1250°C. True density was ~ 5 and ~ 6 g/cm^3 for samples sintered at 1100°C and 1250°C and higher, so residual porosity was above 20% and below 5% at these temperatures, respectively.

Similar to PN samples, as-prepared oxides with the nominal $PrNiO_{3\pm\delta}$ composition are comprised of Pr_6O_{11} and NiO phases with a very little amount of admixture phase identified according to PDF # 0370805 (P.G.hex *P63/mmc*). For $PrNi_{1-x}Co_xO_{3-\delta}$ (PNC, x = 0.4–0.6) samples sintered at T up to 1100°C, the main phase was orthorhombic perovskite (Figure 3) with the admixture of cubic *Fm3m* Pr_6O_{11} phase, its content decreasing with increasing sintering temperature. $Pr_4(Ni,Co)_3O_{10+\delta}$ RP homologue admixture is present in samples sintered at T = 1100°C and higher. For samples sintered at 1300°C, it was the main phase, perovskite phase keeps its orthorhombic symmetry with Ni(Co)O segregation apparently driven by B-site cations' reduction at high temperatures. In a similar way, $PrNiO_{3\pm\delta}$ sample sintered at 1300°C consists of $Pr_2NiO_{4+\delta}$ homologue and NiO (Sadykov et al. 2014a, b, 2015a).

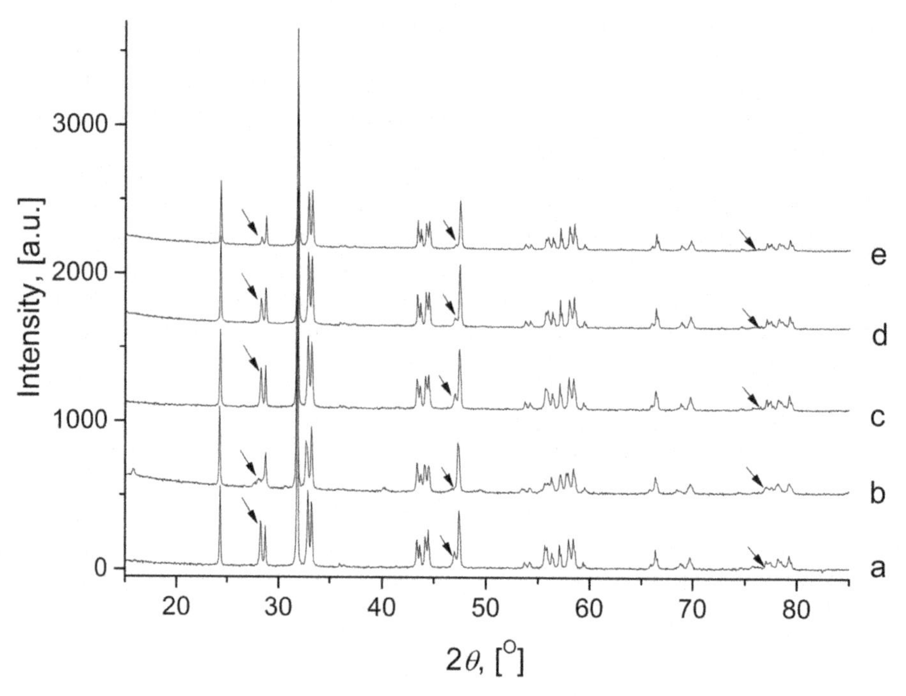

Figure 2. X-ray diffraction patterns of samples with R-P type structure: $Pr_2NiO_{4+\delta}$ sintered at 1100°C in air (a) and nitrogen (b) atmosphere, $Pr_{1.9}NiO_{4+\delta}$ (c), $Pr_{1.8}NiO_{4+\delta}$ (d) and $Pr_{1.7}NiO_{4+\delta}$ (e) sintered at 1100°C in air. ↓ – Pr_6O_{11}.

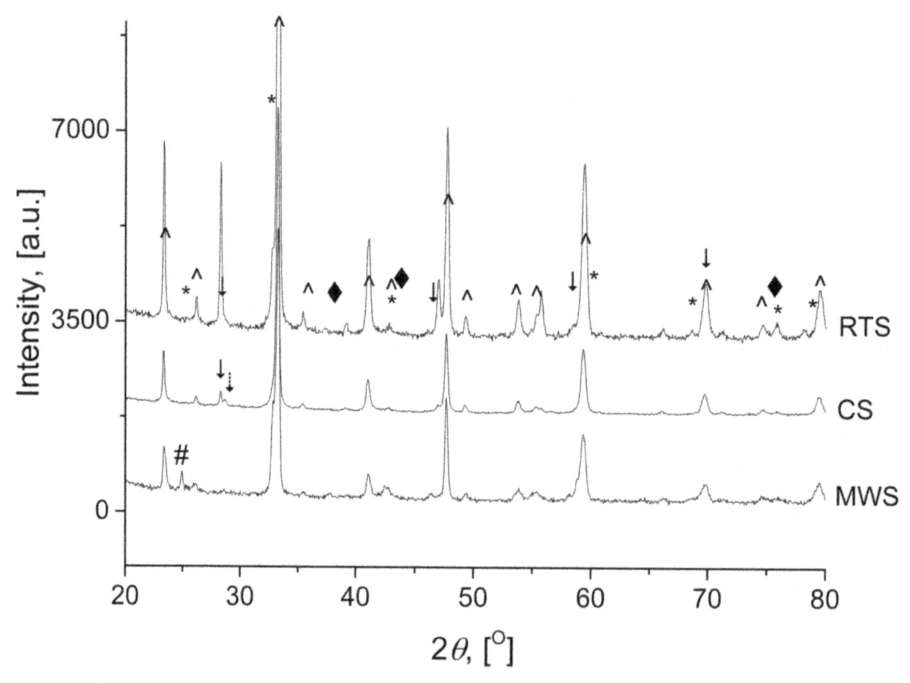

Figure 3. X-ray diffraction patterns of $PrNi_{0.5}Co_{0.5}O_{3-\delta}$ samples sintered at 1000°C using convention (CS), microwave radiation (MWS) and radiation-thermal (by e-beam, RTS) sintering techniques. ^ – $PrNi_{0.5}Co_{0.5}O_{3-\delta}$ perovskite, ↓ – Pr_6O_{11}, ⇣ – PrO_2, * – $Pr_4(Ni,Co)_3O_{10+\delta}$, ♦ – Ni(Co)O, # – plasticine.

Sm-doped PNC analogues show almost the same tendencies. For $Pr_{1-x}Sm_xNi_{0.5}Co_{0.5}O_{3-\delta}$ (PSNC, x = 0.1–0.3) sintered at $T \leq 1000°C$, the main phase is orthorhombic P with the lattice parameter ~ 5.450 Å. Admixture of Pr-Sm-O phase is present in these samples, its content decreasing with the increase of sintering temperature. PSNC sintered at higher T have no $Pr_{1-x}Sm_xO_{2-\delta}$ admixture but the content of $Pr_{4-x}Sm_xNi_{1-y}Co_yO_{10+\delta}$ RP homologue dramatically increases being the main phase for samples sintered at 1300°C.

PNC–YDC and PSNC–YDC composites are bi-phasic samples with traces of NiO at high sintering T. The most prominent feature is incorporation of Pr cations into doped ceria phase apparently disordering both perovskite-like and fluorite-like phases. Domain sizes of both phases in samples sintered at 1100°C are in the range of 50–100 nm. At higher (1300°C) sintering temperature, redistribution of elements in nanocomposites results in transformation of orthorhombic perovskite structure into disordered cubic one. Doping by Sm stabilizes the orthorhombic structure of P phase even at such high temperatures (Sadykov et al. 2016b).

Specific surface area of samples sintered at 1100°C and 1300°C was ~ 1 and ~ 0.5 m^2/g (or lower), respectively. True density was ~ 5–5.5 g/cm^3. Residual porosity was ~ 30% for samples sintered at 1000°C and ~ 25% for those sintered at 1100°C.

To provide a low residual porosity at not so high temperatures in the range of perovskite phase stability, one way to solve this problem is using strongly non-equilibrium processing by advanced sintering techniques. Microwave sintering was applied with variation of sintering temperature, heating rate and time of keeping in the isothermal mode. It was found that PNC samples contained Pr_6O_{11} and/or $Pr_4(Ni,Co)_3O_{10+\delta}$ admixtures as well, but their content was lower (for some samples much lower) compared to CS ones. Additional Pr_6O_{11} fluorite-like phase forms due to strong cation redistribution for PNC–YDC nanocomposites. MWS technique was successful in providing residual porosity ~ 14% and < 5% for sintered at 1000°C PNC and PNC–YDC samples, respectively. More details upon MWS Pr nickelates-cobaltites and their nanocomposites can be found in our recent study (Sadykov et al. 2016a). Phase composition of PNC samples depending on sintering technique is given in Figure 3.

Conductivity

PN samples present bell-shaped total electric conductivity curve temperature dependence with maximum 50–100 S/cm at 400–450°C. This was shown in our study (Sadykov et al. 2013b) agreeing with the data presented by other authors (Kovalevsky et al. 2007a). The effective activation energy of total conductivity is 10–12 kJ/mole at low temperatures and ~ 3 kJ/mole in the intermediate temperature range. The best conductivity is for samples with small Pr deficiency ($Pr_{1.9}NiO_{4+\delta}$ and $Pr_{1.8}NiO_{4+\delta}$). This can be due to decreasing the content of Pr_6O_{11} which is known to be a poor electronic (and good ionic) conductor (Biswas et al. 1997). A higher Pr deficiency decreases conductivity probably due to partial fragmentation of RP structure. This fragmentation may occur at the domain or subdomain scale, since no peak distortion or broadening was observed in XRD patterns. Arrhenius plot for conductivity of $Pr_{1.9}NiO_{4+\delta}$ is given in Figure 4, curve a.

For "$PrNiO_3$" (in fact $Pr_2NiO_{4+\delta}$–NiO) sample, the temperature dependence of total conductivity has the same features as those for PN samples. Doping by Co in B-site dramatically increases electric conductivity. PNC samples have conductivity ~ 200–300 S/cm at 600°C comparable with that for LSM, LSFN and LSFC (see Figure 4, curves b, c and d) (Murray et al. 1998, Minh 1993, Basu 2007, Sadykov et al. 2014a). The total conductivity increases with Co content. The same tendency was shown for Sr-doped Pr nickelates-cobaltites $Pr_{0.5}Sr_{0.5}Ni_{1-x}Co_xO_{3-\delta}$ (x = 0.2–0.6) (Kostogloudis and Ftikos 1998). The effective activation energy increases with temperature from ~ 10 kJ/mole to ~ 25 kJ/mole. Such a low E_a value may be explained by $Pr_4(Ni,Co)_3O_{10+\delta}$ phase effect. PNC–YDC electric conductivity was shown to be noticeably (approximately by one and half order of magnitude) lower with its effective activation energy being ~ 5 kJ/mole lower compared to PNC; however, it remains at the level sufficient for application of such composites as cathodes (Sadykov et al. 2014a).

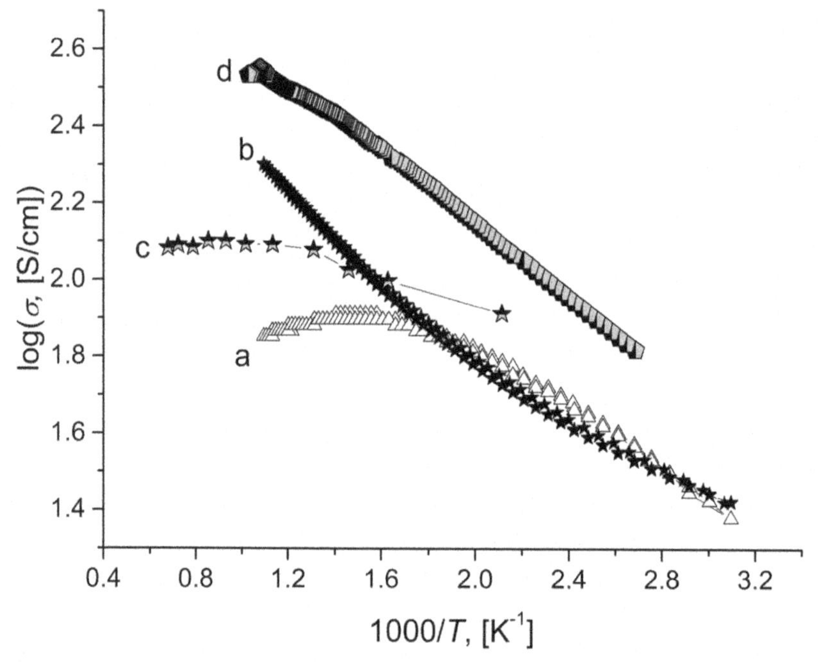

Figure 4. Arrhenius plots for the total conductivity (σ) of $Pr_{1.9}NiO_{4+\delta}$ (a) and $PrNi_{0.4}Co_{0.6}O_{3-\delta}$ (b) samples compared to LSM (c) and LSFN (d) (Murray et al. 1998, Minh 1993, Basu 2007, Sadykov et al. 2014a).

Temperature-programmed desorption of oxygen

O_2 TPD curves of PN samples have 4 peaks corresponding to α, β_1, β_2 and γ oxygen forms. The first three peaks are situated at 450–600°C, with the last one—at 880°C. α-, β- and γ-oxygen are types of bulk oxygen classified by bond strength with the activation energy of desorption ~ 40, 80–120 and 160–400 kJ/mol, respectively (Kuznetsova et al. 2005). All the peaks have almost Gauss shape. The total amount of oxygen desorbed in TPD run is 60–100 monolayers (ML), thus demonstrating a high bulk oxygen mobility. This can be explained by cooperative mechanism of oxygen transport described in (Sadykov et al. 2013b). The rate of desorption and the amount of desorbed oxygen are ~ 2 orders of magnitude higher compared to $La_2NiO_{4+\delta}$ agreeing with well-known tendency of oxygen mobility increasing while substituting La by Pr explained by a higher oxygen excess due to the presence of Pr cations in both 3+ and 4+ states (Kovalevsky et al. 2007b) and elevated content of highly mobile oxygen (Sadykov et al. 2013b) as well as a lower barrier for cooperative migration route via interstitial oxygen jump through triangle of small Pr^{3+} and Pr^{4+} cations (Li and Benedek 2015). The first three desorption peaks can be attributed to removing weakly bound interstitial oxygen from the rock salt layers accompanied by rearrangement of coordination spheres of Pr cations. The fourth desorption peak situated at 880°C can be assigned to perovskite layers. This peak is specific for Ni-containing perovskites corresponding to Ni cations reduction from 3+ to 2+ (Sadykov et al. 2010a).

Such a high oxygen mobility and surface reactivity revealed by TPD O_2 studies might be explained by the cooperative mechanism of oxygen migration. As stated above, this mechanism involves interstitial oxygen of rock salt layers and vacancies of perovskite layers. This expressly characterizes $Pr_{2-x}NiO_{4+\delta}$ RP oxides due to Pr cations ability to change their charge from 4+ to 3+ and back. Rock salt layers can accumulate weakly bound interstitial oxygen atoms. Also, perovskite layers, especially in the case of Pr deficiency, have a lot of oxygen vacancies. Synergetic effect of highly mobile oxygen and oxygen vacancies widens bulk oxygen migration pathway with this process being characterized by cooperative conjugation. This mechanism was confirmed by modeling performed by using quantum mechanics calculation for $Pr_{2-x}NiO_{4+\delta}$ systems (Sadykov et al. 2013b, Li et al. 2015).

PNC samples conventionally sintered (CS) at 1000°C have 4 peaks as well at ~ 400, 500, 550 and 880°C corresponding to α, β, γ_1 and γ_2 oxygen forms, respectively (see Table 2). The peaks situated in the intermediate temperature range are less intense compared to the high-temperature one. These low-temperature peaks can be assigned to the presence of some extended defects (stacking faults, domain boundaries, etc.) (Sadykov et al. 2010a) with the local coordination environment of Pr cations similar to that in R-P type structure. The total amount of oxygen desorbed is less than in the case of RP Pr nickelates being not higher than 10–15 ML, thus corresponding to some defects. For PNC–YDC nanocomposites, much more intense desorption in the intermediate-temperature range is observed, while the total amount of oxygen desorbed in the run increases up to 20 ML. The most distinctive feature of TPD O_2 curves for PNC–YDC is domination of the intermediate-temperature desorption peak not observed for perovskite-fluorite nanocomposites used as traditional SOFC cathode materials (LSFN(C)–GDC, LSM–GDC, etc. (Sadykov et al. 2010a)). Doping by Sm slightly decreases desorption rate due to the decreasing content of $Pr^{3+/4+}$ cations and associated oxygen excess (Sadykov et al. 2016b).

TPD O_2 curves for microwave sintered (MWS) PNC samples reveal only α- and β-oxygen forms, with the total amount of oxygen desorbed up to 20 ML (see Table 2). For MWS PNC–YDC composites, only β- (the most intense) and γ_2-peaks are observed, with the total amount of desorbed oxygen up to 40 ML for conventionally sintered (CS) samples and up to 70 ML for MWS samples. Such a high bulk oxygen mobility is explained by developed interface between P and F nanodomains as well as cations' redistribution between phases (Sadykov et al. 2014b, 2016a).

These features can be explained by domination of slow diffusion channel for PNC and PSNC and fast diffusion channel for PNC–YDC and PSNC–YDC samples, which will be discussed below while analyzing data of the oxygen isotope exchange techniques and calculating such characteristics of oxygen mobility and surface reactivity as oxygen tracer diffusion coefficients and surface exchange constants.

Table 2. Amount of oxygen desorbed q_{O2} and effective activation energy of oxygen desorption $E_{a, TPD}$ for samples sintered conventionally and by microwave heating.

Peak No.	T_{max}, [°C]	q_{O_2}, [monolayers]	q_{O_2}, [%]	$E_{a, TPD}$, [kJ/mol]
$PrNi_{0.5}Co_{0.5}O_{3-\delta}$, 1000°C (conventional sintering)				
1	390	0.6	6	50
2	470	0.3	3	140
3	550	0.7	7	190
4	880*	7.9	84	150
Total desorbed		9	100 (1% of overall oxygen)	
$PrNi_{0.5}Co_{0.5}O_{3-\delta}$, 1000°C (microwave sintering)				
1	451	5	25	64
2	880*	15	75	110
Total desorbed		21	100 (1% of overall oxygen)	
$PrNi_{0.5}Co_{0.5}O_{3-\delta}$ – YDC, 1100°C (conventional sintering)				
1	470	24	57	110
2	880*	23	43	25
Total desorbed		47	100 (3% of overall oxygen)	
$PrNi_{0.5}Co_{0.5}O_{3-\delta}$ – YDC, 930°C (microwave sintering)				
1	428	53	78	81
2	530			–
3	880*	15	22	47
Total desorbed		68	100 (5% of overall oxygen)	

*Peak maximum in isothermal conditions (T = 880°C, t = 70 min).

Oxygen isotope exchange

According to $^{18}O_2$ oxygen isotope exchange data for PN samples, a part of exchangeable oxygen atoms (so called dynamic extent of exchange x_S) is up to 750 ML corresponding to ~ 80% of the total bulk oxygen, thus evidencing its high mobility. Exchange process goes via R^2 type of mechanism involving two surface oxygen atoms per one act and is limited by the surface exchange. Oxygen tracer diffusion coefficients D_O are not so high (~ 10^{-10} cm²/s at 700°C) but the surface exchange constants (k_{ex} ~ 10^{-7} cm/s at the same temperature) are comparable, or even higher, compared to LSM, LSFN and LSFC (Sadykov et al. 2013b). For the sample sintered at 1250°C, D_O and k_{ex} values are ~ 3 orders of magnitude higher, which exceed LSFN and LSFC by ~ 2 orders of magnitude (Tai and Nasrallah 1995). Figure 5 shows $D_O(T)$ dependences obtained by temperature programmed isotope exchange of oxygen with $C^{18}O_2$ data for the materials involved in this work compared to LSFN and LSFC (Tai and Nasrallah 1995).

Oxygen tracer diffusion coefficient and surface exchange constant values almost do not vary with Pr deficiency. Due to cooperative mechanism of oxygen transport discussed above, leading to fast exchange of interstitial and regular oxygen, all oxygen atoms in layered structure become kinetically equivalent, which was shown in modeling isotope kinetic curves.

For PN–GDC nanocomposites, $^{18}O_2$ exchange process can be generally described by R^2 mechanism type (contribution of R^1 type involving one surface oxygen atom per exchange act is 5–10%) and is limited by the surface exchange too. They have D_O by ~ 1 order of magnitude lower due to hampering the cooperative mechanism of oxygen transfer caused by cations' redistribution, namely, by Ce^{4+} cations' incorporation into rock salt layers in RP structure.

For undoped Pr nickelates-cobaltites and their nanocomposites with Y doped ceria, oxygen isotope exchange process with $^{18}O_2$ is described by R^2 type of mechanism, while exchange with $C^{18}O_2$—by R^1 type of mechanism. This is naturally explained by the rapid exchange of CO_2 with the surface oxygen species via fast and easy formation of carbonates CO_3^{2-} complexes. For Sm-doped samples, the process is limited by bulk diffusion, so the surface exchange mechanism and specific exchange rate could not be determined (Sadykov et al. 2015a, 2016b). P(S)NC, especially P(S)NC–YDC composites, have a high oxygen mobility and surface reactivity. Two types of bulk oxygen with D_O differing by ~ 2 orders of magnitude are present. For undoped PNC samples, D_{fast} ~ 10^{-7} cm²/s and D_{slow} ~ 10^{-8} cm²/s at 700°C, that is by 1–2 orders of magnitude higher compared to LSFN and LSFC (Tai and Nasrallah 1995). Comparison of oxygen tracer diffusion coefficient values is given in Figure 5. A narrow (~ 10% of total oxygen) channel of fast diffusion is observed to be suppressed doping by Sm leads to its complete disappearance. PNC–YDC are characterized by even higher bulk oxygen mobility (D_{fast} ~ 10^{-6} cm^{-2}/s at 700°C). This fast diffusion channel involves ~ 60% of the total oxygen, so this fraction increases while doping by Sm. MWS samples have characteristics comparable with CS ones (Sadykov et al. 2016a).

For perovskites, presence of fast channel of oxygen diffusion involving up to 10% of total bulk oxygen can be explained by the existence of surface/bulk extended defects described above as well as by a small content of Pr_6O_{11} ($Pr_{1-x}Sm_xO_{2-\delta}$ in the case of Sm-doped samples), since this channel could not be observed for MWS samples (Sadykov et al. 2015a, 2016a). The oxygen tracer diffusion coefficient at 700°C for Pr_6O_{11} is ~ 10^{-7} cm²/s being close to the one for PNC (see Figure 6). The fast oxygen diffusion channel in nanocomposites can be assigned to F domains and P–F interfaces. It appears due to cation redistribution and developed perovskite-fluorite interface. Cation redistribution slightly distorts P and F cells weakening M–O bonds strength. Important role can be attributed to Pr and Sm cations migration into Y-doped ceria causing dramatic increase of its oxygen mobility and surface reactivity. This model is supported by a high value of D_O ~ 10^{-5} cm²/s at 700°C for specially prepared Pr-doped F sample with composition $Ce_{0.65}Pr_{0.25}Y_{0.1}O_{2-\delta}$ (see Figure 6) corresponding to the average F phase composition in PNC–YDC nanocomposites. Enhanced oxygen mobility and surface reactivity in Pr-doped ceria were observed in many works (Borchert et al. 2005, Sinev et al. 1996, Sadykov et al. 2007, 2014b, Shuk et al. 1999). This is the result of Pr cations being able to easily change their charge and, therefore, to vary the oxygen content in its coordination sphere. A highly mobile oxygen of F phase along with partial P and F cells distortion as well as developed P–F interface are responsible for the presence of broad channel of fast oxygen diffusion.

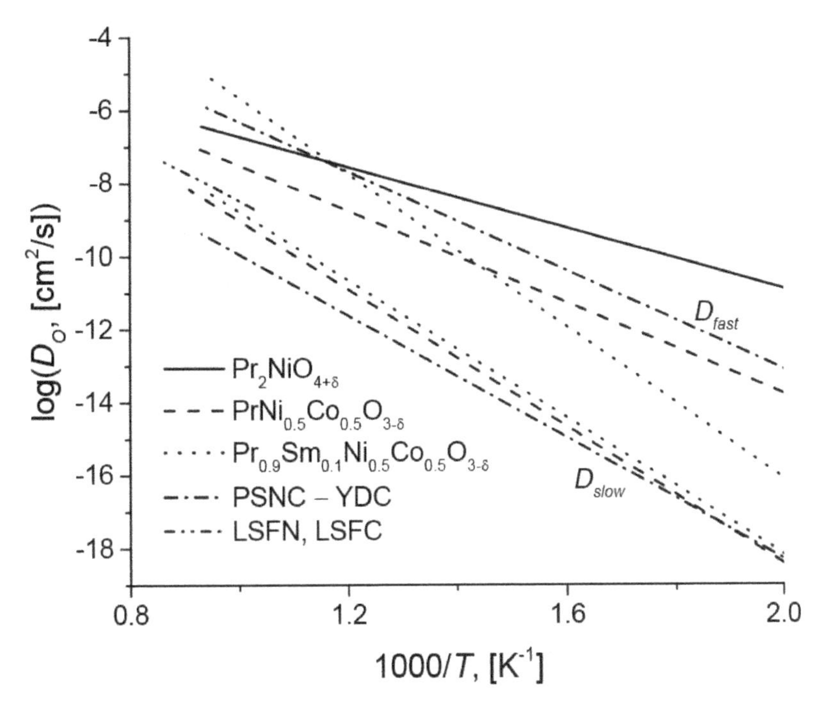

Figure 5. Arrhenius plots of oxygen tracer diffusion coefficients (D_O) for the samples involved in the current work compared to LSFN and LSFC (Tai and Nasrallah 1995).

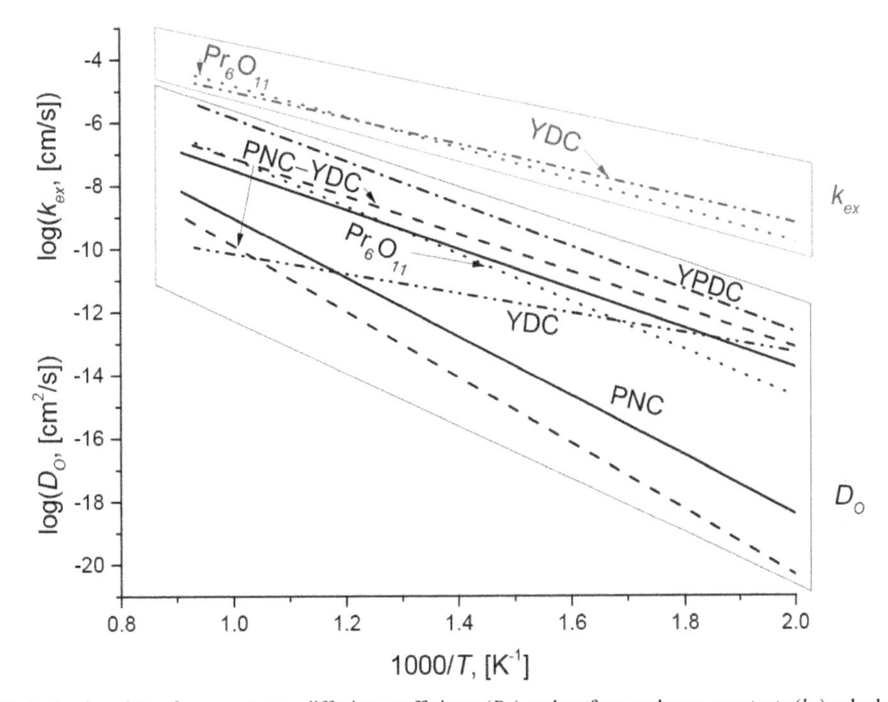

Figure 6. Arrhenius plots of oxygen tracer diffusion coefficients (D_O) and surface exchange constants (k_{ex}) calculated from temperature programmed isotope exchange of oxygen with $C^{18}O_2$ data for $PrNi_{0.5}Co_{0.5}O_{3-\delta}$ (PNC), $PrNi_{0.5}Co_{0.5}O_{3-\delta}$ (50 wt. %) – $Ce_{0.9}Y_{0.1}O_{2-\delta}$ (PNC–YDC) compared to Y-doped ceria (YDC), Y- and Pr-doped ceria (YPDC) and Pr_6O_{11}. Reprinted with permission from Sadykov, V.A., S.N. Pavlova, Z.S. Vinokurov, A.N. Shmakov, N.F. Eremeev, Yu.E. Fedorova et al. 2016c. Physics Procedia 84: 397. Copyright@Sadykov, V.A. et al., open access article under the CC BY-NC-ND license.

Chemical diffusion and surface exchange

Chemical diffusion coefficient (D_{chem}) and chemical exchange constant (k_{chem}) values were estimated for $Pr_{2-x}NiO_{4+\delta}$ and their nanocomposites with Gd-doped ceria using mass (MR), electric conductivity (ECR) and unit cell volume relaxation (UCVR) techniques (Sadykov et al. 2013a, 2014c). According to ECR data, $D_{chem} = 3 \cdot 10^{-4}$ cm²/s and $k_{chem} = 8 \cdot 10^{-4}$ cm/s at 700°C for $Pr_2NiO_{4+\delta}$, which is comparable with the data reported in other works (Boehm et al. 2005). Amongst the PN samples row, according to MR data, the highest D_{chem} and k_{chem} values were observed for $Pr_{1.9}NiO_{4+\delta}$, being $5 \cdot 10^{-4}$ cm²/s and $2 \cdot 10^{-3}$ cm/s at 500°C, respectively, and were 1–1.5 orders of magnitude higher compared to $Pr_{1.8}NiO_{4+\delta}$ and $Pr_{1.7}NiO_{4+\delta}$. D_{chem} and k_{chem} values for $Pr_{1.9}NiO_{4+\delta}$–GDC nanocomposite are a few times lower compared to $Pr_{1.9}NiO_{4+\delta}$, apparently due to cooperative mechanism of oxygen transport hampering (Sadykov et al. 2013a, b). These values for $Pr_{1.9}NiO_{4+\delta}$ exceed the ones for traditional and state-of-the-art cathode materials such as LSM, LSFN и LSFC (Tietz et al. 2006, Murray et al. 1998, Minh 1993, Basu 2007, Steinberger-Wilckens et al. 2009). According to UVCR data, the total effect of the structure relaxation results in the expansion of $Pr_{2-x}NiO_{4+\delta}$ lattice caused by the oxygen removal accompanied by Ni^{3+} and Pr^{4+} cations' reduction to Ni^{2+} and Pr^{3+} state, respectively, with the increase of ionic radii (Sadykov et al. 2014c, Yoo and Lee 2009). The value of $\log(k_{chem}/[\text{cm/s}]) = -5.0 \pm 0.1$ found for $Pr_{1.8}NiO_{4+\delta}$ is close to that obtained by MR technique: $\log(k_{chem}/[\text{cm/s}]) = -5.2 \pm 0.1$ at 400°C. Preliminary estimation of $\log(D_{chem}/[\text{cm}^2/\text{s}])$ gives the value ~ -4.8 ± 0.3 at 470°C, being close to that estimated by MR technique (Sadykov et al. 2014c).

In MR experiments carried out for PNC and their nanocomposites with YDC, relaxation curves are highly distorted with the relaxation itself going sufficiently fast. These factors complicate numeric modeling of the relaxation process. According to UCVR data, PNC samples (both CS and MWS) do not lose oxygen actively, so k_{chem} and D_{chem} values could not be estimated (Sadykov et al. 2016c). k_{chem} calculated for PNC–YDC is $1.2 \cdot 10^{-7}$ cm/s for CS sample and $2.0 \cdot 10^{-5}$ cm/s for MWS one at 600°C (Sadykov et al. 2016c, see Figure 7a). Since the process is limited by the surface exchange, only the lower limit of D_{chem} can be found (e.g., $D_{chem} \geq 8 \cdot 10^{-8}$ cm²/s for MWS sample, see Figure 7a). In comparison with traditional SOFC cathode materials, such values are ~ 2 orders magnitude higher compared to LSM (Belzner et al. 1992) and comparable with LSC (Sadykov et al. 2015b, Kan and Wachsman 2010). These

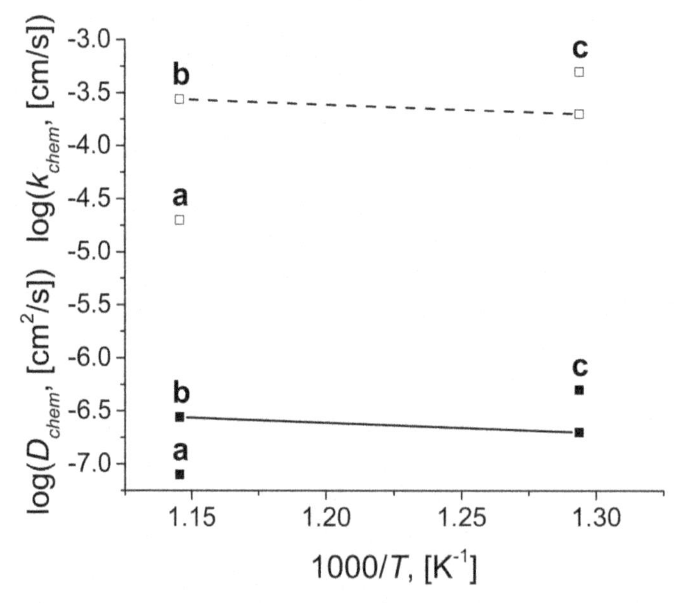

Figure 7. Arrhenius plots of chemical diffusion coefficients (D_{chem}) and exchange constants (k_{chem}) for $PrNi_{0.5}Co_{0.5}O_{3-\delta}$ (50 wt. %)–$Ce_{0.9}Y_{0.1}O_{2-\delta}$ (a), $Ce_{0.65}Pr_{0.25}Y_{0.1}O_{2-\delta}$ (b) and Pr_6O_{11} (c). Reprinted with permission from Sadykov, V.A., S.N. Pavlova, Z.S. Vinokurov, A.N. Shmakov, N.F. Eremeev, Yu.E. Fedorova et al. 2016c. Physics Procedia 84: 397.

data are in agreement with the ones based upon the oxygen isotope exchange showing thermodynamic factor $\gamma = D_{chem}/D_O \approx k_{chem}/k_{ex}$ being ~ 10^2, which is close to usual values for these systems (Lane et al. 1999).

Fast oxygen diffusion channel in PNC shown by TPIE data involving small amount of oxygen can be attributed to admixture of Pr_6O_{11} ($D_{chem} = 1.1 \cdot 10^{-4}$ cm/s and $k_{chem} = 4.7 \cdot 10^{-7}$ cm²/s at 500°C) along with the surface/bulk extended defects. Transport in PNC–YDC nanocomposites is determined by the fluorite phase(s) and interfaces, which is also in agreement with TPIE data showing fast oxygen diffusion channel domination. Like TPIE data showing high k_{ex} and D_O values of the F phases (which are given above), UVCR data demonstrated high D_{chem} and k_{chem} values for YPDC and Pr_6O_{11} (the second Pr_6O_{11}-like fluorite phase was revealed for MWS PNC–YDC nanocomposite). For YPDC, $k_{chem} = 1.6\ 10^{-4}$ cm/s and $D_{chem} \geq 3 \cdot 10^{-7}$ cm²/s at 600°C (see Figures 7b and c).

Cell testing

Working characteristics of single button SOFCs with functionally graded $Pr_{1.9}NiO_{4+\delta}$ | $Pr_{1.9}NiO_{4+\delta}$ – GDC and with simple active $Pr_{1.9}NiO_{4+\delta}$ – GDC sublayer cathodes were similar. At 700°C, the maximum power density value (P_{max}) was 0.4 W/cm² for the first one and 0.5 W/cm² for the second one (Sadykov et al. 2014a). P_{max} for SOFC with functionally graded PNC based cathode was shown to be close to these values, being about 0.4 W/cm². Nevertheless, as shown by comparison, transition from the $PrNi_{0.5}Co_{0.5}O_{3-\delta}$ to PNC–YDC nanocomposite increases P_{max}, which can be explained by the increase in overall oxygen mobility and surface reactivity, according to the studies described above. Comparison with a standard LSCF–based cathode (Forschungszentrum Jülich) (Steinberger-Wilckens et al. 2009) shows that a composite cathode based on praseodymium nickelates-cobaltites even without microstructure optimization allows obtaining a higher power. Moreover, such a cathode allows obtaining close or even higher power density values than those obtained for nanocomposite LSM–ScCeSZ based cathodes with the optimized composition and microstructure (Sadykov et al. 2014a).

For comparison, functionally graded LSM cathode with LSF40–GDC (Fuel Cell, USA) active sublayer was deposited upon the same anode half-cell. The cell obtained showed $P_{max} = 0.5$ W/cm² at 700°C. So, there was no significant difference between cell working characteristics at 700°C, although cells with PN- and PNC-based cathodes shown even better characteristics compared to the ones with cathodes based upon traditional materials at 600°C (see Figure 8).

A small difference in the performance of single button SOFCs can be explained by domination of ohmic losses. Polarization curves are linear showing ohmic loss due to non-uniformity of the anode and not optimized procedure of sintering the cathode. Cross section SEM images revealed non-uniformity in the used anode half-cells after their reduction (Figure 9). Thin net-like YSZ layer deposited on the anode working surface to prevent anode half-cell plate bending under sintering can provide additional active resistance of the anode. Also, linear shape of $U(I)$ curves for all the samples make evidence of ohmic loss. Ohmic voltage loss (overpotential) is known to have linear dependence on the current density with its slope being determined by resistance of cell components (Kopasakis et al. 2008).

In addition, it is to be noted that microstructure of the cathodes was not optimized, so the data obtained are preliminary. Hence, these tests of single button solid oxide fuel cells demonstrate results promising for the practical application of such Sr-free materials stable to carbonization. Further optimization of cathode layer microstructure is expected to increase the power density of fuel cells with these promising materials (Sadykov et al. 2014a).

Membrane testing

Working characteristics of oxygen separation membrane with $PrNi_{0.5}Co_{0.5}O_{3-\delta}$ (50 wt. %)–$Ce_{0.9}Y_{0.1}O_{2-\delta}$ (PNC–YDC) and $MnFeO_3$–$Ce_{0.9}Gd_{0.1}O_{2-\delta}$ (MF–GDC) layers with Pt (1.4 wt. %)–$Sm_{0.15}Pr_{0.15}Ce_{0.35}Zr_{0.3}O_{2-\delta}$ (Pt/SmPrCeZrO) catalytic layer were successfully estimated in processes of methane partial oxidation and oxi-dry reforming of natural gas.

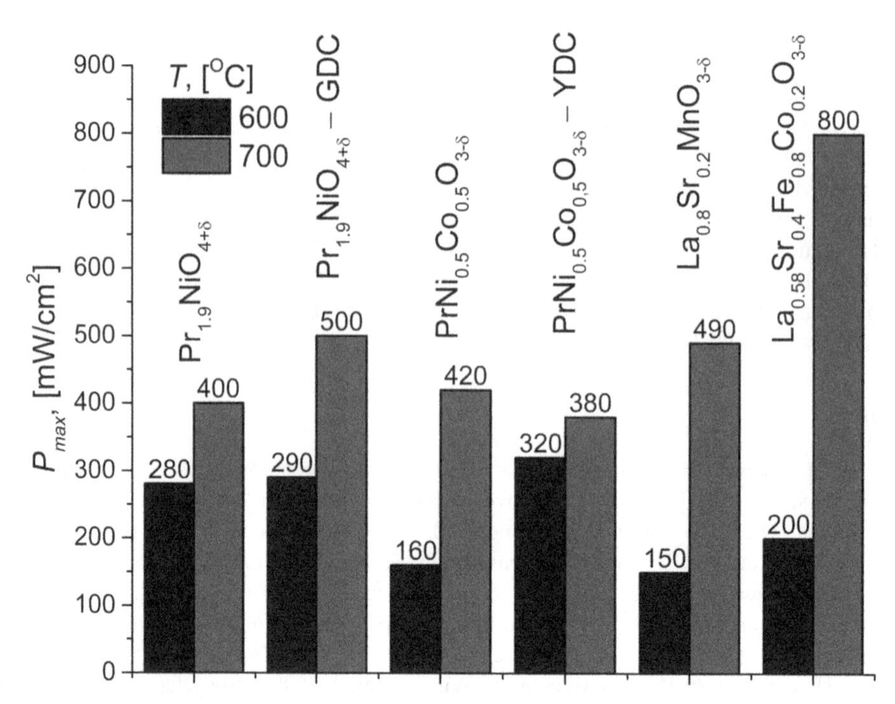

Figure 8. Comparison of maximum power density (P_{max}) of single button SOFCs with the cathode based upon the materials indicated.

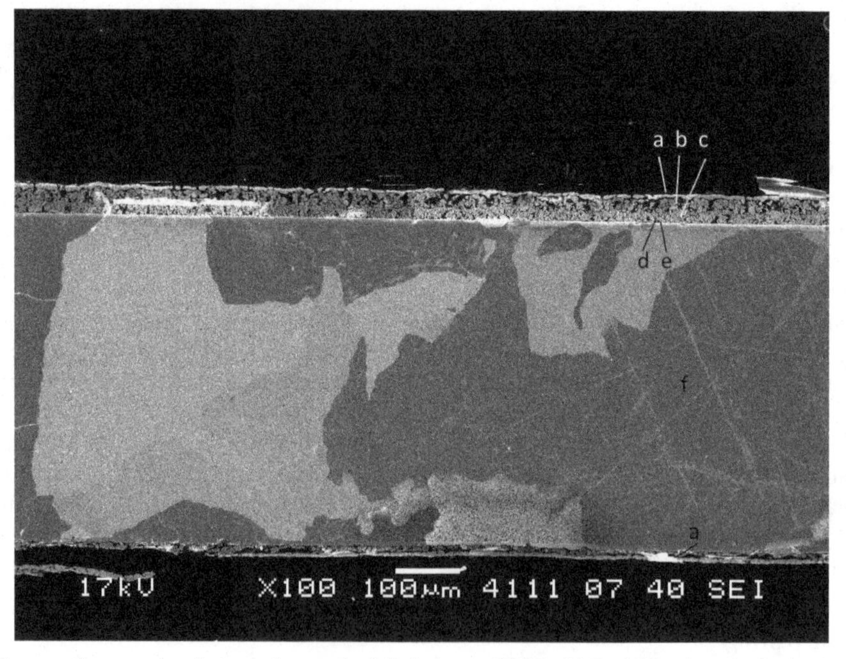

Figure 9. Cross section scanning electron micrograph of single-button SOFC with $Pr_{1.9}NiO_{4+\delta}$-based cathode. (a) Ag current collector, (b) $Pr_{1.9}NiO_{4+\delta}$ cathode, (c) $Pr_{1.9}NiO_{4+\delta}$–$Ce_{0.9}Gd_{0.1}O_{2-\delta}$ sublayer, (d) $Ce_{0.9}Y_{0.1}O_{2-\delta}$ buffer layer, (e) $Zr_{0.84}Y_{0.16}O_{2-\delta}$ electrolyte, (f) Ni/$Zr_{0.84}Y_{0.16}O_{2-\delta}$ anode.

As shown in Figure 10, in the partial oxidation of CH_4 its conversion only slightly depends upon the inlet concentration, while syngas selectivity increases with its increase, while selectivity for combustion products—H_2O and CO_2– decreases. This is explained by decreasing the coverage of the catalytic particles' surface by reactive oxygen species with increasing CH_4 concentration (Shelepova et al. 2016).

At constant feed composition and feed rate, syngas selectivity and methane conversion increase with temperature (Figure 11), which is typical for the process of methane's partial oxidation into syngas.

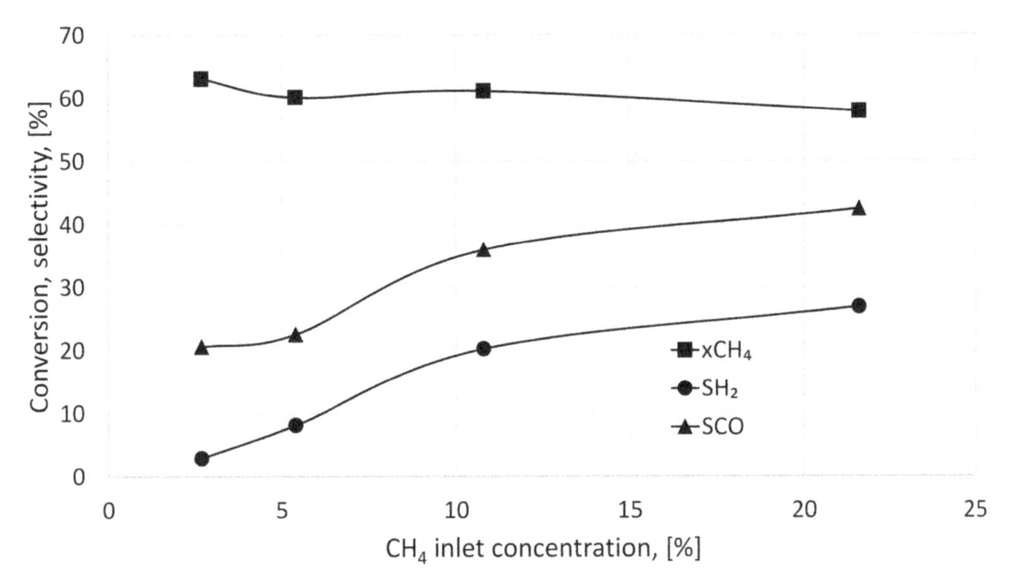

Figure 10. Effect of CH_4 inlet concentration in Ar on degree of CH_4 conversion and products selectivities in membrane reactor. 950°C, feed flow rate 2 L/h, air flow rate 2 L/h. Reprinted with permission from Sadykov, V.A., N.F. Eremeev, Z.S. Vinokurov, A.N. Shmakov, V.V. Kriventsov, A.I. Lukashevich et al. 2017. J. Ceram. Sci. Tech. 08: 129. Copyright@ Göller Verlag.

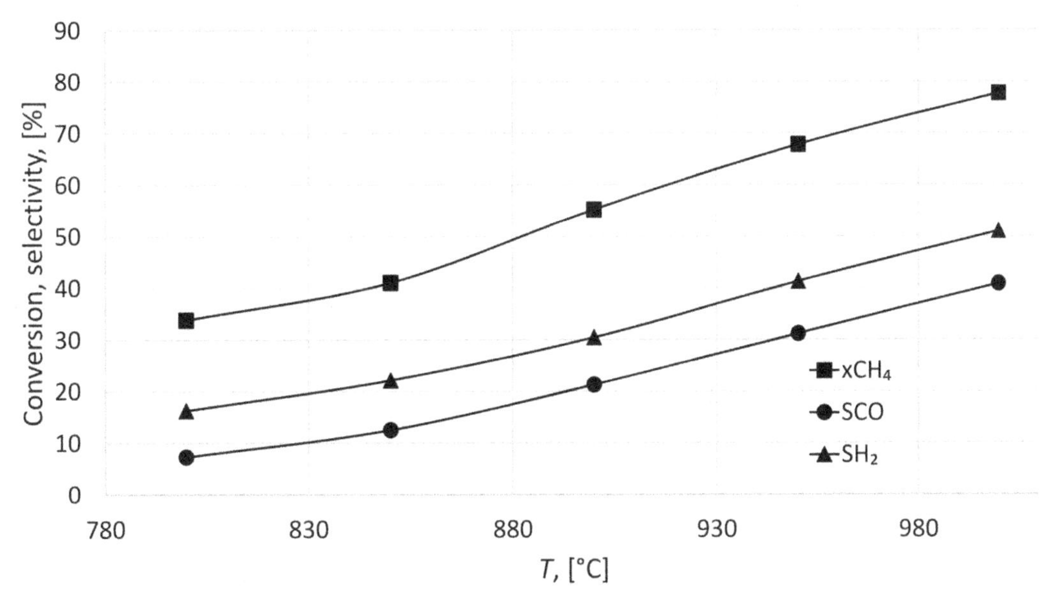

Figure 11. Temperature dependence of CH_4 conversion and products' selectivities in membrane reactor. Feed 21% CH_4 in Ar, flow rate 2 L/h, air flow rate 2 L/h.

The oxygen flux also increases with temperature and inlet CH_4 concentration achieving values up to 2.5 mLO_2/cm^2min at 950°C for 21% CH_4 in feed and up to ~10 mLO_2/cm^2min for feeds with 50–100% CH_4. Hence, the oxygen flux in designed membrane is close to values required for the practical application (Sadykov et al. 2010a).

For natural gas (NG) oxi-dry reforming, CH_4 conversion and syngas yield are higher in oxi-dry reforming than in partial oxidation increasing with temperature (Figure 12) and contact time (Figure 13).

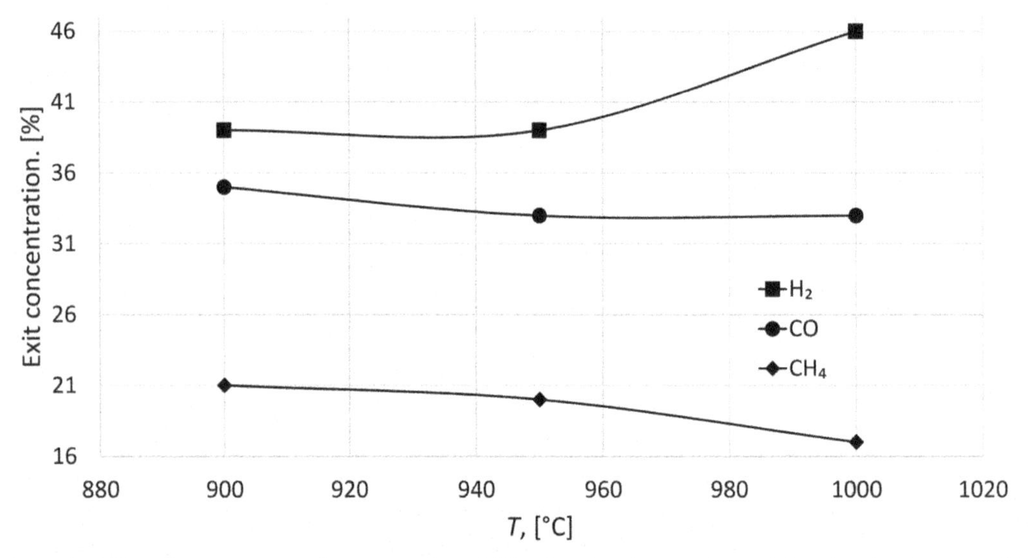

Figure 12. Effect of temperature on exit concentrations of CH_4, H_2 and CO in natural gas oxi-dry reforming in reactor with PNC–YDC | MF–GDC | Pt/SmPrCeZrO membrane. Feed 46% CO_2 + 48% NG + N_2, contact time 0.18. Reprinted with permission from Sadykov, V.A., N.F. Eremeev, Z.S. Vinokurov, A.N. Shmakov, V.V. Kriventsov, A.I. Lukashevich et al. 2017. J. Ceram. Sci. Tech. 08: 129. Copyright@ Göller Verlag.

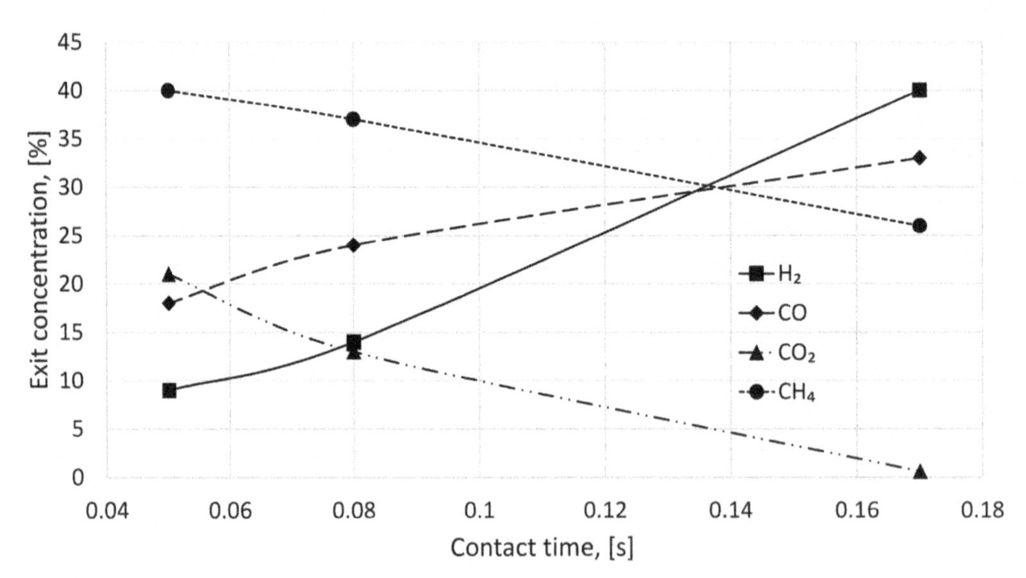

Figure 13. Effect of relative contact time (related to empty space volume ~ 1 mm before membrane surface covered with catalytic layer) on exit concentrations of reagents and products in natural gas oxi-dry reforming in reactor with PNC–YDC | MF–GDC | Pt/SmPrCeZrO membrane at 950°C. Feed 46% CO_2 + 48% NG + N_2. Reprinted with permission from Sadykov, V.A., N.F. Eremeev, Z.S. Vinokurov, A.N. Shmakov, V.V. Kriventsov, A.I. Lukashevich et al. 2017. J. Ceram. Sci. Tech. 08: 129. Copyright@ Göller Verlag.

Since at all temperatures and contact time 0.18 s H_2 concentration is higher than that of CO, this demonstrates that the share of the process of partial oxidation of methane by oxygen transferred through membrane is at least comparable with that of dry reforming. At much shorter contact time (Figure 13), the ratio of CO/H_2 in products is ~ 2. This suggests that at high feed rates, methane conversion and products selectivities are mainly determined by fast reactions of methane dry reforming and reverse water gas shift $CO_2 + H_2 = CO + H_2O$ in the porous catalytic layer, while the oxygen flux through membrane is not sufficiently fast to affect the product composition.

From the practical point of view, H_2/CO ratio in syngas can be tuned by variation of the inlet concentration of CO_2 in the feed even at rather short contact times (Figure 14).

In both reactions, stability of membrane's performance was demonstrated for up to 200 h time-on-stream.

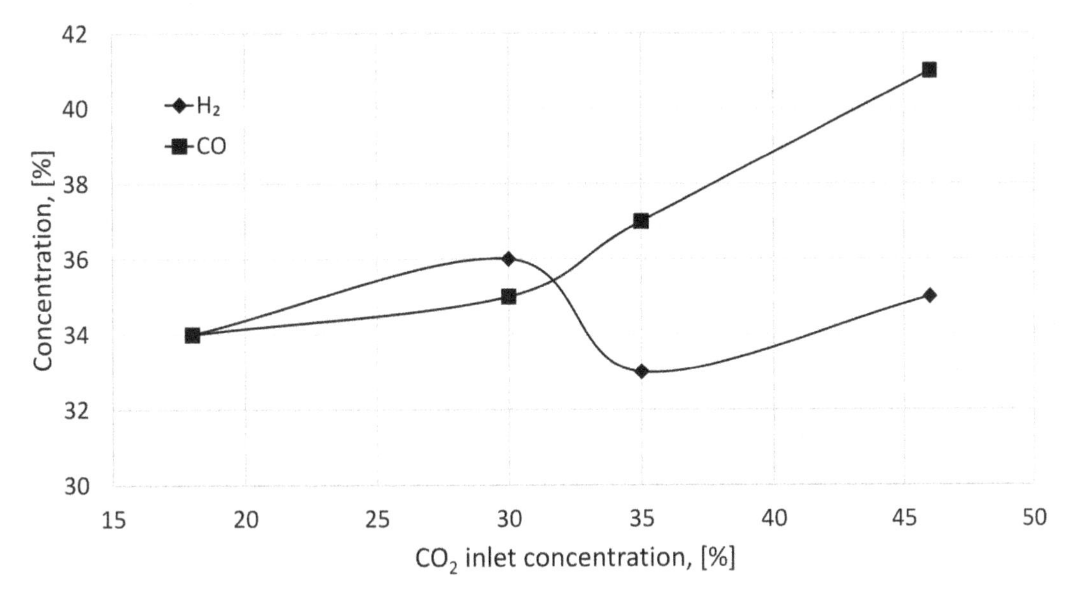

Figure 14. Effect of CO_2 content in the feed x% CO_2 + (99-x)% NG +1% N_2 on exit concentrations of CO and H_2 in natural gas oxi-dry reforming in reactor with PNC–YDC | MF–GDC | Pt/SmPrCeZrO membrane at 950°C and contact time 0.1 s.

Conclusions

Procedures for synthesis and controlled sintering of perovskite-like oxides based on praseodymium nickelates and their nanocomposites with doped ceria were developed and optimized, and their real structure/microstructure and transport properties (conductivity, oxygen mobility) were studied in details. Best systems demonstrate very high lattice oxygen mobility and surface reactivity due to disordering caused by cations' redistribution between nanosized domains of perovskite-like and fluorite-like phases. This allows providing promising performance as materials for intermediate temperature SOFC cathodes and functional layers of oxygen separation membranes along with a high stability to carbonization due to the absence of alkaline-earth dopants.

Different parts of this work are supported by the Russian Science Foundation (Project 16-13-00112), the Ministry of Education and Science of Russia (in frames of Top 5-100 Program), BIOGO FP7 Project, German-Russian Project N_CATH (Contract No. 14.740.12.1357), the Russian Academy of Sciences and Federal Agency of Scientific Organizations (Project V.45.3.8), Complex Program of Siberian Branch of RAS (Project 0303-2015-0001) and Zamaraev Foundation.

References

Ananyev, M.V. 2011. ROSPATENT author registration certificate of "ECRPro" software. Patent # 2,011,614,003.

Atkinson, A., S. Barnett, R. Gorte, J. Irvine, A. Mcevoy, M. Mogensen et al. 2004. Advanced anodes for high-temperature fuel cells. Nature 3: 17–27.

Basu, R.N. 2007. Materials for solid oxide fuel cells. pp. 286–331. *In*: S. Basu (ed.). Recent Trends in Fuel Cell Science and Technology. Anamaya Publishers, New Delhi, India.

Belzner, A., T.M. Giir and R.A. Huggins. 1992. Oxygen chemical diffusion in strontium doped lanthanum manganites. Solid State Ionics 57: 327–337.

Biswas, R.G., A. Hartridge, K.K. Mallick and A.K. Bhattcharaya. 1997. Preparation, structure and electrical conductivity of $Pr_{1-x}La_xO_{2-\delta}$ (x = 0.05, 0.1, 0.2). J. Mater. Sci. Lett. 16: 1089–1091.

Boehm, E., J. Bassat, P. Dordor, F. Mauvy, J. Grenier and P. Stevens. 2005. Oxygen diffusion and transport properties in non-stoichiometric $Ln_{2-x}NiO_{4+\delta}$ oxides. Solid State Ionics 176: 2717–2725.

Borchert, H., Y.V. Frolova, V.V. Kaichev, I.P. Prosvirin, G.M. Alikina, A.I. Lukashevich et al. 2005. Electronic and chemical properties of nanostructured cerium dioxide doped with praseodymium. J. Phys. Chem. B 109.12: 5728–5738.

Duval, S.B.C., P. Holtappels, J.P. Ouweltjes and B. Rietveld. 2010. Evaluation of the perovskite $(La_{0.8}Sr_{0.2})_{0.95}Fe_{0.8}Ni_{0.2}O_{3-\delta}$ as SOFC cathode. pp. 1047–1056. *In:* P. Connor (ed.). Proceedings of 9th European Solid Oxide Fuel Cell Forum. Chapter 10. Lucerne, Switzerland.

Fan, L., M. Chen, C. Wang and B. Zhu. 2012. Pr_2NiO_4–Ag composite cathode for low temperature solid oxide fuel cells with ceria-carbonate composite electrolyte. Int. J. Hydrogen Energy 37: 19388–19394.

Ferchaud, C., J.C. Grenier, Y. Zhang-Steenwinkel, M.M.A. van Tuel, F.P.F. van Berkel and J.M. Bassat. 2011. High performance praseodymium nickelate oxide cathode for low temperature solid oxide fuel cell. J. Power Sources 196: 1872–1879.

Hjalmarsson, P. and M. Mogensen. 2011. $La_{0.99}Co_{0.4}Ni_{0.6}O_{3-\delta}$–$Ce_{0.8}Gd_{0.2}O_{1.95}$ as composite cathode for solid oxide fuel cells. J. Power Sources 196: 7237–7244.

Huang, S., Q. Lu, S. Feng, G. Li and C. Wang. 2012. $PrNi_{0.6}Co_{0.4}O_3$–$Ce_{0.8}Sm_{0.2}O_{1.9}$ composite cathodes for intermediate temperature solid oxide fuel cells. J. Power Sources 199: 150–154.

Irvine, J.T.S., D. Neagu, M.C. Verbraeken, C. Chatzichristodoulou, C. Graves and M.B. Mogensen. 2016. Evolution of the electrochemical interface in high-temperature fuel cells and electrolysers. Nature Energy 1: 1–13.

Janney, M.A., C.L. Calhoun and H.D. Kimrey. 1992. Microwave sintering of solid oxide fuel cell materials: I. Zirconia—8 mol% yttria. J. Am. Ceram. Soc. 75: 341–346.

Kan, C.C. and E.D. Wachsman. 2010. Isotopic-switching analysis of oxygen reduction in solid oxide fuel cell cathode materials. Solid State Ionics 181: 338–347.

Kopasakis, G., T. Brinson and S. Credle. 2008. A theoretical solid oxide fuel cell model for system controls and stability design. J. Fuel Cell Sci. Tech. 5: 041007.

Kostogloudis, G. Ch. and Ch. Ftikos. 1998. Structural, thermal and electrical properties of $Pr_{0.5}Sr_{0.5}Ni_{1-y}Co_yO_{3-\delta}$ perovskite-type oxides. Solid State Ionics 109: 43–53.

Kovalevsky, A.V., V.V. Kharton, A.A. Yaremchenko, Y.V. Pivak, E.V. Tsipis, S.O. Yakovlev et al. 2007a. Oxygen permeability, stability and electrochemical behavior of $Pr_2NiO_{4+\delta}$-based materials. J. Electroceram. 18: 205–218.

Kovalevsky, A.V., V.V. Kharton, A.A. Yaremchenko, Y.V. Pivak, E.N. Naumovich and J.R. Frade. 2007b. Stability and oxygen transport properties of $Pr_2NiO_{4+\delta}$ ceramics. J. Eur. Ceram. Soc. 27: 4269–4272.

Kuznetsova, T.G., V.A. Sadykov, V.A. Matyshak, L.Ch. Batuev and V.A. Rogov. 2005. Catalysts based on complex oxides with fluorite and perovskite structures for soot removal from the exhaust gas and diesel engines. Chem. Sustain. Dev. 13: 779–785.

Lane, J.A., S.J. Benson, D. Waller and J.A. Kilner. 1999. Oxygen transport in $La_{0.6}Sr_{0.4}Co_{0.2}Fe_{0.8}O_{3-\delta}$. Solid State Ionics 121: 201–208.

Li, X. and N.A. Benedek. 2015. Enhancement of ionic transport in complex oxides through soft lattice modes and epitaxial strain. Chem. Mater. 27: 2647–2652.

Minervini, L., R.W. Grimes, J.A. Kilner and K.E. Sickafus. 2000. Oxygen migration in $La_2NiO_{4+\delta}$. J. Mater. Chem. 10: 2349–2354.

Minh, N.Q. 1993. Ceramic fuel cells. J. Amer. Chem. Soc. 76: 563–588.

Murray, E.P., T. Tsai and S.A. Barnett. 1998. Oxygen transfer processes in $(La,Sr)MnO_3/Y_2O_3$-stabilized ZrO_2 cathodes: an impedance spectroscopy study. Solid State Ionics 110: 235–243.

Nishimoto, Sh., S. Takahashi, Y. Kameshima, M. Matsuda and M. Miyake. 2011. Properties of $La_{2-x}Pr_xNiO_4$ cathode for intermediate-temperature solid oxide fuel cells. J. Ceram. Soc. Jpn. 11: 246–250.

Ormerod, M.R. 2003. Solid oxide fuel cells. Chem. Soc. Rev. 32: 17–28.

Pikalova, E.Yu. and A.A. Kolchugin. 2016. Influence of the substituting element (M = Ca, Sr, Ba) in $La_{1.7}M_{0.3}NiO_{4+\delta}$ on the electrochemical performance of the composite electrodes. Euras. Chem. Technol. J. 18: 3–11.

Sadovskaya, E., V. Goncharov, Yu.K. Gulyaeva, G.Y. Popova and T.V. Andrushkevich. 2010. Kinetics of the $H_2{}^{18}O/H_2{}^{16}O$ isotope exchange over vanadia–titania catalyst. J. Mol. Catal. A: Chem. 316: 118–125.

Sadykov, V., N. Mezentseva, G. Alikina, A. Lukashevich, V. Muzykantov, R. Bunina et al. 2007. Doped nanocrystalline Pt-promoted ceria-zirconia as anode catalysts for IT SOFC: Synthesis and properties. Mat. Res. Soc. Symp. Proc. 1023: 1023-JJ02-07.

Sadykov, V.A., S.N. Pavlova, T.S. Kharlamova, V.S. Muzykantov, N.F. Uvarov, Yu.S. Okhlupin et al. 2010a. Perovskites and their nanocomposites with fluorite-like oxides as materials for solid oxide fuel cells cathodes and oxygen-conducting membranes: Mobility and reactivity of the surface/bulk oxygen as a key factor of their performance. pp. 67–178. *In*: M. Borovski (ed.). Perovskites: Structure, Properties and Uses. Nova Science Publishers, New York, USA.

Sadykov, V., V. Zarubina, S. Pavlova, T. Krieger, G. Alikina, A. Lukashevich et al. 2010b. Design of asymmetric multilayer membranes based on mixed ionic–electronic conducting composites supported on Ni–Al foam substrate. Catal. Today 156: 173–180.

Sadykov, V., N. Mezentseva, V. Usoltsev, E. Sadovskaya, A. Ishchenko, S. Pavlova et al. 2010c. SOFC composite cathodes based on perovskite and fluorite structures. pp. 104–1016. *In*: P. Connor (ed.). Proceedings of 9th European Solid Oxide Fuel Cell Forum. Chapter 10. Lucerne, Switzerland.

Sadykov, V., N. Mezentseva, V. Usoltsev, E. Sadovskaya, A. Ishchenko, S. Pavlova et al. 2011a. Solid oxide fuel cell composite cathodes based on perovskite and fluorite structures. J. Power Sources 196: 7104–7109.

Sadykov, V., N. Mezentseva, G. Alikina, R. Bunina, V. Pelipenko, A. Lukashevich et al. 2011b. Nanocomposite catalysts for steam reforming of methane and biofuels: design and performance. pp. 909–947. *In*: B. Reddy (ed.). Advances in Nanocomposites—Synthesis, Characterization and Industrial Applications. InTech, New York, USA.

Sadykov, V., V. Usoltsev, Yu. Fedorova, N. Mezentseva, T. Krieger, N. Eremeev et al. 2012. Advanced sintering techniques in design of planar IT SOFC and supported oxygen separation membranes. pp. 121–140. *In*: A. Lakshmanan (ed.). Sintering of Ceramics—New Emerging Techniques. InTech., New York, USA.

Sadykov, V., V. Usoltsev, N. Yeremeev, N. Mezentseva, V. Pelipenko, T. Krieger et al. 2013a. Functional nanoceramics for intermediate temperature solid oxide fuel cells and oxygen separation membranes. J. Eur. Ceram. Soc. 33: 2241–2250.

Sadykov, V.A., N.F. Eremeev, V.V. Usol'tsev, A.S. Bobin, G.M. Alikina, V.V. Pelipenko et al. 2013b. Mechanism of oxygen transfer in layered lanthanide nickelates $Ln_{2-x}NiO_{4+\delta}$ (Ln = La, Pr) and their nanocomposites with $Ce_{0.9}Gd_{0.1}O_{2-\delta}$ and $Y_2(Ti_{0.8}Zr_{0.2})_{1.6}Mn_{0.4}O_{7-\delta}$ solid electrolytes. Russ. J. Electrochem. 49: 645–651.

Sadykov, V.A., N.F. Eremeev, E.M. Sadovskaya, A.S. Bobin, Y.E. Fedorova, V.S. Muzykantov et al. 2014a. Cathodic materials for intermediate-temperature solid oxide fuel cells based on praseodymium nickelates-cobaltites. Russ. J. Electrochem. 50: 669–679.

Sadykov, V.A., N.F. Eremeev, G.M. Alikina, E.M. Sadovskaya, V.S. Muzykantov, V.V. Pelipenko et al. 2014b. Oxygen mobility and surface reactivity of $PrN_{1-x}Co_xO_{3+\delta}$–$Ce_{0.9}Y_{0.1}O_{2-\delta}$ cathode nanocomposites. Solid State Ionics 262: 707–712.

Sadykov, V., Y. Okhlupin, N. Yeremeev, Z. Vinokurov, A. Shmakov, V. Belyaev et al. 2014c. *In situ* X-ray diffraction studies of $Pr_{2-x}NiO_{4+\delta}$ crystal structure relaxation caused by oxygen loss. Solid State Ionics 262: 918–922.

Sadykov, V., N. Eremeev, E. Sadovskaya, A. Bobin, A. Ishchenko, V. Pelipenko et al. 2015a. Oxygen mobility and surface reactivity of $PrNi_{1-x}Co_xO_{3-\delta}$ perovskites and their nanocomposites with $Ce_{0.9}Y_{0.1}O_{2-\delta}$ by temperature-programmed isotope exchange experiments. Solid State Ionics 273: 35–40.

Sadykov, V.A., V.S. Muzykantov, N.F. Yeremeev, V.V. Pelipenko, E.M. Sadovskaya, A.S. Bobin et al. 2015b. Solid oxide fuel cell cathodes: Importance of chemical composition and morphology. Catal. Sustain. Energy 2: 57–70.

Sadykov, V.A., E.M. Sadovskaya and N.F. Uvarov. 2015c. Methods of isotopic relaxations for estimation of oxygen diffusion coefficients in solid electrolytes and materials with mixed ionic-electronic conductivity. Russ. J. Electrochem. 51: 529–539.

Sadykov, V.A., N.F. Eremeev, V.A. Bolotov, Yu.Yu. Tanashov, Yu.E. Fedorova, D.G. Amanbayeva et al. 2016a. Oxygen mobility in microwave sintered praseodymium nickelates-cobaltites and their nanocomposites with yttria-doped ceria. Solid State Ionics 288: 76–81.

Sadykov, V.A., N.F. Eremeev, D.G. Amanbayeva, T.A. Krieger, Yu.E. Fedorova, A.S. Bobin et al. 2016b. Sm-doped praseodymium nickelates-cobaltites and their nanocomposites with Y-doped ceria as promising cathode materials. Integrated Ferroelectrics 173: 71–81.

Sadykov, V.A., S.N. Pavlova, Z.S. Vinokurov, A.N. Shmakov, N.F. Eremeev, Yu.E. Fedorova et al. 2016c. Application of SR methods for the study of nanocomposite materials for Hydrogen Energy. Phys. Proc. 84: 397–406.

Sadykov, V.A., N.F. Eremeev, Z.S. Vinokurov, A.N. Shmakov, V.V. Kriventsov, A.I. Lukashevich et al. 2017. Structural studies of Pr nickelate-cobaltite – Y-doped ceria nanocomposite. J. Ceram. Sci. Tech. 08: 129–140.

Sadykov, V.A., E.M. Sadovskaya, E.Yu. Pikalova, A.A. Kolchugin, E.A. Filonova, S.M. Pikalov et al. 2018. Transport features in layered nickelates: correlation between structure, oxygen diffusion, electrical and electrochemical properties. Ionics 24: 1181–1193.

Sammes, N. and Y. Du. 2005. Intermediate-temperature SOFC electrolytes. pp. 19–34. *In*: N. Sammes, A. Smirnova and O. Vasylyev (eds.). Full Cell Technologies: State and Perspectives. Springer, Netherlands, Dordrecht.

Shelepova, E., A. Vedyagin, V. Sadykov, N. Mezentseva, Y. Fedorova, O. Smorygo et al. 2016. Theoretical and experimental study of methane partial oxidation to syngas in catalytic membrane reactor with asymmetric oxygen-permeable membrane. Catal. Today 268: 103–110.

Shuk, P. and M. Greenblatt. 1999. Hydrothermal synthesis and properties of mixed conductors based on $Ce_{1-x}Pr_xO_{2-\delta}$ solid solutions. Solid State Ionics 116: 217–223.

Sinev, M.Yu., G.W. Graham, L.P. Haack and M. Shelef. 1996. Kinetic and structural studies of oxygen availability of the mixed oxides $Pr_{1-x}M_xO_y$ (M = Ce, Zr). J. Mater. Res. 11.08: 1960–1971.

Steinberger-Wilckens, R., L. Blum, H.P. Buchkremer, L.G.J. de Haart, M. Pap, R.W. Steinbrech et al. 2009. Recent results in solid oxide fuel cell development at Forschungszentrum Juelich. ECS Trans. 25: 213–220.

Tai, L.W., M.M. Nasrallah, H.U. Anderson, D.M. Sparlin and S.R. Sehlin. 1995. Structure and electrical properties of $La_{1-x}Sr_xCo_{1-y}Fe_yO_3$. 1. The system $La_{0.8}Sr_{0.2}Co_{1-y}Fe_yO_3$. Solid State Ionics 76: 259–271.

Tietz, F., I. Arul Raj, M. Zahid and D. Stöver. 2006. Electrical conductivity and thermal expansion of $La_{0.8}Sr_{0.2}$(Mn,Fe,Co)O_{3-y} perovskites. Solid State Ionics 177: 1753–1756.

Yoo, H.-I. and C.-E. Lee. 2009. Conductivity relaxation patterns of mixed conductor oxides under a chemical potential gradient. Solid State Ionics 180: 326–337.

12

Recent Advances in Polymer Nanocomposite Coatings for Corrosion Protection

Subramanyam Kasisomayajula, Niteen Jadhav* and *Victoria Johnston Gelling*

Introduction

One of the primary properties of protective coatings in providing adequate anti-corrosion performance is their barrier protection to metal substrate under corrosive environment. Polymer Nanocomposite Coatings (PNCs) have exhibited improved barrier performance due to the presence of nanomaterials in them because of higher aspect ratio and larger surface area to volume ratio of nanomaterials (Makhlouf et al. 2016). In addition to barrier protection, other types of protection mechanisms such as cathodic and anodic protections are also included into the coatings in order to extend their lifespan. Cathodic and anodic protection mechanisms involve electrochemical reactions. Generally, corrosion is a combination of electrochemical reactions between metal substrate and corrosive species such as O_2 and water. PNCs containing especially electrochemically active materials have been found to interact with metals and corrosive species electrochemically to resist corrosion processes at the metal-coating interface. Electrochemically active materials such as conductive polymers are capable of conducting electron transfer (electrical conductivity) and participating in electrochemical reactions (Deshpande et al. 2014, Federica De Riccardis et al. 2014). Several researchers have reported the application of PNCs consisting of electrochemically active materials for corrosion protection. Even though considerable degree of improvement in corrosion protection performance has been noticed with the presence of PNCs, there is still ambiguity in understanding the value of various PNCs in terms of their electrochemical properties. In this chapter, it is intended to help the readers to obtain up-to-date knowledge on PNCs for corrosion protection as well as to be able to distinguish the performance level and benefits of different PNCs based on their electrochemical behavior.

North Dakota State University, Department of Coatings and Polymeric Materials, Fargo, ND, USA, 58108.
Email: venkats.kss@gmail.com
* Corresponding author

This chapter starts with the basics of corrosion and electrochemical techniques such as potentiodynamic polarization (PP) and electrochemical impedance spectroscopy (EIS) that are most commonly used to evaluate corrosion protection. Subsequent sections will discuss the classification of PNCs based on the electrochemical properties. This will help to correlate the concepts with the literature findings discussed in the following sections. Finally, this chapter will conclude with a summary of the research findings relevant to a few applications for aerospace, automotive, marine, and infrastructure fields.

Basics of corrosion and electrochemical techniques

Corrosion is an electrochemical process that involves electrons transfer via chemical reactions. As shown in Figure 1, it is a combination of two half-cell reactions, namely oxidation and reduction reactions. At the anodic site, metal undergoes oxidation by losing electrons and converts into oxides or hydroxides. At the cathodic site, any species that is electrochemically more positive than metal can consume those electrons to undergo reduction. The most common reduction reactions at the cathodic site are oxygen reduction and hydrogen evolution (McCafferty et al. 2010). The following three conditions are necessary for corrosion to occur:

1. An anodic reaction

$$M(s) \longrightarrow M^{n+}(aq) + ne^-$$

2. A cathodic reaction

$$O_2(g) + 2H_2O\,(aq) + 4e^- \longrightarrow 4OH^-\,(aq)$$

and/or

$$2H^+\,(aq) + 2e^- \longrightarrow H_2\,(g)$$

3. A conductive path between anodic site and cathodic site (In atmospheric corrosion, an electrolyte with dissolved oxygen and ions serves as a conductive path).

In addition to oxygen and hydrogen ion reduction reactions, the reduction of a metal cation can also occur when two electrochemically different metals are in contact with each other in the presence of an electrolyte solution. The cations of electrochemically more positive metal are reduced while other metal gets oxidized to convert into cations. This is called galvanic corrosion. The electrochemically more positive metal acts as cathode and the corroding metal becomes anode. Table 1 shows electrochemical series of selected half-cell reactions with corresponding standard potentials (vs. standard hydrogen electrode (SHE)). As it can be seen in the series, iron is more electrochemically positive than aluminum. Thus, aluminum will corrode preferentially when it is in contact with iron.

An electrolyte is an aqueous solution that contains many different polar (e.g., H_2O), non-polar, and ionic species. The presence of ionic species is the source for electrolyte to conduct electricity, whereas in a metal substrate, electrons are delocalized over a large number of molecular orbitals of closely packed metal atoms. These electrons can move freely from one atom to another due to overlap between metal orbitals. As these electrons travel through metal substrate, metal atoms exist in the form of positive ions capable of creating positive charge/negative at the surface depending on whether electrons are withdrawn or supplied to the metal (McCafferty et al. 2010).

A phenomenon called electrical double layer occurs at the metal-electrolyte interface as a result of the interactions between ions present in both metal and electrolyte. While Figure 2(a) depicts the basic model, Figure 2(b) is the complex model of electrical double layer, the Bockris-Devanathan-Muller (BDM) model, which is the modified version of previous models, the Stern and the Gouy-Champan models. According to the BDM model, if a positive charge develops on metal surface, it will be neutralized by the combination of negatively charged anions and water molecules with oxygen side (negative side

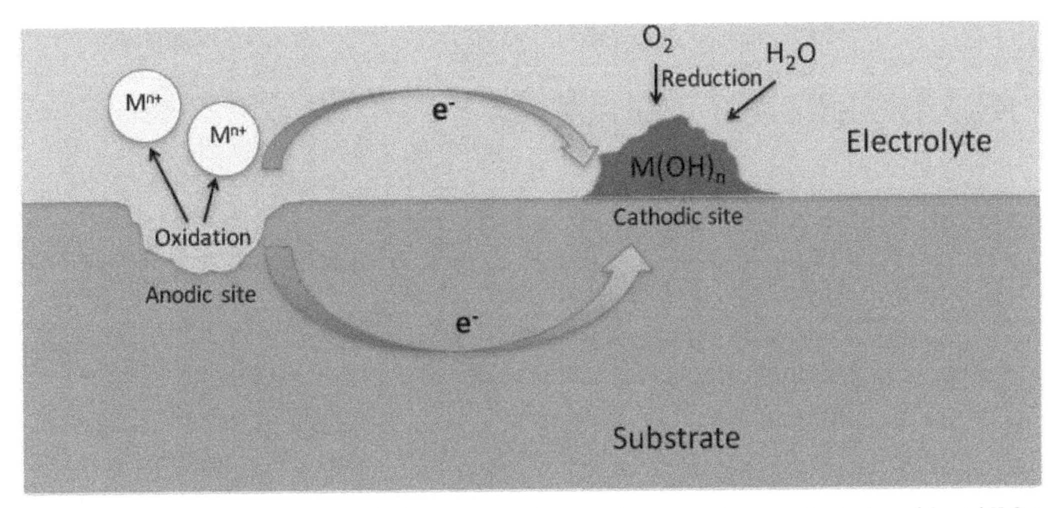

Figure 1. Corrosion process involving oxidation of metal substrate (M) at the anodic site and reduction of O_2 and H_2O at the cathodic site.

Table 1. Electrochemical series of selected half-cell reactions.

Half-cell reaction	E vs. SHE (V)
$Al^{3+} + 3e^- \longrightarrow Al(s)$	-1.662
$2H_2O\ (l) + 3e^- \longrightarrow H_2\ (g) + 2OH^-\ (aq)$	-0.827
$Zn^{2+} + 2e^- \longrightarrow Zn\ (s)$	-0.762
$Fe^{2+} + 2e^- \longrightarrow Fe\ (s)$	-0.447
$Fe^{3+} + 3e^- \longrightarrow Fe\ (s)$	-0.037
$Cu^{2+} + 2e^- \longrightarrow Cu\ (s)$	+0.342
$O_2 + 2H_2O + 4e^- \longrightarrow 4OH^-\ (aq)$	+0.401

of dipole) facing toward metal surface. The layer containing adsorbed negatively charged ions in simple model (Figure 2(a)) or the layer containing both adsorbed negatively charged ions and adsorbed water molecules in the BDM model (Figure 2(b)) is known as Stern layer or Inner Helmholtz plane. Since this layer is formed right at the interface as a result of attraction from positive ions of metal substrate, it creates a very high field of strength up to 1×10^7 V/cm. In the subsequent layer, negatively charged ions adsorbed to metal surface in Stern layer are counterbalanced by opposite ions and these counter-ions are further counterbalanced by their opposite ions in electrolyte solution. This second layer continues to exist in the solution and diminishes gradually with increase in distance away from surface of substrate. This layer is called the Gouy-Chapman diffuse layer.

Formation of electrical double layer causes a potential difference at the interface. This phenomenon controls the rates of oxidation and reduction reactions because the movement of metal ions and corrosive species is controlled by this potential barrier at the interface. The potential barrier of electrical double layer is generally represented by a circuit model using the combination of a capacitor and a resistor as shown in Figure 2(c). The capacitance of capacitor part of electrical double layer is called double layer capacitance (Cdl) and the resistance part is called charge transfer resistance (R_{tr}) or polarization resistance (R_p). Higher the value of R_{tr} or R_p, lower is the corrosion rate and therefore, better is the corrosion resistance of metal substrate at the metal-electrolyte interface (McCafferty et al. 2010).

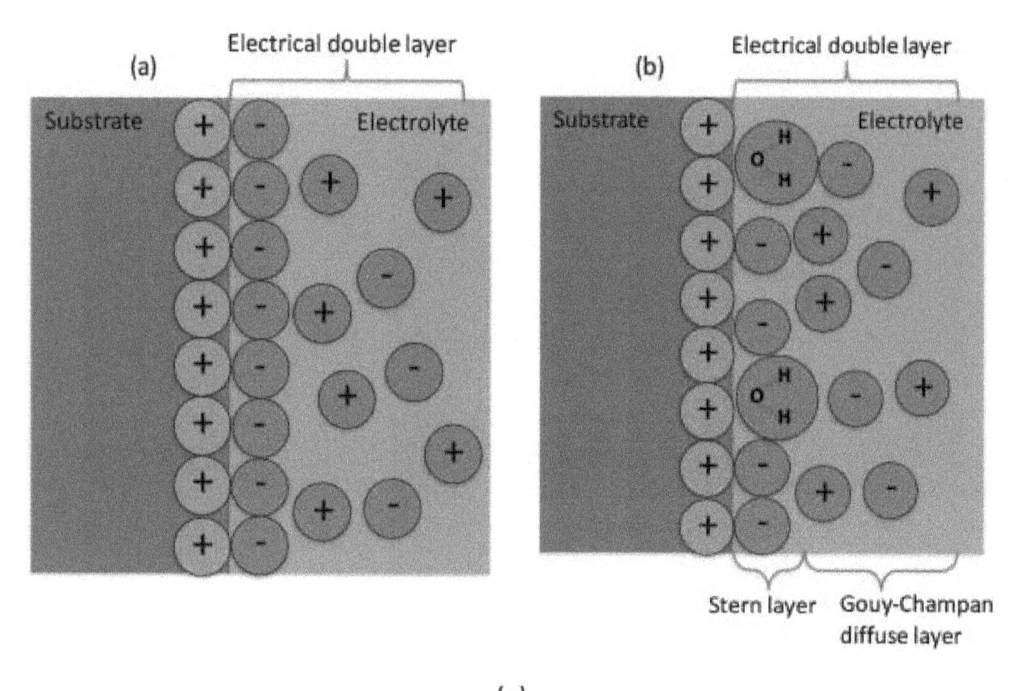

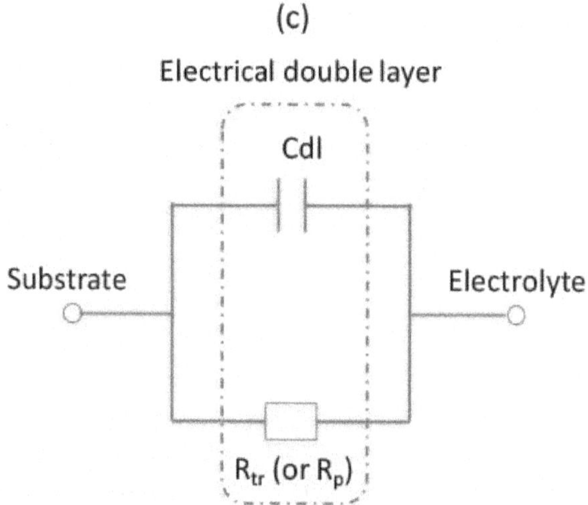

Figure 2. (a) Simple model of electrical double layer, (b) the Bockris-Devanathan-Muller (BDM) model, (c) representative circuit model.

The corrosion rate can be determined by many different methods including weight loss/gain method, chemical analysis of electrolyte solution, thickness measurement via profilometry, and electrochemical methods. In this chapter, the electrochemical methods, particularly potentiodynamic polarization (PP) and electrochemical impedance spectroscopy (EIS), will be discussed as these methods have been most commonly used to investigate PNCs for corrosion protection applications.

Potentiodynamic polarization (PP)

As previously discussed in Figure 1, corrosion of metal substrate involves two reactions: an anodic reaction in which metal is oxidized (oxidation) and a cathodic reaction in which corrosive species is reduced (reduction). The corrosion rate is determined by the kinetics of anodic and cathodic reactions. The kinetics of these reactions measures the rates at which electrons are released and absorbed in the process. They depend on various factors including pH of electrolyte solution/environment, the concentration of corrosive species available in the environment, and potential difference at the metal-electrolyte interface.

Behavior of a metal substrate under steady corrosive environment (keeping concentration of corrosive species constant) can be plotted in relation to the changes in pH and potential difference. These plots are called Pourbaix diagrams. As shown in Figure 3(a), Pourbaix diagram of a metal shows regions of corrosion, immunity, and passivity. However, Pourbaix diagrams do not provide any information on corrosion kinetics. In order to obtain the information about corrosion kinetics, polarization curves are

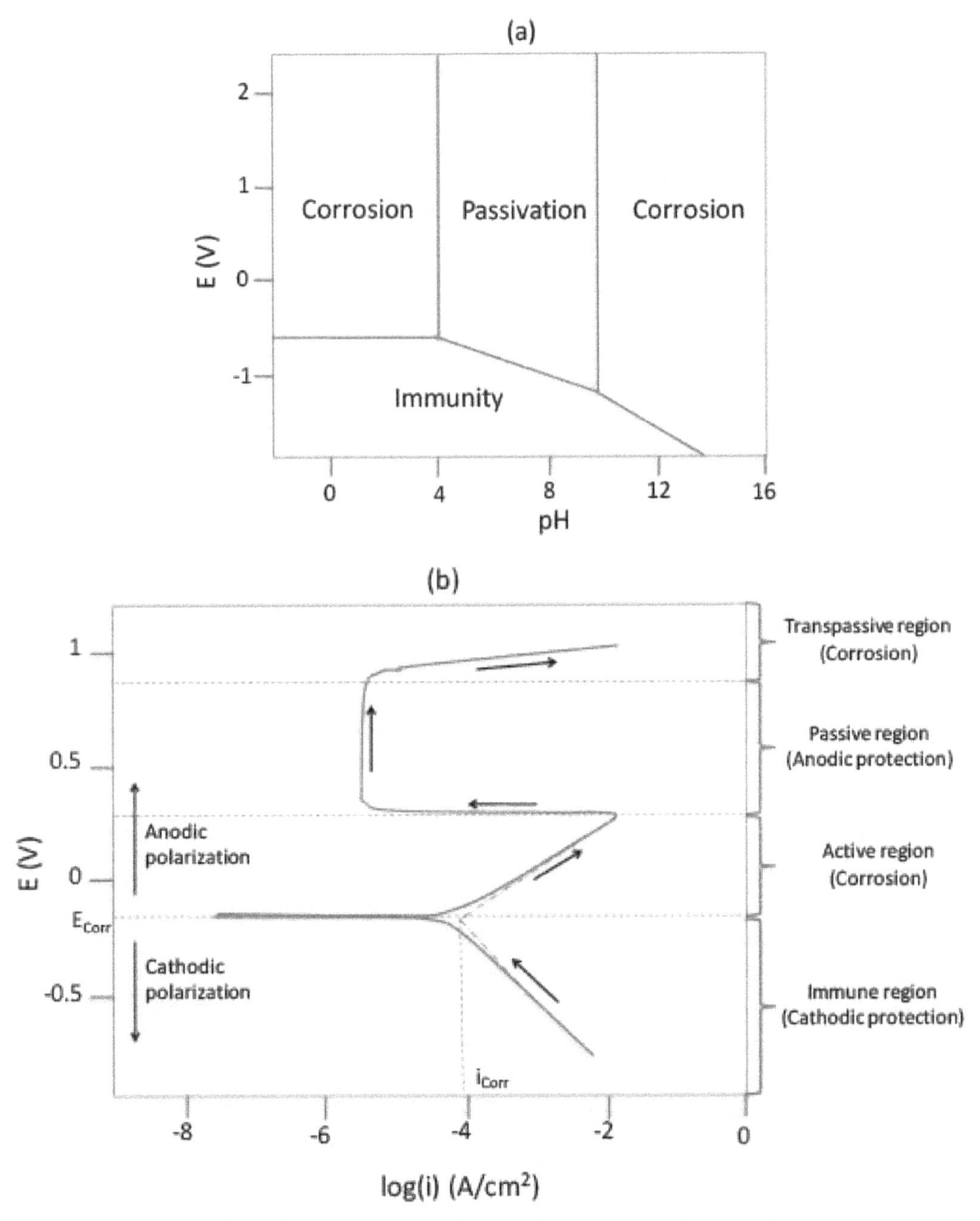

Figure 3. (a) Pourbaix diagram (potential vs. pH), (b) Potentiodynamic polarization curve.

acquired by plotting the relationship between potential and current density measurements (McCafferty et al. 2010).

PP is one of the direct current (DC) polarization methods in which the potential of a working electrode is scanned over a potential range with respect to a standard electrode (e.g., SHE) at a constant scan rate and current through electrolyte solution is measured using a potentiostat/galvanostat (Kelly et al. 2003). The PP curves can be obtained in many different ways depending on the type of corrosion characteristics to be studied. Generally, PP curves provide the information on corrosion potential, corrosion current density, and corrosion resistance. Corrosion potential (E_{corr}) (also known as mixed potential or rest potential) is the potential at which metal corrodes at natural corrosion rate when no polarization (external voltage difference) is applied. At this potential, the system is in equilibrium at which electrons are released at constant rate from anodic reactions and consumed at constant rate by cathodic reactions. Both anodic current (i_a) and cathodic current (i_c) at this potential are equal to current density, which is also known as corrosion current density or corrosion rate (i_{corr}). It is measured using Tafel extrapolation (not discussed in this chapter) as shown in Figure 3(b). Typically, the potential scan starts at about 250 mV negative of open circuit potential (or corrosion potential), sweeps through zero, and ends at 250 mV positive of open circuit potential. However, in order to determine some of the other characteristics such as passivation and pitting potential, scanning is continued toward very high positive potential values. Sometimes, anodic polarization part of the curve does not appear as straight as shown in Figure 3(b) due to the effect of passive layer formation or changes in surface profile. In such cases, the corrosion rate is determined from the intersection of Tafel extrapolation of cathodic part and open circuit potential line.

At any point on cathodic line in the polarization curve, metal substrate is cathodically protected as current density in this region is mainly due to supply of electrons from the potentiostat. As the applied voltage is increased from the initial potential on cathodic part towards the positive direction (called forward scan or anodic polarization), current density will decrease as the number of electrons supplied from potentiostat decreases. The region just above the corrosion potential is the active region where electrons are withdrawn from metal substrate. Further increase in potential leads to passivation of metal indicating the formation of a protective oxide layer, which will reduce the corrosion current density. At breakaway potential, passive layer breaks down mainly due to pitting corrosion giving rise to higher current density. Above this potential (transpassive region), passive oxide layer can no longer protect metal surface. The reduction of oxygen occurs with the absorption of electrons released from the corrosion of metal surface (Kelly et al. 2003).

Electrochemical impedance spectroscopy (EIS)

Electrochemical Impedance Spectroscopy (EIS) is one of the most useful electrochemical techniques that utilize alternating current (AC) to study corrosion processes by measuring corrosion related parameters. This method involves the application of a small perturbation (~ 10–20 mV) to a system of interest and measuring response, usually current, of the system with respect to this perturbation. This technique has been widely used to investigate various systems containing surface-pretreatments, oxide films, metal-electrolyte interfaces, and protective coatings. One of the major advantages of EIS over other electrochemical techniques is that EIS is considered a non-destructive method since only a small voltage is applied to the system of interest (Loveday et al. 2004).

In EIS, the applied voltage is represented by a complex form that comprises of amplitude (E_0) and angular frequency (ω). The amplitude of this voltage is the maximum magnitude and the frequency represents the rate at which the magnitude of voltage changes with time. When this voltage is applied to a system, the response of the system measured in current will also be in complex form. However, there will be a time difference between the application of voltage and the response received. This time difference is measured in terms of phase angle depending on the responsive properties of the system. The phase angle is 0° when voltage and current are in the same phase. This is the characteristic of a resistor.

The phase angle is –90° when voltage and current are in opposite phases. This is the characteristic of a capacitor. In addition, the ability to resist the passage of current is called the resistance (for direct current) or the impedance (for AC current) (Su et al. 2013, Loveday et al. 2004). The most common methods used to present EIS data are: (1) the Bode plots, in which the magnitude of impedance and the phase angle are plotted as a function of angular frequency, (2) the Nyquist plot, which depicts the relationship between the imaginary part (on y-axis) and the real part (on x-axis) of impedance. Since both impedance and frequency are measured over several orders of magnitude, they are plotted in logarithmic scale (e.g., log(|Z|) vs log (ω)) in the Bode plot.

Although the Nyquist plot does not explicitly provide the information about frequency, it allows determining likely mechanisms involved in distinct processes, for example the interactions at the metal-electrolyte interface and diffusion of various ionic species through a coating or an electrolyte solution. The EIS results are fitted with circuit models to calculate parameters including electrolyte resistance (R_S), coating resistance (also called pore resistance (R_{pore})), coating capacitance (C_C), double layer capacitance (C_{dl}), and charge transfer resistance (R_{tr}) (Loveday et al. 2005).

Figure 4 shows the typical Bode plots and Nyquist plots and their corresponding equivalent circuit models of some of the common phenomena including metal-electrolyte interface, an intact organic coating, coating with low pore resistance and a damaged coating. Figure 4(a) depicts the representative EIS plots and model of a metal-electrolyte interface. As previously discussed, a metal surface, in the absence of any protective coating, forms an electrical double layer at the interface with the electrolyte solution. This phenomenon can also be observed in a damaged coating as shown in Figure 4(d). The electrolyte travels across the damaged areas of the coating and reaches the metal surface. The degradation process of a protective coating occurs through the stages starting from an intact coating (Figure 4(b)) through the development of pores (Figure 4(c)), and eventually loss of adhesion due to penetration of electrolyte solution reaching the interface (Figure 4(d)) (Loveday et al. 2005).

Introduction to polymer nanocomposite coatings (PNCs)

This has already been discussed in the beginning of the chapter. PNCs are generally considered one of the fields in the broad area of composites where one or more of filler materials with nanometer dimensions are dispersed in a polymeric matrix. Most commonly investigated nanomaterials are nanoparticles, nanotubes, nanorods, nanofibers, nanowires, and nanolayered sheets. Materials at nanoscale have shown to exhibit distinct physical and chemical properties compared to their micro and macro scale counterparts, apparently due to large surface area to volume ratio of nanomaterials. For a given volume, the total surface area of all the particles of a material increases with the reduction in size of each particle. When these particles are dispersed as filler material in a polymer matrix, molecular interactions such as hydrogen bonding, eletrostatic and vander wall forces between the particles and the matrix significantly increase as the particle size decreases from micro to nanoscale. PNCs have been used in various forms depending on the end-use applications such as coatings, composites, and adhesives (Hung et al. 2011).

Recent review articles have discussed different fabrication approaches, microstructure of PNCs and the characterization for corrosion protection. Various factors such as size, shape, dispersion, and orientation of nanofillers embedded within the polymer matrix have shown to significantly affect the mechanical, physical, and chemical properties of PNCs. In addition, the interfacial interactions and bonding between polymer matrix and nanofillers incorporated within the matrix also play an important role in defining the properties of PNCs. Furthermore, PNCs consisting of same composition but prepared through different fabrication methods are found to have distinct microstructures leading to differences in properties (Hintze-Bruening et al. 2012).

In literature, the classification of PNCs has often been mentioned in terms of microstructure of nanocomposites and the structural characteristics of nanomaterials. However, the classification of PNCs in terms of electrochemical properties, especially electronic conductivity and redox phenomena, can

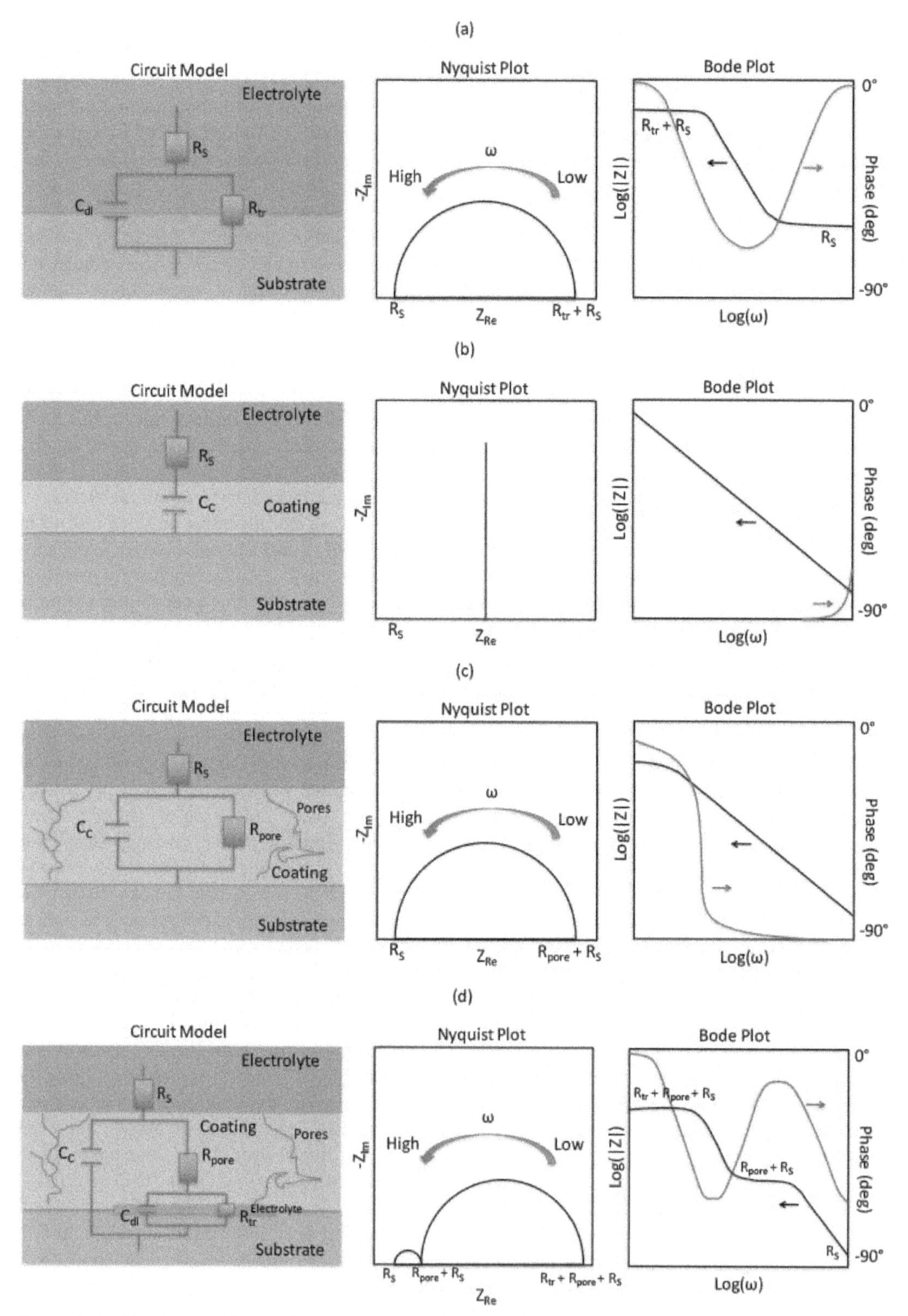

Figure 4. (a) Circuit model, Nyquist plot, and Bode plot of substrate-electrolyte interface, (b) Circuit model, Nyquist plot, and Bode plot of intact coating, (c) Circuit model, Nyquist plot, and Bode plot of coating with low pore resistance, (d) Circuit model, Nyquist plot, and Bode plot of damaged coating.

be beneficial as corrosion is an electrochemical process and the electrochemical interactions of PNCs with metal and corrosive species can alter the corrosion process. For instance, electrochemically active materials such as conductive polymers, when compared to non-electrochemically active materials, exhibit distinct interactions with metal substrate and corrosive species in providing corrosion protection to the substrate.

As shown in Figure 5, both polymer matrix and nanofiller are divided into two different groups: conductive and non-conductive. Depending on the group of nanofiller incorporated into the group of polymer matrix, four categories of PNCs can be fabricated as follows: (1) conductive nanofiller incorporated into conductive polymer matrix, (2) non-conductive nanofiller incorporated into conductive polymer matrix, (3) conductive nanofiller incorporated into non-conductive polymer matrix, and (4) non-conductive nanofiller incorporated into non-conductive polymer matrix.

Several PNCs have been evaluated as anti-corrosion coatings for metals and alloys used in the aerospace, automotive, marine, infrastructure and many other industries. This chapter focuses on recent advances in PNCs for corrosion protection of various metal substrates most commonly used in the above-mentioned industries. In each section, the use of each category of PNCs will be discussed with case studies as examples. Table 2 summarizes the typical roles of conductive and non-conductive polymer matrices, and conductive and non-conductive nanofillers used in PNCs for corrosion protection.

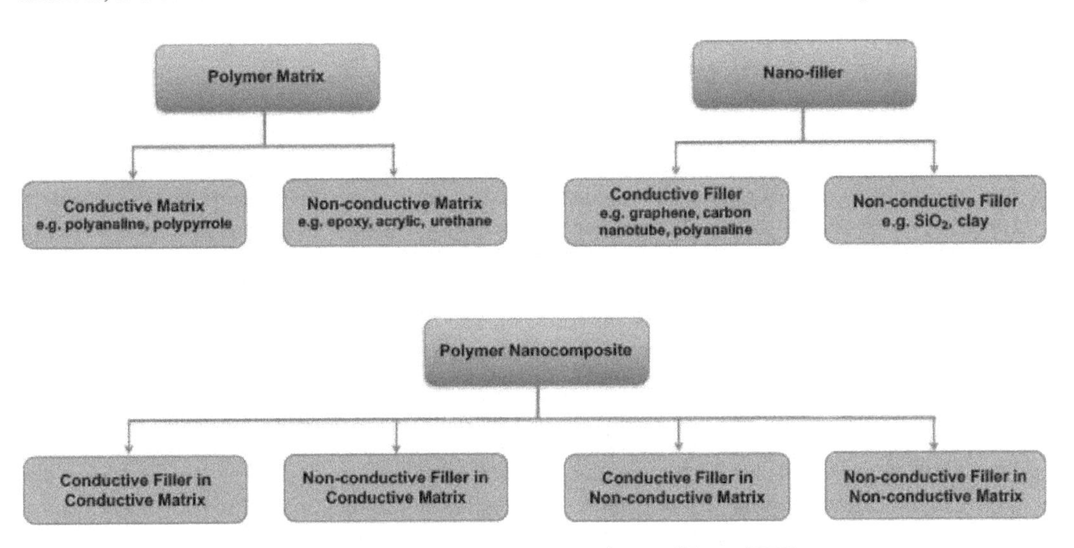

Figure 5. Classification of polymer matrix, nanofiller, and PNCs.

Table 2. Typical roles of polymer matrix (conductive and non-conductive) and nanofiller (conductive and non-conductive) in providing corrosion protection to the metal substrate.

Component	Typical Role
Conductive polymer matrix	• Oxidizes metal substrate to form a strongly adhered protective passive layer • Provides anodic protection to metal substrate • Inhibits oxygen by reducing it at the coating-air interface • Transfers and discharges the charge at the coating-substrate interface
Non-conductive polymer matrix	• Provides barrier protection • Improves adhesion
Conductive nanofiller	• Reduces permeability to corrosive species • Increases hydrophobicity, e.g., graphene • Increases connectivity between polymer chains, especially in the case of conductive polymer matrix
Non-conductive nanofiller	• Improves barrier protection of polymer matrix • Self-healing/passive layer formation

Recent advances in PNCs for corrosion protection

Aerospace

Corrosion has always been one of the most common failure modes for major causalities in the aerospace industry. Various metals and alloys used for the fabrication of different parts in the aircraft construction are susceptible to corrosion attacks when they are exposed to severe environmental conditions. When corrosion develops in the weaker areas of structural components, it results in failure of structural integrity. In order to prevent from catastrophic failures, aircraft components are manufactured through proper selection of materials such as metals, alloys, and composites based on their intrinsic properties to resist fatigue and corrosion. In addition, they are further protected by the application of pre-treatments and corrosion protective coatings to increase the service-life of an aircraft.

Majority of aircraft structural parts are composed of aluminum alloys such as AA2024 and AA7075 because of their high strength to weight ratio. Basically, these alloys are prepared by adding certain elements to aluminum and processing through specific heat treatments. While AA2024 is alloyed with copper (4–5%), magnesium (1–1.5%), and manganese (0.5–0.6%), AA7075 is alloyed with zinc (5–6%), magnesium (2–2.5%), and copper (1–1.5%). The presence of these alloying elements in small quantities can result in significant improvement in the mechanical properties of aluminum alloy (600 MPa Yield Strength) compared to pure aluminum (90 MPa Yield Strength). Due to their excellent fatigue resistance and ultimate tensile strength, they can withstand large magnitudes of stresses and are therefore considered the most suitable materials for aerospace applications. However, the addition of other elements such as copper and zinc reduces the corrosion resistance of alloys, due to the fact that these elements are more electrochemically positive than aluminum. These metals can form corrosion cells with aluminum by serving as cathodic sites in corrosive environments leading to corrosion of aluminum. The development of corrosion in the aircraft structural components leads to the reduction in fatigue resistance suffering sudden and cataclysmic events (Musa et al. 2007, Montemor et al. 2014).

Hexavalent chromium-containing pre-treatments and coatings have been heavily used in the aerospace industry as they provide excellent corrosion resistance to aluminum alloys. However, due to their carcinogenic nature, there is an enormous pressure on the aerospace and coatings industries by the environmental and health organizations such as the Occupational Safety and Health Administration (OSHA) to find a replacement for these hexavalent chromium-containing materials (Spinks et al. 2002).

Although there has been significant development in achieving comparable performance to that of hexavalent chromium materials, the confidence on the alternatives to replace hexavalent chromium materials in the industry is still very low due to other technical challenges such as ease of application and costs associated with the alternative technologies. A proper replacement should have the capability to perform similar to chromate coatings in adhesion, chemical resistance, mechanical properties, ease of application combined with curing under most of the conditions, compatibility with several topcoats and pre-treatments, and last but not the least, long term corrosion protection. When considering corrosion protection alone, the replacement coating should possess excellent barrier properties to prevent diffusion of corrosive species such as H_2O, O_2, and various ions, and be able to inhibit corrosion process in the damaged areas. There is always a huge demand for innovative coatings in the aerospace industry that are not only environment friendly but also can exhibit improved efficiency (Spinks et al. 2002).

A typical aircraft protective coating system shown in Figure 6 contains 3 layers: (1) surface pre-treatment, (2) primer coat, and (3) top-coat. Generally, a chromate conversion coating such as Alodine 1200S with a thickness less than 2 μm is applied as surface pre-treatment on the surface of aluminum alloy to increase adhesion between primer coat and substrate. A primer coat is usually an epoxy-polyamide coating that may contain corrosion inhibitors such as strontium chromate in the form of pigments. In addition to providing corrosion protection to metal substrate, the primer coat is also intended to help to maintain adhesion between the surface pre-treatment and topcoat. Even though the key purpose of a topcoat, generally a polyurethane coating, is to bestow an aesthetic appearance to the overall coating system, it is sometimes designed to add protective functionality as well.

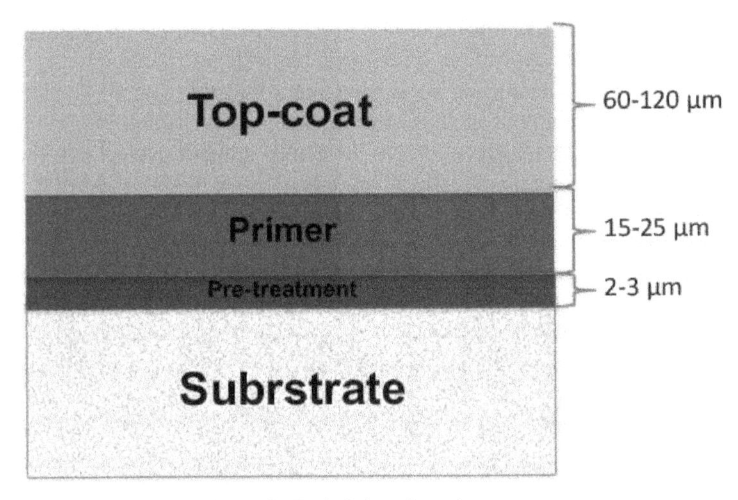

Figure 6. Typical aircraft coating system.

Among many materials that have been investigated as potential replacement of chromates for the corrosion protection of aluminum alloys AA2024 and AA7075, PNCs have exhibited promising performance. They have been typically evaluated for pre-treatments or primers. Among four categories of PNCs mentioned earlier, it has been noticed that the application of PNCs prepared with non-conductive polymer matrix has been widely studied for the corrosion protection of aluminum alloys AA2024 and AA7075. While only a very few research articles were found on the use of PNCs prepared with conductive matrix and non-conductive nano-filler, there was apparently no report found on the use of PNCs with conductive nanofillers incorporated into conductive matrix for the corrosion protection of these two alloys.

(a) PNCs containing non-conductive nanofillers in conductive matrix

For the past few years, several research reports have been published on the use of conductive polymer coatings such as polyaniline (PANI) and polypyrrole (PPY) in providing corrosion protection to Al2024-T3 alloy with partial to considerable success. Subsequently, researchers have begun investigating the effect of nanofillers in conducting polymer matrix on barrier and self-healing properties of PNCs. Generally, conducting polymer coatings are found to exhibit high porosity due to their inadequate mechanical properties and the challenges associated with processing and application. In several cases, the incorporation of nanofillers into conducting polymer coatings has led to the reduction in porosity and significant improvement in mechanical properties.

In Zubillaga's research work, PANI containing either TiO_2 or ZrO_2 nanoparticles was synthesized on anodized aluminum alloy AA2024-T3 via electrochemical polymerization in the presence of oxalic acid (Zubillaga et al. 2009). Using PP evaluation, the performance of PANI with and without nanoparticles was compared to the performance of the alloy anodized in chromic acid (A2-CAA). In this work, PANI-TiO_2 coated anodized alloy (A2-PA-T1) showed significant decrease in passive current density to 10^{-5}–10^{-6} A/cm^2; however, it was still two orders magnitude higher than that of A2-CAA (Figure 7). The improvement in corrosion performance of PANI-TiO_2 coating, compared to pure PANI coating (A2-PA-0) and PANI-ZrO_2 coating (A2-PA-Z1), was attributed to considerable increase in barrier protection of PANI with the incorporation of TiO_2 nanoparticles.

Mert et al. also reported similar corrosion current densities in the case of TiO_2 doped PPY coating on Al 7075–T6 (Mert et al. 2015). The author mentioned that the improvement in corrosion resistance was due to the formation of p-n junction between PPY (p-type semiconductor) and TiO_2 (n-type semiconductor) leading to the decrease in charge transfer. Even though the barrier properties of the nanocomposite were considerably low ($5 \times 10^3 \Omega$-cm^2), the impedance was observed to increase with the immersion time from 2 hours to 168 hours as a result of deposition of corrosion products into pores.

In a recent work of Singh-Beemat et al. it was found that the composite of PANI-PPY could significantly perform better than any of the individual coatings of PANI and PPY (Singh-Beemat et al. 2012). Based on this finding, Kartsonakis et al. has developed a hybrid nanocomposite coating system consisting of two layers in which the bottom layer was a copolymer of PANI and PPY and the top layer was a sol-gel coating for the corrosion protection of aluminum alloy Al2024-T3 (Kartsonakis et al. 2012). The copolymer of PANI (PANI) and polypyrrole (PPY) was filled with CeO_2 nanocontainers containing a corrosion inhibitor, 2-mercaptobenzothiazole (MBT). In their work, the effect of each component in the hybrid system on corrosion performance was studied using EIS and potentiodynamic polarization (PP). The corrosion tests were performed in 0.05 M NaCl solution for 72 hours. As presented in Figure 8, the PANI-PPY composite coating (Combo) was able to slightly reduce corrosion current as well as corrosion potential as compared to that of bare Al2024-T3. According to the PP characterization, it appeared that the sol-gel coating (Coat) could make the system nobler and further reduce corrosion current. However, the combination of Combo and Coat could not considerably improve the performance as compared to that of sol-gel coating (Coat). Nevertheless, with the incorporation of nanocontainers in the PANI-PPY composite, the performance was significantly improved with respect to all the coatings tested.

As per the EIS results, the coatings Coat-Combo-CeO_2-MBT and Coat exhibited higher impedance at low frequency range (0.1 Hz) after 72 hours of immersion. However, the coating Coat suffered sudden drop at higher frequencies while the coating Coat-Combo-CeO_2-MBT maintained its protective ability at higher frequencies indicating its enhanced barrier properties. This enhancement was explained by the

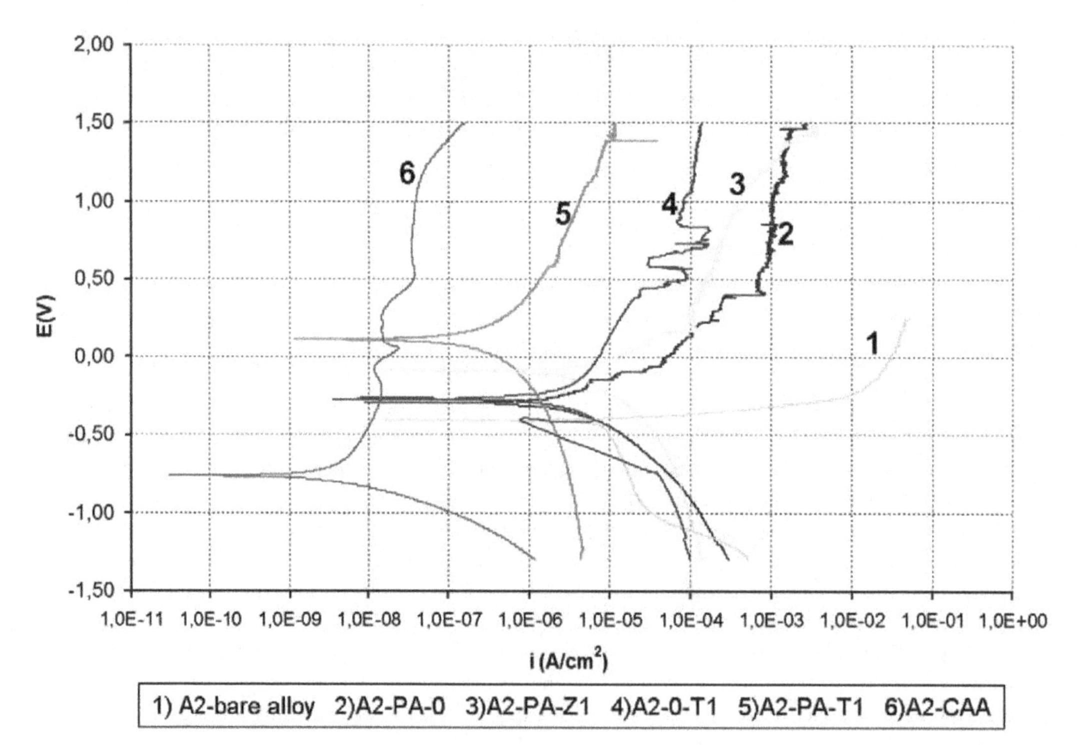

Figure 7. Potentiodynamic polarization curves recorded in 5 mM NaCl and 0.1 M Na_2SO_4 solution for the AA2024T3 alloy with films A2-PA-0, A2-PA-T1, A2-PA-Z1, A2-CAA (chromic acid anodised AA2024T3 alloy) and bare alloy. (Reprinted with permission from Zubillaga, O., F.J. Cano, I. Azkarate, I.S. Molchan, G.E. Thompson and P. Skeldon. 2009. Anodic films containing polyaniline and nanoparticles for corrosion protection of AA2024T3 aluminium alloy. Surface and Coatings Technology 203(10-11): 1494–1501. Copyright@Elsevier.)

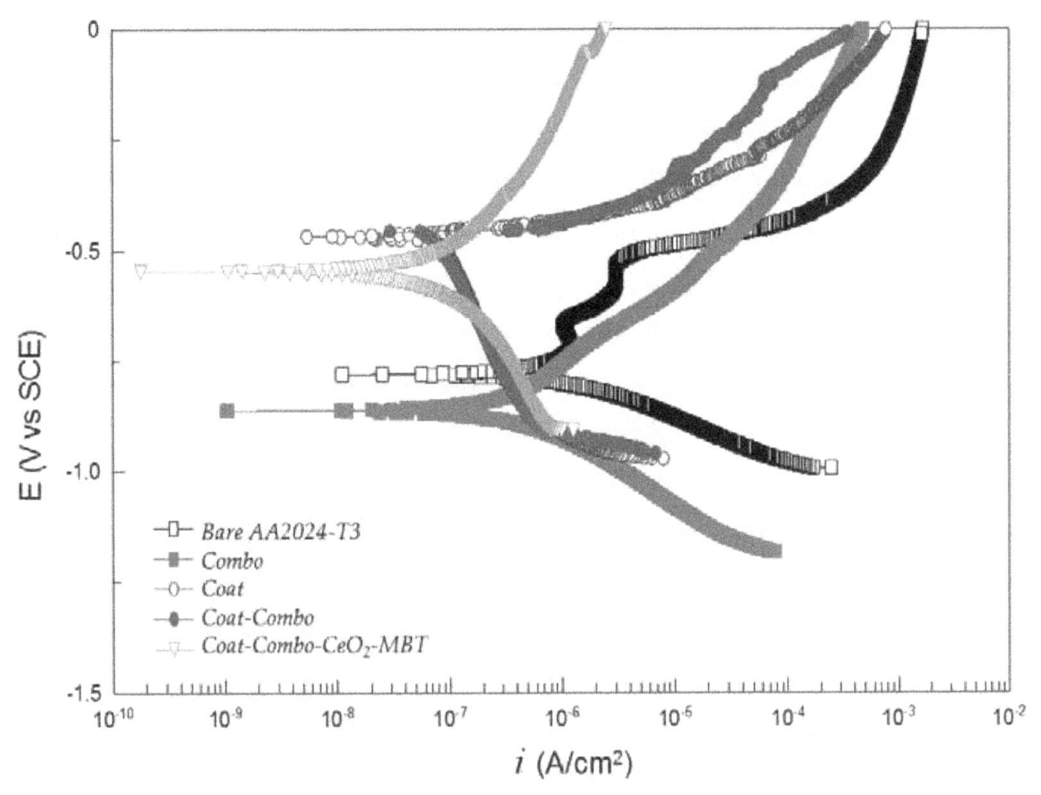

Figure 8. Potentiodynamic polarization curves for bare AA2024-T3, Combo (a copolymer of PANI-PPY electrodeposited), Coat (a sol-gel coating via dip coating process), Coat-Combo (two layers-a copolymer of PANI-PPY and a sol-gel coating), Coat-Combo-CeO_2-MBT (two layers-a copolymer of PANI-PPY containing CeO_2 nanocontainers loaded with MBT and a sol-gel coating after exposure to 0.05 M NaCl solution at room temperature for 72 h. (Reprinted with permission from Kartsonakis, I.A., E.P. Koumoulos, A.C. Balaskas, G.S. Pappas, C.A. Charitidis and G.C. Kordas. 2012. Hybrid organic–inorganic multilayer coatings including nanocontainers for corrosion protection of metal alloys. Corrosion Science 57: 56–66. Copyright@ Elsevier.)

reduction in crack development with the addition of nanocontainers in the coating, demonstrating the ability of nanofiller to exhibit both barrier and self-healing phenomenon.

(b) PNCs containing conductive nanofillers in non-conductive matrix

It has been actively debated whether the use of conductive fillers such as graphene and carbon nanotubes (CNTs) in PNCs is beneficial for the corrosion protection of metals and alloys as they can serve as electron transferring mediators in propagating localized galvanic corrosion.

Monetta et al. prepared a hybrid coating on AA2024-T3 by incorporating graphene flakes (1 wt%) into a water based epoxy resin (Monetta et al. 2015). The EIS results indicated that the incorporation of graphene flakes into the epoxy coating significantly improved the barrier properties of the coating. Due to hydrophobic nature of graphene flakes, water absorption into the epoxy coating was minimized. Generally, the capacitance of a coating increases with the absorption of water. The trends in the capacitance of epoxy coating with and without graphene flakes, as estimated from EIS analysis, were compared. Lower capacitance of coating containing graphene flakes was correlated with the improvement in corrosion resistance and the reduction in blistering. From the EIS analysis in this work, it can be understood that the incorporation of graphene flakes at lower loadings (1 wt%) improves the hydrophobicity of the coating but does not cause increase in charge transfer through the coating.

In the research work of Gkikas et al. the effect of varying concentrations (0 wt%, 0.5 wt%, and 1 wt%) of multi-walled CNTs incorporated into a commercial epoxy adhesive was investigated for the adhesion and corrosion protection properties of the adhesive on anodized Al2024-T3 using electrochemical characterization (Gkikas et al. 2012). In this study, it was demonstrated that the incorporation of CNTs in the adhesive caused the reduction of corrosion current density and a slight increase in corrosion potentials making the substrate nobler indicating anodic protection of the metal substrate. Although the increase in concentration of CNTs in epoxy increased the conductivity of the film, the comparison in their polarization behavior revealed that the corrosion current (i_{corr}) was decreased in the following order: neat epoxy film (0.006 mA/cm^2), 0.5 wt% CNT loaded film (0.002 mA/cm^2), and 1 wt% CNTs loaded film (0.001 mA/cm^2). The forward polarization curves of the films containing CNTs showed a flat anodic polarization behavior, similar to that of non-coated anodized Al2024, indicating localized pitting corrosion at the substrate and coating interface. As shown in Figure 9, among the four films, Al2024 anodized exhibited largest flat region followed by NE+0.5 wt% CNTs, NE+1 wt% CNTs, and the neat epoxy (NE) films. This lower protective ability of the films containing CNTs against pitting corrosion was due to water and ions uptake with the increase in conductivity leading to the degradation of the matrix. However, the ability to resist pitting corrosion appeared to improve from 0.5 wt% CNTs to 1 wt% CNTs, implying that there might be a balance between the increase in hydrophobic nature of the film in resisting water uptake and the increase in conductivity of the film causing the diffusion of water. The authors suggested that the improvement in bonding between CNTs and the matrix at interfacial region could minimize the diffusion of water while retaining hydrophobicity of the film.

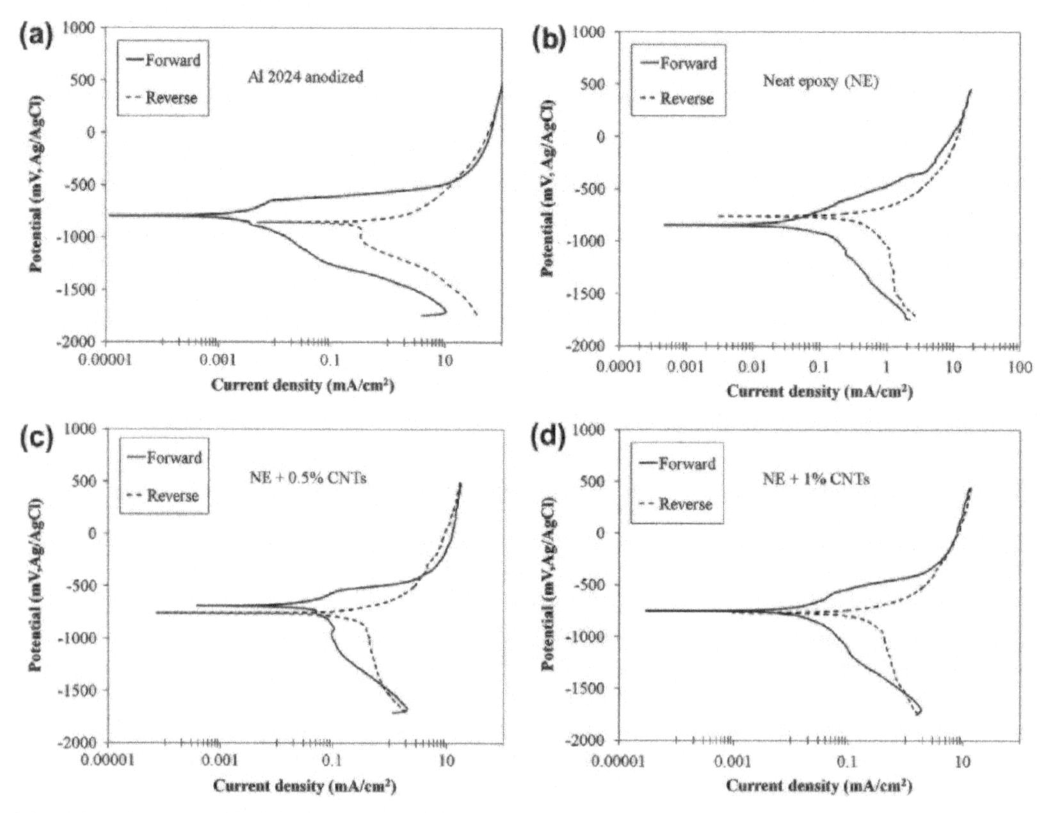

Figure 9. Polartization curves for (a) uncoated anodised Al2024 T3, (b) neat epoxy film on anodised Al2024 T3, (c) 0.5 wt.% CNT doped epoxy film on anodised Al2024 T3 and (d) 1 wt.% CNT doped epoxy film on anodised Al2024 T3 (aerated 3.5% NaCl, R.T.) (Reprinted with permission from Gkikas, G., D. Sioulas, A. Lekatou, N.M. Barkoula and A.S. Paipetis. 2012. Enhanced bonded aircraft repair using nano-modified adhesives. Materials and Design 41: 394–402. Copyright@Elsevier).

In another research study by Khun et al. an epoxy PNC impregnated with multi-walled carbon nanotubes was evaluated for the corrosion protection of AA2024-T3 substrate in the presence of 0.5 M NaCl solution as a function of immersion time for 45 hours (Khun et al. 2014). The results obtained from the EIS data modeling illustrated that the incorporation of carbon nanotubes into the epoxy coating mainly improved the pore resistance and reduced the coating capacitance. It was explained by the fact that the multi-walled carbon nanotubes can form 3-dimensional networks into the epoxy coating impeding the diffusion of corrosive species. Although there was significant improvement in charge transfer resistance as a result of the incorporation of carbon nanotubes into the epoxy coating during the initial immersion times, the maintenance did not last longer.

Tallman et al. reported the application of polyester-urethane coating filled with PPY coated alumina nanoparticles for the corrosion protection of aluminum alloy Al2024-T3 (Tallman et al. 2008). The nanoparticles were coated with PPY in the presence of an adhesion promoter, 1,2-dihydroxybenzene, via chemical oxidative polymerization. The impedance of the coatings with and without nanoparticles as well as with and without PPY coating on the nanoparticles was studied using EIS characterization. Figure 10 shows the comparison of impedance of the coatings (neat coating, coating with 0.5 wt% PPY coated nanoparticles, coating with 1 wt% PPY coated nanoparticles, coating with 2 wt% PPY coated nanoparticles, and coating with 1 wt% nanoparticles without PPY) at low frequency (0.1 Hz) between 1 hour and 168 hours of immersion in diluted Harrison solution (DHS) (3.5 wt% $(NH_4)_2SO_4$ and 0.5wt% NaCl aq. solution).

While the coatings containing 0 wt%, 0.5 wt%, and 1 wt% PPY coated nanoparticles exhibited some reduction in the mangnitude of impedance, the coating containing 2 wt% PPY coated nanoparticles gained an increase (10.3%) in its impedance value presumably due to dedoping and swelling of PPY with water absorption leading to the reduction in pore permeability. The connectivity between the substrate and PPY leads to the oxidation of metal to form passive layer at the interface and the reduction of PPY to release dopant ion.

Gupta et al. were able to present substantial improvement in the corrosion protection of AA2024-T3 with the application of epoxy nanocomposite coating containing 5 wt% of lignosulfonate doped PANI nanoparticles (Gupta et al. 2013). They revealed that the nanoparticles incorporated coating exhibited enhanced barrier properties as well as self-healing capability. They proposed that the protection of aluminum surface occurs through the reduction of PANI and the formation of passive layer via Al^{3+} and sulfonate complex deposition. Even though the nanocomposite coating at optimized concentration of nanoparticles was able to demonstrate self-healing property for smaller defects, it suffered deterioration as a result of corrosion developed under larger defects after 30 days immersion due to insufficient concentration of PANI.

(c) PNCs containing non-conductive nanofillers in non-conductive matrix

Sol-gel coatings containing nanomaterials have recently become one of the most popular PNCs for corrosion protection due to their excellent adhesion and barrier properties. In Shchukin's work, a PNC based on SiO_2-ZrO_2 sol-gel chemistry was developed for the corrosion protection of aluminum alloy 2024. The corrosion performance was further improved by the incorporation of halloysite nanotubes loaded with corrosion inhibitor (2-mercaptobenzothiazole) (MBT), which upon damage of the coating is released to inhibit corrosion process (Shchukin et al. 2008).

In another work of Kartsonakis et al. MBT was encapsulated in cerium molybdate containers and incorporated into an epoxy primer coating for AA2024-T3. The effect of encapsulated MBT on corrosion protection was evaluated using EIS. It was found that the PNC containing cerium molybdate containers with MBT could improve the performance by at least one order magnitude as compared to the PNCs containing either empty containers or just MBT. Similarly, nanoporous silica particles were also used as nanocontainers to encapsulate cerium based corrosion inhibitors and incorporated into sol-gel coating for the corrosion protection evaluation on both AA2024 and AA7075 (Kartsonakis et al. 2011).

Chen et al. first investigated epoxy nanocomposite with layered silicate nanosheets as nanofillers for corrosion protection of AA2024-T3 (Chen et al. 2003). In this work, they found that there was minimal

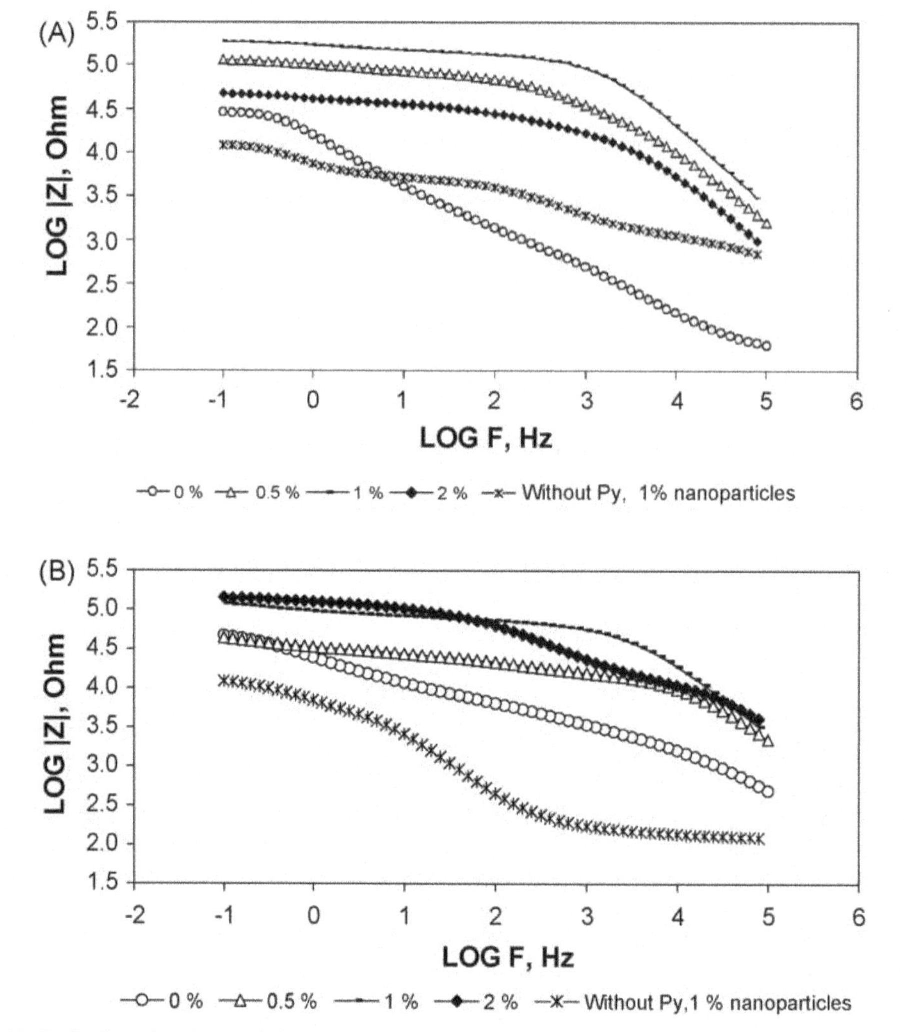

Figure 10. Bode plots of PNCs containing PPY coated alumina nanoparticles at varying amounts (0, 0.5, 1, and 2 wt%) and uncoated alumina nanoparticles (1%). (Reprinted with permission from Tallman, D.E., K.L. Levine, C. Siripirom, V.G. Gelling, G.P. Bierwagen and S.G. Croll. 2008. Nanocomposite of polypyrrole and alumina nanoparticles as a coating filler for the corrosion protection of aluminium alloy 2024-T3. Applied Surface Science 254(17): 5452–5459. Copyright@Elsevier.)

improvement in anti-corrosion properties of the nanocomposite over epoxy coating when the nanosheets were exfoliated, and no improvement when the nanosheets were intercalated. The improvement in performance was attributed to better dispersion of nanosheets in the case of exfoliated nanocomposite. In general, while encapsulated corrosion inhibitors systems enable sol-gel coatings to be active corrosion protective system, nanoparticles such as ZrO_2 improve the passive corrosion protection of sol-gel coatings by enhancing barrier properties.

Automotive industry

Corrosion is a major problem in the automotive industry as it causes degradation of materials such as metals and alloys used in manufacturing vehicles. The selection of materials in the manufacturing process depends on various factors including the properties of materials and the criteria formed by the legislatives and regulatory laws to address environmental and safety issues. Alloys of iron, aluminum, and magnesium are the most commonly used materials in the fabrication of structural parts in the automotive industry. A wide variety of steel materials with different grades produced according to the specifications constitute large percentage (currently more than 60% of automobile body) in structural components of vehicles. In addition to a broad range of versatile properties that they can offer, they have low cost to mass ratio as compared to the alloys of aluminum and magnesium. However, due to increased pressure on the automotive industry to reduce vehicle's weight, thinner steel materials and lightweight alloys of aluminum and magnesium have been introduced. This modification has aggravated corrosion problems as these materials are more prone to corrosion attacks under severe corrosive conditions (Partnership et al. 1999, Ghassemieh et al. 2011).

As shown in Figure 11(a), four types of corrosion are prevalent in the structural components of vehicle underbody: (1) pitting, (2) crevice, (3) galvanic, and (4) cosmetic. As discussed previously, pitting corrosion is a localized corrosion caused by chloride ion attacks. Crevice corrosion occurs when small quantities of electrolyte solution are trapped in crevices and joints. Galvanic corrosion is also called sacrificial/preferential corrosion that occurs when two electrochemically dissimilar metals are in contact with each other in an electrolyte solution. For example, when aluminum is connected to steel,

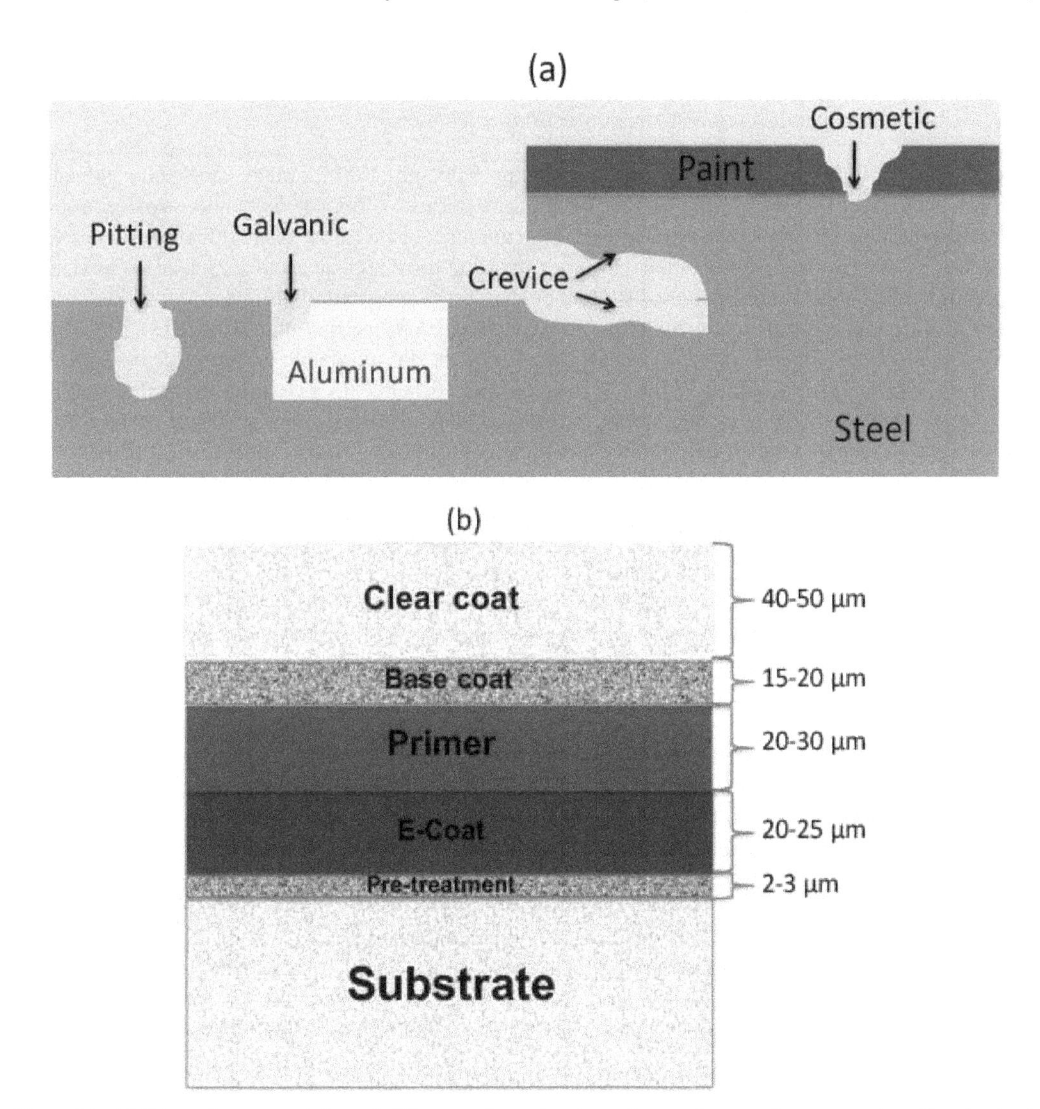

Figure 11. (a) Schematic representation of pitting, crevice, galvanic, and cosmetic corrosion on an automotive structural component, (b) a typical automotive coating system.

aluminum will preferentially corrode as it is electrochemically more active than steel. Cosmetic corrosion is normally seen underneath protective coating when it is damaged due to scratches.

Structural components are coated with multiple layers of coating to provide protection for the substrate (mild steel is the most commonly used in the industry) against corrosion. A typical automotive coating system is shown in Figure 11(b). The first three layers of the coating system including pre-treatment, E-coat, and primer serve the purpose of corrosion protection. While a zinc phosphate conversion coating is normally employed as pre-treatment, both E-coat and primer are epoxy-based coatings—but E-coat is deposited from a water suspension by means of voltage (Partnership et al. 1999).

The research on investigating PNCs for corrosion protection of automotive parts began in 1980s with the pioneering work of Lagaly et al. on PNCs based on clay particles. Later, Toyota's R&D introduced the use of nanocomposite containing Montmorillonite layered sheets embedded in polyamide matrix. Following this work, many different PNCs have been developed in order to improve the corrosion resistance of pre-treatment and E-coat, and the impact resistance of primer layers on substrates such as mild steel, hot-dip galvanized steel, aluminum, and magnesium (Hintze-Bruening et al. 2012). In this chapter, a few examples in each category of PNCs have been reviewed analyzing their performance on these substrates.

(a) PNCs containing conductive matrix and conductive nanofiller

Zinc rich coatings have been in use for the corrosion protection of steel for many years. Zinc galvanized steel has been widely utilized by the automotive industry because of its efficient anti-corrosive property. In order to sacrificially protect the steel surface, high amounts of Zn (more than critical pigment volume concentration) are required in the coating. However, the addition of high quantity of Zn leads to weakening of mechanical and adhesion properties of the coating. Olad et al. prepared a hybrid nanocomposite coating system in which Zn nanoparticles were loaded into an epoxy/PANI matrix (Olad et al. 2012). They varied the concentrations of Zn and epoxy in the system to study the anti-corrosive performance on AISI 1006 carbon steel in acidic conditions. While PANI being electrochemically active improves the sacrificial protection of Zn, epoxy enhances the barrier, mechanical and adhesion properties of the coating. In this work, they demonstrated that at optimum concentrations of Zn (4 wt%) and epoxy (3–7 wt%), the anti-corrosion performance was significantly improved as compared to pure PANI coating.

Kumar et al. synthesized PANI nanocomposite coating containing multi-walled carbon nanotubes (MWCNTs) on mild steel via electrochemical polymerization in order to study the corrosion protection in 3.5 wt% NaCl solution (Kumar et al. 2015). They were able to achieve good dispersion of MWCNTs in PANI through functionalization of MWCNTs with sodium dodecyl sulfate (SDS).

The corrosion performance of the PNC was investigated using electrochemical characterization such as PP and EIS. As shown in Figure 12(a), the PP results of the coatings illustrated that the incorporation of MWCNTs into PANI coating caused a positive shift in corrosion potential and a significant reduction in corrosion current density. Moreover, it was also reported in this work that the improvement in corrosion performance was further enhanced with the increase in percentage of MWCNTs in the PANI coating. The EIS results presented in Figure 12(b) also indicated similar trend in corrosion protection performance by the coatings evaluated in terms of barrier properties (impedance at higher frequency) as well as corrosion protection properties (impedance at lower frequency) at the coating-substrate interface. The EIS data modeling revealed that the coatings containing MWCNTs, especially PCNT3, exhibited considerable increase in pore resistance (from 6698 in PANI to 549,924 in PCNT3) and charge transfer resistance (from 16,731 in PANI to 901,239 in PCNT3) as compared to pure PANI coating and uncoated substrates. Ionita et al.'s work further substantiated the corrosion protective ability of PANI-CNTs nanocomposites for carbon steel (Ionita et al. 2010).

(b) PNCs containing non-conductive filler in conductive matrix

PPY coatings containing ZnO nanorods were prepared by Hosseini et al. (2011) to study their corrosion protective ability for mild steel (Makhlouf and Hosseini 2016a). In the work of Hosseini et al. (2011), it

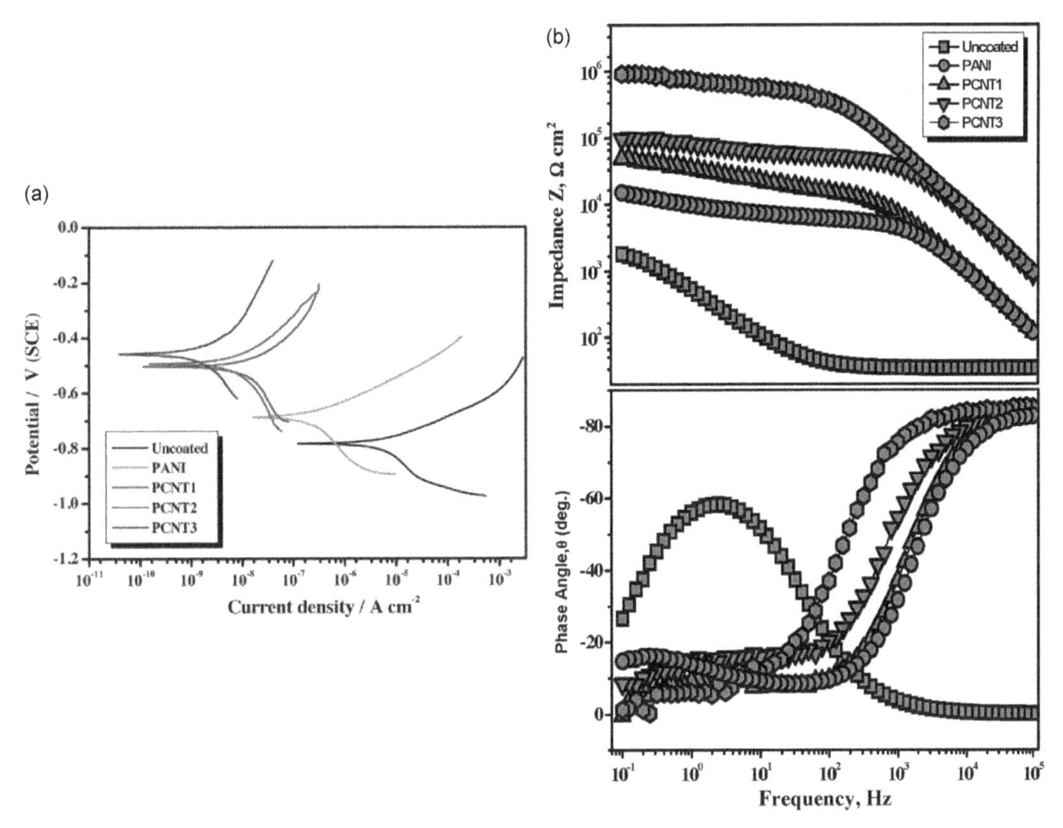

Figure 12. (a) Potentiodynamic polarization curves of uncoated, PANI coated, and PNC coated metal substrates, (b) Bode plots of uncoated, PANI coated, and PNC coated metal substrates (Reprinted with permission from Madhan Kumar, A. and Z.M. Gasem. 2015. *In situ* electrochemical synthesis of polyaniline/f-MWCNT nanocomposite coatings on mild steel for corrosion protection in 3.5% NaCl solution. Progress in Organic Coatings 78: 387–394. Copyright@Elsevier).

was demonstrated that the deposition of ZnO nanorods into PPY coatings was able to improve corrosion protective ability for mild steel by enhancing the barrier and electrochemical properties of the coating. This improvement was attributed to the increase in diffusion path of corrosive species as well as the reduction in porosity due to the deposition of products formed as a result of electrochemical reactions. While ZnO nanorods with large aspect ratio and surface area increases diffusion path, the reduction of PPY releases dopant ions to bind with metal cations to form metal-dopant complex.

Other non-conductive nanofillers such as zeolites type materials were also evaluated for corrosion protection by incorporating them into conductive matrices. In the work of Olad et al. nanocomposite of PANI and clinoptilolite was prepared with varying amounts (1, 3, and 5 wt%) of clinoptilolite on low carbon steel (Olad et al. 2010). It was shown in this work that the key aspect in improving corrosion efficiency of nanocomposites containing zeolites was a proper encapsulation of conductive matrix into their channels. Corrosion studies indicated that these nanocomposites, especially with 3 wt% clinoptilolite, were able to provide corrosion protection to iron substrates, but only in acidic environments.

(c) PNCs containing conductive nanofiller in non-conductive matrix

In the work of Chen et al. PNC prepared by dispersing PANI/partially phosphorylated poly(vinyl alcohol) (PANI/P-PVA) nanoparticles in a waterborne epoxy matrix was investigated for the corrosion protection of mild steel (Chen et al. 2011). It was indicated in this work that the improvement in performance of the

PNC was mainly because of two reasons: (1) excellent dispersibility of the (PANI/P-PVA) nanoparticles in epoxy leading to enhanced barrier properties and better adhesion to the metal substrate, (2) corrosion inhibition by PANI causing the formation of passive layer at the metal-coating interface.

PANI was further utilized in conjunction with other electroactive materials such as multi walled carbon nanotubes (MWCNTs) for the corrosion protection of low carbon/mild steel. Deshpande et al. demonstrated the improved performance of epoxy coating with the incorporation of PANI coated MWCNTs on low carbon steel in 3.5 wt% NaCl solution (Deshpande et al. 2013). The corrosion rate was reduced to about 5.2 times lower than that of uncoated substrate and 3.6 times lower than pure epoxy coated substrate. They concluded that the performance could further be improved by the optimization of PANI/MWCNTs' content in the nanocomposite.

The challenges in obtaining long-term corrosion protection performance of zinc rich coatings were addressed in the work of Gergely et al. Although zinc rich coatings appear to perform effectively in the beginning of service-life of the coatings, their long-term protective capability weakens due to porosity and severe galvanic corrosion under damaged conditions (Gergely et al. 2011). Gergely et al. prepared efficient zinc rich coating by incorporating highly dense, structured, and dispersed PPY coated alumina nanoparticles into the coating. The incorporation of nanoparticles controlled the galvanic function of Zn and also significantly increased the barrier properties.

Similar to Zinc, magnesium (Mg) can also provide sacrificial cathodic protection to iron, as can be seen in Table 1; Mg is also more electrochemically active (negative) than iron. Dennis et al. investigated the sacrificial corrosion protection of nanostructured Mg composite containing Mg nanoplatelets dispersed in poly(ether imide) on low alloy steel in 3.5 wt% NaCl solution. The nanocomposite exhibited better corrosion protection when compared to Zn galvanized steel and Zn-rich primer coating of similar thickness (Dennis et al. 2014).

(d) PNCs containing non-conductive nanofiller in non-conductive matrix

In Ramezanzadeh et al.'s work, the incorporation of ZnO nanoparticles was found to improve the corrosion resistance and hydrolytic degradation resistance of an epoxy coating when exposed to 3.5 wt% NaCl solution. In another study by the same authors, the addition of silane modified or functionalized ZrO_2 nanoparticles into the epoxy coating was also assessed in terms of corrosion protection for hot-dip galvanized steel substrate (Ramezanzadeh et al. 2011a and b).

Montmorillonite (MMT), a layered silicate clay material, was surface-functionalized with urea and incorporated into an epoxy matrix to employ as a protective coating for carbon steel. In this work, Hosseini and Makhlouf 2016 utilized PP and EIS characterizations to determine corrosion rate and impedance of the coated samples in two different corrosive conditions (3.5 wt% NaCl solution at room temperature and 80°C). The trend in corrosion rate at room temperature followed a gradual decrease with the increase in percentage of modified clay from 0 wt% to 5 wt% in the nanocomposite. However, for the tests at 80°C, only 1 wt% and 2 wt% nanocomposites exhibited lower corrosion rates than pure epoxy. Using EIS results and modeling, the charge transfer resistance, which is an indicative of corrosion resistance at the metal-coating interface, was calculated for each coating tested at room temperature and 80°C. According to them, the percentages of clay for achieving optimum resistance of the nanocomposite coating were 3 wt% at room temperature and 1–2 wt% at 80°C.

Heidarian et al. reported a similar observation for polyurethane/organo clay nanocomposites as well. The corrosion performance of polyurethane coating was improved by the incorporation of organically modified montmorillonite clay as nanofillers (Heidarian et al. 2010). As per the polarization behavior comparison of the coatings shown in Figure 13(a), the inclusion of clay made the coating nobler and reduced the corrosion current density as its concentration increased to 3 wt%. Further, as shown in Figure 13(b), the analysis of the results obtained from the EIS tests indicated that the coating containing 3 wt% clay sustained the highest coating resistance (3000 times higher than pure polyurethane coating) as well as the lowest capacitance among the coatings tested over a period of 69 days exposure.

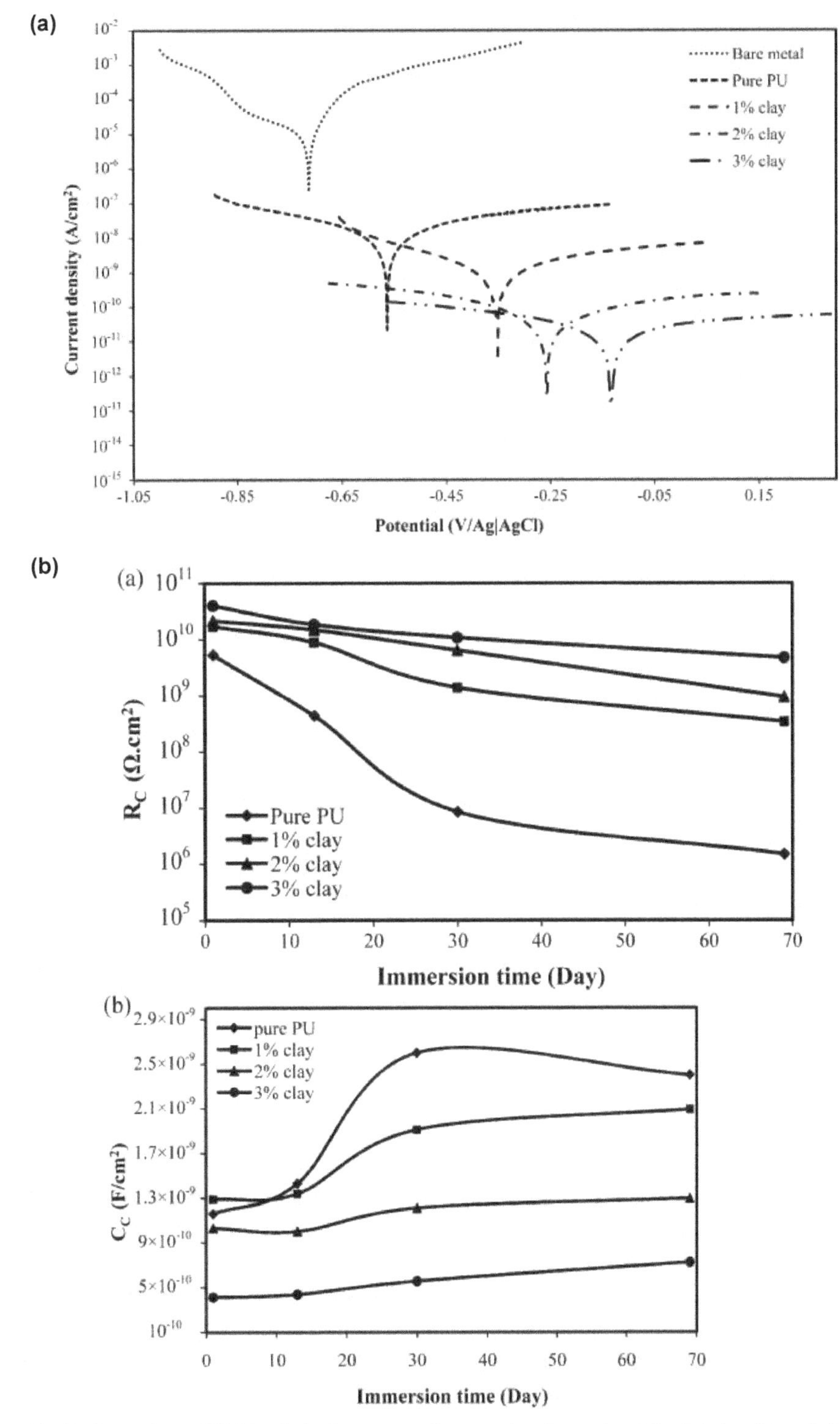

Figure 13. (a) Comparison of Tafel polarization curves of uncoated and coated metal substrates, (b) evolution of coating resistance (R_c) and coating capacitance (C_c) as obtained from EIS (Reprinted with permission from Heidarian, M., M.R. Shishesaz, S.M. Kassiriha and M. Nematollahi. 2010. Characterization of structure and corrosion resistivity of polyurethane/organoclay nanocomposite coatings prepared through an ultrasonication assisted process. Progress in Organic Coatings 68(3): 180–188. Copyright@Elsevier).

Marine

The total corrosion costs including installation and maintenance costs of marine structures incurred by the industry are ever increasing due to the fact that the demand for offshore oil and gas exploration is constantly growing. Additionally, the corrosion costs for the maintenance of ships are enormous as large-scale transportation of goods still occurs via seawater (Scotto et al. 1985, Ivanov et al. 2016). Since the majority of structural components in both ships and marine structures are mainly constructed using various metals and alloys, these structures are greatly affected by the corrosive marine environment as they are submerged in the seawater for longer periods of time. Moreover, the presence of aggressive corrosive species combined with the threat from marine organisms makes seawater, relatively, much severe. Therefore, the selection of materials for the construction and protection of ships and marine structures is rather quite different from other applications. Materials of stainless steel (SS) have been extensively used for marine applications due to its excellent strength, toughness, and corrosion resistance. However, in chloride ion environments, SS components in marine structures suffer from pitting and crevice corrosion. Several methods including cathodic protection and the use of protective coatings based on organic and inorganic chemistries have been implemented for the corrosion protection of SS. Currently, it is the combination of both cathodic protection and protective coatings that offers the best performance (Tezdogan et al. 2014).

The major problem for the structures and equipment in marine environment is fouling. It causes accelerated corrosion and significant load on structures leading to disastrous incidents. Conventional marine protective coatings utilize toxic ingredients containing metal-based compounds such as tri-n-butyltin to fight against fouling. However, the use of this type of ingredients has been restricted internationally due to their harmful nature to marine atmosphere. There has been a great demand in the marine industry for novel, environment friendly and efficient protective coatings for ships and marine structures. As a part of search for an appropriate coating system, several PNCs have been evaluated using various corrosion tests and electrochemical techniques.

(a) PNCs containing conductive nanofiller in conductive matrix

Jafari et al. reported an improved efficiency of corrosion inhibition by PANI-graphene nanocomposite on stainless steel substrate in a chloride environment (Jafari et al. 2016). The performance of the nanocomposite was compared to bare substrate and pure PANI coating using the results obtained from the PP and EIS tests. The corrosion current density (i_{corr}) was reduced from 5.2 μA/cm^2 for base SS (blank) to 1.4 μA/cm^2 when coated with pure PANI and to 0.15 μA/cm^2 when coated with PANI-graphene nanocomposite. This resulted in reduction of corrosion rate about 116 times as compared to bare SS. The EIS results also indicated similar improvement in performance of the nanocomposite. The size of semicircle in Nyquist plot increased from bare SS to PANI coating and further increase in the case of the nanocomposite, indicating the increase in charge transfer resistance. As compared to bare SS, the nanocomposite exhibited significant improvement (32 times) in charge transfer resistance, which is inversely related to corrosion rate. Overall, the improvement with the nanocomposite was attributed to the anodic protection of SS substrate by PANI-graphene synergic interactions and enhanced barrier properties with the incorporation of graphene.

(b) PNCs containing non-conductive nanofiller in conductive matrix

Ganash et al. published the use of conductive PNC composed of poly(o-phenylenediamine) (PoPD) and ZnO nanoparticles as a corrosion protective coating for Type 304 stainless steel (SS) in chloride conditions (Ganash et al. 2014). The coating was prepared on the substrate via electrodeposition by the oxidation of o-phenylenediamine in the presence of ZnO nanoparticles. The coating of pure PoPD on 304 SS increased the corrosion potential and reduced the current density indicating that the metal substrate was anodically protected due to the formation of passive layer. The interactions between PoPD and SS substrate formed a galvanic couple, which caused reduction/undoping of PoPD and oxidation of

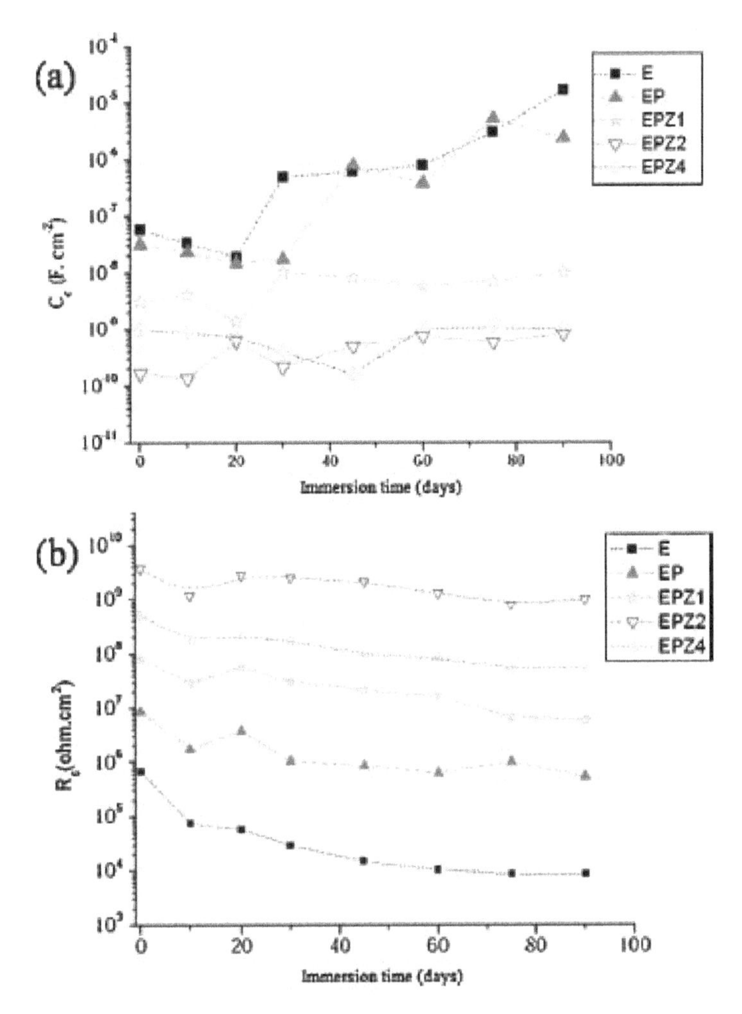

Figure 14. Evolution of coating resistance (R_c) and coating capacitance (C_c) from EIS (Reprinted with permission from Mostafaei, A. and F. Nasirpouri. 2014. Epoxy/polyaniline-ZnO nanorods hybrid nanocomposite coatings: Synthesis, characterization and corrosion protection performance of conducting paints. Progress in Organic Coatings 77(1): 146–159. Copyright@Elsevier).

SS substrate. The undoping process of PoPD would release dopant anion, which could bind with metal cation to form a passive layer. The incorporation of ZnO nanoparticles into the PoPD polymer matrix improved the interactions between the conductive polymer and the substrate by increasing the surface area leading to further increase in corrosion potential toward positive direction and decrease in corrosion current density.

(c) PNCs containing conductive nanofiller in non-conductive matrix

The reports published by Mostafaei et al. discussed the improvement in anticorrosion and antifouling properties of epoxy coating with the incorporation of PANI coated ZnO nanorods (Mostafaei et al. 2013). They evaluated the performance of the nanocomposite on marine grade carbon steel for 9 months in the Caspian Sea and Persian Gulf. As shown in Figure 14, they also used EIS to study the anticorrosion properties of the nanocomposite with varying concentrations of PANI/ZnO in epoxy coating on carbon steel ST37 in 3.5% NaCl solution at 25°C for 0, 30 and 90 days. From the Bode plots of the coatings after

90 days, the nanocomposite containing 4 wt% PANI/ZnO (EPZ4) exhibited substantial improvement in barrier properties. Further, after fitting the curve with models, the evolutions obtained for coating capacitance (C_c) and resistance (R_c) were correlated with the diffusion of electrolyte and corrosive species through the coatings. As depicted in Figure 15, the proposed mechanism for corrosion protection with PANI follows via PANI reduction, dopant anion release, and complex formation between metal cation and dopant anion.

(d) PNCs containing non-conductive nanofiller in non-conductive matrix

Behzadnasab et al. prepared a nanocomposite by treating ZrO_2 nanoparticles with amino propyl trimethoxy silane (APS) and incorporating these nanoparticles into an epoxy coating for the evaluation of corrosion protection on mild steel in chloride solutions (Behzadnasab et al. 2011). The surface treatment of ZrO_2 nanoparticles with APS improved their dispersion in the target epoxy coating. The effect of varying concentrations of ZrO_2 nanoparticles (0, 1, 2, and 3 wt%) on corrosion performance was investigated using EIS. Improved barrier properties and ohmic resistance was noticed in the case of nanocomposites containing 2 and 3 wt% nanoparticles. Although the rate of decrease in coating resistance over 120 days immersion time considerably reduced with the increase in concentration of nanoparticles, the difference in coating capacitance between 0 wt% and 3 wt% at the end of 120 days immersion time was significantly less indicating that even higher concentration of nanoparticles could not resist water intake.

In this category of PNCs, sol-gel method has been extensively used to prepare various PNCs and deposit them on marine steel substrates for corrosion protection evaluation. Among them, TiO_2 based sol-gel coatings have become very popular due to their excellent chemical stability, hardness, and barrier properties. Curkovic et al. observed increase in corrosion performance of TiO_2 sol-gel coating in 3.5 wt% NaCl solution as the number of layers increases (Ćurković et al. 2013). Cheraghi et al. also noticed similar phenomenon in the case of TiO_2-NiO nanocomposite preparation process (Cheraghi et al. 2012). Significant reduction in corrosion current density was noticed with the number of layers (tested

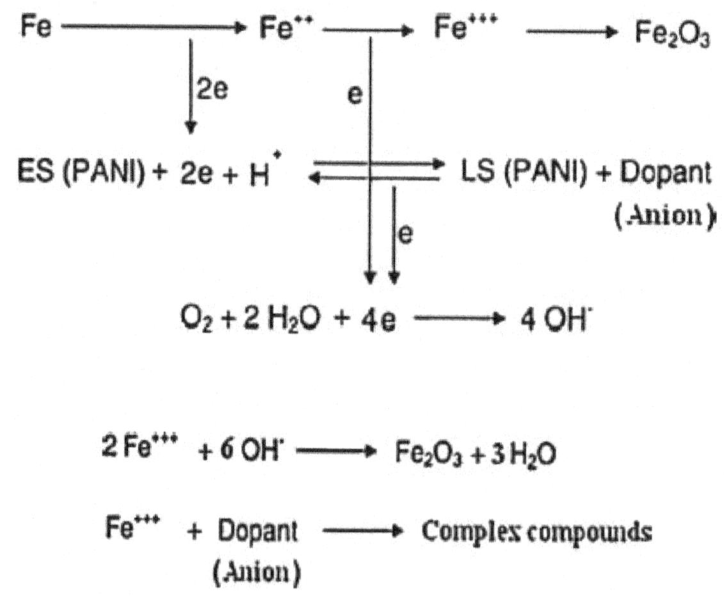

Figure 15. Proposed mechanism of corrosion protection of metal substrate in the presence of PANI-ZnO nanocomposite (Reprinted with permission from Mostafaei, A. and F. Nasirpouri. 2014. Epoxy/polyaniline-ZnO nanorods hybrid nanocomposite coatings: Synthesis, characterization and corrosion protection performance of conducting paints. Progress in Organic Coatings 77(1): 146–159. Copyright@Elsevier).

up to 6 dipping times) formed on the metal substrate. In addition to TiO_2 sol-gel coatings, organically modified sol-gel coatings based on 3-glycidoxypropyltrimethoxysilane (GPTMS) were also effectively implemented for the corrosion protection of stainless steel in marine environments. Further, in order to improve the efficiency of these sol-gel coatings, corrosion inhibitors such as cerium nitrate ($Ce(NO)_3$) were incorporated into them as observed in the work of Zandi Zand et al. (Zandi Zand et al. 2011).

Layered materials such as hexagonal boron nitride nanosheets (h-BN) and cloisite 15A nanoclay have been found to be useful as nanofillers in PNCs in providing corrosion protection to marine stainless steel substrates. Husain et al. demonstrated an effective corrosion protective ability of PNC containing h-BN for type 316L SS substrate (Husain et al. 2013). Similarly, Jeeva Jothi et al. utilized cloisite 15A nanoclay as nanofiller to improve barrier properties of the maleic anhydride grafted low-density polyethylene (LDPE) and ultra-high-molecular weight polyethylene (UHMWPE) matrix (Jeeva Jothi et al. 2014). The addition of layered sheets, when dispersed homogenously throughout the entire coating, can extend the corrosion performance of the coating by increasing the diffusion path of corrosive species.

Infrastructure

Generally, metal-based facilities and infrastructure such as bridges, pipelines, and buildings are vulnerable to corrosion due to their continuous exposure to corrosive atmosphere. According to the United States Federal Highway Administration, the current direct costs associated with the damages due to corrosion in the infrastructure in the United States alone are estimated to be more than $300B per year, which represents approximately 3% of nation's Gross Domestic Product (International and Paper 2012) (Fhwa et al. 2012). Without proper corrosion prevention and control measures, it can cause sudden structural failures leading to substantial loss of life and severe damage to environment. Corrosion prevention measures include appropriate selection of materials, design, quality assurance, manufacturing, maintenance, and protective technologies. Most commonly used materials in the construction of infrastructure are stainless steel, galvanized steel, carbon steel, copper, brass, aluminum alloys, and composite materials. Steel materials constitute the highest percentage of structural components in the construction industry. Since stainless steel and galvanized steel are more corrosion resistant than carbon steel, they relatively account for lower costs of protection and maintenance. However, in chloride environments, they suffer from localized corrosion such as pitting, crevice, and stress corrosion cracking. Despite its lower corrosion resistance, carbon steel is primarily used in the construction of infrastructure due to its lower cost to strength ratio and excellent wear resistance. Depending on physical, mechanical, chemical, and electrochemical properties of these structural metals and alloys, they are used for specific parts in the infrastructure industry and protected from corrosive environment by employing different protective technologies. According to American Petroleum Institute, large quantity of underground pipelines for the transportation of oil and natural gas is made up of carbon steel (Figueira et al. 2016). As far as bridges, buildings, and large structures are concerned, different types of steel such as carbon steel, weathering steel, and stainless steel, and other metals such as aluminum and copper are used for various structural elements.

Similar to the applications in the areas discussed previously in this chapter, the coatings technology has also been used to protect metallic substrates in the infrastructure industry for many years (Lyon et al. 2016). Traditionally, a conversion coating such as phosphate on steel and chromate on aluminum is applied as pre-treatment to form a passive oxide layer of the metal substrate. Primer coatings that contain corrosion inhibitors such as chromates or phosphates, or sacrificial fillers such as Zn particles for steel substrates, are applied on top of pre-treatment to provide corrosion protection and sufficient adhesion. Intermediate coating and topcoat are applied on primers to improve mechanical strength and UV resistance of the complete coating system (Figueira et al. 2016).

In recent years, PNCs have been thoroughly studied to meet changing requirements in the infrastructure industry and environmental regulations passed by the U.S. federal agencies. Below are some of the case studies of the applications of PNCs for the metals and alloys used in the infrastructure industry.

(a) PNCs containing conductive nanofiller in conductive matrix

Chang et al. reported the application of PNCs based on graphene for the corrosion protection on steel. Graphene sheets were functionalized with 4-aminobenzoyl groups and subsequently incorporated into polyaniline coating via chemical oxidation process. The corrosion protection of polyaniline/graphene nanocomposite coating on steel substrate in corrosive medium containing 3.5 wt% NaCl aqueous solution was compared to the corrosion protection of polyaniline coating as well as polyanaline/clay nanocomposite coating using potentiodynamic polarization method. It was shown in this study that the incorporation of graphene in polyaniline coating improved the barrier properties of the coating by increasing the diffusion path of O_2 and H_2O, retarding their movement through the coating, and thereby reducing their concentrations at the metal surface. In addition, when compared with the well-established polyanaline/clay nanocomposite, polyanaline/graphene nanocomposite coating exhibited higher corrosion protection probably due to larger filler matrix interfacial area and aspect ratio in the case of graphene.

(b) PNCs containing conductive nanofiller in non-conductive matrix

In the research work of Pour-Ali et al. it was found that the nanocomposite coating prepared by the incorporation of PANI (PANI)-camphorsulfonate (CSA) nanoparticles in epoxy exhibited long term corrosion protection for reinforcing steel rebars in chloride-laden concrete environment (Pour-Ali et al. 2015). The improvement in performance of the nanocomposite coating was analyzed using EIS measurements while samples were immersed in 3.5 wt% NaCl solution for 12 months. The comparison of open circuit potentials (OCPs) among different coatings indicated that the OCP decreased from ca. –300 mV to –500 mV except for the nanocomposite coating where the OCP decreased slightly and remained constant. This shows that the presence of PANI-CSA particles can maintain passivity at the substrate-coating interface for longer periods of time. Figures 16(a) and (b) show the evolutions of coating resistance (R_{layer}) and coating capacitance (C_{layer}) over a year as obtained from EIS measurements. As it can be seen, the nanocomposite coatings on both normal concrete steel bars (NC) and self-compacting concrete steel bars demonstrated higher resistances and lower capacitances than epoxy coatings or bare steel bars.

In another similar work by Valenca et al. polypyrrole (PPy) coated ZnO nanoparticles were incorporated into epoxy coating and tested to protect structural steel (SAE 1020 carbon steel) from corrosion. EIS and OCP tests revealed that the best corrosion protection could be obtained when the concentration of corrosion inhibiting ZnO/PPy nanoparticles was 0.2% w/w (Valenca et al. 2015).

Recently, Yuan et al. reported the fabrication of superamphiphobic and electroactive nanocomposite of PANI/fCNTs incorporated into ethylenetetrafluoroethylene (ETFE) matrix as self-cleaning, anti-wear, and anti-corrosion coating on the etched aluminum substrate (Yuan et al. 2016). The corrosion resistance of the nanocomposite coating was investigated using PP and EIS under immersion conditions in 3.5 wt% NaCl solution for 90 days. The PP results displayed in Figure 17(a) showed that the corrosion rate was significantly reduced from 152 μm/year in the case of uncoated (curve a) to 0.004 μm/year in the case of PANI-fCNTs –6 wt% nanocomposite (curve e). As shown in Figure 17(b), the EIS results also indicated that the incorporation of PANI-fCNTs into the matrix could improve the barrier properties of the coating as well as lead to passivation of the metal surface.

(c) PNCs containing non-conductive nanofiller in non-conductive matrix

Silane functionality plays an important role in sol-gel coatings when it comes to adhesion to metal substrate and water absorption properties. The alkoxy groups (Si-O-R) present in silanes can hydrolyze in the aqueous conditions during the dipping process and convert into silanol (Si-O-H) groups, which can react with the hydroxyl groups on the metal surface to form a strong (M-O-Si) bond at the metal-coating interface. This allows sol-gel coatings to exhibit excellent adhesion properties.

Corrosion resistance of sol-gel coatings is generally further increased by the addition of corrosion inhibitors. Sol-gel coatings based on tetraethoxysilane (TEOS) and 3-glycidoxypropyl-triethoxysilane containing SiO_2 nanoparticles have been used as barrier coatings for carbon steel. Santana et al. included

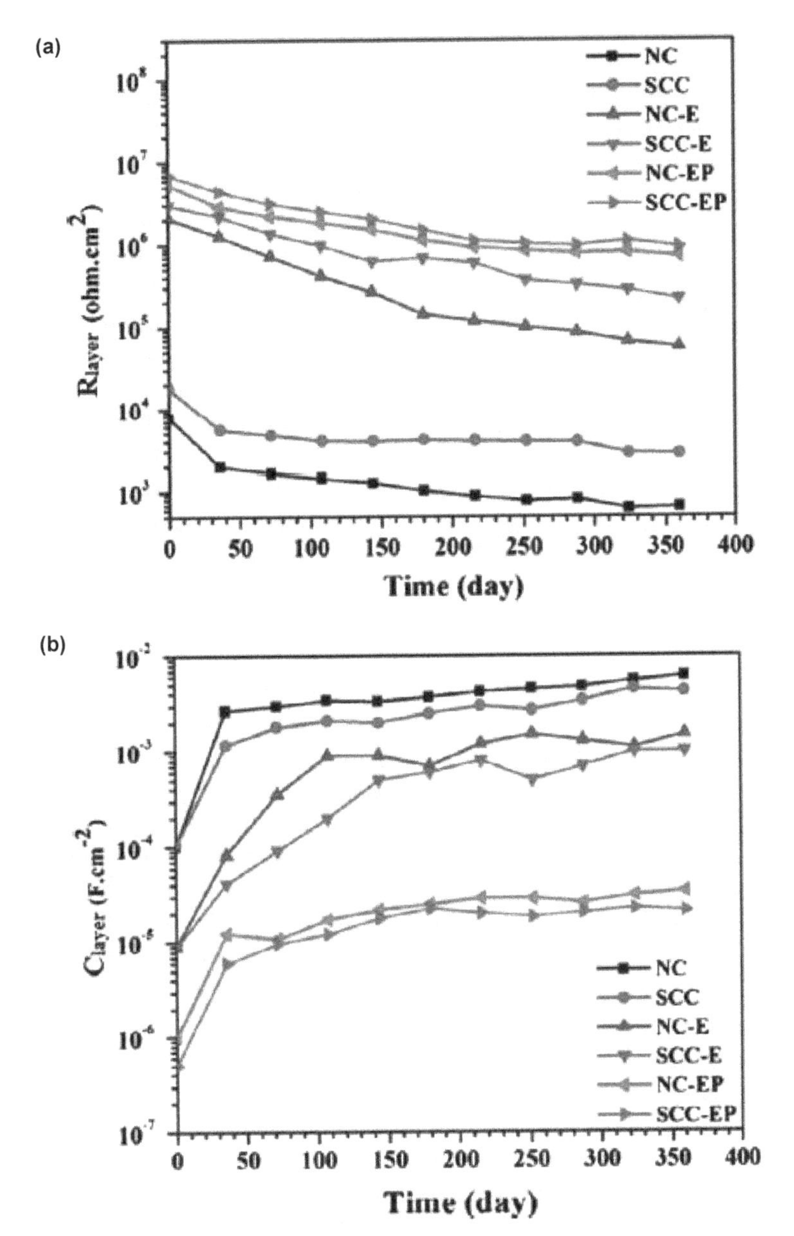

Figure 16. (a) Evolution of coating resistance (R_{layer}) and (b) coating capacitance (C_{layer}) as obtained from EIS (Reprinted with permission from Pour-Ali, S., C. Dehghanian and A. Kosari. 2015. Corrosion protection of the reinforcing steels in chloride-laden concrete environment through epoxy/polyaniline-camphorsulfonate nanocomposite coating. Corrosion Science 90: 239–247. Copyright@Elsevier).

corrosion inhibitors such as cerium nitrate ($Ce(NO_3)_3$) into SiO_2 nanoparticles to provide corrosion inhibition (Santana et al. 2015). The corrosion resistance of the nanocomposite was evaluated using PP and EIS in 0.35% NaCl solution for 24 h immersion and was found to increase with the increase in the concentration of nanoparticles containing corrosion inhibitor. In order to further enhance barrier and adhesion properties of sol-gel coatings, Joncouc-chabrol et al. investigated talc-like phyllosilicates nanosheets having silanol groups as nanofillers. The authors mentioned that the presence of silanol groups enabled nanosheets to covalently bond with the matrix as well as the metal surface. Better dispersion

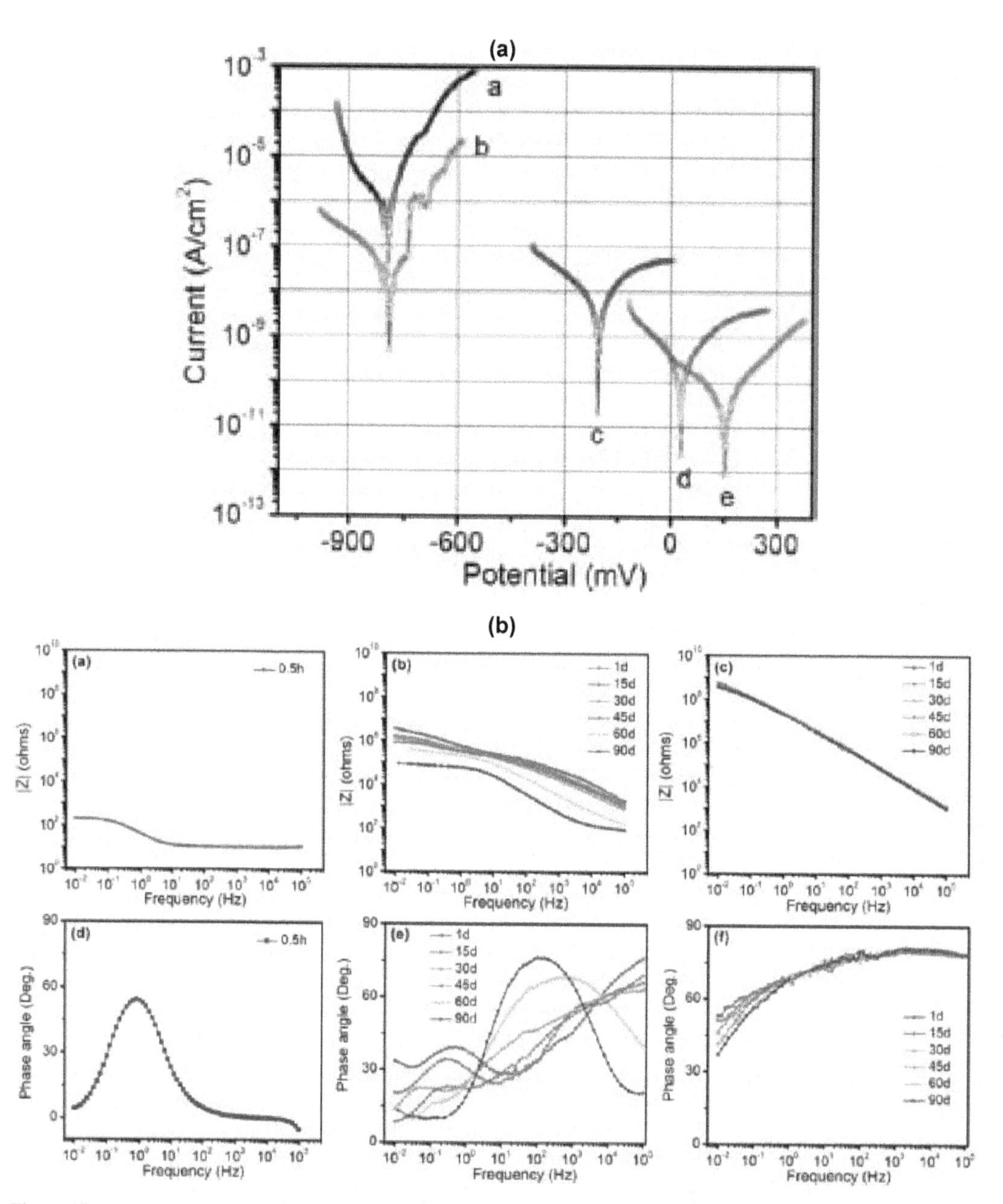

Figure 17. (a) Potentiodynamic polarization curves of uncoated Al substrate after 0.5 h immersion (curve a), ETFE coating after 1 day (curve c) and 90 days (curve b) immersion, and ETFE-PANI/fCNTs-6 after 1 day (curve e) and 90 days (curve d) immersion, (b) Bode plots of Al substrate, ETFE coated Al substrate, and ETFE-PANI/fCNTs-6 nanocomposite as a function of time (Reprinted with permission from Yuan, R., S. Wu, P. Yu, B. Wang, L. Mu, X. Zhang et al. 2016. Superamphiphobic and electroactive nanocomposite toward self-cleaning, antiwear, and anticorrosion coatings. ACS Applied Materials and Interfaces 8(19): 12481–12493. Copyright@American Chemical Society).

and proper distribution of the nanosheets caused the increase in tortuous pathways leading to improved barrier properties of the coating.

In Dong et al.'s work, the effect of layered double hydroxide (LDH) and montmorillonite (MMT) incorporated in a nanocomposite coating was analyzed for corrosion protection performance on steel pipes that are normally used for oil and gas exploration (Dong et al. 2013). In order to assess the effectiveness of the nanocomposite coating that needs to sustain in chloride environment at high

Table 3. Summary of recent research work on application of PNCs for corrosion protection of substrates used in aerospace, automotive, marine, and infrastructure industries.

Industry	PNC Category	Description of PNC	Substrate	Electrochemical Analysis	Reference
Aerospace	Conductive nanofiller in non-conductive matrix	Graphene in epoxy matrix	Al 2024 – T3	• EIS measurement from immersion in 3.5 wt% NaCl solution for 10 days • Impedance (|Z|) after 10 days: 10^6–10^7 Ohm-cm^2	(Monetta et al. 2015)
		Multi-walled CNTs in epoxy matrix	Al 2024 – T3	• EIS measurement from immersion in 0.5 M NaCl solution for 45 hours • Impedance (|Z|) after 45 hours: 4×10^5 Ohm-cm^2	(Khun et al. 2014)
		GFRP/multi-walled CNTs in epoxy matrix	Al 7075	• Corrosion rate of metal increased to 1.375 mm/yr	(Ireland et al. 2012)
		CNTs in epoxy matrix	Al 2024 – T3	• PP measurement from immersion in 3.5 wt% NaCl solution for 24 hours • Corrosion rate: 1×10^{-6} A/cm^2	(Guadagno et al. 2014)
		PPY coated Al_2O_3 in polyester-urethane matrix	Al 2024 – T3	• EIS measurement from immersion in DHS solution (3.5 wt% $(NH_4)_2SO_4$ and 0.5 wt% NaCl) solution for 168 hours • Impedance (|Z|) after 168 hours: 10^5–$10^{5.5}$ Ohm-cm^2	(Tallman et al. 2008)
		PANI/lignosulfonate in epoxy coating matrix	Al 2024 – T3	• PP measurement from immersion in 0.6 M NaCl solution for 30 days • Corrosion rate: 7.62×10^{-10} A/cm^2 • Impedance (|Z|) from EIS after 30 days: 10^8 Ohm-cm^2	(Gupta et al. 2013)
		TiO_2/multiwalled CNTs in Bis-[triethoxysilypropyl] tetrasulphide (BTESPT) matrix	Al 2024 – T3	• PP measurement from immersion in 0.6 M NaCl solution for 30 days • Corrosion rate: 5.6×10^{-8} A/cm^2 • Impedance (|Z|) from EIS after 6 days: 10^6 Ohm-cm^2. But, impedance decreased to 10^4 Ohm-cm^2 after 7 days	(Zhang et al. 2016)
	Non-conductive nanofiller in conductive matrix	TiO_2 or ZrO_2 in PANI matrix	Al 2024 – T3	• PP measurement from immersion in 5 mM NaCl and 0.1 M Na_2SO_4 solution • Corrosion rate: 10^{-5}–10^{-6} A/cm^2	(Zubillaga et al. 2009)
		TiO_2 in PPY matrix	Al 7075 – T6	• PP measurement from immersion in 5 mM NaCl and 0.1 M Na_2SO_4 solution • Corrosion rate: ≈ 10^{-6} A/cm^2 • Impedance (|Z|) from EIS after 6 days: 5×10^3 Ohm-cm^2	(Mert 2015)

Table 3 contd. ...

...Table 3 contd.

Industry	PNC Category	Description of PNC	Substrate	Electrochemical Analysis	Reference
		1st Layer: CeO_2/MBT nanocontainers in PANI-PPY composite coating **2nd Layer:** Sol-gel coating	Al 2024 – T3	• PP measurement from immersion in 0.0.5 M NaCl solution for 72 hours • Corrosion rate: 4.36×10^{-8} A/cm^2 • Impedance (|Z|) from EIS after 6 days: 1.05×10^5 Ohm-cm^2	(Ioannis A. Kartsonakis et al. 2011)
	Non-conductive nanofiller in non-conductive matrix	Halloysite nanotubes/MBT in sol-gel matrix	Al 2024 – T3	• Impedance (|Z|) from EIS after 2 weeks: 2×10^6 Ohm-cm^2	(Shchukin et al. 2008)
		SiO_2 nanoparticles in epoxy matrix via sol-gel method	Al 2024 – T3	• PP measurement from immersion in 0.0.5 M NaCl solution for 168 hours • Corrosion rate: 5.9×10^{-9} A/cm^2	(Pirhady Tavandashti et al. 2009)
		Ce(III)$(NO_3)_3$ in sol-gel matrix	Al 2024 – T3	• PP measurement from immersion in 3.5 wt% NaCl solution for 168 hours • Corrosion rate: 6×10^{-9} A/cm^2 • Impedance (|Z|) from EIS after 168 hours: 2×10^5 Ohm-cm^2	(Lakshmi et al. 2013)
		$CeMoO_4$/MBT in epoxy matrix	Al 2024 – T3	• Impedance (|Z|) from EIS after 28 days of immersion in 0.05 M NaCl Ohm-cm^2	(Kartsonakis et al. 2011)
		Cloisite 15A clay in epoxy ester	Al 2024 – T3	• PP measurement from immersion in 3.5 wt% NaCl solution for 84 days • Corrosion rate: 1×10^{-6} A/cm^2	(Singh-Beemat et al. 2012)
Marine	Conductive nanofiller in conductive matrix	Graphene in PANI matrix	Type 310 Stainless Steel	• PP measurement from immersion in 3.5 wt% NaCl solution for 168 hours • Corrosion rate: 1.5×10^{-7} A/cm^2 • Impedance (|Z|) from EIS after 168 hours: 1.3×10^4 Ohm-cm^2	(Jafari et al. 2016)
	Conductive nanofiller in non-conductive matrix	PANI/ZnO in epoxy matrix	Carbon Steel ST 37	• Impedance (|Z|) from EIS after 28 days of immersion in 3.5 wt% NaCl for 90 days: 1×10^9 Ohm-cm^2	(Mostafaei et al. 2013)
	Non-conductive nanofiller in conductive matrix	ZnO in Poly(o-phenylenediaimne) matrix	Type 304 Autenestic stainless steel	• PP measurement from immersion in 3.5 wt% NaCl solution • Corrosion rate: 34.21×10^{-9} A/cm^2 • Impedance (|Z|) from EIS: 4.98×10^5 Ohm-cm^2	(Ganash et al. 2014)

Table 3 contd. ...

...Table 3 contd.

Industry	PNC Category	Description of PNC	Substrate	Electrochemical Analysis	Reference
	Non-conductive nanofiller in non-conductive matrix	TiO_2-NiO sol-gel coating via dip-coating method	Type 316 L Autenestic stainless steel	• PP measurement from 1 hour immersion in 3.5 wt% NaCl solution • Corrosion rate: 34.21×10^{-9} A/cm^2 • Impedance (|Z|) from EIS: 4.98×10^5 Ohm-cm^2	(Cheraghi et al. 2012)
		TiO_2 sol-gel coating	Type 304 Autenestic stainless steel	• PP measurement from 1 hour immersion in 3.5 wt% NaCl solution • Corrosion rate: 7.77×10^{-9} A/cm^2 • Impedance (|Z|) from EIS: 6.8×10^4 Ohm-cm^2	(Ćurković et al. 2013)
		TiO_2-Al_2O_3 (80–20 wt%) sol-gel coating	Type 316 L Autenestic stainless steel	• PP measurement from 1 hour immersion in 0.5 M NaCl solution • Corrosion rate: 12×10^{-9} A/cm^2	(Vaghari et al. 2011)
		Hexagonal boron nitride nanoflakes in poly(vinyl alcohol) matrix	Type 316 L Autenestic stainless steel	• PP measurement from immersion in 3.5 wt% NaCl solution • Corrosion rate: 5.18×10^{-8} A/cm^2 • Impedance (|Z|) from EIS: 2.1×10^5 Ohm-cm^2	(Husain et al. 2013)
		Cloisite 15A nanoclay in low density polyethylene/ultra-high molecular weight polyethylene matrix	Mild steel	• PP measurement from immersion in 0.5 M NaCl solution • Corrosion rate: 2.48×10^{-6} A/cm^2	(Jeeva Jothi et al. 2014)
		TEOS-MTES based sol-gel coating	AISI 304 Stainless steel	• PP measurement from immersion in 0.5 M NaCl solution • Corrosion rate: 9×10^{-11} A/cm^2	(Pepe et al. 2006)
		$Ce(NO_3)_3.6H_2O$ in sol-gel coating of GPTMS + BPA	AISI 304 stainless steel	• PP measurement from immersion in 3.5 wt% NaCl solution • Corrosion rate: 2.3×10^{-8} A/cm^2 • Impedance (|Z|) from EIS: 7.44×10^5 Ohm-cm^2	(Zandi Zand et al. 2012)
Infrastructure	Conductive nanofiller in non-conductive matrix	CNTs in PMMA-silica matrix	Carbon steel A1020	• Impedance (|Z|) from EIS @25C immersion in 3.5 wt% NaCl until pitting develops: 1×10^9 Ohm-cm^2 after 53 days	(Harb et al. 2016)
		PANI/ Camphorsulfonate in epoxy	Mild steel	• Impedance (|Z|) from EIS immersion in 3.5 wt% NaCl: 1×10^6 Ohm-cm^2	(Pour-Ali et al. 2015)

Table 3 contd. ...

...Table 3 contd.

Industry	PNC Category	Description of PNC	Substrate	Electrochemical Analysis	Reference
		ZnO/PPY in epoxy matrix	Carbon steel A1020	• Impedance (|Z|) from EIS in 3.5 wt% NaCl: 5.21×10^8 Ohm-cm^2 zero immersion time	(Valença et al. 2015)
		PANI nanoparticles in epoxy ester matrix	Carbon steel A1020	• Impedance (|Z|) from EIS in 3.5 wt% NaCl: 1×10^{10} Ohm-cm^2 after 77 days immersion time	(Arefinia et al. 2012)
	Non-conductive nanofiller in non-conductive matrix	$Ce(NO_3)_3.6H_2O$ in sol-gel matrix	Carbon steel	• Impedance (|Z|) from EIS in 3.5 wt% NaCl: 7.2×10^4 Ohm-cm^2 after 24 hours immersion time	(Santana et al. 2015)
		LDH/MMT in epoxy matrix	Mild steel	• Impedance (|Z|) from EIS in 3.5 wt% NaCl: 1×10^{11} Ohm-cm^2 zero immersion time	(Dong et al. 2013)
Automotive	Conductive nanofiller in non-conductive matrix	PANI nanoparticles in phorylated poly(vinyl alcohol) matrix	Mild steel	• Impedance (|Z|) from EIS in 3 wt% NaCl: 1×10^7 Ohm-cm^2 at zero immersion time	(Chen et al. 2011)
		PANI/MWCNTs in epoxy matrix	Low carbon steel	• PP measurement after 144 hours immersion in 3.5 wt% NaCl solution • Corrosion rate: 1.26×10^{-5} A/cm^2 • Impedance (|Z|) from EIS: 1.313×10^3 Ohm-cm^2	(Deshpande et al. 2013)
		PPY coated alumina nanoparticles in zinc rich epoxy matrix	Cold rolled steel	• Not available	(Gergely et al. 2011)
		Mg nanoparticles in poly (ether imide)	Galvanized steel or low alloy steel	• PP measurement immediately after immersion in 3.5 wt% NaCl solution • Corrosion rate: 3.6×10^{-11} A/cm^2	(Dennis et al. 2014)
	Non-conductive nanofiller in non-conductive matrix	ZnO nanoparticles in epoxy matrix coating on conversion coating	Hot-dip galvanized steel	• Impedance (|Z|) from EIS after 120 days immersion in 3.5 wt% NaCl: 1×10^6 Ohm-cm^2	(Ramezanzadeh et al. 2011a)
		Organically modified MMT in polyurethane matrix	Cold rolled carbon steel	• PP measurement immediately after immersion in 3.5 wt% NaCl solution • Corrosion rate: 0.139×10^{-9} A/cm^2 • Impedance (|Z|) from EIS: 1×10^{10} Ohm-cm^2	(Heidarian et al. 2010)
		Silane treated ZrO_2 in epoxy matrix	Mild steel	• Impedance (|Z|) from EIS immediately after immersion in 3.5 wt% NaCl solution: 5×10^8 Ohm-cm^2	(Behzadnasab et al. 2011)

Table 3 contd. ...

...Table 3 contd.

Industry	PNC Category	Description of PNC	Substrate	Electrochemical Analysis	Reference
		Modified ZrO_2 in epoxy matrix	Mild steel	• Impedance (|Z|) from EIS immediately after immersion in NaCl solution: 3.9 × 10^7 Ohm-cm^2	(Haddadi et al. 2015)
		Ceramic nanocontainers containing 2-MBT in epoxy matrix	Hot-dip galvanized steel	• Impedance (|Z|) from EIS immediately after 31 days immersion in 5 mM NaCl solution: 2 × 10^7 Ohm-cm^2	(Kartsonakis et al. 2012)
		Clay Na-MMT in sol-gel coating	Hot-dip galvanized steel	• Impedance (|Z|) from EIS after 24 hours immersion in 3.5 wt% NaCl: 7.4 × 10^4 Ohm-cm^2	(Poelman et al. 2015)
		TiO_2/MgO/Al_2O_3 in polyetherimide matrix	High strength steel	• No positive results	(Rout et al. 2015)
		1 wt% modified clay in polyester-amide hyperbranched polymer matrix	Mild steel	• Impedance (|Z|) from EIS immediately after 50 days immersion in 3.5 wt% NaCl solution: 1 × 10^8 Ohm-cm^2	(Ganjaee et al. 2015)
		ZnO nanoparticles in epoxy-polyamide matrix	Hot-dip galvanized steel	• Impedance (|Z|) from EIS immediately after 7 days immersion in 3.5 wt% NaCl solution: 1 × $10^{8.5}$ Ohm-cm^2	(Ramezanzadeh et al. 2011)
	Non-conductive nanofiller in conductive matrix	Clinoptilolite in PANI matrix	Low alloy steel	• PP measurement immediately after immersion in 1 M HCl and 3.5 wt% NaCl solutions • Corrosion rates: 3.16 × 10^{-7} A/cm^2 and 1.04 × 10^{-6} A/cm^2	(Olad et al. 2010)
		ZnO nanorods in PPY matrix	Mild steel	• PP measurement in 3.5 wt% NaCl solution • Corrosion rate: 8.45 × 10^{-3} A/cm^2 • Impedance (|Z|) from EIS in 3.5 wt% NaCl solution: 1.7 × 10^4 Ohm-cm^2	(Hosseini et al. 2011)
	Conductive nanofiller in conductive matrix	Epoxy/Zn nanoparticles in PANI matrix	Mild steel/ Galvanized steel	• PP measurement after 1 week immersion in 3.5 wt% NaCl solution • Corrosion rate: 1 × 10^{-8} A/cm^2	(Olad et al. 2012)
		MWCNTs in PANI matrix	Mild steel	• PP measurement in 3.5 wt% NaCl solution • Corrosion rate: 3.4 × 10^{-9} A/cm^2 • Impedance (|Z|) from EIS in 3.5 wt% NaCl solution: 1 × 10^6 Ohm-cm^2	(Madhan Kumar et al. 2015)

temperatures (90°C), it was evaluated using EIS by immersing the coatings in 3.5 wt% NaCl solution at 90°C for 25 days. According to the EIS results, the incorporation of nanoparticles into epoxy coating improved the corrosion performance. Moreover, the best corrosion protection performance was obtained when the combination of LDH and MMT was used in comparison to the nanocomposites containing either of them. Table 3 provides the summary of the literature discussed in this chapter including various PNCs and their performances.

Conclusions

Corrosion has continuously been one of the major concerns in various industries where metals and alloys are used as one of the primary structural components. Several new technologies have recently evolved in addressing the challenges that arise due to constantly changing requirements in industries and regulatory laws by the legislative agencies. In recent times, polymer nanocomposite coatings (PNCs) have shown promising results when compared to conventional coatings in providing protection to metals and alloys from corrosion.

In this chapter, the fundamental concepts of corrosion and the electrochemical techniques such as potentiodynamic polarization (PP) and electrochemical impedance spectroscopy (EIS) that are used to evaluate corrosion have been discussed. Since corrosion is an electrochemical event in which metal is degraded/oxidized by losing electrons, the need to understand the effect of electrochemical properties of PNCs on their corrosion protection performance has been emphasized. PNCs have been classified into four categories on the basis of electrochemical properties of the components such as electrical conductivity and redox activity. While non-conductive materials used in making PNCs serve as physical barriers against the passage of corrosive species to the metal surface, conductive materials interact with the metal surface and corrosive species electrochemically to resist corrosion process at the interface. For example, sol-gel coatings have proven to provide excellent barrier protection and good adhesion to the substrate because of their hydrophobic nature and the ability to form strong covalent bonds with metals such as Fe-O-Si and Al-O-Si. Conductive materials such as PANI and PPY have been identified to demonstrate corrosion resistance at the metal-coating interface by either causing metal to form passive layer or inhibiting oxygen. Ideally, a system, which can do both these jobs effectively, would be considered the best as far as long-term protection of the metal substrate is concerned.

This chapter has provided a general review on the performance of various PNCs developed at the academic level for corrosion protection of different metals and alloys analyzing them to recognize their potential use at the industrial level. The prospective applications of PNCs for corrosion protection of most commonly used metals and alloys in the industries including aerospace, automotive, marine, and infrastructure have been focused. As it can be realized from this review, several researchers at the academic level have reported successful results with the application of PNCs for the corrosion protection of metal alloys, especially aluminum and iron, used in various industries. However, the most challenging task will be the process that involves delivering any promising technology developed at the academic level to the industrial scale with an equal degree of quality and efficiency. Evaluation and execution of some of these technologies for real-time applications have already begun in some cases and it is believed that they will, someday in near future, significantly benefit the society and improve the quality of life of humankind.

Acknowledgements

The authors would like to gratefully acknowledge the support from US Army Research Laboratory under Grant Nos. W911NF-09-2-0014, W911NF-10-2-0082, and W911NF-11-2-0027.

References

Arefinia, R., A. Shojaei, H. Shariatpanahi and J. Neshati. 2012. Anticorrosion properties of smart coating based on polyaniline nanoparticles/epoxy-ester system. Progress in Organic Coatings 75(4): 502–508.

Behzadnasab, M., S.M. Mirabedini, K. Kabiri and S. Jamali. 2011. Corrosion performance of epoxy coatings containing silane treated ZrO_2 nanoparticles on mild steel in 3.5% NaCl solution. Corrosion Science 53(1): 89–98.

Chen, C., M. Khobaib and D. Curliss. 2003. Epoxy layered-silicate nanocomposites. Progress in Organic Coatings 47(3-4): 376–383.

Chen, F. and P. Liu. 2011. Conducting polyaniline nanoparticles and their dispersion for waterborne corrosion protection coatings. ACS Applied Materials and Interfaces 3(7): 2694–2702.

Cheraghi, H., M. Shahmiri and Z. Sadeghian. 2012. Corrosion behavior of TiO_2-NiO nanocomposite thin films on AISI 316L stainless steel prepared by sol-gel method. Thin Solid Films 522: 289–296.

Ćurković, L., H.O. Ćurković, S. Salopek, M.M. Renjo and S. Šegota. 2013. Enhancement of corrosion protection of AISI 304 stainless steel by nanostructured sol-gel TiO_2 films. Corrosion Science 77: 176–184.
Dennis, R.V., L.T. Viyannalage, J.P. Aldinger, T.K. Rout and S. Banerjee. 2014. Nanostructured magnesium composite coatings for corrosion protection of low-alloy steels. Industrial & Engineering Chemistry Research 53(49): 18873–18883.
Deshpande, P.P., N.G. Jadhav, V.J. Gelling and D. Sazou. 2014. Conducting polymers for corrosion protection: A review. Journal of Coatings Technology Research 11(4): 473–494.
Deshpande, P., S. Vathare, S. Vagge, E. Tomšík and J. Stejskal. 2013. Conducting polyaniline/multi-wall carbon nanotubes composite paints on low carbon steel for corrosion protection: electrochemical investigations. Chemical Papers 67(8): 1072–1078.
Dong, Y., L. Ma and Q. Zhou. 2013. Effect of the incorporation of montmorillonite-layered double hydroxide nanoclays on the corrosion protection of epoxy coatings. Journal of Coatings Technology Research 10(6): 909–921.
Federica De Riccardis, M. and Virginia Martina. 2014. Hybrid conducting nanocomposites coatings for corrosion protection. Chapter 13: 271–317.
Fhwa. 2012. Steel Bridge Design Handbook Corrosion Protection of Steel Bridges, 19(November).
Figueira, R., I. Fontinha, C. Silva and E. Pereira. 2016. Hybrid sol-gel coatings: Smart and green materials for corrosion mitigation. Coatings 6(1): 12.
Ganash, A. 2014. Anticorrosive Properties of Poly (o-phenylenediamine)/ZnO Nanocomposites Coated Stainless Steel, 2014.
Ganjaee Sari, M., B. Ramezanzadeh, M. Shahbazi and A.S. Pakdel. 2015. Influence of nanoclay particles modification by polyester-amide hyperbranched polymer on the corrosion protective performance of the epoxy nanocomposite. Corrosion Science 92: 162–172.
Gergely, A., É. Pfeifer, I. Bertóti, T. Török and E. Kálmán. 2011. Corrosion protection of cold-rolled steel by zinc-rich epoxy paint coatings loaded with nano-size alumina supported polypyrrole. Corrosion Science 53(11): 3486–3499.
Ghassemieh, E. 2011. Materials in automotive application, state of the art and prospects. New Trends and Developments in Automotive Industry, 365–394.
Gkikas, G., D. Sioulas, A. Lekatou, N.M. Barkoula and A.S. Paipetis. 2012. Enhanced bonded aircraft repair using nano-modified adhesives. Materials and Design 41: 394–402.
Guadagno, L., M. Raimondo, V. Vittoria, L. Vertuccio, C. Naddeo, S. Russo et al. 2014. Development of epoxy mixtures for application in aeronautics and aerospace. Royal Society of Chemistry 15474–15488.
Gupta, G., N. Birbilis, A.B. Cook and A.S. Khanna. 2013. Polyaniline-lignosulfonate/epoxy coating for corrosion protection of AA2024-T3. Corrosion Science 67: 256–267.
Haddadi, S.A., M. Mahdavian and E. Karimi. 2015. Evaluation of the corrosion protection properties of an epoxy coating containing sol-gel surface modified nano-zirconia on mild steel. RSC Advances 5(36): 28769–28777.
Harb, S.V., F.C. dos Santos, S.H. Pulcinelli, C.V. Santilli, K.M. Knowles and P. Hammer. 2016. Protective coatings based on PMMA–silica nanocomposites reinforced with carbon nanotubes. Carbon Nanotubes—Current Progress of their Polymer Composites (September).
Heidarian, M., M.R. Shishesaz, S.M. Kassiriha and M. Nematollahi. 2010. Characterization of structure and corrosion resistivity of polyurethane/organoclay nanocomposite coatings prepared through an ultrasonication assisted process. Progress in Organic Coatings 68(3): 180–188.
Hintze-Bruening, H. and F. Leroux. 2012. Nanocomposite based multifunctional coatings. New Advances in Vehicular Technology and Automotive Engineering, 55.
Hosseini, M.G., R. Bagheri and R. Najjar. 2011. Electropolymerization of polypyrrole and polypyrrole-ZnO nanocomposites on mild steel and its corrosion protection performance. Journal of Applied Polymer Science 121(6): 3159–3166.
Hosseini, M. and A.S.H. Makhlouf (eds.). 2016. Industrial Applications for Intelligent Polymers and Coatings, Chapter 18.
Hung, W., K. Chang, Y. Chang and J. Yeh. n.d. Advanced Anticorrosive Coatings Prepared from Polymer-Clay Nanocomposite Materials. Structure.
Husain, E., T.N. Narayanan, J.J. Taha-Tijerina, S. Vinod, R. Vajtai and P.M. Ajayan. 2013. Marine corrosion protective coatings of hexagonal boron nitride thin films on stainless steel. ACS Applied Materials & Interfaces 5(10): 4129–35.
International, N. and W. Paper. 2012. Corrosion control plan for bridges (November).
Ionita, M., I.V. Branzoi and L. Pilan. 2010. Multiscale molecular modeling and experimental validation of polyaniline-CNTs composite coatings for corrosion protecting. Surface and Interface Analysis 42(6-7): 987–990.
Ireland, R., L. Arronche and V. La Saponara. 2012. Electrochemical investigation of galvanic corrosion between aluminum 7075 and glass fiber/epoxy composites modified with carbon nanotubes. Composites Part B: Engineering 43(2): 183–194.
Ivanov, H. 2016. Corrosion Protection Systems in Offshore Structures.
Jafari, Y., S.M. Ghoreishi and M. Shabani-Nooshabadi. 2016. Electrochemical deposition and characterization of polyaniline-graphene nanocomposite films and its corrosion protection properties. Journal of Polymer Research 23(5).
Jeeva Jothi, K., A.U. Santhoskumar, S. Amanulla and K. Palanivelu. 2014. Thermally sprayable anti-corrosion marine coatings based on MAH-g-LDPE/UHMWPE nanocomposites. Journal of Thermal Spray Technology 23(8): 1413–1424.
Kartsonakis, I.A., A.C. Balaskas and G.C. Kordas. 2011. Influence of cerium molybdate containers on the corrosion performance of epoxy coated aluminium alloys 2024-T3. Corrosion Science 53(11): 3771–3779.
Kartsonakis, I.A., A.C. Balaskas, E.P. Koumoulos, C.A. Charitidis and G.C. Kordas. 2012. Incorporation of ceramic nanocontainers into epoxy coatings for the corrosion protection of hot dip galvanized steel. Corrosion Science 57: 30–41.

Kelly, R.G., J.R. Scully, D.W. Shoesmith and R.G. Buchheit. 2003. Electrochemical techniques in corrosion science and engineering. World Wide Web Internet and Web Information Systems (Vol. 18).
Khun, N.W., B.C.R. Troconis and G.S. Frankel. 2014. Effects of carbon nanotube content on adhesion strength and wear and corrosion resistance of epoxy composite coatings on AA2024-T3. Progress in Organic Coatings 77(1): 72–80.
Lakshmi, R.V., G. Yoganandan, K.T. Kavya and Bharathibai J. Basu. 2013. Effective corrosion inhibition performance of Ce_{3+} doped sol–gel nanocomposite coating on aluminum alloy. Progress in Organic Coatings 76: 367–374.
Loveday, D., P. Peterson and B. Rodgers. 2004. Evaluation of organic oatings with electrochemical impedance spectroscopy. Part 2: Application of EIS to coatings. JCT Coatings Tech. October (10): 88–93.
Loveday, D., P. Peterson and B. Rodgers. 2005. Evaluation of organic coatings with electrochemical impedance spectroscopy; Part 3: Protocols for testing coatings with EIS. JCT Coatings Tech. 2(13): 22–27.
Lyon, S.B., R. Bingham and D.J. Mills. 2016. Advances in corrosion protection by organic coatings: What we know and what we would like to know. Progress in Organic Coatings. http://doi.org/10.1016/j.porgcoat.2016.04.030.
Madhan Kumar, A. and Z.M. Gasem. 2015. *In situ* electrochemical synthesis of polyaniline/f-MWCNT nanocomposite coatings on mild steel for corrosion protection in 3.5% NaCl solution. Progress in Organic Coatings 78: 387–394.
Makhlouf, A.S.H. and M. Hosseini. 2016a. Industrial Applications for Intelligent Polymers and Coatings, 511–535.
Makhlouf, A.S.H. and M. Hosseini. 2016b. Industrial Applications for Intelligent Polymers and Coatings.
McCafferty, E. 2010. Introduction to corrosion science. Introduction to Corrosion Science, 1–575.
Mert, B.D. 2015. Corrosion protection of aluminum by electrochemically synthesized composite organic coating. Corrosion Science 103: 88–94.
Monetta, T., A. Acquesta and F. Bellucci. 2015. Graphene/epoxy coating as multifunctional material for aircraft structures. Aerospace 2(3): 423–434.
Montemor, M.F. 2014. Functional and smart coatings for corrosion protection: A review of recent advances. Surface and Coatings Technology 258: 17–37.
Mostafaei, A. and F. Nasirpouri. 2013. Preparation and characterization of a novel conducting nanocomposite blended with epoxy coating for antifouling and antibacterial applications. Journal of Coatings Technology Research 10(5): 679–694.
Mostafaei, A. and F. Nasirpouri. 2014. Epoxy/polyaniline-ZnO nanorods hybrid nanocomposite coatings: Synthesis, characterization and corrosion protection performance of conducting paints. Progress in Organic Coatings 77(1): 146–159.
Musa, A.Y. 2007. Corrosion protection of Al alloys: Organic coatings and inhibitors. Recent Researches in Corrosion Evaluation and Protection, 51–66.
Olad, A. and B. Naseri. 2010. Preparation, characterization and anticorrosive properties of a novel polyaniline/clinoptilolite nanocomposite. Progress in Organic Coatings 67(3): 233–238.
Olad, A., M. Barati and S. Behboudi. 2012. Preparation of PANI/epoxy/Zn nanocomposite using Zn nanoparticles and epoxy resin as additives and investigation of its corrosion protection behavior on iron. Progress in Organic Coatings 74(1): 221–227.
Partnership, S. 1999. A guide to corrosion protection. Anti-Corrosion Methods and Materials 35(8).
Pepe, A., P. Galliano, M. Aparicio, A. Durán and S. Ceré. 2006. Sol-gel coatings on carbon steel: Electrochemical evaluation. Surface and Coatings Technology 200(11): 3486–3491.
Pirhady Tavandashti, N., S. Sanjabi and T. Shahrabi. 2009. Corrosion protection evaluation of silica/epoxy hybrid nanocomposite coatings to AA2024. Progress in Organic Coatings 65(2): 182–186.
Poelman, M., M. Fedel, C. Motte, D. Lahem, T. Urios, Y. Paint et al. 2015. Influence of formulation and application parameters on the performances of a sol-gel/clay nanocomposite on the corrosion resistance of hot-dip galvanized steel. Part I. Study of the sol preparation parameters. Surface and Coatings Technology 274: 1–8.
Pour-Ali, S., C. Dehghanian and A. Kosari. 2015. Corrosion protection of the reinforcing steels in chloride-laden concrete environment through epoxy/polyaniline-camphorsulfonate nanocomposite coating. Corrosion Science 90: 239–247.
Ramezanzadeh, B. and M.M. Attar. 2011a. An evaluation of the corrosion resistance and adhesion properties of an epoxy-nanocomposite on a hot-dip galvanized steel (HDG) treated by different kinds of conversion coatings. Surface and Coatings Technology 205(19): 4649–4657.
Ramezanzadeh, B. and M.M. Attar. 2011b. Studying the effects of micro and nano sized ZnO particles on the corrosion resistance and deterioration behavior of an epoxy-polyamide coating on hot-dip galvanized steel. Progress in Organic Coatings 71(3): 314–328.
Rout, T.K. and A.V. Gaikwad. 2015. *In-situ* generation and application of nanocomposites on steel surface for anti-corrosion coating. Progress in Organic Coatings 79(C): 98–105.
Santana, I., A. Pepe, E. Jimenez-Pique, S. Pellice, I. Milošev and S. Ceré. 2015. Corrosion protection of carbon steel by silica-based hybrid coatings containing cerium salts: Effect of silica nanoparticle content. Surface and Coatings Technology 265(3): 106–116.
Scotto, V., R. Di Cintio and G. Marcenaro. 1985. The influence of marine aerobic microbial film on stainless steel corrosion behaviour. Corrosion Science 25(3): 185–194.
Shchukin, D.G., S.V. Lamaka, K.A. Yasakau, M.L. Zheludkevich, M.G.S. Ferreira and H. M??hwald. 2008. Active anticorrosion coatings with halloysite nanocontainers. Journal of Physical Chemistry C 112(4): 958–964.
Singh-Beemat, J. and J.O. Iroh. 2012. Characterization of corrosion resistant clay/epoxy ester composite coatings and thin films. Progress in Organic Coatings 74(1): 173–180.

Spinks, G.M., A.J. Dominis, G.G. Wallace and D.E. Tallman. 2002. Electroactive conducting polymers for corrosion control: Part 2. Ferrous metals. Journal of Solid State Electrochemistry 6(2): 85–100.

Su, Y., I. Zhitomirsky, M. Olivier, M. Poelman, Y. Su and I. Zhitomirsky. 2013. Use of Electrochemical Impedance Spectroscopy (EIS) for the evaluation of electrocoatings performances. Journal of Colloid and Interface Science 399(1): 1–27.

Tallman, D.E., K.L. Levine, C. Siripirom, V.G. Gelling, G.P. Bierwagen and S.G. Croll. 2008. Nanocomposite of polypyrrole and alumina nanoparticles as a coating filler for the corrosion protection of aluminium alloy 2024-T3. Applied Surface Science 254(17): 5452–5459.

Tezdogan, T. and Y.K. Demirel. 2014. An overview of marine corrosion protection with a focus on cathodic protection and coatings. Brodogradnja/Shipbuilding 65(2): 49–59.

Vaghari, H., Z. Sadeghian and M. Shahmiri. 2011. Investigation on synthesis, characterisation and electrochemical properties of TiO_2-Al_2O_3 nanocomposite thin film coated on 316 L stainless steel. Surface and Coatings Technology 205(23-24): 5414–5421.

Valença Alves, K.G.B., C.P. De Melo and N.D.P. Bouchonneau. 2015. Study of the efficiency of polypyrrole/ZnO nanocomposites as additives in anticorrosion coatings. Materials Research 18(Suppl 2): 273–278.

Yuan, R., S. Wu, P. Yu, B. Wang, L. Mu, X. Zhang et al. 2016. Superamphiphobic and electroactive nanocomposite toward self-cleaning, antiwear, and anticorrosion coatings. ACS Applied Materials and Interfaces 8(19): 12481–12493.

Zandi Zand, R., K. Verbeken and A. Adriaens. 2011. The corrosion resistance of 316L stainless steel coated with a silane hybrid nanocomposite coating. Progress in Organic Coatings 72(4): 709–715.

Zandi Zand, R., K. Verbeken and A. Adriaens. 2012. Corrosion resistance performance of cerium doped silica sol-gel coatings on 304L stainless steel. Progress in Organic Coatings 75(4): 463–473.

Zhang, Y., H. Zhu, C. Zhuang, S. Chen, L. Wang, L. Dong et al. 2016. TiO_2 coated multi-wall carbon nanotube as a corrosion inhibitor for improving the corrosion resistance of BTESPT coatings. Materials Chemistry and Physics 179: 80–91.

Zubillaga, O., F.J. Cano, I. Azkarate, I.S. Molchan, G.E. Thompson and P. Skeldon. 2009. Anodic films containing polyaniline and nanoparticles for corrosion protection of AA2024T3 aluminium alloy. Surface and Coatings Technology 203(10-11): 1494–1501.

13

Recent Advances in the Design of Nanocomposite Materials via Laser Techniques for Biomedical Applications

Monireh Ganjali,[1,*] *Mansoureh Ganjali,*[2] *Somayeh Asgharpour*[2] and *Parisa Vahdatkhah*[2,3]

Introduction

There are two main approaches to synthesis of nanomaterials and the fabrication of nanostructures: the "Bottom-up" and "Top-down" approaches. The bottom-up approach includes the miniaturization of material components with further self-assembly processes which results in the formation of nanostructures (up to the atomic level). This approach leads to time-consuming developed atomic or molecular-scaled products in nano-fabrication processes. In contrast, the Top-down fabrication reduces the large pieces of materials all the way down to the nanoscale. The second approach requires larger amounts of materials and is more probable to result in substance waste.

Various bottom up- or top down-based methods have been developed like sol-gel processing (Moncada et al. 2007), high-energy ball milling (Zeng et al. 2012), laser ablation (Xu et al. 2013, Nikov et al. 2012), sonication (Narongdej and Boonroung 2010), chemical vapor deposition (CVD) (Veith et

[1] Bioengineering Research Group, Nanotechnology and Advanced Materials Department, Materials and Energy Research Center (MERC), Emam Khomeini Blvd, Meshkin Dasht, Karaj, Iran 31787-316.
[2] Noure Zoha Materials Engineering Research Institute, Shadmehr Street, Tehran, Iran, 14566-55945.
[3] Sharif University of Technology, Department of Materials Science and Engineering, Azadi Avenue, Tehran, Iran, 11365-11155.
Emails: ganjali.m@gmail.com; asgharpour.somayeh@yahoo.com; parisa.vahdatkhah@gmail.com
* Corresponding author: monireh.gan@gmail.com

al. 2010), hydrothermal process (Mishara et al. 1999), chemical vapor condensation (CVC) (Yang et al. 2002), plasma or flame spraying synthesis (Kavitha et al. 2007), laser pyrolysis (Reau et al. 2007, D'Amato et al. 2013), precipitation (Rashad and Ibrahim 2012) and reduction process (Pastoriza-Santos and Liz-Marzán 1999).

The laser assisted methods have been proved to be promising in Nano-composite synthesis due to its easy, rapid, precise, pure, versatile and cost-effective nature. This chapter will discuss the recent efforts and key research challenges in applying the laser-assisted methods for producing nano-materials.

Laser methods for synthesis of nanocomposites

Laser Ablation Method

The early uses of Laser Ablation Method (LAM) go back to 1960s when the laser technology was found to be highly attractive among the scientists. The first experimental paper about the applications of laser ablation was published in 1963, though the method was not employed for synthesizing nanomaterials with gas sensing purposes until mid-1990s (Williams and Coles 1999), when it was used for removing the unwanted materials from a specific solid using the laser beam in both pulsed and continuous modes. Numerous researches took place up to 1985 and resulted in successful different applications in science such as medical, bio-technology, quantitative and qualitative spectroscopy analysis and thin film growth functions. The LAM was used as early as 20 years after the invention of the laser technology for analyzing the majority of chemical elements which is called Inductively coupled plasma mass spectrometry (ICP-MS) (Devos et al. 2000, Mank and Mason 1999, Horn et al. 2000, Watmough et al. 1997, Li et al. 2000, Bi et al. 2000). Another device based on laser ablation of materials is the Laser Induced Breakdown Spectroscopy (LIBS) that was developed to perform the job both in qualitative and quantitative modes. These nondestructive techniques which are less limited in detection in comparison with other methods are used in several fields such as archaeology, geology, environment, and forensic investigation and semiconductor industries.

The ablation was applied through different wavelengths and pulse durations to fabricate different materials (Miller 2008, Russo et al. 2002a, Gunther et al. 2003, Gonzalez et al. 2007, Koch et al. 2004). Far from the early machine which was called the Ruby laser, the Nd:YAG (NIR at 1064 nm) laser machine is one of the best types of laser to use for ablation in different analytical sciences due to having cost-effective price, easy to handle set-up and better maintenance conditions in comparison with other machines. As the wavelength gets shorter, more particles will be ablated from the surface which is the ruling relationship in the LAM technique.

Principle of laser ablation

The laser ablation technique is in need of having one or more laser beams, the subject material, and a specific inert gas to be applied in liquid and gas medium. Also, magnetic stirrer in case of laser ablation in liquids is needed. Based on the wavelength and the pulse duration, the amount of the ablated micro- or nano-particles varies inside the medium. Inert gases, specially the Argon and Helium, are used to prevent the particles to be oxidized. According to Kosuke et al. (2015), the XRD analysis would approve the oxidization of micro- or nano-particles during the process in case the researcher decides not to use the inert gases (Kosuke et al. 2015).

There are two main mechanisms for ablating the particles which are both dependent on the wavelength directly: the thermal and non-thermal processes. As the laser light is absorbed by the subject's surface, the energy will result in breakdown, melting, vaporization and finally the ablation in atomic scales. The mentioned series would usually happen in case the pulse duration is as short as a nano-second and is called a thermal process. On the other hand, in cases of using ultra-short pulse durations, the breakdown, melting and vaporization time periods will be ignored as the ablation would happen without any heating process. This is called a non-thermal process where the absorbed energy level is much higher than the bonding energy. Additionally, during the increase in intensity of laser. Before an extended vapor plume created, increasing the power density of laser may abundantly excite electrons layer caused both, ionization and multiphoton absorption near the sample. In other words, the laser intensity is able to break the atomic lattices and eject ions and atoms without any heating process. Regarding the independence of the particles removed in thermal features of target components, this mechanism called non-thermal ablation may also take part in decreasing fractionation (Russo et al. 2002b). In fact, the non-thermal process is an endeavor to increase the energy level of surface electrons which results in ion creation. These ions stay between the laser beam and the surface in a medium called the "vapor plume" (Russo et al. 2002a).

The principle of laser ablation is that plasma plums, shockwaves and thermal pressure lead to ablate particles in nanosizes while powerful, very short, rapid and moving laser pulses scan various target surfaces. This approach is to focus laser beam precisely for vaporization of the target coating to create compositions based on metal or ceramic (see Figure 1; Ganjali et al. 2011).

Furthermore, this process is able to optimize the laser beam which produces a maximum reaction with the target material for speed while, simultaneously, does so safely and without any harm to the base composites. Laser beam power density is to accurately and easily provide impossible products in nanoscales with high quality and purity (Pillai and Kamat 2004, Brust et al. 1994, Pastoriza-Santos et al. 1999, Bonet et al. 1999).

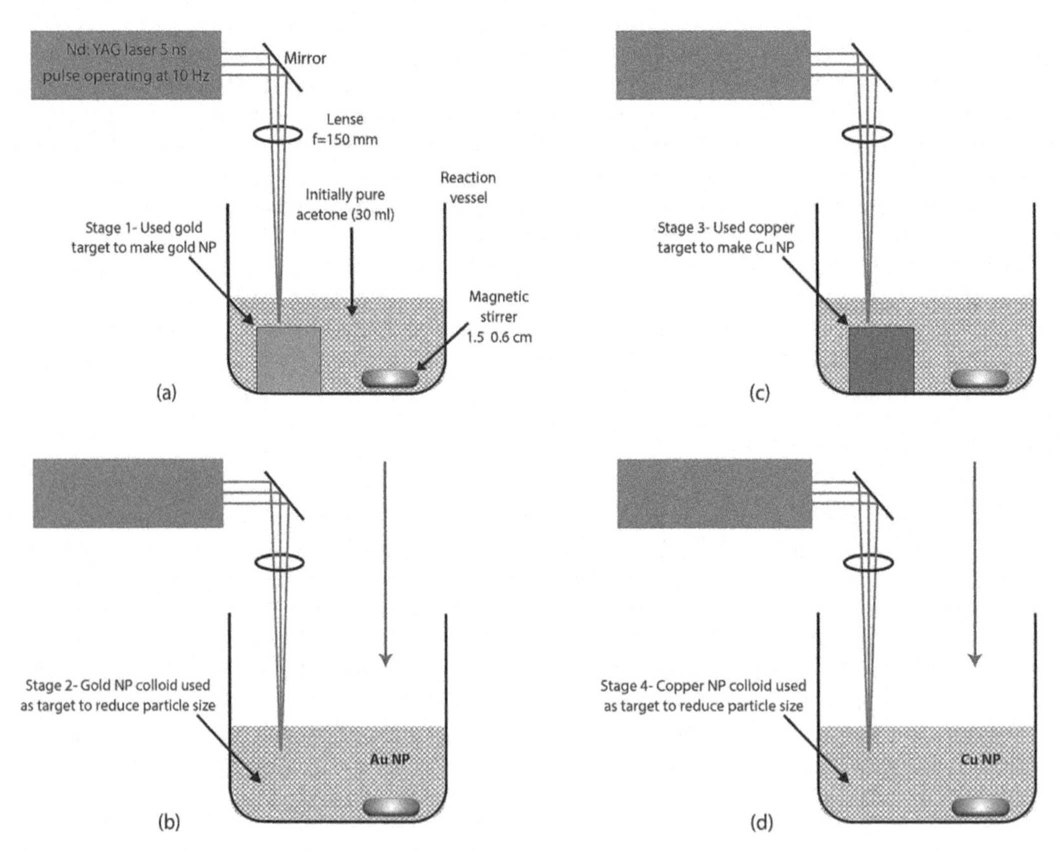

Figure 1. Schematic of laser ablation method in liquid. Reprinted with permission from Ganjali, Man., Mo. Ganjali, S. Khoby and M.A. Meshkot. 2011. Nano-Micro Lett. 3: 256–263. Copyright@Springer.

Advantages and disadvantages of laser ablation

Comparing with chemical synthesis, LA has been demonstrated as a simple, compatible, promising, reliable, relatively low-cost and green approach with absence of chemical toxic and reagents, which may pose low environmental and biological risks, be proficient to ablate different type of materials such as metals, ceramic and polymer considering the ultra-high energy density, control over the process parameters like pulse width, energy density, wavelength, scanning rate and so forth (Parashar et al. 2009, Tsuji et al. 2002, Mafuné et al. 2000a and b, Yamamoto and Nakamoto 2003). Gram per hour can be production rates using laser–synthesized nanoparticles whereas their concentration depends on laser parameters and solvent (Bärsch et al. 2009, Sajti et al. 2010).

Comparing to chemical methods, the laser synthesized nanomaterials have obviously had certain particularities considering both stabilization of the particles and agents. Consequently, such method considerably enables to gain contrasting images in biomedicine. The detection includes even single particles with high resolution (Klein et al. 2010).

The ablation process with laser involves several important disadvantages that are unavoidable. First of all, in comparison to chemical and other conventional methods the sample consumption is significantly decreased via LAM and that is why this approach is considered microchemical whereas dissolution ones are exposed to a bulk analyses in quantitative/qualitative characterizing of samples. Nonetheless, fractionation can also appear because of lower laser intensities and longer pulse widths (Russo et al. 2002).

Laser ablation in liquids

There are various aspects of creating nanosize materials using different phases such as solid, liquid and gas targets in a liquid, gas and even vacuum as media in this regard. Based on reports by various scientists, the most effective collection of synthesized nanostructures have been achieved via the LAM in a liquid phase with specific focus on their purity, size, shape and stability (Amendola et al. 2007, Amendola and Meneghetti 2009).

The potential of functionalized issues in nanosizes and their supreme features have not fully been achieved. To integrate these features into production, this procedure in aqueous solution is an appropriate technique to versatilely produce highly pure and functionalized nanomaterials in polymer and biomolecules. These versatile nanomaterials are widely applied in medicine and engineering. For instance, scratch resistance of surface in ceramics, antibacterial properties in silver and medical diagnoses and therapies in gold or iron can only be examples for up to date application possibilities.

Laser ablation for synthesis of nanocomposite materials

As the recent published efforts, nanocomposites have become of significant interest to bio-scientific researchers in terms of their applications in science and technology with novel optical, chemical and physical properties.

Nanomaterial phases such as nanocomposites and nanoalloys with free-agglomeration simple lead to available and unique productions having the most common advantages in medicine, biotechnology and other relevant industries (Barcikowski et al. 2009).

The layered nanocomposites are synthesized and analyzed using LAM devoted to magnetism effect, efficient catalysis and high ion-exchange capacity, enhancement of the mechanical and thermal stability of polymers, and applying of the optoelectronic device lasers as well as sensors (Fujita and Awaga 1997, Okazaki et al. 2000, Newman and Jones 1998, Kandare et al. 2006, Usui et al. 2005, Van der Molen et al. 2007, Kumar et al. 2006, Yang et al. 2007).

Noble-metal/oxide material as a nanocomposite with high surface energy and high specific surface area is another important and promising type of nanocomposite which has been applied in photocatalysis, chemical catalyst, modified biotechnological functions, naked-eye colorimetric detection, therapeutics, diagnosis and rapid improvement in analytical techniques on surface and nanoscale materials (Liu et al. 2016, Lin et al. 2010, Folarin et al. 2011).

It is also possible that since composite only includes a heterogeneous mixture, an alloy may be defined as a specific type of composite with heterogeneous and homogeneous mixtures. So in this part, authors provide pieces of evidence in which bimetallic alloys have successfully been synthesized using LA method in liquid with coated form on a substrate. Multipurpose sensors, optical markers, microelectronic devices, and filters are among many other applications with unique properties. The size, stoichiometry (atomic arrangement) and their compositions remarkably play a great role to combine these alloys (Ganjali et al. 2011, 2014, Sinfelt 1983, Ferrando et al. 2008, Wang et al. 2010, Johnston et al. 1992).

Laser assisted Self-propagated High-temperature Synthesis (LSHS)

Principle of self-propagated high-temperature synthesis

During the recent decades, a lot of attention has been focused on the production of Nano-materials and Nano-powders because they show novel electronic, optical, magnetic, photochemical, electrochemical and mechanical properties (Yang and Riehemann 2001). The researches display that an effective and efficient process to produce powders is Self-propagated High-temperature Synthesis, SHS, which has been taken as an important goal for the future of technology in the composites (Mukasyan and Dinka 2007).

The use of SHS in ceramic technology, due to production of sinterable and fine-sized or high purity micro powders, has been significantly developed (Vincenzini et al. 2010). SHS is an exothermic process in which the reaction occurs between two or more solids or gaseous reactants and this reaction occurs with a self-propagating regime during the formation of its solid product (McCauley and Puszynski 2007). The SHS process has two basic steps: ignition of reaction and propagation of reaction. The ignition is first initiated by an external heating source such as laser beam irradiation, radiant flux, resistance heating, spark or chemical oven, and then induced by the chemical reaction inside the heated materials (Shen et al. 2000). Combustion synthesis may occur in two different modes: In wave propagation mode, the reaction is started from a small but sufficient large volume of the reactant sample by the localized heat and is expanded through the pressed mixture like a combustion wave. But in explosion mode, the whole of pressed mixture is simultaneously heated to reach the initial reaction temperature. In this mode, the reactions occur spontaneously and simultaneously in the whole of sample (see Figure 2) (Pacheco et al. 2008).

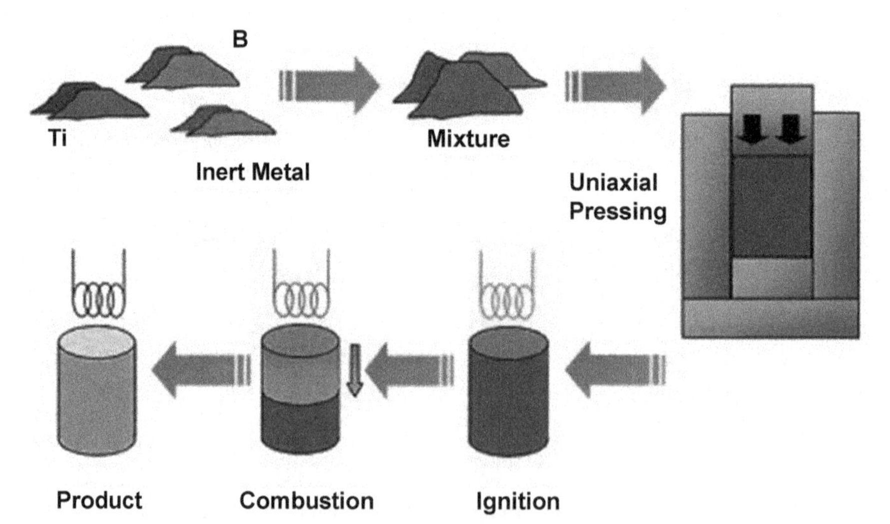

Figure 2. Scheme of the basic steps during the combustion synthesis process. Reprinted with permission from Martinez pacheco, M., R.H.B. Boumar and I. katgerman. 2008. Applied Physics A 90: 159–163. Copyright@Springer.

Methods of composite synthesis

Mechanical Alloying (Mukasyan and Dinka 2007), Plasma, Physical or Chemical Vapor Condensation (Yang et al. 2002), Spray Pyrolysis (Mishra et al. 1999), Sol-/Gel (Chen et al. 1995), Laser Ablation (Mukasyan and Dinka 2007) and Combustion Synthesis are various methods to synthesise composite. In Table 1, some advantages and disadvantages of these methods are given.

Some more significant advantages of SHS include: the high temperature of reaction causes the evaporation of contaminations which have low boiling temperature. So the final product has high purity. The low time consumption of reactions causes the reduction of the costs. The exothermic and simple nature of SHS process provides for the reaction to occur without needing some expensive facilities and on the other hand, the energy is saved (McCauley and Puszynski 2007).

Table 1. Various methods of different composites' synthesis.

Various Methods of Composite Synthesis	**Advantage**	**Disadvantage**
Mechanical Alloying	Simple	Impurity and taking time
Plasma	High purity and grain size < 10 nm	Expensive
Physical or Chemical Vapor	High purity, appropriate for thin film	Inability to produce on industrial scale
Condensation		------
Spray Pyrolysis	High purity	Unability to produce industrials scale
Sol-/Gel	Same size products	Takes time
Laser Ablation	High purity	Expensive

New approach

Today, taking maximum advantage of the material properties is not possible without utilizing the benefits of nanotechnology (Abbasi et al. 2014, Ramezanalizadeh and Heshmati-Manesh 2012). On the other hand, the product of SHS procedures is fine enough, but some provisions must be made to optimize the synthesis of nanostructures. One of the common methods used to achieve nanostructures is "Mechanical Alloying" (MA); however, this method has proven its own ability to achieve stable compounds with nanostructure; the researchers have also faced some problems. For example, the synthesis of a broad pattern of main phases with intermediate phase, the entrance of impurity through balls and cups, use of too much energy and excessive length of the process (Suryanarayana 2001).

Recently, combining milling and SHS process has led to faster and better combustion processes (Gras et al. 2001). It is important to mention that the SHS itself imposing very high temperatures (although in short time) usually does not have the ability to develop compounds with nanostructures and should be combined with mechanical activation methods (MASHS)[1] as an auxiliary procedure. It is well recognized that a combustion based technique as Self propagating High-temperature Synthesis (SHS) is an effective energy and time saving method for synthesis of a variety of advanced materials (Yang and Riehemann 2001). Various types of laser as igniter were utilized such as Neodymium-doped glass laser, ruby laser, cesium bromide laser (Barzykin 1992), diode laser (Biffi et al. 2017), pulsed and continuous wave CO_2 laser (Volpi et al. 1998, Chen et al. 2002, Farley et al. 2011, Masanta et al. 2016). By using laser beam as igniter, the process is more controllable (Gras et al. 2001).

[1] Mechanical Activated SHS.

Some researches were done via SHS in order to fabricate some materials which can be used in medicine as artificial bone graft substitutes and cell tissue carriers (Amosov et al. 2013, Masanta et al. 2016).

In the recent research, continuous wave of CO_2 (CW-CO_2) laser beam was used as an external heating source for ignition and to approach nano scale materials, a short duration milling was used. By considering unique properties of composite Al_2O_3-ZrO_2 such as high toughness, high wear resistance and relative low thermal expansion, this binary system was chosen as a model system for its many applications in toughened ceramics, wear resistant ceramics, thermal barrier coating in gas turbine, oxygen sensor, dental implant and the femoral head of hip replacement (Yang and Riehemann 2001). The results of X-ray analysis are equal for both samples which are synthesized in furnace and via laser beam radiation. But the SEM images, as you can see in Figure 3, are different.

The white part mostly consisted of ZrO_2 as the EDAX analysis showed and the black part mostly consisted of Al_2O_3. Because the mode of propagation is explosion mode, particles do not have enough time to replace a specific order so they have formed in dendrite shape but via laser, in wave propagation, particles find appropriate time to form a circular shape. One of the most important advantages of synthesizing via laser is time conservation. Also, this process doesn't need high-tech facilities and it obviously is a simple method with low-energy consumption.

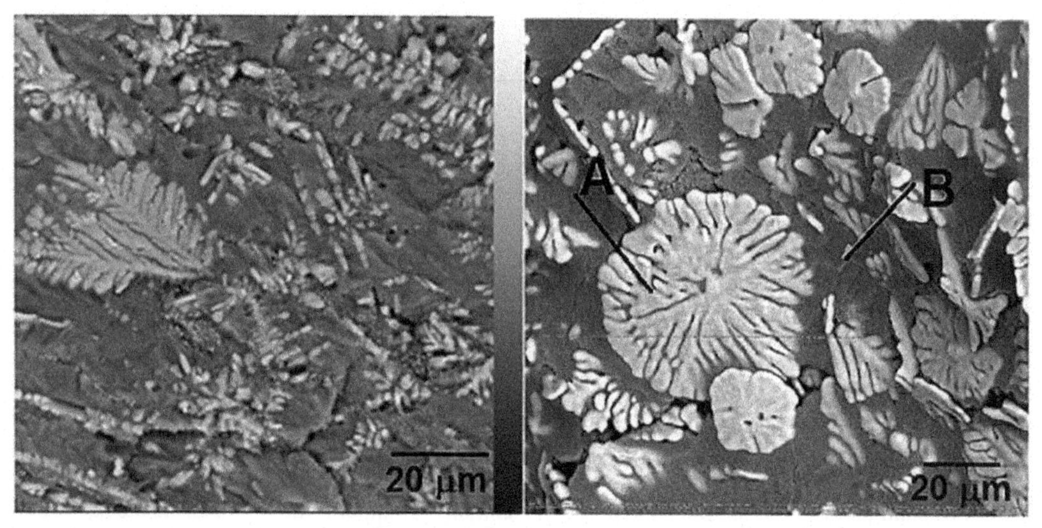

Figure 3. The image in the left: sample was synthesized in furnace and the right one was synthesized via laser. With permission from Ganjali, M., M. Vaezi, S. Tayebifard and S. Asgharpour. 2013. Int. J. Eng. Trans. A 27: 615–620. Copyright@ Materials and Energy Research Center (MERC).

Laser cladding

Laser cladding process

Laser Cladding (LC) is a processing technique used for melting a substrate surface via a focused laser beam in which cladded layer is formed by feeding materials into the melt pool continuously. In this process, coating layer is chemically different from the substrate (Toyserkani et al. 2004, Pawlowski 1999). Schematic of laser cladding is shown in Figure 4. Feeding materials can be in the form of wire, strip, or powder, where the powder form is most widely used due to ease of control for lower feed rates. The powder can be transported by an inert carrier gas (dynamic blow method) or preplaced (pre-placed method) into the melt pool. This laser cladding process is emerging as a strategic technique for improving wear resistance, corrosion resistance and bioactivity, and repairing critical parts (Capello et al. 2005, Brandt et al. 2009, Yakolev et al. 2004).

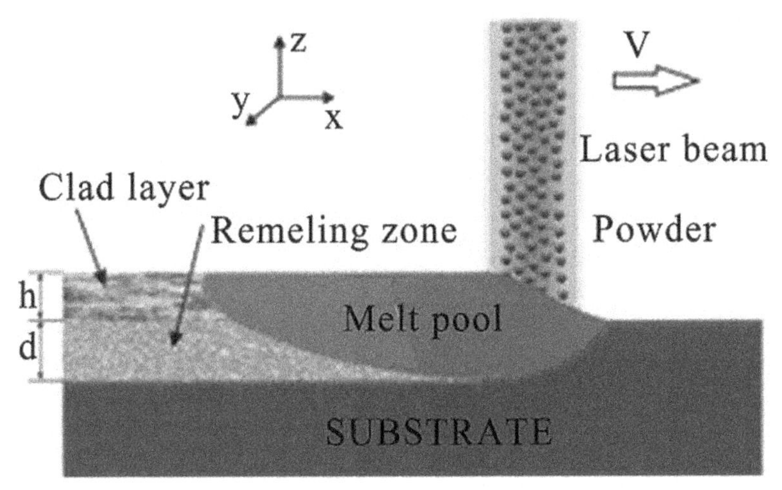

Figure 4. Schematic of laser cladding. Reprinted with permission from Niziev, V.G., F.Kh. Mirzade, V.Ya. Panchenko, M.D. Khomenko, R.V. Grishaev, S. Pityana and C.V. Rooyen. 2013. Modeling and Numerical Simulation of Material Science 3: 61–69. Copyright@Scientific reseach.

New approach

Recently, laser cladding has been applied for depositing bioactive Glass (Comesaña et al. 2010, Fei et al. 2014), Calcium phosphate (Hydroxyapatite, Fluorapatite and Tricalcium phosphate) (Krishna et al. 2008, Chien et al. 2012, Wang et al. 2008) on metallic surface to improve corrosion resistance and biocompatibility of these materials. However, the mechanical properties of these materials, including their low fracture toughness and ductility, high elastic modulus and brittleness, cannot meet the demands of the load-bearing applications (Kokubo et al. 2003, Cheng et al. 2005). So, one way to solve this problem is to use composite materials. For example, the mixture of $CaHPO_4.2H_2O$ and CaCO and little ceria powder were used to coat thin plate of the titanium alloy (Ti–6Al–4V) by laser cladding (Zheng et al. 2008). The XRD spectra of surface coating identified that the HA and β-tricalcium phosphate as bioactive phases were synthesized on the surface coating. Furthermore, the methyl thiazolyl tetrazolium (MTT) assay of cell proliferation revealed that cell growth distinctly increased on the laser-cladded bioceramic coating compared with the untreated substrate.

In another work, carbon nanotube/HA nanocomposite on pure titanium was coated by laser cladding technique (Pei et al. 2011) and improved bone osseointegration, the hardness of composite coating and bonding strength of metal implants. The micro-hardness of CNTs/HA nanocomposite coatings gradually increased with the increase of thickness of coating. The results of *in vitro* biocompatibility tests on MC3T3-E1 osteoblast-like cell line indicated that the biocompatibility of coating containing CNTs/HA was comparable with that of pure HA coating due to its high surface area and porous structure of the CNTs-containing coatings (Lin 2008).

A novel nanocomposite layer of nano-silver-containing hydroxyapatite (Ag-HA) was prepared on the surface of biomedical Ti6Al4V by laser processing (Liu 2017). In this study, different contents of Ag were added to HA powder and prepared Silver-hydroxyapatite (Ag-HA) composite powder using a reduction method. Then Ag-HA composite powders were mixed with polyvinyl alcohol (PVA) and painted with a brush on the Ti6Al4V samples and laser cladding was performed using a CW 2kW Nd:YAG laser. SEM images show that this method is an effective method for preventing the coatings from peeling during long-term implantation and deposited Ag-HA coatings are tightly bonded with the substrate by laser melting process. 5% Ag-HA was immersed in SBF solution for different periods for up to 21 days. The results indicated the formation of different HA nucleation and growth on surface coating, which is also confirmed by the XRD patterns acquired from the immersed 5% Ag-HA sample.

Laser pyrolysis

Laser pyrolysis process

Among other synthesis techniques, research tends to focus on pyrolysis, which means chemical decomposition by heat supplied by flame combustion (Kavitha et al. 2007), laser (Reau et al. 2007, D'Amato et al. 2013) or another general heat source. The laser interacts with gas precursor and due to high temperature gradient and fast reaction time, reagents decompose in the form of atomic radical which recombine together to form particles with uniform and controllable particle size distribution. In Figure 5, a schematic general-purpose laser pyrolysis installation is depicted. Laser pyrolysis is a potential strategy to produce powders with quality, high capacity and good reliability with appropriate robustness.

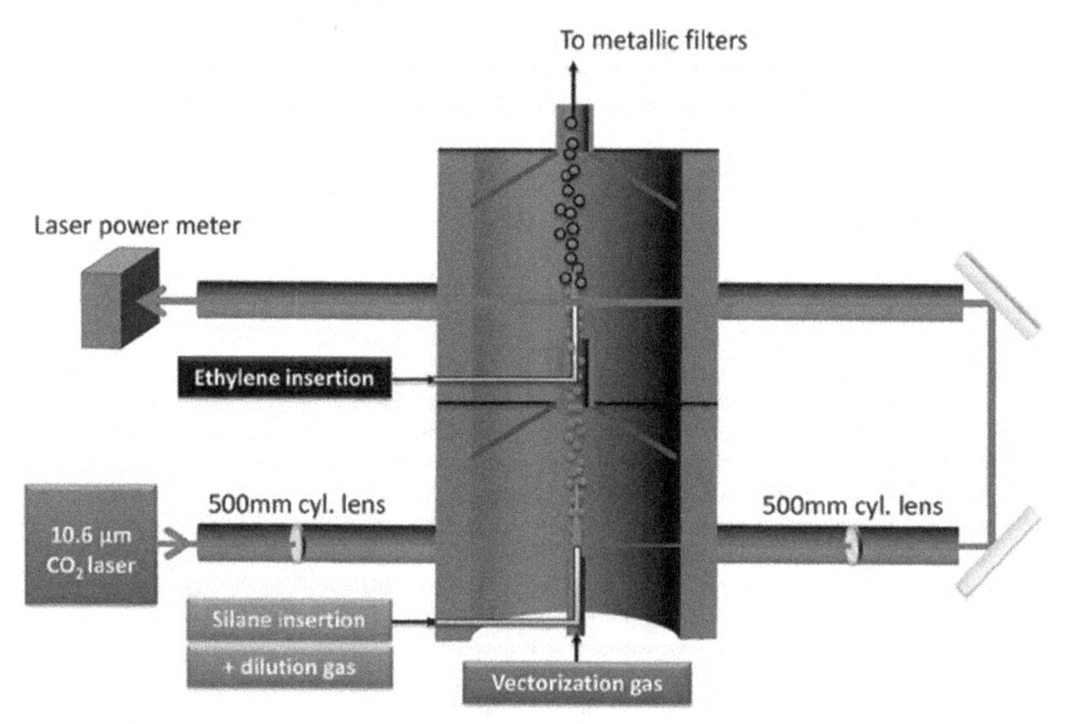

Figure 5. Schematic diagram of the laser pyrolysis reactor in the two stages configuration. Reprinted with permission from Sourice et al. 2015. ACS Appl. Mater. Interfaces 7: 6637–6644.

New approach

This method has been used for the synthesis of a wide range of iron and carbon-based oxide semiconductors, metal-based polymer nanocomposites, etc. (Borsella et al. 2008, 1997). These powders have been widely used for the preparation of colloidal dispersions in various biomedical applications due to their non-cytotoxicity (Bomatí-Miguel et al. 2010).

Iron nanoparticles, due to their high magnetic response, are widely studied because of their potential application in various fields such as in microelectronics (magnetic recording and magneto-optics, data storage devices) (Raj and Moskowitz 1990, Piramanayagam and Chong 2011), bio-medicine (magnetic resonance imaging, magnetic carriers for drug delivery, a.c. magnetic field-assisted cancer therapy) (Jain et al. 2005, Patel et al. 2008, Davaran et al. 2014) or catalysis (Genuino et al. 2013). Some studies were performed on synthesis and magnetic properties of iron based nanocomposites using laser pyrolysis. Iron nanocomposites consisting mainly of iron cores embedded in a carbon matrix have been directly synthesized by the laser induced pyrolysis (Schinteie et al. 2013, Morjan et al. 2012). The results show that the magnetic measurements of the composite nanoparticles indicate an almost direct correlation between the residence time in the laser beam and the saturation magnetization.

In another work, magnetic nanoparticles encapsulated in a silica matrix were prepared by laser pyrolysis technique (Sabino et al. 2007). The advantages of the silica composites prepared by laser pyrolysis are not only the synthesis procedure itself, that is a continuous method with no need for size selection process after synthesis, but also the fact that particles are obtained directly coated with silica, which is easily functionalized with different biomolecules and have a high resistance to de gradation in a biological environment.

Selected Laser Melting (SLM)

Selected Laser Melting (SLM) process

Another well-known method used to create nanocomposites for medical application is Selective Laser Melting (SLM). This technique provides very well-fitted implants surrounded with supportive tissues. In most cases, SLM technique is a special form of Selective Laser Sintering (SLS). In this process, the laser with high power density is used to melt and fuse layers of metallic powders. This layer-to-layer process continues to complete the fabricated composite.

New approach

Shishkovskii et al. (2011) investigated the layer-by-layer synthesis of full density 3D parts nitinol and hydroxyapatite using SLS/SLM process. A used SLS/SLM setup in their work consists of several parts: a laser (operation mode: continuous); deflectors of laser radiation; a computer and its relevant software to control the process and to design the layer-by-layer synthesis; a roller for distributing the powder mixture; and two pistons where the first one contains the powder mixture and another one is intended for a 3D layer-by-layer synthesis. When the first layer is sintered, the left platform goes up and then the powder mixture is distributed over the right platform by roller. This process continues until the completion of a 3D composite part. Also, they studied the effect of laser parameters on the phase composition and structure of composite. No remarkable decomposition of HA under laser irradiation is observed which indicates that its pharmacological properties are preserved. By developing cracks in SLM process, which results in full density 3D parts, the specimens can be used as scaffolds in tissue engineering applications.

Recently, Wei et al. (2015) manufactured layer-by-layer stainless-steel/nano-hydroxyapatite implant by using SLM as a solid free-form fabrication technique. Their study set out to fabricate the load-bearing bone implant with high strength and uniform metallurgical bonding of metal and ceramic compositions. To improve the mechanical properties of the implants, another phase with higher strength was added to HA coating. Nanocomposite fabrication was first carried out by mixing commercial gas atomization 316L SS and nHA powders with average particle sizes of 28 μm and less than 100 nm, respectively, in ball grinder. Wei et al. (2015) revealed the influence of 5, 10 and 15 vol% ratios of nHA to SS and also SLM processing parameters on the nanocomposite quality.

The in-house HRPM-type SLM equipment (shown in Figure 6) using a fiber laser at a wavelength of 1090 ± 5 nm with an output power of 200 W was employed to fabricate the SS/nHA composites' specimen under argon atmosphere. The fabrication conditions were laser beam spot size = 50–80 μm, laser power = 110 W, layer thickness = 0.02 mm, scanning spacing = 0.08 mm and scanning speed = 250, 300, 350 and 400 mm/s. By changing the scanning speed, the effect of laser energy density on the prepared composite was investigated.

In all specimens, nHA were uniformly distributed. Then, they adhered to the surfaces of SS sphere and flowability of the composite powders did not change after adding nHA powders. Due to lower density of nHA relative to SS, during SLM process, nHA and its decomposed products floated up and, therefore, nHA aggregation generated cracks around the "track-track" melt pool boundaries (MPBs) (Figure 7). Wei et al. (2015) demonstrated that the regular and fine columnar grains were surrounded by MPBs.

Figure 6. The in-house developed HRPM-type SLM equipment. Reprinted with permission from ref. Wei, Q., S. Li, C. Han, W. Li, L. Cheng, L. Hao and Y. Shi. 2015. J. Mater. Process. Tech. 222: 444–453. Copyright@ Elsevier BV.

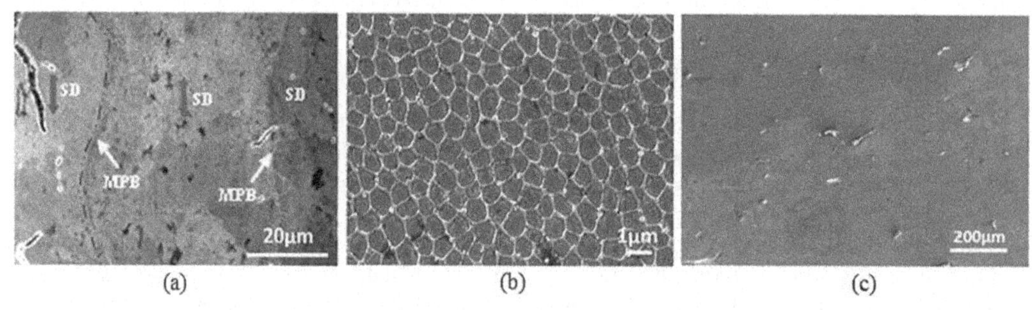

Figure 7. (a), (b) SEM morphologies of melt pool overlap as well as a magnified view of grain structure (with nHA content of 15 vol.% at scanning speed of 400 mm/s): Observed in the X–Y reference plane (perpendicular to the build direction), BD—building direction; MPB—melt pool boundary; SD—scanning direction of laser beam; and (c) Low magnification SEM morphologies of the cross sections parallel to the laser scanning plane (X–Y plane) with nHA contents (C) = 5 vol.%, and processed at scanning speeds (V) = 350 mm/s. Reprinted with permission from ref. Wei, Q., S. Li, C. Han, W. Li, L. Cheng, L. Hao and Y. Shi. 2015. J. Mater. Process. Tech. 222: 444–453. Copyright@ Elsevier BV.

In addition, due to the uniform distribution of Ca and P elements on the composite surfaces, the bone osseointegration was improved more than the pure metal implants.

As laser scanning speed increased, the degree of cracking was reduced. This, in turn, reduced the density of laser energy which led to formation of pores. As a result, the optimal condition with minimum cracks and pores belonged to the sample with 5 vol. % nHA content and 350 mm/s scanning speed which exhibited high tensile strengths in the range 130 MPa to 590 MPa. The identified mechanical properties such as elastic modulus and hardness showed that SS/nHA porous scaffolds could be used for bioactive and load-bearing bone applications.

Hybrid laser technology

It should be noted that various hybrid techniques such as pulsed laser deposition (PLD) + magnetron deposition, PLD + RF discharges, double or triple PLD and laser chemical vapor deposition (LCVD) can also be used for coating the implants. In a recent work, Jelinek et al. (2016) developed the hybrid methods to fabricate doped biocompatible diamond-like carbon coating using PLD + magnetron and double PLD. They found that using various techniques can control different film parameters. Their study showed that hybrid techniques such as pulsed laser deposition (PLD) + magnetron deposition can be possibly used to fabricate nanocomposites for biomedical applications.

New approach

Veith et al. (2010) assessed the cell compatibility of Al/Al_2O_3 biphasic nanowires, which was used as an implant, by normal human dermal fibroblast (NHDF) as a cell model. Al/Al_2O_3 biphasic nanowires were synthesized by chemical vapor deposition and pulsed laser treatment. First, $(tBuOAlH_2)_2$ was decomposed on metal, silicon and glass substrates within a CVD chamber. The diameters and lengths of the synthesized nanowires were about 25 nm and several micrometers, respectively. Veith et al. (2010) found that the nanowires were composed of Al core and Al_2O_3 shell. Then, nanosecond Nd:YAG pulsed laser was used to heat treatment of Al/Al_2O_3 nanowires at a central wavelength of 532 nm and a constant laser fluence of 0.2 J cm^{-2}. Due to laser treatment in air, the melting and, subsequently, oxidation of the Al-core to Al_2O_3 occurred.

Veith et al. (2010) showed that pulsed laser can prepare multiscale features. The mixture of nano- and microstructures is necessary for interaction between cell and surface of implant because the extracellular matrix has a variety of micro- and nanostructures as well. The results showed that treating nanowires with one and two laser pulses improves the cell compatibility. However, by increasing the number of laser pulses, the result was reversed and sphere-like microstructures were formed (Figure 8). Therefore, to control cellular adhesion better, the post-laser treatment of Al/Al_2O_3 composite nanowires was proved useful in modifying the surface.

Laser spinning technique

Laser spinning technique: new approach

Quintero et al. (2007) developed the laser spinning technique for creating glass nanofibers of several centimeters in length using the supersonic gas jet to clear the cut edge in microseconds. Some of the features of the laser spinning technique are its high speed, high product content, adjustable chemical compositions, and no need for post heat treatments and chemical additives (Boccaccini et al. 2010). The function of the frictional drag action of the supersonic gas jet is rapid stretching and cooling of the molten material (Quintero et al. 2007). Due to the high cooling speed in this technique, the resulting fibers are amorphous. The length to diameter ratio of a fiber is the function of the process time (Boccaccini et al. 2010). Quintero et al. (2009) proposed the production of bioactive glass nanofibers in the 45S5 Bioglass® and 52S4.6 compositions by the laser spinning technique. They found that the obtained nanofibers can be used as a scaffold, as a synthetic bone graft, or as reinforcement in bioactive nanocomposites. Energy dispersive X-ray spectroscopy (EDS), selected area electron diffraction (SAED) TEM and bioactivity tests confirmed the bioactivity of these glass nanofibers (Figure 9).

Laser printing

Laser printing process

Although the location and shape of the cell pattern is a key issue in tissue engineering, laser-assisted bioprinting (LAB) can be used for fabrication of two- and three-dimensional biological materials. In this

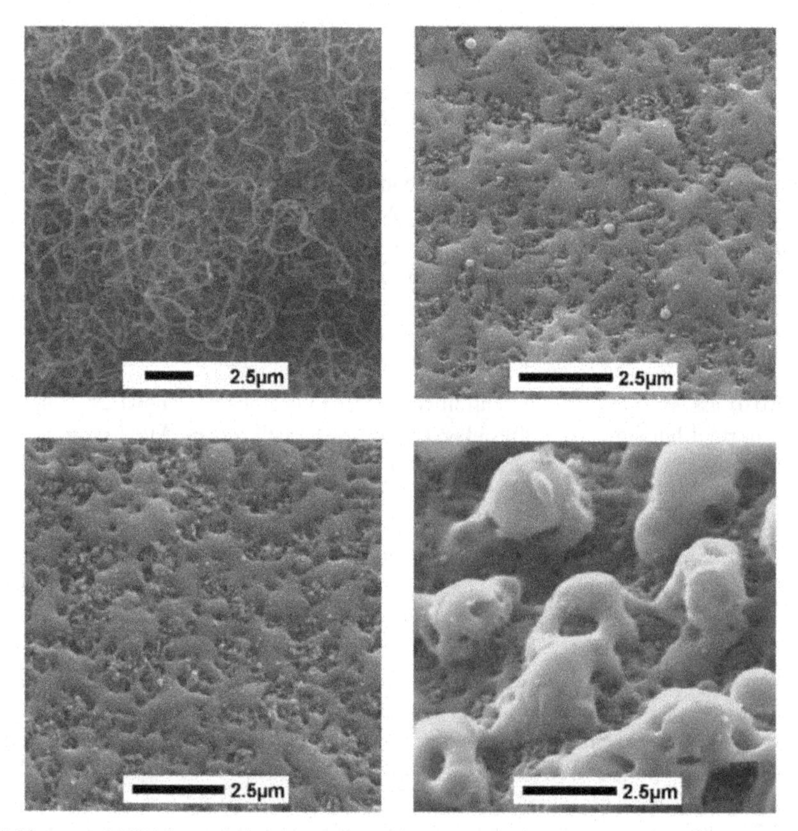

Figure 8. SEM images of Al/Al_2O_3-nanowires: (a) as-deposited; (b) after one laser pulse; (c) after two laser pulses and (d) after three laser pulses. Reprinted with permission from ref. Veith, M., O.C. Aktas, W. Metzger, D. Sossong, H.U. Wazir, I. Grobelsek, N. Pütz, G. Wennemuth, T. Pohlemann and M. Oberringer. 2010. Biofabrication. 2(3): 035001. Copyright@ Institute of Physics.

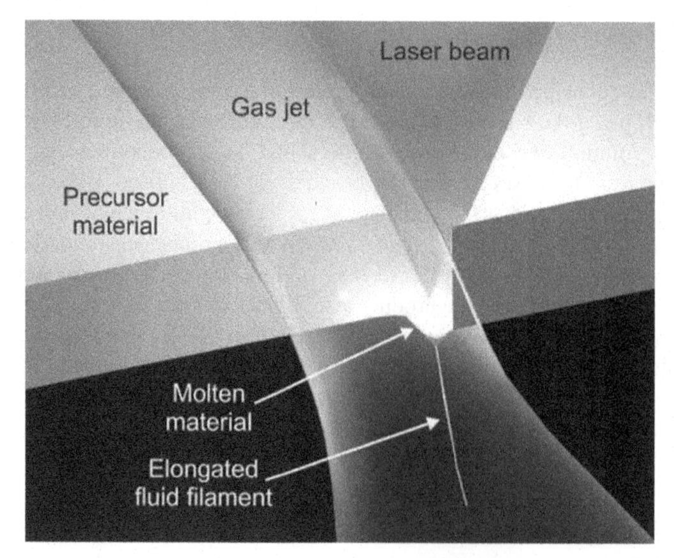

Figure 9. The laser-spinning process involves the use of a high-power laser to melt a very small volume of the precursor material. A high speed supersonic gas jet causes the rapid elongation and cooling of the melt. Reprinted with permission from ref. Quintero, F., J. Pou, R. Comesaña, F. Lusquiños, A. Riveiro, A.B. Mann, R.G. Hill, Z.Y. Wu and J.R. Jones. 2009. Adv. Funct. Mater. 19(19): 3084–3090. Copyright@ Wiley.

process, optical forces are used to direct materials to substrate and attach them through van-der Waals interactions.

New approach

Catros et al. (2011) reported two and three dimensional laser assisted bioprinting of hydroxyapatite (nHA) and living components (human osteoprogenitors: HOPs). In order to obtain the suitable patterns of nHA and HOPs for bone repair applications, Catros et al. (2011) investigated the laser printing parameters and experimental conditions. The LAB set-up comprised a Nd:YAG infrared laser focusing on a quartz ribbon, which was coated with a thin absorbing interlayer of titanium and a thick layer of biological (nHA) printing material (bioink) (Figure 10). A dedicated software was used to control the scanning system in the LAB set-up.

The two and/or three dimensional sequential printing of different biological materials can be done with this set-up. The nHA was prepared by chemical precipitation and was characterized using TEM, FTIR and XRD. The Live/Dead assay and osteoblastic phenotype markers (alcaline phosphatase and osteocalcin) were used to characterize the HOP cells after printing. The physico-chemical properties of nHA as well as the adhesion, proliferation activity, and osteoblast phenotype of HOPs did not change up to 15 days. The results showed that the LAB method is an alternative technology for biofabrication of 2D and 3D composite materials.

In a similar study, Guillemot et al. (2010) investigated the LAB of a sodium alginate (biopolymer), nano-sized hydroxyapatite (nHA) and human endothelial cells (EA.hy926) for 3D tissue engineering materials. They evaluated the laser fluence, focalization condition and writing speed in the LAB process. Bioprinting efficiency coefficient (BEC) is also studied in terms of the writing speed, the volume fraction of deposited biological component, the resolution, and the integrability of the biological laser printing method.

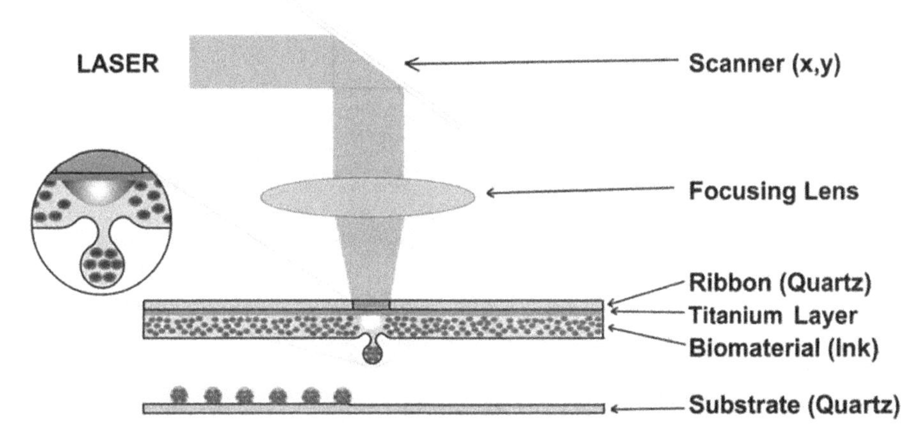

Figure 10. Laser-assisted bioprinting (LAB) set-up. Reprinted with permission from ref. Catros, S., J.C. Fricain, B. Guillotin, B. Pippenger, R. Bareille, M. Remy, E. Lebraud, B. Desbat, J. Amédée and F. Guillemot. 2011. Biofabrication. 3(2): 025001. Copyright@ Institute of Physics.

References

Abbasi, B.J., M. Zakeri and S. Tayebifard. 2014. High frequency induction heated sintering of nanostructured Al_2O_3–ZrB_2 composite produced by MASHS technique. Ceram. Int. 40: 9217–9224.

Amosov, A., D. Andriyanov, A. Samboruk and D. Davydov. 2013. SHS of porous cermets from Ti-B-C preforms, XII International Symposium on Self-propagating High-Temperature Synthesis (SHS 2013) 266–268.

Amendola, V., S. Polizzi and M. Meneghetti. 2007. Free silver nanoparticles synthesized by laser ablation in organic solvents and their easy functionalization. Langmuir 23: 6766–6770.

Amendola, V. and M. Meneghetti. 2009. Laser ablation synthesis in solution and size manipulation of noble metal nanoparticles. Phys. Chem. Chem. Phys. 11: 3805–3821.
Barzykin, V.V. 1992. Initiation of SHS processes. Pure Appl. Chem. 64: 909–918.
Barcikowski, S., F. Devesa and K. Moldenhauer. 2009. Impact and structure of literature on nanoparticle generation by laser ablation in liquids. J. Nanopart. Res. 11: 1883–1893.
Bärsch, N., J. Jakobi, S. Weiler and S. Barcikowski. 2009. Pure colloidal metal and ceramic nanoparticles from high-power picosecond laser ablation in water and acetone. Nanotechnology 20: 445603-1–445603-9.
Biffi, C.A., P. Bassani and ZahraSajedi, A.Tuissi. 2017. Laser ignition in self-propagating high temperature synthesis of porous NiTinol shape memory alloy. Materials Letters 193: 54–57.
Bi, M., A.M. Ruiz, I. Gornushkin, B.W. Smith and J.D. Winefordner. 2000. Appl. Surf. Sci. 158: 197.
Boccaccini, A.R., M. Erol, W.J. Stark, D. Mohn, Z. Hong and J.F. Mano. 2010. Polymer/bioactive glass nanocomposites for biomedical applications: a review. Compos. Sci. Technol. 70: 1764–1776.
Bomatí-Miguel, O., X. Zhao, S. Martelli, P. Di Nunzio and S. Veintemillas-Verdaguer. 2010. Modeling of the laser pyrolysis process by means of the aerosol theory: Case of iron nanoparticles. J. Appl. Phys. 107: 014906-1–014906-8.
Bonet, F., V. Delmas, S. Grugeon, R.H. Urbina, P. Silvert and K. Tekaia-Elhsissen. 1999. Synthesis of monodisperse Au, Pt, Pd, Ru and Ir nanoparticles in ethylene glycol. Nanostruct. Mater. 11: 1277–1284.
Borsella, E., S. Botti and S. Martelli. 1997. Nano-powders from gas-phase laser driven reactions: Characteristics and applications. Materials Science Forum 235: 261–266, Trans. Tech. Publ.
Borsella, E., S. Botti, L. Caneve, L. De Dominicis and R. Fantoni. 2008. IR multiple-photon excitation of polyatomic molecules: a route towards nanostructures. Phys. Scr. 78: 058112-1–058112-9.
Brandt, M., S. Sun, N. Alam, P. Bendeich and A. Bishop. 2009. Laser cladding repair of turbine blades in power plants: From research to commercialisation. International Heat Treatment & Surface Engineering 3: 3.105.
Brust, M., M. Walker, D. Bethell, D.J. Schiffrin and R. Whyman. 1994. Synthesis of thiol-derivatised gold nanoparticles in a two-phase liquid–liquid system. J. Chem. Soc. Chem. Commun. 801–802.
Capello, E., D. Colombo and B. Previtali. 2005. Repairing of sintered tools using laser cladding by wire. Journal of Materials Processing Technology 990: 164–165.
Catros, S., J.-C. Fricain, B. Guillotin, B. Pippenger, R. Bareille, M. Remy et al. 2011. Laser-assisted bioprinting for creating on-demand patterns of human osteoprogenitor cells and nano-hydroxyapatite. Biofabrication 3: 025001-1–025001-12.
Chen, Y.-W., T.-M. Yen and C. Li. 1995. Preparation of alumina-zirconia materials by the sol-gel method from metal alkoxides. J. Non-Cryst. Solids 185: 49–55.
Chen, S., Y. Chi, Y. Shi and S.-h. Huang. 2002. Ignition time of self-propagating high-temperature synthesis by laser. Trans. Nonferrous Met. Soc. China-Eng. Ed. 12: 49–53.
Cheng, G.J., D. Pirzada, M. Cai, P. Mohanty and A. Bandyopadhyay. 2005. Bioceramic coating of hydroxyapatite on titanium substrate with Nd-YAG laser. Materials Science and Engineering: C 25: 4.541–547.
Chi-Sheng, Chien, Tze-Yuan, Liao, Ting-Fu, Hong, Tsung-Yuan, Kuo, Chih-Han, Chang, Min-Long, Yeh et al. 2012. Surface microstructure and bioactivity of hydroxyapatite and fluorapatite coatings deposited on Ti-6Al-4V substrates using Nd-YAG laser. Journal of Medical and Biological Engineering 34: 2: 109–11.
Comesaña, R., F. Quintero, F. Lusquiños, M.J. Pascua, M. Boutinguiza, A. Durán et al. 2010. Laser cladding of bioactive glass coatings. Acta biomaterialia 6: 3.953–961.
D'Amato, R., M. Falconieri, S. Gagliardi, E. Popovici, E. Serra, G. Terranova et al. 2013. Synthesis of ceramic nanoparticles by laser pyrolysis: From research to applications. J. Anal. Appl. Pyrolysis 104: 461–469.
Davaran, S., S. Alimirzalu, K. Nejati-Koshki, H.T. Nasrabadi, A. Akbarzadeh, A.A. Khandaghi et al. 2014. Physicochemical characteristics of Fe. Asian Pac. J. Cancer Prev. 15: 49–54.
Devos, W., M. Senn-Luder, C. Moor, C. Salter and J. Fresenius'. 2000. Anal. Chem. 366: 873.
Farley, C., T. Turnbull, M.L. Pantoya and E.M. Hunt. 2011. Self-propagating high-temperature synthesis of nanostructured titanium aluminide alloys with varying porosity. Acta Materialia 59: 2447–2454.
Ferrando, R., J. Jellinek and R.L. Johnston. 2008. Nanoalloys: from theory to applications of alloy clusters and nanoparticles. Chem. Rev. 108: 845–910.
Folarin, O.M., E.R. Sadiku and A. Maity. 2011. Polymer-noble metal nanocomposites: review. Int. J. Phys. Sci. 6: 4869–4882.
Fujita, W. and K. Awaga. 1997. Reversible structural transformation and drastic magnetic change in a copper hydroxides intercalation compound. J. Am. Chem. Soc. 119: 4563–4564.
Ganjali, Man., Mo. Ganjali, S. Khoby and M.A. Meshkot. 2011. Synthesis of Au-Cu nano-alloy from monometallic colloids by simultaneous pulsed laser targeting and stirring. Nano-Micro Lett. 3: 256–263.
Ganjali, Man., Mo. Ganjali and P. Sangpour. 2014. Synthesis of bimetallic nanoalloy layer using simultaneous laser ablation of monometallic targets. J. Appl. Spectrosc. 80: 991–997.
Ganjali, M., M. Vaezi, S. Tayebifard and S. Asgharpour. 2013. Synthesis of Al_2O_3-ZrO_2 nanocomposite by mechanical activated self-propagating high temperature synthesis (MASHS) and ignited via laser. Int. J. Eng. Trans. A 27: 615–620.
Genuino, H., N. Mazrui, M. Seraji, Z. Luo and G. Hoag. 2013. Green synthesis of iron nanomaterials for oxidative catalysis of organic environmental pollutants. Elsevier: Amsterdam. pp. 41–61.
González-Castillo, J.R., E. Rodriguez, D. Jimenez-Villar, I. Rodríguez, F. Salomon-García, T. Gilberto de Sá et al. 2015. Synthesis of Ag@Silica nanoparticles by assisted laser ablation. Nanoscale Research Letters 10: 399.

Gonzalez, J., C. Liu, S. Wen, X. Mao and R.E. Russo. 2007. Metal particles produced by laser ablation for ICP–MS measurements. Talanta 73: 567–576.

Gras, C., D. Vrel, E. Gaffet and F. Bernard. 2001. Mechanical activation effect on the self-sustaining combustion reaction in the Mo–Si system. J. Alloys Compd. 314: 240–250.

Guillemot, F., A. Souquet, S. Catros, B. Guillotin, J. Lopez, M. Faucon et al. 2010. High-throughput laser printing of cells and biomaterials for tissue engineering. Actabiomaterialia 6: 2494–2500.

Gunther, D., B. Hattendorf and C. Latkoczy. 2003. Laser ablation-ICPMS. Anal. Chem. 341A–347A.

Horn, I., R.L. Rudnick and W.F. Mcdonough. 2000. Precise elemental and isotope ratio determination by simultaneous solution nebulization and laser ablation-ICP-MS: application to U–Pb geochronology. Chem. Geol. 164: 281–301.

Jain, T.K., M.A. Morales, S.K. Sahoo, D.L. Leslie-Pelecky and V. Labhasetwar. 2005. Iron oxide nanoparticles for sustained delivery of anticancer agents. Mol. Pharmaceutics 2: 194–205.

Jelinek, M., L. Bacakova, J. Remsa, T. Kocourek, J. Miksovsky, P. Pisarik et al. 2016. Hybrid laser technology for creation of doped biomedical layers. J. Chem. Eng. Mater. Sci. 4: 98.

Johnston, G.P., R. Muenchausen, D.M. Smith, W. Fahrenholtz and S. Foltyn. 1992. Reactive laser ablation synthesis of nanosize alumina powder. J. Am. Ceram. Soc. 75: 3293–3298.

Johnston, R. and R. Ferrando. 2008. Nanoalloys: from theory to application. Preface Farad. Discuss. 138: 9.

Kandare, E., G. Chigwada, D. Wang, C.A. Wilkie and J.M. Hossenlopp. 2006. Nanostructured layered copper hydroxy dodecyl sulfate: A potential fire retardant for poly (vinyl ester) (PVE). Polym. Degrad. Stabil. 91: 1781–1790.

Kavitha, R., S. Meghani and V. Jayaram. 2007. Synthesis of titania films by combustion flame spray pyrolysis technique and its characterization for photocatalysis. Mater. Sci. Eng. B 139: 134–140.

Klein, S., S. Petersen, U. Taylor, D. Rath and S. Barcikowski. 2010. Quantitative visualization of colloidal and intracellular gold nanoparticles by confocal microscopy. J. Biomed. Opt. 15: 036015-036015-036011.

Koch, J., A. Von Bohlen, R. Hergenroder and K. Niemax. 2004. Particle size distributions and compositions of aerosols produced by near-IR femto- and nanosecond laser ablation of brass. J. Anal. At. Spectrom. 19: 267–272.

Kokubo, T., H.-M. Kim and M. Kawashita. 2003. Novel bioactive materials with different mechanical properties. Biomaterials 24: 2161–2175.

Kosuke Kawasoe, Yoshie Ishikawa, Naoto Koshizaki, Tetsuji Yano, Osamu Odawara and Hiroyuki Wada. 2015. Preparation of spherical particles by laser melting in liquid using TiN as a raw material. Appl. Phys. B 119: 3.475–483.

Kumar, N., A. Dorfman and J.-i. Hahm. 2006. Ultrasensitive DNA sequence detection using nanoscale ZnO sensor arrays. Nanotechnology 17: 2875–2881.

Li, F.H., M.K. Balazs and R. Pong. 2000. Total dose measurement for ion implantation using laser ablation ICP-MS. J. Anal. At. Spectrom. 15: 1139.

Lin, C., H. Han, F. Zhang and A. Li. 2008. Electrophoretic deposition of HA/MWNTs composite coating for biomaterial applications. J. Mater. Sci. Mater. Med. 19: 2569–2574.

Lin, F., J. Yang, S.-H. Lu, K.-Y. Niu, Y. Liu, J. Sun et al. 2010. Laser synthesis of gold/oxide nanocomposites. J. Mater. Chem. 20: 1103–1106.

Liu, S., M.D. Regulacio, S.Y. Tee, Y.W. Khin, C.P. Teng, L.D. Koh et al. 2016. Preparation, functionality, and application of metal oxide-coated noble metal nanoparticles. Chem. Rec. 16: 1965–1990.

Mafuné, F., J.-y. Kohno, Y. Takeda, T. Kondow and H. Sawabe. 2000a. Structure and stability of silver nanoparticles in aqueous solution produced by laser ablation. J. Phys. Chem. B 104: 8333–8337.

Mafuné, F., J.-y. Kohno, Y. Takeda, T. Kondow and H. Sawabe. 2000b. Formation and size control of silver nanoparticles by laser ablation in aqueous solution. J. Phys. Chem. B 104: 9111–9117.

Mank, A.J.G. and P.R.D.J. Mason. 1999. A critical assessment of laser ablation ICP-MS as an analytical tool for depth analysis in silica-based glass samples. Anal. At. Spectrom. 14: 1143–1153.

Martinez Pacheco, M.M., R. Bouma and L. Katgerman. 2008. Combustion synthesis of TiB_2-based cermets: modeling and experimental results. Appl. Phys. A 90: 159–163.

Masanta, M., S. Shariff and A.R. Choudhury. 2016. Microstructure and properties of TiB_2–TiC–Al_2O_3 coating prepared by laser assisted SHS and subsequent cladding with micro-/nano-TiO_2 as precursor constituent. Mater. Des. 90: 307–317.

McCauley, J.W. and J.A. Puszynski. 2008. Historical perspective and contribution of US researchers into the field of self-propagating high-temperature synthesis (SHS)/combustion synthesis (CS): personal reflections. Int. J. Self Propag. High Temp. Synth. 17: 58–75.

Miller, J.C. 1993. A brief history of laser ablation. AIP Conference Proceedings 288: 619.

Mishra, R.S., V. Jayaram, B. Majumdar, C. Lesher and A. Mukherjee. 1999. Preparation of a ZrO_2–Al_2O_3 nanocomposite by high-pressure sintering of spray-pyrolyzed powders. J. Mater. Res. 14: 834–840.

Mishra, D., S. Anand, R. Panda and R. Das. 2000. Hydrothermal preparation and characterization of boehmites. Mater. Lett. 42: 38–45.

Moncada, E., R. Quijada and J. Retuert. 2007. Nanoparticles prepared by the sol–gel method and their use in the formation of nanocomposites with polypropylene. Nanotechnology 18: 1–7.

Morjan, I., F. Dumitrache, R. Alexandrescu, C. Fleaca, R. Birjega, C. Luculescu et al. 2012. Laser synthesis of magnetic iron–carbon nanocomposites with size dependent properties. Adv. Powder Technol. 23: 88–96.

Mukasyan, A. and P. Dinka. 2007. Novel approaches to solution-combustion synthesis of nanomaterials. Int. J. Self Propag. High Temp. Synth. 16: 23–35.

Narongdej, T. and T. SR Boonroung. 2010. The effectiveness of a calcium sodium phosphosilicate desensitizer in reducing cervical dentin hypersensitivity. J. Am. Dent Assoc. 41: 995–999.

Newman, S.P. and W. Jones. 1998. Synthesis, characterization and applications of layered double hydroxides containing organic guests. New. J. Chem. 22: 105–115.

Nikov, R., A. Nikolov, N. Nedyalkov, I. Dimitrov, P. Atanasov and M. Alexandrov. 2012. Stability of contamination-free gold and silver nanoparticles produced by nanosecond laser ablation of solid targets in water. Appl. Surf. Sci. 258: 9318–9322.

Niziev, V.G., F.Kh. Mirzade, V.Ya. Panchenko, M.D. Khomenko, R.V. Grishaev, S. Pityana et al. 2013. Numerical study to represent non-isothermal melt-crystallization kinetics at laser-powder cladding. Modeling and Numerical Simulation of Material Science 3: 61–69.

Okazaki, M., K. Toriyama, S. Tomura, T. Kodama and E. Watanabe. 2000. A monolayer complex of $Cu_2(OH)_3C_{12}H_{25}SO_4$ directly precipitated from an aqueous SDS solution. Inorg. Chem. 39: 2855–2860.

Pacheco, M., R.H.B. Bouma and L. Katgerman. 2008. Combustion synthesis of TiB_2-based cermets: modeling and experimental results. Applied Physics A 90: 159–163.

Parashar, U.K., P.S. Saxena and A. Srivastava. 2009. Bioinspired synthesis of silver nanoparticles. Dig. J. Nanomater. Biostruct. 4: 159–166.

Pastoriza-Santos, I. and L.M. Liz-Marzán. 1999. Formation and stabilization of silver nanoparticles through reduction by N,N-Dimethyl formamide. Langmuir 15: 948–951.

Patel, D., J.Y. Moon, Y. Chang, T.J. Kim and G.H. Lee. 2008. Poly (D, L-lactide-co-glycolide) coated superparamagnetic iron oxide nanoparticles: Synthesis, characterization and *in vivo* study as MRI contrast agent. Colloids Surf. A 313: 91–94.

Pawlowski, L. 1999. Thick laser coatings: A review. Journal of Thermal Spray Technology 8: 2.279–287.

Pei, X., J. Wang, Q. Wan, L. Kang, M. Xiao and H. Bao. 2011. Functionally graded carbon nanotubes/hydroxyapatite composite coating by laser cladding. Surf. Coat. Technol. 205: 4380–4387.

Pillai, Z.S. and P.V. Kamat. 2004. What factors control the size and shape of silver nanoparticles in the citrate ion reduction method? J. Phys. Chem. B 108: 945–951.

Piramanayagam, S. and T.C. Chong. 2011. Developments in Data Storage: Materials Perspective. John Wiley & Sons.

Quintero, F., A. Mann, J. Pou, F. Lusquiños and A. Riveiro. 2007. Rapid production of ultralong amorphous ceramic nanofibers by laser spinning. Appl. Phys. Lett. 90: 153109-1–153109-3.

Quintero, F., J. Pou, R. Comesaña, F. Lusquiños, A. Riveiro and A.B. Mann. 2009. Laser spinning of bioactive glass nanofibers. Adv. Funct. Mater. 19: 3084–3090.

Raj, K. and R. Moskowitz. 1990. Commercial applications of ferrofluids. J. Magn. Magn. Mater. 85: 233–245.

Ramezanalizadeh, H. and S. Heshmati-Manesh. 2012. Preparation of $MoSi_2$–Al_2O_3 nano-composite via MASHS route. Int. J. Refract. Met. Hard Mater. 31: 210–217.

Rashad, M. and I. Ibrahim. 2012. Structural, microstructure and magnetic properties of strontium hexaferrite particles synthesised by modified coprecipitation method. Materials Technology 27: 308–314.

Reau, A., B. Guizard, C. Mengeot, L. Boulanger and F. Ténégal. 2007. Large scale production of nanoparticles by laser pyrolysis. Materials Science Forum 534: 85–88, Trans. Tech. Publ.

Remsa, J., J. Mikšovský and M. Jelínek. 2012. PLD and RF discharge combination used for preparation of photocatalytic TiO_2 layers. Applied Surface Science 258: 9333–9336.

Russo, R.E., X. Mao, H. Liu, J. Gonzalez and S.S. Mao. 2002a. Laser ablation in analytical chemistry—a review. Talanta 57: 425–451.

Russo, R.E., Mao Xianglei and S. Samiel. 2002b. The physics of laser ablation in microchemical, analytical chemistry. Lawrence Berkeley National Laboratory 70A–77A.

Sajti, C.L., R. Sattari, B.N. Chichkov and S. Barcikowski. 2010. Gram scale synthesis of pure ceramic nanoparticles by laser ablation in liquid. J. Phys. Chem. C 114: 2421–2427.

Schinteie, G., V. Kuncser, P. Palade, F. Dumitrache, R. Alexandrescu and I. Morjan. 2013. Magnetic properties of iron–carbon nanocomposites obtained by laser pyrolysis in specific configurations. J. Alloys Compd. 564: 27–34.

Shen, P., Z. Guo, J. Hu, J. Lian and B. Sun. 2000. Study on laser ignition of Ni-33.3 at% Al powder compacts. Scr. Mater. 43: 893–898.

Sinfelt, J.H. 1983. Bimetallic Catalysts: Discoveries, Concepts, and Applications. Wiley-Interscience.

Suryanarayana, C. 2001. Mechanical alloying and milling. Prog. Mater Sci. 46: 1–184.

Sourice, J., A. Quinsac, Y. Leconte, O. Sublemontier, W. Porcher, C. Haon et al. 2015. One-step synthesis of Si@C Nanoparticles by laser pyrolysis: High-capacity anode material for lithium-ion batteries. ACS Appl. Mater. Interface. 7: 6637–6644.

Tsuji, T., K. Iryo, N. Watanabe and M. Tsuji. 2002. Preparation of silver nanoparticles by laser ablation in solution: influence of laser wavelength on particle size. Appl. Surf. Sci. 202: 80–85.

Toyserkani, E., S. Corbin and A. Khajepour. 2004. Laser Cladding. Boca Raton, FL: CRC Press.

Usui, H., T. Sasaki and N. Koshizaki. 2005. Ultraviolet emission from layered nanocomposites of $Zn(OH)_2$ and sodium dodecyl sulfate prepared by laser ablation in liquid. Appl. Phys. Lett. 87: 063105–063108.

Van der Molen, K.L., A.P. Mosk and A. Lagendijk. 2007. Quantitative analysis of several random lasers. Opt. Commun. 278: 110–113.

Veintemillas-Verdaguer, S., Y. Leconte, R. Costo, O. Bomati-Miguel, B. Bouchet-Fabre and M.P. Morales. 2007. Continuous production of inorganic magnetic nanocomposites for biomedical applications by laser pyrolysis. J. Magn. Magn. Mater. 311: 120–124.

Veith, M., O. Aktas, W. Metzger, D. Sossong, H.U. Wazir and I. Grobelsek. 2010. Adhesion of fibroblasts on micro- and nanostructured surfaces prepared by chemical vapor deposition and pulsed laser treatment. Biofabrication 2: 035001-1–035001-10.

Vincenzini, P., R. Riedel, A.G. Merzhanov and G.E. Chang-Chun. 2010. Composites produced by SHS method-current development and feature trends, 12th International Ceramics Congress Part B: Advances in Science and Technology 63: 263–272.

Volpi, A., C. Zonotti, P. Giulioni, F. Pqssoretti and E. Olzi. 1998. Preliminary study on self-sustained high-temperature synthesis of magnesium compound. Metallurgical Science and Technology 16(1-2): 107–110.

Wang, D.G., C.Z. Chen, J. Ma and G. Zhang. 2008. *In situ* synthesis of hydroxyapatite coating by laser cladding. Colloids and Surfaces B: Biointerfaces 66: 155–162.

Wang, R., X. Zi, L. Liu, H. Dai and H. He. 2010. The study and application of core-shell structure bimetallic nanoparticles. HuaxueJinzhan 22: 358–366.

Watmough, S.A., T.C. Hutchinson and R.D. Evans. 1997. Environ. Sci. Technol. 31: 114.

Weng, Fei, Chen Chuanzhong and Yu Huijun. 2014. Research status of laser cladding on titanium and its alloys: a review. Materials and Design 58: 412–425.

Wei, Q., S. Li, C. Han, W. Li, L. Cheng, L. Hao et al. 2015. Selective laser melting of stainless-steel/nano-hydroxyapatite composites for medical applications: Microstructure, element distribution, crack and mechanical properties. J. Mater. Process. Technol. 222: 444–453.

Williams, G. and G.S. Coles. 1999. The gas-sensing potential of nanocrystalline tin dioxide produced by a laser ablation technique. MRS Bulletin 24: 25–29.

Wongpisutpaisan, N., P. Charoonsuk, N. Vittayakorn and W. Pecharapa. 2011. Sonochemical synthesis and characterization of copper oxide nanoparticles. Energy Procedia 9: 404–409.

Xibo, Pei, Jian, Wang, Qianbing, Wan, Lijuan, Kang, Minglu, Xiao and Hong Bao. 2011. Functionally graded carbon nanotubes/hydroxyapatite composite coating by laser cladding. Surface & Coatings Technology 205: 4380–4387.

Xu, H., B.W. Zeiger and K.S. Suslick. 2013. Sonochemical synthesis of nanomaterials. Chem. Soc. Rev. 42: 2555–2567.

Yakovlev, A., P. Bertrand and I. Smurov. 2004. Laser cladding of wear resistant metal matrix composite coatings. Thin Solid Films 133: 453–454.

Yamamoto, M. and M. Nakamoto. 2003. Novel preparation of monodispersed silver nanoparticles via amine adducts derived from insoluble silver myristate in tertiary alkylamine. J. Mater. Chem. 13: 2064–2065.

Yang, G. 2007. Laser ablation in liquids: applications in the synthesis of nanocrystals. Prog. Mater. Sci. 52: 648–698.

Yang, X.-C. and W. Riehemann. 2001. Characterization of Al_2O_3–ZrO_2 nanocomposite powders prepared by laser ablation. Scr. Mater. 45: 435–440.

Yang, X., W. Riehemann, M. Dubiel and H. Hofmeister. 2002. Nanoscaled ceramic powders produced by laser ablation. Mater. Sci. Eng. B 95: 299–307.

Zeng, H., X.W. Du, S.C. Singh, S.A. Kulinich and S. Yang. 2012. Nanomaterials via laser ablation/irradiation in liquid: A review. Advanced Functional Material 22: 1333–1353.

14

Film Formation from PS Latex/Al_2O_3 Composites Prepared by Dip-Drawing Method

Şaziye Uğur[1,*] and *Önder Pekcan*[2]

Introduction

Polymer colloids, often referred to as latex dispersion, are used in a broad range of applications ranging from adhesives, inks, paints, coatings, drug delivery systems, and films to cosmetics (Fitch 1997, Keddie 1997). In many of these applications, latexes are used as the building blocks of larger structures and are utilized as thin polymer films on a substrate surface. Particle identity is usually lost upon film formation and a homogenous film is formed. Colloidal particles with glass transition temperatures (T_g) above the drying temperature are referred to as hard latex (high-T_g) particles, and colloidal particles with T_g below the drying temperature are known as soft latex (low-T_g) particles. Film formation from soft and hard latex dispersions can occur in several stages. In both cases, the first stage corresponds to the initial wet stage. Evaporation of the solvent leads to the second stage in which the particles form a close packed array, and if the particles are soft they are deformed to polyhedrons. Hard latex, however, stays undeformed at this stage. The annealing of soft particles causes diffusion across particle–particle boundaries, which leads to a homogeneous continuous material. In contrast, in the annealing of hard latex, deformation of particles first leads to void closure (Sperry et al. 1994, Mazur 1995, Mackenzie and Shutlewort 1946, Vanderhoff 1970) and then after the voids disappear, diffusion across particle–particle boundaries starts; this means that the mechanical properties of hard latex films evolve during annealing, after all the solvent has evaporated and all voids have disappeared (Pekcan et al. 1990, Wang and Winnik 1993, Canpolat and Pekcan 1996, Pekcan and Arda 1999).

[1] Istanbul Technical University, Department of Physics, Maslak, Istanbul, Turkey, 34469.
[2] Kadir Has University, Faculty of Arts and Science, Cibali, Istanbul, Turkey, 34320.
Email: pekcan@khas.edu.tr
* Corresponding author: saziye@itu.edu.tr

Organization of monodispersed colloidal particles like latex and silica microspheres into higher-order microstructures is attracting growing interest (Wang and Mohwald 2004, Dinsmore et al. 2002), since it provides unique structures suitable for various advanced devices and functional materials such as photonic crystals (Chomski and Ozin 2000) and porous polymers (Moon et al. 2004). Colloidal crystals, consisting of three dimensional ordered arrays of monodispersed spheres, represent novel templates for the preparation of highly ordered macroporous inorganic solids, exhibiting precisely controlled pore sizes and highly ordered three-dimensional porous structures. This macroscale templating approach typically consists of three steps. First, the interstitial voids of the monodisperse sphere arrays are filled with precursors of various classes of materials, such as ceramics, semiconductors, metals, and monomers. In the second step, the precursors condense and form a solid framework around the spheres. Finally, the spheres are removed by either calcination or solvent extraction. Therefore, arrangements formed by latex microspheres have been extensively used as a target on which to template advanced materials. Ordered arrays of polymer (e.g., polystyrene or poly(methyl methacrylate)) or silica nanospheres have been extensively studied in recent years for photonic crystal applications (Zakhidov et al. 1998, Holland et al. 1998, Vlasov et al. 1999). Such systems can be used as the "host" for chemically or electrochemically immobilizing semiconductor particles. Thus, the pores and voids of the ordered matrix can be filled with a metal, semiconductor, or both, which act as the "guest" material. The guest follows the symmetry layout of the voids or pores in the host matrix by self-organization, resulting in the formation of a three-dimensional array nanoarchitecture. Importantly, the macroporous films retain the periodicity of the templates and exhibit strong photonic band-gaps that can be tuned by varying the template diameter. Photonic crystals (i.e., spatially periodic structures of dielectric materials with different refractive indices) have extensively been investigated worldwide. Because the lattice constant of photonic crystals is in the visible or infrared wavelength range, they can control the propagation of photons in a way similar to the way a semiconductor does for electrons. Many studies have been carried out to predict and produce the 3D complete photonic band gap structures because of their wide potential applications in optics (Joannopoulos et al. 1995). Recently, they have attracted renewed interest, mainly because they provide a much simpler, faster, and cheaper approach than complex semiconductor nanolithography techniques to create three-dimensional photonic crystals working in the optical wavelength range (Vlasov et al. 2001, Blanco et al. 2000).

In this chapter, based on steady-state fluorescence (SSF) and UV-vis (UVV) data and SEM micrographs, the effect of dip drawing rate and time, Al_2O_3 content and latex particle size on the structure and film formation properties of PS/Al_2O_3 films is discussed. In the first part, the effect of dip drawing rate and time on the film formation of PS/Al_2O_3 composites is studied. For this purpose, films were prepared first by casting PS dispersion on clean glass substrates which creates a close-packed array of small PS sphere (SmPS: 203 nm) templates. These templates were then covered with Al_2O_3 utilizing the dipdrawing method for various dip-drawing rates and dipping times in Al_2O_3 sol. The film formation of these composites was studied by annealing them at a temperature range of 100°C to 270°C and monitoring the scattered light (I_{sc}), fluorescence (I_P), and transmitted light (I_{tr}) intensities after each annealing step. The results demonstrated that the film formation behavior and morphology of composites depended mainly on dipping time, and no dependence on the dip-drawing rate was observed. The optical results indicated that PS/Al_2O_3 films undergo complete film formation independent of the dip-drawing rate and dipping time. After completion of film formation, PS polymers were extracted to obtain porous Al_2O_3 thin films. Highly ordered porous structures were observed for long dipping time in Al_2O_3 sol but no change was observed for different dip-drawing rates, confirming the optical data. In the second part, in order to study the effect of latex particle size on film formation behavior of PS/Al_2O_3 composites, two film series (SmPS/Al_2O_3 and LgPS/Al_2O_3) were prepared covering PS sphere (SmPS: 203 nm; LgPS: 382 nm) templates with various layers of Al_2O_3 by dip coating method for a constant dip-drawing rate and dipping time. The film formation behaviors of these composites were studied as in the first part by annealing them at a temperature range of 100–250°C and monitoring the I_{sc}, I_P and I_{tr} intensities after each annealing step. The results indicate that LgPS/Al_2O_3 films underwent complete film formation independent of Al_2O_3 content while no film formation occurred above a certain Al_2O_3 content for SmPS/Al_2O_3. Extraction of PS produced ordered porous structures for high Al_2O_3 content in both series. Scanning electron microscopy images showed that the pore size and porosity could be easily tailored by varying PS particle size and Al_2O_3 content.

Experimental part

Materials

Synthesis of polystyrene (latex) spheres

Fluorescent PS latexes were produced via emulsion polymerisation process (Liu et al. 1994). The polymerisation was performed batchwise, using a thermostated reactor equipped with a condenser, thermocouple, mechanical stirring paddle and nitrogen inlet. Water (50 mL), styrene monomer (3 g; 99% pure from Janssen) and 0.014 g of fluorescent 1-Pyrenylmethyl methacrylate (PolyFluor 394) were first mixed in the polymerisation reactor where the temperature was kept constant at 70°C. The water soluble radical initiator potassium persulphate (1.6% wt/wt over styrene) which dissolved in a small amount of water (2 mL) was then introduced in order to induce styrene polymerisation. Different surfactant sodium dodecyl sulphate concentrations (0.03 and 0.12% wt/vol) were added in the polymerisation recipe to change the particle size keeping all other experimental conditions the same. The polymerisation was conducted under 400 rev.min^{-1} agitation during 12 h under nitrogen atmosphere at 70°C. The particle size was measured using Malven Instrument NanoZS. The mean diameter of these particles is 203 nm (SmPS) and 382 nm (LgPS). The weight average molecular weights (M_w) of individual PS chain (M_w) were measured by gel permeation chromatography and found to be 90×10^3 g.mol^{-1} for both 203 nm (SmPS) and 382 nm (LgPS). Glass transition temperature (T_g) of the PS latexes were determined using differential scanning calorimeter and found to be around 105°C.

Al_2O_3 solution

Al_2O_3 sol was prepared in the following way: a total of 2 mL aluminium-tri-sec-butoxide (Aldrich; 97%) was dissolved in 45 cc water at 70°C. The solution was stirred for 30 min. A small amount of acid was continuously added as catalyst, until the solution became transparent and this was stirred for another 2 h. Oxide networks are formed upon hydrolytic condensation of alkoloxide precursors. Finally, a uniform and transparent Al_2O_3 sol was obtained for film fabrication.

Preparation of PS/Al_2O_3 films

Firstly, the glass substrates (0.8×2.5 cm^2) were cleaned ultrasonically in acetone and deionised water. PS aqueous suspensions were dropped on clean glass substrates using the casting method and dried at room temperature. Upon slow drying at room temperature, PS sphere templates (films) were produced. The film thickness of these powder films was determined to be 5 µm on average. Then, the dip-drawing method was used to fill the Al_2O_3 sol into the spaces among the close-packed PS sphere templates.

In the first part, as our aim was to study the effects of dip-drawing rate and dipping time in Al_2O_3 sol on the film formation behavior of PS/Al_2O_3 composites, we prepared two series of films. In the first series, PS films prepared from SmPS (203 nm) dispersion were dipped into the Al_2O_3 sol at the rate of 106 and 60 mm/min and then drawn out from the sol at the same rates and dried in an oven at 100°C for 10 min. In the second series, the PS covered glass substrates were placed vertically in the Al_2O_3 sol with a dipping rate of 106 mm/min and kept for 0, 5, 10, 20, 40, and 60 min in the sol. Then, the films were drawn out at the same rate and dried at 100°C in an oven for 10 min. When the templates were immersed in the Al_2O_3 sol, the Al_2O_3 precursor could permeate the close-packed arrays of PS by capillary force and form a solid skeleton around the PS spheres. In order to study the film formation behavior of PS/Al_2O_3 composites, the films were separately annealed above the T_g of PS at 105°C for 10 min at temperatures ranging from

100°C to 270°C. After each step in the annealing process, the films were removed from the oven and allowed to cool to room temperature. Lastly, after the film formation process of the PS latexes was complete, the PS/Al_2O_3 composite films were dissolved in toluene to remove the PS templates and obtain the porous structure of the Al_2O_3 thin films.

In the second part, in order to study the particle size effect of PS latex and Al_2O_3 content on film formation of PS/Al_2O_3 composites, we prepared two series of films:

series 1: LgPS and Al_2O_3 (LgPS/Al_2O_3), series 2: SmPS and Al_2O_3 (SmPS/Al_2O_3).

The LgPS and SmPS covered glass substrates are placed vertically into the Al_2O_3 sol for 2 min, and then taken out and dried at 100°C for 10 min. Here the Al_2O_3 content in the films could be adjusted by dipping cycle and therefore, to investigate the effect of Al_2O_3 content, consecutive dipping was performed. Six different films for each series of films were produced with 0, 3, 5, 8, 10 and 15 layers (dipping cycle) of Al_2O_3. The produced films were then separately annealed at temperatures ranging from 100 to 250°C for 10 min. After each annealing step, films were removed from the oven and cooled down to room temperature.

Methods

Fluorescence measurements

After annealing, each sample was placed in the solid surface accessory of a Perkin–Elmer model LS-50 fluorescence spectrometer. Pyrene was excited at 345 nm and scattering and fluorescence emission spectra were detected between 300 and 500 nm. All measurements were carried out in the front face position at room temperature. Slit widths were kept at 8 nm during all SSF measurements.

Photon transmission measurements

Photon transmission experiments were carried out using a Carry-100 Bio UV-Visible (UVV) scanning spectrometer. The transmittances of the films were detected at 500 nm. A glass plate was used as a standard for all UVV experiments, and measurements were carried out at room temperature after each annealing processes.

Scanning electron microscopy (SEM) measurements

Scanning electron micrographs of the PS/Al_2O_3 films were taken at 10–20 kV in a JEOL 6335F microscope. A thin film of gold (10 nm) was sputtered onto the surface of samples using a Hummer-600 sputtering system to help image the PS/Al_2O_3 films against the glass background.

Results and discussion

Effect of dip-drawing rates and dipping time in Al_2O_3 sol on film formation of PS/Al_2O_3 composites

In this first part, the effect of dip-drawing rates on the film formation behavior of PS/Al_2O_3 was studied (Ugur and Kislak 2012). Figure 1 illustrates the fluorescence (I_P), scattered (I_{sc}) and transmitted (I_{tr}) light intensities versus annealing temperatures for pure PS (with no Al_2O_3) film and PS/Al_2O_3 composite film for 60 mm.min^{-1} dip-drawing rate. For both film samples, upon annealing, the transmitted light intensity I_{tr} started to increase above a certain onset temperature which is known as the minimum film formation temperature, T_0. Scattered light intensity, I_{sc}, showed a sharp increase at the single temperature named as the void closure temperature, T_v. At this point, it should be noted that I_{sc} is scattered from below the surface as well as from the surface of the latex film; however, I_{tr} goes completely through the film.

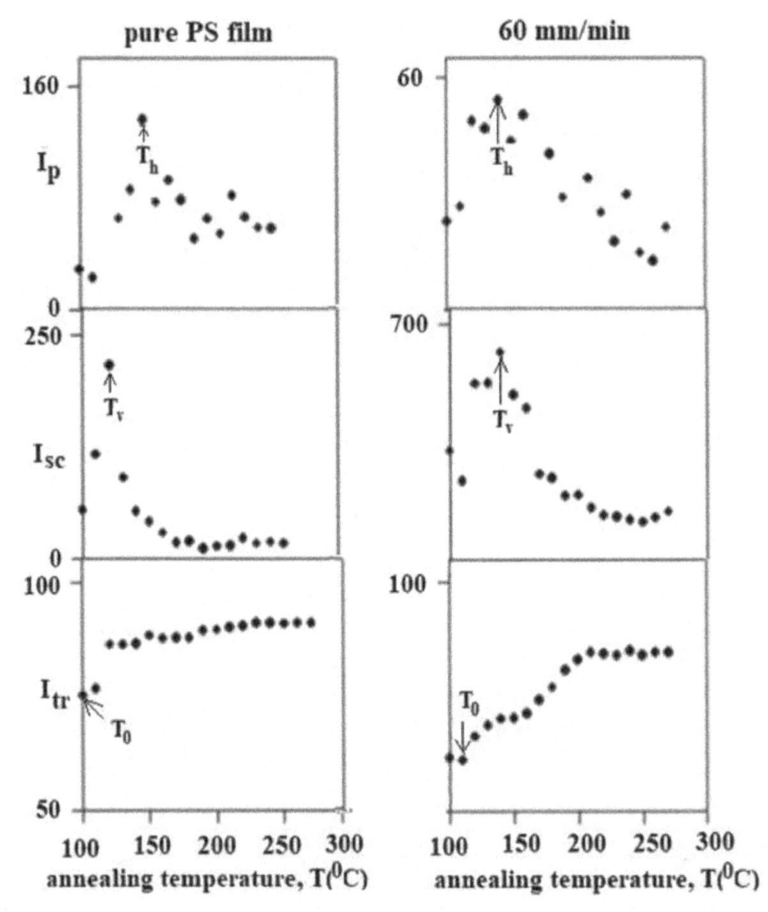

Figure 1. Plot of I_p, I_{sc}, and I_{tr}, intensities versus annealing temperature, T for pure PS film and PS/Al_2O_3 composite film prepared at 60 mm.min^{-1} dip-drawing rate.

The fluorescence intensity, I_p, of all film samples first increased, reached a maximum, and then decreased as the annealing temperature increased (Canpolat and Pekcan 1996, Ugur et al. 2003, 2006, 2007). The temperature at which I_p reached the maximum is the healing temperature, T_h. The increase in I_{tr} above T_0 can be explained through an evaluation of the transparency of the composite films upon annealing. Since we measured the transmittances of the films at 500 nm (above the absorption wavelength of both pyrene and Al_2O_3), the absorption of the films is negligible at this wavelength. Most probably, an increase in the I_{tr} in the visible region corresponds to the void closure process (Canpolat and Pekcan 1996), i.e., polystyrene begins to flow upon annealing and the voids between particles are then filled. Since a higher I_{tr} corresponds to a higher clarity of the composite, an increase in I_{tr} indicates that the microstructure of these films changes considerably when they are annealed, meaning that the transparency of these films develops upon annealing. The PS starts to flow as the result of the annealing process, and the voids between particles can then be filled due to the viscous flow. Further annealing at higher temperatures caused healing and interdiffusion processes (Canpolat and Pekcan 1996, Ugur et al. 2003, 2006, 2007), resulting in a more transparent film. The sharp increase in I_{sc} occurred at T_v, which intersects with the inflection point on the I_{tr} curve. Below T_v, light scattered isotropically because of the rough surface of the PS films. Annealing of the film at T_v created a flat surface on the film, which behaved like a mirror. As a result, light was reflected toward the photomultiplier detector of the spectrometer. Further annealing rendered the PS film completely transparent to light, and the I_{sc} decreased to its minimum.

On the other hand, the increase in I_P above T_0 presumably corresponds to the void closure process up to the T_h point where the healing process takes place. The decrease in I_P above T_h can be understood as interdiffusion processes between the polymer chains (Wu et al. 1995).

In the second experiment, the PS array templates on glass substrates were settled vertically into the Al_2O_3 sol at a rate of 106 mm/min and kept for 0 to 60 min to understand the effect of dipping time in Al_2O_3 sol on the film formation process (Ugur and Kislak 2012). After this, they were drawn out from the sol at the same rate and dried in an oven at 100°C for 10 min to form a solid skeleton around the PS spheres. These films were then annealed in the same temperature range of 100 to 270°C. Figure 2 shows the I_p, I_{sc}, and I_{tr} curves of the PS/Al_2O_3 composite films kept in Al_2O_3 sol for 0 and 40 min. The curves for all dipping times are similar to those for different dip-drawing rates, indicating that these films also underwent complete film formation (Canpolat and Pekcan 1996, Ugur et al. 2003, 2006, 2007).

Minimum film formation (T_0), void closure (T_v), and healing (T_h) temperatures are important characteristics related to the film formation properties of latexes. T_0 is often used to indicate the lowest possible temperature for particle deformation sufficient to decrease interstitial void diameters to sizes well below the wavelength of light (Keddie et al. 1995). Below this critical temperature, the dry latex is opaque and powdery. However, at and/or above this temperature, a latex cast film becomes a continuous and clear film (Eckersley and Rudin 1990). Therefore, T_0 has been used in this study as the temperature above which I_{tr} starts to increase. Here, T_v is the lowest temperature at which I_{sc} reaches its highest point, and is defined as the maxima of the I_{sc} curve. T_v is thus the minimum temperature at which I_{sc} becomes

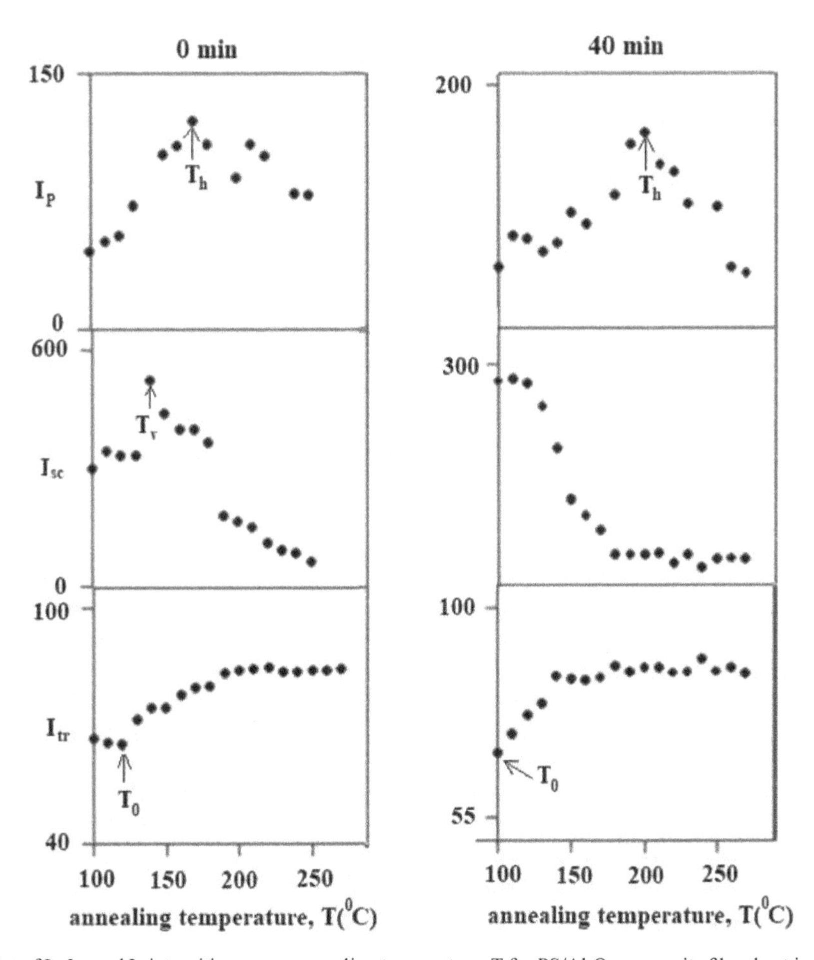

Figure 2. Plot of I_p, I_{sc}, and I_{tr} intensities versus annealing temperature, T for PS/Al_2O_3 composite films kept in Al_2O_3 sol for 0 and 40 min.

the highest. The healing temperature (T_h) is the minimum temperature at which the latex film becomes continuous and free of voids. The healing point indicates the onset of the particle–particle adhesion (Eckersley and Rudin 1990). Here, T_h is defined as the maxima of the I_p curves versus temperature. The T_0, T_v, and T_h temperatures measured for all of the films are reported in Tables 1 and 2 for both film series, respectively. As can be seen from Table 1, the T_0, T_v, and T_h temperatures change only slightly with dip-drawing rates. This behavior clearly indicates that the dip-drawing rate of the films does not affect the film formation behavior of PS/Al_2O_3 films. In Table 2, it should be noted that the T_0 and T_v values for composite films kept in Al_2O_3 for different times change only slightly with an increase in dipping time, indicating that the void closure process is not affected by dipping time. However, T_h shifts to a higher temperature region, demonstrating that the healing process is strongly influenced by dipping time.

To support these findings, SEM micrographs of both composite series were taken after annealing at 100°C and 270°C. Figure 3 shows SEM graphs of pure PS (with no Al_2O_3) and PS/Al_2O_3 composite film for dip-drawing rate of 60 mm/min annealed at 100°C (see Figure 3(a,b)) and 270°C (see Figure 3(c,d)) for 10 min, respectively. Figure 3(a,b) indicates that the structure of PS colloidal templates is well preserved after the heat treatment at 100°C. The particles are in a typically ordered close-packed array, and no deformation of particles was found. This confirms that particle coalescence and film formation have not yet occurred after annealing at 100°C.

When the annealing temperature is 270°C (see Figure 3(c,d)), considerable changes occur; the particle–particle boundaries completely disappear and the film surfaces appear flat, confirming our propositions regarding the optical data for different dip-drawing rates.

Figure 4 shows SEM micrographs of PS/Al_2O_3 composite films kept for 0 and 40 min in Al_2O_3 sol annealed at 100°C (see Figure 4(a,b)) and 270°C (see Figure 4(c,d)), respectively. Similar behavior can be observed for these composite films when Figure 4 is compared with Figure 3. However, some spherical nanoparticles are seen in these figures, which can be understood to be the PS particles encapsulated by Al_2O_3. This indicates that these highly encapsulated particles cannot be destroyed and contribute to the film formation process even when annealing occurs at high temperatures. Both the optical (I_p and I_{tr}) and SEM data suggest that the film formation process of PS latexes are not affected by the dip-drawing rate, while it is strongly affected by the dipping time. The SEM images also indicate that the latex particles in these films are completely covered with Al_2O_3, and this depends on dipping time which affects the film formation process of PS particles. This confirms the shift in the T_h temperature of PS/Al_2O_3 composites kept in Al_2O_3 sol for varying dipping times (see Table 2).

Table 1. T_0, T_v and T_h temperatures of PS/Al_2O_3 films for various dip-drawing rates.

Dipping Speed (mm.min^{-1})	T_0 (°C)	T_v (°C)	T_h (°C)
Pure PS film	100	120	140
106	120	140	170
60	110	140	140

Table 2. T_0, T_v and T_h temperatures of PS/Al_2O_3 films kept in Al_2O_3 sol for various dipping times (rate: 106 mm.min^{-1}).

Dipping Time (min)	T_0 (°C)	T_v (°C)	T_h (°C)
0	120	140	170
5	110	120	140
10	120	130	140
20	120	130	180
40	100	–	200
60	110	130	190

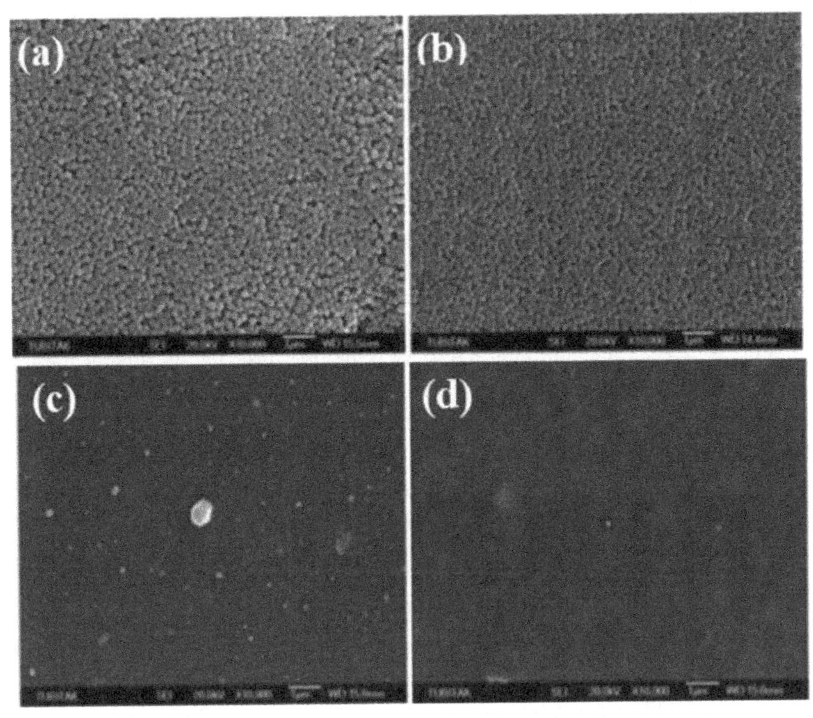

Figure 3. SEM images of pure PS film and PS/Al_2O_3 composite film prepared with dip-drawing rate of 60 mm/min annealed at 100°C (a,b) and 270°C (c,d). In every case, the scale bar is 1 μm.

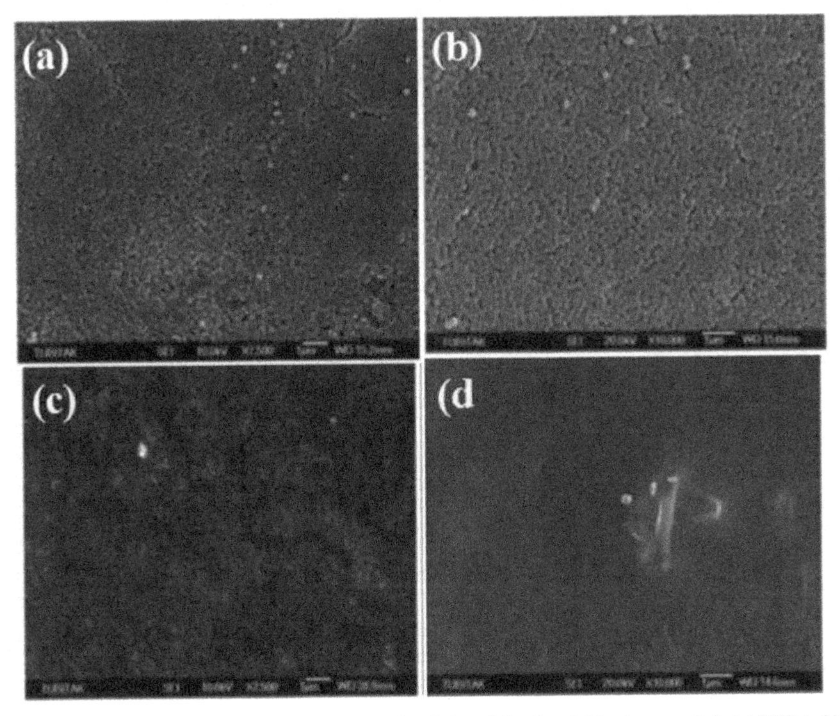

Figure 4. SEM images of PS/Al_2O_3 composite films kept in Al_2O_3 sol for 0 and 40 min annealed at 100°C (a,b) and 270°C (c,d). In every case, the scale bar is 1 μm.

To determine the extent of film formation, the films in both series were dissolved in toluene for 24 h to completely dissolve the PS template after film formation was completed. After dissolution of the PS/Al_2O_3 films, for the dip-drawing rate of 60 mm/min some pores were observed, as seen in Figure 5. These pores have a well-pronounced circular shape and must belong to the Al_2O_3 encapsulated replica of PS latexes. This can be explained by removal of PS from the surface of the Al_2O_3 covered latex particles during the dissolution process. In other words, the film formation from PS particles occurred on top of the Al_2O_3 covered PS particles during annealing and, during dissolution, the PS material completely dissolved showing the microstructure of PS particles covered by Al_2O_3. This indicates that the dip-drawing rate has no considerable effect on the morphology of PS/Al_2O_3 films. In addition to the pores, there are also large voids that may be left by the interconnected PS aggregated spheres.

The SEM images of PS/Al_2O_3 composites for 0 and 40 min dipping times after extraction of PS are given in Figure 6(a,b), respectively. In Figure 6(a) for 0 dipping time, the PS spheres highly coated with Al_2O_3 can be seen clearly. However, some spherical pores which must belong to the Al_2O_3 encapsulated replica of the PS latexes are also observed. These pores have a well-pronounced circular shape and are isolated from each other.

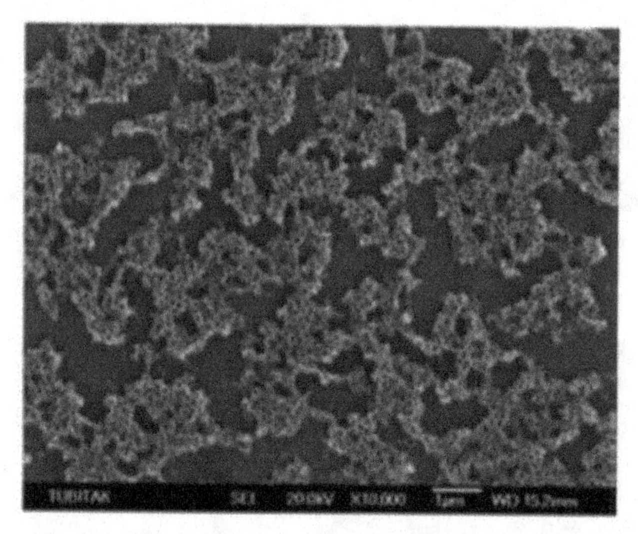

Figure 5. SEM images of PS/Al_2O_3 composite film prepared with dip-drawing rate of 60 mm/min after extraction of PS template with tolüene. The scale bar is 1 μm.

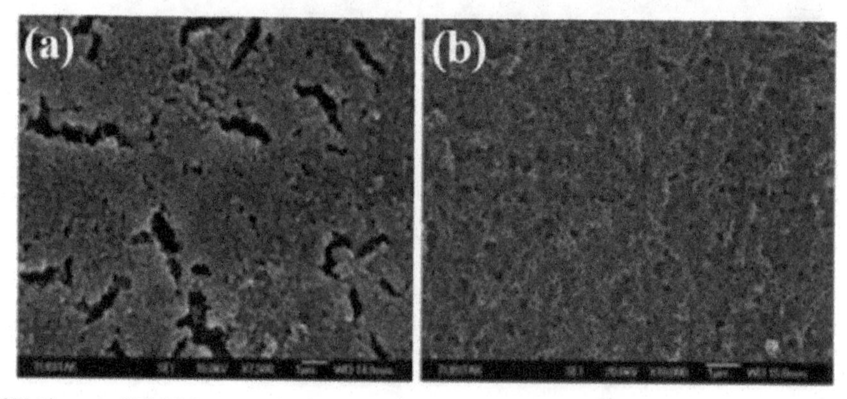

Figure 6. SEM images of PS/Al_2O_3 composite films kept in Al_2O_3 sol for different dipping times; (a) 0 min and (b) 40 min after extraction of PS template with toluene. In every case, the scale bar is 1 μm.

However, the SEM image of the PS/wall film kept for 40 min in inorganic sol (Figure 6(b)) provides nice hole depictions. From this image, it can also be seen that the level of order and the uniformity of the inorganic wall improve with an increase of dipping time. The SEM image in Figure 6(b) shows an interconnected and open porosity with average pore size diameter of 203 nm, corresponding approximately to the PS template diameter. It can be inferred that a longer dipping time (meaning higher wall content) leads to a porous material after the extraction of the PS template.

The effects of immersion time and dip-drawing rate on the structure of polymer-inorganic oxide films have been extensively studied (Brinker et al. 1991, Li et al. 2011, Kuaia et al. 2003). These studies noted that besides immersion time, drawing rate also dramatically affects surface structures. Kuaia et al. reported that excessively long immersion times may result in a thick inorganic (titania) surface layer, which would considerably deteriorate the optical properties; therefore, an appropriate immersion time of 5 min was chosen in their experiment (Kuaia et al. 2003). They also found that at high drawing rates the surface titania layer is thick and heavy, and that it peeled up easily during the template removal process; in contrast, at low drawing rates the surface titania layer is thin and fragile, and may be seriously damaged during the repeated dipping and drawing process. A smooth and integrated titania surface can be obtained only via an intermediate rate.

In our study, as indicated in the SEM images, the level of order and the uniformity of porous inorganic structures improves even at very long dipping times (60 min). It can be inferred that longer dipping times cause the wall material to penetrate into the voids of the PS template and this results in a high coverage of PS particles. After the film formation process, the extraction of the PS template creates a highly ordered porous structure for longer times. On the other hand, contrary to the results reported in the literature (Brinker et al. 1991, Li et al. 2011, Kuaia et al. 2003), we did not find any dependence on dip-drawing rates as confirmed by the optical data. In other words, the film formation from the PS particles occurs on top of the Al_2O_3 covered particles during annealing and, during dissolution, the PS material is completely dissolved showing the microstructure of the PS particles covered by Al_2O_3 layer.

Film formation mechanisms

Voids closure

In order to quantify the behavior of I_p below T_h and I_{tr} above T_0 in Figures 1 and 2, a phenomenological void closure model can be introduced. Latex deformation and void closure between particles can be induced by shearing stress which is generated by surface tension of polymer, i.e., polymer-air interfacial tension. The void closure kinetics can determine the time for optical transparency and latex film formation (Keddie et al. 1996). In order to relate the shrinkage of spherical void of radius, r to the viscosity of surrounding medium, η an expression was derived and given by the following relation (Keddie et al. 1996).

$$\frac{dr}{dt} = -\frac{\gamma}{2\eta}\left(\frac{1}{\rho(r)}\right) \tag{1}$$

where γ is the surface energy, t is time and ρ(r) is the relative density. It has to be noted that here the surface energy causes a decrease in void size and the term ρ(r) varies with the microstructural characteristics of the material, such as the number of voids, the initial particle size and packing. If the viscosity is constant in time, integration of Eq. 1 gives the relation as

$$t = -\frac{2A}{\gamma}\exp\left(\frac{\Delta H}{kT}\right)\int_{r_0}^{r}\rho(r)dr \tag{2}$$

where ΔH is the activation energy of viscous flow, i.e., the amount of heat which must be given to one mole of material for creating the act of a jump during viscous flow. Here A represents a constant for the related parameters which do not depend on temperature.

In order to quantify the above results, Eq. 2 can be employed by assuming that the interparticle voids are in equal size and number of voids stay constant during film formation (i.e., $\rho(r) \propto r^{-3}$), then integration of Eq. 2 gives the relation

$$t = \frac{2AC}{\gamma} \exp\left(\frac{\Delta H}{kT}\right)\left(\frac{1}{r^2} - \frac{1}{r_o^2}\right) \quad (3)$$

As we stated before, decrease in void size (r) causes an increase in I_p. If the assumption is made that I_p is inversely proportional to the 6th power of void radius, r then Eq. 3 can be written as

$$I_p(T) = S(t) \exp\left(-\frac{3\Delta H}{k_B T}\right) \quad (4)$$

where $S(t) = (\gamma t/2AC)^3$ and C is a constant related to relative density $\rho(r)$.

As it was already argued above, the increase in I_p originated due to the void closure process, then Eq. 4 was applied to I_p below its maxima for all film samples. Figure 7 presents the Ln (I_p) versus T^{-1} and Figure 8 presents Ln (I_{tr}) versus T^{-1} plots for both film series from which ΔH^P and ΔH^{tr} activation energies were obtained. The measured ΔH^P and ΔH^{tr} energies are listed in Tables 3 and 4 for both film series, respectively. It is seen that the ΔH^P and ΔH^{tr} values do not change much as the result of an increase in both of the dip-drawing rate and time. This indicates that the amount of heat that was required by one mole of polymeric material to accomplish a jump during viscous flow does not change by varying the dip-drawing rate and time.

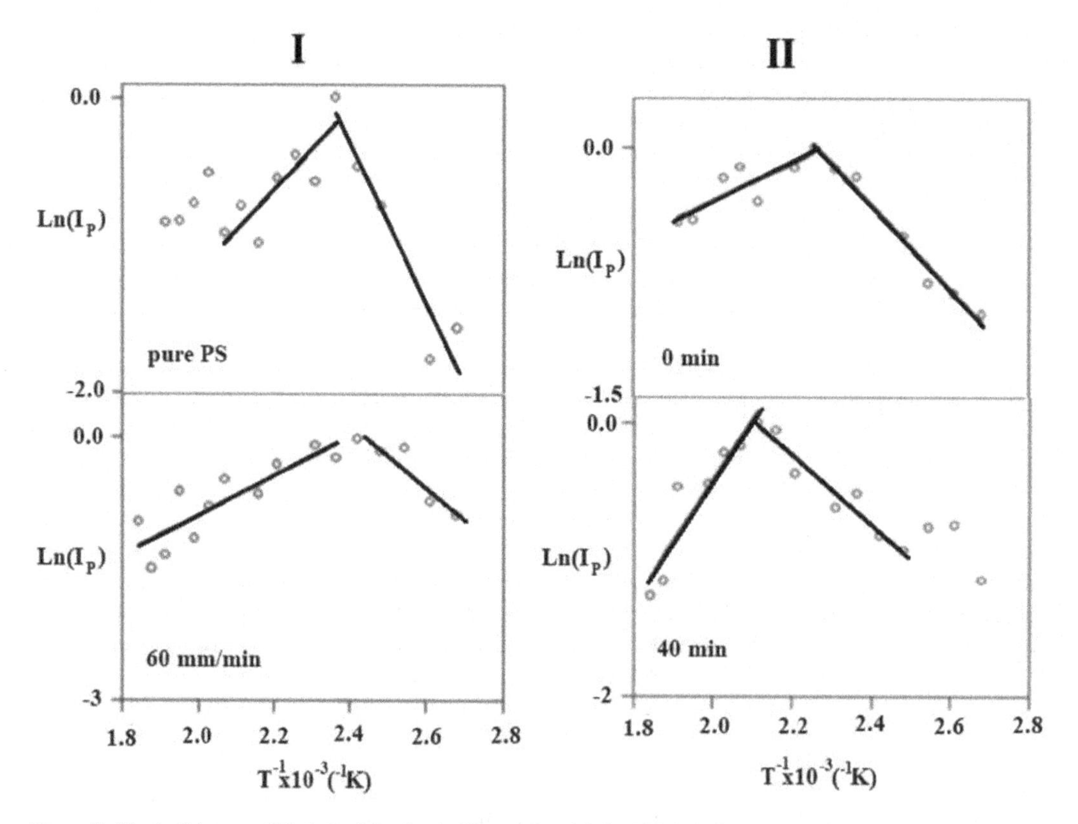

Figure 7. The ln (I_p) versus T^{-1} plots of the data in Figures 1 and 2 for PS/Al_2O_3 composite films (I) prepared at different dip-drawing rates and (II) kept in Al_2O_3 sol for different dipping times. The slope of the straight lines on the right and left hand side of the graph produce ΔH^P and ΔE activation energies, respectively.

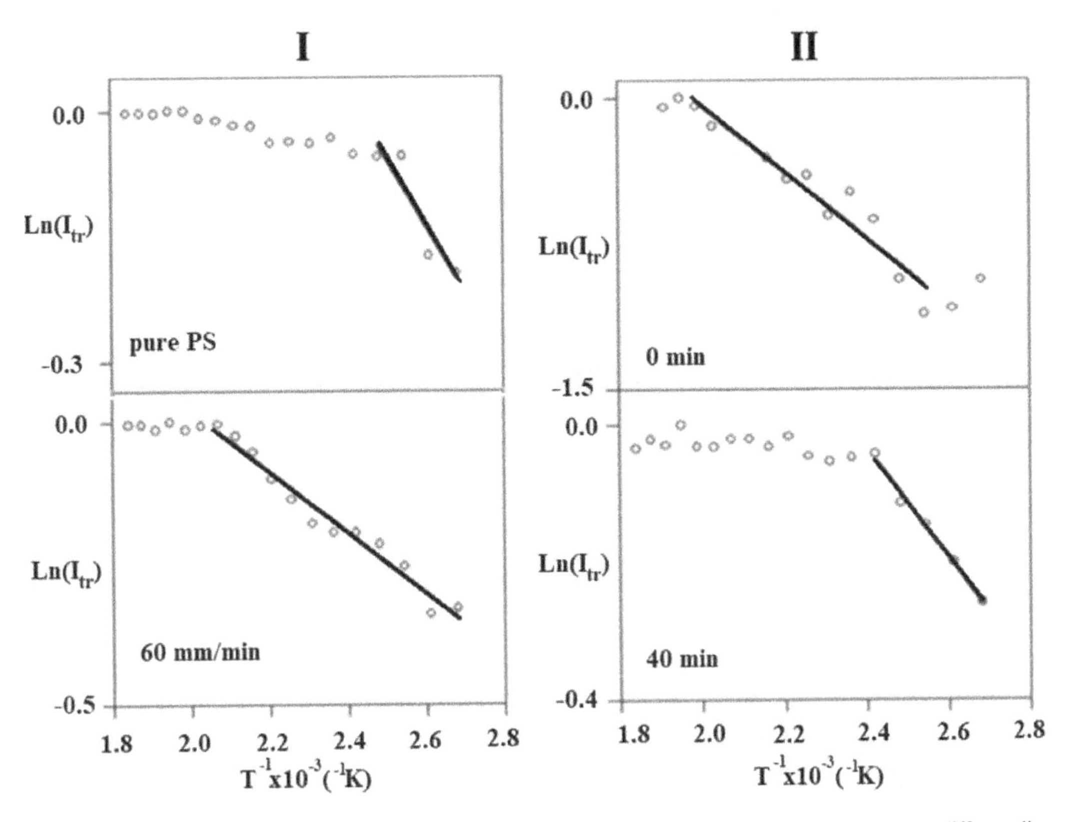

Figure 8. The ln (I_{tr}) versus T^{-1} plots of the data in Figures 1 and 2 for PS/Al_2O_3 composite films prepared at different dip-drawing rates and (II) kept in Al_2O_3 sol for different dipping times. The slope of the straight lines produces ΔH^{tr}.

Table 3. Experimentally produced activation energies for various dip-drawing rates.

Dip-drawing Rate (mm.min^{-1})	**ΔH_{tr} (kcal.mol^{-1})**	**ΔH_p (kcal.mol^{-1})**	**ΔE (kcal.mol^{-1})**
Pure PS film	0.37	2.50	7.51
106	0.25	1.15	3.15
60	0.25	1.66	6.14
Average	0.29	1.77	5.60

Table 4. Experimentally produced activation energies for various dipping times (rate: 106 mm.min^{-1}).

Dipping Time (min)	**ΔH_{tr} (kcal.mol^{-1})**	**ΔH_p (kcal.mol^{-1})**	**ΔE (kcal.mol^{-1})**
0	0.25	1.15	3.15
5	0.22	1.66	3.15
10	0.16	3.10	4.45
20	0.18	1.74	5.80
40	0.37	1.16	12.36
60	1.08	1.90	8.86
Average	0.52	1.79	

When comparing the activation energies of both series, it can be seen that the average ΔH value of films in the first series is larger than the ΔH value produced for the second series. This difference can be explained by the greater coverage of PS latex with Al_2O_3 in the second series which prevents the PS latex from flowing.

Healing and interdiffusion

As the annealing temperature is increased, some part of the polymer chains might cross the junction surface and particle boundaries disappear, and as a result I_p decreases due to transparency of the film. In order to quantify these results, the Prager-Tirrell (PT) model (Prager and Tirrell 1981, Wool et al. 1989) for the chain crossing density can be employed. The total "crossing density" σ(t) (chains per unit area) at junction surface was then calculated from the contributions $\sigma_1(t)$ due to chains still retaining some portion of their initial tubes, plus the remainder $\sigma_2(t)$, i.e., contribution comes from chains which have relaxed at least once. In terms of reduced time $\tau = 2\nu t/N^2$, the total crossing density can be written as (Prager and Tirrell 1981).

$$\sigma(\tau)/\sigma(\infty) = 2\pi^{-1/2}\tau^{1/2} \tag{5}$$

where ν and N are the diffusion coefficient and number of freely jointed segment of polymer chain (Zakhidov et al. 1998).

In order to compare our results with the crossing density of the PT model, the temperature dependence of σ(τ)/σ(∞) can be modeled by taking into account the following Arrhenius relation for the linear diffusion coefficient, which gives the following useful relation as

$$\sigma(\tau)/\sigma(\infty) = R_o \exp(-\Delta E / 2kT) \tag{6}$$

Here ΔE is defined as the activation energy for backbone motion depending on the temperature interval and $R_o = (8\nu_o t/\pi N^2)^{1/2}$ is a temperature independent coefficient. The decrease in I_p in Figures 1 and 2 above maximum is already related to the disappearance of particle-particle interface. As annealing temperature increased, more chains relaxed across the junction surface and as a result, the crossing density increases. Now, it can be assumed that I_p is inversely proportional to the crossing density σ(T) and then the phenomenological equation can be written as

$$I_P(\infty) = R_0^{-1} \exp(\Delta E / 2k_B T) \tag{7}$$

The activation energy of the backbone motion, ΔE, is produced by least-squares fitting the data in Figure 7 (the left hand side) to Eq. 7, and is listed in Tables 3 and 4. When comparing the ΔE energies of both series, it is seen that the ΔE values for the first series change only slightly while those for the second series increase with increasing dipping time. This implies that the interdiffusion process is significantly affected by dipping time. Since long dipping time means higher Al_2O_3 content, the film forming ability of PS latexes in composite films is limited by Al_2O_3 for long dipping times in Al_2O_3 sol. PS chains are not completely mixed in these composite films, where interpenetration is inevitably limited by Al_2O_3 particles. As a result, in the case of long dipping time (or an increase in Al_2O_3 content), the interpenetration of PS chains requires more energy to achieve motion due to the physical restrictions of the Al_2O_3 particles. Furthermore, the ΔE values are larger than the void closure activation energies for both series. This result is understandable because a single chain needs more energy to execute diffusion across the polymer-polymer interface than the viscous flow process.

Effect of latex size on film formation from PS Latex/Al_2O_3 nanocomposites

Figures 9 and 10 show fluorescence (I_p), scattered (I_{sc}) and transmitted (I_{tr}) light intensities versus annealing temperatures for both pure SmPS and LgPS latex films and (SmPS/Al_2O_3 and LgPS/Al_2O_3)

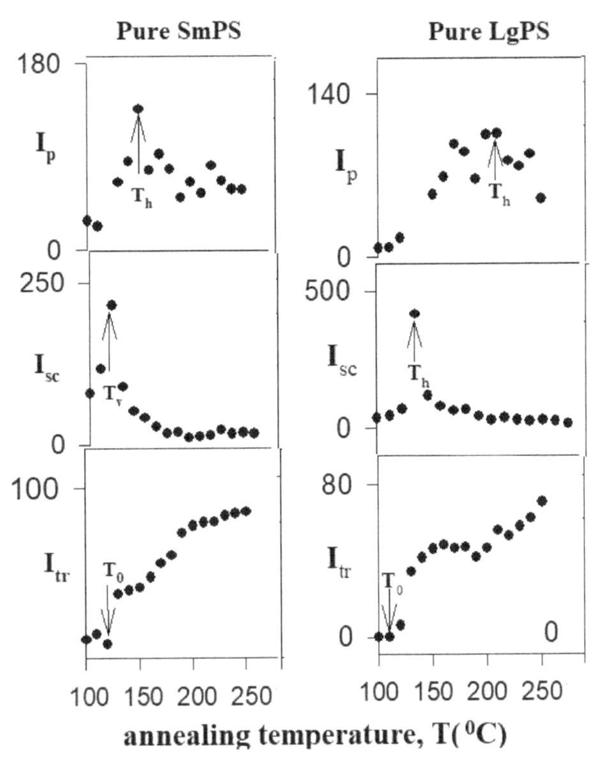

Figure 9. Plot of I_p, I_{sc} and I_{tr} intensities versus annealing temperature, T for pure SmPS and LgPS films.

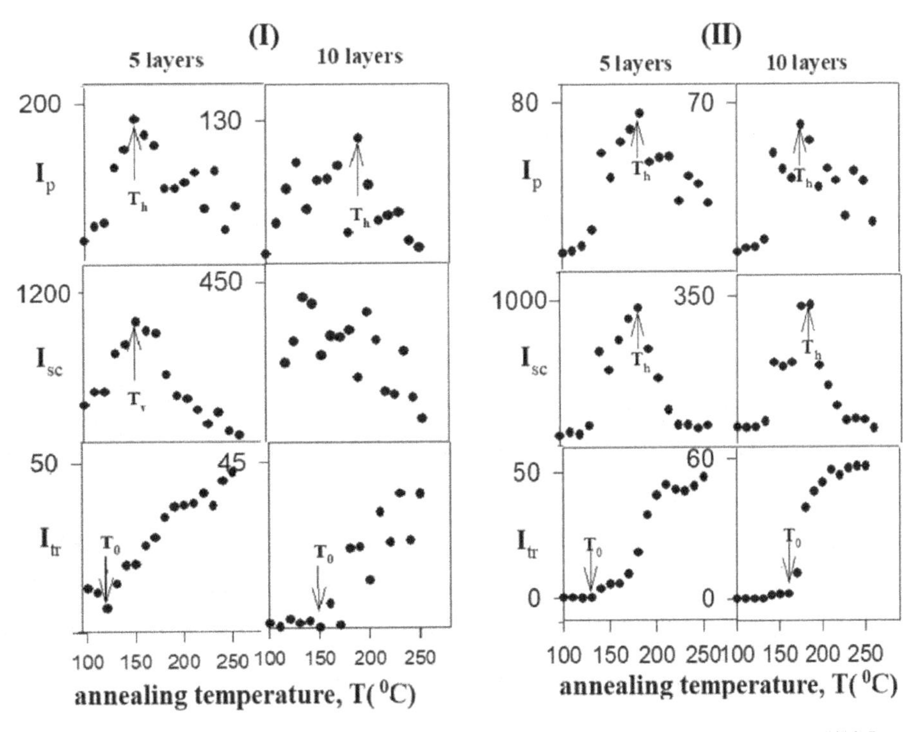

Figure 10. Plot of I_p, I_{sc} and I_{tr} intensities versus annealing temperature, T for (I) SmPS/Al_2O_3 and (II) LgPS/Al_2O_3 composite films with 0, 5 and 10 layer of Al_2O_3. Numbers on each curve shows Al_2O_3 content.

composite film series (Ugur et al. 2014). It is clear that all curves of SmPS/Al_2O_3 and LgPS/Al_2O_3 films show similar behaviors indicating that these films undergo complete film formation (Ugur et al. 2006, 2007). The behavior of T_0, T_v and T_h supports these findings.

T_0, T_v and T_h, temperatures measured for two series are reported in Table 5. From Table 5, it should be noticed that T_0 and T_v values increase with increasing Al_2O_3 content. This behavior of T_0 and T_v clearly indicates that the existence of Al_2O_3 delays the latex film formation process. However, the healing process is almost not affected by the presence of Al_2O_3, which is not surprising. As a result, film formations of PS latexes were strongly influenced by both the Al_2O_3 content and PS particle size.

In order to support these findings, SEM micrographs of both composite series with 0, 5 and 10 layers of Al_2O_3 were taken after annealing at 100°C and 250°C. Figures 11(a) and 11(b) show SEM graphs of SmPS/Al_2O_3 films with 0, 5 and 10 layer of Al_2O_3 annealed at 100°C and 250°C for 10 min, respectively. Here it has to be noted that the pure SmPS latex film with no Al_2O_3 in Figure 11(a) presents perfect order whereas inclusion of Al_2O_3 destroys the order in the latex system. As the Al_2O_3 content is increased, SmPS spheres seem to attach to each other resulting in the agglomeration of particles. When the annealing temperature is 250°C (see Figure 11(b)), considerable change occurs. Particle–particle boundaries almost disappear and the film surface appears flat confirming our arguments related with

Table 5. Minimum film formation (T_0), void closure (T_v) and healing (T_h) temperatures for films.

	SmPS/Al_2O_3			LgPS/Al_2O_3		
Al_2O_3 Layer	**T_0 (°C)**	**T_v (°C)**	**T_h (°C)**	**T_0 (°C)**	**T_v (°C)**	**T_h (°C)**
0	120	120	150	120	130	180
3	120	170	170	140	180	180
5	120	150	150	130	180	180
10	170	120	160	160	180	170
15	170	190	190	160	170	170

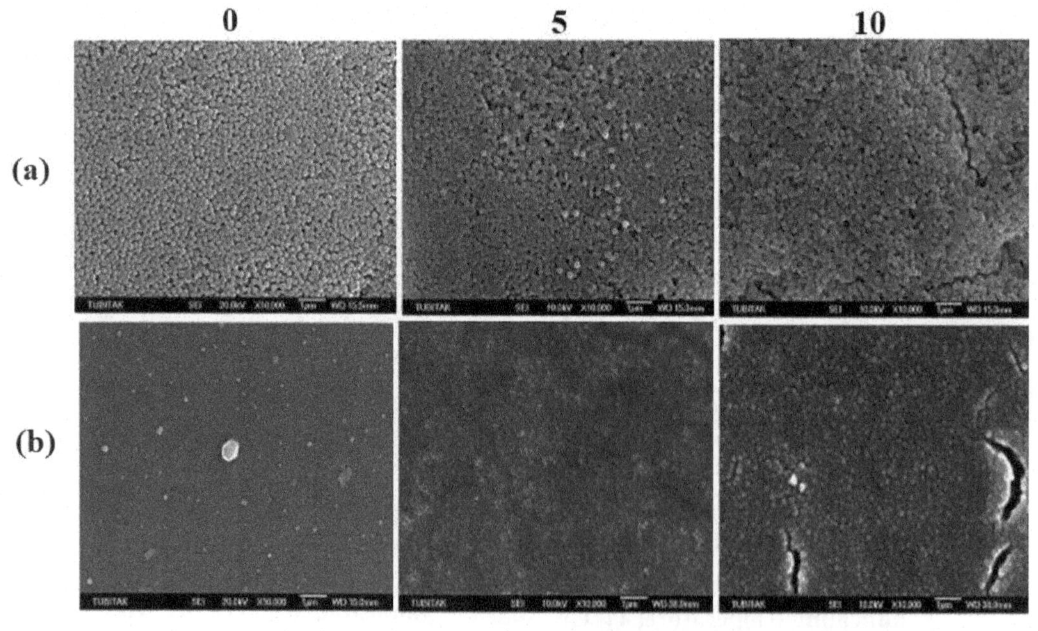

Figure 11. SEM images of SmPS/Al_2O_3 composite films with 0, 5 and 10 layer of Al_2O_3 annealed at (a) 100°C and (b) 250°C. In every case, scale bar is 1 μm.

I_P and I_{tr} data. However, in Al_2O_3 content films, some spherical nanoparticles are seen which can be attributed to the SmPS particles encapsulated by Al_2O_3. These highly encapsulated particles cannot be destroyed and contribute to the film formation process. Very similar behavior can be observed for LgPS/Al_2O_3 composite films when Figure 12 is compared with Figure 11. Nevertheless, the nanoparticles in Figure 12(b) seem larger in size than those in Figure 11(b), indicating the dependence on PS particle size used as template.

After extraction of PS (see Figure 13), the morphology of SmPS/Al_2O_3 films with 5 layer of Al_2O_3 does not change significantly. In these images, SmPS spheres, highly coated with Al_2O_3, are clearly seen. However, some spherical pores which must belong to the Al_2O_3 encapsulated replica of SmPS latexes are also observed. These pores have a well pronounced circular shape and are isolated from each other. Scanning electron microscopy images of the film prepared with 10 layers of Al_2O_3 in Figure 13 show an interconnected and open porosity with average pore size diameter of 203 nm, corresponding approximately to the SmPS template diameter. In all images, besides the pores there are also voids that

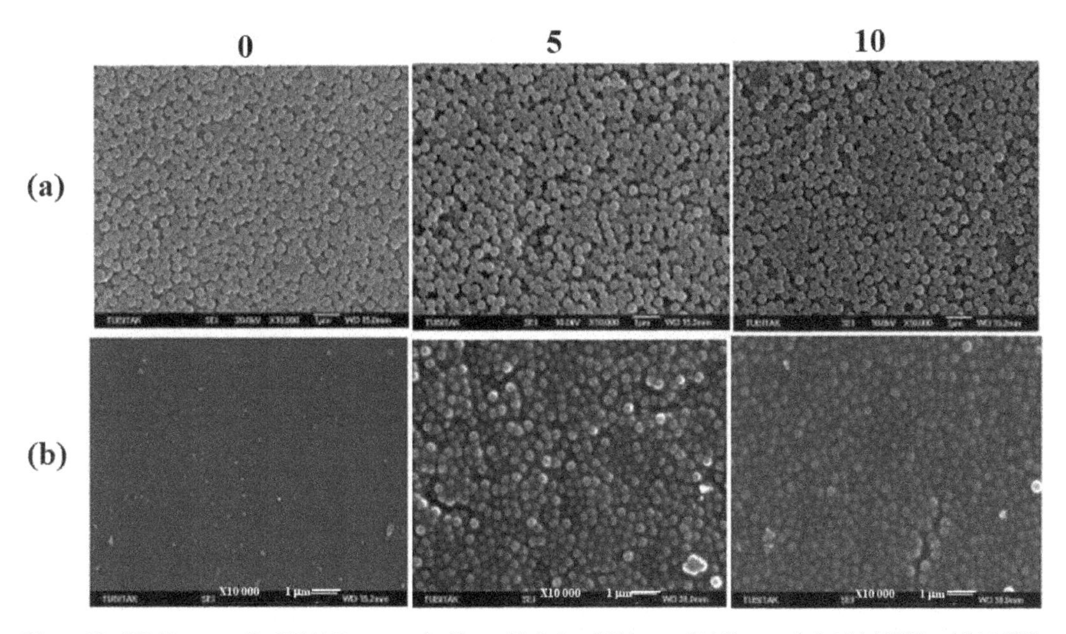

Figure 12. SEM images of LgPS/Al_2O_3 composite films with 0, 5 and 10 layer of Al_2O_3 annealed at (a) 100°C and (b) 250°C. In every case, scale bar is 1 μm.

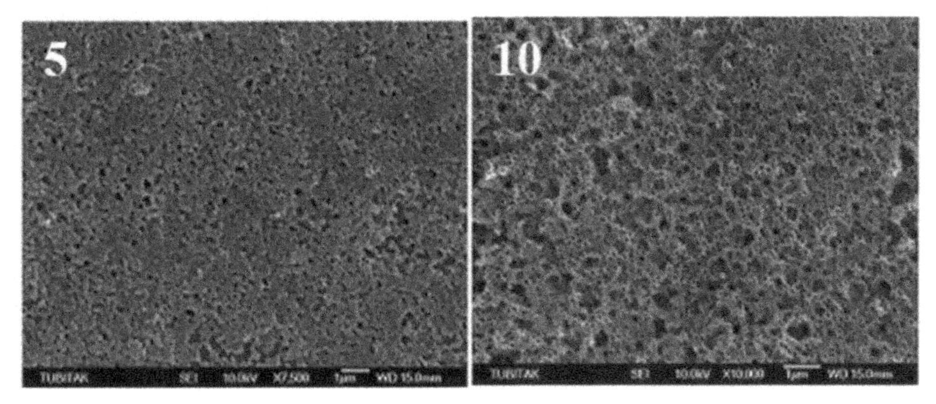

Figure 13. SEM images of SmPS/Al_2O_3 films with 5 and 10 layers of Al_2O_3 after extraction of PS template with toluene. In every case, scale bar is 1 μm.

might have been left by the interconnected SmPS aggregated spheres. It is understood that higher Al_2O_3 content and small PS size created a porous, disordered material after extraction of template.

On the other hand, SEM images of LgPS/Al_2O_3 composites given in Figure 14 presents highly porous structures after the extraction process. The pores are uniformly distributed in space, but random pore morphology in the film with 5 layer of Al_2O_3 destroy their spherical shape. These films show a poorly ordered pore structure and heterogeneous pore size distribution indicating that the Al_2O_3 particles might not be sufficient to fully cover the surface of the LgPS template. However, it could be seen that film with 10 layers of Al_2O_3 shows well defined spherical ordered pores. The pores are uniformly distributed in the sample and show an ordered connected porous structure. The homogenous distribution of the pores in the Al_2O_3 framework indicates that the porous structure retains the periodicity of the LgPS template.

In conclusion, SEM images of both film series showed that by varying the latex particle diameter, it is possible to systematically control the pore size in the final porous structure. There is also a close relationship between morphology and the Al_2O_3 content. Well defined open structure and interconnected porosity were obtained when Al_2O_3 content was increased. This behavior can be explained by the removal of PS from the surface of the Al_2O_3 covered latex particles during the dissolution process. In other words, the film formation from SmPS and LgPS particles has occurred on top of the Al_2O_3 covered particles during annealing and, during dissolution, PS material is completely dissolved showing the microstructure of PS particles covered by the Al_2O_3 layer.

As it was already explained above, the increase in both I_P and I_{tr} originated due to the void closure process, then Eq. (4) was applied to I_{tr} above T_0 and to I_P below T_h for all film samples in two series. Figure 15 presents the ln (I_P) versus T^{-1} and Figure 16 presents ln (I_{tr}) versus T^{-1} plots for both film series from which ΔH^P and ΔH^{tr} activation energies were obtained. The measured ΔH^P and ΔH^{tr} activation energies are listed in Table 6. It is seen that ΔH^P values, except for pure SmPS and LgPS films, for both series do not change much with increasing Al_2O_3 layers. In addition, ΔH^{tr} values of both film series also do not change much.

When comparing the activation energies of both series, it is seen that ΔH values of LgPS/Al_2O_3 series are larger than those of SmPS/Al_2O_3 series. This implies that the viscous flow process is significantly affected by the PS particle size. With smaller diameter (i.e., 203 nm), the SmPS particles have larger surface area or surface free energy.

The driving force for film formation is proportional to the inverse of the particle size, according to the descriptions of film formation driven by capillary forces (Wool et al. 1989). The greater curvature and higher surface area of small particles are expected to encourage film formation. The specific surface area or the total surface energy of SmPS particles (diameter 203 nm) is much larger than that of LgPS particles (diameter 382 nm). As their total surface energy is much less than that of SmPS particles, LgPS particle requires higher energy to complete the viscous flow process.

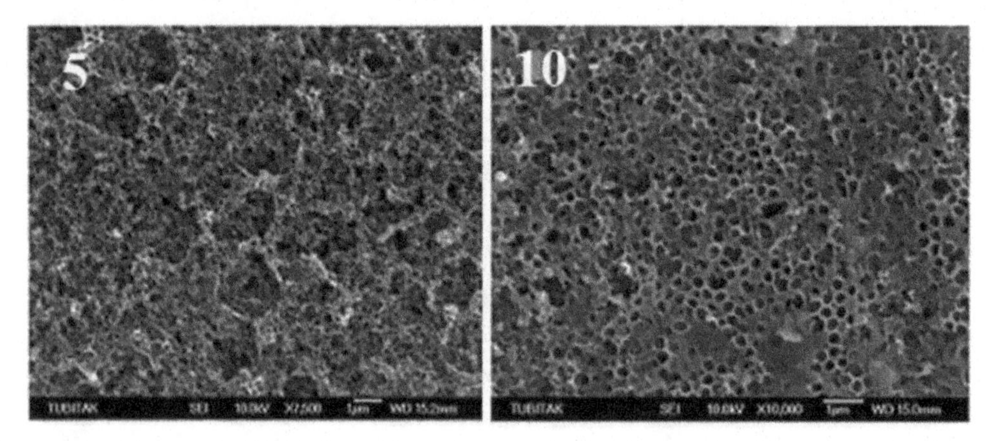

Figure 14. SEM images of LgPS/Al_2O_3 films with 5 and 10 layer of Al_2O_3 after extraction of PS template with toluene. In every case, scale bar is 1 μm.

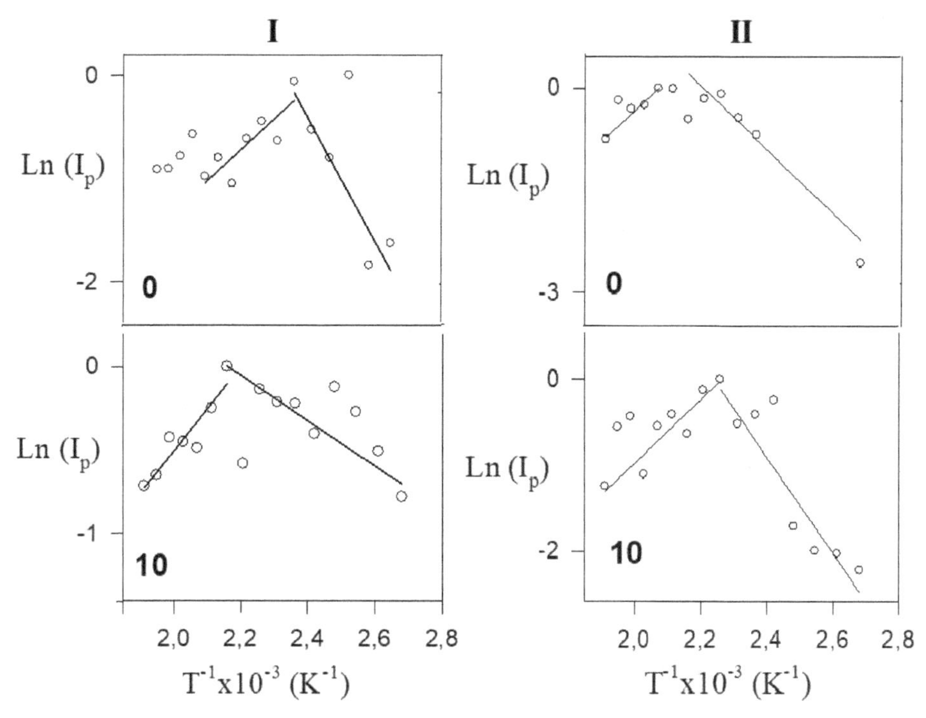

Figure 15. ln (I_p) versus T^{-1} plots of data in Figures 9 and 10 for (I) SmPS/Al_2O_3 and (II) LgPS/Al_2O_3 composite films contain 0, 5 and 10 layer of Al_2O_3. Slope of straight lines on right and left hand side of graph produce ΔH^P and ΔE activation energies, respectively.

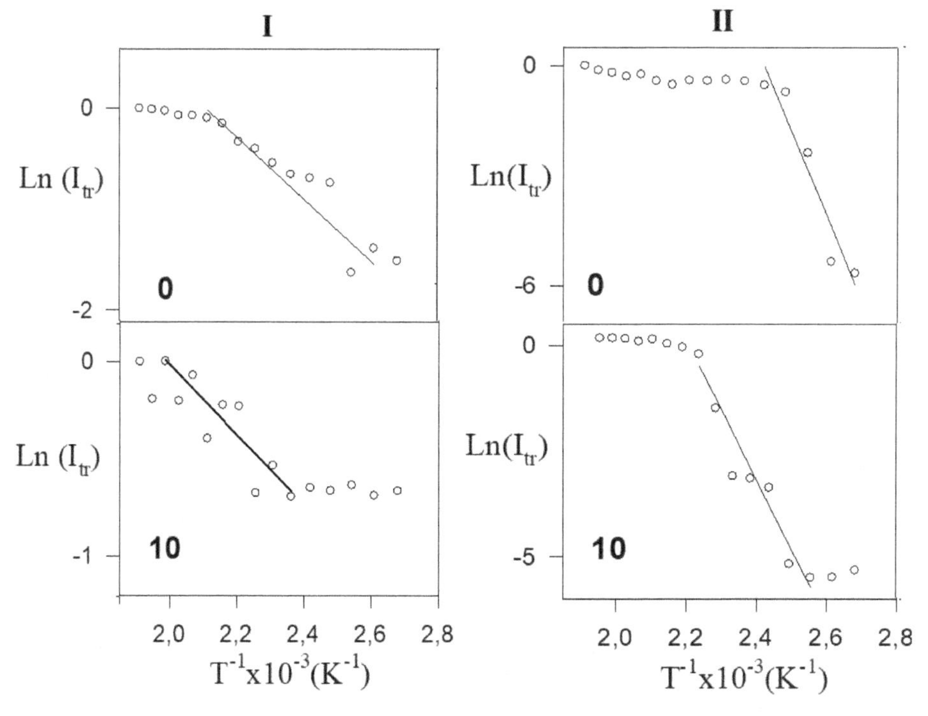

Figure 16. ln (I_{tr}) versus T^{-1} plots of data in Figures 9 and 10 for (I) SmPS/Al_2O_3 and (II) LgPS/Al_2O_3 composite films contain 0, 5 and 10 layer of Al_2O_3. Slope of straight lines produces ΔH^{tr}.

Table 6. Experimentally produced activation energies of both film series for various Al_2O_3 layers.

	$SmPS/Al_2O_3$			$LgPS/Al_2O_3$		
Al_2O_3 Layer	ΔH_p(kcal.mol^{-1})	ΔH_{tr} (kcal.mol^{-1})	ΔE (kcal.mol^{-1})	ΔH_p(kcal.mol^{-1})	ΔH_{tr} (kcal.mol^{-1})	ΔE (kcal.mol^{-1})
0	2.5	2.2	7.5	2.2	10.6	12.6
3	1.7	0.6	3.4	2.2	4.9	5.8
5	2.2	1.0	7.7	3.0	6.1	7.1
10	0.9	0.8	10.0	2.2	6.7	8.7
15	1.5	1.1	7.8	2.4	8.3	5.1

The activation energy of back bone motion, ΔE is produced by least squares fitting the data in Figure 15 (the left hand side) to Eq. (7) and these are listed in Table 6. The ΔE values for each series almost does not seem to change with increasing Al_2O_3 content showing that the interdiffusion process is not affected by Al_2O_3 content. Furthermore, ΔE values for $LgPS/Al_2O_3$ series are slightly larger than that of $SmPS/Al_2O_3$ series. The polymer chains contain more free volume and less interactions between segments in SmPS particles leading to higher conformational energy and less interaction of polymer chains (Wu et al. 1995, Qu et al. 2001). Polymer chains in the SmPS particle are in a highly confined state because of the spatial limitation compared to that of the random coil state (Wu et al. 1995) in LgPS particles. This is the major reason for the SmPS particles' need for less energy to accomplish the interdiffusion process in comparison with LgPS particles in composite films.

Conclusions

In this chapter, we discussed the effects of dip-drawing rate and dipping time in sol and PS particle size on film formation and morphology of PS/Al_2O_3 nanocomposites based on SSF technique in conjunction with UVV and SEM techniques. From optical data, it was observed that classical latex film formation occurred for all films independent of dip-drawing rate and dipping time and particle size. This demonstrates that the film formation process of PS/Al_2O_3 composite films in both series was unaffected by the Al_2O_3 content, since the film formation occurred on top of the Al_2O_3 covered particles during annealing. However, the activation energy values of the PS/Al_2O_3 films for different dipping times were found to differ slightly from the films prepared with different dip-drawing rates. This shows that dipping time plays an important role in the film forming process. It can then be inferred that as dipping time increases, more Al_2O_3 sol can penetrate the voids between PS particles, resulting in a higher coverage of PS particles. After extraction of the PS, these composite films produced highly ordered porous structures for longer dipping times. The measurement obtained from the SEM showed that the porosity could be easily tailored simply by varying the dipping time. We found that it is only dipping time that dramatically affects the surface structures, and that there is no relevance to dip-drawing rates.

However, activation energy values of the $LgPS/Al_2O_3$ series were found to be slightly larger than for the $SmPS/Al_2O_3$ series which can be explained by PS size effect. The surface morphology of the films was found to vary with the PS particle size and Al_2O_3 content. Extraction of PS produced highly ordered porous structures for high Al_2O_3 content in both film series. The measurement obtained from the SEM showed that the pore size and porosity could be easily tailored by varying the PS particle size and the Al_2O_3 content. The results also showed that there is a good correspondence between the optical data and SEM images.

In summary, first, these findings provide insight into the principle mechanism of latex film formation in inorganic oxide-based systems and the kinetics of film formation in composite systems. Second, we found that changing the PS particle size, dip-drawing rate and dipping time made it possible for us to tailor the porosity with the same thickness, resulting in different pore size and pore fraction. Third, the high transmission of PS/Al_2O_3 films is promising for their use in antireflective applications. In addition, this approach provides a general and effective approach to prepare ordered porous inorganic materials, which can be used for various other materials.

References

Blanco, A., A. Chomski, S. Grabtchak, M. Ibisate, S. John, S.W. Leonard et al. 2000. Large-scale synthesis of a silicon photonic crystal with a complete three-dimensional bandgap near 1.5 micrometers. Nature 405: 437–440.

Brinker, C.J., G.C. Frye, A.J. Hurd and C.S. Ashley. 1991. Fundamentals of sol-gel dip coating. Thin Solid Films 201: 97–108.

Canpolat, M. and O. Pekcan. 1996. Healing and photon diffusion during sintering of high-T latex particles. J. Polym. Sci. Polym. Phys. Ed. 34: 691–698.

Chomski, E. and G.A. Ozin. 2000. Panoscopic silicon-a material for "all" length scales. Adv. Mater. 12: 1071–1078.

Dinsmore, A.D., M.F. Hsu, M.G. Nikolaides, M. Marquez, A.R. Bausch and D.A. Weitz. 2002. Colloidosomes: Selectively permeable capsules composed of colloidal particles. Science 298: 1006–1009.

Eckersley, S.T. and A. Rudin. 1990. Mechanism of film formation from polymer latexes. J. Coat. Technol. 62: 89–99.

Fitch, R.M. 1997. Polymer Colloids: A Comprehensive Introduction. Academic Press, New York-USA.

Holland, B.T., C.F. Blanford and A. Stein. 1998. Synthesis of macroporous minerals with highly ordered three-dimensional arrays of spheroidal voids. Science 281: 538–540.

Joannopoulos, J.D., R.D. Meade and N. Winn. 1995. Photonic Crystals, Molding the Flow of Light. Princeton University Press, Princeton-USA.

Keddie, J.L., P. Meredith, R.A.L. Jones and A.M. Donald. 1995. Kinetics of film formation in acrylic latices studied with multiple-angle-of-ıncidence ellipsometry and environmental SEM. Macromolecules 28: 2673–2682.

Keddie, J.L., P. Meredith, R.A.L. Jones and A.M. Donald. 1996. Film formation in waterborne coatings. *In*: T. Provder, M.A. Winnik and M.W. Urban (eds.). ACS Symp. Ser. Amer. Chem. Soc. 648: 332–348.

Keddie, J.L. 1997. Film formation of latex. Mater. Sci. Eng. R 21: 101–170.

Kuaia, S.L., X.F. Hub and V.V. Truonga. 2003. Synthesis of thin film titania photonic crystals through a dip-infiltrating sol–gel process. J. Cryst. Growth 259: 404–410.

Li, H., Z. Yi, C. Lv and M. Dang. 2011. Templated synthesis of ordered porous TiO_2 films and their application in dye-sensitized solar cell. Mater. Lett. 65: 1808–1810.

Liu, J.S., J.F. Feng and M.A. Winnik. 1994. Study of polymer diffusion across the interface in latex films through direct energy transfer experiments. J. Chem. Phys. 101: 9096–9103.

Mackenzie, J.K. and R. Shutlewort. 1949. A phenomenological theory of sintering. Proc. Phys. Soc. 62: 833–852.

Mazur, S. 1995. Coalescence of polymer particles. *In*: N. Rosenweig (ed.). Polymer Powder Processing. Wiley, New York-USA.

Moon, J.H., S. Kim, G.R. Yi, Y.H. Lee and S.M. Yang. 2004. Fabrication of ordered macroporous cylinders by colloidal templating in microcapillaries. Langmuir 20: 2033–2035.

Pekcan, O., M.A. Winnik and M.D. Croucher. 1990. Fluorescence studies of coalescence and film formation in poly (methyl methacrylate) nonaqueous dispersion particles. Macromolecules 23: 2673–2678.

Pekcan, O. and E. Arda. 1999. Void closure and interdiffusion in latex film formation by photon transmission and fluorescence methods. Colloids Surf. A 153: 537–549.

Prager, S. and M.J. Tirrell. 1981. The healing-process at polymer-polymer interfaces. Chem. Phys. 75: 5194–5198.

Qu, X., Y. Tang, L. Chen and X. Jin. 2001. Novel sintering behavior of polystyrene nano-latex particles in filming process. Chin. Sci. Bull. 46: 991–995.

Sperry, P.R., B.S. Synder, M.L. O'Dowd and P.M. Lesko. 1994. Role of water in particle deformation and compaction in latex film formation. Langmuir 10: 2619–2628.

Ugur, S., A. Elaissari and O. Pekcan. 2003. Void closure and interdiffusion processes during latex film formation from surfactant-free polystyrene particles: a fluorescence study. J. Colloid Interface Sci. 263: 674–683.

Ugur, S., M.S. Sunay, A. Elaissari, F. Tepehan and O. Pekcan. 2006. Film formation from nano-sized polystrene latex covered with various TiO_2 layers. Polym. Compos. 27: 651–659.

Ugur, S., M.S. Sunay, F. Tepehan and O. Pekcan. 2007. Film formation from TiO_2-polystyrene latex composite: a fluorescence study. Compos. Interfaces 14: 243–260.

Ugur, S. and Y. Kislak. 2012. Film formation from PS/Al_2O_3 nanocomposites prepared by dip-drawing method. Polym. Compos. 33: 1274–1287.

Ugur, S., M.S. Sunay and O. Pekcan. 2015. Fluorescence study of effect of particle size in PS Latex/Al_2O_3 nanocomposite films. Plastics, Rubber and Composites 44: 129–141.

Vanderhoff, J.W. 1970. Mechanism of film formation of latexes. Br. Polym. J. 2: 161–173.

Vlasov, Y.A., N. Yao and D.J. Norris. 1999. Synthesis of photonic crystals for optical wavelengths from semiconductor quantum dots. Adv. Mater. 11: 165–169.

Vlasov, Y.A., X.Z. Bo, J.C. Sturm and D.J. Norris. 2001. On-chip natural assembly of silicon photonic bandgap crystals. Nature 414: 289–293.
Wang, Y. and M.A. Winnik. 1993. Energy transfer study of polymer diffusion in melt-pressed films of poly(Methyl Methacrylate). Macromolecules 26: 3147–3150.
Wang, D. and H. Mohwald. 2004. Template-directed colloidal self-assembly, the route to 'top-down' nanochemical engineering. J. Mater. Chem. 14: 459–468.
Wool, R.P., B.L. Yuan and O.J. McGarel. 1989. Welding of polymer interfaces. J. Polym. Eng. Sci. 29: 1340–1367.
Wu, C., K.K. Chan, K.F. Woo, R. Qian, X. Li, L. Chen et al. 1995. Characterization of pauci-chain polystyrene microlatex particles prepared by chemical initiator. Macromolecules 28: 1592–1597.
Zakhidov, A.A., R.H. Baughman, Z. Ighal, C. Cui, I. Khayrullin, S.O. Dantas et al. 1998. Carbon structures with three-dimensional periodicity at optical wavelengths. Science 282: 897–901.

15

Nanocomposite Electrolytes for Advanced Fuel Cell Technology

Rizwan Raza,[1,3,*] *Ghazanfar Abbas,*[1] *M. Saleem,*[1] *Saif Ur Rehman,*[1] *Asia Rafique,*[1] *Naveed Mushtaq*[1,2] and *Bin Zhu*[3]

Introduction

Among the different types of fuel cells, SOFCs have gained attention because of their high energy conversion efficiency and fuel flexibility (Hibino et al. 2000, Park et al. 2000, Liu et al. 2002, McIntosh and Gorte 2004a, Sasaki et al. 2004). Different kinds of fuels, such as gases (e.g., hydrogen, syngas, and bio-gas), liquids (e.g., methanol, ethanol and glycerol) and solids (e.g., carbon and lignin), can be used in SOFCs (Liu et al. 2002, McIntosh and Gorte 2004a, Sasaki et al. 2004). The current SOFC technology is still expensive and requires new materials that can work more efficiently at lower temperatures. SOFCs operate at high temperatures and employ ceramics as functional elements of the cell. Each cell is composed of an anode (e.g., Ni) and a cathode (e.g., LSCF, BSCF, etc.) separated by a solid impermeable electrolyte (e.g., YSZ, SDC, GDC, etc.). During operation, the electrolyte conducts oxygen ions from the cathode to the anode, where they react chemically with the fuel. The electric charge induced by the passage of the ions may then be collected and conducted away from the cell (Steele and Heinzel 2001).

The electrolyte is the most critical material of the fuel cell. The performance of the cell is dependent on the conductivity of the electrolyte materials. The electrolyte has to be dense to prevent gas permeability (Liu et al. 2002, McIntosh and Gorte 2004a, Sasaki et al. 2004, Steele and Heinzel 2001) to opposite sides and have a large area to minimise the bulk resistance. It should be an electronic insulator but a good ionic conductor. For example, for state-of-the-art SOFCs with YSZ electrolytes, temperatures above 700–800°C are needed to ensure adequate ionic conductivity to guarantee sufficient power outputs. An extensive range of research has been devoted to the development of new alternative electrolyte materials

[1] Clean Energy Lab, Department of Physics, COMSATS Institute of Information Technology, 54000, Lahore, Pakistan.
[2] Hubei Collaborative Innovation Center for Advanced Materials, Faculty of Physics and Electronic Science, Hubei University, Wuhan, Hubei 430062, P.R. China.
[3] Department of Energy Technology, Royal Institute of Technology (KTH), 10044, Stockholm, Sweden.
* Corresponding author: razahussaini786@gmail.com

for SOFCs that operate at low temperatures, e.g., ion-doped ceria (GDC) (Shaorong et al. 2000) (SDC) (Park et al. 2000) (YDC) (Hartmanov 2005), perovskite-type oxides ($La_{1-x}Sr_xGa_{1-y}Mg_yO_3$) (Joshi et al. 2005) mixed-ion conductors ($BaZr_{0.1}Ce_{0.7}Y_{0.2-x}Yb_xO_{3-\delta}$) (Yang et al. 2009) and complex materials such as $La_2Mo_2O_9$ (Tealdi et al. 2004) and apatite-type oxides (Islam et al. 2003). Ceria-carbonate composite electrolytes have attracted significant attention during the last decade for their application in LTSOFCs. Such composite electrolytes are considered to be a new class of ionic conductors because of their high conductivity, which occurs via interfaces and gives these materials promising potential applications in LTSOFCs (Zhu and Mellander 1994, Zhu and Tao 2000, Zhu et al. 2001, 2003, Zhu and Mat 2006, Huang et al. 2007a). A general description of a composite electrolyte (ceria-carbonate) is given in Table 1.

Table 1. Fuel cells with the ceria-carbonate composite electrolytes overview.

Ceria-carbonates	Fuel/ Oxidant	Conductivity (S/cm)	Power Density (mW/cm^2)	Working Temp. (°C)	References
GDC-salt composites	H_2/air	0.01–1	200–800	400–660	(Zhu et al. 2003)
Co-doped ceria-carbonate	H_2/air	0.01–1	900	560	(Raza et al. 2010a)
GYDC-40 wt.% $LiKCO_3$	H_2/air	–	70–300	480–530	(Zhu et al. 2003)
SDC-Na_2CO_3 nanocomposite	H_2/air	–	500–900	450–580	(Wang et al. 2008)
YDC-22 wt.% $LiNaCO_3$	H_2/air	0.01–0.78	200–700	400–660	(Zhu et al. 2003)
SDC-10 wt.% $LiNaCO_3$	H_2/air	0.001–0.03	430	400–625	(Huang et al. 2007b)
GDC-30 wt.% $LiKCO_3$	H_2/air	0.002–0.09	–	300–700	(Benamira et al. 2011)
SDC-25 wt.% $LiNaCO_3$	H_2/air	–	100–1100	400–600	(Hou et al. 2008)
SDC-30 wt.% $LiNaCO_3$	H_2/air	0.01–0.2	200–1000	500–650	(Huang et al. 2008)
Ceria-oxide nanocomposite	H_2/air	–	200–700	450–580	(Zhu et al. 2003)
SDC-Na_2CO_3	H_2/air	0.2	1000	550	(Raza et al. 2010b)

What is NANOCOFC science?

The development of nanocomposite materials for fuel cells (NANOCOFC) is a science and an approach for developing functional nanocomposite materials for low-temperature SOFCs. This LTSOFC is named as ASOFC due to below NANOCOFC theory. In some cases, these nanocomposite materials can be used for other solid electrochemical devices (Zhu et al. 2001, 2002, 2003, 2008, Zhu 2006c, 2009, Zhu and Mat 2006).

The development of advanced nano-architectures requires wide flexibility and feasibility. The two-phase material (TPM) nanocomposite exhibits multiple functions and unique characteristics (Zhu et al. 2001, 2002, 2003, 2008, Zhu 2006c, 2009, Zhu and Mat 2006).

(i) TPM composites consist of two phases with interfaces. The material functionalities are created through the two-phase interfacial regions (typically a core–shell structure). Such interfacial functionalities can break conventional structural limits.

(ii) The ionic transport phenomena differ from the conventional structure/bulk effects, which do not play a major role in two-phase composite electrolytes. Instead, the interfacial mechanisms or fast super-ionic conduction (SIC) through the interfaces determine the electrical properties and other functionalities of the composite material.

(iii) The interfacial SIC and the two source ions (O^{2-} and H^+) can significantly enhance SOFC power output at 300–600°C.

(iv) The interfacial phenomena and mechanism provide a wide range of possibilities for the development of functional materials and even for the development of functional materials from non-functional materials as long as functional interfaces can be created.

(v) Interfacial and surface redox reactions could provide new opportunities for the further development of functional nanocomposites for advanced fuel-cell technology.

Experimental

Nanocomposite electrolytes

Firstly, different electrolytes have been synthesized (LN-SDC, NSDC, CSDC-carbonate, and YSDC) using a simple co-precipitation method.

Synthesis of ceria-carbonate composites electrolyte (two-steps) (Raze et al. 2009)

LN-SDC composite electrolyte was synthesized by co-precipitation method. 0.5 M Ce $(NO_3)_3.6H_2O$ (Sigma-Aldrich, USA) solution was mixed with Sm $(NO_3)_3.6H_2O$ (Sigma-Aldrich, USA) using 1:4 molar ratios. An appropriate amount of ammonia (33% NH_3, Sigma-Aldrich) was added in order to co-precipitate the cations, e.g., Sm^{3+} and Ce^{3+} for SDC precursor in the hydroxide state, having pH value of 10. The precipitate was sluiced thrice in deionized water, followed by washing with ethanol several times in order to remove water from the particle surfaces. The resultant precipitates were dried in an oven at 100°C overnight and then grinded in a mortar, after which the powder was sintered at 700°C for 2 hours to obtain the SDC powder.

The composites of SDC and carbonates were then prepared. Subsequently, the SDC and $LiNaCO_3$ powders were mixed according to suitable weight ratio. Finally, the powder was sintered at 700°C in the furnace for 30 minutes and two steps LN-SDC composite electrolyte was achieved.

Improved ceria-carbonate (SDC-Na_2CO_3, single step) (Raza et al. 2010b)

Nanocomposite electrolyte was synthesized by a co-precipitation process (Figure 1a). In the synthesis of ceria carbonate composite, the following raw chemicals were used for 1.0 M solutions, $Ce(NO_3)_3.6H_2O$ (Sigma-Aldrich, USA) and Sm $(NO_3)_3.6H_2O$ (Sigma-Aldrich, USA). According to desired molar ratios (4:1), the solution of Ce $(NO_3)_3.6H_2O$ was mixed with Sm $(NO_3)_3.6H_2O$ solution. According to "metal ion :carbonate ion = 1: 2" in molar ratio, a pertinent amount of Na_2CO_3 solution (1.0 M) was gradually added at a rate of 10 ml/min to complete the ceria-carbonate composites within a wet-chemical co-precipitation process. After this process, the mixture was filtered by suction filtration method. The precipitate was dried overnight in the oven at 80°C. Finally, the dried solid powder was crushed in a mortar with pestle, and sintered at 800°C for 2 hours and SDC-Na_2CO_3 (NSDC) composites were obtained in nano-scale, a so-called nanocomposite.

Co-doped SDC-based composite electrolyte ($Ce_{0.8}Sm_{0.2-x}Ca_xO_{2-\delta}$ - Na_2CO_3) (Raza et al. 2011)

A co-doped SDC-based electrolyte (CSDC–Na_2CO_3) was synthesized by a co-precipitation method. Initial ingredients Ce $(NO_3)_3.6H_2O$ (Sigma-Aldrich, USA), Sm $(NO_3)_3 \cdot 6H_2O$ (Sigma-Aldrich, USA) and Ca $(NO_3)_2 \cdot 4H_2O$ (Sigma-Aldrich, USA) were dissolved in de-ionized water with an optimal molar ratio of Ce^{3+}:Sm^{3+}:Ca^{2+} = 4:1:1 to form a 0.1 mol/L solution. Na_2CO_3 solution of 0.2 mol/L was gradually introduced in droplets to this solution, which was stirred for 30 minutes, giving a white precipitate. The precipitate was washed, filtered and dried in the oven for 12 h at 80°C and sintered at 800°C for 4 h in the air atmosphere. Subsequently, the powder was grinded in a mortar prior to the measurements.

Composite electrolyte preparation (Ceria-Oxide) (Raza et al. 2010a)

The oxide based composite electrolytes [($Ce_{0.8}Sm_{0.2}O_{2-\delta}$ – Y_2O_3) = YSDC)] were synthesized using a co-precipitation method (Figure 1b). The initial ingredients Ce $(NO_3)_3 \cdot 6H_2O$ and Sm $(NO_3)_3 \cdot 6H_2O$ were

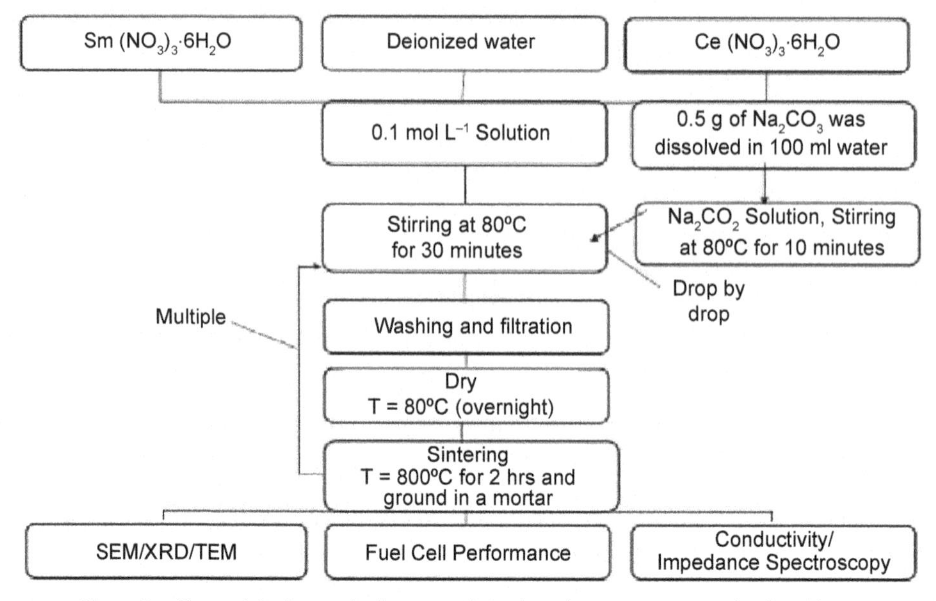

Figure 1a. Co-precipitation synthesis process for ceria-carbonate nanocomposite electrolyte.

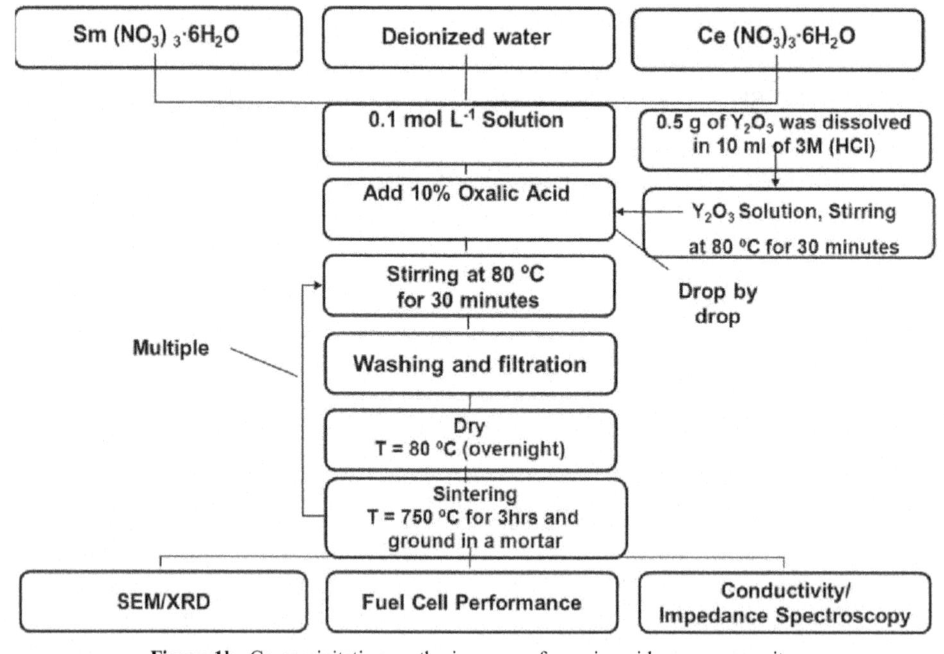

Figure 1b. Co-precipitation synthesis process for ceria-oxide nanocomposite.

dissolved in de-ionized water with an optimal molar ratio of Ce^{3+}:Sm^{3+} = 4:1 to prepare a 0.1 mol/L solution. Oxalic acid (10% of SDC) was added to the SDC solution as the precipitation agent to prepare the precursor. The precipitate was washed three times in de-ionized water. Another solution of yttrium oxide (Y_2O_3) was prepared separately; 0.5 g of Y_2O_3 was dissolved in 10 ml of 3 M hydrochloric acid (HCl) to form a solution, heated and stirred at 80°C for 30 minutes. Subsequently, Y_2O_3 solution was added drop by drop to the first SDC solution and stirred for 30 minutes, giving a white precipitate. After washing and filtering, the precipitate was dried in an oven overnight at 80°C and sintered at 750°C for 3 hrs in atmospheric conditions. Ultimately, the homogenous nanocomposite electrolyte was obtained.

Results and discussion—nanocomposite electrolytes (CERIA)

In this part, the prepared ceria-carbonate based nanocomposite electrolytes for ASOFCs are presented.

Enhancement of conductivity in ceria-carbonate nanocomposites

The purpose of this work is to study the SDC-carbonate (two steps) composite microstructures with emphasis on interfacial regions between the SDC and LiNa-carbonates in order to determine the interfaces and provide proofs of the existence of the interfaces, as well as the interfacial mechanism for superionic conduction.

Interfacial regions between the SDC and carbonates

The two-phase ranges are further observed by the high resolution TEM analysis as shown in Figure 2(a), where there is a bright-field image of the particle of (Ce, Sm) O_2 + LiNa-carbonate. Figure 2(b) shows a rectangle particle which is likely $(Sm,Ce)O_2$ and the FFT diffraction pattern (inset in Figure 2(b)) clearly shows two sets of reflections from two phases.

It is obviously observed that the SDC-carbonate is a two-phase composite. The materials consist of grains of SDC mixed with carbonate salt. An identify confirmation of the interfaces between the SDC and carbonate, where the carbonates seem to be coated on the SDC particle can also be seen from Figure 2(a). The coated layer is ranged around 5 nm thick. From Figure 2(a), we can see nano-particle SDC ranges 10–30 nm, while the SDC-carbonate composite particles range from 50–100 nm due to two-phase composite particle where the carbonate seems to be coated on the SDC particles.

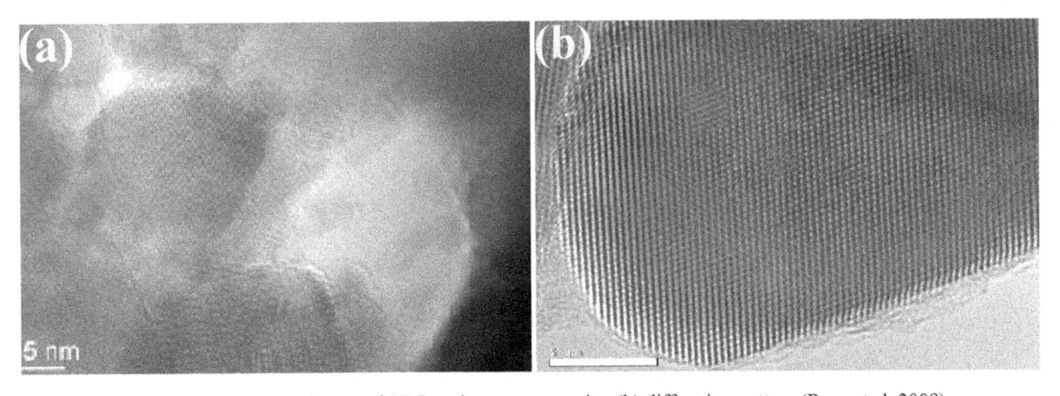

Figure 2. (a) HRTEM image of SDC-carbonate composite, (b) diffraction pattern (Raza et al. 2009).

Interfacial superionic conduction

Unlike the conventional superionic conduction in a single-phase material, it takes place from low conductive phase transferred to the superionic conduction one accompanying phase structure changes. In the ceria-carbonate two-phase systems, the superionic conduction occurs at the interfacial regions between the two constituent phases. It is thus determined by the interfaces, i.e., the change in the interfacial properties without involving individual phase structural changes. The conductivity is then strongly dominated by the interactions or coupling effect between the constituent phases. Therefore, the two-phase co-existence can create the interfacial effects and thus multi-functions, typically the superionic conduction.

The SDC-oxide and carbonate composite approach can offer a careful control of the carbonate amount in a level not only to form the percolative network and continuous ion conducting channels/framework, but also to maintain good mechanical strength. In our experiments, 20 wt% carbonate can perform these functions. In this case, the carbonates are in stick or attached tightly with the SDC-oxide without flowing freedom causing strong mechanical strength; in the same time, they can optimize the interfaces and interfacial interactions to facilitate the oxygen ion transport, especially when temperature near or above the carbonate melting points, the high mobility of carbonates, especially the fast rotation of carbonate anion groups may enhance the ceria (SDC) surface oxygen ion mobility, resulting in interfacial superionic conduction. The superionic conduction often takes place accompanying the carbonate (salt) melting. It is questioned if the conductivity enhancement is due to the molten salt behaviour, and not the interfacial mechanism. As pointed out by Schober (2005): "Conceivably, the newly formed phase between the two constituent phases could have its melting point at the transition".

Improved ceria carbonates composite electrolytes

The microstructure of sintered SDC–Na_2CO_3 (NSDC) was examined by scanning electron microscopy (SEM), as shown in Figure 3(a). The image reveals that prepared NSDC composites' nanoparticles' morphology is tetrahedron shaped in nano scale between 30 and 100 nm. The size and morphology of

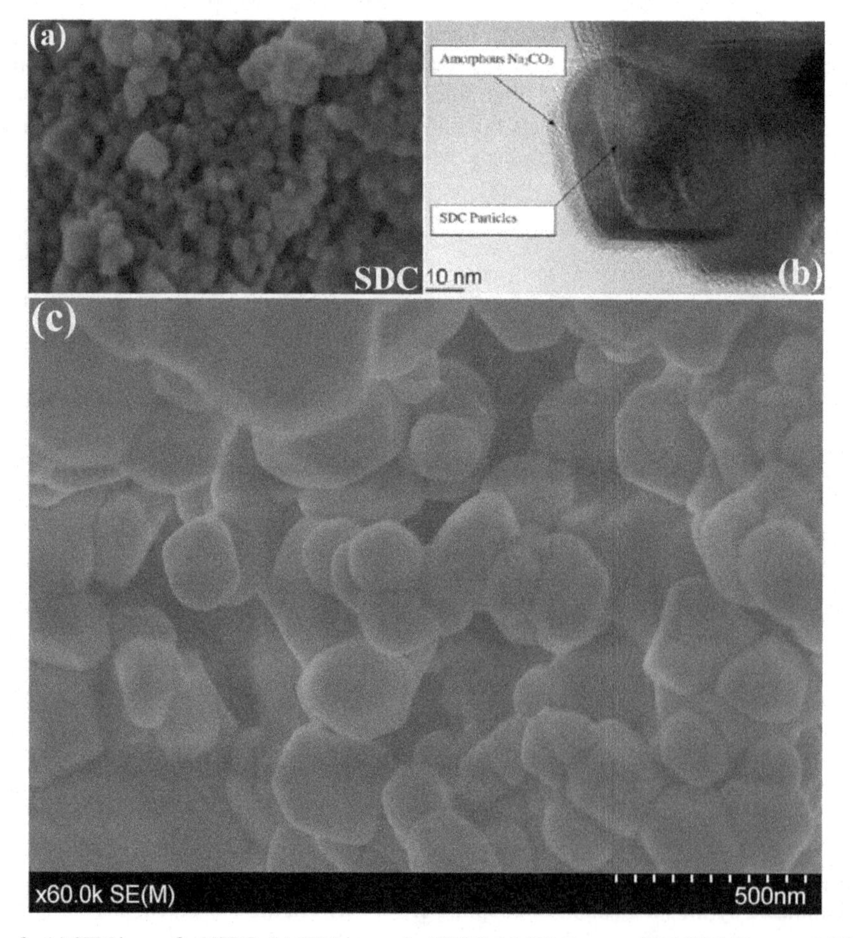

Figure 3. (a) SEM image for NSDC, (b) TEM image for NSDC, (c) SEM image of LN-SDC (Raza et al. 2010b).

nanocomposite particles can be controlled by the preparation skills, sintering temperature and time as well. On the other hand, the conventional two-step prepared SDC–$LiNaCO_3$ composites consist of much large particle size in micro meter level (Huang et al. 2008, Banerjee et al. 2007).

The NSDC nanocomposite and conventional LN-SDC composite electrolytes have exhibited very different morphologies, especially, when we used the NANOCOFC approach to develop the materials. The crystallographic and more details of nanostructure of the NSDC nanocomposites were investigated by transmission electron microscopy (TEM) as shown in Figure 3(b). The microstructure of LN-SDC is depicted in Figure 3(c).

The carbonate is coated on the SDC particle surface in a core shell structure, as observed clearly by TEM. The carbonate shell is amorphous. This amorphous nature can facilitate ionic conduction.

Calcium and samarium co-doped ceria based nanocomposite electrolytes

Microstructure/morphology

Figure 4(a) shows the microstructure of the co-doped ceria-Na_2CO_3 (CSDC-carbonate) composites, obtained by SEM. The morphology of the sample is largely homogeneous, signifying a rather dense. In the co-doped powder, homogeneity and chemical composition were determined by EDX as shown in Figure 4(b). The results are in good agreement with the experimental stoichiometric indexes and the nominal composition of CSDC-carbonate.

There is another way to develop ceria-host structure to form the ceria-carbonate nanocomposites, e.g., co-doping. Co-doping can stabilize the ceria properties, e.g., against the ceria reduction in H_2 to cause electron conduction (Banerjee et al. 2007). The CSDC-carbonate electrolytes have also achieved the same or even better fuel cell performance. The cell power density P_{max}, with H_2 as a fuel, reached a maximum of 980 mW/cm^2 at 560°C. The open-circuit voltage (OCV) was 1.05 V at 560°C and indicates that the electrolyte membrane is sufficiently dense. The performance is very encouraging at very low temperatures with the two phase co-doped nanocomposite electrolyte. This may be attributed from improved chemical surface properties of particles prepared using the co-doping technique, which appears to be useful and effective for fuel cell performance at lower temperatures.

The obtained performance at lower temperature with co-doped ceria electrolyte is comparatively better with conventional electrolytes YSZ and SDC based on the previous reported results by the researchers. Particularly, Huang et al. obtained maximum power density 131 mW/cm^2 at very low temperature 350°C using YSZ electrolytes (Huang et al. 2007d); Hibino et al. obtained 101 mW/cm^2 at 350°C using ceria based electrolytes (Hibino et al. 2000); Shao and Haile were reported 1010 mW/cm^2

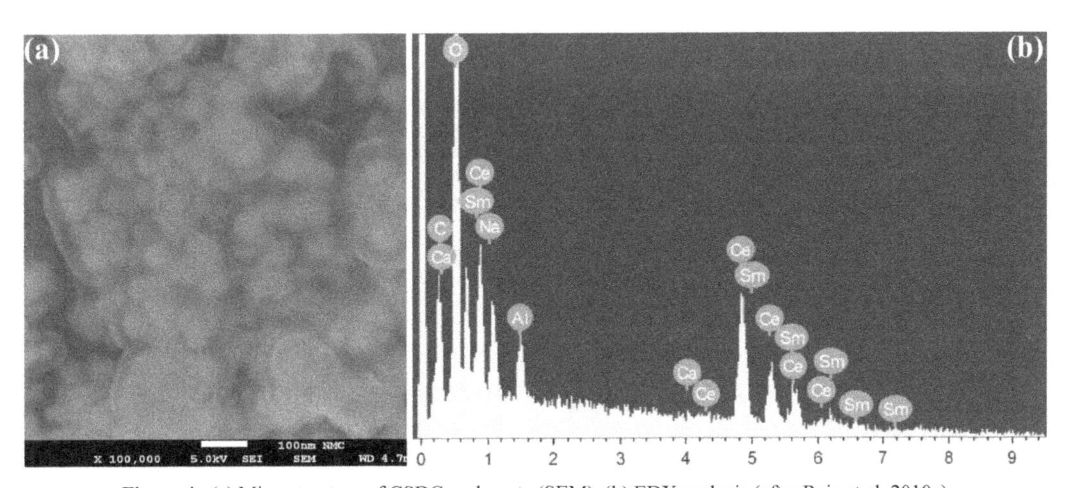

Figure 4. (a) Microstructure of CSDC-carbonate (SEM), (b) EDX analysis (after Raja et al. 2010a).

and 402 mW/cm^2 at 600°C and 500°C using single phase SDC electrolyte (Shao and Haile 2004); and Suzuki et al. demonstrated 1000 mW/cm^2 at 600°C using zirconia based electrolytes (Suzuki et al. 2009).

There is no significant electrode polarization process from Figures 3–6, because the I–V characteristics are linear. The less IR drop from electrolyte ohmic behaviour may be the reason for a higher performance at such low temperature. Due to the calcium co-doped two phase electrolyte, it has a higher conductivity which also makes the performance excellent. By viewing high performance SOFCs in literature (Shao and Haile 2004, Suzuki et al. 2009, Fan et al. 2011), it is a common fact that the high performance corresponds to linear I–V characteristics. In our work, when we used CSDC-carbonate mixed with the anode and cathode, the reaction at electrodes would be enhanced because of the high ionic conductivity of CSDC-carbonate. Then, electrode polarization resistance may be decreased for this development.

The Arrhenius plots of the ionic conductivity for different composite electrolytes have been shown in Figure 5. Figure 5 describes the oxide ion conductivity.

It can be seen that the CSDC-carbonate exhibits the best ionic conductivity as shown in Figures 3–7a. There are two effects: one is co-doping and the other is composite, where the interface is built between ceria and carbonate. The co-doping can enhance the oxygen ion concentrations and interface between ceria and carbonate facilitates the ions oxide/proton mobilities where the super-ionic conduction pathway takes place. The conductivity of all materials is high at higher temperature due to the conductivity which largely depends on the oxygen/proton ions' mobile concentration and has higher mobility at high temperature. The activation energy of the CSDC-carbonate (0.238 eV) is obtained from the Arrhenius relation. It is lower than that of the co-doped ceria electrolyte as reported by (Banerjee et al. 2007). This is attributed to the interfaces of the composite material.

Figure 5 shows proton ion conductivity of composite electrolyte at 500°C. The proton conduction in bulk oxide, extensive proton exchange and transfer along the interface of the two percolating phases contributes to the enhancement of conductivity.

Another important parameter that was investigated was the stability of such nanocomposite electrolytes' conductivity due to their major contribution in the performance of the cell. Therefore, we have measured the conductivity for the duration of 21 hrs in 3 days (7 hrs per day) at 550°C as shown in Figure 6.

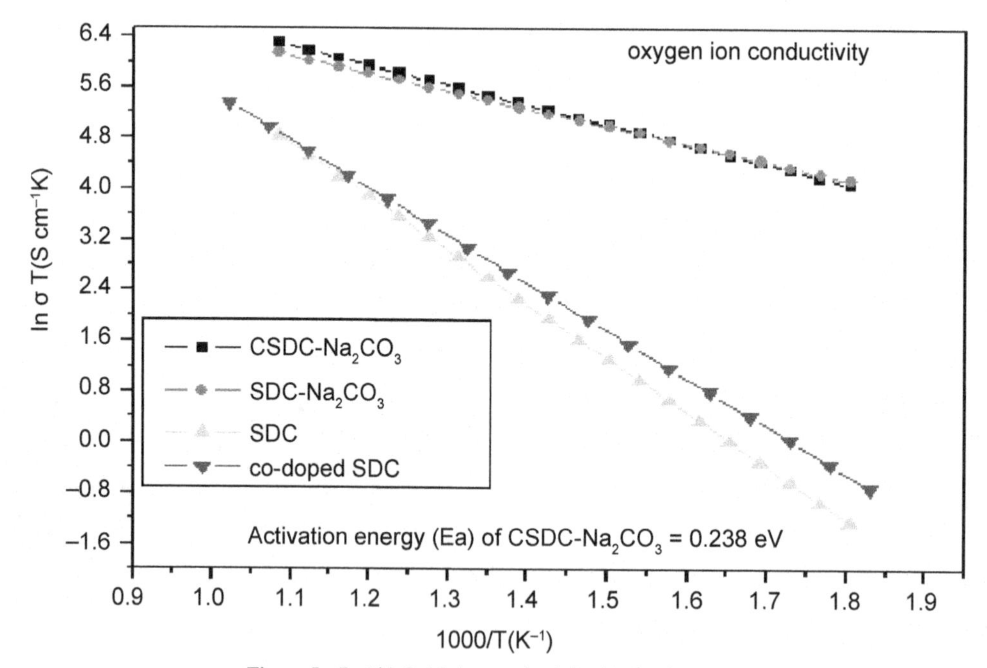

Figure 5. (In Air) Oxide ion conductivity (Arrhenius plot).

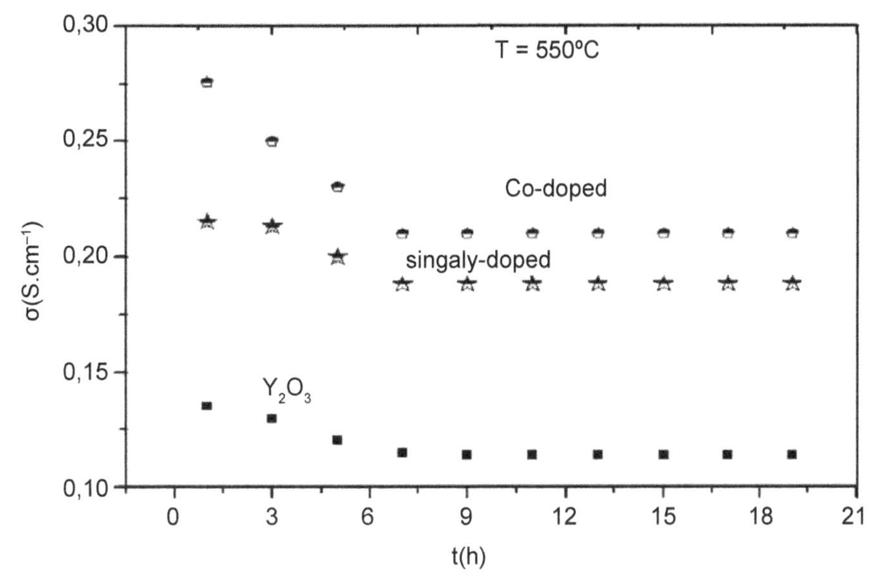

Figure 6. Stability of ionic and protonic conductivity @550°C of CSDC-carbonate (Raja et al. 2010a).

Composite electrolyte based on samaria-doped ceria and containing yttria as a second phase

Figure 7(a) shows a comparison of the electrolyte samples with different compositions. The diffractogram is indexed and the crystalline structure of the sample is cubic fluorite. The average crystallite size of SDC and Y_2O_3 was calculated by Debye-Scherrer formula and values were 40 and 20 nm, respectively. In Figure 7(a), it can be seen that the composite electrolyte consists of doped ceria and yttria. All peaks can be indexed with two crystal reference. The small peaks show that some tiny ones are crystalline form of the Y_2O_3. No individual phase of samaria was found, i.e., Samaria formed a solid solution with the ceria in the SDC. In addition, there is no lattice change from SDC to Y_2O_3-SDC composites. This implies that yttrium did not dope the ceria and thus didn't make the SDC host structural change. The intensities of the peaks related to yttrium oxide indicate different content in each sample. The XRD results prove the yttrium oxide-SDC electrolyte as a two-phase composite system. In addition, the diffraction patterns are also the proof of a nanocrystalline structure of the sample.

The density of the sintered electrolyte (YSDC) is 6.4939 g/cm^3 and the relative density was about 91%. The highest densification is obtained at 40% concentration of yttria in the SDC electrolyte.

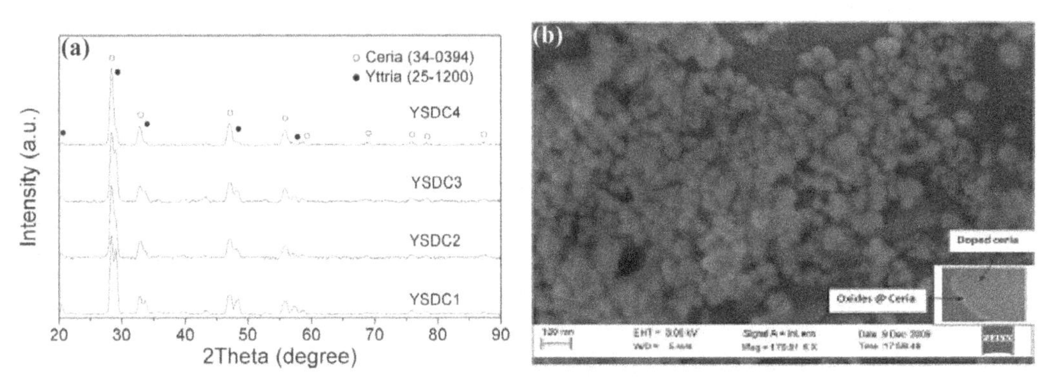

Figure 7. (a) Crystal structure (XRD), (b) microstructure of YSDC (SEM) (Ref: after (Raza et al. 2011)).

Figure 7(b) shows the morphology and nanostructure of the Y_2O_3-SDC electrolyte of SEM. Different particle sizes are seen in the figure in the range of 30 nm–80 nm. The prepared Y_2O_3-SDC composite electrolyte is denser than single doped ceria materials. It has been observed that for the smaller particle size, the conductivity will be higher because more and more ions can pass through the interface of the materials. The interfacial region between the yttrium oxide coating and samaria doped ceria thus plays a key role in the material conductivity and cell performance. These results also support the J. Maier theory published in the Nature journal that the interfaces are increasing with decreasing of the size of the particle (Maier 2005).

The core shell structure of the Y_2O_3-SDC was observed at an optimized sintering temperature of 700°C. A nanoparticle of the synthesized electrolyte is shown in the inset of Figure 7(b) as a two-phase composite consisting of a coating layer (yttrium oxide) named as a surface region, an inner part named as core of the SDC particle and the region between these two parts called interface.

IV/IP Characteristics (Fuel Cell Performance)

Figure 8 shows the I–V characteristics for comparison of three fuel cells using the co-doped ceria-carbonate composite, singly doped ceria (NSDC) composite, and singly doped ceria-oxide (YSDC) composite as electrolytes. The maximum fuel cell power density reached about 1000 mW/cm^2, 800 mW/cm^2, 700 mW/cm^2 at temperatures 550°C as shown in Figure 8 for co-doped (CSDC–Na_2CO_3), singly doped ceria (SDC–Na_2CO_3), and SDC–Y_2O_3, respectively. The open circuit voltage (OCV) is 0.98 V at 550°C for CSDC while 1.018 V at 500°C for NSDC electrolytes. The improved ceria–carbonate electrolyte (one step) displays significantly higher OCV than that of the conventional one. The lower OCV in the conventional composite (two-step) electrolyte fuel cell may be due to the low density of the electrolyte which caused some gas penetration directly through the electrolyte to make electrical voltage losses. This confirms that the nanocomposite NSDC electrolyte (one step) membrane is dense. We observed that the performance of fuel cell using the NSDC nanocomposite electrolyte was enhanced significantly at lower temperatures, 300–500°C, even functioned well between 200 and 300°C than those

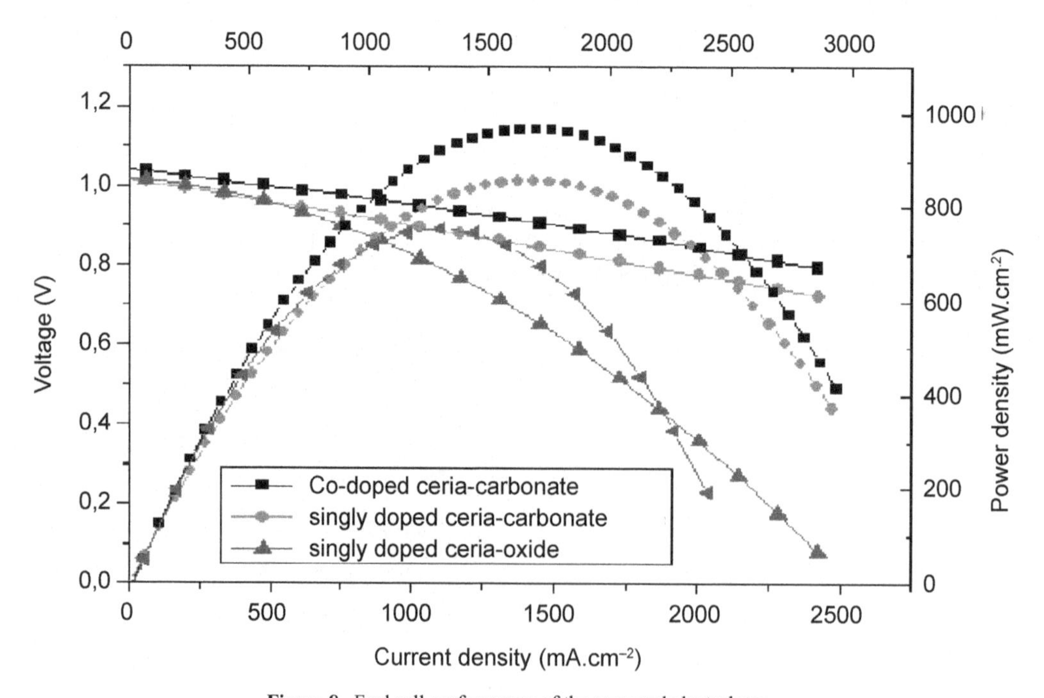

Figure 8. Fuel cell performance of the prepared electrolytes.

using the conventional composite electrolytes. It may be explained that due to an interfacial superionic conduction mechanism in the two-phase composites first reported by Schober (2005), large amount of surfaces of nanoparticles ceria and interfaces in the SDC nanocomposites can significantly enhance and improve the ionic conductivity.

The best performance of about 1000 mW/cm^2 was reported by Huang et al. for the conventional LN-SDC composite electrolyte fuel cell at 600°C (Huang et al. 2008, 2007c, d), while a better fuel cell performance of 1000 mW/cm^2 was obtained at temperature 500°C in this work. The advantages of the nanocomposites are obvious. The excellent fuel cell performances achieved so far for the nanocomposite electrolytes show that ceria–carbonates nanocomposite electrolyte is appropriate for the low temperature (300–500°C, even 200–300°C) fuel cells. The stability of the fuel cell can be expected to improve by only involving the solid carbonate and also at much lower temperatures in fuel cell operations. In the meantime, it also requires developments of high catalytic electrodes which we will report elsewhere.

Ion transportation mechanism

The oxide/carbonate was introduced in a doped ceria electrolyte to create a 2nd phase as a core-shell structure. J. Maier reported that there are highly conducting paths near phase boundaries due to space charge zones and it is much higher than in bulk (Maier). The ion transportation mechanism is shown in Figure 9. The two-phase composite may form a large interface region for ion conduction paths between the SDC and oxide/carbonate. It is called "superionic highway", as reported before by Schober and Zhu, for ceria-carbonate and ceria-oxide composite electrolytes (Zhu 2009, Suzuki et al. 2009), so as to greatly enhance the material conductivity. This interface between the host SDC and yttrium oxides has no bulk structural limit for creation of high concentration of mobile ions, and might be greatly disordered. It means that the interfaces have the capability to increase the mobile ion concentration than that inside the bulk. The electric field distribution in the interfaces between two phases is the key to realize the interfacial superionic conduction, allowing ions to move on particle's surfaces or interfaces through high conductivity pathways (Zhu 2009, Schober 2005).

The performance of cell with Y_2O_3 40% content has the highest power density, see Figure 8. The OCV and performance of our material is relatively better than that of pure SDC and YDC at ≤ 600°C, as reported (Huang et al. 2007c, d, Shao and Haile 2004, Suzuki et al. 2009). It means that the introduction

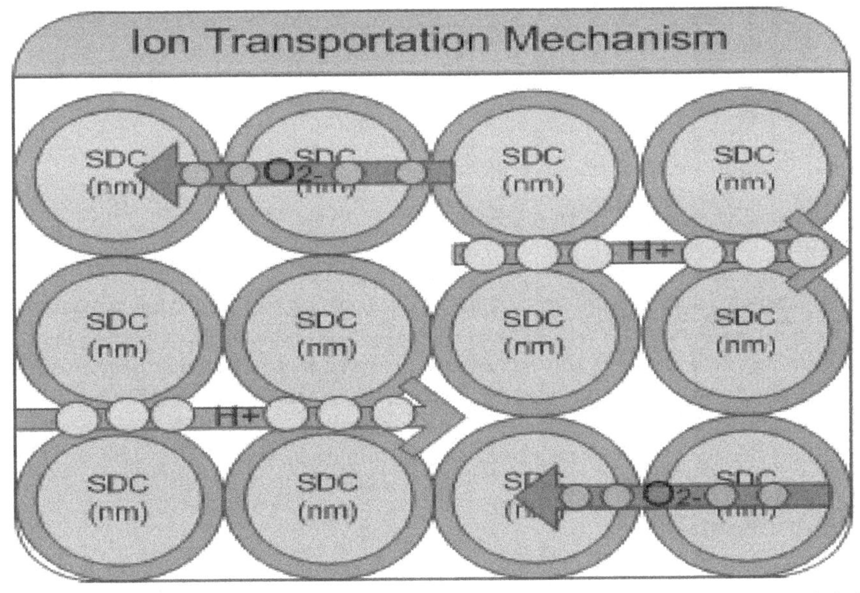

Figure 9. Ion transport mechanism in two phase composite electrolytes (where core is doped ceria and shell is oxide/carbonate) (Ref: after (Raza et al. 2011)).

of the second phase can extract the electronic conductivity of the SDC-based solid electrolyte. The electrolytes are stable in the fuel cell environment (Zhu and Mat 2006, Zhu 2006c).

Typical EIS spectra are shown in Figure 10. These spectra were obtained in air at 400–600°C. The impedance spectra have been modelled with an equivalent circuit, shown inset. From the equivalent circuit, L (inductance) is the effect of the stainless tube from the testing device. The R_1 (ohm resistance) predominately includes the electrolyte resistance. The electrochemical resistance, R_2, is probably associated with the electrode processes' combined cathode and anode reactions. Q (constant phase element) might be due to the interfaces between the electrodes and the electrolyte. It can be seen from EIS spectra at different temperatures that R_{ohm} decreases with the increase in temperature and, in parallel, the sample ionic conductivity increases.

It can also be seen from Figure 10 that there is only one arc observed which may be related to the electrode diffusion process. The arc corresponding to the bulk conductivity cannot be measured due to frequency a limit that is not sufficiently high. However, it is visible from the spectra that the intersections of the EI spectra with the Zre axis are not at the origin of the axis. A semi-circle from the origin of the impedance plot can be simulated as presented in Figure 10.

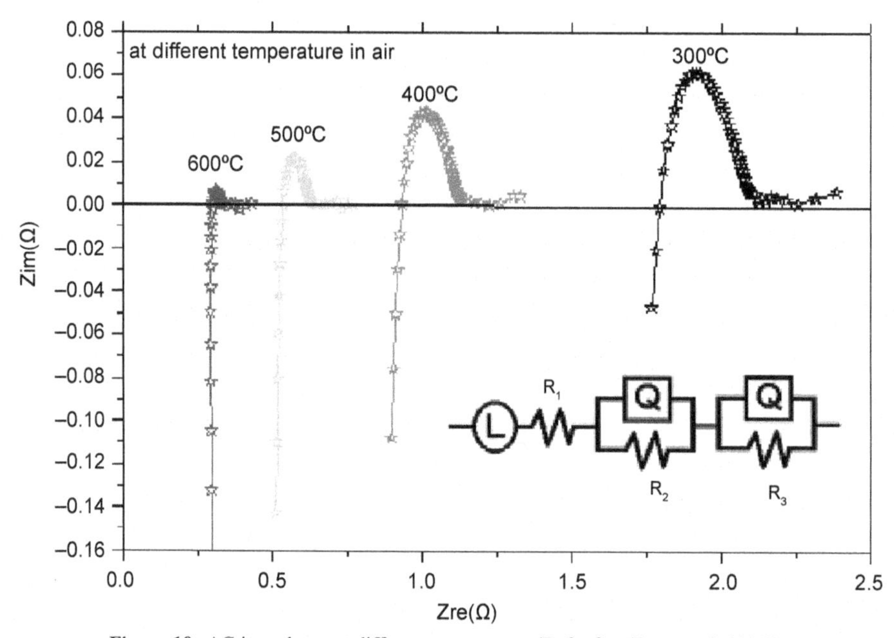

Figure 10. AC impedance at different temperature (Ref: after (Raza et al. 2011)).

Conclusions

It has been concluded that the different homogenous nanocomposite electrolyte with an enhanced conductivity of 0.1 S/cm, power density of 200 mW/cm^2 @350°C and the highest power density of about 1 watt/cm^2 was achieved @550°C. It has been also observed that the melting point of carbonate was lowered by the presence of SDC which caused lowering of the temperature of the FC.

It can be concluded that ASOFCs can provide a significantly high energy density, which is chemically stable and is cost effective with such nanocomposite electrolytes.

Acknowledgements

This work is supported by the Swedish Innovation System (VINNOVA), Swedish Research Council (VR)/Swedish agency for international cooperation development (Sida). The Higher Education

Commission (HEC) of Pakistan and COMSATS Institute of Information Technology, Lahore Pakistan is also acknowledged.

References

Banerjee, S., P.S. Devi, D. Topwal, S. Mandal and K. Menon. 2007. Enhanced ionic conductivity in $Ce_{0.8}Sm_{0.2}O_{1.9}$: Unique effect of calcium Co-doping. Advanced Functional Materials 17(15): 2847–2854.

Benamira, M., A. Ringuedé, V. Albin, R.N. Vannier, L. Hildebrandt et al. 2011. Gadolinia-doped ceria mixed with alkali carbonates for solid oxide fuel cell applications: I. A thermal, structural and morphological insight. Journal of Power Sources 196(13): 5546–5554.

Fan, L., C. Wang, M. Chen, J. Di and J. Zheng. 2011. Potential low-temperature application and hybrid-ionic conducting property of ceria-carbonate composite electrolytes for solid oxide fuel cells. International Journal of Hydrogen Energy. In Press, n., pp. DOI:10.1016/j.ijhydene.2011.1005.1055.

HartmanovÃ, M., E. Lomonova, V. NavrÃTil, P.Å. Utta and F. Kundracik. 2005. Characterization of yttria-doped, ceria prepared by directional crystallization. Journal of Materials Science 40(21): 5679–5683.

Hibino, T., A. Hashimoto, T. Inoue, J. Tokuno and S. Yoshida. 2000. A low-operating-temperature solid oxide fuel cell in hydrocarbon-air mixtures. Sciences 288(5473): 2031–2033.

Hou, Z., X. Wang, J. Wang and B. Zhu. 2008. Structural studies on ceria-carbonate composite electrolytes. 368–372 PART 1: 278–281.

Huang, J.B., L.Z. Yang and Z.Q. Mao. 2007a. High performance low temperature ceramic fuel cell with zinc doped ceria-carbonates composite electrolyte. Key Engineering Materials 336: 413–416.

Huang, J., Z. Mao, Z. Liu and C. Wang. 2007b. Development of novel low-temperature SOFCs with co-ionic conducting SDC-carbonate composite electrolytes. Electrochemistry Communications 9(10): 2601–2605.

Huang, H., M. Nakamura, P. Su, R. Fasching and Y. Saito. 2007c. High-performance ultrathin solid oxide fuel cells for low-temperature operation. Journal of the Electrochemical Society 154(1): B20–B24.

Huang, J., Z. Mao, Z. Liu and C. Wang. 2007d. Development of novel low-temperature SOFCs with co-ionic conducting SDC-carbonate composite electrolytes. Electrochemistry Communications 9(10): 2601–2605.

Huang, J., Z. Mao, Z. Liu and C. Wang. 2008. Performance of fuel cells with proton-conducting ceria-based composite electrolyte and nickel-based electrodes. Journal of Power Sources 175(1): 238–243.

International. 2006. Journal of Electrochemical Science 1: 383–402.

Islam, M.S., J.R. Tolchard and P.R. Slater. 2003. An apatite for fast oxide ion conduction. Chemical Communications (13): 1486–1487.

Joshi, A.V., J.J. Steppan, D.M. Taylor and S. Elangovan. 2005. Solid electrolyte materials, devices and applications. ChemInform. (13): 619–625.

Liu, J., B.D. Madsen, Z.Q. Ji and S.A. Barnett. 2002. A fuel-flexible ceramic-based anode for solid oxide fuel cells. Electrochem. Solid State Letters 5(6): A122–A124.

Maier, J. 2005. Nanoionics: ion transport and electrochemical storage in confined systems. Nat Mater. 4(11): 805–815.

McIntosh, S. and R.J. Gorte. 2004a. Direct hydrocarbon solid oxide fuel cells. Chemical Reviews 104(10): 4845–4866.

Park, S., J.M. Vohs and R.J. Gorte. 2000. Direct oxidation of hydrocarbons in a solid-oxide fuel cell. Nature 404(6775): 265–267.

Raze, R., X. Wang, Y. Ma, Y. Huang and B. Zhu. 2009. Enhancement of conductivity in ceria-carbonate nanocomposites for LTSOFCs. Journal of Nano Research 6: 197–203.

Raza, R., X. Wang, Y. Ma and B. Zhu. 2010a. Study on calcium and samarium co-doped ceria based nanocomposite electrolytes. Journal of Power Sources 195(19): 6491–6495.

Raza, R., X. Wang, Y. Ma, X. Liu and B. Zhu. 2010b. Improved ceria-carbonate composite electrolytes. International Journal of Hydrogen Energy 35(7): 2684–2688.

Raza, R., G. Abbas, X. Wang, Y. Ma and B. Zhu. 2011. Electrochemical study of the composite electrolyte based on samaria-doped ceria and containing yttria as a second phase. Solid State Ionics 188(1): 58–63.

Sasaki, K., K. Watanabe, K. Shiosaki, K. Susuki and Y. Teraoka. 2004. Multi-fuel capability of solid oxide fuel cells. Journal of Electroceramics 13(1): 669–675.

Schober, T. 2005. Composites of ceramic high-temperature proton conductors with inorganic compounds. Electrochemical and Solid-State Letters 8(4): A199–A200.

Shao, Z. and S.M. Haile. 2004. A high-performance cathode for the next generation of solid-oxide fuel cells. Nature 431(7005): 170–173.

Shaorong, W., K. Takehisa, D. Masayuki and H. Takuya. 2000. Electrical and ionic conductivity of Gd-doped ceria. Journal of the Electrochemical Society 147(10): 3606–3609.

Steele, B.C.H. and A. Heinzel. 2001. Materials for fuel-cell technologies. Nature 414(6861): 345–352.

Suzuki, T., Z. Hasan, Y. Funahashi, T. Yamaguchi and Y. Fujishiro. 2009. Impact of anode microstructure on solid oxide fuel cells. Science 325(5942): 852–855.

Tealdi, C., G. Chiodelli, L. Malavasi and G. Flor. 2004. Effect of alkaline-doping on the properties of La2Mo2O9 fast oxygen ion conductor. Journal of Materials Chemistry 14(24): 3553–3557.

Wang, X., Y. Ma, R. Raza, M. Muhammed and B. Zhu. 2008. Novel core-shell SDC/amorphous Na_2CO_3 nanocomposite electrolyte for low-temperature SOFCs. Electrochemistry Communications 10(10): 1617–1620.
Yang, L., S. Wang, K. Blinn, M. Liu and Z. Liu. 2009. Enhanced sulfur and coking tolerance of a mixed ion conductor for SOFCs: $BaZr_{0.1}Ce_{0.7}Y_{0.2}YbxO_3$. Science 326(5949): 126–129.
Zhu, B. and M. Mat. 2006. Studies on dual phase ceria-based composites in electrochemistry.
Zhu, B. and B. Mellander. 1994. Ionic conduction in composite materials containing one molten phase. Solid State Phenomena 39(40): 19–22.
Zhu, B. and S. Tao. 2000. Chemical stability study of Li_2SO_4 in a H_2S/O_2 fuel cell. Solid State Ionics 127(1-2): 83–88.
Zhu, B., C.R. Xia, X.G. Luo and G. Niklasson. 2001. Transparent two-phase composite oxide thin films with high conductivity. Thin Solid Films 385(1-2): 209–214.
Zhu, B., X. Liu, P. Zhou, X. Yang and Z. Zhu. 2002. Innovative solid carbonate—Ceria composite electrolyte fuel cells. Fuel Cells Bulletin 2002(1): 8–12.
Zhu, B., X. Yang, J. Xu, Z. Zhu and S. Ji. 2003. Innovative low temperature SOFCs and advanced materials. Journal of Power Sources 118(1-2): 47–53.
Zhu, B. 2006c. Next generation fuel cell R&D. International Journal of Energy Research 30(11): 895–903.
Zhu, B., S. Li and B.E. Mellander. 2008. Theoretical approach on ceria-based two-phase electrolytes for low temperature (300–600°C) solid oxide fuel cells. Electrochemistry Communications 10(2): 302–305.
Zhu, B. 2009. Solid oxide fuel cell (SOFC) technical challenges and solutions from nano-aspects. International Journal of Energy Research 33(13): 1126–1137.

16

Carbonaceous Nanostructured Composites for Electrochemical Power Sources

Fuel Cells, Supercapacitors and Batteries

Sethu Sundar Pethaiah,[1,*] *J. Anandha Raj*[2] and *Mani Ulaganathan*[2,*]

Introduction

Elemental carbon plays a significant role in the development of alternative, clean and sustainable energy technologies; they have been used in many potential applications in day-to-day life activities, for example: abrasives, carbon fibers as reinforcement material in tennis racquets, activated carbons for water purification, carbon blacks for inks, carbon membrane switches in electronic gadgets, graphite electrodes in electrochemical industries, carbon arc electrodes in projectors, leads in pencils, etc. (Prem Kumar 2009). One of the important applications is as electrode materials in electrochemical energy storage (supercapacitors (SCs) and batteries) and energy conversion devices (fuel cells). Electrochemical energy storage devices have potential applications in electric and hybrid electric vehicles, low-energy industrial equipment, consumer electronics and power grid. There are various kinds of active electrode materials that have been used to obtain an improved performance of such devices. Most commonly, carbonaceous materials with different structures (particularly porosity) represent a very important family of electrode materials for electrochemical energy storage and conversion applications. Different forms of the carbon materials with various significant characteristics, such as high surface area, pore volume, pore size, good electrical conductivity, chemical stability, and low cost are playing a key role in the improvement of electrode performances which are used in supercapacitors, batteries, and fuel cell applications.

[1] GasHubin Engineering Pte. Ltd., Singapore 536200.
[2] CSIR-Central Electrochemical Research Institute, Karaikudi, Tamil Nadu, India-630003.
Email: janandharaj@gmail.com
* Corresponding authors: sundar@fuelcell.sg; nathanphysics@gmail.com

A series of carbonaceous materials, including activated carbon, carbon nanotubes (CNTs), graphene-based nanostructures and their composites were prepared, characterized and evaluated in the energy storage (SCs and batteries) and conversion (fuel cells) applications. In addition, carbon materials have been used as an additive in the electrode preparation to enhance the electronic conductivity of the composite electrodes. Hence, in the present chapter, recent developments of the carbonaceous materials in different electrochemical energy storage and conversions devices are discussed in detail.

Fuel cells

Fuel cells are energy conversion devices which convert the chemical energy of the fuel directly into electrical energy, heat and water as a by-product. Generally, fuel cells are classified based on the type of electrolyte and fuel used, which includes Alkaline Fuel Cells (AFCs), Low Temperature Polymer Electrolyte Membrane Fuel Cells (LT-PEMFCs), High Temperature Polymer Electrolyte Membrane Fuel Cells (HT-PEMFCs), Phosphoric Acid Fuel Cells (PAFCs), Molten Carbonate Fuel Cells (MCFCs) and Solid Oxide Fuel Cells (SOFCs). Further classifications includes direct ethanol fuel cell, biological fuel cells, composite solid oxide/molten carbonate fuel cells, direct ammonia fuel cells and direct carbon fuel cell, which differs from above mentioned types. The electric-energy converting efficiency of fuel cells typically varies from 40 to 60% and about 30–40% of the energy is available as heat (Giddey et al. 2012). Among the above different types of fuel cells, Polymer Electrolyte Membrane Fuel Cells (PEMFC) have large potential for commercialization and its applications range from portable and transportation to large-scale stationary power systems (Wee 2007).

Nanostructured carbonaceous composites as a structural component in fuel cells

Carbon materials are playing a significant role in the development of electrochemical applications due to their high electrical conductivity, acceptable chemical stability and low cost (Dicks 2006).

Various types of carbonaceous materials have been employed in PEMFCs and play a different role such as conducting agent, supporting catalyst, etc.; in particular, the carbon composite materials are mainly used as major structural component in the fuel cell (Figure 1), which includes

- Bipolar plates
- Gas diffusion layer
- Electrocatalyst support
- Used as a fuel in direct carbon fuel cell (DCFC)

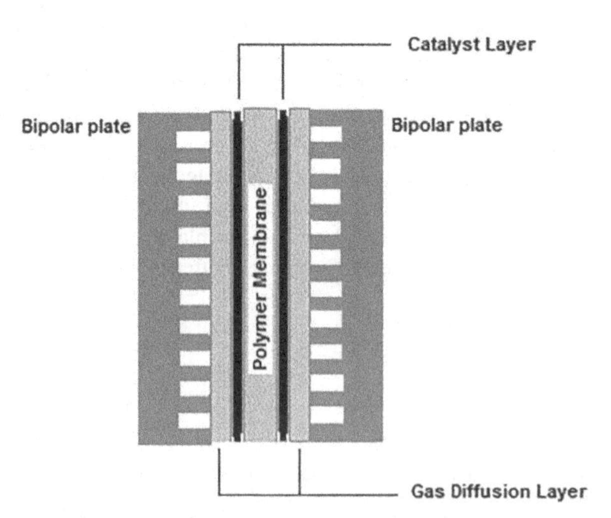

Figure 1. Basic structure and components of PEMFCs.

Hence, in this section, the roles of the carbonaceous nanostructure and their composites have been explained, particularly in PEMFC applications.

Bipolar plate (BP)

Bipolar plates play a vital role in the determination of the performance characteristics of the PEMFCs. As we know, in PEMFC BP act as current collector and fuel distributor (through flow channels); hence, it should have good electrical conductivity, gas impermeability, low density, and corrosion resistance. In addition, BP might also have a good mechanical strength and thermal stability in order to fabricate the PEMFC in a compact design and transfer the heat which evolved during the cell operation, respectively (Adloo et al. 2016). Department of energy (DOE), USA targets the PEMFC characteristics towards the successful transportation applications and the targets fixed to achieve by 2020 are given in Table 1 (https://energy.gov/eere/fuelcells/doe-technical-targets-polymer-electrolyte-membrane-fuel-cell-components (Assessed as on 02/08/217)).

Mostly, non-metal, metal and composite materials have been used as BP in fuel cells (Figure 2), based on the thermal and electrical conductivity and corrosion resistance (Hermann et al. 2005). In recent years, various research works have been focused on the development of various composite materials to replace the pure graphite BP, due to their attractive features which include acceptable corrosion resistance, light weight, etc. (Antunes et al. 2010). However, low electrical conductivity, poor mechanical properties, and high gas permeability are the major concern in composite plate. Recently, various reports have been focused on different composite BPs to replace the graphite BPs—Bipolar plates (Adloo et al. 2016, Mamunya et al. 2002, Antunes et al. 2011, Dhakate et al. 2010, Pethaiah et al. 2011). Adloo et al. (2016), studied the addition of different electrically conductive additives such as graphite, graphene, and high structure nano-carbon black on the manufacturing of polypropylene bipolar plates. The best composition of 23% polypropylene, 65% graphite, 7% high structure carbon black, 5% polypropylene maleic anhydride possessed the electrical conductivity of 104.63 $S.cm^{-1}$ and flexural strength of 44.28 MPa. Mamunya et al. (2002), explained the conduction mechanism of composites by the classical percolation theory (Figure 3).

The composite behaves as an insulator when the volume fraction of the conductive filler is below the percolation threshold (φ_c). The maximum electrical conductivity (m) is achieved only as the filler content (F) increases above the percolation threshold and below percolation threshold value, the composite conductivity is determined by polymer matrix (Antunes et al. 2011). Dhakate et al. (2010) investigated different vol.% of Multiwall Carbon Nanotubes (MWCNTs) in graphite—polymer composite bipolar plate and found that adding of 1 vol.% of MWCNTs in graphite composite plate, the electrical and thermal conductivity of nanocomposite increased by 100%. This enhancement is due to the orientation

Table 1. Technical targets: bipolar plates for transportation applications. https://energy.gov/eere/fuelcells/doe-technical-targets-polymer-electrolyte-membrane-fuel-cell-components (Assessed as on 02/08/217).

Characteristic	Units	2015 Status	2020 Targets
Cost	$\$/kW_{net}$	7	3
Plate weight	kg/kW_{net}	< 0.4	0.4
Plate H_2 permeation coefficient	Std $cm^3/(sec\ cm^2Pa)$	0	$< 1.3 \times 10^{-14}$
Corrosion, anode	@ 80°C, 3		
Corrosion, cathode	atm, 100%	no active peak	< 1 and no active peak
Electrical	RH	< 0.1	< 1
conductivity	$\mu A/cm^2$	> 100	> 100
Areal specific	$\mu A/cm^2$	0.006	< 0.01
resistance	S/cm	> 34 (carbon plate)	> 25
Flexural strength	ohm cm^2	20–40	40
Forming elongation	MPa		
	%		

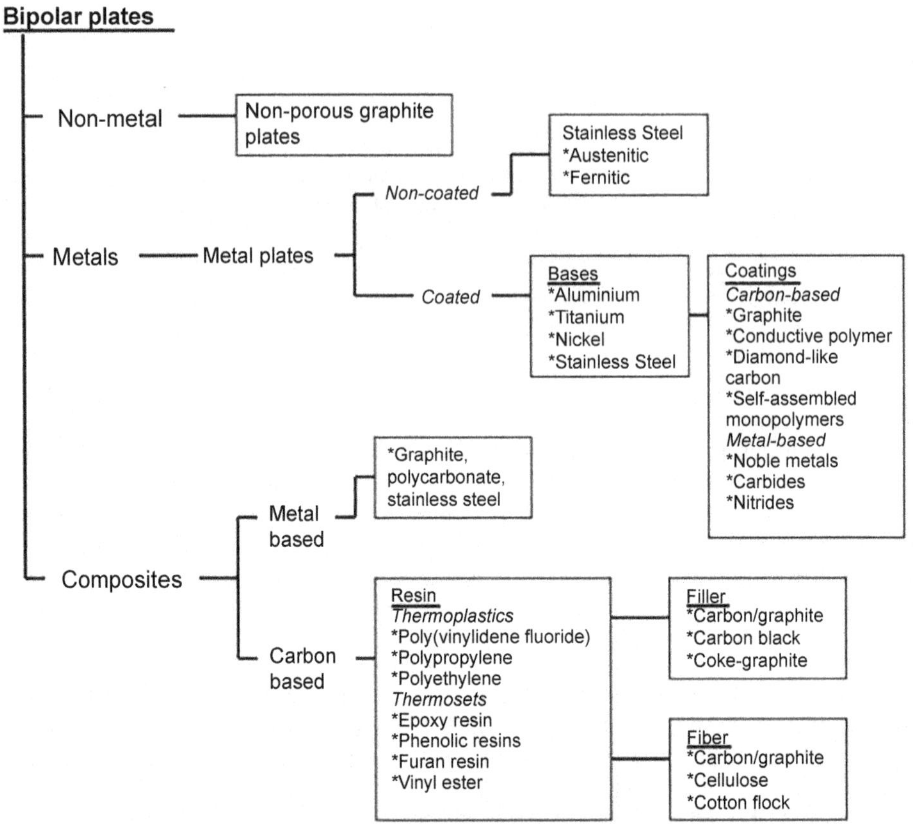

Figure 2. Classification of various materials used for Bipolar plates. Reprinted with permission from ref. Hermann, A., T. Chaudhuri and P. Spagnol. 2005. Int. J. Hydrogen Energy 30: 1297. Copyright @ Elsevier.

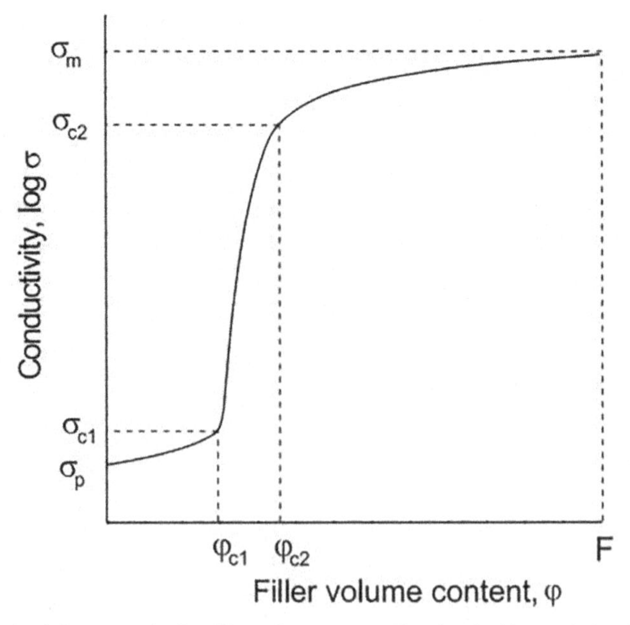

Figure 3. Electrical conductivity on conductive filler volume content. Reprinted with permission from ref. Antunes, R.A., M.C.L. de Oliveira, G. Ett, V. Ett. 2011. J. Power Sources 196: 2945. Copyright @ Elsevier.

of MWCNTs in all the directions of composite, positive synergistic effect of MWCNTs and heat transfer along the axis directions. The improvement in conductivity due to an increase in the electron transfer ability within the composite plate leads to the enhancement of the I–V performance of fuel cell.

Generally, carbon will undergo electrochemical oxidation in the PEMFC environment (Pethaiah et al. 2011); thus, due to the chemical instability of such carbon material, a high corrosion rate will be raised. de Oliveira et al. (2013) investigated MWCNTs, synthetic graphite particles and acrylonitrile-butadiene-styrene composite BP for PEMFC applications. It was reported that 2 wt.% of MWCNTs provided the best compromise between through-plane electrical conductivity and corrosion resistance. However, higher MWCNT loadings decrease the corrosion resistance and also, there is no significant gain in the electrical conductivity. Kakati et al. 2010, also studied above similar corrosion resistance of a composite bipolar plate prepared from a mixture of phenolic resin, graphite, carbon black and carbon fibers; it was reported that the corrosion current density increased with the addition of the conductive carbon black and carbon fiber fillers. Attaining good electrical conductivity and corrosion resistance which can improve the performance of the BPs is always a major challenge in the development of the composite BPs. Comparison of the results of different matrixes and fillers, and the variation in the electrical conductivity with respect to the various processings are given in Table 2 (Antunes et al. 2011).

In recent years, carbon coated metal plates are also proposed by various research groups (Mirzaee Sisan et al. 2014, Chung et al. 2008, Pech-Rodríguez et al. 2014) due to its good gas impermeability, high mechanical strength and low cost (for mass production). Sisan et al. (2014), coated carbon on SS-316L plate and they found 200 nm coating thickness provides best corrosion resistance. Chung et al. (2008) reported that the carbon coating on the metallic compound enhanced the performance of the PEMFC. Pech-Rodríguez et al. (2014) investigated polypyrrole-Vulcan XC-72 corrosion protection coatings on SS-304 BP plates by using Electrophoretic deposition techniques and confirmed that polypyrrole-Vulcan XC-72

Table 2. Electrical conductivity composite bipolar plates with different matrix and filler (Antunes et al. 2011).

Matrix	Filler	Filler Loading	σ ($S.cm^{-1}$)	Reference
Epoxy	Graphite	70 vol.%	153.8	
	Carbon Black	5 vol.%		
Epoxy	Graphite	73 vol.%	254.7	(Du et al. 2010a)
	MWCNT	2 vol.%		
Epoxy	Graphite	68 vol.%	237	
	Carbon Fiber	7 vol.%		
Epoxy	Expanded Graphite	69 wt.%.	121	
Epoxy	Expanded Graphite	60 wt.%.	500	(Du and Jana 2007)
Compressed expanded graphite	Epoxy	40 vol.%	175	
Phenolic resin (novolac type)	Expanded Graphite	50 vol.%	280	(Du et al. 2010b)
Phenolic resin (novolac type)	Expanded Graphite	50 vol.%	285	
	Carbon Black	5 vol.%		
Phenolic resin (novolac type)	Natural Graphite	65 wt.%.	178, 30.0	(Dhakate et al. 2009)
	MWCNT	1 wt.%.		(Dhakate et al. 2008)
Phenol formaldehyde resins (novolac type)	Carbon Fiber	5 wt.%.	281.97	
	Carbon Black	5 wt.%.	79.84	
	Natural Graphite	60 wt.%.		(Dhakate et al. 2010)
	Graphene	1.5 wt.%.	319.42	
Phenol formaldehyde resins (novolac type)	Carbon Fiber	5 wt.%.	113.87	
	Carbon Black	5 wt.%.		
	Natural Graphite	58.5 wt.%.		(Ghosh et al. 2014)
	Exfoliated Graphite	35 wt.%	374.42 and 97.32	
Phenolic Resin	Carbon Black	5 wt.%		(Ghosh et al. 2014)
	Graphite Powder	3 wt.%		(Gautam and Kar 2016)

coated on SS BP plates is a promising candidate for PEMFCs due to their high electrical conductivity and enhanced corrosion resistance. Recently, Hao et al. (2016) studied magnesium phosphate cement (MPC) based composite as the construction material for bipolar plates of fuel cells. They used graphite, carbon black, carbon fiber, and MWCNTs to construct the conductive phase; using their optimize ration, they achieved the target fixed by DOE, USA-2015.

Gas-diffusion layer in fuel cell

Gas diffusion layer (GDL) is an integral part of a membrane electrode assemble (MEA) of fuel cell, and its principal functions are to efficiently transport the reactants and the products to and from the reaction sites as well as to conduct heat and current. Integration of the microporous layer (MPL) and gas diffusion medium (GDM) is called Gas diffusion layer. MPL comprises of carbon for electrical conductivity and PTFE for hydrophobicity (Jayakumar et al. 2015, Cindrella et al. 2009, Park et al. 2015, Park et al. 2012).

Generally, carbon-fiber-based products, such as carbon cloth or nonwoven carbon paper are widely used as GDM due to its high gas permeability, and good electronic and thermal conductivity. Scanning electron microscope (SEM) surface image of carbon cloth and carbon paper is shown in Figure 4. More details about the carbon-fiber products can be found in (Mathias et al. 2010). Carbon nanotube bucky papers have also been used in GDL due to its excellent conductivity, good durability and ease of production. So far, a variety of carbon-based materials have been used in MPL due to good conductivity and hydrophobicity, which includes 1. Vulcan XC-72, 2. Black Pearl 2000 carbon, 3. Shawinigan

(a) (b)

Figure 4. SEM image of (a) Carbon Cloth and (b) Carbon Paper.

acetylene black (SAB), 4. Asbury 850, 5. Mogul L, 6. Carbon Nanotubes, 7. Acetylene Black carbon, 8. Platelet carbon nanofiber, etc. (Jayakumar et al. 2015, Park et al. 2012, Schweiss et al. 2012, Kitahara et al. 2015, Kim et al. 2013).

Electrocatalyst support

Platinum (Pt) or Pt alloy nanoparticles, supported on porous materials based electrocatalysts, are mostly used in fuel cell application. Since Pt catalyst is very expensive, carbon based nano materials have been employed as supporting catalyst in fuel cell applications for improving the performances. A good electrocatalyst support might have the following properties: (1) high specific surface area and high porosity, (2) high electrochemical stability and easy Pt recycling, and (3) high electric conductivity (Candelaria et al. 2012). Vulcan XC-72 carbon black is the most commonly used catalyst support; however, the activity and durability has still become a challenge for fuel cell commercialization. Hence, in recent years, many research and development activities have mainly focused on carbon nanotubes (CNT), mesoporous carbon and graphene supported catalyst towards achieving the good performance of PEMFCs.

CNT based electrocatalyst delivered superior performance than conventional carbon black in PEMFC applications; this is mainly due to high durability and good electrical conductivity (Candelaria et al. 2012, Shao et al. 2006, Niu and Wang 2008, Zhou et al. 2010). In addition, CNT has less impurity than Vulcan XC-72 carbon. Wang (2008) studied platinum loaded on activated carbon nanotubes and reported activated carbon nanotubes (ACNT) with high Pt-loading of 50 wt.% which showed good electrochemical activity not only in H-adsorption/desorption but also in methanol oxidation compared to commercial Pt/C (Table 3). This is due to high BET surface area with plenty of mesopores and defects, which leads to high Pt utility and, consequently, to excellent electrochemical activity of the ACNTs.

Very few types of nano tubes such as single wall carbon nanotubes (SWCNT) and double wall carbon nanotubes (DWNT) supported by Pt alloy electrocatalysts are studied as catalyst support in fuel cell applications (Shao et al. 2006). Wang et al. (2006) compared the surface oxidation of carbon black Vulcan XC-72 and MWCNT. The durability study demonstrated that MWCNT as catalyst support exhibits less Pt surface area loss without sacrificing catalytic activity, suggesting that MWCNT is a promising catalyst support for PEMFC application. In addition, chemically inert nature of CNTs lowers the effective reaction sites for attachment of metal nanoparticles. To overcome this, Gupta et al. (2016) investigated the modification of the surface of CNTs by chemical functionalization. From their report, the peak power density of 156 mW cm^{-2} has been achieved with catalyst prepared in alkaline medium, an increase of > 110% as compared to 72 mW cm^{-2} catalyst prepared in acidic medium and tested under similar conditions.

To achieve higher dispersion of Pt nanoparticles, CNTs generally require functionalization, which may significantly reduce their electrical conductivity. In contrast to CNTs, graphene can offer edge plane anchor sites for the Pt catalyst nanoparticles (Pham et al. 2016), which enhances activity and durability as fuel cell catalyst support. Moreover, the unique planar geometry of graphene provides an opportunity to create promising three phase zone, which leads to improve the performance of PEM fuel cell (Figure 5). Hence, in recent years, interest in the synthesis of graphene based catalyst support has increased drastically.

Şanlı et al. (2016) demonstrated synthesis, characterization and fuel cell performance of Pt/graphene electrocatalysts. It is observed that the Pt nanoparticles were anchored onto the graphene oxide (GO), graphene nanoplatelets (GNP) and thermally reduced GO (TRGO) using several impregnation reduction methods including ethylene glycol reflux, sodium borohydride reduction and ascorbic acid reduction. From the report, better results in terms of Pt dispersion, particle size, and electro catalytic activity were obtained with ethylene glycol reflux and GO as the support material. The maximum power density of 320 mW cm^2 (40% higher than that of Pt/Vulcan XC-72) was achieved for Pt/r-GO with a Pt loading of 0.25 mg cm^2.

Mesoporous carbon (MC) is one of the important catalyst support material that has received much attention in recent years. The physical properties such as surface area, pore sizes, surface groups, etc have shown to improve the catalytic effect, nanoparticles' distribution across the surface, anchoring of the nanoparticles to the MC support and improvement of the reactant mass transport (Viva et al. 2014). Shrestha et al. (2011) developed Pt/MC and Pt/Vulcan XC-72 by a pH-controlled modified polyol process and found Pt/MC performs lower specific activity but higher mass-specific activities compared to Pt/Vulcan XC-72 carbon. Hasché et al. (2012) synthesised a mesoporous nitrogen doped carbon supported

Table 3. Performance comparison of various carbon support with different Pt loading (Niu and Wang 2008).

Samples	**50 wt.% Pt/C (Vulcan XC- 72)**	**50 wt.% Pt/CNT**	**30 wt.% Pt/ ACNT**	**40 wt.% Pt/ACNT**	**50 wt.% Pt/ACNT**
Electro active specific surface area (m^2/g)	14.0	~ 24.6	~ 17.7	~ 33.6	~ 41.1
Pt utility ratio (%)	32.3	~ 30.8	~ 10.0	~ 30.2	~ 41.7
Effective accessible surface area (m^2/g)	51.4	~ 117.6	~ 120.2	~ 200.6	~ 190.0

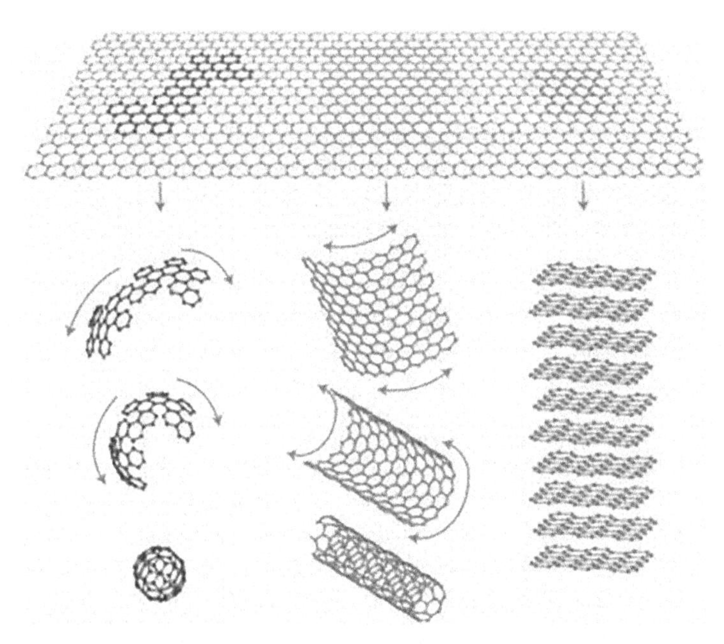

Figure 5. Schematic representation of graphene, which is the starting material for bucky balls, carbon nanotubes, and graphite. Reprinted with permission from ref. Hou, J., Y. Shao, M.W. Ellis, R.B. Moore and B. Yi. 2011. Phys. Chem. Chem. Phys. 13: 15384. Copyright @ Royal Society of Chemistry.

platinum catalyst and compared with commercial high surface area carbon supported platinum catalyst. It was found that the Pt supported with N-doped mesoporous carbon electrochemical active surface area was significantly enhanced when compared to the commercial catalyst.

Fuel in direct carbon fuel cell (DCFC)

The basic structure and operation principle of direct carbon fuel cell is same as other fuel cells and is classified based on electrolyte used (Figure 6). DCFC directly converts the chemical energy in carbon (Fuel) into electrical energy through its direct electrochemical oxidation (Giddey et al. 2012).

Generally, various types of fuels such as acetylene black, coal derived carbon, furnace oil carbon black, graphite particles, heat treated petroleum coke, etc. have been investigated in fuel cells; Giddey et al. (2012) and Cooper (2007) reported that the peak current density achieved was 110 $mA.cm^{-2}$ at 0.8 V and at 800°C (96 $mA.cm^{-2}$ power density, 80% efficiency) with furnace black. Similar studies were also carried by Hackett et al. (2007) and reported that the graphite rods produced open-circuit voltages of up to 0.788 V and current densities up to 230 $mA.cm^{-2}$ while coal-derived rods produced open-circuit voltages of up to 1.044 V and only 35 $mA.cm^{-2}$ in current density. The main advantage of DCFC is that the fuel utilization and its theoretical efficiency are almost equal to 100%, which leads to electric efficiency of DCFC to nearly 80%.

Supercapacitors (SCs)

For the past few decades, ***global warming and environmental pollution*** have forced people to move from tradition fossil fuel based energy system to green and sustainable energy resources. As we know, sustainable energy sources such as wind and solar could supply enormous amount energy to the entire world. For these renewable and environment-friendly energy resources, a proficient energy storage system is extremely important to meet the requirements of the smart grid management systems. In

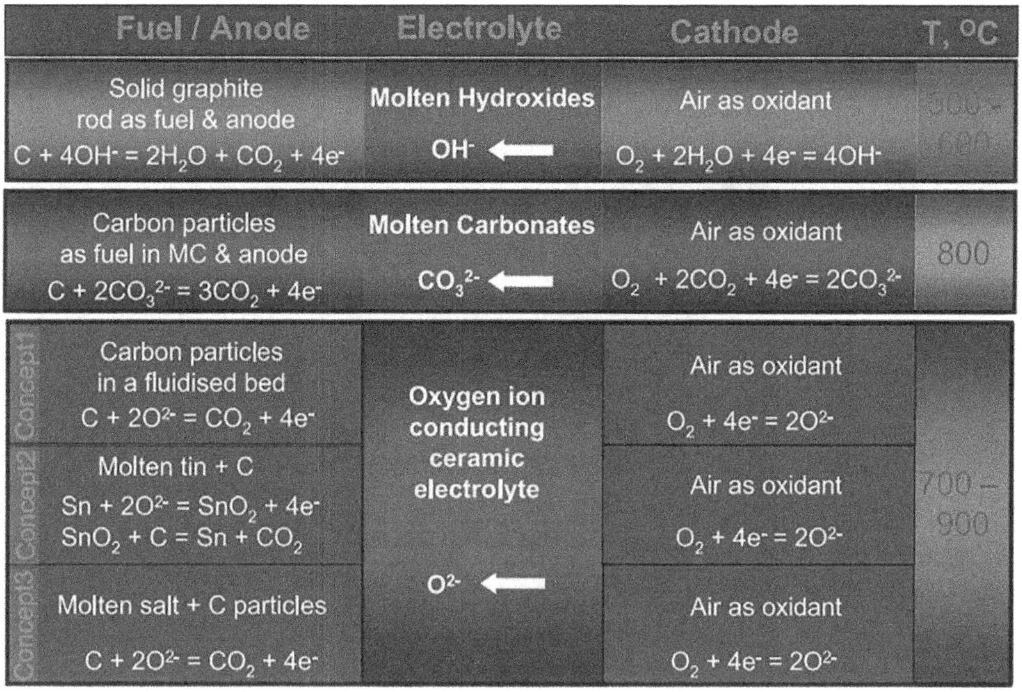

	Fuel / Anode	Electrolyte	Cathode	T, °C
	Solid graphite rod as fuel & anode $C + 4OH^- = 2H_2O + CO_2 + 4e^-$	Molten Hydroxides OH^- ⟵	Air as oxidant $O_2 + 2H_2O + 4e^- = 4OH^-$	500–600
	Carbon particles as fuel in MC & anode $C + 2CO_3^{2-} = 3CO_2 + 4e^-$	Molten Carbonates CO_3^{2-} ⟵	Air as oxidant $O_2 + 2CO_2 + 4e^- = 2CO_3^{2-}$	800
Concept1	Carbon particles in a fluidised bed $C + 2O^{2-} = CO_2 + 4e^-$	Oxygen ion conducting ceramic electrolyte O^{2-} ⟵	Air as oxidant $O_2 + 4e^- = 2O^{2-}$	700–900
Concept2	Molten tin + C $Sn + 2O^{2-} = SnO_2 + 4e^-$ $SnO_2 + C = Sn + CO_2$		Air as oxidant $O_2 + 4e^- = 2O^{2-}$	
Concept3	Molten salt + C particles $C + 2O^{2-} = CO_2 + 4e^-$		Air as oxidant $O_2 + 4e^- = 2O^{2-}$	

Figure 6. Main types of direct carbon fuel cells and fuel cell reactions. Reprinted with permission from ref. Giddey, S., S.P.S. Badwal, A. Kulkarni and C. Munnings. 2012. Prog. Energy Combust. Sci. 38: 360. Copyright @ Elsevier.

general, rechargeable batteries dominate such energy storage requirements. However, they could not sustain all conditions, particularly high power requirement and long cycle life when used in the grid system. In this case, supercapacitor could be considered as suitable candidate for providing huge power density and long cycle life (Miller and Simon 2008). As we know, electrochemical capacitors differ from an ordinary capacitor. In a conventional capacitor, the two conducting plates are separated by a relatively thick dielectric medium which includes ceramic, thin plastic film and/or air. When the potential is applied between the plates (often called charging), electric charges are induced on the plates and generate an electric field between them. Thus, the electric field polarizes the molecules or atoms of the dielectric and makes them aligned in the opposite direction to the field.

In SCs, there is no dielectric used as in conventional capacitor. Instead, both the conducting plates are soaked in an electrolyte and separated by a very thin insulator film. When the plates are charged up, an opposite charge forms on either side of the separator, creating charge layers called an electric double-layer. In comparison, SCs store more energy due to the formation of the thin layer of the charge between two plates; hence, SCs are often referred to as double-layer capacitors (DLC), also called electric double-layer capacitors (EDLCs). Owing to the high power density (kW kg^{-1}) and moderate energy density (Wh kg^{-1}), SCs are considered as alternative energy storage devices to batteries (Li\Na) or fuel cells and conventional capacitors (Simon and Gogotsi 2008, Kötz and Carlen 2000, Pech et al. 2010). The Ragone plot which compared the power and energy density of various electrochemical power sources is shown in Figure 7.

There are two different mechanisms are involved in the charge storage process of the SCs. (1) the electrical double layer (EDL) capacitance from the pure electrostatic charge accumulation at the electrode interface (non-Faradic reactions) and (2) the pseudo-capacitance due to fast and reversible surface redox processes at characteristic potentials (Faradic reaction). Based on the storage mechanism involved in the electrode materials, the current SCs are classified as given below (Figure 8).

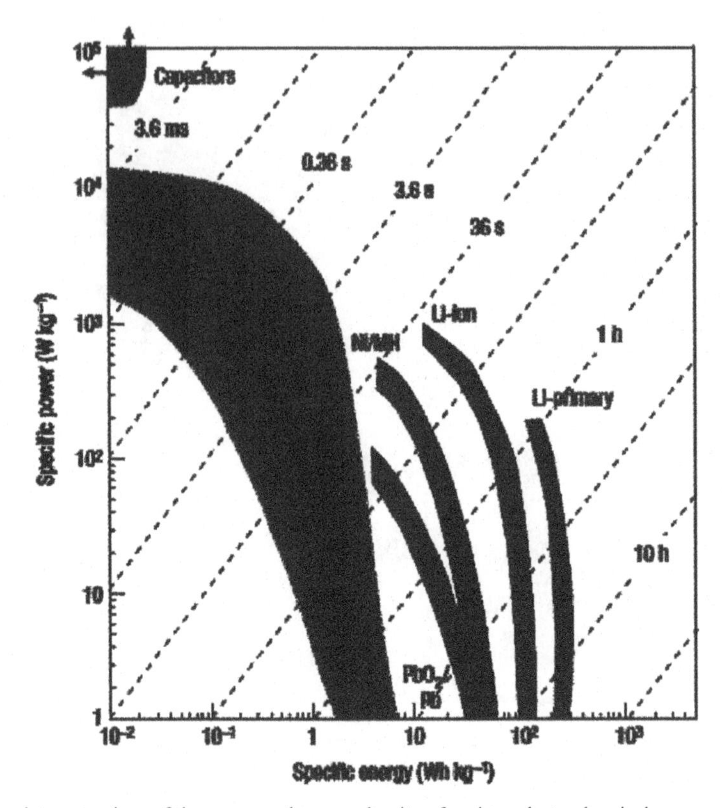

Figure 7. Ragone plot comparison of the power and energy density of various electrochemical power sources (permission with reprint (Simon and Gogotsi 2008)).

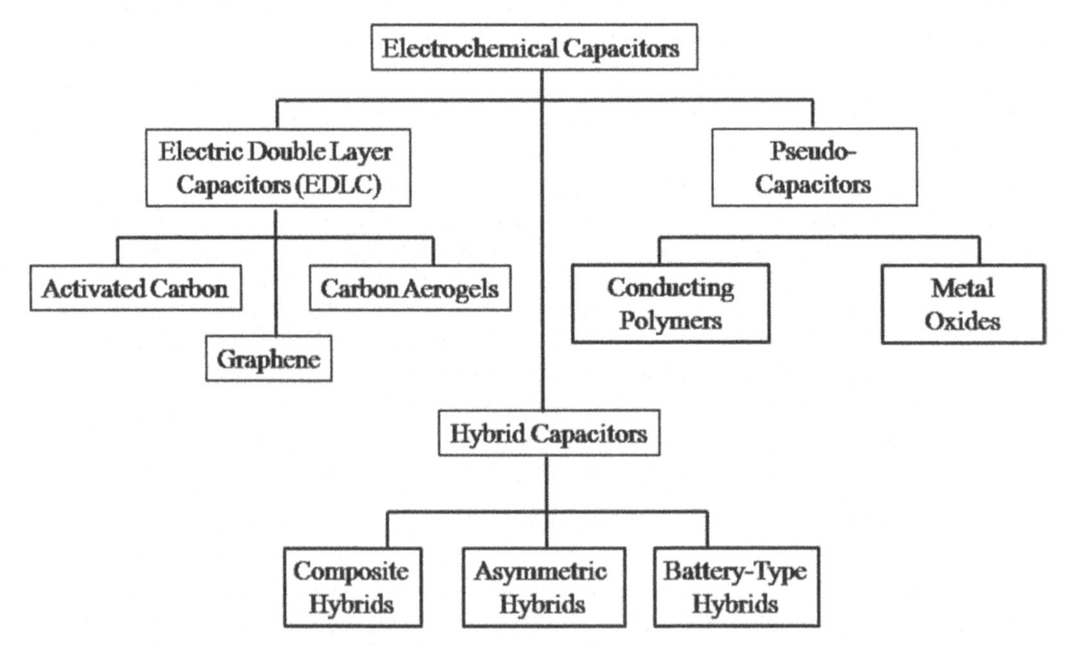

Figure 8. Classification of the electrochemical capacitors, taking into account the charge storage mechanism and active material in each electrode.

Charge storage mechanism

EDLC stored the charges electrostatically (non-faradic) by adsorbing the ions of the electrolyte onto the surface of the active electrode materials. This reaction is highly reversible (adsorption and desorption) and electrochemically stable. The charge separation occurs at the electrode-electrolyte interface as explained by Helmholtz in 1853. The Helmholtz double layer model states that two layers of opposite charge form at the electrode-electrolyte interface and are separated by an atomic distance (Figure 9a).

This model was further modified by Gouy and Chapman 1910; they stated that presence of diffusion layer in the electrolyte is due to the accumulation of ions close to the electrode surface (Figure 9b). Gouy and Chapman stated that the capacitance of two separated arrays of charges increases inversely with the charge separation distance. Later, Stern 1924 combined both the models and clearly recognized that there are two ion distribution regions that occur: first (inner) one is called compact layer or stern layer or Helmholtz layer (Figure 2c) and the second one is diffusion layer (Figure 9c). The first compact layer is formed by adsorbed ions and non-specifically adsorbed counter ions on the electrode and these two types of adsorbed ions are distinguished as inner Helmholtz plane (IHP) and outer Helmholtz plane (OHP). The second layer is called diffusion layer, which is the same as what the Gouy–Chapman model described (Zhang and Zhao 2009, Conway 1999, Zhao et al. 2011).

(i) EDLC capacitor

The capacitance in the EDL (C_{dl}) can be treated as a combination of the capacitances of the two regions, the Stern type of Helmholtz double layer capacitance (C_H) and the diffusion region capacitance (C_{diff}) (Figure 10). Thus, C_{dl} can be expressed by the following equation:

$$1/C_{dl} = 1/C_H + 1/C_{diff} \tag{1}$$

For the EDL type of supercapacitor, the specific capacitance, C, (F g^{-1}) of each electrode is generally assumed to follow that of a parallel-plate capacitor:

The charge storage on the capacitor is defined as follows:

$$C = (A \times \varepsilon_o.\varepsilon_r)/d \tag{2}$$

where ε_o is the permittivity of free space (8.854 10^{-12} F.m^{-1}) and ε_r is the relative dielectric constant of the electrolyte as liquid or solid and depends on the ionic concentration of the electrolyte; A is the surface area of the electrodes (m^2) and d is the effective thickness of the electric double layer (EDL) or Debye length (DL) (nm).

(ii) Pseudocapacitors

Charge storage process in pseudocapacitors is mainly due to the reversible redox processes called Faradaic reaction, in which the valence electrons of electroactive materials are crossed over the electrode/ electrolyte interface, simply due to the charge acceptance (Q) and a change in potential (V).

$$C = Q/V \tag{3}$$

Unlike redox processes in a battery, Faradaic processes in a pseudo-capacitor are subject to a thermodynamic change of potential during charge accumulation and have better reversibility. Consequently, redox supercapacitors can be identified of as having both battery-like behaviors with Faradaic reactions occurring across the double layer and electrostatic capacitor-like behavior with high reversibility and high power. While pseudo-capacitance is higher than EDL capacitance, it suffers from the drawbacks of low power density (due to poor electrical conductivity), and poor stability.

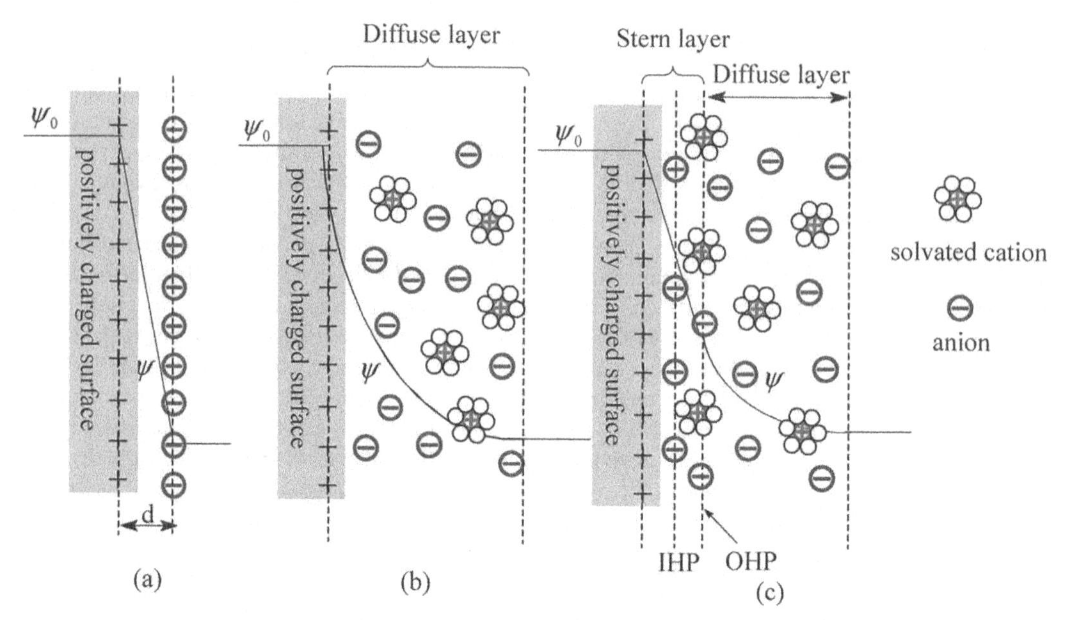

Figure 9. Models of the electrical double layer at a positively charged surface: (a) the Helmholtz model, (b) the Gouy–Chapman model, and (c) the Stern model, showing the inner Helmholtz plane (IHP) and outer Helmholtz plane (OHP). The IHP refers to the distance of closest approach of specifically adsorbed ions (generally anions) and OHP refers to that of the non-specifically adsorbed ions. The OHP is also the plane where the diffuse layer begins. d is the double layer distance described by the Helmholtz model. J0 and j are the potentials at the electrode surface and the electrode/electrolyte interface, respectively (Reprinted with permission (Zhang and Zhao 2009)).

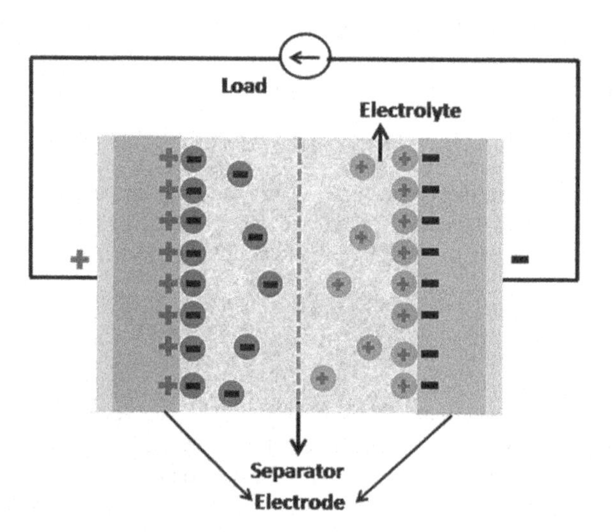

Figure 10. Schematic representation of the EDLC capacitor.

Electrode materials for supercapacitors

The charge storage mechanism in SC mainly depends on the types of the electrode materials used. There are different kinds of electrode materials that have been extensively used in SCs; each of them has their own electrochemical behaviors. Electrode materials used in the SCs might have the following properties for obtaining better performances (Zhi et al. 2013):

- High specific surface area: To improve the electrolyte accessibility or to accommodate more number of electrolyte ions
- Uniform pore size distribution: Significantly influences the specific capacitance and the rate capability of the electrode
- High electronic conductivity: Affects the rate capability and the power density of the electrode, and reduces the equivalent serial resistance (ESR)
- Good electrochemically active sites: For the contribution of pseudocapacitance
- Good chemical and cyclic stability: For practical applications
- Low cost: To aid manufacturing

Eco-friendly carbonaceous materials such as AC, CNT and graphene have been used by different manufactures due to their excellent features such as good electrical conductivity, high specific surface area, chemical stability, high thermal stability, controlled pore size distribution, excellent long term stability upon charge and discharge process, and low cost, etc. (Kötz and Carlen 2000, Conway, Bose et al. 2012, Levi et al. 2009).

Activated carbon

Carbonaceous materials have always been regarded as a significant candidate for EDLC applications due to their high specific surface area, excellent chemical and electrochemical stability, porous nature, and electrical conductivity (Zhang and Zhao 2009, Zhai et al. 2011, Ghosh and Lee 2012). AC can be derived from non-renewable resources such as petroleum coke, tar pitches, and coal; however, the cost of the extraction and the demands of such sources restrict the mass production of the activated carbon.

Choice of EDLC materials has been expanded from activated carbon (Balathanigaimani et al. 2008), biomass-derived carbon, mesoporous carbon, carbon nanotubes/fibers, carbide-derived carbon (Korenblit et al. 2010, Jost et al. 2013), to the most recent graphene (or reduced graphene oxide (rGO)) (Liu et al. 2010). So far, plenty of bio materials have been investigated as starting material for the preparation of porous activated carbons. Biomass wastes offer the combined advantages of waste utilization, reduction in the use of non-renewable resources and contribution to make a green environment. To date, various precursors have been investigated for the preparation of porous AC. Among these, biomass has drawn huge attention due to its low-cost, abundance and environment friendly renewable resource. Utilization of biomass, particularly its waste, offers the combined advantages of recycling, reduction in the use of non-renewable resources and contribution towards creating a green environment. Studies have utilized various biomass as precursor materials for porous AC synthesis such as coconut shell (Jayaraman et al. 2017, Ulaganathan et al. 2015, Jain et al. 2013, Aravindan et al. 2014), bamboo waste (Kim et al. 2006), sunflower seed shell (Li et al. 2011), Prosopis juliflora wood (Sennu et al. 2016a, b), Jackfruit peel (Sennu et al. 2016c), sorghum-pith (Senthilkumar et al. 2011), durian shell (Chandra et al. 2009), seaweeds (Raymundo-Piñero et al. 2009), coffee ground (Rufford et al. 2009, 2008, Jisha et al. 2009), corn grains (Balathanigaimani et al. 2008), banana fibers (Subramanian et al. 2007), sugar cane bagasse (Rufford et al. 2010), dead neem leaves (Chhatre et al. 2015), rice husk (Guo et al. 2003), orange peel (Maharjan et al. 2017), teak wood (Jain et al. 2017), etc. (Wei and Yushin 2012, Yan et al. 2014).

As mentioned earlier, performance of carbon-based supercapacitor is attributed to high specific surface area coupled with well-developed pore structure (Chmiola et al. 2006, Raymundo-Piñero et al. 2006, Sevilla et al. 2007, Sevilla and Mokaya 2014) and electrode surface chemistry (Raymundo-Piñero et al. 2006). In high power applications, smaller pores of materials create hindrance to efficient ionic motion and results in the reduced rate of energy delivery (Sevilla et al. 2007, Kim et al. 2004, Fuertes et al. 2005, Salitra et al. 2000, Eliad et al. 2005), whereas too large pores reduce the discharging time with lower capacitance (Raymundo-Piñero et al. 2006, Chmiola et al. 2006, Barbieri et al. 2005, Jain et al. 2015). In contrast, studies have demonstrated that ultra-fast ion transport is also possible with microporous structure, provided the micropores have straight shapes without bottlenecks (Chmiola et al. 2006, Kajdos et al. 2010, Wei et al. 2011).

A better capacitive performance was obtained for highly microporous AC with regular pores despite its lower specific surface than that of the mesoporous activated carbon with irregular and different void sizes. However, researchers have reported that porous carbon with the co-existence of optimal mesopores with micropores in carbons permit rapid transport of the solvated ions which serve better storage behavior in supercapacitor applications (Sevilla and Mokaya 2014, Lv et al. 2012, Pech et al. 2010, Miller et al. 2010, Sevilla et al. 2007, Portet et al. 2007, Frackowiak and Béguin 2001). Porous structure and typical cyclic voltammetry curve of the biomass derived activated carbon are shown in the below Figure 11a and 11b.

High surface area and mesoporous structure of the activated carbons are favored to the excellent performance of the EDL supercapacitors. This reflects in the CV curve, which shows typical rectangular shape in 1 M H_2SO_4 solution.

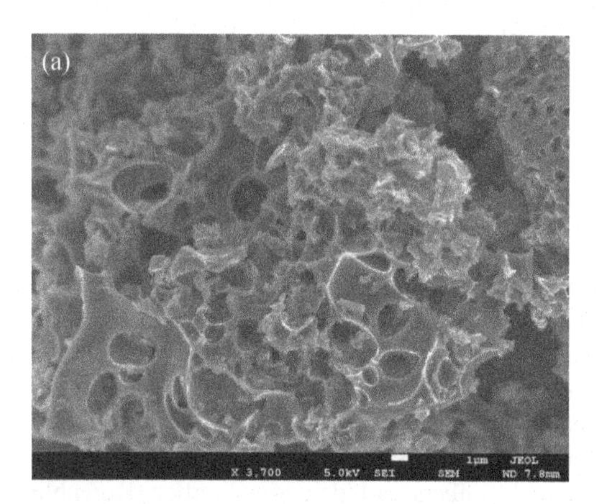

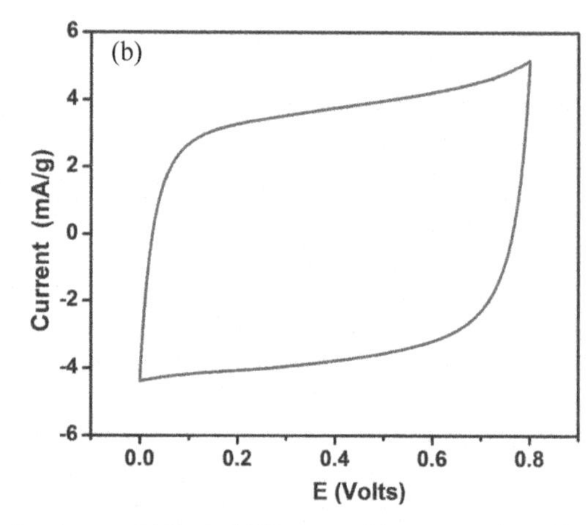

Figure 11. Biomass derived activated carbon: (a) FESEM surface image; (b) Typical CV curve of the symmetric cell in 1 M H_2SO_4.

Graphene

Graphene-based materials (graphene sheets, flakes, and their hybrids) have been demonstrated as one of the most promising and attractive electrode materials for SCs because of their excellent electrical conductivity in plane (10^6 S/cm), thermal properties, high theoretical specific surface area (SSA) (~2620 m^2/g), exceptional intrinsic double-layer capacitance (~21 $\mu F/cm^2$) and high theoretical capacitance (~ 550 F/g), as well as high mechanical strength and chemical stability comparable with or even better than CNT. Owing to its unique properties, graphene have attracted much attention for a wide range of potential applications, specifically in supercapacitors.

Graphene-based ultracapacitors have been developed and tested using chemically modified graphene (CMG) as an electrode. Performances of the CMG-based ultracapacitor were tested indifferent in electrolyte medium which includes 5.5 M KOH (aqueous electrolyte) and TEABF4 in acetonitrile (AN) solvent and TEA-BF4 in propylene carbonate (PC) solvent (organic electrolytes). It was noted that cell showed specific capacitances of 135 in aqueous electrolytes and was higher than the organic electrolyte system. In addition, the cell showed excellent cell performances even at high scan rate which is due to the high electrical conductivity of the CMG (Stoller et al. 2008).

Graphene was also synthesized and activated using KOH using microwave exfoliated oxides and thermally exfoliated graphite oxides, where the specific surface area of the activated graphene was about 3100 m^2/g. The electrode testing showed specific capacitance of 200 F/g in organic electrolyte solutions (Zhu et al. 2011).

The one-dimensional (1D) CNTs and two-dimensional (2D) graphene sheets are considered as ideal 'building blocks' for the bottom-up strategy by combining them to create three-dimensional (3D) nano networks. Recent theoretical studies predicted that the 3D pillared CNT-graphene network nanostructure possesses desirable out-of-plane properties while maintaining in-plane properties, attractive for numerous innovative applications, including supercapacitors.

Composite materials of graphene and CNT have been described in recent years to establish synergistic effects between these two different carbon nanostructures both having unique electronic, thermal and mechanical properties. Another advantage of this hybrid structure is that CNT can work as the space provider between graphene sheets preventing the re-stack of graphene, and thus enhance the performance of electrodes based on carbon nanostructures (Tung et al. 2009).

Pham et al. (2015) synthesized the carbon nanotube (CNT)-bridged graphene 3D building blocks via the Coulombic interaction between positively charged CNTs grafted by cationic surfactants and negatively charged graphene oxide sheets, followed by KOH activation. In that specially structured composite, the CNTs were intercalated into the nanoporous graphene layers to build pillared 3D structures. It will be helped for faster ion diffusion during the charge and discharge process. The supercapacitors fabricated using the obtained composite film exhibit an outstanding electrochemical performance in an ionic liquid electrolyte with a maximum energy density of 117.2 Wh/L or 110.6 Wh/kg at a maximum power density of 424 kW/L or 400 kW/kg, which is based on thickness or mass of total active material.

Gr-SWCNT composite film was prepared and tested in supercapacitor applications (Cheng et al. 2011). Specific capacitances of 290.6 F/g and 201 F/g were obtained for a single electrode in aqueous and organic electrolytes, respectively. The graphene/CNT electrodes exhibited high energy density of 155.6 Wh/kg in ionic liquid at room temperature. The cell also showed a good cycle life of over 10000 cycles in an ionic liquid electrolyte medium. SWCNTs also acted as a conductive additive, spacer, and binder in the graphene/CNT supercapacitors.

Graphene–CNT composite with lamellar structure by amidation reaction was prepared and tested in supercapacitor applications (Jung et al. 2013). It was noted that the complementary bonding between the graphene oxide (GO) and CNTs prevented restacking between graphene layers and the agglomeration of CNTs attached to the edges and surface of the graphene. It improved the increase of the electrolyte-accessible surface area, but also provided 3D electrical conduction path to the ion diffusion. In this work, the authors reported the highest volumetric capacitance of 165 $F.cm^{-3}$ for the binder free composite electrode.

In addition, there are plenty of research activities with improved supercapacitor performances which have also been evaluated by different research groups in different electrolyte media (Wu et al. 2013, Yoo et al. 2011, DeYoung et al. 2014, Huang et al. 2015). 3D nanohybrid architectures consisting of MnO_2 nanoneedles, carbon nanotubes (CNTs) and graphene sheets are fabricated by varying the percentage of the CNT content in the composite. The prepared electrode was tested in 1 M Na_2SO_4 aqueous electrolyte medium at room temperature. The prepared system at 10 wt.% CNT showed higher capacitance (228.92 F/g) than other studied composites (Kim et al. 2014).

1-dimensional (1D) CNTs and 2-dimensional (2D) graphene sheets are an effective way to build hybrid carbon architectures with fascinating new properties. Cui et al. (2015) synthesized binder-free reduced graphene oxide/carbon nanotube (rGO/CNT) hybrid film and the supercapacitor performances were evaluated in both positive and negative potential window using 1 M Na_2SO_4 aqueous electrolyte medium. A small amount of CNTs into rGO sheets showed an excellent specific capacitance of 272 F g^{-1} at a scan rate of 5 mV s^{-1} in negative potential window of −0.8 to 0 V. However, it showed poor capacitance in the positive window at the same scan rate. The higher value of capacitance in the negative potential window is mainly due to the strong cation adsorption at the oxidized surface of rGO sheets (Cui et al. 2015).

$NiCo_2S_4$ nanoparticles coated on the graphene fiber (GF/$NiCo_2S_4$) composites were prepared and their performance was evaluated in three and two electrode system level. Electrochemical characterizations showed 100% higher capacitance of the GF/$NiCo_2S_4$ as compared to that of pure graphene fibers. The fabricated device achieves high energy density, up to 12.3 mWh/cm with a maximum power density of 1600 mW/cm (Cai et al. 2016).

Graphene and SWCNT composites are explored as the electrodes for supercapacitors by coating polyaniline (PANI) nano-cones onto the graphene/CNT composite to obtain graphene/CNT–PANI composite electrode (Cheng et al. 2013). The composite electrode assembled with graphene/CNT electrode into an asymmetric pseudocapacitor showed a maximum energy density of 188 Wh/kg and maximum power density of 200 kW/kg in 1 M KCl aqueous electrolyte solution. Over all, graphene and graphene based composites showed better performances than the other studied electrodes due to its good electrical surface rated properties.

Carbon nanotubes (CNTs) and their composite materials

Since the discovery by Iijima in 1991 (Iijima 1991), CNTs have attracted much attention in electronic and electrochemical applications such as Li-ion secondary batteries, photovoltaic cells and hydrogen storage in fuel cells. The unique properties of CNTs, such as structure (one-dimensional geometry), high surface area, remarkable chemical stability, high electric conductivity, as well as being light weight, have great use in energy generation and storage applications. CNTs could be viewed as a graphene sheet rolled up into a nanoscale tube form to produce the SWCNTs. There may be additional graphene tubes around the core of an SWNT to form MWCNTs. As compared to activated carbon, CNTs provide novel properties of high electrical conductivity, high specific surface area, high charge transport capability, high mesoporosity, and high electrolyte accessibility; hence, CNTs are used as potential electrode materials in high-performance supercapacitors.

CNT was suggested by Niu et al. (1997) for using in supercapacitor applications. MWCNT was functionalized and used in supercapacitor applications. The functionalized CNT showed a specific area of 430 m^2/g and delivered a gravimetric capacitance of about 102 F/g in a sulfuric acid medium. The storage capacity behavior is attributed to both EDLC and pseudocapacitance. The cell also delivered a power density of > 8 kW/kg.

It is well accepted that the unique mesoporosity of the CNT is favored for high storage capacitance; it further improves electrolyte accessibility which can enhance the storage capacitance. An et al. (2001), tested supercapacitor performance of the SWCNT and achieved 180 F/g. It also showed promising power density of 20 kW/kg with maximum energy density of ~ 10 Wh/kg in KOH electrolyte medium.

In recent years, aligned CNTs have been studied for supercapacitor applications. It is highly desirable to have aligned/patterned structures of CNTs so that their structure/property could be easily accessed; hence, they can be effectively incorporated into devices (Dai et al. 2003). Recent research demonstrated that vertically aligned CNTs are advantageous over their randomly entangled counterparts for supercapacitor applications. Vertically aligned MWCNTs were grown directly on commercially available metallic alloy substrate, and the supercapacitor performance of the CNT was investigated with respect to the length of the CNT. It was reported that as the length of CNT increased, the specific capacitance also increased, from 10.75 to 21.51 F/g with an energy density from 2.3 to 5.4 Whkg and power density of 315 kWkg (Rakesh shah et al. 2009). Kim et al. (2010) have also synthesized vertically aligned CNTs (dia. 6 nm) on carbon paper using Al/Fe as a catalyst. It was reported that the SC showed 200 F/g^- at 20 Ag with 20 Wh/kg in 1 M H_2SO_4; the same electrode in organic electrolyte delivered a maximum energy density of 100 Wh/kg.

In general, CNTs deliver a relatively low energy density. In order to improve the energy density of EDLC, many attempts were made using composites or additives in electrode and electrolytes. CNT-ZnO composite was prepared and reported by Zhang et al. (2009a); the specific capacitance of 323.9 F/g was reported when evaluated in poly (vinyl alcohol)-phosphomolyhydric based gel electrolyte. MoO_2-SWCNT composite electrode was prepared via electrochemically induced deposition method which delivered a high specific capacitance of 597 F/g at 10 mV/s in 0.1 M H_2SO_4 (Gao et al. 2010). RuO_2-CNT composite electrode was synthesized and the supercapacitor performances were evaluated. The composite electrode showed capacitance value of 203 F/g over 20,000 charge/discharge cycles (Liu et al. 2009). Very recently, 3D CNT/graphene composite was synthesized by Chemical vapour deposition CVD method (Fan et al. 2010); this composite showed enhanced specific capacitance to 385 F/g in 6 M KOH at a scan rate of 10 mV/s. Recently, aligned CNT and the composite with polydiacetylene was studied and reported (Ulaganathan et al. 2016). The as prepared composite electrode showed high capacitance value (1111 F/g) than the bare aligned CNT.

Carbon/conducting polymer composites

In recent years, conducting polymers are employed in supercapacitors owing to their relatively higher electrical conductivity, larger pseudo-capacitance, faster doping/de-doping rate during the charge–discharge process and low equivalent series resistance (ESR). Polypyrrole (Wang et al. 2016), polyaniline (PANI) (Wei et al. 2016), and poly-(3,4-ethylenedioxythiophene) (PEDOT) (Perez-Madrigal et al. 2016) are the most commonly used conducting polymers. However, low mechanical strength, poor conductivity, and low porosity are the major drawbacks, which restricts them to be used as SCs electrode as an independent content. Hence, in many cases, the polymer is complexed with metal oxides and CNT for improving their performances in SCs.

Wang et al. (2016) synthesized the functionalized 3D porous nitrogen-doped carbon nanosheet frameworks (FT-PNCNFs). Their charge storage behavior was tested in aqueous electrolyte medium. The as prepared FT-PNCNFs electrode was tested in symmetric configuration using KOH electrolyte; the symmetric configuration showed superior energy storage behavior at an extremely high scan rate of 3000 mV/s; the cell delivered capacitances of ~ 247 F/g at 10 mV/s and 146 F/g at 3000 mV/s; the cell also showed good cycling stability even after 8000 cycles with the capacitance retention of 95%.

Cai et al. (2013) prepared CNT-PANI composites by electrochemical polymerization process and studied the SC behavior. MWCNT/PANI composite was also prepared using an *in situ* electrochemical polymerizations onto stainless-steel from an aniline solution containing small content of MWCNT (Zhang et al. 2009b). The prepared composite electrodes showed a specific capacitance of 500 F/g which has 0.8 wt% MWCNT on the total weight of the composites.

Li/Na ion batteries

Carbon based materials play an important role in the development of Li/Na-ion batteries. Before the discovery of graphite anode materials, Li metal was used as main candidate in Li-ion batteries; however, the safety related issues due to the dendrite formation of the lithium metal is a major concern. Graphite is proposed as an ion intercalating electrode materials and till date is being used as the anode in commercial products. Graphite is attractive due to its high in-plane electron conductivity due to the p-bond and weak interaction with Li ions, giving rise to high Li ion storage capacity and fast Li-ion diffusion.

The first lithium ion secondary battery (LIB) technology was pioneered by Sony Corporation (1991) and was commercialized as lithium coin-type cells. The characteristics of the lithium ion batteries are:

1) high energy density (both gravimetric and volumetric),
2) high operating voltage,
3) low self-discharge rate,
4) no memory effect,
5) high drain capability,
6) quick response upon charging,
7) wide operation temperature.

The charge and discharge mechanism is as follow:

$$\text{LiaC} = \text{Lia-dxC} + \text{dxLi}^{+} + dxe$$

As mentioned in equation above, graphite has an orderly layered structure and lithium can intercalate only into the spacings between the layers to form LiC_6, and the lithium doping capacity of graphite is subject to the stoichiometry of LiC_6 (372 mAh/g). It is well known that the d_{002} spacing of graphite is 0.335 nm and when lithium intercalates between the layers of graphite, d_{002} spacings expand to 0.372 nm, then shrink back to 0.336 nm again by lithium deintercalation (Nishi 2000). In the case of the graphite anode, significant deformation of the electrode is brought about by the expansion/shrinkage of d_{002} spacings during charge/discharge cycles, which results in poor cyclability of the graphite anode cell (Nishi 2000).

At the same time, graphite is not favorable to Na ions due to its smaller inter spacings when compared to the Na-ion size. Though it accommodates the Na-ions, it destroyed the structure due to larger anion size of the Na. As a result, it exhibited poor performance in Na-ion batteries. However, hard carbon has been employed as anode successfully in Na-ion batteries and is commercially available.

Carbon nanotubes (CNTs) have been extensively investigated as anode materials for Li-ion batteries due to their mesoporosity, high chemical stability, low resistance, strong mechanical strength, and high activated surfaces (that enhance the electrolyte accessibility); the different forms of SWCNT and MWCNT showed a higher li-ion storage capacity than the graphite. However, the storage capacity mainly depends on the specific alignment and diameter of the CNTs. In order to improve the performance of the CNT in batteries, it can be used as composite with other metal oxides which can give better performances in terms of rate capability and cycle life.

Summary and future perspectives

Nano structured carbonaceous materials have been extensively used in energy storage and energy conversion application in many forms. The carbonaceous nanostructures' materials determined the specific properties of the devices; carbon based materials have been used in both the electrodes in the form of composite, either as binary or as ternary with polymer/metal oxide materials. The selection of carbon materials for supercapacitor electrodes is mainly based on the important properties such as high surface area, controlled pore size matching the electrolyte ions, high electrical conductivity, good electrolyte accessibility, interconnected pore structure and the presence of electrochemically stable surface functionalities, etc. In order to develop high energy supercapacitors, the electrolyte is also an very important factor. To obtain the super capacitors > 4 V, ionic liquids are the best choice; however, the cost of that is the major worrying factor. At the same time, it is important to achieve the energy density without compromising the rate capability and the cycle life of the supercapacitors.

One of the major issues in developing supercapacitor devices is the cost factor. Hence, it is important that the future research activities should be concentrated in the direction of fabricating carbonaceous materials based supercapacitors with high capacity and long cycle life in a cost effective manner. In that case, it can be considered a one-step synthesis process for obtaining the high density carbonaceous material synthesis which will be beneficial to design high power and energy sources.

In terms of battery, there are various carbonaceous materials which have been investigated so far, but none of them have been commercialized due to their poor rate performance and poor cycle life. However, Graphite based cells have been used as anode in the commercial batteries though it showed a poor cycle life. Recently, graphene has been investigated as a superior anode material in batteries and has showed an excellent performance. Hard carbon has been identified as good anode material for Na-ion battery anodes; however, the performance of the hard carbon still needs to improve.

Certainly, it is also expected that new generation of supercapacitors will replace batteries in some applications where high power, high level of reliability and high efficiency is required.

Carbonaceous materials have also been employed as an electrode material in the form of Bipolar plates and as supporting catalyst in GDL and MPL. However, the heavy weight and poor mechanical stability of the graphite bipolar plate is a major drawback; these can be improved by preparing the composite bipolar plate without compromising the exciting performances. In the preparation of the composites, different structures of carbonaceous materials with good mechanical stability and electrical conductivity will be used for further improvement in the performance of the fuel cells.

Thus, it is no doubt that the low cost, eco-friendly carbonaceous materials will play a major role in the future energy market devices and to store the renewable energy sources because of their outstanding electrochemical properties. Research on the carbonaceous materials is expected to continue to expand rapidly in the future.

References

Adloo, A., M. Sadeghi, M. Masoomi and H.N. Pazhooh. 2016. Renewable Energy 99: 867.
An, K.H., W.S. Kim, Y.S. Park, J.M. Moon, D.J. Bae, S.C. Lim, Y.S. Lee and Y.H. Lee. 2001. Advanced Functional Materials 11: 387.
Antunes, R.A., M.C.L. Oliveira, G. Ett and V. Ett. 2010. International Journal of Hydrogen Energy 35: 3632.
Antunes, R.A., M.C.L. de Oliveira, G. Ett and V. Ett. 2011. Journal of Power Sources 196: 2945.
Aravindan, V., J. Sundaramurthy, A. Jain, P.S. Kumar, W.C. Ling, S. Ramakrishna et al. 2014. ChemSusChem. 7: 1858.
Balathanigaimani, M.S., W.-G. Shim, M.-J. Lee, C. Kim, J.-W. Lee and H. Moon. 2008. Electrochemistry Communications 10: 868.
Barbieri, O., M. Hahn, A. Herzog and R. Kötz. 2005. Carbon 43: 1303.
Bose, S., T. Kuila, A.K. Mishra, R. Rajasekar, N.H. Kim and J.H. Lee. 2012. Journal of Materials Chemistry 22: 767.
Cai, W., T. Lai, J. Lai, H. Xie, L. Ouyang, J. Ye et al. 2016. Scientific Reports 6: 26890.
Cai, Z., L. Li, J. Ren, L. Qiu, H. Lin and H. Peng. 2013. Journal of Materials Chemistry A 1: 258.
Candelaria, S.L., Y. Shao, W. Zhou, X. Li, J. Xiao, J.-G. Zhang et al. 2012. Nano Energy 1: 195.
Chandra, T.C., M.M. Mirna, J. Sunarso, Y. Sudaryanto and S. Ismadji. 2009. Journal of the Taiwan Institute of Chemical Engineers 40: 457.
Chapman, D.L. Philos. Mag. 6: 475.
Cheng, Q., J. Tang, J. Ma, H. Zhang, N. Shinya and L.-C. Qin. 2011. Physical Chemistry Chemical Physics 13: 17615.
Cheng, Q., J. Tang, N. Shinya and L.-C. Qin. 2013. Journal of Power Sources 241: 423.
Chhatre, S., V. Aravindan, D. Puthusseri, A. Banerjee, S. Madhavi, P.P. Wadgaonkar et al. 2015. Materials Today Communications 4: 166.
Chmiola, J., G. Yushin, Y. Gogotsi, C. Portet, P. Simon and P.L. Taberna. 2006. Science 313: 1760.
Chung, C.-Y., S.-K. Chen, P.-J. Chiu, M.-H. Chang, T.-T. Hung and T.-H. Ko. 2008. Journal of Power Sources 176: 276.
Cindrella, L., A.M. Kannan, J.F. Lin, K. Saminathan, Y. Ho, C.W. Lin et al. 2009. Journal of Power Sources 194: 146.
Conway, B.E.
Cooper, J.F. 2007. *In*: S. Basu (ed.). Recent Trends in Fuel Cell Science and Technology. Springer New York: New York, NY, p. 248.
Cui, X., R. Lv, R.U.R. Sagar, C. Liu and Z. Zhang. 2015. Electrochimica Acta 169: 342.
Dai, L., A. Patil, X. Gong, Z. Guo, L. Liu, Y. Liu et al. 2003. ChemPhysChem. 4: 1150.
de Oliveira, M.C.L., G. Ett and R.A. Antunes. 2013. Journal of Power Sources 221: 345.
DeYoung, A.D., S.-W. Park, N.R. Dhumal, Y. Shim, Y. Jung and H.J. Kim. 2014. The Journal of Physical Chemistry C 118: 18472.
Dhakate, S.R., S. Sharma, M. Borah, R.B. Mathur and T.L. Dhami. 2008. Energy & Fuels 22: 3329.
Dhakate, S.R., R.B. Mathur, S. Sharma, M. Borah and T.L. Dhami. 2009. Energy & Fuels 23: 934.
Dhakate, S.R., S. Sharma, N. Chauhan, R.K. Seth and R.B. Mathur. 2010. International Journal of Hydrogen Energy 35: 4195.
Dicks, A.L. 2006. Journal of Power Sources 156: 128.
Du, C., P. Ming, M. Hou, J. Fu, Q. Shen, D. Liang et al. 2010a. Journal of Power Sources 195: 794.
Du, C., P. Ming, M. Hou, J. Fu, Y. Fu, X. Luo et al. 2010b. Journal of Power Sources 195: 5312.
Du, L. and S.C. Jana. 2007. Journal of Power Sources 172: 734.
Eliad, L., E. Pollak, N. Levy, G. Salitra, A. Soffer and D. Aurbach. 2005. Applied Physics A 82: 607.
Fan, Z., J. Yan, L. Zhi, Q. Zhang, T. Wei, J. Feng et al. 2010. Advanced Materials 22: 3723.
Frackowiak, E. and F. Béguin. 2001. Carbon 39: 937.
Fuertes, A.B., G. Lota, T.A. Centeno and E. Frackowiak. 2005. Electrochimica Acta 50: 2799.
Gao, F., L. Zhang and S. Huang. 2010. Materials Letters 64: 537.
Gautam, R.K. and K.K. Kar. 2016. Fuel Cells 16: 179.
Ghosh, A. and Y.H. Lee. 2012. ChemSusChem 5: 480.
Ghosh, A., P. Goswami, P. Mahanta and A. Verma. 2014. Journal of Solid State Electrochemistry 18: 3427.
Giddey, S., S.P.S. Badwal, A. Kulkarni and C. Munnings. 2012. Progress in Energy and Combustion Science 38: 360.
Gouy, G.J. Phys. Chem. A 4: 457.
Guo, Y., J. Qi, Y. Jiang, S. Yang, Z. Wang and H. Xu. 2003. Materials Chemistry and Physics 80: 704.
Gupta, C., P.H. Maheshwari and S.R. Dhakate. 2016. Materials for Renewable and Sustainable Energy 5: 1.
Hackett, G.A., J.W. Zondlo and R. Svensson. 2007. Journal of Power Sources 168: 111.
Hao, W., H. Ma, Z. Lu, G. Sun and Z. Li. 2016. RSC Advances 6: 56711.
Hasché, F., T.-P. Fellinger, M. Oezaslan, J.P. Paraknowitsch, M. Antonietti and P. Strasser. 2012. ChemCatChem 4: 479.
Hermann, A., T. Chaudhuri and P. Spagnol. 2005. International Journal of Hydrogen Energy 30: 1297.
Hou, J., Y. Shao, M.W. Ellis, R.B. Moore and B. Yi. 2011. Physical Chemistry Chemical Physics 13: 15384.
https://energy.gov/eere/fuelcells/doe-technical-targets-polymer-electrolyte-membrane-fuel-cell-components (Assed as on 02/08/217)
Huang, J., J. Wang, C. Wang, H. Zhang, C. Lu and J. Wang. 2015. Chemistry of Materials 27: 2107.
Iijima, S. 1991. Nature 354: 56.
Jain, A., V. Aravindan, S. Jayaraman, P.S. Kumar, R. Balasubramanian, S. Ramakrishna et al. 2013. Scientific Reports 3: Art 3002.

Jain, A., C. Xu, S. Jayaraman, R. Balasubramanian, J.Y. Lee and M.P. Srinivasan. 2015. Microporous and Mesoporous Materials 218: 55.
Jain, A., S. Jayaraman, M. Ulaganathan, R. Balasubramanian, V. Aravindan, M.P. Srinivasan et al. 2017. Electrochimica Acta 228: 131.
Jayakumar, A., S.P. Sethu, M. Ramos, J. Robertson and A. Al-Jumaily. 2015. Ionics 21: 1.
Jayaraman, S., A. Jain, M. Ulaganathan, E. Edison, M.P. Srinivasan, R. Balasubramanian et al. 2017. Chemical Engineering Journal 316: 506.
Jisha, M.R., Y.J. Hwang, J.S. Shin, K.S. Nahm, T. Prem Kumar, K. Karthikeyan et al. 2009. Materials Chemistry and Physics 115: 33.
Jost, K., D. Stenger, C.R. Perez, J.K. McDonough, K. Lian, Y. Gogotsi et al. 2013. Energy & Environmental Science 6: 2698.
Jung, N., S. Kwon, D. Lee, D.-M. Yoon, Y.M. Park, A. Benayad et al. 2013. Advanced Materials 25: 6854.
Kajdos, A., A. Kvit, F. Jones, J. Jagiello and G. Yushin. 2010. Journal of the American Chemical Society 132: 3252.
Kakati, B.K., D. Sathiyamoorthy and A. Verma. 2010. International Journal of Hydrogen Energy 35: 4185.
Kim, B., H. Chung and W. Kim. 2010. The Journal of Physical Chemistry C 114: 15223.
Kim, M., Y. Hwang and J. Kim. 2014. Physical Chemistry Chemical Physics 16: 351.
Kim, Y.-J., Y. Horie, Y. Matsuzawa, S. Ozaki, M. Endo and M.S. Dresselhaus. 2004. Carbon 42: 2423.
Kim, Y.-J., B.-J. Lee, H. Suezaki, T. Chino, Y. Abe, T. Yanagiura et al. 2006. Carbon 44: 1592.
Kim, Y.-S., D.-H. Peck, S.-K. Kim, D.-H. Jung, S. Lim and S.-H. Kim. 2013. International Journal of Hydrogen Energy 38: 7159.
Kitahara, T., H. Nakajima and K. Okamura. 2015. Journal of Power Sources 283: 115.
Korenblit, Y., M. Rose, E. Kockrick, L. Borchardt, A. Kvit, S. Kaskel et al. 2010. ACS Nano 4: 1337.
Kötz, R. and M. Carlen. 2000. Electrochimica Acta 45: 2483.
Levi, M.D., G. Salitra, N. Levy, D. Aurbach and J. Maier. 2009. Nat. Mater. 8: 872.
Li, X., W. Xing, S. Zhuo, J. Zhou, F. Li, S.-Z. Qiao et al. 2011. Bioresource Technology 102: 1118.
Liu, C., Z. Yu, D. Neff, A. Zhamu and B.Z. Jang. 2010. Nano Letters 10: 4863.
Liu, X., T.A. Huber, M.C. Kopac and P.G. Pickup. 2009. Electrochimica Acta 54: 7141.
Lv, Y., F. Zhang, Y. Dou, Y. Zhai, J. Wang, H. Liu et al. 2012. Journal of Materials Chemistry 22: 93.
Maharjan, M., M. Ulaganathan, V. Aravindan, S. Sreejith, Q. Yan, S. Madhavi et al. 2017. Chemistry Select 2: 5051.
Mamunya, Y.P., V.V. Davydenko, P. Pissis and E.V. Lebedev. 2002. European Polymer Journal 38: 1887.
Mathias, M.F., J. Roth, J. Fleming and W. Lehnert. 2010. In Handbook of Fuel Cells; John Wiley & Sons, Ltd.
Miller, J.R. and P. Simon. 2008. Science 321: 651.
Miller, J.R., R.A. Outlaw and B.C. Holloway. 2010. Science 329: 1637.
Mirzaee Sisan, M., M. Abdolahi Sereshki, H. Khorsand and M.H. Siadati. 2014. Journal of Alloys and Compounds 613: 288.
Nishi, Y. 2000. Molecular crystals and liquid crystals science and technology. Section A. Molecular Crystals and Liquid Crystals 340: 419.
Niu, C., E.K. Sichel, R. Hoch, D. Moy and H. Tennent. 1997. Applied Physics Letters 70: 1480.
Niu, J.J. and J.N. Wang. 2008. Electrochimica Acta 53: 8058.
Park, J., H. Oh, T. Ha, Y.I. Lee and K. Min. 2015. Applied Energy 155: 866.
Park, S., J.-W. Lee and B.N. Popov. 2012. International Journal of Hydrogen Energy 37: 5850.
Pech, D., M. Brunet, H. Durou, P. Huang, V. Mochalin, Y. Gogotsi et al. 2010. Nat. Nano 5: 651.
Pech-Rodríguez, W.J., D. González-Quijano, G. Vargas-Gutiérrez and F.J. Rodríguez-Varela. 2014. International Journal of Hydrogen Energy 39: 16740.
Perez-Madrigal, M.M., F. Estrany, E. Armelin, D.D. Diaz and C. Aleman. 2016. Journal of Materials Chemistry A 4: 1792.
Pethaiah, S.S., G.P. Kalaignan, M. Ulaganathan and J. Arunkumar. 2011. Ionics 17: 361.
Pham, D.T., T.H. Lee, D.H. Luong, F. Yao, A. Ghosh, V.T. Le et al. 2015. ACS Nano 9: 2018.
Pham, K.-C., D.S. McPhail, C. Mattevi, A.T.S. Wee and D.H.C. Chua. 2016. Journal of the Electrochemical Society 163: F255.
Portet, C., G. Yushin and Y. Gogotsi. 2007. Carbon 45: 2511.
Prem Kumar, T., T.S.D. Kumari and M.A. Stephan. 2009. Journal of the Indian Institute of Science 89: 393.
Rakesh, S., Z. Xianfeng and T. Saikat. 2009. Nanotechnology 20: 395202.
Raymundo-Piñero, E., K. Kierzek, J. Machnikowski and F. Béguin. 2006. Carbon 44: 2498.
Raymundo-Piñero, E., M. Cadek and F. Béguin. 2009. Advanced Functional Materials 19: 1032.
Rufford, T.E., D. Hulicova-Jurcakova, Z. Zhu and G.Q. Lu. 2008. Electrochemistry Communications 10: 1594.
Rufford, T.E., D. Hulicova-Jurcakova, E. Fiset, Z. Zhu and G.Q. Lu. 2009. Electrochemistry Communications 11: 974.
Rufford, T.E., D. Hulicova-Jurcakova, K. Khosla, Z. Zhu and G.Q. Lu. 2010. Journal of Power Sources 195: 912.
Salitra, G., A. Soffer, L. Eliad, Y. Cohen and D. Aurbach. 2000. Journal of the Electrochemical Society 147: 2486.
Şanlı, L.I., V. Bayram, B. Yarar, S. Ghobadi and S.A. Gürsel. 2016. International Journal of Hydrogen Energy 41: 3414.
Schweiss, R., M. Steeb, P.M. Wilde and T. Schubert. 2012. Journal of Power Sources 220: 79.
Sennu, P., H.-J. Choi, S.-G. Baek, V. Aravindan and Y.-S. Lee. 2016a. Carbon 98: 58.
Sennu, P., V. Aravindan, M. Ganesan, Y.-G. Lee and Y.-S. Lee. 2016b. ChemSusChem. 10.1002/cssc.201501621.
Sennu, P., V. Aravindan and Y.-S. Lee. 2016c. Journal of Power Sources 306: 248.
Senthilkumar, S.T., B. Senthilkumar, S. Balaji, C. Sanjeeviraja and R. Kalai Selvan. 2011. Materials Research Bulletin 46: 413.
Sevilla, M., S. Álvarez, T.A. Centeno, A.B. Fuertes and F. Stoeckli. 2007. Electrochimica Acta 52: 3207.
Sevilla, M. and R. Mokaya. 2014. Energy & Environmental Science 7: 1250.

Shao, Y., G. Yin, J. Wang, Y. Gao and P. Shi. 2006. Journal of Power Sources 161: 47.
Shrestha, S., S. Ashegi, J. Timbro, C.M. Lang, W.E. Mustain. 2011. ECS Transactions 41: 1183.
Simon, P. and Y. Gogotsi. 2008. Nat Mater 7: 845.
Stern, O.Z. Electrochem. 30: 508.
Stoller, M.D., S. Park, Y. Zhu, J.H. An and R.S. Ruoff. 2008. Nano Lett. 8: 3498.
Subramanian, V., C. Luo, A.M. Stephan, K.S. Nahm, S. Thomas and B. Wei. 2007. The Journal of Physical Chemistry C 111: 7527.
Tung, V.C., L.-M. Chen, M.J. Allen, J.K. Wassei, K. Nelson, R.B. Kaner et al. 2009. Nano Letters 9: 1949.
Ulaganathan, M., A. Jain, V. Aravindan, S. Jayaraman, W.C. Ling, T.M. Lim et al. 2015. Journal of Power Sources 274: 846.
Ulaganathan, M., R.V. Hansen, N. Drayton, H. Hingorani, R.G. Kutty, H. Joshi et al. 2016. ACS Applied Materials & Interfaces 8: 32643.
Viva, F.A., M.M. Bruno, E.A. Franceschini, Y.R.J. Thomas, G. Ramos Sanchez, O. Solorza-Feria. 2014. H. R. International Journal of Hydrogen Energy 39: 8821.
Wang, H., Y. Zhang, W. Sun, H.T. Tan, J.B. Franklin, Y. Guo et al. 2016. Journal of Power Sources 307: 17.
Wang, X., W. Li, Z. Chen, M. Waje and Y. Yan. 2006. Journal of Power Sources 158: 154.
Wee, J.-H. 2007. Renewable and Sustainable Energy Reviews 11: 1720.
Wei, C., C. Cheng, L. Ma, M. Liu, D. Kong, W. Du et al. 2016. Dalton Transactions 45: 10789.
Wei, L., M. Sevilla, A.B. Fuertes, R. Mokaya and G. Yushin. 2011. Advanced Energy Materials 1: 356.
Wei, L. and G. Yushin. 2012. Nano Energy 1: 552.
Wu, Z.S., K. Parvez, X. Feng and K. Müllen. 2013. Nature Communications 4: 2487.
Yan, J., Q. Wang, T. Wei and Z. Fan. 2014. Advanced Energy Materials 4: 1300816.
Yoo, J.J., K. Balakrishnan, J. Huang, V. Meunier, B.G. Sumpter, A. Srivastava. 2011. Nano Letters 11: 1423.
Zhai, Y., Y. Dou, D. Zhao, P.F. Fulvio, R.T. Mayes and S. Dai. 2011. Advanced Materials 23: 4828.
Zhang, J., L.-B. Kong, B. Wang, Y.-C. Luo and L. Kang. 2009b. Synthetic Metals 159: 260.
Zhang, L.L. and X.S. Zhao. 2009. Chemical Society Reviews 38: 2520.
Zhang, Y., X. Sun, L. Pan, H. Li, Z. Sun, C. Sun et al. 2009a. Solid State Ionics 180: 1525.
Zhao, X., B.M. Sanchez, P.J. Dobson and P.S. Grant. 2011. Nanoscale 3: 839.
Zhi, M., C. Xiang, J. Li, M. Li and N. Wu. 2013. Nanoscale 5: 72.
Zhou, Y., K. Neyerlin, T.S. Olson, S. Pylypenko, J. Bult, H.N. Dinh et al. 2010. Energy & Environmental Science 3: 1437.
Zhu, Y., S. Murali, M.D. Stoller, K.J. Ganesh, W. Cai, P.J. Ferreira et al. 2011. Science 332: 1537.

17

Nanostructured Semiconducting Polymer Inorganic Hybrid Composites for Opto-Electronic Applications

*Sudha J. Devaki** and *Rajaraman Ramakrishnan*

List of Abbreviations

SCPICs	Semiconducting Polymer-Inorganic Nanocomposites	μ	Carrier mobility
PA	Polyacetylene	n	Charge carrier concentration
PVK	Poly (N-vinylcarbazole)	ZnO	Zinc oxide
PANI	Polyaniline	TiO_2	Titanium dioxide
PT	Polythiophene	0D	Zero dimensional
PEDOT	Poly(3,4-ethylenedioxythiophene)	1D	One dimensional
PPy	Polypyrrole	2D	Two dimensional
CV	Cyclic voltammetry	3D	Three dimensional
PS-*b*-PMMA	Polystyrene-*b*-polymethyl methacrylate	I–V	Current-voltage
CS-AFM	Current-sensing atomic force microscopy	WORM	Write once read many times
HOMO	Highest occupied molecular orbital	DRAM	Dynamic random access memory
VB	Valence band	TE	Thermoelectric materials
LUMO	Lowest unoccupied molecular orbital	OFET	Organic field effect transistor
CB	Conduction band	SWNT	Single walled nanotube

Chemical Sciences and Technology Division, CSIR-National Institute for Interdisciplinary Science and Technology, Thiruvananthapuram 695019, India.
* Corresponding author: sudhajd2001@yahoo.co.in

Introduction

Development of Semiconducting Polymer-Inorganic Hybrid Nanocomposites (SCPIHCs) having unique opto-electronic properties are receiving a lot of attention in recent years due to its intriguing properties and potential applications (Agranovich et al. 2011). These SCPIHCs are endowed with combined advantages of inorganic particles (e.g., metal oxides, metal nanoparticles and layered materials) and organic counterparts (e.g., conjugated oligomers and polymers). Recently, researchers have taken greater efforts towards enhancing the properties of hybrid nanocomposites by accomplishing excellent percolation of inorganic nanoparticles within the polymer matrix (Janáky and Rajeshwar 2015, Kao et al. 2013, Holder et al. 2008). Apart from achieving excellent percolation, enhancement in charge carrier mobility and charge carrier concentration is also a great challenge for the development of efficient opto-electronic devices. During the processing of hybrid nanocomposites, an intimate contact between polymer-inorganic nanoparticles is to be established which may help to increase the interfacial area between the junction and controlled distribution of the nanoparticles within the polymeric matrix. The advantage of nanocomposite materials is to facilitate the transfer of charge carriers across the interfaces which results in an increase in the charge carrier mobility and also enhances the electrical conductivity (Holder et al. 2008). In addition, self organization, ordered structure, crystallinity, well defined morphologies and interaction between the semiconducting polymer-nanoparticle in the nano level also play pivotal role in the properties of hybrid nanocomposites (Romero 2001). SCPICs play a crucial role in the development of advanced functional nanomaterials which are expected to revolutionize the emerging opto-electronic technology such as flexible display panels, solar cells, volatile and non-volatile memory device, organic field effect transistors, thermoelectric devices and photocatalysts applications as shown in Figure 1.

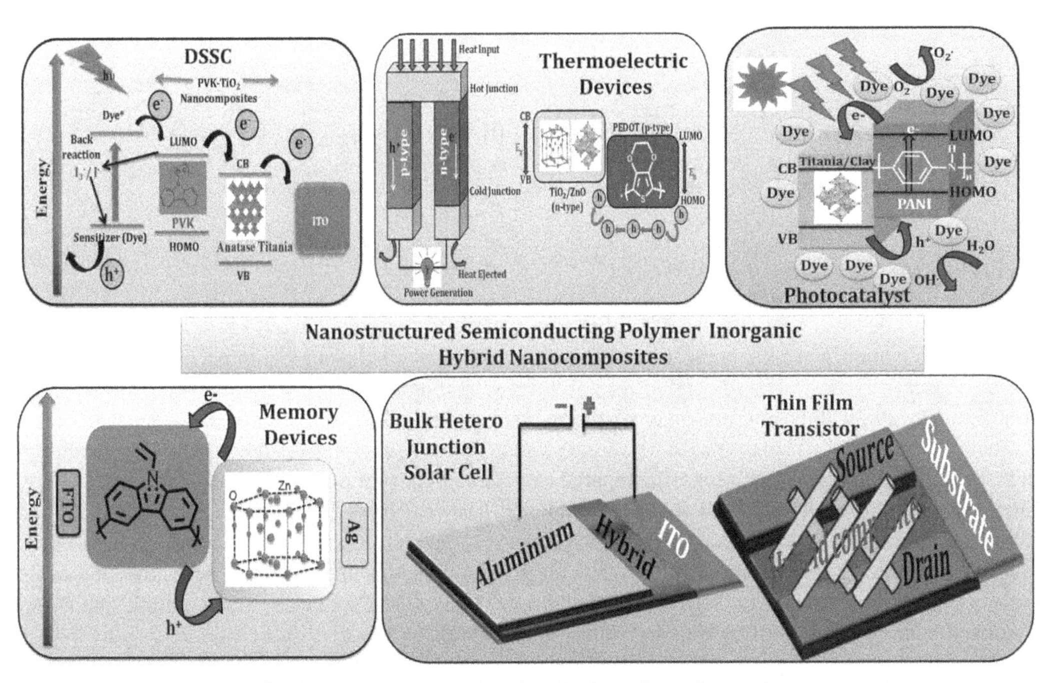

Figure 1. Applications of nanostructured semiconducting polymer-inorganic nanocomposites.

Conducting polymers

Conducting polymers are conjugated organic compounds with an extended π-orbital system through which electrons can move from one end of the polymer to the other. With the invention of conductive polyacetylene in the 1970s, conducting polymers have received significant attention from both science and engineering communities. The Nobel Prize in chemistry in the year 2000 was awarded to Alan J. Heeger, Alan G. MacDiarmid and Hideki Shirakawa for the discovery of conducting polymers (Chiang et al. 1977). For a polymer to be conducting, it should possess conjugated double bonds, and the basic repeating units must be planar and should possess non-even number of electrons. When an electric field is applied, the π electrons move rapidly along the polymer backbone, making it conducting. Since the discovery of intrinsically conducting polymers explored their unusual opto-electronic properties for a wide range of applications because of its good processability, light weight, flexible, thermal stability, tunable conductivity and low production costs (Virji et al. 2006, Song and Choi 2013). Figure 2 represents the molecular structures of extensively studied conducting polymers such as polyacetylene (PA), poly (N-vinylcarbazole) (PVK), polyaniline (PANI), polythiophene (PT), poly(3,4-ethylenedioxythiophene) (PEDOT) and polypyrrole (PPy).

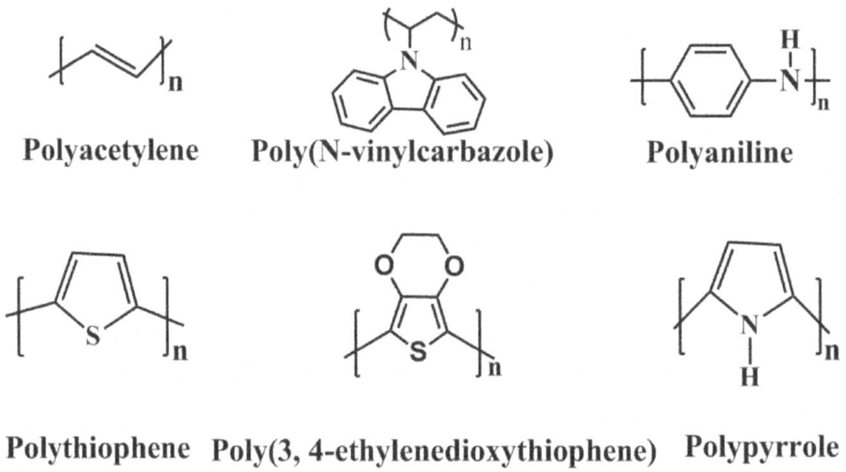

Polyacetylene **Poly(N-vinylcarbazole)** **Polyaniline**

Polythiophene **Poly(3, 4-ethylenedioxythiophene)** **Polypyrrole**

Figure 2. Molecular structures of extensively studied conducting polymers.

Nanostructured conducting polymers

In the past two decades, tremendous efforts have been made towards the development of nanostructured conducting polymers since the realization of these materials can display a great impact on the properties at the nanoscale. Nanostructured conductive polymers with tunable size, shape, composition and spatial arrangement exhibit improved mechanical properties for strain accommodation, large surface areas, shortened pathways for charge/mass/ion transport and also new exciting features including flexibility, light weight and processability for the prepared nanostructures (Hu et al. 1999, Bhadraa et al. 2009, Chen et al. 2013). A wide variety of conductive polymer nanostructures have been developed and applied for a range of applications, such as energy conversion, storage, electronic devices, sensors, field-effect transistors, electrochromic display devices, supercapacitors, actuators and so forth. These novel materials are a fascinating playground for scientists and engineers to explore the effects of low dimensionality on a material's intriguing properties and their multifunctional applications.

Nanostructured conducting polymer can be synthesized by template based methods. The templates can be either hard template or soft template. The most commonly employed templates are anodic aluminium oxide, block co-polymers, porous silicate and mesoporous zeolites. The conventional hard template method is pioneered by Martin's group and has become a powerful tool for controllable synthesis of nanostructured conductive polymer (Martin 1995). Variety of nanostructured conducting polymer such as PANI, PPy, PHT and PEDOT have been chemically or electrochemically synthesized with tunable shape and size by conducting polymerization inside the pores of these templates. The main disadvantage of this method is removal of the template causing damage to the nanostructured conducting polymer during the post processing step. Kim et al. (2011) reported electrochemically generated dense array of vertically aligned PPy nanorods (~ $10^{11}/cm^2$) in a porous polystyrene-*b*-polymethyl methacrylate membrane template (pore diameter of ~ 25 nm) as shown in Figure 3 (Lee et al. 2008). The height of the nanorods was controlled by the time used for the electro-polymerization under a constant voltage. It was also observed that the self-supporting PPy nanorods were oriented normally to the substrate after removal of the template by washing with toluene. They exhibited a higher conductivity than a homogeneous PPy thin-film, as characterized by a technique of current-sensing atomic force microscopy. This study demonstrates a useful hard-template method to construct the ultrahigh density arrays of conducting polymer nanorods with good electrical conductivity and current-voltage characteristics.

Nanostructured conductive polymer can also be prepared using the soft template approach. The soft templates are usually made from mesophase structures such as surfactant micelles, liquid crystals, amphiphilic dopants and block copolymers, etc. (Brinker et al. 1999, Ramanathan et al. 2013, Liao et al. 2010, Devaki et al. 2014). The morphology of resulting products is controlled by the nature and chain length of the surfactants as well as the concentrations of both monomers and surfactants. The dimension and shape of the formed nanostructured polymer can be controlled by tuning the size and shape of the micelle. Amphiphilic dopants, such as dodecyl benzene sulphonic acid and camphor sulphonic acid, form liquid crystalline phases which can act as dual role of dopant cum template. One advantage of this strategy is that dopant template will be a part of the nanostructured polymer and post processing step can be avoided. The template-free method developed by Wan is a simple self-assembly method without an external template (Wan 2008). The self-assembly method is induced by non-covalent forces between polymer chains, such as π-π stacking, dipole–dipole, hydrophobic, van der Waals forces, hydrogen bonding, electrostatic and ion–dipole interactions of the building blocks to spontaneously form anisotropic aggregates or oriented structures under suitable conditions (Kim et al. 2011, Huang and Kaner 2004).

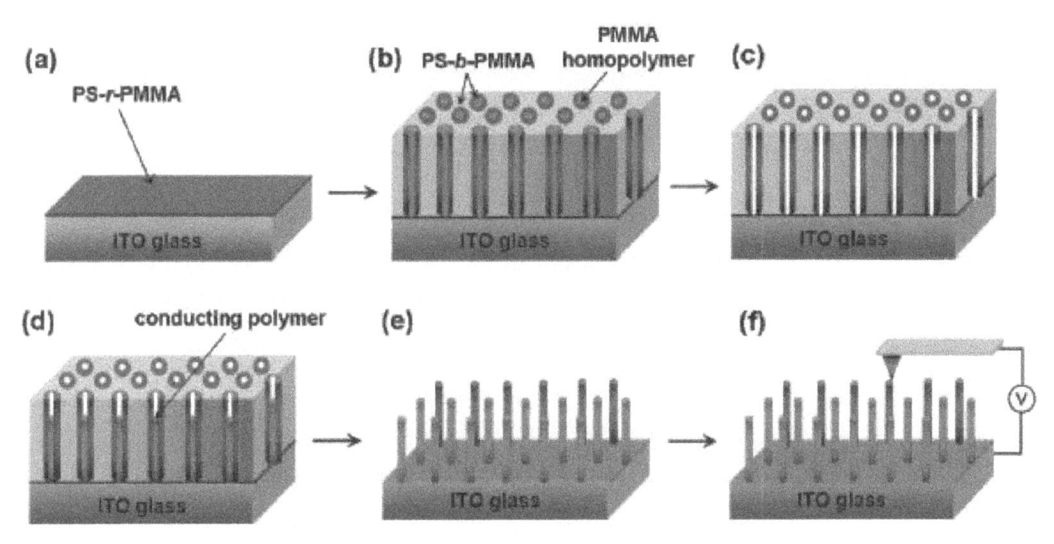

Figure 3. Schematic representation of the hard template approach for the synthesis of ultrahigh density arrays of PPy nanorods using nanoporous block copolymer membrane as hard template [Reprinted with permission from Ref. (Lee et al. 2008)].

Conduction mechanism of conducting polymers

Doping

Conducting polymers are intrinsically conducting and the conductivity can be improved by the process of doping. The process of doping involves either removal of electrons or addition of electrons from the π-orbitals so as to produce charge carriers. Doping is achieved by the formation of charge carriers by electron donors such as sodium or potassium (n-doping, reduction) or by electron acceptors such as I_2, AsF_5, or $FeCl_3$ (p-doping, oxidation) (Huang et al. 1986). As a result of this process, the doped polymer backbone becomes negatively or positively charged with the dopant forming oppositely charged ions (Na^+, K^+, I_3^-, I_5^-, AsF_6, $FeCl_4^-$) (Kaneko et al. 1993). Application of an electric potential results in motion of counter ions in and out which enables to switch the polymer between the doped conductive state and the undoped insulating state. The charge transport mechanism in the conducting polymers is mainly due to the formation of polarons and bipolarons in conducting molecules. This concept is also used for the explanation of the drastic colour changes produced during the doping process. The electronic energy diagrams and doping process in PPy is illustrated in Figure 4. Figure 4a represents the energy diagram of PPy in neutral form. Upon oxidation (removal of π-electrons from the valence band) of neutral PPy, there occurs the local relaxation of the benzoid structure toward a quinoid-like, which creates radical cations (polarons). Formation of polarons induces two new energy levels (Figure 4b) that are symmetrically positioned within the bandgap. Thus, two new electronic transitions at longer wavelengths emerge. Further oxidation results in the formation of bipolarons which are the charge carriers of the coupled cations (dication). Since they have lower energy state empty, bipolarons are signified by broad, low energy absorptions due to the transitions from the top of the valence band (Figure 4c). As the polymer is oxidized further, the bipolaronic energy state overlaps and forms intermediate band structures which are given in Figure 4d. Doping level determines number of polaron and bipolarons, low doping levels gives rise to polarons, whereas higher doping levels produce bipolarons (Camurlu 2014).

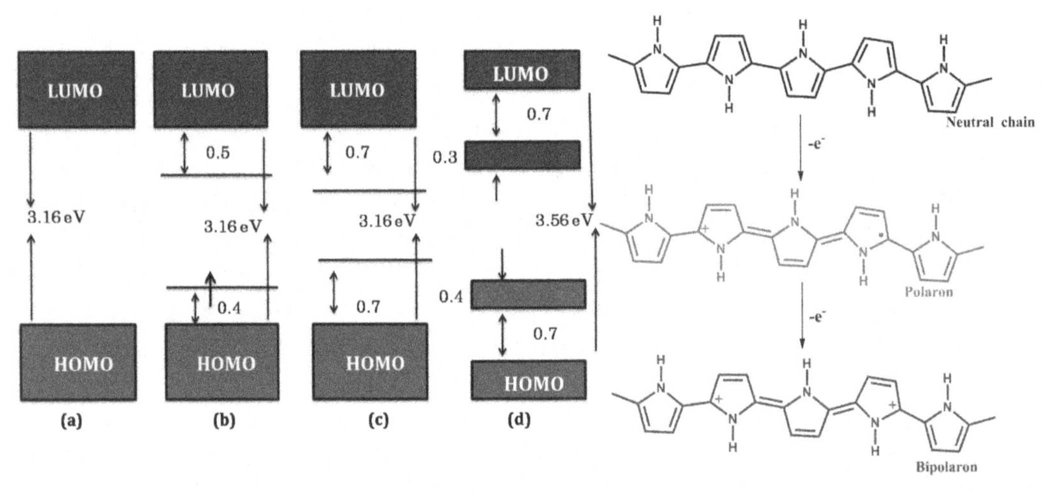

Figure 4. Electronic energy diagrams and structures for (a) neutral polypyrrole (b) polaron (c) bipolaron and (d) fully doped pyrrole [Reprinted with permission from Ref. (Camurlu 2014)].

Band theory

According to molecular orbital theory, atomic orbital from one atom overlaps with atomic orbital of another atom resulting in the formation of two molecular orbitals that are delocalized over both atoms. One of the molecular orbital is bonding (π) which has energy lower than both the combining atomic orbitals while the other is anti bonding (π*) with high energy. The band formed from the highest occupied molecular orbital (HOMO) will be completely filled and is called valence band (VB) while the band formed from the lowest unoccupied molecular orbital (LUMO) will be entirely empty and is called conduction band (CB). The forbidden energy gap value determines the electrical conductivity of the molecules. In case of insulators and semiconductors, this energy gap separating the two bands is high (> 1.5 eV for semiconductor and > 9 eV for insulator) whereas in the case of conductors, two bands overlap to facilitate maximum conduction.

Conductivity is attributed to the flow of mobile charge carriers. Since metals have a large number of mobile electrons, they are good conductors and the conductivity is in the order 10^6 S/cm. The conductivity (σ) is directly proportional to the product of carrier mobility (μ), its charge (Q) and number of carrier or the concentration (n) as shown in Eq. 1,

$$\sigma = \mu . Q . n \quad (1)$$

The electrical conductivity of organic material includes ionic conduction, hopping conduction, excitonic conduction, quantum mechanical conduction and tunneling between metallic domains. Figure 5 illustrates an overview of a broad conductivity range of insulators, semiconductors and metallic conductors.

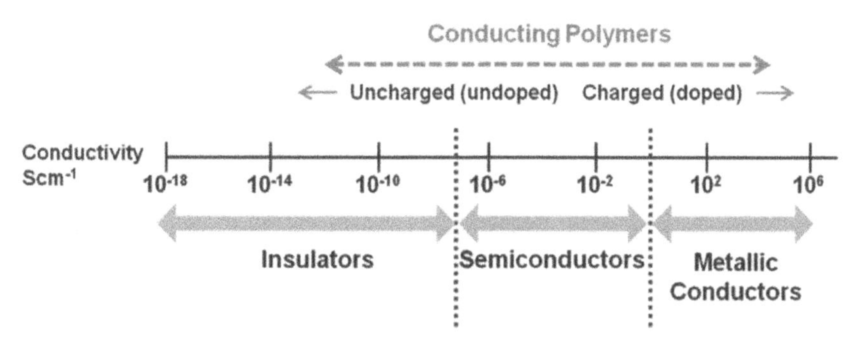

Figure 5. Conductivity chart of insulators, semiconductor and metallic conductors.

Nanostructured semiconducting metal oxides

Research and development on nanostructured semiconducting metal oxide has been growing largely around the world in the past few decades (Ozin et al. 2008, Burda et al. 2005). This is considered as a class of materials that can be distinguished on the basis of tailoring their properties from bulk to nano size. Nanostructures are defined as the materials with at least one dimension between 1 and 100 nm. The physical and chemical properties of a material can also be tuned significantly when the size of the particles is reduced to the nanometer scale because of quantum confinement effect (Alivisatos 1996). Owing to the unique and beneficial properties of these nanomaterials which exhibit excellent chemical and thermal stability, non-toxicity, and low cost, they find applications in the area of energy saving and harvesting devices such as solar cells, transistors/FETs, light emitting devices and lithium ion batteries (Colodrero et al. 2009, Ye et al. 2005, Wood et al. 2009, Poizot et al. 2000) and also as photocatalysts used for the degradation and adsorption of organic/inorganic pollutants present in water (Allam and Grimes 2007). Apart from these, they also have tremendous applications in biological and medical sciences such as drug delivery, cancer treatments, fluorescent imaging, bio-labeling and bio tagging, etc. (Chen et al. 2006, Stoeva et al. 2006, Bae et al. 2006, Aernandez et al. 2007, Suzuki et al. 2008). Generally, in metal oxides, the S-shells of positive metallic ions are always fully filled by electrons and the d-shells are left with vacancy. This characteristic feature endows them with various unique properties, which involve electronic transitions, high dielectric constants, tunable band gaps, good electrical characteristics and so on. Therefore, nanostructured metal oxides are considered to be one of the most fascinating functional materials. They can be prepared in a choice of morphologies such as nanoparticles, hexagonal discs, nanorods, nanowires, tapes, belts and flowers using various physical and chemical routes (Patra and Gedanken 2004, Xia et al. 2009, Yu et al. 2009, Hu et al. 2004).

Strategies for the preparation of nanostructured metal oxides

In the past two decades, tremendous efforts have been made towards the development of synthetic strategies for the preparation of nanostructured materials with well controlled size, shape, and composition.

In general bottom up and top down approach are used for the synthesis of metal oxide nanomaterials which are shown in Figure 6.

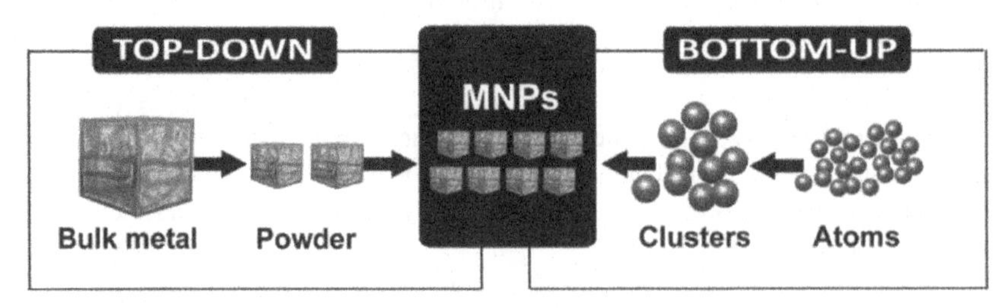

Figure 6. Schematic representation of Bottom-up and Top-down technique.

Top-down approach

This approach begins with large (macroscopic) and complex structures of bulk materials which are sliced or reduced to get the objects with nano-scale order. The top-down approach is essentially a miniaturization technique and leads to the bulk production of nanomaterials. This approach introduces internal stress and imperfections of surface structure and it will continue to play an important role in the synthesis of nanostructures. Lithography, attrition or milling, quenching, etching, and successive cutting are typical examples of top-down approach for the synthesis of metal oxide nanomaterials (Lu et al. 2002).

Bottom-up approach

In "bottom-up" processes, fundamental building blocks or primary units (atoms or molecules) self-associate, interact with each other in a coordinated way, and form ordered, complex and integrated 0-D, 1-D, 2-D and 3-D nanostructures with macroscopic nanoscale objects. During self-assembly or organization of basic units into ordered nanostructures, several non-covalent interactions such as hydrophobic, aromatic stacking, or electrostatic interactions, hydrogen bonds, and physical forces controlling at nanoscale are used to aggregate the building blocks into nanoscale stable structures. Fabrication of metal oxide nanostructures by this synthetic approach is much less expensive. Most of the existing strategies suffer from different drawbacks such as expensive and externally controlled equipments, prolonged reaction times, high temperatures, formation of undesirable by products, complex chemicals or complicated reaction conditions to produce metal oxides in the nanoregime. Therefore, the bottom-up strategy is able to create such nanoscale products at much cheaper rates than the top-down approach, which is normally used for the high yield production with size and shape controllable metal oxide nanostructures (Yang et al. 2000). A number of physical, chemical, hybrid and biological synthetic techniques have been adopted to control the shape, size, homogeneity, and crystalline phases in these metal oxide nanostructures. Nanomaterials have been produced by physical processes such as spray pyrolysis, laser ablation, vapour depositions, arc-discharge, molecular beam epitaxy and sputtering. Several chemical methods including precipitation, hydrothermal, solvothermal, sol-gel, electrochemical and sonochemical methods (Li et al. 2006, Wu et al. 2005, Jokanovic et al. 2004, Tian et al. 2003, Bak et al. 2008) are widely used for the preparation of metal oxide nanomaterials by bottom-up strategy.

Zinc oxide and titanium dioxide

Among the various metal oxides, titania (TiO_2) and zinc oxide (ZnO) nanoparticles based n-type semiconductors have attracted significant attention over the last two decades due to their attractive opto-electrical properties, nontoxicity, long-term stability, catalytic property and low cost. They can be prepared using various methods such as sonochemical, hydrothermal, and sol-gel reactions from their precursors. They have been extensively used in diverse applications such as solar cells, photocatalysts, sensors, photochromic devices, as white pigment in paints, as small additive in food and cosmetics, anti-inflammatory agents and also as a UV-absorbing material in sunscreen applications.

ZnO is transparent in its pure form with a band gap of 3.37 eV, high ionization potential ~ 8 eV, electron affinity ~ 4.7 eV, electron effective mass 0.26, carrier mobility 130–200 $cm^2V^{-1}S^{-1}$, electron diffusion coefficient 1.7×10^{-4} and large exciton binding energy ~ 60 meV at room temperature (Djurisic and Leung 2006). This enables it to exhibit efficient near band edge excitonic emission. The large exciton binding energy leads to a close distance between the electron and hole pairs. As a result, even ZnO quantum dots are fluorescent. Depending on the crystallizing condition, ZnO has three crystal structures: wurtzite, zinc blende and rock salt as schematically shown in Figure 7. The wurtzite structure is a thermodynamically stable crystal structure for ZnO under ambient conditions. If ZnO is grown on the surface of a cubic crystal, a zinc blende structure is expected, whereas the rock salt structure can be only achieved under high pressure (~ 10 GPa). The ZnO wurtzite crystal structure has a hexagonal unit cell with lattice parameters equal to a = 0.324 nm and c = 0.521 nm. The ZnO structure consists of a number of zinc (Zn) and oxygen (O) surfaces stacked alternatively along the c-axis. Each Zn cation is surrounded by four O anions coordinated at the edges of a tetrahedron and the mechanical deformation of the tetrahedral structure leads to polarization, i.e., the formation of electric dipole at the microscopic scale, which in turn results in the piezoelectric property of ZnO.

Titanium dioxide is a wide band gap semiconductor (3.22 eV), which has electron effective mass 9, carrier mobility 0.1–4 $cm^2V^{-1}S^{-1}$ and electron diffusion coefficient 4.3×10^{-4} with similar physical properties as those of ZnO. Titania exhibited three different polymorphs, i.e., anatase (tetragonal), rutile (tetragonal), and brookite (orthorhombic), as shown in Figure 8 (Pan et al. 2013). Tetragonal structure of anatase and rutile are more ordered than the orthorhombic brookite structure. Rutile titania has a tetragonal crystal structure and contains six atoms per unit cell. Rutile is the most thermodynamically stable polymorph of titania at all temperatures, exhibiting lower total free energy than metastable phases of anatase and brookite. Anatase titania has a crystalline structure that corresponds to the tetragonal system but the distortion of the TiO_6 octahedron is slightly larger for the anatase phase. The anatase structure is preferred over other polymorphs for solar cell applications because of its potentially higher conduction band edge energy and lower recombination rate of electron–hole pairs.

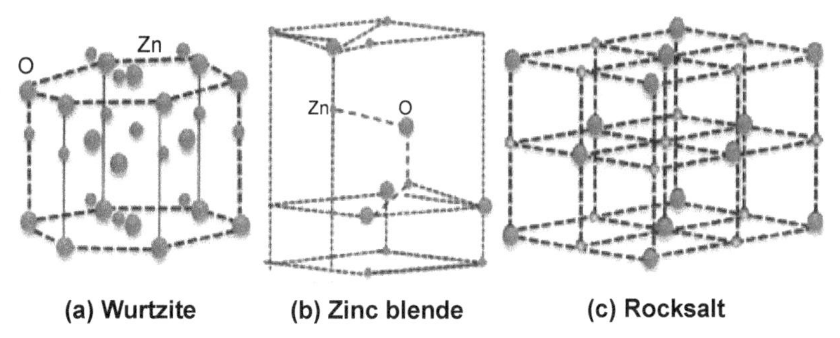

Figure 7. Crystal structures of zinc oxide (a) Wurtzite (b) Zinc blende and (c) Rocksalt.

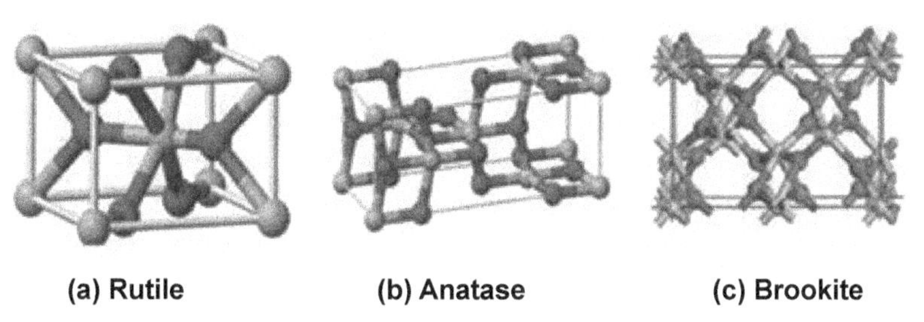

Figure 8. Crystal structures of titanium dioxide (a) Rutile (b) Anatase and (c) Brookite.

Defects in metal oxide nanoparticles

A defect is defined as an imperfection in the regular pattern of the atomic arrangement of a material due to the dislocation of an atom or more from their original positions. Such dislocation can be triggered by foreign atom (impurity) associated during the growth. This is referred to as an *extrinsic defect.* A defect might involve a host atom, in which case one speaks about an intrinsic defect (Janotti and Van de Walle 2009, Lai et al. 2012). Intrinsic defects can be classified into four categories depending on their geometrical structure:

- Zero dimensional (0D) defects or Point defects consist of an isolated atom in localized regions in the host crystal (e.g., vacancies).
- One dimensional (1D) defects or line defects consist of a row of atoms (e.g., a straight dislocation).
- Two dimensional (2D) defects or area defects consist of an area of atoms (e.g., twins).
- Three dimensional (3D) defects or volume defects consist of a volume of atoms (e.g., voids).

The 0D-defects (point defects) are the natural intrinsic defects created inside the metal oxides during the growth mechanism regardless of the growth techniques. Such defects can be vacancies, interstitials, anti-sites and substitution defects. In the case of ZnO, the vacancy defects are created when oxygen or zinc atoms are missing in one of the lattice site. Oxygen vacancy (V_o) acts as deep donor and are located at ~ 2 eV below the conduction band while zinc vacancy (V_{Zn}) is a shallow acceptor located at ~ 0.31 eV above the valence band as shown in Figure 10. If the atom occupies a position where there is usually no atom, then it is called interstitial oxygen. The interstitial defects can originate from oxygen (O_i) or zinc (Zn_i) atoms. The interstitial oxygen (O_i) atoms are deep acceptors while zinc interstitials (Zn_i) are shallow donors. Another type of the point defects is the anti-site defects where an oxygen atom occupies a position normally taken by a zinc atom in the zinc lattice (or vice versa). Both zinc anti-sites (Zn_O) and oxygen anti-sites (O_{Zn}) have high formation energies and are created under non equilibrium condition such as ion implantation. Oxygen anti-site defects are deep acceptors; however, zinc anti-site defects are shallow donors. Finally, substitution defects can be created during ZnO growth process. In such case, an impurity atom has to replace an oxygen atom or a zinc atom in the ZnO lattice. The schematic diagram showing the illustration of formation of various defects and the value of energy levels is depicted in Figure 9 (Petkovich and Stein 2013).

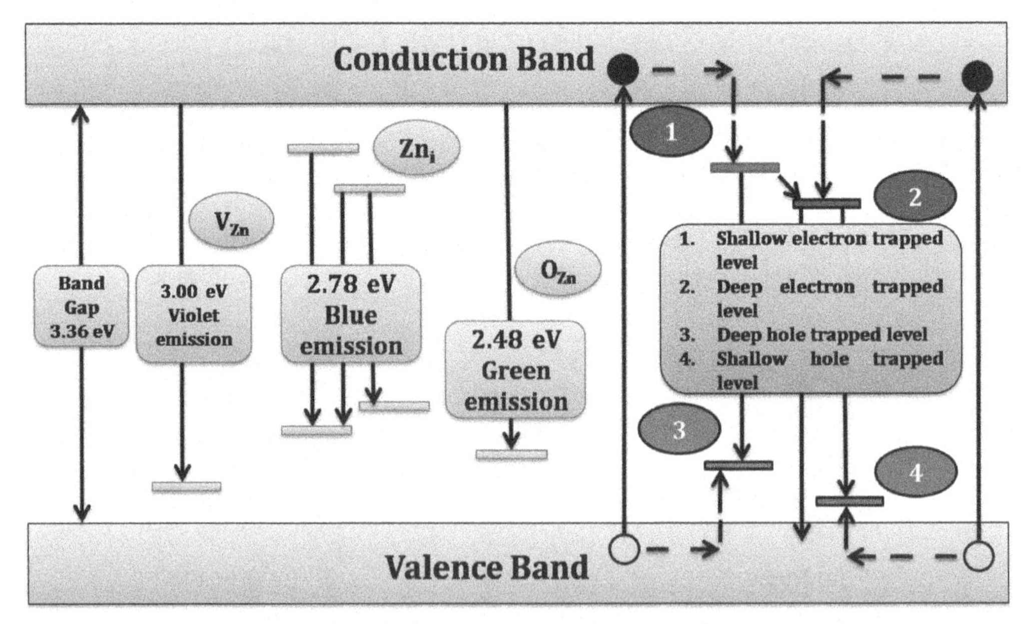

Figure 9. Schematic illustration of various defect energy level of ZnO.

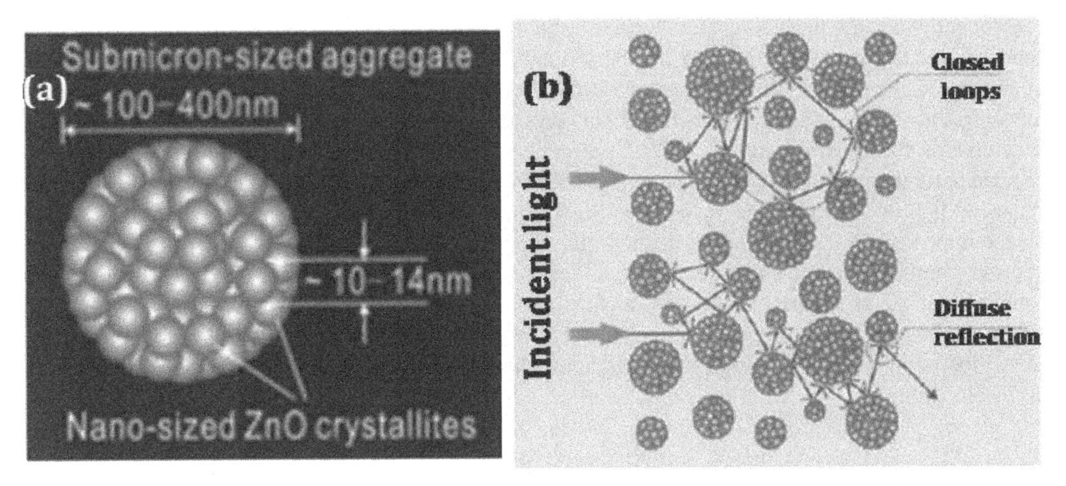

Figure 10. (a) Cartoon showing the hierarchical structure of ZnO and (b) effect of light scattering and photon localization within a film consisting of submicrometer sized aggregates [Reprinted with permission from Ref. (Wang et al. 2006)].

Hierarchical structure

In general, nanostructures possess four different morphologies: zero-dimensional (0D), one-dimensional (1D), two-dimensional (2D) and three dimensional (3D) nanostructures. Among them, 3D hierarchical structures (nanocrystalline aggregates) are an exciting new frontier for materials' research. The advantages of hierarchical materials are good mesoporous and microporous structure that help to impart a high surface area and pore volume. It provides numerous reactive sites and a substantial interfacial area that can improve an enhanced surface-to-volume ratio and reduced transport lengths for both mass and charge transport (Huang et al. 2001, Wang et al. 2004). Owing to their structure, hierarchical materials provide the substantial performance enhancement in numerous applications such as energy storage, energy conversion, catalysis and sensing (Hong et al. 2010, Jing et al. 2013, Baxter and Aydil 2005, Wang et al. 2006).

Hierarchical structure of ZnO and TiO_2 are widely used as active materials in DSSCs and photocatalytic application. In DSSCs, the major restricting problem of improving the higher conversion efficiencies is a dynamic competition between the charge carrier generation and recombination of photoexcited carriers (Law et al. 2005, Paulose et al. 2006). One dimensional nanostructures are able to provide a direct pathway for the rapid collection of photogenerated electrons and thus reduce the degree of charge recombination. However, such one dimensional nanostructures seem to have insufficient internal surface area which limits their energy conversion efficiency at a relatively low level, for example, 1.5% for ZnO nanowires and 4.7% for titania nanotubes (Nishimura et al. 2003, Halaoui et al. 2005). Another way to increase the light harvesting capability of the photo electrode films is utilizing optical enhancement effects. This can be achieved by means of light scattering via introducing scattering particles into the photo electrode film (Zhang et al. 2009, Chou et al. 2007). The following criteria are essential for improving the power conversion efficiency of solar cell. It should allow the complete light absorption in the spectral range of the dye, increase the light scattering of the absorbing layer for enhancing the time spent by light inside the sensitized film and improving light absorption, and inhibition of back electron transfer between the conducting layer at the anode and the electrolyte (Guillen et al. 2013, Zhang et al. 2008). Figures 10(a and b) show the hierarchical structure of ZnO and the effect of light scattering and photon localization within a film consisting of submicrometer sized aggregates.

However, the major drawback is that the introduction of larger particles into nanocrystallline films will unavoidably lower the internal surface area of the photoanode film and therefore counteract the enhancement effect of light scattering on the optical absorption spectra. Also, the incorporation of a layer of titania photonic crystal may lead to an undesirable increase in the electron diffusion length and consequently increase the recombination rate of photogenerated carriers (He et al. 2010). Cao et al. 2007 demonstrated the dye sensitized solar cell performance with polydisperse and monodisperse aggregates of zinc oxide hierarchical films (Chen et al. 2010). Recently, we reported a facile bioanchoring strategy for controlling the crystal growth process of ZnO crystals during calcination to form hierarchical multiple structures as shown in Figure 11. Further performed its application as photoanode in dye sensitized solar cells for improving the power conversion efficiency (PCE) (~ 5.3%) through high dye loading and enhanced light absorption edge via inherent light reflection mechanism (Tian et al. 2014). The reason for the improved PCE for hierarchical ZnO might be arising from the excellent optical absorption and improved molar extinction coefficient. Hierarchical ZnO consists of nano and submicron sized spherical particles that are aggregated to form hierarchical polydisperse hollow structures with a crystal size of 19 nm. These polydispersed hollow structures induce the multiple reflection of incident photon inside the closed loops and lead to the absorption of light at a higher wavelength as well as improve the electron diffusion length.

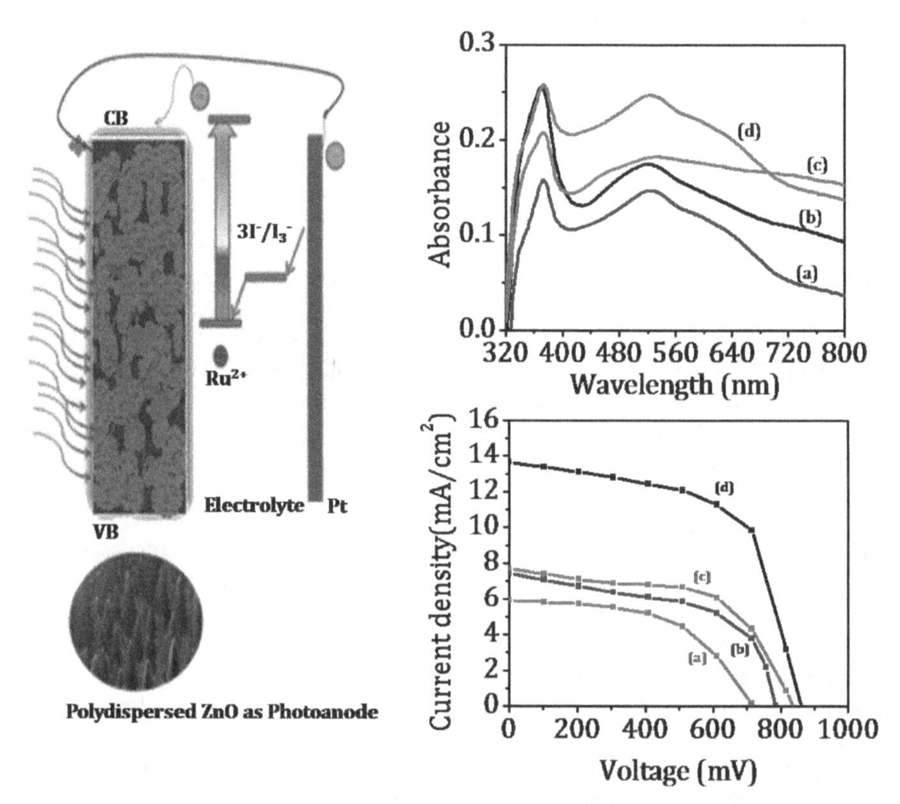

Figure 11. Hierarchical mutiple structures ZnO as Photoanode in DSSC [Reprinted with permission from Ref. (Tian et al. 2014)].

Semiconducting polymer-inorganic nanocomposites

Nanocomposites of conjugated polymers and inorganic nanoparticles (NPs) have emerged as an area of interest motivated by potential applications of these materials in electronics, optics, catalyst and biomedical area. Nanocomposites, in which materials are self assembled or organized in the nanoscale, will help to tailor the optical as well as electronic behaviour of both the NPs and conjugated materials. Additionally, these materials offer an effective route for the stabilization of inorganic nanoparticles and allow accessing of their fascinating physical and chemical properties in a simple and facile way. It has been observed that organic conducting polymers have been shown to be excellent hosts for trapping nanoparticles of metals and semiconductors because of their ability to act as stabilizers or surface capping agents (Barlier et al. 2009). The properties of nanocomposites strongly depend on the composition and nanoscale morphology, most importantly interfacial area and contact between the components. To prepare the nanocomposites materials, several approaches have been employed such as physical mixing, sol-gel technique, *in situ* chemical polymerization in aqueous solution in the presence of monomer and inorganic particles, emulsion technology, sonochemical process and ϒ-radiation technique. Based on the method of preparation, nanocomposites can be classified into three main categories: (i) *ex situ* (sequestered) synthesis, (ii) *in situ* (sequential) synthesis, and (iii) one-pot synthesis.

Ex situ synthesis

In *ex situ* polymerization method, conducting polymer and inorganic species are synthesized separately and then hybridization is achieved during a subsequent step by simple blending of two or more components in which interfacial interaction between the different components determines the major properties of the resulting nanocomposites. Figure 12 shows the preparation of hybrid nanocomposites by *ex situ* polymerization method. These procedures are simple and highly suitable for solution based processing, which is a milestone for enabling mass production and processing via roll-to-roll printing. Another benefit of these methods is the well established synthetic procedures because each component is manufactured separately using a known process. As for the hybridization process, there are various valuable strategies available, ranging from very simple methods such as mechanical mixing of the two components, to more sophisticated methods such as ligand exchange or layer-by-layer fabrication procedures. Finally, physical infiltration of conducting polymers into inorganic nanostructures can also be achieved, resulting in highly ordered nanoarchitectures. However, these are not direct processes because the polymeric guest materials may be prevented from infiltrating the pores of the nanostructured host due to high interfacial tension as a result of physical limitations such as the hydrodynamic radius of the polymer, leading to incomplete pore filling (Liu and Lee 2008).

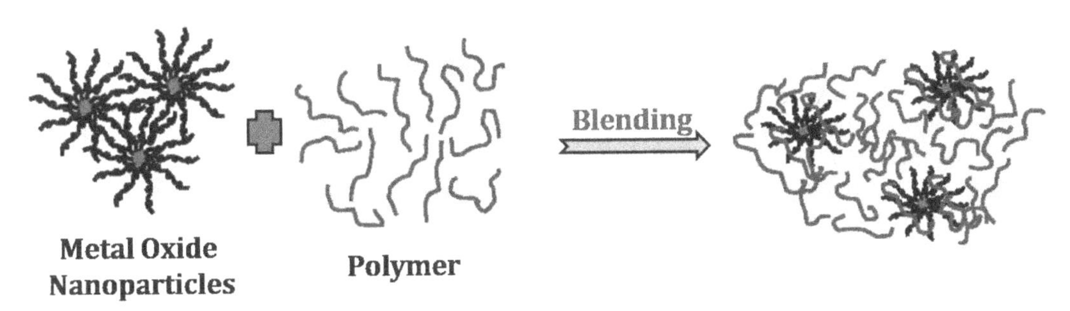

Figure 12. *Ex situ* preparation of hybrid nanocomposites.

In situ synthesis

In this strategy, the conducting polymer is synthesized by the *in situ* polymerization of the monomer in the presence of the semiconductor nanostructures. However the opposite sequence is also possible, namely the semiconductor component is generated *in situ* inside the conducting polymer, either through chemical or electrochemical methods. It is apparent that nanohybrids obtained by *in situ* methods may have several advantages over their *ex situ* prepared counterparts because the organic/inorganic interface can be better controlled at the molecular level. Under this method, two distinctly different strategies can be followed for the preparation of hybrid composites, either chemical polymerization or electro-polymerization method. Specifically, in the case of the electrochemical *in situ* route, heterogeneous components can be incorporated into conducting polymer that grow in the form of a film on the electrode surface. One benefit of such an approach is that the structure and properties of the hybridized material can be controlled by changing the critical variables including deposition time and current density. Figure 13 shows the one step method of synthesizing MnO_2/PEDOT coaxial nanowires by coelectrodeposition in a porous alumina template (Zhou et al. 2010). For chemical *in situ* synthesis, the formation of the nanohybrid is rather straight forward because conducting polymer coexists with nanoparticles and is self assembled to form hybrid composites. In this method, conducting polymers are formed in the presence of the inorganic species; however, the opposite is also possible where the inorganic component is synthesized *in situ* within conducting polymer. The most important advantage of this method is very good control over the structure and morphology of the resultant composite material.

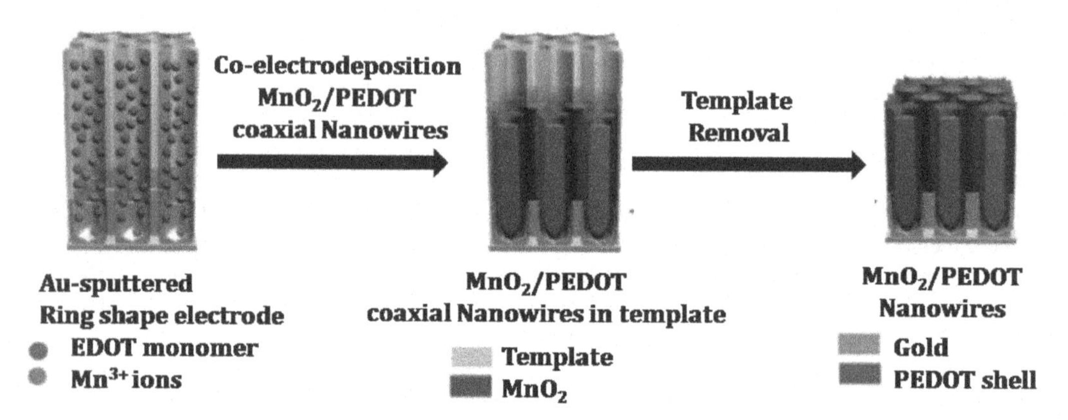

Figure 13. Schematic representation of electro-chemical polymerization route for the semiconducting polymer-inorganic nanocomposite [Reprinted with permission from Ref. (Zhou et al. 2010)].

Synergistic opto-electronic properties in nanocomposites

Semiconducting polymer inorganic nanomaterial is expected to exhibit unassuming high performance arising from the synergistic effects of both nanoparticles and conducting polymer which can affect the optical, electrical and mechanical properties of the hybrid nanocomposites. The property of nanocomposites depends on the (i) percolation of nanoparticles on polymer matrix, (ii) Interfacial interaction between nanoparticle and polymer (iii) effective charge-energy transfer between nanoparticles and conducting polymers (Shen et al. 2007, Kaur et al. 2015, He et al. 2013).

In the case of well dispersed nanoparticles, they can provide better optical, electrical and mechanical properties than bulk nanoparticles because of intimate contact between the polymers and nanoparticles. In the second case, polymer-nanoparticle interactions can control the chain anchoring, carrier mobility, carrier concentration, morphology and interfacial area of conducing polymers and nanoparticle. Depending on the nature of interactions, it may be (i) nanoparticles strongly coupled with conducting polymer (ii) nanoparticle adsorbed on the surface of polymer or vice versa, (iii) remain mobile at the interface. The strong interactions can lead to formation of a well defined network that can provide an enhancement of opto-electronic properties. The electrical properties of the material can be enhanced further by the formation of a percolating network structure that can provide a direct charge/electron transfer (Reiss et al. 2011). The electrical properties of these nanocomposites are strongly dependent on the film morphology, chemical and physical structure, which can be strongly modified via a variety of post treatments such as solvent and thermal treatment. In nanocomposites, large interface between the nanoparticles and conducting polymer facilitates charge and energy transfer between the two components which can occur through a number of potential routes depending upon the electronic energy levels of both the nanoparticle and semiconducting polymer. Figure 14 shows formation of conductive network in (a) insulative polymer-filler composites and (b) semiconducting polymer-inorganic composites.

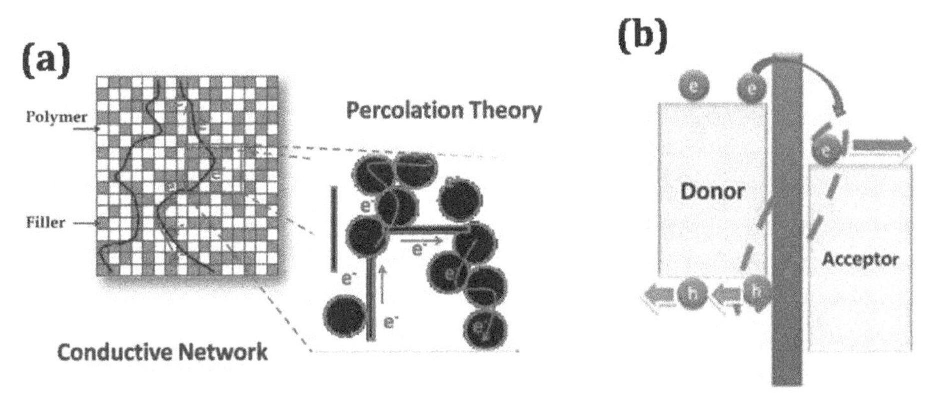

Figure 14. Formation of conductive network in (a) insulative polymer-filler composites and (b) semiconducting polymer-inorganic composites.

Application of semiconducting polymer inorganic nanocomposites

Memory devices

Memory devices are simple metal-insulator-metal structure with an active layer sandwiched between two electrodes and the resistance of the active layer is changed by applying a bias, constituting the set and reset states of the device as shown in Figure 15(a). Ideally, the device at the low conductivity state (OFF) is switched by a threshold voltage to a high conductivity state (ON) and the two states of bistable device differ in their conductivity by several orders in magnitude, retaining a remarkable stability. More importantly, the high and low conductivity states of the device can be precisely controlled by applying a positive voltage pulse to write or a negative voltage pulse to erase, respectively (Heremans et al. 2011, Zhang et al. 2012). Based on the ability to retain information, the electrical switching characteristics of organic memory devices are categorized in two classes, volatile and non-volatile as shown in Figure 15(b). Volatile switching requires periodic refreshing of data as a result of the loss of stored information similar to Dynamic Random Access Memory (DRAM). Conversely, electrically programmed organic memory devices with non-volatile switching characteristics can retain data for extended periods of time. This is similar to conventional flash memory. Non-volatile switching is often classified into three types based on current-voltage (I–V) curves: write once read many times (WORM), and unipolar and bipolar switching memory (II, III, and IV in Figure 15b). WORM type memory devices show electrically irreversible switching characteristics and the original state is never recovered. These devices can be used as storage components for radio frequency identification tags. Both unipolar and bipolar memory systems exhibit electrically reversible switching. Unipolar memory devices use the same voltage polarity to write and erase, while bipolar memory devices require different voltage polarities (Cho et al. 2009).

The origin of this change in conductivity is still unambiguous. Each conductive state has been thoroughly described using the following well established conduction mechanisms: charge transfer, conformational change, space charge limit current and filament formation. The filament formation in conduction mechanism is due to anionic (oxygen vacancies) and cationic (electro metallization) migration of ions. Gao et al. 2012 reported hybrid nanocomposite of P3HT-PCBM enabling the realization of reproducible electrical switching which involves metallic filament formation in memory systems (Gao et al. 2012). When a voltage bias is applied to the Cu electrode, mobile copper ions formed by electrochemical oxidation of the Cu top contact can migrate within the active polymer. From our group, we reported role of zinc oxide in the electrical switching of hybrid memory device as shown in Figure 16. These hierarchical ZnO structures (n-type) will be expected to play a significant role in improving the ON/OFF current ratio of device due to large surface area which has a major role in memory switching. Apart from this, the presence of PVK in the nanocomposites is expected to prevent the aggregation of ZnO particles,

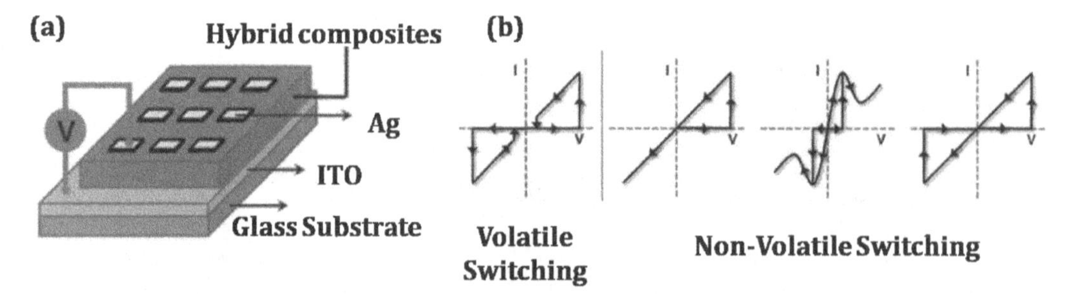

Figure 15. (a) Configuration of hybrid nanocomposite based memory device and (b) Typical I–V curve of resistive memory devices; (I) DRAM, (II) WORM, (III) unipolar and (IV) bipolar switching behaviour.

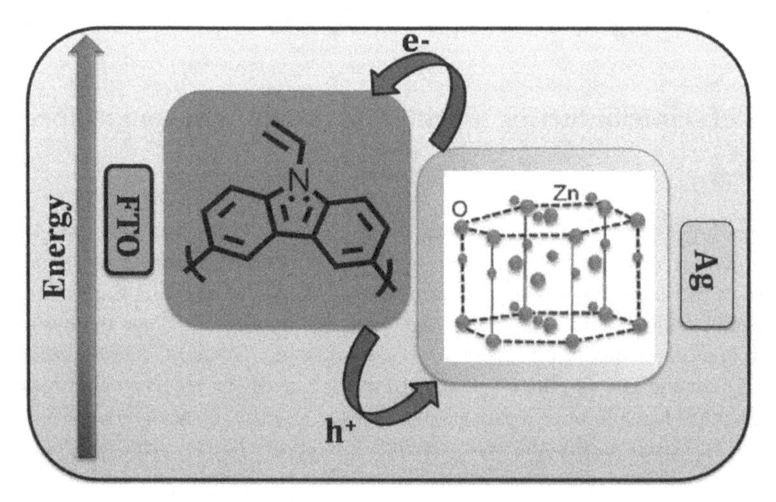

Figure 16. Hybrid composites of polyvinylcarbazole-ZnO based non-volatile memory device [Reprinted with permission from Ref. (Ramakrishnan et al. 2016a)].

reduce the charge loss, current leakage, and increase processability. Also, the formation of p–n junction will be expected to improve the switching and ON/OFF resistance state with narrow sweeping voltage. Studies showed that electrical bistability and conduction mechanism of the devices depends upon the amount of ZnO nanoparticles present in the nanocomposite (Ramakrishnan et al. 2016a).

Thermo electric devices

The direct conversion of electricity from heat source using thermoelectric (TE) materials is an emerging area of research and it offers a promising route towards the development of power generation and refrigeration without moving parts (Tritt et al. 2008, Yu et al. 2011). The energy conversion efficiency of this technology is simple, automatic and eco-friendly. TE devices, quantified by the dimensionless figure-of-merit (ZT), is shown in Eq. (2),

$$ZT = \alpha.\ \sigma.\ T/k \tag{2}$$

where σ is the electrical conductivity, α is the Seebeck coefficient (also called the thermo power), k is the thermal conductivity and T is the absolute temperature. The power factor (PF) is calculated from the measured electrical conductivity and Seebeck coefficient, $PF = \alpha.\sigma^2$. Sketch of a thermoelectric module composed of p-type and n-type hybrid composites of carbon nanotubes coated by PEDOT:PSS particles is shown in Figures 17(a and b).

A high-performance TE material requires a low thermal conductivity to prevent thermal shorting, a high electrical conductivity to reduce Joule heating and a high Seebeck coefficient to promote the energy conversion of heat to electricity or electricity to cooling. Examples of polymers that have been researched for thermoelectric applications are polyacetylene, polypyrrole, polyaniline, polyethylenedioxythiophene, polythiophene, and poly(carbazole)s. It is found that the molecular weights of the polymers have a substantial effect on the electron mobility and consequently affect the electrical conductivity through electron hopping along the polymer backbone (Kline et al. 2003). A high molecular weight polymer will promote the charge carriers to move longer distances before hopping to another chain, and can result in increased charge carrier mobility. The electrical conductivity of device can be tuned by doping; however, it has been found that a higher doping causes decrease in the Seebeck Coefficient. This is because as the Fermi energy is pushed inside the conduction band due to increased number of charge carriers and the number of electronic states above and below, the Fermi energy becomes more equal (Kim et al. 2014, Du et al. 2014, Yao et al. 2010). From our group, we reported hybrid nanocomposites based on ZTO–PEDOT for thermoelectric device application as shown in Figure 18. We have studied the effect of ZTO on the electrical conductivity, carrier concentration and carrier mobility. Finally, we demonstrated its application as a thermoelectric device having configuration Cu/PZTs/Cu and measured Seebeck coefficient, power factor and figure of merit of PZTs. The nanocomposites showed maximum ZT values of 4.8×10^{-3} at 303 K and 1.49×10^{-2} at 383 K (Ramakrishnan et al. 2016b).

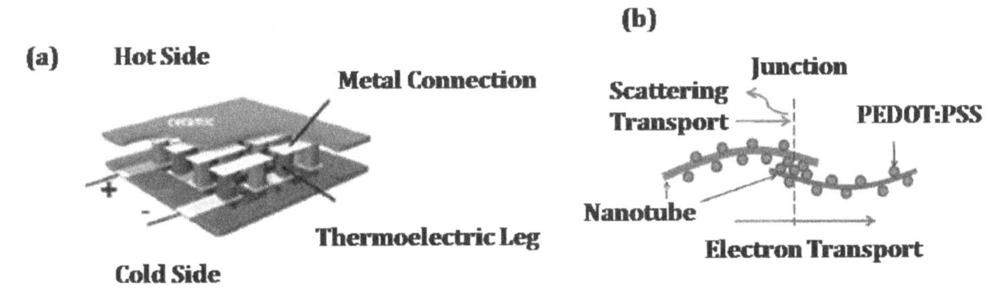

Figure 17. (a) Sketch of a thermoelectric module composed of p-n type legs and (b) hybrid nanocomposites of nanotubes are coated by PEDOT:PSS [Reprinted with permission from Ref. (Yu et al. 2011)].

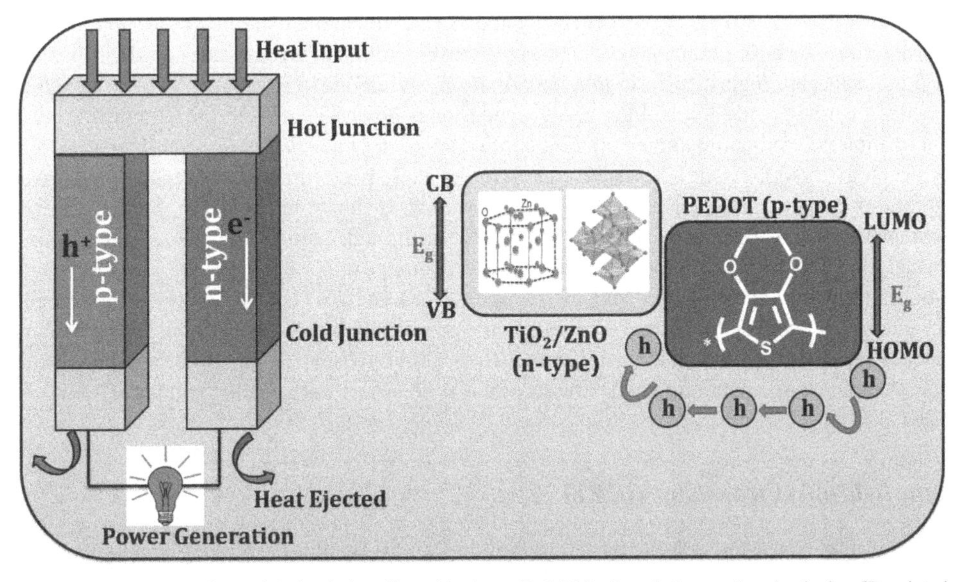

Figure 18. Hybrid composites of Polyethylenedioxythiophene-ZnO/TiO_2 based thermoelectric device [Reprinted with permission from Ref. (Kim et al. 2014)].

Photocatalysts

Photocatalysts are materials which utilize light energy and convert it to chemical energy through a series of electronic processes and surface reactions. Organic dyes are an important source of environmental contamination as they are toxic and mostly non-biodegradable. The key to the success of photocatalyst is the development of high performance materials having well matched photo absorption with the solar spectrum, an efficient photoexcited charge separation to prevent electron–hole recombination and an adequate energy of charges that carry out the desired chemical reactions (Serpone and Emeline 2012). Semiconductor photo-catalysts such as TiO_2 and ZnO are the most studied photocatalysts because of their good stability, low cost, band gap and non-toxicity (Zhang et al. 2007). However, the applications of these metal oxides are limited because of weakly absorbed visible light and they can only collect UV light owing to a large band gap (3.20 eV). Additionally, high recombination efficiency of the photogenerated electron–hole pairs also results in poor photocatalytic property. Therefore, many approaches have been developed to improve the photocatalytic performance of TiO_2 and ZnO under visible light, including non-metal doping, noble metal deposition, semiconductor coupling and conducting polymer sensitized nanoparticles (Ghosh et al. 2015). Among them, conducting polymer sensitized metal nanoparticles have drawn great interest in recent years because metal nanoparticles immobilize on a polymeric support. An ideal photocatalyst should have a narrow bandgap for absorbing solar light, a perfect conjugated structure for fast transfer of charge carriers, good adhesion with the photocatalyst, high adsorption capability toward the reaction species, high chemical inertness, good mechanical stability and a large specific surface area for adsorbing target pollutants. In the past decade, several kinds of conducting polymers, including PANI, P3HT, PPy, PEDOT and others have been used in the preparation of photo-catalysts based on polymer/semiconductor nanocrystal nanocomposites. Upon excitation with light, conducting polymer absorbs light to induce catalyst absorbs light to induce π–π^* electron transition. The excited electrons in the π^*-orbital of CP (lowest unoccupied molecular orbitals, LUMO) activate the adsorbed oxygen molecules to superoxide radicals ($O_2^{\bullet-}$) for oxidizing pollutants. The holes in the highest occupied molecular orbitals (HOMO) of CP can also directly oxidize the pollutants through the formation of hydroxyl radical by interacting with water in the environment. Photocatalytic performance of hybrid material is arising from the effective solar sensitization and suppression of electron-hole recombination (Zhou and Shi 2016, Su et al. 2014).

Recently, we reported photocatalytic properties of rutile titania nanocubes in the visible region which was prepared by a bio-capping strategy and demonstrated its applicability in photocatalytic degradation of dyes (Aashish et al. 2015). The excellent photocatalytic efficiency exhibited by titania is due to the synergistic effect of both photocatalytic and photosensitized oxidation mechanism which is attributed to the low band gap, higher surface area and small crystallite size. Guo et al. 2014 prepared hybrid composites of micro scale hierarchical three-dimensional flowerlike TiO_2/PANI composite by *in situ* polymerization method and demonstrated its application as a photocatalyst for the degradation of dye molecules under both UV and UV-visible light (Guo et al. 2014). Similarly, we reported composites of polyaniline-polytitanate-clay for photocatalytic application as shown in Figure 19. Studies showed that nanocomposites form electrically conductive nanotubes through template/self-assembly process. The photocatalytic effect of nanocomposites towards methyl orange and methylene blue is due to the photo-oxidation and photosensitizing mechanisms between PANI and PHTC. The interface formed between PHTC and PANI transfers the excited electrons to the conduction band of PHTC and hole transferring to the valency band of PANI which in turn increases the yield of hydroxyl, super oxide and positive carbon radicals. A rapid charge separation and slow recombination occurs in the nanocomposite (Ramakrishnan et al. 2012).

Organic field effect transistor (OFET)

Organic thin film transistors are the basic building blocks for flexible integrated circuits and displays. A schematic structure of OFET and representative gate/source current-voltage graph are shown in

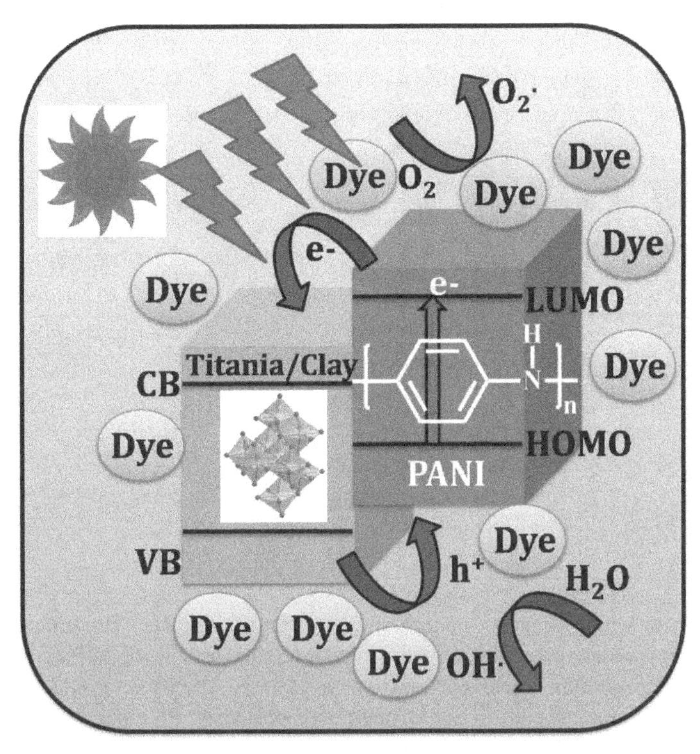

Figure 19. Mechanism for the photodegradation of dye in the presence of PPTC [Reprinted with permission from Ref. (Ramakrishnan et al. 2012)].

Figures 20(a and b). It finds potential applications in organic integrated circuits such as radio frequency identification (RFID) tags, smart cards, and organic active matrix displays (Di et al. 2009). The field effect mobility of OFETs is now comparable to that of devices based on amorphous silicon. Recent attention has been devoted for improving device performance and stability to reduce the power consumption, flexible and lightweight devices for developing simple fabrication techniques compared to typical inorganic semiconducting materials (Torsi et al. 2013). During the operation of the transistor, a gate electrode is used to control the current flow between the drain and source electrodes. Typically, a higher applied gate voltage leads to a higher current flow between the drain and the source electrodes. The current flow between the drain and the source electrode is low when no voltage is applied between the gate and the drain electrodes. This state at which the gate voltage is zero is called the "off" state of a transistor. When a voltage is applied to the gate, charges can be induced into the semiconducting layer at the interface between the semiconductor and dielectric layer. As a result, the drain-source current increases due to increased number of charge carriers, and this is called the "on" state of a transistor. Therefore to construct a FET, the materials ranging from insulating (dielectric material), semiconducting to conducting are required. From the point of a device, its performance is dominated by the properties of its components as well as the properties of the interfaces. In fact, it is well accepted that interface modification is an excellent way to achieve high performance OFETs, since it is an effective approach for improving mobility, device stability, reducing the operating voltage, etc. (Sirringhaus 2014, Fu et al. 2016). Khondaker et al. 2011 fabricated OFETs by directly growing P3HT crystalline nanowires on solution processed aligned array single walled carbon nanotubes (Sarker et al. 2011). The device exhibits high mobility and high current on-off ratio with a maximum of 0.13 cm^2/Vs and 3.1×10^5, respectively. The reason for improvement of device performance can be attributed to improvement of the contact via π-π^* interaction between SWNT with the crystalline P3HT nanowires. Thus, interface engineering has become a general way to fabricate OFETs with excellent device characteristics.

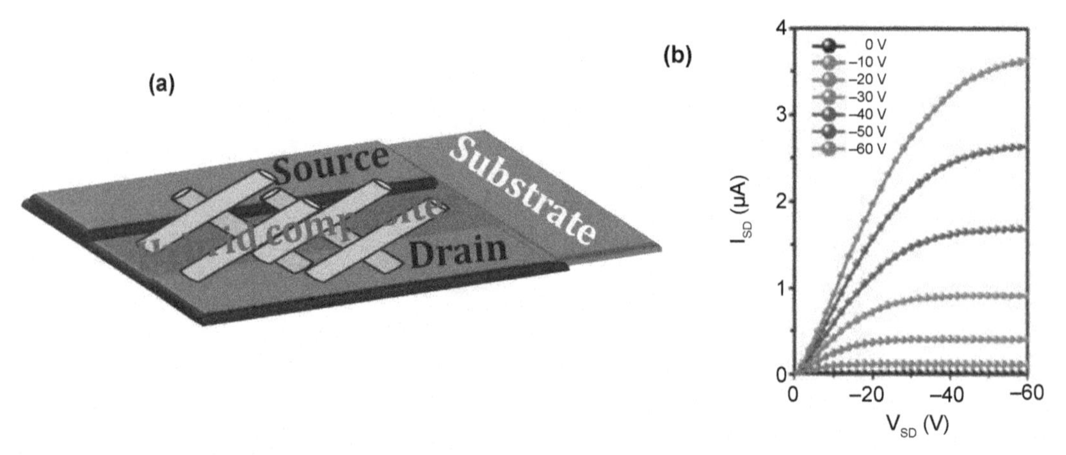

Figure 20. (a) Schematic displays of thin film field effect transistor and (b) representative gate/source current-voltage graph [Reprinted with permission from Ref. (Fu et al. 2016)].

Solar cell

Solar cell is the device which converts the light into electricity. At present, the active materials used for the fabrication of solar cells are mainly inorganic materials, such as silicon, gallium-arsenide, cadmium-telluride, and cadmium-indium-selenide (Kamat et al. 2010). The large production costs for these inorganic materials based solar cells is one of the major obstacles. Polymer solar cells have attracted considerable attention in the past few years owing to their potential for providing environmentally safe flexible, light weight, inexpensive, and efficient solar cells (Jørgensen et al. 2012, Beek et al. 2006, Zeng et al. 2011). Especially, bulk heterojunction solar cells consisting of a mixture of conjugated polymer with a methanofullerne acceptor are considered as a promising approach. In the last five years, there has been an enormous increase in the understanding and performance of polymer-fullerene bulk heterojunction solar cells (Chan et al. 2011). Comprehensive insights have been obtained in crucial material parameters in terms of morphology, energy levels, charge transport, and electrode materials. To date, the power conversion efficiencies close to 5% are routinely obtained and some laboratories have reported power conversion efficiencies of ~ 5–7% and are now aiming at the efficiency to be increased to 10%. By combining synthesis, processing, and materials science with device physics and fabrication, we can improve the performance of the solar cell. In general, for a successful solar cell four important processes have to be optimized to obtain a high conversion efficiency of solar energy into electrical energy: absorption of light, charge transfer, separation of the opposite charges, charge transport and charge collection (Bredas et al. 2009). Schematic diagram showing the bulk heterojunction solar cell and working principle are given in Figures 21(a and b), respectively. For an efficient collection of photons, the absorption spectrum of the photoactive organic layer should match the solar emission spectrum and the layer should be sufficiently thick to absorb all incidents' light. New combinations of materials that are being developed in various laboratories focus on improving the three parameters which determine the energy conversion efficiency of a solar cell, i.e., the open-circuit voltage (V_{oc}), the short-circuit current (J_{sc}), and the fill factor that represents the curvature of the current density-voltage characteristic (Jaegermann et al. 2009). For ohmic contacts, the open-circuit voltage of bulk-heterojunction polymer solar cells is governed by the energy levels of the HOMO and the LUMO of donor and acceptor, respectively. Novel molecular chemistry and materials offer hope for revolutionary, rather than evolutionary, breakthroughs in future device efficiencies.

We reported nanocomposites of polyvinylcarbazole-titania for dye sensitized solar cell application as shown in Figure 22. The hybrid composite was prepared by oxidative polymerization of vinylcarbazole and titania. Finally, the prepared nanocomposites were used as photoanode for the fabrication of DSSC and observed maximum power conversion efficiency of 3.03%.

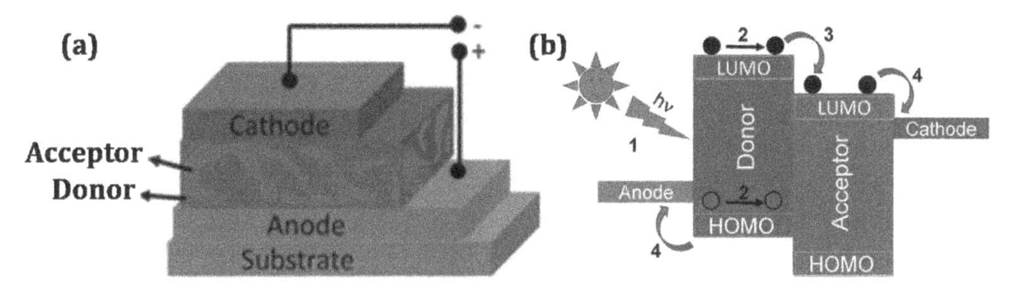

Figure 21. (a) Schematic diagram of bulk heterojunction solar cell and (b) working principle of solar cell [Reprinted with permission from Ref. (Beek et al. 2006)].

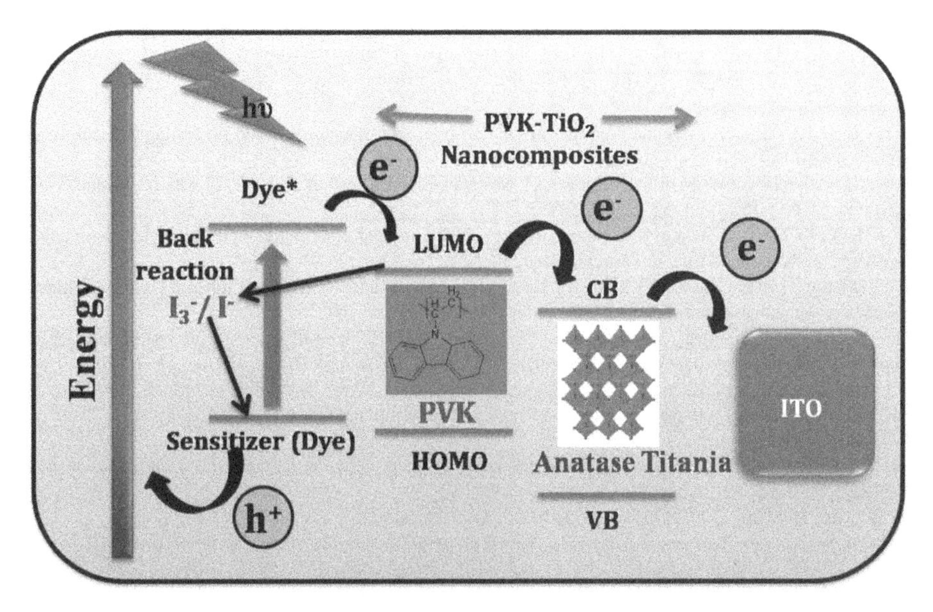

Figure 22. Schematic hybrid composite of polyvinylcarbazole-titania for dye sensitized solar cell application [Reprinted with permission from Ref. (Aashish et al. 2016)].

References

Aashish, A., R. Ramakrishnan, J.D. Sudha, M. Sankaran and G.K. Priya. 2015. Nanocubes of rutile titania for enhanced visible light photocatalytic applications. Mater. Chem. Phys. 157: 31–38.

Aashish, A., R. Ramakrishnan, J.D. Sudha, M. Sankaran and G. Krishnapriya. 2016. Self-assembled hybridpolyvinylcarbazole–titania nanotubes as an efficient photoanodeforsolarenergyharvesting. Sol. Energ. Mat. Sol. Cells 151: 169–178.

Aernandez, M.T.F., A. Yakovlev, R.A. Sperling, C. Luccardini, S. Gaillard, A.S. Medel. 2007. Synthesis and characterization of polymer-coated quantum dots with integrated acceptor dyes as FRET-based nanoprobes. Nano Lett. 7: 2613–2617.

Agranovich, V.M., Yu.N. Gartstein and M. Litinskaya. 2011. Hybrid resonant organic-inorganic nanostructures for optoelectronic applications. Chem. Rev. 111: 5179–5214.

Alivisatos, A.P. 1996. Perspectives on the physical chemistry of semiconductor nanocrystals. J. Phys. Chem. 100: 13226–13239.

Allam, N.K. and C.A. Grimes. 2007. Formation of vertically oriented TiO_2 nanotube arrays using a fluoride free HCl aqueous electrolyte. J. Phys. Chem. C. 111: 13028–13032.

Bae, S., S.W. Lee and Y. Takemura. 2006. Applications of $NiFe_2O_4$ nanoparticles for a hyperthermia agent in biomedicine. Appl. Phys. Lett, 89: 252503–252508.

Bak, T., M.K. Nowotny, L.R. Sheppard and J. Nowotny. 2008. Effect of prolonged oxidation on semiconducting properties of titanium dioxide. J. Phys. Chem. C. 112: 13248–13257.

Barlier, V., V.B. Legare, G. Boiteux, J. Davenas, A. Slazak, A. Rybak et al. 2009. Photo generation and photovoltaic effect in poly(N-vinylcarbazole):TiO_2 bulk-heterojunction elaborated by hydrolysis–condensation reactions of TiO_2 precursors. Synth. Metals 159: 508–512.

Baxter, J.B. and E.S. Aydil. 2005. Nanowire-based dye-sensitized solar cells. Appl. Phys. Lett. 86: 053114–053117.
Beek, W.J.E., M.M. Wienk and R.A.J. Janssen. 2006. Hybrid solar cells from regioregular polythiophene and ZnO nanoparticles. Adv. Funct. Mater. 16: 1112–1116.
Bhadraa, S., D. Khastgir, N.K. Singha and J.H. Lee. 2009. Progress in preparation, processing and applications of polyaniline. Prog. Polym. Sci. 34: 783–810.
Bredas, J.L., J.E. Norton, J. Cornil and V. Coropceanu. 2009. Molecular understanding of organic solar cells: The challenges. Acc. Chem. Res. 42: 1691–1699.
Brinker, C.J., Y. Lu, A. Sellinger and H. Fan. 1999. Evaporation-induced self-assembly: Nanostructures made easy. Adv. Mater. 11: 579–585.
Burda, C., X.B. Chen, R. Narayanan and M.A.E. Sayed. 2005. Chemistry and properties of nanocrystals of different shapes. Chem. Rev. 105: 1025–1102.
Camurlu, P. 2014. Polypyrrole derivatives for electrochromic applications. RSC Adv. 4: 55832–55845.
Chan, S.H., C.S. Lai, H.L. Chen, C. Ting and C.P. Chen. 2011. Highly efficient P_3HT: C_{60} solar cell free of annealing process. Macromolecules 44: 8886–8891.
Chen, H., R.C. McDonald, S. Li, N.L. Krett, S.T. Rosen and T.V. O'Halloran. 2006. Lipid encapsulation of arsenic trioxide attenuates cytotoxicity and allows for controlled anticancer drug release. J. Am. Chem. Soc. 128: 13348.
Chen, J.S., Y.L. Tan, C.M. Li, Y.L. Cheah, D. Luan, S. Madhavi et al. 2010. Constructing hierarchical spheres from large ultrathin anatase TiO_2 nanosheets with nearly 100% exposed (001) facets for fast reversible lithium storage. J. Am. Chem. Soc. 132: 6124–6130.
Chen, S., Y. Lia and Y. Li. 2013. Architecture of low dimensional nanostructures based on conjugated polymers. Polym. Chem. 4: 5162–5180.
Chiang, C.K., C.R. Fincher, Y.W. Park, A.J. Heeger, H. Shirakawa, E.J. Louis et al. 1977. Electrical conductivity in doped polyacetylene. Phys. Rev. Lett. 39: 1098–1101.
Cho, B., T.W. Kim, M. Choe, G. Wang, S. Song and T. Lee. 2009. Unipolar nonvolatile memory devices with composites of poly(9-vinylcarbazole) and titanium dioxide nanoparticles. Org. Electron. 10: 473–477.
Chou, T.P., Q. Zhang, G.E. Fryxell and G. Cao. 2007. Hierarchically structured ZnO film for dye-sensitized solar cells with enhanced energy conversion efficiency. Adv. Mater. 19: 2588–2592.
Colodrero, S., A. Mihi, L. Haggman, M. Ocana, G.A. Boschloo, H. Hagfeldt et al. 2009. Porous one-dimensional photonic crystals improve the power-conversion efficiency of dye-sensitized solar cells. Adv. Mater. 21: 764–770.
Devaki, S.J., N.K. Sadanandhan, R. Sasi, H.J.P. Adler and A. Pich. 2014. Water dispersible electrically conductive poly(3,4-ethylenedioxythiophene) nanospindles by liquid crystalline template assisted polymerization. J. Mater. Chem. C. 2: 6991–7000.
Di, C.A., Y. Liu, G. Yu and D. Zhu. 2009. Interface engineering: An effective approach toward high-performance organic field-effect transistors. Acc. Chem. Res. 42: 1573–1583.
Djurisic, A.B. and Y.H. Leung. 2006. Optical properties of ZnO nanostructures. Small 2: 944–961.
Du, Y., K.F. Cai, S. Chen, P. Cizek and T. Lin. 2014. Facile preparation and thermoelectric properties of Bi_2Te_3 based alloy nanosheet/PEDOT:PSS composite films. ACS Appl. Mater. Interfaces 6: 5735–5743.
Fu, B., J. Baltazar, Z. Hu, A.T. Chien, S. Kumar, C.L. Henderson et al. 2016. Ordering of poly(3-hexylthiophene) in solutions and films: Effects of fiber length and grain boundaries on anisotropy and mobility. Chem. Mater. 28: 3905–3913.
Gao, S., C. Song, C. Chen, F. Zeng and F. Pan. 2012. Dynamic processes of resistive switching in metallic filament-based organic memory devices. J. Phys. Chem C. 116: 17955–17959.
Ghosh, S., N.A. Kouame, S. Remita, L. Ramos, F. Goubard, P.H. Aubert et al. 2015. Visible-light active conducting polymer nanostructures with superior photocatalytic activity. Sci. Rep. 5: 18002–18011.
Guillen, E., E. Azaceta, A.V. Poot, J. Idígoras, J. Echeberría, J.A. Anta et al. 2013. ZnO/ZnO core–shell nanowire array electrodes: Blocking of recombination and impressive enhancement of photo voltage in dye-sensitized solar cells. J. Phys. Chem. C. 117: 13365–13373.
Guo, N., Y. Liang, S. Lan, L. Liu, J. Zhang, G. Ji et al. 2014. Microscale hierarchical three-dimensional flowerlike TiO_2/PANI composite: Synthesis, characterization, and its remarkable photocatalytic activity on organic dyes under UV-Light and sunlight irradiation. J. Phys. Chem. C. 118: 18343–18355.
Halaoui, L.I., N.M. Abrams and T.E. Mallouk. 2005. Increasing the conversion efficiency of dye-sensitized TiO_2 photoelectrochemical cells by coupling to photonic crystals. J. Phys. Chem. B. 109: 6334–6342.
He, C.X., B.X. Lei, Y.F. Wang, C.Y. Su, Y.P. Fang and D.B. Kuang. 2010. Sonochemical preparation of hierarchical ZnO hollow spheres for efficient dye-sensitized solar cells. Chem. Eur. J. 16: 8757–8761.
He, M., F. Qiu and Z. Lin. 2013. Toward high-performance organic–inorganic hybrid solar cells: Bringing conjugated polymers and inorganic nanocrystals in close contact. J. Phys. Chem. Lett. 4: 1788–1796.
Heremans, P., G.H. Gelinck, R. Muller, K.J. Baeg, D.Y. Kim and Y.Y. Noh. 2011. Polymer and organic nonvolatile memory devices. Chem. Mater. 23: 341–358.
Holder, E., N. Tesslerb and A.L. Rogachc. 2008. Hybrid nanocomposite materials with organic and inorganic components for opto-electronic devices. J. Mater. Chem. 18: 1064–1078.
Hong, Y.L., X.S. Chen, X.B. Jing, H.S. Fan, B. Guo, Z.W. Gu et al. 2010. Preparation, bioactivity, and drug release of hierarchical nanoporous bioactive glass ultrathin fibers. Adv. Mater. 22: 754–758.

Hu, J.T., T.W. Odom and C.M. Lieber. 1999. Chemistry and physics in one dimension: Synthesis and properties of nanowires and nanotubes. Acc. Chem. Res. 32: 435–445.
Hu, X.L., Y.J. Zhu and S.W. Wang. 2004. Sonochemical and microwave-assisted synthesis of linked single-crystalline ZnO rods. Mater. Chem. Phys. 88: 421–426.
Huang, J. and R.B. Kaner. 2004. A general chemical route to polyaniline nanofibers. J. Am. Chem. Soc. 126: 851–855.
Huang, W.S., B.D. Humphery and A.G. MacDiarmid. 1986. Polyaniline, a novel conducting polymer morphology and chemistry of its oxidation and reduction in aqueous electrolytes. J. Chem. Soc. Faraday Trans. 82: 2385–2400.
Huang, Y., X.F. Duan, Q.Q. Wei and C.M. Lieber. 2001. Directed assembly of one-dimensional nanostructures into functional networks. Science 291: 630–633.
Jaegermann, W., A. Klein and T. Mayer. 2009. Interface engineering of inorganic thin-film solar cells—materials-science challenges for advanced physical concepts. Adv. Mater. 21: 4196–4206.
Janáky, C. and K. Rajeshwar. 2015. The role of (photo) electrochemistry in the rational design of hybrid conducting polymer/semiconductor assemblies: From fundamental concepts to practical applications. Prog. Polym. Sci. 43: 96–135.
Janotti, A. and C.G. Van de Walle. 2009. Fundamentals of zinc oxide as a semiconductor. Rep. Prog. Phys. 72: 126501–126530.
Jing, L., W. Zhou, G. Tian and H. Fu. 2013. Surface tuning for oxide-based nanomaterials as efficient photocatalysts. Chem. Soc. Rev. 42: 9509–9549.
Jokanovic, V., A.M. Spasic and D. Uskokovic. 2004. Designing of nanostructured hollow TiO_2 spheres obtained by ultrasonic spray pyrolysis. J. Colloid. Interface Sci. 278: 342–352.
Jørgensen, M., K. Norrman, S.A. Gevorgyan, T. Tromholt, B. Andreasen and F.C. Krebs. 2012. Stability of polymer solar cells. Adv. Mater. 24: 580–612.
Kamat, P.V., K. Tvrdy, D.R. Baker and J.G. Radich. 2010. Beyond photovoltaics: Semiconductor nanoarchitectures for liquid-junction solar cells. Chem. Rev. 110: 6664–6688.
Kaneko, H., T. Ishiguro, A. Takahashi and J. Tsukamoto. 1993. Magneto resistance and thermoelectric power studies of metal-nonmetal transition in iodine-doped polyacetylene. Synth. Met. 57: 4900–4905.
Kao, J., K. Thorkelsson, P. Bai, B.J. Rancatore and T. Xu. 2013. Toward functional nanocomposites: taking the best of nanoparticles, polymers, and small molecules. Chem. Soc. Rev. 42: 2654–2678.
Kaur, G., R. Adhikari, P. Cass, M. Bown and P. Gunatillake. 2015. Electrically conductive polymers and composites for biomedical applications. RSC Adv. 5: 37553–37567.
Kim, F.S., G. Ren and S.A. Jenekhe. 2011. One-dimensional nanostructures of π-conjugated molecular systems: Assembly, properties, and applications from photovoltaics, sensors, and nano-photonics to nano-electronics. Chem. Mater. 23: 682–732.
Kim, G.H., L. Shao, K. Zhang and K.P. Pipe. 2014. Engineered doping of organic semiconductors for enhanced thermoelectric efficiency. Nat. Mater. 12: 719–722.
Kline, R.J., M.D. McGehee, E.N. Kadnikova, J. Liu and J.M.J. Frechet. 2003. Controlling the field-effect mobility of regioregular polythiophene by changing the molecular weight. Adv. Mater. 15: 1519–1522.
Lai, X., J.E. Halpert and D. Wang. 2012. Recent advances in micro-/nano-structured hollow spheres for energy applications: From simple to complex systems. Energy Environ. Sci. 5: 5604–5618.
Law, M., L.E. Greene, J.C. Johnson, R. Saykally and P.D. Yang. 2005. Nanowire dye-sensitized solar cells. Nat. Mater. 4: 455–459.
Lee, J.I., S.H. Cho, S.M. Park, J.K. Kim, J.K. Kim, J.W. Yu et al. 2008. Highly aligned ultrahigh density arrays of conducting polymer nanorods using block copolymer templates. Nano Lett. 8: 2315–2320.
Li, X.L., Q. Peng, J.X. Yi, X. Wang and Y.D. Li. 2006. Near monodisperse TiO_2 nanoparticles and nanorods. Chem. Eur. J. 12: 2383–2391.
Liao, Y., X.G. Li and R.B. Kaner. 2010. Facile synthesis of water-dispersible conducting polymer nanospheres. ACS Nano 4: 5193–5202.
Liu, R. and S.B. Lee. 2008. MnO_2/Poly(3,4-ethylenedioxythiophene) coaxial nanowires by one-step coelectrodeposition for electrochemical energy storage. J. Am. Chem. Soc. 130: 2942–2943.
Lu, Z.L., E. Lindner and H.A. Mayer. 2002. Applications of sol–gel-processed interphase catalysts. Chem. Rev. 102: 3543–3578.
Martin, C.R. 1995. Template synthesis of electronically conductive polymer nanostructures. Acc. Chem. Res. 28: 61–68.
Nishimura, S., N. Abrams, B.A. Lewis, L.I. Halaoui, T.E. Mallouk, K.D. Benkstein. 2003. Standing wave enhancement of red absorbance and photocurrent in dye-sensitized titanium dioxide photoelectrodes coupled to photonic crystals. J. Am. Chem. Soc. 125: 6306–6310.
Ozin, G.A., A.C. Arsenault and L. Cademartiri. 2008. Nanochemistry: A Chemical Approach to Nanomaterials, RSC, Cambridge.
Pan, X., M.Q. Yang, X. Fu, N. Zhang and Y.J. Xu. 2013. Defective TiO_2 with oxygen vacancies: synthesis, properties and photocatalytic applications. Nanoscale 5: 3601–3614.
Patra, C.R. and A. Gedanken. 2004. Rapid synthesis of nanoparticles of hexagonal type In_2O_3 and spherical type Tl_2O_3 by microwave irradiation. New J. Chem. 28: 1060–1065.
Paulose, M., K. Shankar, O.K. Varghese, G.K. Mor and C.A. Grimes. 2006. Application of highly-ordered TiO_2 nanotube-arrays in heterojunction dye-sensitized solar cells. J. Phys. D: Appl. Phys. 39: 2498–2503.
Petkovich, N.D. and A. Stein. 2013. Controlling macro- and meso-structures with hierarchical porosity through combined hard and soft templating. Chem. Soc. Rev. 42: 3721–3739.

Poizot, P., S. Laruelle, S. Grugeon, L. Dupont and J.M. Tarascon. 2000. Nano-sized transition-metal oxides as negative-electrode materials for lithium-ion batteries. Nature 407: 496–498.
Ramakrishnan, R., J.D. Sudha and V.L. Reena. 2012. Nanostructured polyaniline-polytitanate-clay composite for photocatalytic applications: preparation and properties. RSC Advances 2: 6228–6236.
Ramakrishnan, R., K.B. Jinesh, S.J. Devaki and M.R. Varma. 2016a. Facile strategy for the fabrication of efficient nonvolatile bistable memory devices based on polyvinylcarbazole—zinc oxide. Phys. Status Solidi A 213: 2414–2424.
Ramakrishnan, R., S.J. Devaki, A. Aashish, S. Thomas, M.R. Varma and K.P.P. Najiya. 2016b. Nanostructured semiconducting PEDOT−TiO_2/ZnO hybrid composites for nanodevice applications. J. Phys. Chem. C 120: 4199–4210.
Ramanathan, M., L.K. Shrestha, T. Mori, Q. Ji, J.P. Hill and K. Ariga. 2013. Amphiphile nano-architectonics: from basic physical chemistry to advanced applications. Phys.Chem. Chem. Phys. 15: 10580–10611.
Reiss, P., E. Couderc, J.D. Girolamo and A. Pron. 2011. Conjugated polymers/semiconductor nanocrystals hybrid materials-preparation, electrical transport properties and applications. Nanoscale 3: 446–489.
Romero, P.G. 2001. Hybrid organic-inorganic materials in search of synergic activity. Adv. Mater. 13: 163–174.
Sarker, B.K., J. Liu, L. Zhai and S.I. Khondaker. 2011. Fabrication of organic field effect transistor by directly grown poly(3 Hexylthiophene) crystalline nanowires on carbon nanotube aligned array electrode. ACS Appl. Mater. Interfaces 3: 1180–1185.
Serpone, N. and A.V. Emeline. 2012. Semiconductor photocatalysis—past, present, and future outlook. J. Phys. Chem. Lett. 3: 673–677.
Shen, Y., Y. Lin, M. Li and C.W. Nan. 2007. High dielectric performance of polymer composite films induced by a percolating interparticle barrier layer. Adv. Mater. 19: 1418–1422.
Sirringhaus, H. 2014. Organic field-effect transistors: The path beyond amorphous silicon. Adv. Mater. 26: 1319–1335.
Song. E. and J.W. Choi. 2013. Conducting polyaniline nanowire and its applications in chemiresistive sensing. Nanomaterials 3: 498–523.
Stoeva, S.I., J.S. Lee, J.E. Smith and C.A. Mirkin. 2006. Multiplexed detection of protein cancer markers with biobar-coded nanoparticle probes. J. Am. Chem. Soc. 128: 8378–8379.
Su, Y.W., W.H. Lin, Y.J. Hsu and K.H. Wei. 2014. Conjugated polymer/nanocrystal nanocomposites for renewable energy applications in photovoltaics and photocatalysis. Small 10: 4427–4442.
Suzuki, N., H. Tanaka and T. Kawai. 2008. Epitaxial transition metal oxide nanostructures fabricated by a combination of AFM lithography and molybdenum lift-off. Adv. Mater. 20: 909–913.
Tian, J., L. Lv, X. Wang, C. Fei, X. Liu, Z. Zhao et al. 2014. Micro sphere light-scattering layer assembled by ZnO nanosheets for the construction of high efficiency (> 5%) quantum dots sensitized solar cells. J. Phys. Chem. C 118: 16611–16617.
Tian, Z.R.R., J.A. Voigt, J. Liu, B. Mckenzie, M.J. Mcdermott, M.A. Rodriguez et al. 2003. Complex and oriented ZnO nanostructures. Nat. Mater. 2: 821–826.
Torsi, L., M. Magliulo, K. Manoli and G. Palazzo. 2013. Organic field-effect transistor sensors: a tutorial review. Chem. Soc. Rev. 42: 8612–8628
Tritt, T.M., H. Boettner and L. Chen. 2008. MRS Bull. 33: 366.
Virji, S., R.B. Kaner and B.H. Weiller. 2006. Hydrogen sensors based on conductivity changes in polyaniline nanofibers. J. Phys. Chem. B. 110: 22266–22270.
Wan, M. 2008. A template-free method towards conducting polymer nanostructures. Adv. Mater. 20: 2926–2932.
Wang, D., F. Qian, C. Yang, Z.H. Zhong and C.M. Lieber. 2004. Rational growth of branched and hyper branched nanowire structures. Nano Lett. 4: 871–874.
Wang, H., C.T. Yip, K.Y. Cheung, A.B. Djurisic, M.H. Xie, Y.H. Leung et al. 2006. Titania-nanotube-array-based photovoltaic cells. Appl. Phys. Lett. 89: 023508–023511.
Wood, V., M.J. Panzer, J.E. Halpert, J.M. Caruge, M.G. Bawendi and V. Bulovic. 2009. Selection of metal oxide charge transport layers for colloidal quantum dot LEDs. Nano Lett. 3: 3581–3586.
Wu, J.M., H.C. Shih and W.T. Wu. 2005. Electron field emission from single crystalline TiO_2 nanowires prepared by thermal evaporation. Chem. Phys. Lett. 413: 490–494.
Xia, J., H. Li, Z. Luo, H. Shi, K. Wang, H. Shu et al. 2009. Microwave-assisted synthesis of flower-like and leaf-like CuO nanostructures via room-temperature ionic liquids. J. Phys. Chem. Solids 70: 1461–1464.
Yang, J., S. Mei and J.M.F. Ferreira. 2000. Hydrothermal synthesis of nanosized titania powders: Influence of peptization and peptizing agents on the crystalline phases and phase transitions. J. Am. Ceram. Soc. 83: 1361–1368.
Yao, Q., L. Chen, W. Zhang, S. Liufu and X. Chen. 2010. Enhanced thermoelectric performance of single-walled carbon nanotubes/polyaniline hybrid nanocomposites. ACS Nano 4: 2445–2451.
Ye, P.D., B. Yang, K.K. Ng, J. Bude, G.D. Wilk, S. Halder et al. 2005. GaN metal-oxide-semiconductor high electron mobility transistor with atomic layer deposited Al_2O_3 as gate dielectric. Appl. Phys. Lett. 86: 063501.
Yu, C., K. Choi, L. Yin and J.C. Grunlan. 2011. Light-weight flexible carbon nanotube based organic composites with large thermoelectric power factors. ACS Nano 5: 7885–7892.
Yu, P., X. Zhang, Y. Chen, Y. Ma and Z. Qi. 2009. Preparation and pseudo-capacitance of birnessite-type MnO_2 nanostructures via microwave-assisted emulsion method. Mater. Chem. Phys. 118: 303–307.
Zeng, T.W., C.C. Ho, Y.C. Tu, G.Y. Tu, L.Y. Wang and W.F. Su. 2011. Correlating interface heterostructure, charge recombination, and device efficiency of poly(3-hexyl thiophene)/TiO_2 nanorod solar cell. Langmuir 27: 15255–15260.

Zhang, H., G.R. Li, L.P. An, T.Y. Yan, X.P. Gao and H.Y. Zhu. 2007. Electrochemical lithium storage of titanate and titania nanotubes and nanorods. J. Phys. Chem. C 111: 6143–6148.
Zhang, Q., T.P. Chou, B. Russo, S.A. Jenekhe and G. Cao. 2008. Polydisperse aggregates of ZnO nanocrystallites: A method for energy-conversion-efficiency enhancement in dye-sensitized solar cells. Adv. Funct. Mater. 18: 1654–1660.
Zhang, Q., C.S. Dandeneau, X. Zhou and G. Cao. 2009. ZnO nanostructures for dye-sensitized solar cells. Adv. Mater. 21: 4087–4108.
Zhang, Q., J. Pan, X. Yi, L. Li and S. Shang. 2012. Nonvolatile memory devices based on electrical conductance tuning in poly(N-vinylcarbazole)–graphene composites. Org. Electron. 13: 1289–1295.
Zhou, Q. and G. Shi. 2016. Conducting polymer-based catalysts. J. Am. Chem. Soc. 138: 2868–5876.
Zhou, Y.F., M. Eck and M. Kruger. 2010. Bulk-heterojunction hybrid solar cells based on colloidal nano-crystals and conjugated polymers. Energy Environ. Sci. 3: 1851–1864.

18

Bismuth Vanadate Based Nanostructured and Nanocomposite Photocatalyst Materials for Water Splitting Application

S. Moscow and *K. Jothivenkatachalam**

Introduction

Ever-increasing environmental issues and consumption of fossil fuels have stimulated extensive research on the utilization of sustainable solar energy (Hoffmann et al. 1995). As the global consumption of fossil fuels grows at an alarming, unsustainable rate has led to the generation of two major problems such as environmental, and energy disputes, which are challenges facing the world in the 21st century. According to Annual Energy Outlook 2016, the energy consumption increases drastically every year and the energy consumption in 2040 will increase nearly two fold from year 2015 (Figure 1) (EIA US 2016).

Achieving this in an economically and environmentally benign manner is important for the conservation of the global atmosphere. The future sustainable development of society relies on alternative energy sources that are renewable and environment friendly. Among various renewable energy resources, solar energy represents the ultimate renewable source because the sun is our largest and cheapest (free) energy resource available. It continuously bombards our planet with solar energy, with 1 h of solar energy equating to more than all of our annual energy consumption (Vayssieries 2009). Investigations on clean energy technology help the researchers to exploit solar energy efficiently and economically to generate, convert, and store electricity (Mao et al. 2007). Solar energy is the major source of renewable energy and also a promising source for renewable hydrogen production.

Department of Chemistry, Anna University – BIT campus, Tiruchirarappalli, Tamil nadu 620024, India.
Email: moscowchem@gmail.com
* Corresponding author: jothivenkat@yahoo.com

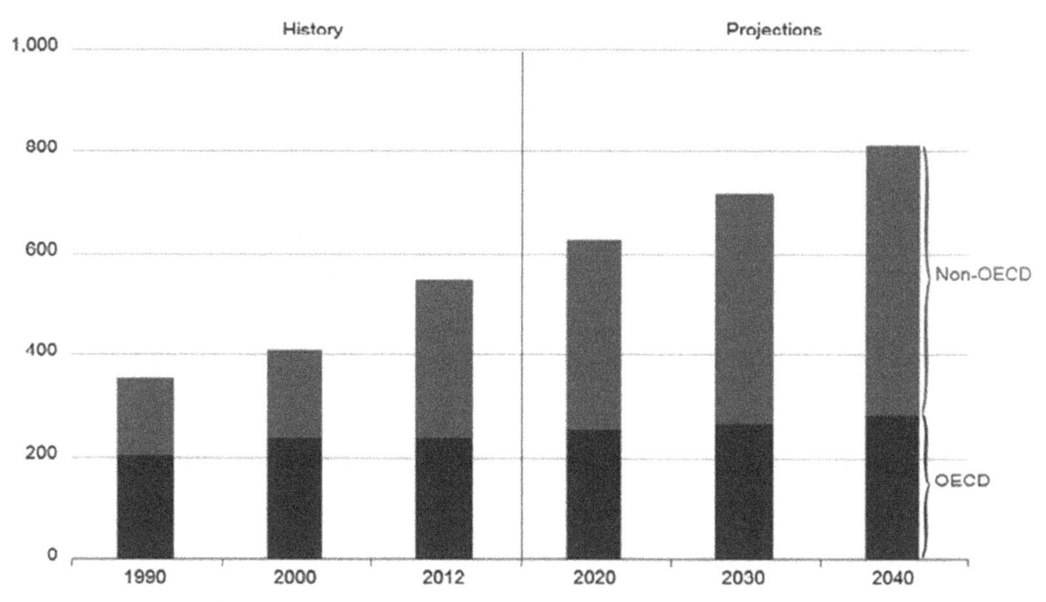

Figure 1. World energy consumption (source: https://www.eia.gov/outlooks/ieo/pdf/0484 (2016)).

Hydrogen is considered as an ideal fuel for the future since it can be produced from clean and renewable energy sources. However, presently, renewable energy contributes only about 5% of the commercial hydrogen production primarily via water electrolysis, while other 95% hydrogen is mainly derived from fossil fuels (Ni et al. 2004). The nanostructured and nanocomposite photocatalytic material offers a promising way for clean, low-cost and environment friendly production of hydrogen by solar energy. Generation of hydrogen (H_2)/oxygen (O_2) by splitting water and using hydrogen as fuel with reduced CO_2 emission conserve the environment. Choosing an appropriate photocatalytic material is the primary requisite to absorb solar energy and to convert it to other useful energies. Most of the conventional photoactive materials are selective in their light absorption, thus limiting the usage of the entire solar spectrum, whereas nanostructured and nanocomposite are more suitable photoactive materials which may fulfil the requirements for photoenergy conversion. There are various studies indicative of surface chemical functionality, local structure and morphological characteristics of catalysts affecting the photocatalytic activity of the water splitting reaction. Surface chemical functionality is modified to protect against corrosion, deactivate destructive surface states, tailor band-edge positions, or selectively extract carriers to improve catalytic activity (Sivula et al. 2016). This chapter concludes with perspectives and an outlook for future efforts aimed at solar water splitting using $BiVO_4$ nanostructured and nanocomposite materials. The realization of practical, efficient and significant development in materials synthesis is useful for water splitting. This chapter is intended to motivate such development in splitting water into H_2 and O_2 using solar light that has drawn wide attention due to the abundance of resources in nature.

Basic principles of photocatalysis

Solar-based water splitting methods and plants can convert and store solar energy as complex molecules, such as carbohydrates and other biomass; however, photosynthesis is not very efficient, and the stored energy in plants and the large amounts of arable land needed to grow them compete with other land uses such as food production. Storing solar energy by splitting water into hydrogen and oxygen has long been considered a promising idea and coal recognized as a possible source to finite supply of hydrogen from electrolysis (Verne 1874). However, there are no economical methods for fulfilling this prediction, partially because of problems of efficiently storing and handling hydrogen. In addition to generating the required H_2 from photocatalytic water splitting, researchers have devised and tested a number of thermochemical cycles that use high temperatures generated in focusing solar concentrators to split water. The discovery of water splitting was announced by Fujishima and Honda (Fujishima and Honda 1972). Photocatalysis is a possible process but achieving considerable effectiveness is still a challenge for the researchers. Moreover, nano sized photocatalyst systems are more advantageous in the area of solar water splitting because of the simplicity and efficiency. So, photocatalytic water splitting is an attractive and ultimate green sustainable technique to solve energy issues resulting in bringing an energy revolution.

The photon energy is converted to chemical energy accompanied by a largely positive change in the Gibbs free energy through water splitting as shown in Figure 2. This reaction is similar to photosynthesis by green plants because these are difficult reactions. Therefore, photocatalytic water splitting is regarded as an artificial photosynthesis and is an attractive and challenging thing in chemistry (Fujishima et al. 2007). The basic processes involved in water splitting are shown in Figure 3. There are three important steps involved in semiconductor photocatalysis: (a) photocatalyst material absorbs photon having energy greater than its band gap and utilizes it to excite an electron from the valence band (VB) of semiconductor to the conduction band (CB) and form an electron/hole (e^-/h^+) pair. (b) Second step involves charge carrier separation into e^-/h^+, and their migration to the surface in order redox reactions should take place. These are the two critical factors in deciding the efficiency of the process because recombination of excited electron and hole pair can occur in femtoseconds in bulk.

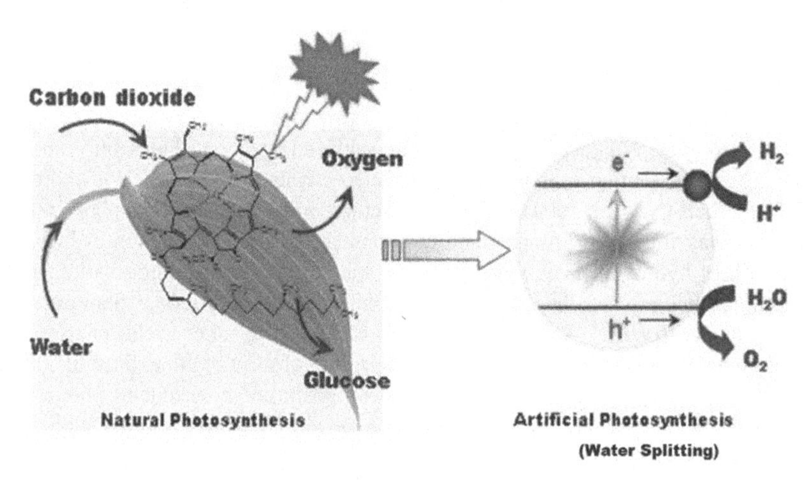

Figure 2. Natural photosynthesis and artificial photosynthesis for photocatalytic water splitting. Source: http://wanglab.fzu.edu.cn/html/RESEARCH/1.html.

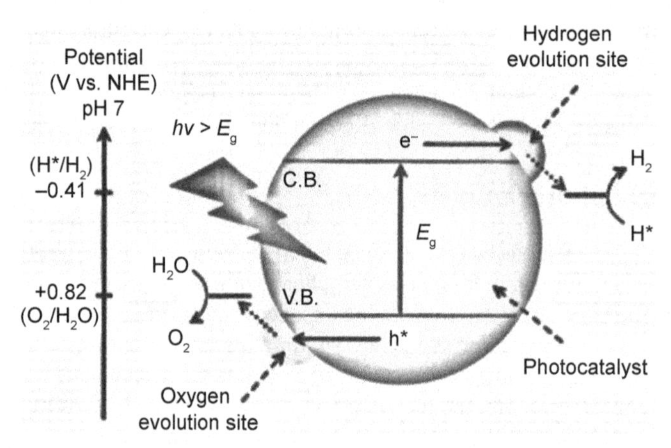

Figure 3. Reaction processes of water splitting on a heterogeneous photocatalyst. Reprinted from ref. 34 Maeda, K. and K. Domen. 2010. J. Phys. Chem. Lett. 18: 2655–2661. Copyright © 2010, American Chemical Society.

Electron migration from bulk to surface without being quenched is assisted by electronic structure of semiconductor, co-catalyst, particle size, crystallinity and other morphological factors in providing appropriate diffusion length of charge carriers (Kudo and Miseki 2009). Surface chemical redox reactions is the final step in which charge carriers are used for the desired purpose, such as the conversion of water molecules into hydrogen and oxygen.

Semiconductors based heterogeneous photocatalysis

The development of new environmental photocatalysts is a great challenge all over the world for researchers, engineers and scientists. Nowadays, many advanced technologies are offered for resolving both the environmental and energy problems. The semiconductors' heterogeneous materials have received more attention and are considered the most potential technique (Qu and Duan 2013). In the past decades, many research groups have investigated the applications of photocatalysis to produce hydrogen from water (Kibria et al. 2013, Xiang et al. 2012, Christians et al. 2014), convert solar energy into electric energy (Zhang et al. 2013) and reduce CO_2 into organic fuels (Roy et al. 2010, Dhakshinamoorthy et al. 2012). Among the many advanced technologies available today, heterogenous semiconductor photocatalyst taking advantage for deficiently utilize solar energy has been identified as one of the most prospective strategies for resolving both the environmental and energy problems and has thus attracted much attention during the recent decades (Li et al. 2014, Jothivenkatachalam et al. 2014). The naturally abundant renewable energy sources such as solar energy can be renewed into chemical or electrical and thermal energies by using semiconductors having persist materials in the process of photocatalysis (Linic et al. 2011, Zhou et al. 2012).

Various semiconductor band edge potentials at NHE scale are shown in Figure 4. Most wide band gap semiconductors are able to provide suitable redox potential for water splitting reaction. However, semiconductors necessarily have to be active in visible light absorption, charge carrier separation, and migration to surfaces; photo- and hydro-stability of the photocatalyst are major issues for sustainable H_2 production. For hydrogen production, the CB level should be more negative than hydrogen production level ($E_{H2/H2O}$) while the VB should be more positive than water oxidation level ($E_{O2/H2O}$) for efficient oxygen production from water by photocatalysis. Band structure of the semiconductor should go hand in hand in deciding light absorption and redox potentials of photogenerated electron and hole (Mapa et al.

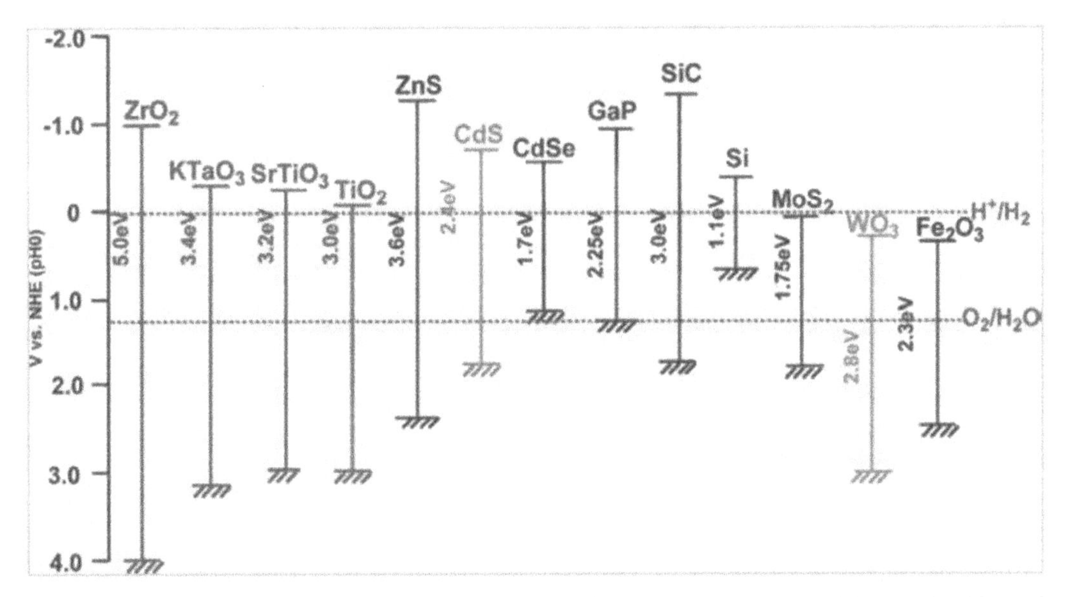

Figure 4. Relationship between band structure of semiconductor and redox potentials of water splitting. Reprinted from ref. 89 Serpone, N. and E. Pelizzetti. 1989. Wiley, New York, NY, USA.

2013, Grätzel 2001). For the water-splitting reaction to be thermodynamically favourable, the band gap of the semiconductor photocatalyst should straddle these redox potentials, that is, the CB should have negative potential rather than the hydrogen-evolution potential and the VB should be positive potential than the oxygen-evolution potential. So electron can be lower their energy by transferred to H^+ in solution and holes to H_2O through a short circuit reaction and balancing the charges transferred to the solution. However, for a single component photocatalyst, it is difficult to possess both wide light–absorption range and strong redox ability concurrently. Besides, in the single-component structure, the photogenerated electrons in the CB can easily return to the VB or remain trapped in the defect state and recombine with the holes, which seriously reduce the utilization efficiency of solar energy (Formal et al. 2014, Zhou et al. 2014). Hence, designing appropriate heterogeneous photocatalytic systems should be an effective way to overcome this problem.

Nano structured and nanocomposite photocatalytic materials for water splitting

Throughout human history, advances in civilization have been associated with the discovery, development, and use of new materials. Hence, materials can be considered as the parent of almost all technologies as most technological breakthroughs have been achieved through the development of new materials (Marquis et al. 2011). We need smart nanomaterials to fully develop effective renewable energy conversion, storage, and utilization. Nanomaterials have also emerged as efficient components of photoelectrochemical cells and photocatalysts. Nano structured photocatalytic materials have greatly attracted interest of the researchers in the last few years since to make use of the tools to produce nanomaterials with small size, defined shape and improved surface properties in order utilised in hydrogen energy generation from water splitting into hydrogen. In this regard, the scientific community is currently focused in relevant direction, which is the synthesis of novel nanostructured and nanocomposite materials capable of converting solar energy with the purpose of turning it into chemical or electrical energy (Serpone and Emeline 2005, Colmenares et al. 2006).

Bismuth vanadate nanostructured photocatalysis for water splitting

Various semiconductor nanomaterials have been widely used to solve energy and environment problems with their excellent photocatalytic properties (Miseki and Miseki 2009). In the past few decades, TiO_2 nanomaterials have attracted tremendous interest to remove organic pollutants and for hydrogen generation owing to their strong oxidation ability, high photo stability and nontoxicity (Hashimoto et al. 2005). However, the TiO_2 nanomaterials have wide band gaps of 3.0–3.2 eV, which means only less than 5% of the entire solar energy can be utilized and its potential as a sustainable technology cannot be entirely fulfilled (Chen and Mao 2007, Moscow and Jothivenkatachalam 2015). Therefore, many researchers have devoted much attention to explore the efficient visible-light-driven photocatalysts compared with single-phase semiconductor photocatalysts. Bismuth metal oxides related nanomaterials such as Bi_2O_3, $BiVO_4$, $BiFeO_3$, Bi_2WO_6, and Bi_2MoO_6 have attracted considerable interest due to high stability, low cost, non-toxicity, their efficient photocatalytic ability in decomposing organic compounds and water splitting (Tang et al. 2004, Bessekhouad et al. 2002). In recent years, extensive efforts have been devoted to fabricate various bismuth metal oxide based nanomaterials to increase the performance of photocatalysts, and to understand the fundamental factors that mediate photocatalytic efficiency (Zhang and Bahnemann 2013, Sun et al. 2013).

Recently, Bismuth Vanadate ($BiVO_4$) has gained increasing attention for its use as a promising candidate under visible light irradiation among the bismuth metal oxide photocatalyst (Pilli et al. 2011, Moscow and Jothivenkatachalam 2016). The $BiVO_4$ photocatalysts are highly promising for different applications such as renewable energy production systems (i.e., solar fuels production from water and sunlight) and to resolve environmental issues. Bismuth vanadate ($BiVO_4$), which is an n-type semiconductor, has been identified as one of the most promising photocatalytic materials. As it is well known, $BiVO_4$ exists in three polymorphs of monoclinic scheelite, tetragonal scheelite, and tetragonal zircon structures, with band gaps of 2.4, 2.34, and 2.9 eV, respectively. It is reported that $BiVO_4$ mainly exists in three crystalline phases: monoclinic scheelite, tetragonal zircon and tetragonal scheelite structure (Lim et al. 1995, Bhattacharya et al. 1997, Luo et al. 2008) (Figure 2). Monoclinic scheelite $BiVO_4$, (~ 2.3 eV band gap) shows both visible-light and UV absorption while tetragonal $BiVO_4$ (~ 2.9 eV band gap) mainly possesses an UV absorption band. The UV absorption observed in both the tetragonal and monoclinic $BiVO_4$ is associated with band transition from O_{2p} to V_{3d}, whereas visible light absorption is due to the transition from a valence band (VB) formed by Bi6s or a hybrid orbital of Bi_{6s} and O_{2p} to a conduction band (CB) of V_{3d} (Ng et al. 2010). The scheelite structure can have a tetragonal crystal system (space group: *I*41/*a* with $a = b = 5.1470$ Å, $c = 11.7216$ Å) or a monoclinic crystal system (space group: *I*2/b with $a = 5.1935$ Å, $b = 5.0898$ Å, $c = 11.6972$ Å, and $b = 90.3871$), while the zircon-type structure has a tetragonal crystal system (space group: *I*41/a with $a = b = 7.303$ Å and $c = 6.584$ Å) (Park et al. 2013).

In addition, the Bi-O bond in monoclinic $BiVO_4$ is distorted, which increases the separation efficiency of photoinduced electrons and holes. In the scheelite structure, four O atoms in a tetrahedral site coordinate with each V ion, and eight O atoms from eight different VO_4 tetrahedral units coordinate with each Bi ion. In zircon type structure, each O atom is coordinated to two Bi centres and one V centre and forms a three-dimensional structure (Figure 5). The only difference between the two structures is that V and Bi ions are more significantly distorted in the monoclinic structure, which removes the four-fold symmetry necessary for a tetragonal system (Wang et al. 2012). Among these, crystalline the monoclinic $BiVO_4$ (m) shows higher activity than the others (Herrmann 1999).

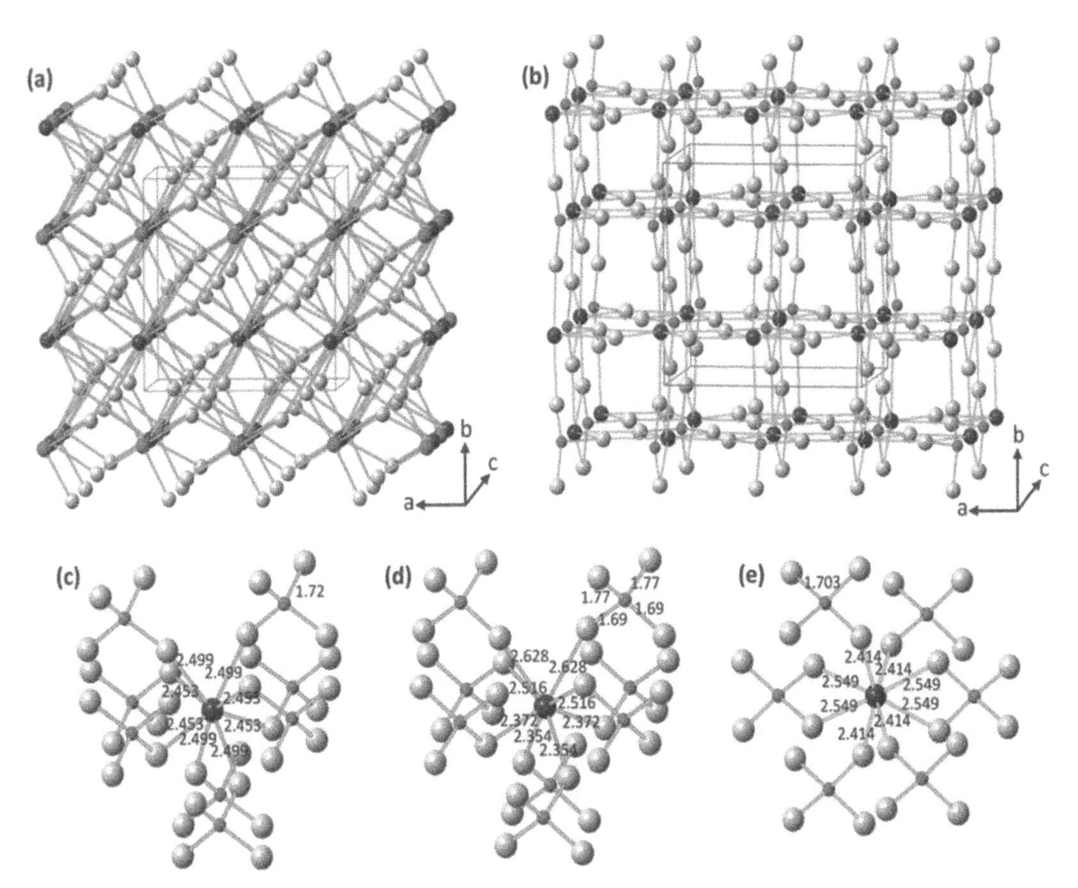

Figure 5. Crystalline structures of (a) tetragonal scheelite and (b) zircon-type $BiVO_4$ (red: V, purple: Bi, and gray: O). Atomic positions of Bi, V, and O. Local coordination of V and Bi ions in (c) tetragonal scheelite, (d) monoclinic scheelite, (e) and zircon–type $BiVO_4$ structure with bond lengths. Reprinted from ref. 56 Tokunaga, S., H. Kato and A. Kudo. 2001. Chem. Mater. 13: 4624–4628. Copyright © (2001) American Chemical Society.

However, the photocatalytic efficiency of the bare $BiVO_4$ is still the bottleneck due to poor adsorptive performance, low mobility and high rate of photogenerated charge carriers before they reach the interfaces (Gotic et al. 2005). Therefore, it is of great interest to find out the technique to enhance photocatalytic activity of $BiVO_4$. Crystalline and morphology control is one of the capable method to improve the light harvesting, charge carrier migration efficiency, assisting the collection and separation of electron-hole pairs at the interface of the materials (Tong et al. 2012, Ma et al. 2014, Kubacka et al. 2011). $BiVO_4$ with controlled morphology techniques have been used for the development of ordered architectures which is much more interesting to improve photocatalytic properties (Eda et al. 2012, He et al. 2014, Sun et al. 2010).

The optical properties of the monoclinic form of $BiVO_4$ make it similar to or even more attractive than other materials in visible light irradiation. It has a band gap in the visible light range (2.4–2.7 eV), a suitable valence band position, crystalline size and controlled morphology for driving water oxidation under illumination. Additionally, the band-edge alignment allows generation of a relatively large photovoltage for water oxidation compared to other metal oxides (Sayama et al. 2006, Walsh et al. 2009). Several research groups have used different methods by the introduction of controlled structure to make efficient $BiVO_4$ photoanodes.

Morphology control is an efficient method to facilitate carrier transportation and light harvesting, accelerate charge movement within the material structure and assist the collection and separation of electron-hole pairs at the interface of the materials (Kubacka et al. 2011). Recently, PEC activity, which improved by various techniques such as thin or thick film deposition, co doped nanostructures, specific crystalline facet, Phase, nano rods, heterojunction, composite and quantum size $BiVO_4$, has been explored.

Berglund et al. (2011) reported that photoelectrochemical water oxidation activity nanostructured Co-$BiVO_4$ films were shown to be several times higher than that of $BiVO_4$ films. They have found that the morphology and film thickness of Co-$BiVO_4$ is, which play role on Oxygen Evolution Reaction (OER) activity of water splitting under illumination. The Co-$BiVO_4$ linear sweep voltammogram (LSV) (Figure 6) of film had an onset potential of about 0.45 V vs. Ag/AgCl under visible light which accounted for more than 50% of the total photocurrent when compared to $BiVO_4$ film. This means that the water splitting properties of the Co-$BiVO_4$ are enriched by the reduction of the over potential for water oxidation or the reduction of the Fermi level (Lopes et al. 2014). The LSV also indicates that the increase in the photocurrent and charge transport is closer to the maximum theoretical efficiency of a 2.3–2.4 eV band

Figure 6. (a and b) The SEM image as synthesised annealed $BiVO_4$ film (c) LSV of $BiVO_4$ film in 0.5 M Na_2SO_4 in dark and visible light and white light from a xenon lamp. Reprint from ref. 1 Berglund, S.P., D.W. Flaherty, N.T. Hahn, A.J. Bard and C.B. Mullins. 2011. J. Phys. Chem. C. 115: 3794–3802. Copyright @ 2011 American Chemical Society.

gap material under AM1.5 illumination. The reason is the addition of Co on the surface morphology (Figure 6). a-b0 of $BiVO_4$ films with equal bismuth and vanadium increased the photocurrent due to improved light absorption and charge transport and achevied the maximum theoretical efficiency. The study indicates the nanostructure film plays a major role and exhibits increasing photocurrent density with the applied potential, whereas the $BiVO_4$ sample shows lower photocurrent.

Chen et al. (2013) reported the synthesis of monoclinic phase $BiVO_4$ thin films by reactive sputtering and investigated the photoactivity for water oxidation with richness of vanadium and bismuth. The highest photocurrents of V excess stoichiometric $BiVO_4$ thin film is ca. 0.8 mA cm^2 at the reversible O_2/H_2O potential with simulated AM 1.5 G illumination. The photoelectrochemical (PEC) performance of the Bi rich stoichiometric $BiVO_4$ films (Figure 7) under anodic bias, both of stoichiometric films exhibited very low photocurrent densities. The conduction band of $BiVO_4$ is lower than the H_2 evolution potential, so this current is not due to water reduction involving $BiVO_4$, but it is due to the phenomenon of an oxygen reduction reaction (Lin et al. 2012). In contrast, the stoichiometric $BiVO_4$ thin film, which possesses a slight excess of V, exhibits a large and dominant photoanodic current density, with little photocathodic current at negative biases. Indeed, the photocurrent density at OER, 0.5 mA /cm^2, is the best conversion efficiency of unmodified $BiVO_4$, and the onset potential of approximately 0.4 V vs. RHE is similar to that reported for $BiVO_4$ thin films deposited via solution-based methods.

This means that photons absorbed deeper by Bi rich stoichiometric $BiVO_4$ thin film have very low probability to generate charges which are able to successfully reach the semiconductor surface, and thus participate in photoreactions, and thus generate low photocurrent. In contrast, V rich stoichiometric $BiVO_4$ thin film shows high photocurrent. Actually, the photogenerated holes recombination of hole with electron is reduced due to their longer diffusion lengths, which is consistent with SEM image (Figure 7 a–b) (Itoh and Bockris 1984).

Nanosized building blocks, such as nanowires, nanorods, nanosheets, and nanotubes possess interesting properties, and their self-assembling into hierarchical architectures is much more interesting and has attracted great attention for water splitting (Kubacka et al. 2011). Pihosh et al. (2014) synthesized nanostructured WO_3 nanorods capped with extremely thin $BiVO_4$ absorber layers, fabricated by sputtering techniques with different morphologies and different phase. The results of the LSV and Amperometric J p-t profiles measurements are presented in Figure 8i and ii. The *J* p behaviour at low bias indicated a good charge separation upon solar illumination. The LSV of prepared samples are shown in Figure 8 (i), where it is observed that the highest photocurrent of 0.95 mA/cm^2 was detected at 1.7 *V* RHE for the WO_3-NRs with the length of 2.5 μm compared with other length of 0.3 μm to 5 μm.

The stability of the photoanodes was characterized by measuring the amperometric J_p Vs *t* profiles under the same solar simulated light and the bias voltage of 1.23 *V* RHE (Figure 8ii). In all cases, Jp-*t*

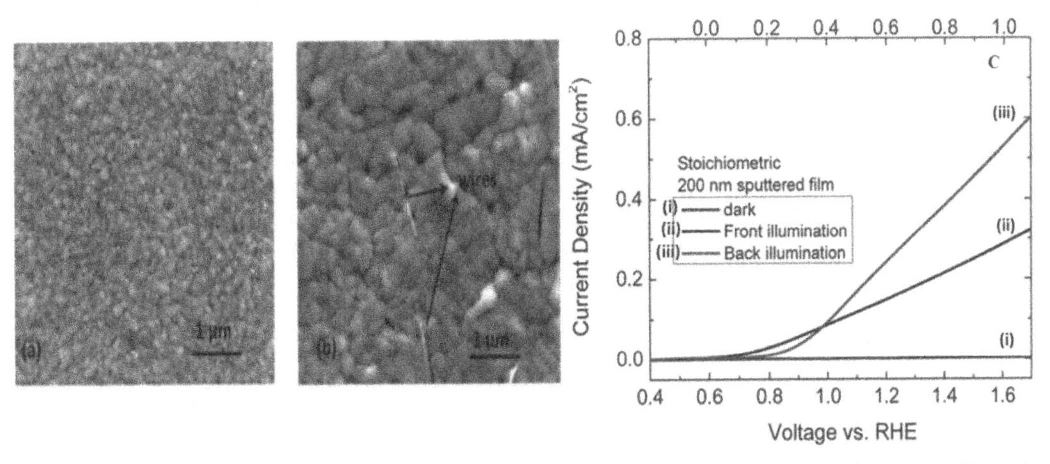

Figure 7. SEM images of sputtered bismuth vanadium oxide thin films with Bi-rich (a), V-rich (b), and stoichiometric ($BiVO_4$) (c) LSV curves under front and back illumination. Reprint from ref. 5 Chen, L., E.A. Lladó, M. Hettick, I.D. Sharp, Y. Lin, A. Javey and J.W. Ager. 2013. J. Phys. Chem. C 117: 21635–21642. Copyright © 2013 American Chemical Society.

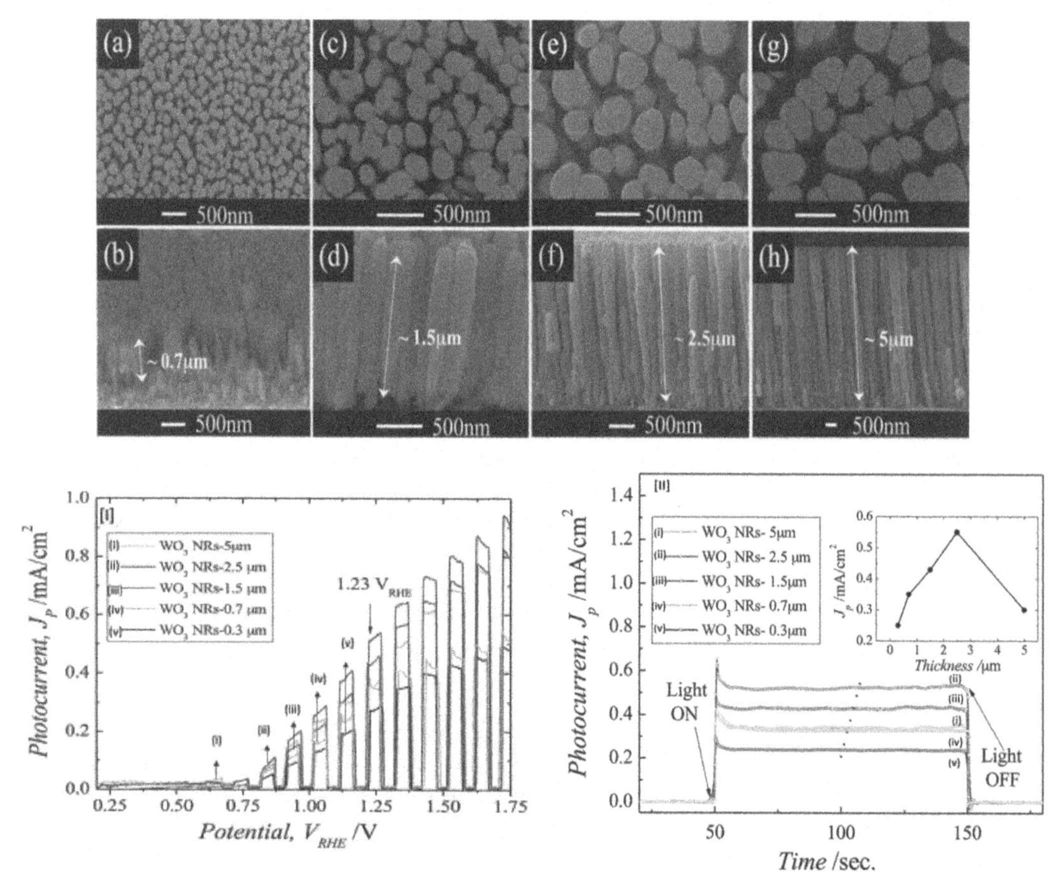

Figure 8. (a–h) Top and cross-sectional SEM images, LSV (i) and Amperometric (ii) *J* p-*t* profiles for various thicknesses of the WO_3 NRs collected in the 0.5 M Na_2SO_4 solution at pH = 7 under the same simulated AM 1.5 solar light at 1.23 *V* RHE. Reprint from ref. 44 Pihosh, Y., I. Turkevych, K. Mawatari, T. Asai, T. Hisatomi, J. Uemura, M. Tosa, K. Shimamura, J. Kubota, K. Domen and T. Kitamori. 2014. Small 10: 3692–3699. © WILEY-VCH Verlag GmbH & Co. KGaA, Weinheim.

profile indicates the excellent stability of the photoanodes. The stable *J* p achieved a maximum value of 0.55 mA/cm^2 at 1.23 *V* RHE for the WO_3-NRs with the length of 2.5 μm (see the inset in Figure 8ii). Apparently, the recombination of charge carriers starts to prevail in the case of longer WO_3-NRs, which in turn decreases the *J* p. The photocurrent enhancement is attributed to the faster charge separation in the electronically thin $BiVO_4$ layer and reduced charge recombination by nanostructured morphology. The enhanced light trapping in the nanostructured WO_3-NRs/$BiVO_4$ photoanode effectively increases the optical thickness of the $BiVO_4$ layer and results in efficient absorption of the incident light. The IPCE spectra revealed that the photocurrent in the optimized photoanode of nanostructured morphology (Figure 8a–h) of the WO_3-NRs/$BiVO_4$ facilitated light trapping that in turn re-established the effective optical thickness of the $BiVO_4$ absorber which offers a promising strategy toward efficient photocatalytic systems for the generation of hydrogen via photocatalytic water splitting.

Apart from morphology of the crystallite size, specific crystal facet determines the surface active sites and even the electronic structure; as a result, crystal facets play a critical role in enhancing the PEC activity (Tokunaga et al. 2001). Thalluri et al. (2016) reported the preparation of W and Mo doped $BiVO_4$ powders with a monoclinic scheelite structure and preferential growth along the {040} facet. This crystalline facet was prepared by green hydrothermal synthesis approach in neutral pH conditions using ammonium carbonate as a structural directing agent. It was also reported that prepared materials demonstrated an extraordinary enhancement of the water oxidation rate under simulated sunlight (i.e.,

942 μmol/g/h O_2), (Figure 9) which is about two-fold and 100 times higher than those of a state-of-art Mo-doped and the non-doped $BiVO_4$ materials.

Moreover, W and Mo dopants play a more important role. It was also observed that the O_2 evolution activity of $BiVO_4$ had a direct correlation with the crystallite size, band gap and the ratio between the {040} and {110} facets. As per report of the FESEM analyses (Figure 9a–c), all the $BiVO_4$ powders have a high exposure to the {040} crystalline surface.

The origin of the doping-induced (W or Mo) photoactivity enhancement on $BiVO_4$ is often explained as being due to increased charge transport efficiency in the $BiVO_4$ bulk. The study predicted that both dopants (W and Mo) are shallow electron donors which can increase the carrier density of $BiVO_4$; it has also been observed that Mo doping generates superior n-type conductivity than W doping in $BiVO_4$ (Parmer et al. 2012).

The metal doped $BiVO_4$ photocatalysts are highly promising for different applications such as renewable energy production systems and to resolve environmental issues. Usai (2013) developed Erbium-yttrium co-doped $BiVO_4$ with a tetragonal structure, synthesized by surfactant free hydrothermal method. The studied photocatalyst shows good O_2 evolution photoactivity under sun-like excitation. From structural and morphological characterization, it has been stated that the presence of lanthanides induces the stabilization of the tetragonal phase (Figure 10a–e).

Figure 9. FESEM images of the (a) $BiVO_4$, (b) W-$BiVO_4$ and (c) Mo-$BiVO_4$ powders. The O_2 evolution by water splitting of pure $BiVO_4$ powder, W-doped (d) and Mo-doped (e) $BiVO_4$ samples under AM 1.5 simulated solar light illumination (100 mW/cm^2). Reprint from ref. 55 Thalluri, S.M., S. Hernández, S. Bensaida, G. Saraccoa and N. Russoa. 2016. Applied Catalysis B: Environmental 180 : 630–636. Copyright @American Chemical Society.

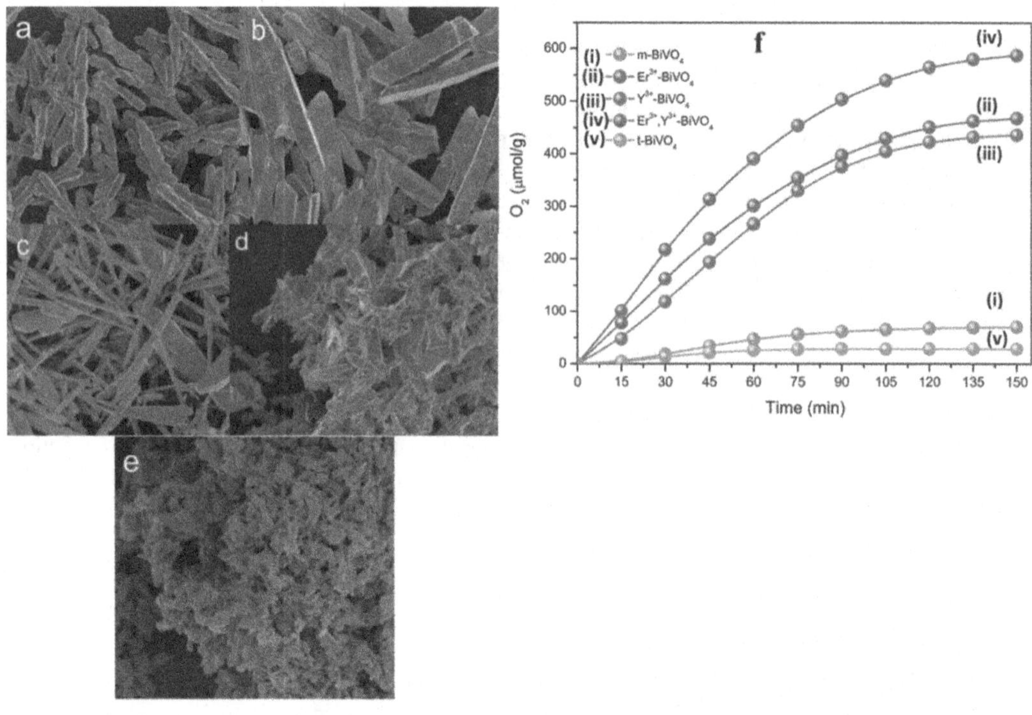

Figure 10. FESEM images of the different $BiVO_4$ systems: (a) bare m-$BiVO_4$; (b) bare t-$BiVO_4$; (c) Er^{3+}-$BiVO_4$; (d) Y^{3+}- $BiVO_4$; (e) Er^{3+},Y^{3+} co-doped $BiVO_4$ (f) Photocatalytic O_2 evolution for different $BiVO_4$ samples as a function of the irradiation time. Reprint from ref. 59 Usai, S., S. Obregón, A.I. Becerro and G. Colón. 2013. J. Phys. Chem. C 117: 24479–24484. Copyright © 2013 American Chemical Society.

Furthermore, Figure 10 shows the photocatalytic water oxidation activity of different $BiVO_4$ samples, from an aqueous solution containing $AgNO_3$ as a sacrificial reagent under sunlight irradiation. The O_2 evolution for different metal-doped $BiVO_4$ catalysts plainly denotes a notably beneficial effect. In all cases, the O_2 evolution rates for these samples are markedly improved with respect to bare m- or t-$BiVO_4$. Moreover, the photocatalytic performance shown by the Er^{3+}, Y^{3+} co-doped $BiVO_4$ appears significantly enhanced compared to the single doped systems (425 μmol h^{-1} g^{-1} vs. ca. 300 μmol h^{-1} g^{-1}) (Figure 10).

Thus, the calculated reaction rate for the co-doped system is 8 times higher than that for the m-$BiVO_4$ as reported in the present work. By observing the reaction rate for t-$BiVO_4$, this enhancement due to the lanthanide dopant led to synergistic effect on bare $BiVO_4$. Also, there is a small contribution of Vis-NIR photons to the UV-active tetragonal co-doped $BiVO_4$ in the overall mechanism, probably due to an energy transfer process from the erbium ions. As a result, the Er^{3+}, Y^{3+} co-doped $BiVO_4$ shows a notably improved photoactivity compared to the m-$BiVO_4$.

The generation of heterojunctions between the nanostructure materials is the key for improving movement and restraining the recombination of photoinduced charge carriers, and finally improving the photocatalytic performance. Recently, Jinzhan et al. (2011) have prepared novel heterojunction WO_3/$BiVO_4$ photoanode for photoelectrochemical water splitting by solvothermal deposition. Compared to planar WO_3/$BiVO_4$ heterojunction films, the nanorod-array films show significantly improved photoelectrochemical properties which are due to the high surface area and improved separation of the photogenerated charge at the WO_3/$BiVO_4$ interface. Figure 11(I a–c) shows that the LSV of WO_3/$BiVO_4$ heterojunction films comprised of a single $BiVO_4$ layer and multiple WO_3 layers, which indicate that with addition of a $BiVO_4$ layer upon the WO_3 base, the onset photocurrent wavelength was shifted from 450 to 525 nm, as the low band gap $BiVO_4$ layer extended the absorption edge to longer wavelengths. The conversion efficiency in the region from 400 to 500 nm for the WO_3/$BiVO_4$ heterojunction films

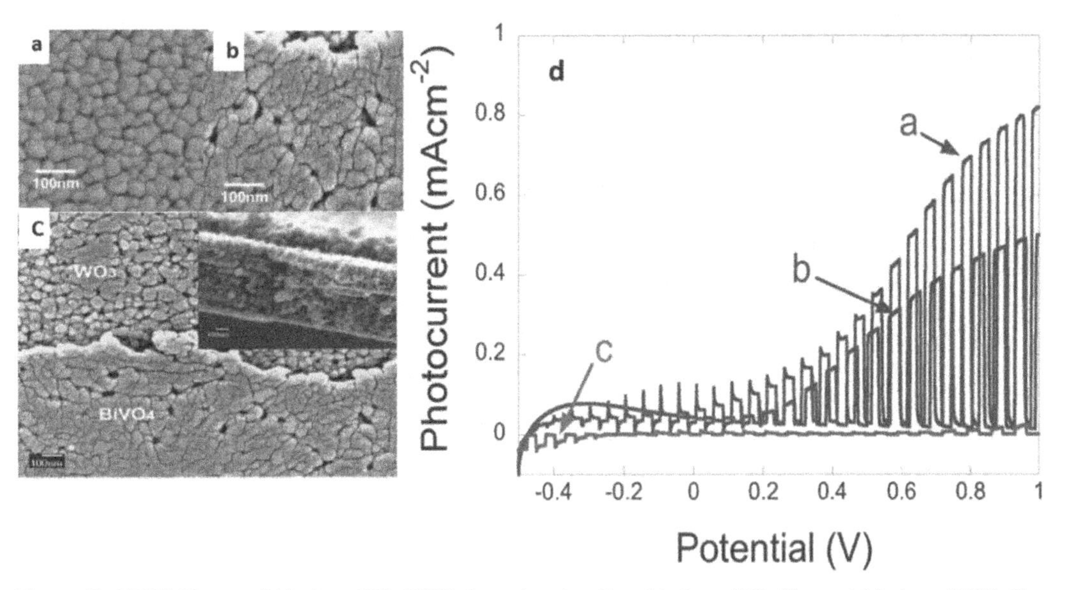

Figure 11. FESEM image of (a) planar $WO_3/BiVO_4$ heterojunction film, (b) planar WO_3 film, and (c) planar $BiVO_4$ film under illumination (100 mW/cm^2) in 0.5 mol Na_2SO_4 (d) LSV plots of photocurrent for planer $WO_3/BiVO_4$, WO_3 and $BiVO_4$ film Reprint from ref. 11 Jinzhan, S., G. Liejin, B. Ningzhong and A.G. Craig. 2011. Nano Lett. 11: 1928–1933. Copyright © 2011, American Chemical Society.

increased for WO_3 layers and photocurrent also increased significantly. The increasing photocurrent was mainly generated in the $BiVO_4$ layer, rather than the WO_3, thus increasing the photocurrent of $WO_3/BiVO_4$. Due to suitable formation of the heterojunction, going by the evidences from SEM image (Figure 11a–c), carrier separation is becoming more efficient in water splitting performance. The results show how one-dimensional nanorod arrays, integrated into a heterojunction structure, promote charge-carrier separation and transfer, as well as photocorrosion stability, offering a promising strategy for improving photoelectrochemical water splitting efficiencies.

$BiVO_4$ is an ideal photocatalytic material for visible light responsive semiconductors, but insufficient for overall water splitting. In addition, it shows significantly increased photocatalytic performance with cocatalyst, which is due to the increased charge separation, visible-light absorbance, specific surface area and reaction sites upon the introduction into pristine. Ding et al. (2013) developed $CoBi/BiVO_4$ photoanode, where it was found that the onset potential is negatively shifted with higher photocurrent, and overall water splitting to H_2 and O_2 is significantly improved. Figure 13 shows the LSV scans of $BiVO_4$ and $CoBi/BiVO_4$ electrodes in different electrolytes under chopped light illumination. As expected for n-type semiconductors, $BiVO_4$ electrode generates anodic photocurrent. Through loading CoBi cocatalyst, the potential needed for the same photocurrent is remarkably lowered and the onset potential is shifted from about –0.23 V to about –0.55 V vs. SCE (Figure 12b), which suggests the decrease of water oxidation over potential. Besides, the photocurrent of the $BiVO_4$ electrode is increased by about 2 folds, both in Na_2SO_4 and borate electrolytes. That is to say, more electrons and holes are used for water oxidation and reduction. This implies that the photocurrent enhancement is mainly from $BiVO_4$ promoted with the cocatalyst.

The cocatalyst can obviously reduce the water oxidation over potential, promote the charge transfer across the semiconductor–electrolyte interface, enhance the stability of $BiVO_4$ photoanode, and thus result in higher PEC water splitting activity. This work revealed the essential role of the cocatalyst and provided a successful demonstration that overall water splitting can be achieved using PEC strategy using a cocatalyst modified $BiVO_4$ photoanode.

Recently, Baek et al. (2012) have demonstrated a simple and efficient approach to prepare $BiVO_4/SnO_2/WO_3$ composites using wet-coating process method. In this method, SnO_2 introducing middle

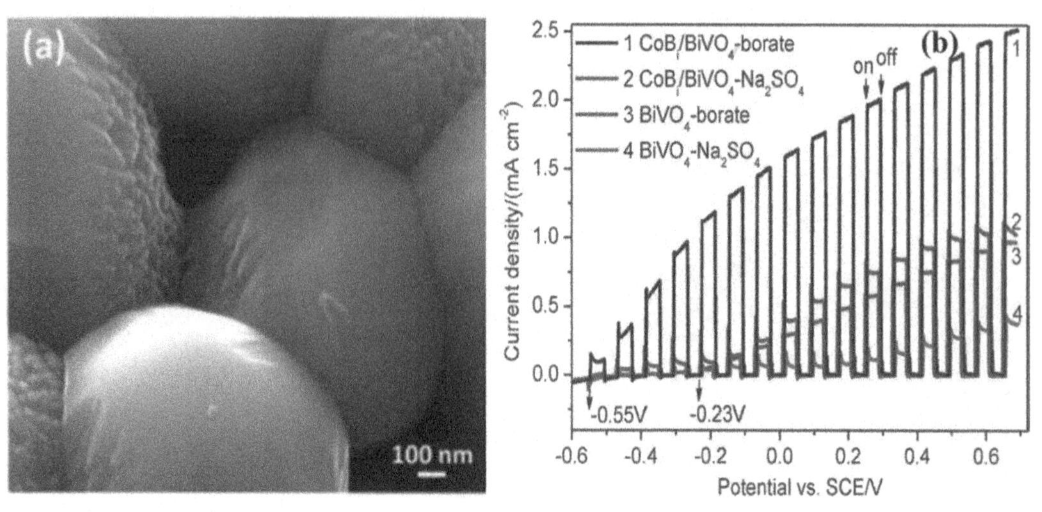

Figure 12. (a) SEM image of $CoBi/BiVO_4$ (b) LSV scans of $BiVO_4$ and $CoBi/BiVO_4$ photoanodes in 0.2 M sodium borate and 0.5 M Na_2SO_4 (pH 9) electrolyte under chopped light illumination. Reprint from ref. 9 Ding, C., J. Shi, Z. Wang and L. Can. 2017. ACS Catal. 7: 675–688. Copyright © 2017, American Chemical Society.

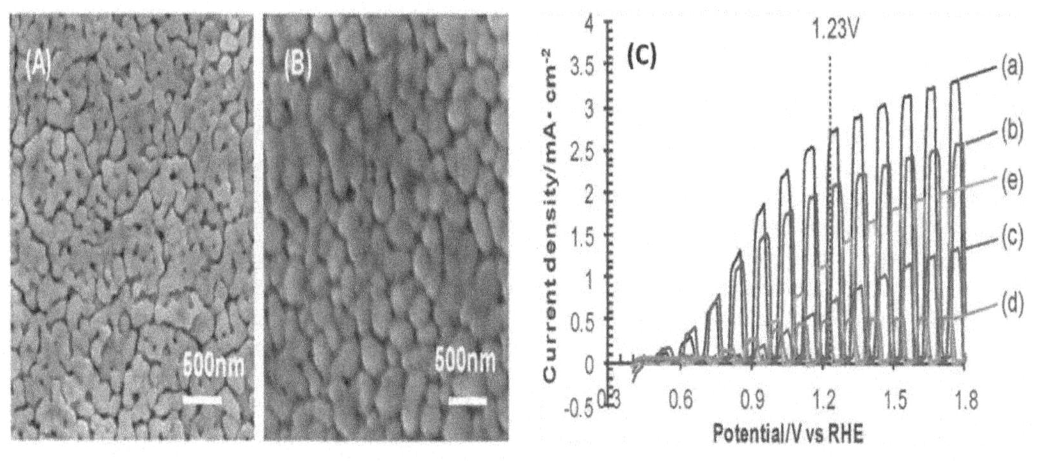

Figure 13. SEM images of the surface of (A) $BiVO_4/WO_3$ and (B) $BiVO_4/SnO_2/WO_3$; I–V curves of multi-composite film, bare $BiVO_4$ and bare WO_3 photoelectrodes. (a) $BiVO_4/SnO_2/WO_3$, (b) $BiVO_4/WO_3$, (c) bare $BiVO_4$ and (d) bare WO_3 in 0.1 M $KHCO_3$ aqueous solution. (e) $BiVO_4/SnO_2/WO_3$ in 0.1 M Na_2SO_4 aqueous solution. Reprint from ref. 20 Hyun, B.J., B.J. Kim, G.S. Han, S.W. Hwang, D.R. Kim, I.S. Cho and H.S. Jung. 2017. ACS Appl. Mater. Interfaces 9: 1479–1487. Copyright © 2016 American Chemical Society.

layer between the $BiVO_4$ and WO_3 layers is enhanced, expands the absorption range, improves the photogenerated electron separation, and increases the active sites and film thickness. The optimized composite sample exhibits outstanding photocatalysis activity compared with pure $BiVO_4$ under simulated sunlight. The photocurrent characteristic of photoelectrodes is known to greatly depend on film thickness. Figure 13C shows the I–V curves of $BiVO_4/SnO_2/WO_3$ (a) and $BiVO_4/WO_3$ (b) LSV scan of $BIVO_4$ and $CoBi/BiVO_4$. The photocurrent of $BiVO_4/SnO_2/WO_3$ (a) was about 1.3 times higher than that of $BiVO_4/WO_3$ (b) at 1.23 V vs. RHE. Therefore, it was concluded that the intrinsic quantum efficiency of the photocurrent generated from excited electrons in $BiVO_4$ was improved in the presence of SnO_2. In addition, there is an electron loss pathway where the electrons in the CB of WO_3 are recombined with the holes in the valence band (VB) of $BiVO_4$, and the SnO_2 middle layer might prevent this unfavourable pathway.

As for the photo-generated holes in the VB of $BiVO_4$, they cannot transfer to VB of WO_3 or SnO_2, and the remaining holes are consumed to oxidize H_2O. Therefore, it is possible that the change in the morphology of the $BiVO_4$ upper layer is one of the factors for the photocurrent improvement. This surface morphology of $BiVO_4$ on the SnO_2/WO_3 thin film favourably functions to improve activated surface reaction for water splitting. It was surmised that the presence of SnO_2 influenced the growth of the $BiVO_4$ coated better because this $BiVO_4/SnO_2/WO_3$ morphology was a little different from $BiVO_4/WO_3$ morphology (Figure 13A–B). Finally, the study talks about the enhancement of the LHE of the photoelectrode, probably due to the electron diffusion length limitation in $BiVO_4$ film using the double-stacked structure and also enhance the water splitting.

Conclusion

Photocatalysis appears to be a promising avenue to solve problems regarding environmental protection and energy production in the near future. Improving the separation and transportation of charge carriers are the main challenges in this area. Hence, the photocatalytic processes involve a complicated sequence of multiple synergistic or competing steps, and the efficient utilization of solar energy. To date, some advances studies have been reported to improve the photocatalytic efficiencies that range from environmental remediation to hydrogen energy production from the water by the utilization of sunlight and led the improvement of the separation/transportation of the charges' carriers. Although great advancements have been made in investigation of nanostructured photocatalytsts, it is still interesting to design more efficient photocatalytic systems. First, the fundamental understandings are necessary to solve bottleneck problems which include improved charge separation, migration efficiency, optical properties with narrow band gap, lowered cost, and toxicity. Second, faceted photocatalysts go on a challenge for the improvement of additive-free synthesis routes which is highly desirable, since most synthesis strategies involve multiple steps in order to obtain clean facets. Therefore, there is an urgent need to explore novel materials, and deepening the knowledge of the designing of nanostructured materials is indispensible to make substantial breakthroughs for practical application of photocatalysts. This chapter has demonstrated the recent research efforts to synthesize visible light driven bismuth vanadate nanostructured materials and nanocomposites.

References

Berglund, S.P., D.W. Flaherty, N.T. Hahn, A.J. Bard and C.B. Mullins. 2011. Photoelectrochemical oxidation of water using nanostructured $BiVO_4$ films. J. Phys. Chem. C 115: 3794–3802.

Bessekhouad, Y., M. Mohammedi and M. Trari. 2002. Hydrogen photoproduction from hydrogen sulfide on Bi_2S_3 catalyst. Solar Energy Materials and Solar Cells 73: 339–350.

Bhattacharya, A.K., K.K. Mallick and A. Hartridge. 1997. Phase transition in $BiVO_4$. Materials Letters 30: 7–13.

Chen, X. and S.S. Mao. 2007. Titanium dioxide nanomaterials: synthesis, properties, modifications, and applications. Chem. Rev. 107: 2891–2899.

Chen. L., E.A. Lladó, M. Hettick, I.D. Sharp, Y. Lin, A. Javey et al. 2013. Reactive sputtering of bismuth vanadate photoanodes for solar water splitting. J. Phys. Chem. C 117: 21635–21642.

Christians, J.A., R.C.M. Fung and P.V. Kamat. 2014. An inorganic hole conductor for organo-lead halide perovskite solar cells. Improved hole conductivity with copper iodide. J. Am. Chem. Soc. 136: 758–764.

Colmenares, J.C., M.A. Aramendía, A. Marinas, J.M. Marinas and F.J. Urbano. 2006. Synthesis, characterisation and photocatalytic activity of different metal doped titania systems. Appl. Catal. A 306: 120–127.

Dhakshinamoorthy, A., S. Navalon, A. Corma and H. Garcia. 2012. Photocatalytic CO_2 reduction by TiO_2 and related titanium containing solids. Energy Environ. Sci. 5: 9217–9233.

Ding, C., J. Shi, Z. Wang and L. Can. 2017. Photoelectrocatalytic water splitting significance of cocatalysts, electrolyte, and interfaces. ACS Catal. 7: 675–688.

Eda, S.I., M. Fujishima and H. Tada. 2012. Low temperature-synthesis of $BiVO_4$ nanorods using polyethylene glycol as a soft template and the visible-light-activity for copper acetylacetonate decomposition. Appl. Catal. B 125: 288–293.

Formal, F.L., S.R. Pendlebury, M. Cornuz, S.D. Tilley, M. Grätzel and J.R. Durrant. 2014. Back electron–hole recombination in hematite photoanodes for water splitting. J. Am. Chem. Soc. 136: 2564–2574.

Fujishima, A. and K. Honda. 1972. Electrochemical photolysis of water at a semiconductor electrode. Nature 238: 37–38.

Fujishima, A., X. Zhang and D.A. Tryk. 2007. Heterogeneous photocatalysis: from water photolysis to applications in environmental cleanup. Int. J. Hydrogen Energy 32: 2664–2672.

Gotic, M., S. Music, M. Ivanda, M. Soufek and A. Popovic. 2005. Synthesis and characterisation of bismuth (III) vanadate. J. Mol. Struct. 744: 535–540.

Grätzel, M. 2001. Photoelectrochemical cells. Nature 414: 338–344.

Hashimoto, K., H. Irie and A. Fujishima. 2005. TiO_2 photocatalysis: A historical overview and future prospects. J. Appl. Phys. 44: 8269–8285.

He, H., S.P. Berglund, A.J.E. Rettie, W.D. Chemelewski, P. Xiao, Y. Zhang et al. 2014. Synthesis of $BiVO_4$ nanoflake array films for photoelectrochemical water oxidation. J. Mater. Chem. A 2: 9371−9379.

Herrmann, J.M. 1999. Heterogeneous photocatalysis: fundamentals and applications to the removal of various types of aqueous pollutants. Catal. Today 53: 115–129.

Hoffmann, M.R., S.T. Martin. W.Y. Choi and D.W. Bahnemann. 1995. Environmental applications of semiconductor photocatalysis. Chem. Rev. 95: 69−96.

Hyun, B.J., B.J. Kim, G.S. Han, S.W. Hwang, D.R. Kim, I.S. Cho et al. 2017. $BiVO_4/WO_3/SnO_2$ double-heterojunction photoanode with enhanced charge separation and visible-transparency for bias-free solar water-splitting with a perovskite solar cell. ACS Appl. Mater. Interfaces 9: 1479–1487.

Itoh, K. and J.O.M. Bockris. 1984. Stacked thin-film photoelectrode using iron oxide. J. Appl. Phys. 56: 874–876.

Jinzhan, S., G. Liejin, B. Ningzhong and A.G. Craig. 2011. Nanostructured $WO_3/BiVO_4$ heterojunction films for efficient photoelectrochemical water splitting. Nano Lett. 11: 1928–1933.

Jothivenkatachalam, K., S. Prabhu, A. Nithyaa and K. Jeganathan. 2014. Facile synthesis of WO_3 with reduced particle size on zeolite and enhanced photocatalytic activity. RSC Adv. 4: 21221–21229.

Kibria, M.G., H.P.T. Nguyen, K. Cui, S. Zhao, D. Liu, H. Guo et al. 2013. One-step overall water splitting under visible light using multiband InGaN/GaN nanowire heterostructures. ACS Nano. 7: 7886−7893.

Kubacka, A., M.F. Garcia and G. Colon. 2011. Advanced nano architectures for solar photocatalytic applications. Chem. Rev. 112: 1555−1614.

Kudo, A. and A. Miseki. 2009. Heterogeneous photocatalyst materials for water splitting. Chem. Soc. Rev. 38: 253–278.

Li, X.J., Q. Wen, J.X. Low, Y.P. Fang and J.G. Yu. 2014. Design and fabrication of semiconductor photocatalyst for photocatalytic reduction of CO_2 to solar fuel. Sci. China Mater. 57: 70−100.

Lim, A.R., S.H. Choh and M.S. Jang. 1995. Prominent ferroelectric domain walls in $BiVO_4$ crystal. Journal of Physics: Condensed Matter 7: 7309–7323.

Lin, Y.J., Y. Xu, M.T. Mayer, Z.I. Simpson, G. McMahon, S. Zhou et al. 2012. Growth of p-type hematite by atomic layer deposition and its utilization for improved solar water splitting. J. Am. Chem. Soc. 134: 5508−5511.

Linic, S., P. Christopher and D.B. Ingram. 2011. Plasmonic-metal nanostructures for efficient conversion of solar to chemical energy. Nat. Mater. 10: 911–921.

Lopes, T., L. Andrade, L. Formal, L. Gratzel, M. Sivula and A. Mendes. 2014. Hematite photoelectrodes for water splitting: evaluation of the role of film thickness by impedance spectroscopy. Phys. Chem. Chem. Phys. 16: 16515–16523.

Luo, H.M., A.H. Mueller, T.M. McCleskey, A.K. Burrell, E. Bauer and Q.X. Jia. 2008. Structural and photoelectrochemical properties of $BiVO_4$ thin films. J. Phys. Chem. C 112: 6099–6102.

Ma, Y., X. Wang, Y. Jia, X. Chen, H. Han and C. Li. 2014. Titanium dioxide-based nanomaterials for photocatalytic fuel generations. Chem. Rev. 114: 9987−10043.

Maeda, K. and K. Domen. 2010. Photocatalytic water splitting: Recent progress and future challenges. J. Phys. Chem. Lett. 18: 2655–2661.

Mao, S.S. and X. Chen. 2007. Selected nanotechnologies for renewable energy applications. International Journal of Energy Research 31: 619–636.

Mapa, M., S. Raja Ambal, S. Chinnakonda and N. Gopinath. 2013. ZnO-based solid solutions for visible light driven photocatalysis. Transactions of the Materials Research Society of Japan 38: 145–158.

Marquis, F.D.S. 2011. The role of nanomaterials systems in energy and environment: renewable energy. Journal of the Minerals Metals and Materials Society 63: 43.

Miseki, A.K. and Y. Miseki. 2009. Heterogeneous photocatalyst materials for water splitting. Chem. Soc. Rev. 38: 253–278.

Moscow, S. and K. Jothivenkatachalam. 2015. Synthesis of $BiVO_4$ nanoparticle by additive assisted microwave hydrothermal method and its photocatalytic performances. J. Environ. Nanotechnol. 4: 31–35.

Moscow, S. and K. Jothivenkatachalam. 2016. Facile microwave assisted synthesis of floral-shaped $BiVO_4$ nano particles for their photocatalytic and photoelectrochemical performances. J. Mater. Sci.: Mater. Electron. 27: 1433–1443.

Ng, Y., H.A. Iwase, A. Kudo and R. Amal. 2010. Reducing graphene oxide on a visible-light $BiVO_4$ photocatalyst for enhanced photoelectrochemical water splitting. J. Mater. Sci. Lett. 1: 2607–2612.

Ni, M., M.K.H. Leung, K. Sumathy and D.Y.C. Leung. 2004. Water electrolysis a bridge between renewable resources and hydrogen. Proceedings of the International Hydrogen Energy Forum 1: 25–28.

Parmer, K.P.S., H.J. Kang, A. Bist, P. Dua, J.S. Jang and J.S. Lee. 2012. Photocatalytic and photoelectrochemical water oxidation over metal-doped monoclinic $BiVO_4$ photoanodes. Chemsuschem. 5: 1926–1934.

Pihosh, Y., I. Turkevych, K. Mawatari, T. Asai, T. Hisatomi, J. Uemura et al. 2014. Nanostructured $WO_3/BiVO_4$ photoanodes for efficient photoelectrochemical water splitting. Small 10: 3692–3699.

Pilli, S.K., T.E. Furtak, L.D. Brown, T.G. Deutsch, J. Turner and A.M. Herring. 2011. Cobalt-phosphate (Co-Pi) catalyst modified Mo-doped $BiVO_4$ photoelectrodes for solar water oxidation. Energy Environ. Sci. 4: 5028–5034.

Qu, Y. and X.F. Duan. 2013. Progress, challenge and perspective of heterogeneous photocatalysts. Chem. Soc. Rev. 42: 2568–258.

Roy, S., C.O.K. Varghese, M. Paulose and C.A. Grimes. 2010. Toward solar fuels: photocatalytic conversion of carbon dioxide to hydrocarbons. ACS Nano 4: 1259–1278.

Sayama, K., A. Nomura, T. Arai, T. Sugita, R. Abe, M. Yanagida et al. 2006. Photoelectrochemical decomposition of water into H_2 and O_2 on porous $BiVO_4$ thin-film electrodes under visible light and significant effect of Ag ion treatment. J. Phys. Chem. B 110: 11352–11360.

Serpone, N. and E. Pelizzetti. 1989. Photocatalysis: Fundamentals and Applications, Wiley, New York, NY, USA.

Serpone, N. and A.V. Emeline. 2015. Modelling heterogeneous photocatalysis by metal-oxide nanostructured semiconductor and insulator materials: factors that affect the activity and selectivity of photocatalysts. Res. Chem. Intermed. 31: 391–432.

Sivula, K. and R. Van de Krol. 2016. Semiconducting materials for photoelectrochemical energy conversion. Nat. Rev. Mater. 70: 15010–15017.

Sun, J., G. Chen, J. Wu, H. Dong and G. Xiong. 2013. Bismuth vanadate hollow spheres: bubble template synthesis and enhanced photocatalytic properties for photodegradation. Appl. Catal. B: Environ. 132: 304–314.

Sun, Y.Y., C. Xie, S. Wu, S. Zhang and S. Jiang. 2010. Aqueous synthesis of mesostructured $BiVO_4$ quantum tubes with excellent dual response to visible light and temperature. Nano Res. 3: 620–631.

Tang, J., Z. Zou and J. Ye. 2004. Efficient photocatalytic decomposition of organic contaminants over $CaBi_2O_4$ under visible-light irradiation. Angew. Chem. Int. Ed. 43: 4463–4466.

Thalluri, S.M., S. Hernándeza, S. Bensaida, G. Saraccoa and N. Russoa. 2016. Green-synthesized W- and Mo-doped $BiVO_4$ oriented along the {040} facet with enhanced activity for the sun-driven water oxidation. Applied Catalysis B: Environmental 180: 630–636.

Tokunaga, S., H. Kato and A. Kudo. 2001. Selective preparation of monoclinic and tetragonal $BiVO_4$ with scheelite structure and their photocatalytic properties. Chem. Mater. 13: 4624–4628.

Tong, H., S. Ouyang, Y. Bi, N. Umezawa, M. Oshikiri and J. Ye. 2012. Nano-photocatalytic materials: possibilities and challenges. Adv. Mater. 24: 229–251.

U.S. energy. Information Administration International Energy Outlook 2016 DOE/EIA-0484 https://www.eia.gov/outlooks/ieo/pdf/0484 (2016).

Usai, S., Obregón, S.A.L. Becerro and G. Colón. 2013. Monoclinic–tetragonal heterostructured $BiVO_4$ by yttrium doping with improved photocatalytic activity. J. Phys. Chem. C 117 : 24479–24484.

Vayssieries, L. 2009. On Solar Hydrogen and Technology. John Wiley and Sons, Singapore.

Verne, J. 1874. The Mysterious Island (available at http :// www.literature-web.net/verne/mysterious island.

Walsh, A., Y. Yan, M.N. Huda, M.M. Al-jassim and S.H. Wei. 2009. Band edge electronic structure of $BiVO_4$: Elucidating the role of the Bi s and V d Orbitals. Chem. Mater. 21: 547–551.

Wang, X., G. Li, J. Ding, J. Peng and H. Chen. 2012. Facile synthesis and photocatalytic activity of monoclinic $BiVO_4$ micro/nanostructures with controllable morphologies. Mater. Res. Bull. 47: 3814–3818.

Xiang, Q.J., J.G. Yu and M. Jaroniec. 2012. Synergetic effect of MoS_2 and graphene as cocatalysts for enhanced photocatalytic H_2 production activity of TiO_2 nanoparticles. J. Am. Chem. Soc. 134: 6575–6578.

Xinchen Wang. 2009. http://wanglab.fzu.edu.cn/html/RESEARCH/1.html.

Zhang, L. and D. Bahnemann. 2013. Synthesis of nanovoid Bi_2WO_6 2D ordered arrays as photoanodes for photoelectrochemical water splitting. Chemsuschem. 6: 283–290.

Zhou, H., Y. Qu, T. Zeid and X. Duan. 2012. Towards highly efficient photocatalysts using semiconductor nano architectures. Energy Environ. Sci. 5: 6732–6743.

Zhou, P., J.G. Yu and M. Jaroniec. 2014. All-solid-state Z-scheme photocatalytic systems. Adv. Mater. 26: 4920–4935.

Index